Peter Clark.
University of St. Andrews.
1979

BOSTON STUDIES IN THE PHILOSOPHY OF SCIENCE

VOLUME XXI

SELECTED PAPERS OF

LÉON ROSENFELD

SYNTHESE LIBRARY

STUDIES IN EPISTEMOLOGY,

LOGIC, METHODOLOGY, AND PHILOSOPHY OF SCIENCE

VOLUME 100

LÉON ROSENFELD (1904-1974)

BOSTON STUDIES IN THE PHILOSOPHY OF SCIENCE

EDITED BY ROBERT S. COHEN AND MARX W. WARTOFSKY

VOLUME XXI

SELECTED PAPERS OF

LÉON ROSENFELD

Edited by

ROBERT S. COHEN AND JOHN J. STACHEL

D. REIDEL PUBLISHING COMPANY

DORDRECHT : HOLLAND / BOSTON : U.S.A.

LONDON : ENGLAND

Library of Congress Cataloging in Publication Data

Rosenfeld, Léon, 1904–1974.
Selected papers of Léon Rosenfeld.

(Boston studies in the philosophy of science; v. 21)
(Synthese library; v. 100)
'Bibliography of the writings of Léon Rosenfeld': p. Includes index.
1. Physics–History–Collected works. 2. Physics–Philosophy–Collected works. 3. Physics–Collected works. 4. Science–History–Collected works. I. Cohen, Robert Sonné. II. Stachel, J.J. III. Title. IV. Series.
Q174.B67 vol. 21 [QC7] 501S [5301.09] 77-17620
ISBN 90–277–0651–4
ISBN 90–277–0652–2 pbk.

The translation of Article 1 in Part II was prepared by Prof. Aage Petersen and edited by RSC and JJS. The translations of Articles 3, 7, 11, 15, 16, 17, and 20 of Part I, 5, 15 and 18 of Part II, 1, 2, and 3 of Part III, and 1 of Part IV were prepared by the AAA Linguistic Service (Boston), Betty Krikorian and Jean Revol and edited by RSC and JJS.

Published by D. Reidel Publishing Company,
P.O. Box 17, Dordrecht, Holland

Sold and distributed in the U.S.A., Canada, and Mexico
by D. Reidel Publishing Company, Inc.
Lincoln Building, 160 Old Derby Street, Hingham,
Mass. 02043, U.S.A.

Printed in The Netherlands

EDITORIAL PREFACE

I

The decision to undertake this volume was made in 1971 at Lake Como during the Varenna summer school of the Italian Physical Society, where Professor Léon Rosenfeld was lecturing on the history of quantum theory. We had long been struck by the unique blend of epistemological, historical and social concerns in his work on the foundations and development of physics, and decided to approach him there with the idea of publishing a collection of his papers. He responded enthusiastically, and agreed to help us select the papers; furthermore, he also agreed to write a lengthy introduction and to comment separately on those papers that he felt needed critical re-evaluation in the light of his current views. For he was still vigorously engaged in both theoretical investigations of, and critical reflections on the foundations of theoretical physics. We certainly did not conceive of the volume as a memorial to a 'living saint', but rather more practically, as a useful tool to place in the hands of fellow workers and students engaged in wrestling with these difficult problems. All too sadly, fate has added a memorial aspect to our labors.

We agreed that in order to make this book most useful for the contemporary community of physicists and philosophers, we should translate all non-English items into English. Selection of the items for inclusion proceeded by correspondence and at the time of Professor Rosenfeld's visit to Boston University in 1973 had reached a stage that enabled us to draw up a nearly final list with him; that list is the basis for this volume. Unfortunately, he had not set to work on the introduction and critical notes at the time of his death. Professor Stefan Rozental, his close friend and co-worker, has kindly supplied us with an introductory tribute; but he joins us in the sense of loss to the volume due to the absence of Rosenfeld's own editorial contributions.

II

Naturally, we cannot dream of filling that gap in these prefatory comments. We should just like to include a few words on some unique features of Professor Rosenfeld's work, which we feel will be of significance to physicists and philosophers, and especially to students.

At the time when Professor Rosenfeld started his scientific career, the number of physicists who concerned themselves with social issues, let alone those who saw any organic relationship between their creative activity and their social concerns, was small. Today, of course, under the impact of the great leap forward in the industrialization of science, and of so many scientific and technological developments upon all aspects of social life – and indeed of possible social death – this number has grown considerably. But often this concern, even among some scientists, takes the form of a rejection of science, or at any rate a rejection of any claim that science gives us a truthful picture of the world. How to reconcile an understanding of the social role of science, and likewise of the role of social forces in the development of science, with a concept of objective scientific truth has become an issue of concern to a growing number of scientists, young and old. Rosenfeld's lifelong example of just such an attempt can serve as an inspiration and example.

III

Professor Rosenfeld was a Marxist. He made no bones about this, as may be seen from the many references to Marx and Engels in this volume, as well as the integral spirit in which he approached such problems as the origins of thermodynamics or the history of electromagnetic theory, to give but two examples. His Marxism was of an undogmatic variety, however, at a time when dogmatism reigned supreme in the dominant currents of several 'official' Marxisms. His work thus provides a valuable example of the attempt by a critical intellect to utilize the categories of Marxist thought in discussing not only the social and historical aspects of the development of physics, but also conceptual evolution within various branches of physics themselves. Indeed, it is precisely in his early refusal to see a sharp separation between these 'external' and 'internal' moments in the evolution of the sciences that we may see a distinctive feature of

Rosenfeld's approach, which he owes to the Marxist tradition, a tradition which his own specific, detailed studies have materially developed and enriched.

This by no means implies that only the committed Marxist can benefit from the reading of those papers most directly concerned with social and historical matters. For while he approaches the issues in each case from his own viewpoint, Rosenfeld does not come armed with a bag of preconceived answers: rather, he displays a fine sense for the uniqueness of the historical event, the role of individual traits and how they blend with, or rub against, the grain of the historical development. We believe readers will time and again find themselves charmed by the concise characterizations of persons and situations, as well as informed about the way they fit into a larger picture of scientific and social development through the clash of individual viewpoints. The Marxist scholar will in addition find numerous models for the tactful and sympathetic application of his method to concrete situations. He will also find some tart comments on what Rosenfeld found to be dogmatic or vulgar Marxist misapplications, notably in the review of Bernal.

IV

Rosenfeld is probably best known to the public of physics for his work on the foundations of quantum theory, especially for his critical development and defense of Bohr's viewpoint. Without going into any discussion of the issues (on which Rosenfeld speaks for himself eloquently enough throughout this volume), it may be helpful to recall that this viewpoint has been, and to some extent still is, under attack by some orthodox Marxists and others as an idealist, or positivist distortion; while at the same time it has been 'defended', under the umbrella labelled the 'Copenhagen' interpretation, by a wide variety of champions who clearly were in fact philosophical idealists. Rosenfeld engaged in a running battle on both fronts. First, he tries to show that Bohr's viewpoint was basically materialist, and that these attacks from the 'left' were being launched from the standpoint of a dogmatically simple or mechanical 'materialism' – which is no true materialism. Second, he argued that views expressed by such 'defenders' of Copenhagen as Heisenberg were really attempts to saddle quantum mechanics with a Platonic ideology. Obviously the debates about the philosophical significance of quantum theory are not over; but those who want to understand some of the genuine issues in these debates must reckon with Rosenfeld's views.

V

Rosenfeld was not, and did not claim to be a philosopher, nor was he a historian. However, he undertook investigations into the philosophical and historical aspects of physics because of two deep motivations. First, he was convinced that a rigorous understanding of the development of ideas would profoundly clarify his own understanding of physics, and that of his contemporaries. He never went so far as to claim that such historical and philosophical studies would *solve* current problems – only *clarify*, which was precious enough in view of the perplexing obscurity of the crises in 20th-century physics.

Second, Rosenfeld's appreciation of the wider social position of scientists, and of their labors, was a component of his general outlook upon society and its history. Hence it was the most natural thing for him to seek understanding of the relationship between scientific ideas and social forces, social problems, social opportunities. It was equally natural for him to apply his mind to contemporary social questions (especially where physics was relevant) as well as to the historical interactions of society and science in classical and post-renaissance times. These interests in the social or 'external' history of physics did not occupy him more than episodically, in no way as extensive a pre-occupation as for some other creative scientists of this century who had similar social sympathies, such as Haldane, Bernal, Needham, and others. Nevertheless, while his research and writing on these social factors of physics was modest, his public efforts to stimulate such interests and concerns among several generations of students and colleagues were far-reaching indeed.

Rosenfeld's work in the history of physics was akin to that combination of historical investigation and logical-conceptual analysis known to us from such giants of the history and logic of physics as Mach and Duhem. Unhappily he produced no treatise in the history of modern physics, and he did not follow through after his dozen superb pioneering papers in the history of physics. Among these we may particularly note:

(1) The Genesis of the Laws of Thermodynamics (1941)
(2) The Velocity of Light and the Evolution of Electrodynamics (1956)
(3) The Evolution of Oersted's Scientific Concepts (1970)
(4) The First Phase in the Evolution of the Quantum Theory (1936)
(5) Max Planck and the Statistical Definition of Entropy (1959)

In addition Rosenfeld continued throughout his scientific life to record and interpret his own historical ambiance, which meant, of course, the history of Niels Bohr, in paper after paper of sometimes repetitive but always thoughtful and provocative writing.

So much for history. As a philosopher of science, Rosenfeld especially was concerned with the classical issues of epistemology. Of first importance perhaps were his joint papers with Bohr on measurement, including the classic paper of 1933. But he went on to explore the central notions, particularly those in quantum physics, and here we would select a noteworthy group:

(1) The Evolution of the Idea of Causality (1942)
(2) Strife about Complementarity (1953)
(3) The Measuring Process in Quantum Mechanics (1965)
(4) On the Foundations of Statistical Thermodynamics (1955)
(5) Questions of Irreversibility and Ergodicity (1960)
(6) Some Reflections on Knowledge (1971)

His last research, in collaboration with Ilya Prigogine and his group, was devoted to elucidating the significance of quantum mechanics for understanding some epistemological features of quantum theory.

At no point was Rosenfeld singled out for intensive study by professional philosophers, nor even by philosophers of science; and the same can be said for his reception by the historians of science. He was seen by both groups mainly as a profound physicist rather than a colleague in their own fields, and to some extent as a profound dilettante. We are not sure he would have minded this description, but we feel sure that his work in the history and philosophy of physics, incomplete though it was, was of first-rate quality. If he had been working with a critical group of logicians and philosophers of science, he might have been stimulated to go further, but it would also be quite difficult to name many other scientists or philosophers who were competent to match him in these matters. The notable exception is Piaget, whose modes of thought and critical exchange were acknowledged by Rosenfeld. At any rate, we believe that Rosenfeld's selected papers in this volume, show a neglected major philosophical historian of modern physics.

VI

In conclusion, we express our hope that the reader will share with us that sense of high spirits which pervaded Rosenfeld's discussion of even the most abstruse topics; one sensed in conversations with him, as in so many of his written works, that he knew that the highest form of human engagement in serious issues must go with a joyfulness of spirit, a good humor in the deepest meaning of these words. It is in this way that we shall remember him.

Boston University Center for the History and Philosophy of Science
October 1977

ROBERT S. COHEN
JOHN J. STACHEL

NOTES

We should like gratefully to acknowledge the help of Mrs. Yvonne Rosenfeld in the preparation of this volume, of Professor S. Rozental in contributing the Introduction, of Professor I. Prigogine for permission to use certain of his papers, and of our good friend G. E. Brown for allowing us to adapt the bibliography of Rosenfeld's writings which he had prepared for *Nuclear Physics*, and some paragraphs from his thoughtful obituary of Rosenfeld.

We are especially pleased that this book contains the classic paper of 1933 on measurement written by Bohr and Rosenfeld (in an accurate translation by Aage Petersen with some emendations by us).

ACKNOWLEDGMENTS

The editors and the publisher are grateful to the following persons and institutions for permission to reprint papers included in this volume:

'On the Method of History of Science', from *Archives Intern. d'Histoire des Sciences* **1** (1947) 1–6, by permission of Académie Internationale d'Histoire des Sciences.

'J.D. Bernal, *Science in History*' (book review), from *Centaurus* **4** (1956) 285–296, by permission of the Board of Editors of *Centaurus*.

'Le problème logique de la définition des nombres irrationnels', from *Isis* **9** (1927) 345–358, by permission of Dr. Robert P. Multhauf, Editor, *Isis*.

'Rationalism in Antiquity', from *Bull. Brit. Soc. Hist. Sc.* **1** (1954) 255–257, by permission of *The British Journal for the History of Science*.

'Atombegrebets forvandlinger gennem tiderne', from *Fysisk Tidsskrift* **67**, nos. 2–3 (1969) 78–91, by permission of N.O. Lassen, Editor, *Fysisk Tidsskrift*.

'Flicker in the Darkness' (book review of *Nicole Oresme and the Medieval Geometry of Qualities and Motions*, ed. M. Clagett), from *Nature* **222** (1969) 197, by permission of *Nature*.

'Marcus Marcis Untersuchungen über das Prisma und ihr Verhaltnis zu Newtons Farbentheorie', from *Isis* **17** (1932) 325–330, by permission of Dr. Robert P. Multhauf, Editor, *Isis*.

'Descartes i Uppsala, Striderna om "nya filosofien" 1663–1689, Rolf Lindborg', from *Isis* **58** (1967) 136, by permission of Dr. Robert P. Multhauf, Editor, *Isis*.

'Newton and the Law of Gravitation', from *Arch. Hist. Exact Sci.* **2** (1965) 365–386, by permission of Prof. C. Truesdell and of Springer-Verlag, Heidelberg.

'Newton's Views on Aether and Gravitation', from *Arch. Hist. Exact Sci.* **6** (1969) 29–37, by permission of Prof. C. Truesdell and of Springer-Verlag, Heidelberg.

'La genèse des principes de la thermodynamique', from *Bull. Soc. Roy.*

Sc. Liege **10** (1941) 199–212, by permission of the Societé Royale des Sciences de Liège.

'Joule's Scientific Outlook', from *Bull. Brit. Soc. Hist. Sc.* **1** (1952) 169–176, by permission of *The British Journal for the History of Science*.

'An Analysis of Joule's Experiments on the Expansion of Air', from *Centaurus* **4** (1956) 311–318, by permission of the Board of Editors of *Centaurus*.

'The Velocity of Light and the Evolution of Electrodynamics', from *Suppl. Nuovo Cim.* **4** (1956) 1630–1669, by permission of the Societa Italiana di Fisica.

'L'évolution des conceptions scientifiques d'Ørsted', from *Overs. Dan. Vid. Selsk.* (1970–71) 1–17, by permission of The Royal Danish Academy of Sciences and Letters.

'La première phase de l'évolution de la theorie des quanta', from *Osiris* **2** (1936) 149–196, by permission of Uitgeverij 'De Tempel'.

'Max Planck et la definition statistique de l'entropie', from *Max-Planck-Festschrift 1958* (1959) 203–211, by permission of Deutscher Verlag der Wissenschaften.

'Matter and Force after Fifty Years of Quantum Theory', from *Physics in the Sixties*, S.K. Runcorn, Editor, (1963), pp. 1–22, by permission of Oliver and Boyd.

'Men and Ideas in the History of Atomic Theory', from *Arch. Hist. Exact Sci.* **7** (1971) 69–90, by permission of Prof. C. Truesdell and of Springer-Verlag, Heidelberg.

'Quantum Theory in 1929: Recollections from the First Copenhagen Conference', from Rhodos (1971), by permission of Niels Blaedel, Rhodos (Copenhagen).

Niels Bohr, An Essay Dedicated to Him on His Sixtieth Birthday 1945 (1945; 2nd. 1961), by permission of North-Holland Publishing Company.

'The Conception of the Meson Field: Some Reminiscences and Epistemological Comments', from *Suppl. Progr. Theor. Phys.* **41** (1968) pp. C1–C7, 257–267, by permission of Prof. Hideki Yukawa, Editor, *Progress of Theoretical Physics*.

'Nuclear Reminiscences', from *Cosmology, Fusion and Other Matters: George Gamow Memorial Volume*, Frederick Reines, Editor, (1972), pp. 289–299, by permission of the Colorado Associated Universities

Press.

'Celestial and Terrestrial Physics in Historical Perspective', from *The Application of Modern Physics to the Earth and Planetary Interiors,* S.K. Runcorn, Editor, (1969), pp. 1–6, by permission of John Wiley & Sons Ltd.

'Zur Frage der Messbarkeit der elektromagnetischen Feldgrössen', from *Mat.-fys. Medd. Dan. Vid. Selsk.* **12**, no. 8 (1933), by permission of the Royal Danish Academy of Sciences and Letters.

'Field and Charge Measurements in Quantum Electrodynamics', from *Phys. Rev.* **78** (1950) 794–798, by permission of The American Physical Society.

'On Quantum Electrodynamics', from *Niels Bohr and the Development of Physics* (1955), pp. 70–95, by permission of Pergamon Press Ltd.

'On Quantization of Fields', from *Nucl. Phys.* **40** (1963) 353–356, by permission of the North-Holland Publishing Company.

'L'évolution de l'idée de causalité', from *Mem. Soc. Roy. Sc. Liège* **6** (1942) 59–87, by permission of the Societé Royale des Sciences de Liège.

'Strife about Complementarity', from *Science Progress* **163** (1953) 393–410, by permission of Blackwell Scientific Publications Ltd.

'Complementarity and Statistics, I–II', from *Det Kgl. Norske Vid. Selsk. Fohr.* **31**, nos. 9–10 (1958), by permission of The Norwegian Academy of Science and Letters.

'Misunderstandings about the Foundations of Quantum Theory', from *Observation and Interpretation*, Colston Paper No. 9, S. Körner, Editor, (1957), pp. 41–61, by permission of Butterworths, London.

'Foundations of Quantum Theory and Complementarity', from *Nature* **190** (1961) 384–388, by permission of *Nature*.

'Le conflit épistémologique entre Einstein et Bohr', from *Revue de métaphysique et de morale* **67** (1962) 147–151, by permission of Paul Ricoeur, Editor, and *Revue de métaphysique et de morale*.

'The Epistemological Conflict between Einstein and Bohr', from *Zeit. für Physik* **171** (1963) 242–245, by permission of *Zeitschrift für Physik*, Max-Planck Institut für Kernphysik.

'Niels Bohr's Contribution to Epistemology' from *Physics Today* **16** (October 1963), © American Institute of Physics.

'The Measuring Process in Quantum Mechanics', from *Suppl. Prog. Theor. Phys., Commemoration Issue for the 30th Anniversary of the*

Meson Theory by Dr. H. Yukawa (1965) 222–231, by permission of Prof. Hideki Yukawa, Editor, *Progress of Theoretical Physics.*

'Statistical Causality in Atomic Theory', from *History of the Interaction Between Science and Philosophy*, Yehuda Elkana, Editor, (The Van Leer Jerusalem Series, Humanities Press, N.Y., 1974), by permission of The Van Leer Jerusalem Foundation.

'General Introduction to Irreversibility', from *Irreversibility in the Many-Body Problem*, J. Biel and J. Rae, Editors, (1972), by permission of Plenum Publishing Corporation.

'The Macroscopic Level of Quantum Mechanics', from *Nature* **240** (1972) 25–27, by permission of *Nature* and Prof. I. Prigogine.

'Quantentheorie und Gravitation', from *Entstehung, Entwicklung und Perspektiven der Einsteinschen Gravitationstheorie* (Akademie-Verlag, Berlin, 1966), by permission of Prof.-Dr. Hans-Jürgen Treder, Akademie der Wissenschaften der DDR.

'Questions of Method in the Consistency Problem of Quantum Mechanics', from *Nucl. Phys.* **A108** (1968) 241–244, by permission of the North-Holland Publishing Company.

'The Method of Physics', a report prepared for UNESCO and reproduced with its permission. © UNESCO 1968.

'Quelques réflections sur la connaissance', from *Mem. Soc. Roy. Sc. Liège*, 6me serie, tome I, fasc. 2 (*Volume Jubilaire Louis d'Or*) (1971), pp. 187–190, by permission of the Societé Royale des Sciences de Liège.

'Erkendelsesteori pa naturvidenskabeligt grundlag', from *Physica Norvegica* **5** (1971) 319–326, by permission of The Norwegian Academy of Science and Letters.

'Condillac's Influence on French Scientific Thought', from *The Triumph of Culture: 18th Century Perspectives*, P. Fritz and D. Williams, Editors, (1972), pp. 157–168, by permission of A. M. Hakkert Ltd.

'Unphilosophical Considerations on Causality in Physics', reprinted from *Perspectives in Quantum Theory: Essays in Honor of Alfred Landé*, edited by Wolfgang Yourgrau and van der Merwe (1971, pp. 219–235), by permission of the MIT Press, Cambridge, Mass.

'Berkeley redivivus' (book review of *Natural Law and the Structure of Matter* by W. Heisenberg), from *Nature* **228** (1970) 479, by permission of *Nature*.

'Sur le tenseur d'impulsion-énergie', from *Mem. Acad. Roy. de Belgique*

18 no. 6 (1940) 1–30, by permission of the Academie Royale des Sciences, des Lettres et des Beaux-Arts de Belgique.

'Sur la définition du spin d'un champ de rayonnement', from *Bull. Acad. Roy. de Belgique* **28** (1942) 562–568, by permission of the Academie Royale des Sciences, des Lettres et des Beaux-Arts de Belgique.

'Sur le comportement d'un ensemble canonique lor d'une transformation adiabatique', from *Proc. Nederl. Akad. van Wetenschappen* **45** (1942) 970–972, by permission of Koninklijke Nederlandse Akademie van Wetenschappen.

'On the Isolated and Adiabatic Susceptibilities', from *Physica* **27** (1961) 67–78, by permission of the North-Holland Publishing Company.

'On the Foundations of Statistical Thermodynamics', from *Acta Physica Polonica* **14** (1955) 3–39, by permission of the Institute of Physics, Polish Academy of Science.

'Questions of Irreversibility and Ergodicity', from *Ergodic Theories*, P. Caldirola, Editor, (*Rendiconti della Scuola Intern. d. Fisica "Enrico Fermi"*, corso 14, Varenna 1960) (Zanichelli, 1962), pp. 1–20, by permission of the Societa Italiana di Fisica.

'Dynamical Theory of Nuclear Resonances' and 'Coupling between Compound and Single-Particle Resonances', from *Spectroscopic and Group Theoretical Methods in Physics, Racah memorial volume* (1968), pp. 203–230, by permission of the North-Holland Publishing Company.

'The Structure of Quantum Theory', from *Revue de Synthese* **3**, nos. 49–52 (1968), by permission of Madame Suzanne Delorme, Centre International de Synthese.

'Aspects techniques et sociaux du développement des organismes européens de recherche scientifique', from *Le Monde Scientifique* **14** (1970) 13–15, by permission of the World Federation of Scientific Workers.

'Sociale og individuelle aspelter af naturvidenskabens udvikling', from *Fysisk Tidsskrift* **69**, nos. 4–5 (1971) 97–106, by permission of N. O. Lassen, Editor, *Fysisk Tidsskrift*.

Special Note: The illustration which appears in 'Irreversibility – A lay Sermon', entitled 'Cast with Proposed Restoration of Three-Sided Relief', a large marble sculpture, from ca. 470–460 B.C. (original piece), H. L. Pierce Fund, 08.205, was reproduced by courtesy of the Museum of Fine Arts, Boston.

PUBLISHER'S NOTE

The reader should note that due to the fact that the articles by Professor Léon Rosenfeld came from various publishers and sources with different styles in denoting symbols, especially with regard to the differential symbol 'd' and to symbols representing vectors, we had to make a choice in order to be consistent throughout the present volume.

For this reason we have used the following notation: the differential symbol 'd' appears in roman upright type; those symbols representing vectors and indicated by letters with arrows over them in some of the papers from which we set, have been substituted by bold roman or italic type.

TABLE OF CONTENTS

PART I / HISTORY OF SCIENCE

PART II / EPISTEMOLOGY

PART III / THEORETICAL PHYSICS

PART IV / SOCIAL RELATIONS OF SCIENCE

INTRODUCTION

The decision to publish this volume was taken nearly two years ago. Rosenfeld himself took part in the preparations, and he planned to write an introduction which would of course have given an authoritative analysis of the articles and their interconnections. The papers he chose for this volume form only a small part of Rosenfeld's rich and varied scientific production, but the choice he made can yield an insight into his way of thinking by showing what he regarded to be really important and representative of his views.

His extensive training in the humanities enabled Rosenfeld to survey many sides of human life and to trace interconnections between various fields of knowledge. Having mastered a number of languages, he could follow developments in many countries. In this respect his ramified international relations also played an important role.

As a student Rosenfeld had already published several articles on mathematical and physical topics in a students' journal in Liège, but his scientific activity began in earnest in the late twenties. It was an exciting time, this 'heroic period' of physics when the newly developed methods of quantum physics, were being tested to their limits. Nearly every issue of the main physical journals – especially the German *Zeitschrift für Physik* – brought theoretical contributions dealing with properties of atomic structure and proving the power of the novel theoretical tools. The rapid progress also gave a beautiful example of international co-operation on a broad scale, and Rosenfeld, at that time on studies abroad, in Göttingen and in Zurich, took a lively part in it. While some of his papers from that time deal with important problems, none of them are included into Part III (Theoretical Physics) where they could have served as examples of his participation in the early stages of the establishment of a consistent system of quantum physics. Perhaps he regarded these contributions rather as an introduction to what he felt to be most important in his life work: the study of the very foundations of the new field of physics. But even his later work on physical problems is represented only to a small extent. As a matter of fact, Rosenfeld had, through the years, made important contributions to nearly every field of modern physics. This can be seen from the full list of publications [See Bibliography, p. 911 – Ed.]. Here mainly such papers are included which touch upon foundations of physics. In particular, Rosenfeld's special interest in statistical physics and thermodynamics is connected with the role which statistical considerations play in quantum theory.

It was at the end of the twenties that Rosenfeld became attached to the circle around Niels Bohr in Copenhagen. Bohr had just formulated his idea of complementarity, and problems of the epistemological implications of quantum theory were then very much in the foreground of interest. Even when he held permanent positions in Belgium, the Netherlands, and Great Britain, and visiting professorships in other countries, Rosenfeld continued to keep up his close connection with the Copenhagen group, until in 1958 he moved to Copenhagen to take a position at Nordita where he was active until his death in March 1974.

The papers presented in Part II (Epistemology) occupy a most central position in Rosenfeld's scientific work. When Rosenfeld arrived in Copenhagen for the first time the discussion on the foundations of quantum theory, a discussion initiated by Bohr, had just started. Rosenfeld became fascinated by the perspectives opened by the new approach. He was interested in many aspects of physical theory, but here he was confronted with a way of presenting problems which appealed to his searching mind. Complementarity made it possible to penetrate much deeper into the fundamental problems of physical knowledge and threw light on the general situation in which a scientist finds himself when trying to bring experimental evidence into a consistent logical scheme. As is well known, the starting point of this development was due to the discovery of the quantum of action which specifies a lower limit to all interactions between physical objects. The apparent contradiction exhibited by the necessity to describe microscopic objects in some cases as particles and in other cases as waves finds its resolution in the fact that the experimental arrangements used in the one case are mutually exclusive, or, in Bohr's terminology, complementary, to those used in the other case. This complementary relationship between those situations for which space-time coordination, characteristic of the particle picture, is envisaged and those situations where we aim at establishing an energy-momentum balance, is essential for Bohr's reasoning.

As a consequence of this analysis one has to replace the deterministic laws, typically used in classical physics, by statistical laws. This forced departure from a way of description of physical phenomena followed by scientists for many centuries gave rise to many, at times violent discussions among physicists, echoes of which still can be heard. Rosenfeld took a leading part in these discussions. They are to be observed in many papers from his hand, some of them reprinted in Part II. It is characteristic of Rosenfeld that the chronologically most early paper on physical problems in the present volume (the still earlier articles I.3 and I.7, from 1927 and 1932 respectively, deal with the history of mathematics and classical optics) is placed as the first one in Section II. It is the classical paper on the measurement of electric and magnetic fields, written together with Bohr in 1933, one of the most fundamental contributions to the discussions of those years.

Most of the other articles in this section contain more general presentations of their subject. Great effort is made and much skill displayed to explain the situation from many angles in order to convince those still reluctant that complementarity is only a logical consequence of experimental experiences obtained in atomic physics. In the first years, some physicists could not get out of "the metaphysical jungle which obscures the view". They could not reconcile themselves with the prospect of abandoning the accustomed idea of causality and replacing it by the more general method of statistical causality. In this connection the analysis of thought experiments (*Gedankenexperimente*), often sophisticated and ingeniously conceived, played an important role. The most famous example is the discussion between Bohr and Einstein which lasted for a quarter of a century (cf. II.10 and II.11). As can be seen from the articles, but perhaps even more from his correspondence with authors of papers sent to him, Rosenfeld took painstaking care in trying to disclose and discuss in detail basic mistakes and misunderstandings in the attempts by various authors to find contradictions in complementarity. His criticism could be slashing and expressed in an extremely sharp way if the author's reasoning was based on some vague philosophical ideas. However, if the manuscript – especially when written by a young scientist – contained a fair presentation and original ideas, Rosenfeld could go far in expressing his appreciation and encouragement and in assisting in the publication of the paper.

It was Bohr who on many occasions pointed to the far-reaching validity of complementarity, extending beyond the domain of quantum theory. Rosenfeld followed this thought in several directions. Thus, in later years he revived his active interest in thermodynamics. He took the view that in this field, although it belongs to classical physics, a statistical approach is unavoidable and that a kind of complementary relationship can be established. This point was taken up in a work in collaboration with I. Prigogine (II.14).

Also disciplines which seem to lie far away from physics came under the searchlight. An experiment is – to use Bohr's expression – something by which we can tell others what we have done and what we have learned. This emphasizes of course the question of the technical side of the experiment but also the importance of language as a tool for communication both on daily life level and in exchange of scientific information (see e.g. II.18). Therefore, a study of the function of language is essential.

Going further in the same direction we are presented with the problem of how the human mind works and this is again highly dependent on how it develops from its very beginning in infancy. It was no accident that Rosenfeld became so passionately interested in the work of Piaget whose research into the evolution of children's thinking through the various periods of their growth led to a reliable picture of changes in coordination of impressions from the outside world. Piaget's work filled a gap, confirming what Helmholtz and Mach anticipated. Shortly before his last illness, Rosenfeld gave an inspiring course of lectures on Piaget's results. The importance he ascribed to Piaget's work can be seen from the fact that when he was entrusted with the organization of a symposium to commemorate the 50th anniversary of the Niels Bohr Institute in Copenhagen he chose as its subject 'Statistical Causality'. In the introduction to the program of the symposium he emphasized that "in order to obtain an all-round survey of the position of statistical causality, it was necessary to include a consideration of the psychological side of the problem". A substantial part of the time was devoted to reports by Piaget and other psychologists working on similar lines, and a fruitful and animated discussion developed.

Rosenfeld's immediate acceptance of the idea of complementarity was no doubt mainly due to his strict and logical way of thinking. Besides, he had for many years a close relation with Niels Bohr and in the day-to-day working together and discussing, Rosenfeld could not but find stimulus for his own lines of thinking. Now, Bohr had a peculiar method of writing papers. Every attempt to formulate sentences, every discussion in order to find the right word or expression, was used to clarify his own ideas. No criticism, even the most superficial, was disregarded; on the contrary it gave Bohr a welcome opportunity to improve the argumentation. This procedure, even if it often became rather time-consuming, must have appealed very much to Rosenfeld's analytical mind. His own manuscripts were always the result of a prolonged and determined work at the improvement of the text.

The way in which the idea of complementarity developed from the recent advances in physics was also in harmony with Rosenfeld's conception of the structure of science. In many of his articles he advocates the materialist view of the development of science. He is materialist in the sense that he regards experience and experiment as the only source of knowledge. All theories based on 'self-evident axioms' or *a priori* ideas are of no use and only apt to obscure things, to "put physics upside down". What matters is only and solely what we can learn from nature by asking it the right questions. All the time we have to be prepared for the possibility that new empirical evidence can force us to revise and even to abandon accustomed ideas.

In the dialectical approach which is the *leitmotif* of the papers presented here, the development of science can, in short, be described as follows. When a certain amount of empirical information is collected, scientists try to create a conceptual framework which allows us to comprehend the known phenomena and predict the results of further investigations. The usefulness of the theory depends entirely on agreement of its predictions with experimental findings. Classical mechanics is an example of such a beautiful and consistent theory. The concepts developed for a certain domain can still be useful when our knowledge grows, perhaps supplemented by some new concepts. But it can also happen that in this development we are brought to the limit of the applicability of the old concepts. For the field lying outside this limit a new conceptual frame has to be conceived which is an antithesis of the old one. What one is aiming at is to create a synthesis, a broader, more general theory in which the old theory will be a special case for a limited domain of application. Examples of such a situation can, in our century, be found in the relativity theory and in quantum physics. In the latter, description by means of deterministic laws had to be replaced by the more general notion of statistical causality. The deterministic description can however still be used in the limit where the quantum of action can be neglected, that is generally in macroscopic physics. There is nothing extraordinary or alarming in such a step. It has however given rise to so many discussions because determinism has been regarded by generations of scientists as a firm and *a priori* inherent basis of all scientific thinking. Rosenfeld's view was, as we have briefly sketched, diametrically opposite. The application of any method in science is justified only as long as it is adapted to the empirical situation. A detailed account of the whole problem is to be found for instance in II.6.

A major part of the present volume, Part I, is devoted to the history of science. One might think that occupation with this subject was for Rosenfeld a kind of a hobby. In the light of what was said earlier it becomes clear however that it is not so. Historical research was for him a serious matter, a necessary consequence of his interest in epistemological problems. It is from the lesson derived from studies of the past that the present can better be understood. When Rosenfeld, a few years ago, gave an extensive course of lectures in Copenhagen on the foundations of physics it turned out to be a harmonious synthesis of a thorough analysis of the present situation in physics and a broad outlook upon various periods in the history of science and philosophy. The lectures were recorded and the transcript was intended to be used as outline materials for a major publication. It is most deplorable that Rosenfeld, who was one of the very few people with the necessary broad knowledge, was not given the time to write the planned book.

Obviously, the borderline between Parts I and II cannot be very sharp. In II one finds extensive use of historical evidence, while in I many of the articles are more or less connected with the problems discussed in II. One important point should especially be noticed. As a natural consequence of his materialist view, Rosenfeld stresses the connection between scientific progress and social conditions (cf. I.1 and 2). He is aware of the fact that the studies in this field are only in their beginnings, but he did his best to encourage them. Actually, to underline the importance of the problem, Part IV, albeit rather short, is introduced as a separate heading: 'Social relations of science'. International co-operation in science should also be mentioned in this context. Rosenfeld not only expressed theoretical views of the paramount importance of such co-operation, but he also tried, in many ways, to implement his views in practice. He served as member of a number of international scientific bodies. He was the founder and editor of a journal which under his guidance became one of the leading journals in its field and brought contributions from all over the world. He also took part in meetings whose purpose was to bridge over difficulties in relations between people from countries with different cultural and political backgrounds, and he never gave up even when the difficulties seemed to be insurmountable.

In the eyes of many people, a materialistic conception of history is unavoidably connected with a certain political attitude and above all with an uniformity of thought which does not allow for divergent opinions. Nothing could be more alien to Rosenfeld. In science, he was devoted to a complete openness to new experience and new ideas, and in general human relations he fought against prejudice and dogmatism. His materialist conception has – as already stressed – to be understood as the belief that a scientific theory can only be based on what we learn from experience. He attacks severely the adherents of the politically coloured 'dialectical materialism' (his quotation marks) which he calls a metaphysical system (I.2, p. 12). It is hardly possible to state his position in sharper terms than he does himself in speaking of a dogmatic attitude, a "contemporary scholasticism which makes its adherents blind to the progress of scientific thought and misleads them into sterile opposition to it" (I.2, p. 13). Complementarity allowed Rosenfeld to realize that mutually excluding approaches to cope with a given situation may both be necessary to ensure a complete description (cf. I.23, p. 329) where he says: "Philosophers will endlessly dispute whether this (duality of matter and force) is a victory for materialism or for idealism. I am afraid I can only say, in the spirit of complementarity, that is neither the one nor the other". The same point is made in a witty travesty of Mark Twain's lines introducing the article on method on physics (II.17).

Niels Bohr used to quote as an example the complementary relationship between seriousness and humour. The more stress is put on the one of them, the more the other disappears, but both are necessary to provide a proper balance in human life. Rosenfeld lived up to this balanced attitude and made many contributions to the humourous picture of physics. He did not refrain – as can be seen in several articles in this volume – from putting witty remarks into a serious scientific text. He also took active part in the organization of joking entertainment in connection with conferences and congresses. They always had some scientific point with a crazy twist. One of the last contributions from his side was the amusing sketch played after the closing dinner at a relativity conference in Copenhagen. It was also meant to celebrate J. A. Wheeler's 60th birthday and contained subtle digs at a number of participants. Rosenfeld disliked, to say the least, those philosophers who created theories at their writing desks with no regard to the information gained by the study of nature (cf. the quotation in II.7 from Morgenstern, a German writer of grotesque poetry, very popular in the 1930's among the physicists in Copenhagen). Their products, he says, are more metaphysics than science. But if real philosophy is defined as an attempt to find common features and laws in the wide variety of human knowledge, without prejudices and above all without preconceived opinions, then Rosenfeld may rightly be called a philosopher *par excellence*.

It would therefore seem appropriate to end this introductory note with a quotation from one of Rosenfeld's articles (See Bibliography [1953e]):

> (It) is a vital requirement for fruitful scientific enquiry that the truth can only impose itself by its own strength and that the greatest disservice which can be done to it is to try to force evidence into any rigid system.

STEFAN ROZENTAL

Niels Bohr Institute
University of Copenhagen
1974

MY INITIATION*

[1945f]

The first message I got from Bohr was a telegram, announcing that the Easter Conference† was to be postponed two days. I was then – in 1929 – in Göttingen and, together with Heitler, had expressed the wish to attend that famous Conference; we had both received from Klein a favourable answer, to which the aforementioned telegram brought the master's eleventh-hour correction. When we arrived in Copenhagen, Bohr informed us of the reason for the postponement: he had to complete ('with Klein's help') a Danish translation of some of his recent papers to be published as a Festskrift of Copenhagen University; he told us all about this venerable Festskrift tradition and added: 'It would have been a catastrophe if that work had not been ready in time!' This struck me as a hyperbolic way of stating the matter. How little I imagined at that moment the tragedy hidden behind this seemingly innocuous procedure of putting the finishing touch to a paper! How little I knew that it was my destiny to play a part in a whole lot of such tragedies!

My sole excuse for the failure to grasp the earnestness of this paper-writing subject is that I was by no means an exception in that respect. In fact, as experience taught me since, people are on the whole distressingly unimaginative on that point. Take, for instance, the case of the Faraday Lecture. Bohr arrived in London for the Faraday celebration with a manuscript of his lecture, which he described as 'practically finished.' There were just a few pages lacking. The plan was to seek the required isolation in the romantic environment of some old English inn, and in a week's time, 'with Rosenfeld's kind help' (he explained to Mr Carr, the secretary of the Chemical Society), the thing would be definitively disposed of. Mr Carr was delighted. After a week's hard labour in a rather crowded and thoroughly unromantic hotel, in which we had to wage a regular war of nerves against an irascible schoolmistress for the possession of the parlour, the ten odd lacking pages had actually been written. We had furthermore gained the insight that a great improvement could be obtained by the mere addition of some twenty more pages. Bohr quite warmed up at this idea, which (he persuaded me) brought us really a good deal nearer to the end. I was accordingly dispatched to Mr Carr to report on the new development. Well, Mr Carr did not at all cheer the prospect; he was just annoyed; he even made no effort to conceal his annoyance. When I alluded to our having worked the whole week without respite, I am sorry to say he looked decidedly incredulous. I was quite downhearted when I left him. Fortunately, I had just then an appointment with Delbrück, whom I found in company of one of his innumerable lady-friends. *He* was a man of feeling and understanding; he comforted me like the true friend he was.

But to return to the scene on the platform in Copenhagen station. What impressed me most about Bohr at this first meeting, was the benevolence radiating from his whole being. There was a paternal air about him, which was enhanced by the presence of some of his sons. These sons of Bohr's were a great matter of speculation to me. When I again saw Bohr at the Institute the next morning, there were also a few sons around him. Different ones, I thought; he must have a host of them. On the afternoon of the same day, however, I was bewildered at the sight of still another son at his side. He seemed to stamp them from the ground or draw them forth from his sleeve, like a conjurer. At length, however, I learned to distinguish the sons from one another and I found out that there was only a finite number of them after all.

I don't know how the Athenian delegates for oracle consultation felt on their return from Delos. But I imagine their feelings must have been akin to mine after I had listened to Bohr's introductory lecture at the Conference. He had begun with a few general considerations calculated, no doubt, to convey to the audience that peculiar sensation of having the ground suddenly removed from under their feet, which is so effective in promoting receptiveness for complementary thinking. This preliminary result being readily achieved, he had eagerly hastened to his main subject and stunned us all (except Pauli) with the non-observability of the electron spin. I spent the afternoon with Heitler pondering on the scanty fragments of the hidden wisdom which we had been able to jot down in our note books. Towards the evening we felt the need for some fortification and proceeded to the Strøg.

The following evening we spent at the cinema, together with some others. Picture theatres have always been institutions of high educational value to young theoretical physicists. So it turned out this time too. There it was that Casimir started his important calculation of the magnetic field exerted by an atomic Dirac electron on the nucleus of the atom. He had to work under very trying circumstances. For as soon as any part of the show started, the lights went out, and poor Casimir had to wait until the lovers had safely got over their troubles and married and all before he could resume his calculations. He did not lose a second either: every time the lights went up, they invariably disclosed our friend bent over odd bits of paper and feverishly filling them with intricate formulae. The way in which he made the best of a desperate situation was truly admirable. It was inspiring to watch him.

On the last day of the Conference I experienced the climax of my Copenhagen initiation. It came about rather unexpectedly in the following way. At the meeting that morning one of the most distinguished guests had developed some views about the vexed question of the 'cut' between system and observer, which seemed to me rather erroneous. Bohr, however, had only opposed them feebly (as I thought); in his rather confused speech, the phrase 'very interesting' recurred insistently; and finally, turning to the speaker, he had concluded by expressing the conviction that 'we agree much more than you think.' I was much worried by this extraordinary attitude, the more so as the highbrow bench seemed to find it all right. I therefore ventured to explain my doubts straight away to Bohr. I began cautiously to state that the speaker's argument did not seem to me quite justified. 'Oh,' said Bohr quickly, 'it is pure nonsense!' So I knew I had been led astray by a mere matter of terminology.

But now the unexpected happened. Bohr summoned me to a little room, in the middle of which stood a rather long table. He manoeuvred me towards that table and as soon as I stood leaning against it, he began to describe around it, at a rather lively pace, a Keplerian ellipse of large eccentricity, of which the place where I was standing was a focus. All the time, he was talking in a soft low voice, explaining to me the broad outlines of his philosophy. He walked with bent head and knit brows; from time to time, he looked up at me and underlined some important point by a sober gesture. As he spoke, the words and sentences which I had read before in his papers suddenly took life and became loaded with meaning. It was one of the few solemn moments that count in an existence, the revelation of a world of dazzling thought, truly an initiation.

It is generally recognized that no initiation can be properly accomplished without being combined with a painful experience of some sort. In that particular also my initiation left nothing to be desired. For since I had to strain my hearing to the utmost to catch the master's words, I was compelled to execute a continuous rotation at the same rate as that of his orbital motion. The true purpose of the ceremony, however, did not occur to me until Bohr ended by emphasizing that you can't even catch a glimpse of complementarity if you don't feel completely dizzy. When I heard that, I realized everything and I could only pay him a silent homage of thankfulness and admiration for such touching solicitude.

NOTE

* [This text was condensed from (1945f) for *A Random Walk in Science*, compiled by R. L. Weber and edited by E. Mendoza (The Institute of Physics, London and Bristol, and Crane, Russak & Co., New York, 1973) – Ed.]

† [For a detailed account by Rosenfeld of the first Copenhagen Conference in 1929 see [1971c] reprinted in this volume on p. 302 – Ed.]

PART I

HISTORY OF SCIENCE

CHAPTER 1

ON THE METHOD OF HISTORY OF SCIENCE

[1947b]

The history of science confronts us with the problem, not only of recording as accurately as possible, but also of 'explaining' the development of scientific thought and discovery, i.e. of reducing this development to some laws or principles of more or less general scope. Of course, even this simple assertion, which just means that history of science should itself be a science, is in itself questionable and is sometimes seriously questioned: some people seem to think that the evolution of scientific discoveries is largely due to 'chance'; they emphasize the fortuitous aspect of the circumstances in which individual discoverers and thinkers have grown to maturity or have been led to their discoveries. Quite apart from its barrenness, this view is, however, definitely disproved by fact. In a large number of instances, it is possible to show quite conclusively that in spite of the capricious interplay of casual events the development of a particular branch of knowledge has followed a general trend, towards which apparently unconnected individual efforts were clearly converging. Also the often-noticed phenomenon of simultaneous discovery cannot be attributed to mere chance, but calls for a closer analysis.

When we look for some general guiding principle in such a casual analysis of the historical development of science, there is a point of view which immediately offers itself and which has in fact extensively been adopted. There is in any scientific doctrine an internal consistency, a logical coherence which makes that the disclosure of a particular point necessarily orients the mind in a definite direction and leads it to gradual unfolding of the whoie structure. As an example of explanation of the historical development of science based on internal consistency, I should like to cite the masterly studies of Prof. F. Enriques on the presocratic cosmogonies: in this work, the fruitfulness of this principle of explanation is especially striking, in that it has for the first time made possible an intelligible interpretation of the obscure fragments which constitute our only material for the reconstruction of this fascinating development.

It is clear, therefore, that the internal consistency of scientific thought furnishes a most powerful instrument for a rational understanding of its historic development. But it is equally clear that this principle alone is not capable of providing an exhaustive explanation. In the first place, the ideas of a scientific theory are not arranged in a unique logical sequence; they are rather interconnected by a whole net of logical relations, so that the point of view of internal consistency by itself does not suggest a single, but many equivalent lines of possible evolution. The choice which has actually been made in the historical course of events must therefore have been determined by other, extraneous circumstances. It has even more than once happened that different lines of approach have developed simultaneously and only at a late stage merged into a connected whole. Moreover, straightforward evolution of scientific thought along some clear logical line is seldom an adequate description of the historical reality. The most conspicuous feature is the occurrence of breaks in the development; at some times a given line of thought seems to exhaust itself – in spite of still unexplored logical potentialities – and makes place for an entirely different one, which is not logically related to the first. Taking again as an example the evolution of Hellenic thought, its magnificent blossoming seems to wither rather suddenly away at the time of Socrates and the whole outlook then becomes entirely different. Just to say, as Whitehead once put it, that Greek philosophy "took a wrong turn" leaves the historical problem as to the causes underlying this transformation entirely unanswered.

The missing principle which ought to help us towards an understanding of such historical phenomena, since it is not contained in the abstract structure of the scientific theory itself, must be sought in the concrete, material substrate on which this structure necessarily rests. The example of Socrates, just alluded to leads our thoughts towards the consideration of the possible influence of human personality. History of science, or, for that matter, all history, would be moulded by the decisive intervention of great men. This romantic view is immediately refuted by the simple observation that the 'greatness' of a scientist is itself an historical phenomenon in need of interpretation; for it is well-known that the recognition of the merit of any scientific discovery requires a certain 'maturity' of contemporaneous thought.

This last remark, pointing to the social aspect of scientific activity, brings us at last on the track of the principle of explanation we are seeking.

Since this principle cannot be found either in the internal structure of scientific thought or in its spontaneous generation in the minds of individual men, it can only lie in the social function of science. In fact, in all cases in which it has been attempted, the analysis of the social implications and reactions of science has thrown a most revealing light on features of the historical development which had so far remained obscure. Once more I may quote as an illustration the history of ancient Greek science, the social relations of which have been investigated by Prof. Farrington. The comparison of this study with that of Enriques, based on purely logical criteria, strikingly demonstrates the wideness of scope and fertility of the new principle. Not only does it enable us to fill the gaps left by the insufficiency of the logical analysis, but it even leads to an essential deepening of our understanding of the logical aspect itself of the evolution.

Notwithstanding the promising outlook already offered by the study of the social relations of science in its historical development, it must be stressed that the work hitherto carried out in this direction leaves much to be desired. The number of serious investigations bearing on the problem on hand is relatively small; no useful purpose is served by uncritical attempts, embracing too vast a subject and consequently liable to unwarranted generalisations. In this as in any other scientific problem, only painstaking, detailed discussion of concrete and limited questions can yield valuable results; in particular, a more precise formulation of the laws of interaction between science and society, as well as a clearer view of the limitations of the social aspect of scientific thought as a general principle of explanation of its historical evolution, can only emerge from a prolonged and extensive study of such concrete problems.

Clearly, progress along these lines requires above all the co-ordination of all converging efforts – a co-ordination which has been sadly lacking up to now. This co-ordination is rendered especially difficult by the very nature of the problem, which necessitates the closest co-operation between those historians who recognize the decisive importance of economic and social forces in the general course of human evolution and the scientists with a sense for history who would be ready to tackle the specific problems of the history of their branch of knowledge in the same spirit. These men should in the first place be given the opportunity of coming together to discuss the broad lines of the problem and eventually to agree upon a definite policy. In my opinion, the first task would be to set up a comprehensive and critical survey of the work already extant or in progress; on the basis of this information, it would then be possible to elaborate the planning of future investigations.

BIBLIOGRAPHY

Enriques, F., *Problems of Science*, tr. by K. Royce (LaSalle, Ill., Open Court, 1943);
The Historic Development of Logic, tr. by J. Rosenthal (New York, Holt, 1929);
Signification de l'histoire de la pensée scientifique (Paris, Hermann, 1934);
La théorie de la connaissance scientifique de Kant à nos jours (Paris, Hermann, 1938).

Farrington, B., *Head and Hand in Ancient Greece: Four Studies in the Social Relations of Thought* (London, Watts & Co., 1947);
Greek Science: Its Meaning for Us (London, Penguin Books, 1953);
Science and Politics in the Ancient World (London, Allen & Unwin, 1946).

CHAPTER 2

SCIENCE IN HISTORY

(Review of J.D. Bernal's *Science in History* [Watts & Co., London, 1954. xxiv + 967 pp.])

[1956f]

I did not open this volume without great expectations. It is well-known that history of science is much hampered by the double obstacle that most scientists have no sense of history, while most historians are ignorant not only of the facts, but of the very spirit of science. Now, Prof. Bernal is one of the few eminent scientists of our time who have shown a keen interest for, and deep understanding of, the historical and social aspects of the development of science. Several shorter essays from his versatile pen are as remarkable for the competence and shrewd judgement exhibited in the analysis of the economic and social background of scientific progress as for the generous sincerity inspiring the author's exposure of social evils detrimental to this progress. He would seem, therefore, the ideal author for such a book as the present one, which aims at a synthetic exposition of the development of science in its broad historical setting and contemporary context. Only a man with Prof. Bernal's rare combination of talents, and with his courage, could be expected to attempt with any chance of success this much needed, but formidable task.

Shall I confess that I am disappointed? This seems a harsh statement, in view not only of the immense labour which has gone into the preparation of this book, but also of the many admirable pages it contains. I am not going to cavil at minor inaccuracies, inevitable in an undertaking of this scope; my chief criticism bears on the one hand on the composition of the book, on the other of Prof. Bernal's method.

This bulky volume has its origin in a course of lectures, the plan of which is still recognisable in the general disposition of the subject matter: a historical survey fills the first half of the book; this is followed by an outline of the present state of science and a discussion of its technological and social relations, in which is included a detailed analysis of sociology in its historical perspective. The author tells us how he was led, in the course of years, to elaborate upon each aspect of this vast enquiry. I am afraid he has been carried away by his enthusiasm for a fascinating subject and has not exercised sufficient firmness in deciding where to stop. At any rate, the final product is a disconcerting hybrid between the racy essay and the encyclopaedic digest, the latter element rather spoiling by its massive intrusion the effect aimed at by the former. The diligent reader is presumed to absorb in bewildering succession a treatise of archaeology, a textbook on the social history of science, a popular survey of modern physics and biology and an ABC of Marxism, in order to prepare himself for the political pamphlet on science in the time of the Cold War, in which the book culminates. No wonder if he now and then, in spite of the solid structural arrangement of the parts, loses the thread of the argument.

However, this lack of balance and unity would be readily forgiven if one felt that the accumulated material had been critically gathered with the view of guaranteeing the accuracy of the picture and providing a firm basis for rational analysis. The general impression is unfortunately quite different; in fact, one wonders whom Prof. Bernal really expects to convince by the dogmatic way in which he has chosen to approach his subject. In his shorter essays, whatever dogmatic tendencies he might have were kept in the background, much to the advantage of the argument; it would seem that in this larger work they have been brought to the fore by some unhappy effort at systematisation. However this may be, the result is the more unfortunate as the message Prof. Bernal intends to convey is a timely and important one. The main theme is familiar: the development of science is primarily conditioned by economic and social factors; history shows how its progress has been fostered by rising social classes and held up by the reactionary elements of society, and thus has its clear lesson for our own times. It is only by a painstaking elaboration of details, however, that such a general view can get substance and meaning, and prove its fruitfulness as a guiding principle. And it is just this critical point that Prof. Bernal, for all his provocative originality, has failed to make out. On almost every page one meets with statements not explicitly supported by any evidence (even though such evidence may be available), and worse still, one constantly gets the impression that the author is building up artificial constructions to suit preconceived ideas, rather than arriving at the proposed interpretation by rigorous inference.

The weakness of Prof. Bernal's method is already noticeable in the historical part of his work. This is especially to be regretted because it is just in this field that the introduction of a sound scientific method of analysis is most urgently needed and could yield extremely valuable results. The excellent section on prehistory, which could rely on Prof. Gordon Childe's fundamental work, and that on modern science (from the XVIIth to the XIXth century), of which Prof. Bernal has obviously made a deeper study, shows what he might have done in the way of a general synthesis, had he not, by sheer impatience, so often shot beside the mark.

Let us look, for instance, at his account of the history of Greek science. It is presented in its entirety, from presocratic to hellenistic times, against the background of a 'slave-holding society'. As a result, Prof. Bernal, who never shrinks from giving the picture an emotional colouring, passes condemnation over it, as the kind of emasculated science one expects from a reactionary class. To be sure, he himself feels that it is a bit hard to squeeze into this scheme Democrit, for instance, but the half-hearted restrictions which he inserts here and there only succeed in blurring the argument and depriving it from whatever superficial plausibility it might have. In point of fact, Prof. Bernal's description does not even remotely tally the evidence of either history or science: the connection between the successive phases of Greek scientific thought and the evolution of the class structure of Greek society is certainly a very real and essential one, but it is much more complex and subtle than the crude opposition of a cultured leisure class and brutish slaves can possibly suggest.

I am not here blaming the author for 'oversimplification' in the pointless way in which historians without scientific education are wont to do. History, like any other science, is essentially a process of simplification; to deny this would be to reduce it to a barren literary exercise. It will not do, however, to tamper with a complicated dialectical process in the hope that after ironing out some of its oscillations the longer waves that remain may still show it in a reasonably truthful perspective. In the case under discussion, Prof. Bernal has ironed out the class of free workers and craftsmen which it is indispensable to take into account in order to understand essential features of scientific development in the hellenistic period. Prof. Bernal has actually fallen into that very error of perspective which renders valueless most of the considerations of literary historians on 'Greek man', the 'Greek ideal' and similar fictions; he debunks Plato instead of glorifying him, but he does not seem to realise that the platonic tradition (culminating in Archimedes) is only one aspect of Greek science, and socially not the most important one, in spite of its technical excellence. He ignores the whole tradition of practical science and technology – algebra, hydraulics, astronomy – carried on by free men of humbler station, but by no means inferior ingenuity. It is true that this tradition has left little literary evidence; but the scanty information we have of such men as Ktesibios, Straton of Lampsachos, Hipparchos or Aristarchos of Samos suffices to show us that they were animated by a much more 'modern' spirit than their platonistic contemporaries.

In fact, the hellenistic times present in many respects a striking similarity with the early XVIth century. Science and technology were much on the same level at the two periods, and they were supported by very similar social structures, composed essentially of free men. Only in agriculture and mining ruthless exploitation of human labour prevailed: and there was little to choose for peasants and miners between the status of serfdom and that of slavery. The real challenge, then, for the scientific historian, is to explain why the hellenistic period was followed by a slow, but irremediable decay, whereas the XVIth century witnessed the birth of capitalism and modern science. From what has just been said it should be clear that the presence or absence of slavery cannot have been a decisive factor. The answer must rather be sought along demographic lines; the essential difference between the two periods is above all one of scale: only in the XVIth century was the 'primitive accumulation' large enough to initiate a capitalist economy. On this new economic and social basis, a convergence and mutual fructification became possible of the two scientific traditions: the theoretical, platonistic one and the practical one, which had become divorced from each other during the long period of decay of antique society. Is it not disconcerting that Prof. Bernal's philosophy, which he so insistently describes as 'Marxist' should have prompted him to by-pass such a striking illustration of the rigorous historical method advocated by Marx and Engels?

When he comes to survey the mediaeval period, the author again misses a most interesting problem, just because he is too anxious to persuade us that no good can come out of the Church. He utterly fails to do justice to a very great XIVth century Churchman, Nicole Oresme, whose thought, emerging from the barbarous darkness of the times, exhibits distinctly modern features: an attempt at quantitative description of phenomena, a discussion of mechanical problems foreshadowing

Galilei, truly Copernican speculations on the motion of the earth, and not least, a penetrating analysis of the nature and role of money (which Prof. Bernal does not even mention). Surely, this is a phenomenon which it is futile to ignore or belittle, but which rather demands a careful analysis. The XIVth century in Western Europe is a period of economic expansion and social evolution which shows an unmistakable trend towards capitalism; it is, so to speak, a prefiguration of the movement which was to be successful two centuries later. The effort, hampered by particularism and other weaknesses, was abortive: it was all over by the middle of the XVth century. But so long as it lasted, it created a climate favourable to science, and Oresme is the most brilliant testimony of this intellectual revival. Oresme belonged to the School, but he did not address himself to his fellow-schoolmen; he wrote in the vernacular (a racy French dialect) for the King and his bourgeois administrators, for the merchants, for the 'mechaniques' (as he calls the craftsmen). His attitude was that of the theoretical scientist trying to help practical men. It is obvious that theoretical science could only be cultivated by clerics, since the Church had the monopoly of education in theoretical thinking; but it would be wrong to imagine that the Church encouraged such intellectual adventures. Roger Bacon's confinement, Giordano Bruno's burning at the stake: that was the regular way of the Church with independent thinkers. If Oresme was enabled to write and teach so freely, it is just because the rise of the bourgeoisie, especially in France, led to a temporary eclipse of the authority of the Pope. So, in the end, our estimate of the relationship of science to the social forces at play on the mediaeval scene is not so different from Prof. Bernal's: but his omission of this dramatic social and intellectual upsurge of the XIVth century (there is just an inadequate mention of Oresme's achievements on pp. 221–3–5 and a stray allusion to the economic aspect on p. 884, but no suggestion of any connection between them!) distorts the picture to an inadmissible extent.

By this lengthy discussion of two typical examples I hope to have made it clear in what respect Prof. Bernal's approach to the historical problems sometimes falls short of a scientific analysis. Nothing is more fatal to the latter than setting about it with the (perhaps hardly conscious) wish to fit the picture into a ready-made frame. The idea that there is an interplay of social and intellectual currents is a powerful guiding principle in this analysis, but it must be kept under control: the form of the interplay, or even its presence in a given case, can only be ascertained at the end of the analysis, never decided beforehand. This remark sounds very pedantic, but it is not superfluous, since even such a shrewd and acute scientific historian as Prof. Bernal may fall a victim to hasty schematising. In fact, 'Marxist' historians would do well to mediate 'Engels' admirable letter to Bloch (21-22 Sept. 1890)[1] before throwing about their 'Marxist interpretations'. Strictly speaking, there is no such thing as a 'Marxist interpretation of history', just as little as there is a 'Maxwellian interpretation of electromagnetism': there is scientific history, of which Marx was an inspiring master, but to this science there is no dogmatic *via regia.*

Scientific history has a great humanist tradition. True historians of this school, Augustin Thierry and Henri Pirenne as well as Karl Marx, were also true humanists: they paid due regard to the importance of the individual thoughts and passions which do no not cease to animate men even when they act collectively under the pressure of economic necessity. Scholastic schematisation banishes this essential human element and degrades the drama of history to a Punch and Judy show.

However disastrous upon historical studies, the effect of dogmatism is worse still in the proper domain of scientific thought, physics or biology, where it numbs critical judgement and diverts thinking into the barren ground of casuistics. It is distressing that we should be reminded of this, not by calling up scenes of mediaeval obscurantism from a forgotten past, but by looking around us now, in the middle of the XXth century, – in fact, I grieve to say, by Prof. Bernal's book itself. Contemporary scholasticism is especially treacherous because it keeps the outward appearances of scientific criticism and usurps a glorious parentage. Adepts of 'dialectical materialism' are offended by the suggestion that the metaphysical system they advocate under this name is no more than a caricature, in fact the dialectical negation, of the original ideas of Marx and Engels.[2] Yet, all they can do to reassure us is to point (as Prof. Bernal does on p. 670) to such events as 'Stalin's contribution to the linguistics controversy in 1950 with its condemnation of imposed orthodoxy in science and its encouragement of free discussion'. What impression is such a statement likely to make on scientists whose tradition of free discussion goes back to the XVIth century?

However, the proof of the pudding is in the eating, and Prof. Bernal gives us enough to taste and make short process of his conception of 'dialectical materialism'. If the development of the theory of atomic phenomena initiated by the discovery of the quantum of action and the elementary constituents of atoms does not represent one of the major dialectical processes in the whole dialectical growth of scientific thought, and if quantum mechanics is not its crowning synthesis, frankly it is hard to say what a dialectic process could well be.[3] Well now, Prof. Bernal does not like the idea at all: according to him, quantum mechanics is just an obscure, incomplete, provisional theory. The widening of scope, the increased power of prediction, the deeper insight into the dialectical relationships between physical concepts, which the new theory affords, mean nothing to him. The replacement of determinism by a more general, statistical, form of causality is not greeted by him as a tremendous step forward in the elaboration of a rational conception of the physical world, but scorned as an 'idealistic' aberration. In such philosophical assessments, he puts physicists just as much as physics upside down: he praises Einstein in the idealistic mood of his later years, and scoffs at Bohr without realising that the latter's common-sense attitude is fundamentally 'materialistic'. All this trouble because the quantum is not dreamt of in his philosophy! What respect does a philosophy deserve (by whatever name it is called) which makes its adherents blind to the progress of scientific thought and misleads them into sterile opposition to it?

Prof. Bernal's account of modern physics is, I am afraid, rendered largely useless by this astonishing error of judgment. Other sections concerned with recent aspects of scientific development suffer less from the straitjacket of 'dialectical materialism' and are very readable; but the whole treatment is marred by the frequent recurrence of obviously biased assertions. This most regrettably casts a shadow of doubt upon the accompanying discussion of the social relations of science, which the author takes up in a most interesting way with his accustomed vigour and fearlessness. In principle, it is certainly right not to separate the social aspects from the purely scientific ones, but the analysis of their interrelations is beset with logical pitfalls, which can only be avoided by utmost caution. Prof. Bernal exercises no such caution, and as a result repeatedly falls into confusion; as, for instance, when he motivates his rejection of some scientific or philosophic idea by the fact that it has been used for socially reactionary purposes. Of course, most of the people against whom he directs his criticism deserve his pungent strictures; every scientist will applaud his exposure of such idealistic muddleheads as Jeans and Eddington, whose widespread popular writings are doing incalculable damage to the cause of rational thinking. It is all the more unfortunate that he should himself sin against this very cause, albeit on a materialistic basis.

Prof. Bernal's most regrettable injustice is his wholesale condemnation of Mach. I hope it is not solely based on Lenin's authority[4]; I am quite certain that it is not based on a serious study of Mach's work. Mach (together with Fechner, another thinker whose merit is not sufficiently recognised) was a pioneer in the psycho-physiology of sensations, and his acute epistemological analyses have lost nothing to-day of their significance. Historically, he initiated the movement of critical reaction against the shallow mechanistic materialism of the XIXth century, a movement which – in spite of occasional lapses into idealism – had a powerful progressive influence on the development of physics. Even Mach's famous 'error' – his fight against atomism – should not be judged (as is too often done) from the naive standpoint that 'the event put him in the wrong'. What he actually criticised was the unwarranted extrapolation by the atomists of mechanical concepts which after all only correspond to a limited class of macroscopic phenomena[5]. Is this so absurd? I will not fall into the opposite extreme of crediting Mach with prophetic vision: but for us, who currently use many-dimensional configuration space in our description of atomic systems, Mach's reflections on the application of spatial representations to such systems have a curiously modern flavour. Anyhow, it is characteristic for Mach's true scientific openmindedness that towards the end of his life (in 1903), after he had been shown a spinthariscope, he exclaimed: "Now I believe in the existence of atoms!"[6]

I am afraid, however, that Prof. Bernal is not really interested in understanding such nuances of individual attitudes to the problems of science and philosophy: for him, Mach is a positivist, and that's the end of it! This is just another example of that ruthless disregard for human personality which seems to be part of the author's philosophical outlook, and which strangely contrasts with his warm sympathy for people's collective fates and aspirations. Both these features become more and more prominent as we proceed throught the last part of the book, leaving the comparatively quiet abodes of science for the stormy seas of economics and sociology. Here, too, the partisan note becomes shriller (which, incidentally, will frighten away just those readers for whom a soberer exposition would have been most instructive). Doctrines and people are presented in highly imaginative black-and-white pictures. A glowing indictment of our decaying world and an idealised vision of the growth of a new civilisation alternate in a medley of brilliant vistas and piquant irrelevancies, so dazzling as to remind us of the author's Irish origin. At the same time, there is in those pages so much sincerity and devotion, so much generous indignation, so much hope and faith, that one's reason silences its irritation and one's heart fills with sympathy and admiration for a forceful (if somewhat reckless) thinker and a lovable and courageous man.

NOTES

[1] See any edition of the correspondence of Marx and Engels.

[2] See e.g., how Engels protests, in the preface to the *Anti-Dühring*, that he does not want to "oppose another system to Herr Dühring's system".

[3] See e.g., L. Rosenfeld, 'Strife about Complementarity', *Science Progress* **163**: 393 (1953) for a discussion of the ideas of quantum mechanics from a dialectical point of view. [This volume, p. 465].

[4] I had a shock recently when a distinguished French scientist, whom I had long before heard indulging in the usual vilification of Mach, told me that he had just started reading the *Mechanik*!

[5] E. Mach, *Die Geschichte und die Wurzel des Satzes von der Erhaltung der Arbeit*, Prag 1872 (2. Aufl. Leipzig 1909, p. 26 seqq.).

[6] S. Meyer, *Z. Naturforschung* **5a**: 407 (1950).

CHAPTER 3

THE LOGICAL PROBLEM OF THE DEFINITION OF IRRATIONAL NUMBERS

[1927e]

A legend records "that the author of the theory of incommensurables was swallowed up in a shipwreck. Thus heaven punished the one who had 'expressed the inexpressible, represented the unfigurable, unveiled that which should have remained forever hidden'".[1]

The discovery of the existence of irrational numbers by the Pythagorean school is, indeed, one of the most significant events in the history of mathematics, as well as an important event in the history of philosophy. By this discovery, the school of Pythagoras committed suicide, so to speak, itself striking the fatal blow against its doctrine of the harmony of numbers: and this is undoubtedly the reason for the legendary curses we have just recalled. But what particularly interests us here is the repercussion of this discovery on the orientation of the mathematical sciences, still in their infancy during this period.

The Pythagoreans had carried the study of mathematics forward in two principal directions: arithmetic, that is the study of the properties of natural numbers, and geometry, which they based especially on the notion of the rational relationship of magnitudes (as we must state it today, in spite of the apparent pleonasm, to indicate that they meant only relationships between commensurable magnitudes). In particular, they invented the well-known algorithm which made it possible to find the greatest common measure of two magnitudes (segments of a line).

They also took up the study of metric relationships between the elements of various geometric figures and found, especially, the famous property of right-angled triangles which today bears the name of Pythagoras. It was in attempting to apply their theories on the relationships of magnitudes and the properties of (rational) numbers to the particular right-angled triangle where two sides are equal that they encountered singular contradictions.

Aristotle[2] has preserved for us the purely arithmetic considerations of the Pythagoreans on this subject. If it be possible, let $\alpha : \beta$ be the ratio, reduced to its simplest form, between the diagonal and the side of the square. Then $\alpha > \beta$ and so $\alpha > 1$. Moreover $\alpha^2 = 2\beta^2$. Thus α^2 and consequently α are even. Since α and β have no common factors, β is necessarily *odd.* Now α being even, suppose $\alpha = 2\gamma$; then $4\gamma^2 = 2\beta^2$, $\beta^2 = 2\gamma^2$, and β^2 is even. It follows that β is *even*, contrary to our first conclusion.

Thus the theory of rational numbers was utterly insufficient to represent relationships among magnitudes in complete generality. Consequent to this conclusion, the Greek mathematicians abandoned the study of arithmetic to devote themselves to geometry, which offered them more solid ground: it is the discovery of $\sqrt{2}$ which gave the Greeks their notion of the preeminence of geometry and their disdain for arithmetic, that is, which helped make Greek mathematics take on its most original and essential characteristics.

But the crisis we are discussing was not only an arithmetic crisis: it was far more serious, threatening to undermine the very foundations of Pythagorean geometry. In effect, in applying their general method to the search for the largest common measure of the diagonal and the side of a square, the Pythagoreans were led to a limitless series of operations: the fundamental concept of their geometry was found wanting, just like that of their arithmetic. It was, as Paul Tannery says, "a veritable logical scandal": all the proofs which depended on the theory of proportions became obsolete and the correctness of many theorems was once again put in question.

The first reaction which followed this crisis was naturally an attempt at 'patching up': they became ingenious at replacing proofs which had become doubtful by others no longer dependent on the theory of proportions and containing a myriad of clever artifices. The echo of these efforts is found in the first four books of Euclid. Other mathematicians, Democritus of Abdera, Theodore of Cyrenia, Theatetus the Athenian, studied certain special classes of the new incommensurable magnitudes, 'inexpressibles' (ἄλογοι) they called them: their results are collected in Book X of Euclid.

It was left to the genius of Eudoxus of Knidos to understand the sterility of these disorganized efforts, to realize that the true manner of resolving the problem was to generalize appropriately the notion of ratio, and finally to discover the principle of that generalization.

We do not know the original theory of Eudoxus of Knidos, but though

Euclid greatly improved its exposition in Book V, it is certain that he changed nothing in the principle of the method, that is, in the new definition of the notion of ratio and the equality of two ratios.

"DEFINITION IV. Two magnitudes are said to have a ratio to each other when it is possible to take multiples of them in such a manner that one becomes greater than the other."

(This definition applies equally well to incommensurable magnitudes as to others: its justification, that is, the proof of the existence of what is defined, requires the important axiom usually given the name of Archimedes.)

"DEFINITION V. Two pairs of magnitudes a,b,c,d are said to have the equal ratios $a : b$, $c : d$ when, if any equal multiples of a and c and any equal multiples of b and d are taken, the first equal multiples are simultaneously greater than, equal to, or smaller than the latter equal multiples taken in the corresponding order."

[Briefly, one has $a/b=c/d$ when one has simultaneously, for any integers m, m',

$$ma > m'b \text{ and } mc > m'd,$$
or
$$ma = m'b \text{ and } mc = m'd,$$
or
$$ma < m'b \text{ and } mc < m'd]$$

"Perhaps the greatest homage to this marvellous definition," said Heath (*op. cit.*), "is its adoption by Weierstrass as the definition of equal numbers".[3]

Such was the first solution to the logical problem which had so brutally confronted the Pythagoreans. As can be seen, it consists essentially in the conception of the most general real number as a ratio of magnitudes.

It has been given its just place of honor again in our time, not only by Weierstrass, but in a slightly different manner, by the brilliant Italian school which, under the guidance of Peano, Enriques and Burali-Forti, made such progress in the criticism of the foundation of the exact sciences. Burali-Forti[4] introduced the extraordinarily fruitful concept of *operator* on magnitudes: the benefits he and his school derived from this are well-known: not only a simple general theory of real numbers in the same spirit as the theory of Eudoxus, but also that elegant and powerful vector analysis[5] which is in the process of becoming an indispensable working tool for the theoretical physicist.

I mention these details to be able to characterize clearly the Eudoxian concept of irrational numbers; in the area of analysis of *geometrical and physical magnitudes*, this concept appears to be the most natural and the most fruitful; it is the theory par excellence of *applied* mathematics. But from the point of view of the theory of knowledge, it is open to serious criticism which we shall discuss later.

II

This situation was not to change until the beginning of the nineteenth century. Earlier, in the medieval lethargy and then in the feverish period of discovery in the seventeenth and eighteenth centuries, "the tangible foundation found for numbers in the geometry of measurement made all doubts disappear completely".[6] The great innovator Cauchy, himself, in his *Cours d'Analyse* (1821), stated, without finding it necessary to prove, the fundamental theorem of the theory of sequences (the convergence criterion). Nevertheless, it should be noted that Bolzano (1817) had already recognized the need for an *arithmetic* theory of irrational numbers as a basis for the theory of sequences. But the ideas of this precursor had no repercussions and it was only much later that this need was clearly felt by everyone. The "arithmetization" of infinitesimal analysis had almost been completed when it was realized that an arithmetic definition of the real number, the fundamental element of the whole theory of functions, was lacking. The second phase of the logical problem which concerns us began at this point: as we have seen, the ancients had resolved it within the framework of geometry; but now that the work of arithmetization begun by Descartes had developed to the point of entirely transforming the conception of mathematics, the logical problem of the definition of irrational numbers reappeared in the framework of arithmetic. It appeared in this form to the moderns just as it had already appeared to Pythagoras; but the moderns were otherwise equipped to solve it than the philosopher of Tarentum. The introduction of analytical geometry and the application of infinitesimal calculus to geometrical problems had familiarized scholars' minds with the concept of *continuity* and the manner in which this concept enters into the science of numbers.

At the start of the second half of the nineteenth century it had been recognized that the problem in question was equivalent to an analysis of the continuum, and several mathematicians proposed, almost simultaneously and independently of each other, solutions resulting from such an analysis. We will review the basic features of the principal ones very quickly and then call attention to one of their common characteristics which especially concerns us.

Meray[7] studied sequences of rational numbers which he termed *variants*: he said that a variant is *convergent* when the absolute value of the difference between two terms became less than any given quantity, beyond a certain point. A variant was said to have a (rational) *limit V*, when the absolute value of the difference between a term of the variant and the rational number became less than any quantity beyond a certain point. Given this, it is easy to see that any variant which has a (rational) limit is convergent, but that the converse is not true, "at least as long as one gives the word *number* or *quantity* the same meaning that we have accorded it thus far" [that is, the meaning of rational number]. However, it is preferable to give our terminology uniformity and picturesqueness through the use of a set of conventions, of which the following are the most important:

"I. When a convergent variant does not tend toward some [rational] limit, it is assigned an *ideal* one called an *incommensurable* [=irrational] number or quantity and represented by the same symbol as if it reaily existed. Then, the convergence of any variant may be expressed by saying that *it tends toward a certain limit* (effective or ideal as the case may be . . ."

In the course of his brilliant investigations on set theory, G. Cantor undertook an analysis of the continuum and applied it to the theory of irrational numbers. The best exposition of his ideas on the latter subject is by Heine.[8] Cantor's view does not differ greatly from that of Meray in its details, but starts from a very different concept, as the following quotation shows: "Ich stelle mich, bei der Definition [der Zahlen] auf den rein formalen Standpunkt, indem ich gewisse greifbare Zeichen Zahlen nenne, so dass die Existenz dieser Zahlen also nicht in Frage steht."*

* "In defining [numbers] I take a purely formal point of view, by calling certain tangible symbols numbers, so that the existence of these numbers is not in question."

Another analysis of the continuum, less profound than that of Cantor,[9] was made by Dedekind.[10] It is summarized in the following *principle of continuity*: "If all points of the straight line fall into two classes such that every point of the first class lies to the left of every point of the second class, then there exists one and only one point which produces this division of all points into two classes, this severing of the straight line into two portions. . . ."

"The assumption of this property of the line is nothing else than an axiom by which we attribute to the line its continuity. . . ."

"And if we knew for certain that space was discontinuous there would be nothing to prevent us, in case we so desired, from filling up its gaps, in thought, and thus making it continuous; this filling up would consist in a creation of new point-individuals and would have to be effected in accordance with the above principle."

Carrying over this analysis into the arithmetic domain, Dedekind defined what he called a *cut* (Schnitt): he applied this term to any division of a series of rational numbers into two classes A_1, A_2. He easily demonstrated that any rational number engenders a cut, but that cuts exist which cannot be engendered by any rational number.

"Whenever, then, we have to do with a cut (A_1, A_2) produced by no rational number, we create a new, an *irrational* number a, which we regard as completely defined by this cut (A_1, A_2); we shall say that the number a corresponds to this cut, or that it produces this cut."

If we compare these various methods, there is a point which strikes us immediately: the new 'numbers' are introduced by a 'convention', by a 'creation'; they are new 'signs' which enter the domain of mathematical considerations. But what is the logical nature of such a way of proceeding? The matter is far from being clear, especially (as we shall show) for the very authors of the theories in question. Thus we are faced with a new crisis in our problem. This crisis is far from abated today and, in fact, we propose to bring forward a contribution to its solution in Section IV of this paper. But first, the present state of the question must be presented.

The current opinion is the following: 'creations' in the manner of Dedekind cannot be considered as definitions of the new numbers, because a definition does not imply the existence of the defined, and thus, from a logical point of view, is not sufficient by itself to justify its introduction. This statement is correct, but another is added which, as we shall demonstrate, is not: the newly 'created' concepts are not equivalent to postulates, because (it is said) one can postulate something only on the basis of concepts whose existence is already assured. This stated, an attempt is made to justify Dedekind's procedure by affirmations of the following sort:[11] "... la creazione stessa è un atto perfettamente legittimo nella matematica pura... è perfettamente logico e consentaneo all'indole della matematica, creare un sistema di simboli che servano in ogni caso a rappresentare il rapporto di due grandezze."*

The last word of this quotation alone would suffice to point up the misunderstanding which has crept into this attempted justification; Natucci adopts the viewpoint of the Italian school and considers numbers as operators on magnitudes. If it were possible to adopt this point of view, his remark would be pertinent. But as we have already said, though the notion of the operator is very useful for applied mathematics, it has no place in pure mathematics; this is not a semantic quarrel, but in my opinion a very important point in the theory of knowledge. The present opinion (as opposed to the Hellenistic opinion) is that pure mathematics consists of the study of the properties of natural numbers and their various *extensions*, that is, the concepts called relative, fractional, irrational, complex, etc., numbers. We will return later to the precise meaning of the word 'extension' applied to the concept of number. Let it suffice now for us to bring out the opposition between this point of view and the theory of operators: in the latter theory the concept of the operator is introduced, an idea which has nothing in common with that of the natural number; the general notion of the real number is deduced from it and then, as a special case, that of the natural number *considered as an operator*. To sum up, the only significance the theory of operators can possibly have is of a *practical* rather than a theoretical order.

III

What is more, B. Russell and Whitehead, taking up the pioneering work of Peano and Frege, demonstrated[12] that it is possible, starting

* Translation of the Italian is as follows: "... creation itself is a perfectly legitimate act in pure mathematics ... it is perfectly logical and in accord with the nature of mathematics to create a system of symbols which can serve in all cases to represent the ratio between two magnitudes."

from the notion of the natural number, to introduce successively, by purely logical means and without any new postulates, the various 'extensions' of the concept of number, and particularly the notion of irrational number.

The following sketch will give an idea of their *constructive* method (as opposed to Dedekind's *creative* method). Let us return to Dedekind's cuts. For these cuts, we can define the relationships $>$, $<$ and the operations $+$, $\times$, etc., that is, create an 'arithmetic of cuts'. This done, a cut which is engendered by a rational number a will be called the 'real-rational number [a]', and a cut (A_1, A_2), which is not engendered by a rational number will be called the 'real-irrational number (A_1, A_2)'. In general, the term 'cut' is replaced by 'real number'. One notes that the *real-rational* numbers enjoy properties completely analogous to those of *rational* numbers, but there is no identity between these two numerical classes, only a simple 'isomorphism'.

Comparing his method to that of Dedekind, B. Russell[13] criticized the former in fairly biting terms:

"The method of 'postulating' what we want has many advantages; they are the same as the advantages of theft over honest toil. Let us leave them to others and proceed with our honest toil."

But the adversaries of the "puritans of logic" (as Natucci said)[14] continued to make criticisms of a very general order, which did not lack interest.[15]

First of all, the entities thus defined, thus constructed, by Russell are not the traditional numbers, that is to say, they do not correspond at all to the intuitive concepts as they were formed historically or psychologically. The theory of operators, on the contrary, is completely adequate for this psychological formation. But we cannot demand that logical constructions always be in agreement with intuitive views, for the latter are not formed according to logical rules, but follow their own psychological rules. A logical analysis always starts from a determined base and the form of its result depends essentially on the choice of this base. The intuitive idea is much richer and, at the same time, much less precise: it contains simultaneously the results of all possible logical analyses and their multiple connections, but it contains them only potentially in the sense that logical analysis is necessary to reveal them clearly. The choice of the point of departure for the logical analysis is guided primarily by reasons of 'economy of thought' (Mach). Moreover, mathematics does not lack theories which are open to analogous objections (if there is an objection): what could be more different from the intuitive idea of continuity than the arithmetic definition of it? What could be more contrary to the order of historical development than the current exposition of the theory of elliptical functions which takes as its point of departure the property of double periodicity? In this discussion, one must beware of the mistake of introducing didactic considerations: the problem we are examining is a logical problem; if the correct solution to this problem is not a good one from a didactic point of view, then, of course, it should not be introduced in teaching (I think that, for teaching, the theory of operators is best). But these considerations are totally foreign to our subject; if I speak of them, it is because certain opponents of the logical method have a tendency to use them as arguments.

Russell's method has been the object of another, apparently more serious, criticism. Undoubtedly, the argument runs, this method does not require existential postulates for the various sorts of numbers, but this only 'shifts the question', because existential postulates are necessary for the very bases of the theory. Undoubtedly, but the important thing is that these postulates are no longer mathematical postulates, but rather logical postulates, i.e., with a much broader significance. It is not at all unimportant to 'shift the question' in this sense.

Finally, the constructive method has been criticized for its complicatedness. We could find no better response to this objection than to refer again to the words of Russell which we quoted earlier.

In summary, the criticisms of a methodological nature which have been directed at Russell's method do not stand up to examination, and consequently I think that it is this method which completely resolves the logical problem of the definition of irrational numbers.

It now remains for us to analyze the logical nature of the creative method of Dedekind.[16] At the same time, this analysis will permit us better to characterize Russell's method in comparison with that of Dedekind.

IV

It is important, first, to define what is to be understood by 'extension of the concept of number': Cipolla gave a very general definition, by introducing the notion of isomorphism of two classes of numbers; but in the particular case which concerns us, we can proceed more simply.

Let us first sketch a few concepts which are necessary to understand what follows. In general, a series is defined by a 'serial' relationship P, possessing certain properties analogous to those of the relationship > among rational numbers. A *cut* in the series P may be defined by means of the relationship P in the same manner that a cut in a series of rational numbers was defined by means of the relationship >. As an example of a serial relationship P other than >, one can cite the relationship 'to the left of' applied, for example, to the series of points in a horizontal straight line. The element of a series P which engenders a given cut in this series, when it exists, will be called the *limit* of the cut. It is said further that two series P, Q are *ordinally similar* when their elements correspond to each other biunivocally and occupy, relative to the relationships P, Q, similar positions in the two series. Finally, any series which is ordinally similar to the series of rational numbers (ranked by order of increasing magnitude) will be called a *series of ratios*.[17]

That established, we shall have realized the desired extension of the series of rational numbers if we find a series P with the following properties:

1. Series P contains a series of ratios as a part;
2. Any cut in the series of ratios contained in P has as its limit an element of the series P;
3. The two preceding properties of series P must not depend on the special nature of the elements of P.

The two first conditions are evident according to the previously mentioned analysis of Dedekind. If we have a series P with these two properties, we will call these elements 'real numbers', and more particularly, the elements of the series of ratios contained in P 'real-rational numbers'; the relationship P itself will be called 'inferior to' and will be the analogue for real numbers of the relationship < for rational numbers; the properties of real-rational numbers will be analogous to those of rational numbers, on the condition, naturally, that the addition of real numbers is also properly defined.

But it is necessary to add the third condition if we want to justify logically the introduction of *any* sort of new symbols which will serve to designate the new numbers we wish to consider. For if we repeat the same considerations for another series Q with the first two properties, we shall indeed obtain another series of real numbers, and we even know that the two series of corresponding real-rational numbers are ordinally similar (*237′4);[18] but we cannot affirm that the two entire series of real numbers are also ordinally similar. From this point on, any rational number is, indeed, univocally determined in series P as well as in series Q, but the same is not necessarily true of an irrational number: the determination of this latter could depend on the particular series P or Q, that is, on the particular symbol which designates it.

What we have just said allows us, at the same time, to transform the statement of the third condition which must be satisfied by series which can define real numbers; if P, Q are two such series, they must be ordinally similar in such a manner that the rational series they contain also correspond ordinally and that the limits of two corresponding cuts also correspond.

At this point, it is easy to see that *continuous* series (in Cantor's sense) possess all the required properties. (The first two result from *275′13; the third is obtained from the demonstration of *271′39.)

The result of our analysis is thus that *any continuous series whatever can be defined as the series of real numbers.*

In other words, as soon as a continuous series is given, one is in a position to define the irrational numbers. *The logical problem of the definition of irrational numbers is thus reduced to the logical problem of the existence of a continuous series.*

Now this last problem is only susceptible to two solutions:

either *postulate* the existence of a continuous series;

or *demonstrate* the existence of a continuous series.

Moreover, there is only one way to demonstrate the existence of a logical individual and that is actually to construct an *example* of this individual.

At present, we are in a position to recognize clearly the status of the method of 'creation' of Dedekind, Meray and others. Having finished no existential demonstration, they necessarily *postulated the existence of a continuous series*: such is the true logical nature of the 'creative' method.

The authors of the various theories of this type do not appear to have recognized this logical problem. In the passage that we reprinted in

Section II, Heine is right to say that the question of the existence of the 'tangible' signs which he introduces does not arise, but he did not see that the question which really does arise is that of the possibility of arranging these signs in a continuous series. As for Dedekind, we have seen that his analysis of the continuum was insufficient in the sense that it led only to conditions 1 and 2, mentioned above, but omitted the important condition 3. Moreover, we have another indication which shows that Dedekind had completely overlooked the logical problem in question: H. Weber observed to him that one could do without the postulate or the principle of continuity by considering the cuts themselves as numbers (as Russell did later); and Dedekind rejected this proposal. It is amusing to note that Dr. Perron,[19] who reports this anecdote, sides with Dedekind, saying: "But this is only a difference in the manner of expression (in der Ausdrucksweise!), and does not touch the heart of the matter."

The results we have obtained also permit us to characterize the difference between Dedekind's method and that of Russell. The first author postulates the existence of a continuous series; the second, by constructing such a series, demonstrates Dedekind's postulate, and he naturally chooses the continuous series he constructed to be the series of real numbers. In the same line of thought, I mention that the Meray-Heine point of view, slightly modified, can be used to give another demonstration of the existence of a continuous series: that is what M. Deruyts has done (though for a purpose completely foreign to the present considerations) by defining certain series of rational numbers which he calls 'gradients' and taking the continuous series of gradients for the series of real numbers.[20]

To sum up, our analysis has led us to a criterion which permits the characterization, from a logical point of view, of any arithmetic theory of irrational numbers.

NOTES

[1] P. Boutroux, *L'idéal scientifique des mathématiciens* (Alcan, Paris, 1920; P.U.F., Paris, 1955), p. 48.

[2] From Heath, *A History of Greek Mathematics*, 2 vols. (Oxford 1921); see the index.

[3] See Natucci, *Il Concetto di Numero e le sue Estensioni* (Turin 1923) pp. 207 ff.

[4] *Ibid.*, pp. 357 ff.

[5] C. Burali-Forti and R. Marcolongo, *Analyse vectorielle générale*, vols. I and II (Mattei, Pavia, 1912–13).

[6] O. Perron, *Irrationalzahlen* (Leipzig 1921).

[7] *Revue des Sociétés Savantes* (2), vol. IV, 1869; *Leçons nouvelles sur l'analyse infintésimale*, vol. I, ch. II, 1894.

[8] 'Die Elemente der Funktionenlehre', *Crelle's Journal*, 1872.

[9] For a comparison of the Cantorian and Dedekindian definitions of the continuum, see B. Russell, *Introduction to Mathematical Philosophy*, ch. X (George Allen and Unwin, London, 1919); B. Russell and A. N. Whitehead, *Principia Mathematica*, vol. III, *275 (Cambridge University Press, Cambridge, England, 1913).

[10] *Stetigkeit und irrationale Zahlen* (Braunschweig 1872). I am quoting from a translation (unpublished) of the principal passages of the book that I made several years ago. [We have quoted the standard English translation: R. Dedekind, *Essays on the Theory of Numbers* (Open Court, Chicago, 1901), pp. 11–12, 15 – Eds.]

[11] Natucci, *op. cit.* (3), p. 259; cf. also Natucci, 'Origine e Sviluppo del Concetto di Numbero irrazionale', *Scientia*, 1925, pp. 293 ff.

[12] *Principia Mathematica*, 3 vols., 1910–1911–1913. – Of course Russell and Whitehead found many much more important things than this.

[13] *Introduction to Mathematical Philosophy*, *op. cit.*, Note 9, p. 71.

[14] *Op cit.*, Note 11

[15] Cf. Natucci, *op. cit.*, Note 3, pp. 446–449; Poincaré. *Science et méthode* [English transl. by G. B. Halsted in *The Foundations of Science* (Science Press, New York, 1913) – Ed.] and *Dernières pensées* (*passim*) (Flammarion, Paris, 1913). [English transl. by J. Bolduc: *Mathematics and Science: Last Essays* (New York, Dover reprint, 1963) – Ed.]

[16] I presented this analysis at the Congress of the Association française pour l'Avancement des Sciences, held in Lyon in 1926.

[17] For more precise definitions, see Russell and Whitehead, *Principia Mathematica*, vols. 2 and 3.

[18] The numbers preceded by asterisks are references to *Principia Mathematica*, vol. 3.

[19] *Op. cit.*, Note 6, p. 57.

[20] Deruyts, Congress of the Association française pour l'Avancement des Sciences, Liège, 1924.

CHAPTER 4

RATIONALISM IN ANTIQUITY*

[1954c]

The growth and development of rationalism in antiquity raises several historical problems which have perhaps not hitherto been sufficiently recognized. The purely philosophical aspect of the remarkable elaboration of a rationalistic view of the world by the preSocratic thinkers has received a great deal of attention and can now be regarded as completely elucidated by the work of Paul Tannery and Enriques. The latter especially put into its right light the great figure of Demokritos, whose atomic system represents the culmination of this evolution.

When we try, however, to follow up the development of rationalism after Demokritos, the situation at once becomes obscure. If we compare the views of Epicurus and Lucretius with what we can reconstruct from those of Demokritos, we are struck by the fact that while the scientific content of the system remains very much the same (with only some signs of decay), stronger and stronger emphasis is put on the more human and social problems. Lucretius' atheism, for example, is not merely an intellectual point of view, but a distinctly anti-clerical attitude which, so far as we know, was at any rate not prominent in Demokritos' views. This then raises the first problem that I wish to discuss: how is this change of attitude of the rationalistic philosophers related to the changes of their social environment? For the answer to this question one has not far to seek when one remembers how the Greek world was shaken to its very foundations by the breakdown of the Athenian democracy and the following conquest of the Persian empire by Alexander. As the Greeks suddenly enlarged their horizon to a world-wide scale, their economy underwent a complete change which confronted them with new political and social problems. A characteristic repercussion of this turmoil is the apparition of a new literary genre, the utopia, which of course is a convenient medium for social reformers and satirists who are unwilling to risk the displeasure of the authorities. None of these utopias has survived, but we know the names of some authors, the most famous of whom is Euhemeros of Messene. The views of Euhemeros have quite a rationalistic flavour. In particular, his attempt to give a rational interpretation of religious myths is so characteristic that the word *Euhemerism* has been coined to denote this type of rationalization of religion. It is significant that one of the utopists who influenced Euhemeros, Hekataios of Abdera, is described as a disciple of Demokritos. The popularity of Euhemeros' book lasted a long time and was even transmitted to the Romans by a translation due to Ennius.

The preoccupation of the utopists with a rational interpretation of religion was not original with them. The question had already been raised by the Sophists and we find echoes of a similar religious crisis in the two great philosophical currents which developed in Hellenistic times, the Stoic and the Epicurean schools. In fact the conflict between popular religion and philosophical views of God became inevitable in all the great schools to which the development of preSocratic philosophy led. The issue came to a crisis when its social implications became more and more prominent in the Hellenistic period. The evolution of rationalism during this period reflects the slow process of gradual decay of ancient society, strikingly illustrated by the contrast between the progressive and optimistic outlook of Demokritos and Lucretius' last pathetic assertion of the power of human reason against the obscurantism which was going to overwhelm it.

Returning now to Demokritos and the evolution leading to him, we may also ask what social preoccupations are apparent among the preSocratics and how they can be related to the economic and social background against which their thinking developed. It is indeed possible to trace back the rationalistic analysis of social phenomena to the Milesian school from which the whole body of preSocratic philosophy originated. We meet here another Hekataios, who developed for the first time a rational view of history in which he endeavoured to interpret the tales about heroes by the same method Euhemeros later applied to the gods. It has often been pointed out, most convincingly by Professor Farrington, that there is a direct causal connection between the rationalist outlook of presocratic philosophers and the industrial and commercial type of culture to which they belonged. Judging from other well-known examples, this connection does seem to be a historical law of wide validity. The present context offers us a further occasion of testing it. Philosophy did not originate in Miletos. As our knowledge of Oriental sources increases we realize more and more how largely the Ionians were indebted to the Orient for the fundamental features of their systems. As the main carriers of this Oriental tradition one thinks in the first place of the Phenicians. Now, if ever a civilization was of the commercial and industrial type it was that of the Phenicians. It is therefore of special interest to enquire of what kind their philosophy was. In this enquiry we are unfortunately hampered by the scarcity of the evidence. About all we know of Phenician cosmogony is the indirect evidence associated with the shadowy figure of Sanchuniathon. A convincing case, however, has recently been made by Eissfeldt, on the basis of the archaeological evidence from Ugarit, for the authenticity of the Sanchuniathon tradition. At any rate our information concerning him is not worse than most of the doxography pertaining to the preSocratics. The rationalistic character of Sanchuniathon's views is very striking indeed, as well as the analogy with those of the Milesian Anaximenes. It therefore seems that we may risk going one more step back towards the origin of rationalism in that commercial civilization which preceded the Ionian one. It would be wrong, however, to suggest that the Ionians just took over Phenician views. The transmission of philosophical and scientific ideas is never a mere imitation. Even though the general trends may remain essentially the same, the details of the picture must be worked out anew under the influence of ever-changing circumstances.

Looking back on the whole evolution we have endeavoured to reconstruct, we see rationalism at a nascent stage among the Phenicians, revived and developed by the preSocratic philosophers, and reaching its acme with Demokritos. The opening-up of the Orient enabled the Greeks to rediscover the Phenician origins of their thinking. Phenician books were actively translated and the ideas they contained were one of the sources of inspiration of the utopists. Together with these, the Epicurean school prolonged the rationalistic tradition, and fought a losing fight for it until the very end of the ancient world.

NOTE

* Abstract of Address delivered by Prof. L. Rosenfeld at the Inaugural Meeting of the Northern Branch of the British Society for the History of Science held at Manchester University on 30th January, 1954. [The draft typescript of the full lecture was not sufficiently complete for publication, in Rosenfeld's judgment; hence we include only the abstract, as originally published. At his request, we have followed his original (German) transliteration of Greek names throughout this paper.–Ed.)

CHAPTER 5

THE TRANSFORMATIONS OF THE ATOMIC CONCEPT THROUGH THE AGES

[1969f]

If one opens a textbook of atomic physics, one will read on the first page that the atomic concept was introduced into science by the Greeks. It is true that the word 'atom' is a Greek word that means 'indivisible'; however, every child knows today that we are perfectly able to split atoms, and that the components of atoms are themselves made up of even smaller particles. From this point of view it is a good thing that children are no longer forced to learn Greek; for the poor Greek children the word 'atom' is perhaps even more confusing, for in modern Greek it denotes a single human being, an individual. This discordance is already a hint that the introduction of the atomic concept did not proceed quite so smoothly: in the course of time this concept has undergone an essential alteration, which has of course been conditioned by our increased knowledge of the structure of matter. If we look more closely at the historical development, we get in fact a very varied picture of continually changing notions about the smallest parts of matter – whether the idea of atom was upheld in some form or other, or was completely rejected. It is not without interest to try to understand the motivations that at various periods led philosophers and scientists to so different opinions: such an analysis sheds light upon the way in which the construction of our scientific concepts is influenced by our knowledge of the phenomena and related to the tasks imposed on us by nature itself or by society.

We shall begin, then, with the Greeks, because we have insufficient knowledge of earlier cultures to be able to say anything definite about their views of the structure of matter. We have only a brief reference to a Phoenician thinker who is said to have developed an atomic theory, and it is possible that the Greeks learnt it from him, but we do not know what it consisted of. On the other hand, we are fairly well informed about the origin of the atomic concept in early Greek philosophy, which distinguished itself, of course, by bold speculations about the creation and constitution of the universe.

Our starting-point must be that singular thinker Parmenides, who tried to carry speculation to the utmost limit pure human reason can reach. He began with the simplest assertion that reason provides, that "being is, and non-being is not". That was how he put it; he must of course have meant something by this enigmatic utterance, and his meaning can be reconstructed from the consequences we see that he derived from it. It turns out that what he meant by 'being' corresponds to our concept of matter, what the world consists of. Then we find a meaning in his statement: everything that is, is matter; where there is no matter, there is nothing. In other words, it is impossible for there to be any place where there is no matter. The whole world must therefore consist of a compact aggregation of matter without the slightest gap or crack, and it is inconceivable and counter to reason that any part of this solid world could move in relation to any other. Since he conceived the world as being finite, the only possible form for it was a large sphere, completely solid and without internal motion. The only movement the sphere could have was a rotation as a whole around some axis. How can one get this strange world-picture to harmonize with the evidence of our senses, which after all give us precisely the opposite impression: that everything in the world moves and is constantly changing? Parmenides' reply is simply that our senses deceive us: sense impressions are illusions upon which we cannot rely; reason is our only source of assured truth. He propounded this philosophy in a very beautiful poem in which he depicts his own ascension to heaven, where he receives the divine revelation of the true basis of wisdom.

So much for Parmenides. His sharp, uncompromising presentation of the dilemma between reason and the senses was a direct challenge to attempt to circumvent it, to find a way out leading to a world-picture that would agree better with our experience and therefore also be more useful. The simplest solution was doubtless to adopt the opposite point of view from Parmenides: if reason leads us to such absurd conclusions, it is not worth much, and all we can rely upon is our senses: only through direct observation do we obtain any knowledge of nature. This purely empirical philosophy gained wide circulation, because it spoke directly to common sense, unlike Parmenides' paradoxical notion of the use of reason. Yet it was much too superficial to further any fruitful scientific thinking.

Among the disciples of Parmenides in Elea in southern Italy a more critical attitude prevailed. The concept of motion was particularly subjected to a very penetrating analysis which illustrated the inadequacy of a logic that operated only with finite concepts and conclusions. Although the Eleatics – as far as we know – did not themselves reach a solution to their ingenious paradoxes, their efforts were of importance as the first step along the road that was soon to lead to the admirable elucidation of the concept of continuity, which still forms the basis of the modern mathematical treatment of physical phenomena.

There was, however, a third school that arose from the teaching of Parmenides, the atomists. These were also disciples of Parmenides and tried to preserve as much as they could of his conclusions without coming into conflict with experience. What was required was to introduce a possibility of motion into the Parmenidic universe, and this possibility would obviously be present if it were assumed that the Parmenidic solid matter might exist as separate pieces with space between them. That a void, a vacuum, might exist was a very bold assumption, in a Parmenidic sense nothing but a denial of reason, but the atomists Leucippus and Democritus had nevertheless the audacity to introduce it. As soon as they allowed this spaciousness in the world, they had room not only for all kinds of motion, but also for much more, and in the building up of a completely atomic world-picture the atomists proceeded with the greatest consistency as well as the greatest imagination. They realized that not only the movements of large bodies, but also the slightest changes that we can observe in phenomena, could be ascribed to the movements of atoms, if it were only assumed that atoms, or at least most of them, are small enough not to be individually accessible to our senses, but that we only perceive the result of the movements of large numbers of atoms. In modern philosophical language this meant that the various qualities attributed to matter on the basis of various sense perceptions are reduced to quantity, i.e. to a few quantitative properties of the atoms that give rise to the quality in question, as for instance their number, shape, size, and motion. In this way, one achieves in principle an enormous simplification of the whole conception of the structure of the world, and furthermore a description of all phenomena that rests on a purely rationalistic basis without the intervention of supernatural forces. The world consists of a heap of atoms which in all eternity are moving in all directions without the slightest order. All the order, all the various qualities and regularities we observe, are simply the result of the chaotic movement of those many atoms.

Both the atomists and their opponents were fully aware that they had achieved a completely materialistic description of the world. Democritus himself lived in an enlightened age in which there was great freedom for discussion of the boldest speculations, but his successors Epicurus and Lucretius lived under completely different social conditions. The latter particularly gave sharp expression to the contradiction he felt between the atomistic conception and his contemporaries' belief in divine powers who were supposed to watch over human societies; but that is an aspect of the matter into which I have no time to go further. If I mention this character of atomism, of being a consistently developed materialistic description of the universe, it is rather to point to its deficiencies, which are just as instructive as its merits.

The atomists' description of phenomena was a purely imaginary picture; they did not care, as we would, to compare in detail the atomistic models of phenomena with experience; they were content with superficial explanations of the various phenomena on the basis of the movement of atoms, without checking more accurately how far these explanations were valid. This great weakness in the attitude of the atomists was an inevitable result of the fact that their speculations were not closely linked to any social need: whether their explanations were correct or not had no practical consequences, their critical alertness was not stimulated by any demand for results of practical use.

As a consequence of this, atomism was not able either to raise itself above other philosophical trends as the one that provided the only possible explanation of phenomena. Thus Anaxagoras, a contemporary of Democritus, developed an equally bold and consistent argument which, however, led him to a notion of the structure of matter that on essential points was just the opposite of the atomistic conception. Far from trying to eliminate qualities, he laid stress on experiences that show that qualities are preserved in matter even when it is broken down into smaller and smaller parts. In addition, he considered that no limit could be set to the divisibility of matter; if a particular body were continually subdivided more and more finely, he thought one would find at every stage of the process the same mixture of qualitatively different components as in the original body. This is certainly the meaning of the main concept of his philosophy, which he called "homeomery". According to tradition, he illustrated this concept by saying that if one could divide a bone into smaller and smaller parts, one would still find that these parts were of the same nature as the bone itself. The idea he was thus trying to express bears a striking resemblance to the modern concept of field, which we use when we are thinking of a certain qualitative property, an electrical or magnetic force for instance, which is continuously distributed in space.

So it does look as if we could discover in Greek philosophy forerunners of the two very fundamental concepts, atom and field, that are part of our own world description. However, one must beware of such superficial comparisons. Certainly, Greek thinkers had carried on speculation about the ultimate constituents of the world so far in all directions that they could not avoid coming upon features in nature that later research has recognized to be essential. But they lived in a different society from ours with a quite different view of the task of science: their motivations, which I have tried to outline, carried them along trains of thought that are quite alien to modern scientific thinking and only show casual points of resemblance to it.

We now make a leap to a completely different society, the Arab empire of the Middle Ages, whose philosophy and science provide a quite different picture – and a fresh lesson for us. In earliest Arab philosophy we immediately encounter a form of atomism that is even more bold and developed than that of the Greeks, but which at the same time is completely unscientific. It arose in the philosophical school that is called "kalam". Originally, "kalam" meant conversation, but it is used here in the sense of controversy, for these philosophers had set themselves the task of defending Islam against infidels: their attitude was apologetic. They wanted to make plain the omnipotence of Allah, and they did it thoroughly. Not only does the world consist of atoms, but even the course of time is atomized: the world that Allah creates in a moment exists only for a very short time, and so it must be created anew, possibly in a somewhat altered form, corresponding to the changes we observe. But whether these repeated creations are continued or not is determined solely by the will of Allah: at every moment the world is lying in the hand of Allah. This is doubtless the only example we have of a philosophy that completely denies any possibility of expecting any coherence in the course of the world; a more radical exclusion of causality in any form can hardly be imagined.

Soon, however, a more rational attitude appeared among the Arabs, and this had its origin in the Greek tradition. The influence that made itself felt most strongly came from the school founded by Aristotle, whose most famous Arab disciple was Ibn Roshd, whom the Latin schools called Averroës. His great commentary on the writings of Aristotle was soon translated into Latin and came to play an important part in the philosophy of the western Middle Ages. Now Aristotle, like Anaxagoras, had rejected the idea of an atomistic division of matter; he could not see any rational cause for the existence of any limit to the divisibility of matter. On the other hand, he was as little able as the Eleatics to solve the difficulties of the concept of infinity – a concept that the Greeks always found it hard to reconcile with their sense of the sharply defined, the sharply limited. What would the continued division of space lead to? To points? But a point has no longer any spatial quality. To escape from this dilemma, Aristotle sought refuge in the distinction between 'potentia' and 'actus' ('dynamis' and 'entelecheia'), which was the basis of his dialectics: matter, he said, is divisible without limit in conception, but not in reality. In fact, there is always a smallest dimension beyond which a particular body would lose its characteristic quality if one tried to proceed further with its division.

Now Aristotle did not stress this argument particularly, and he did not make much use of the idea of the 'smallest parts' of matter ('elachista'), probably because he felt it did not fit in particularly well with his general philosophy. Ibn Roshd, however, made the most of it. When we ask why, we encounter for the first time a new motivation that was going to dominate all further development. Right from Hellenistic times a considerable chemical industry had been growing up, which continued to flourish in the Arab empire. These chemists had their own theoretical problems: they were seeking guidance in their groping attempts with various transformation processes – not least the transmuting of base metals into gold and silver. What happens when a chemical reaction takes place – how are the qualities of the constituents mingled to form the new qualities that characterize the product of the reaction? As soon as this question is asked, it almost inevitably leads to some kind of atomic conception. However, as the Democritic atoms were banished from Aristotelian philosophy, Ibn Roshd seized upon the next best thing he could find there, viz. the smallest parts into which materials could be divided without losing their qualitative properties. Thus, this concept acquired in Arabic natural philosophy (and thereby also under the name 'minima naturalia' in the scholastic philosophy of the West) a significance and a position that Aristotle had never imagined. The doctrine of 'minima naturalia' cannot stand comparison, in point of rigour or profundity, with the old atomic theory, but the latter's clear outline of the problems had by now dropped into oblivion. Then, too, if the concept of the smallest parts was not nearly so pure as the atomic concept proper, neither were the methods of the chemists pure enough for them to feel the need of more sophisticated theoretical interpretations. In any case, we have here an instance of an atomic conception not being derived from philosophical reflections, but forced upon people by a problem of their experience; at the same time we see how radically this simple shift of motivation could alter a concept like that of the smallest parts of matter.

Let us now proceed to the seventeenth century, the period in which modern science emerges. Here we immediately observe what a tremendous broadening there has been of the empirical basis on which the question of the structure of matter is now being re-examined: along with chemical problems there is now a whole multiplicity of physical properties of bodies to be considered; at the same time more is being required of the theoretical explanation of phenomena, greater stress is being laid on quantitative rather than qualitative relations. A new spirit reigns among the investigators of nature: they face problems with completely fresh and receptive minds. They have read the ancient works, either willingly or as students, but they make no use of them. What they read about Epicurus (Democritus is hardly known) or Aristotle conveys nothing to them; they attack the problems with a new and powerful method that they have worked out for themselves. This rigorous inductive method, as it is described in the preface to the second edition of Newton's *Principia*, has come into being and been refined in the course of the research work itself. But the new spirit that animates these investigators they owe to their predecessor, Descartes, the great thinker who paved the way for the renewal of science without entering upon it himself.

Descartes proceeded from the conception that matter and space were identical; from this he concluded that nature could hardly tolerate a vacuum, but in order to allow for the possibility of motion, he imagined the filling of space with matter in a less uncompromising form than Parmenides. Descartes allowed that 'gross' matter, of which the bodies we observe directly are composed, could consist of individual pieces; all the spaces between these pieces would then be filled with 'subtle matter', itself consisting of pieces of much smaller size in contact with each other without the slightest interspace. Such a filling of space permits the 'gross' matter only one kind of motion: a cyclic movement along a closed path, which arises by pieces of 'gross' matter being carried along by the surrounding 'subtle' matter in such a way that any vacuum that the shifting of a piece of matter would leave is immediately filled by 'subtle' matter flowing into it. Thus the motions of the planets around the sun are explained by imagining that the sun produces vortices in the subtle matter, which carry the planets along with them.

To the great scientists of the age, Huygens and Newton, Descartes' purely speculative arguments, to which he himself attached the greatest value, had no longer any meaning, but they retained the main ideas of the Cartesian world-picture in so far as these proved to be compatible with the regularities deduced from experience: once more we see a new generation with new motivations appropriating the ideas of previous ages in modified forms. The idea of subtle matter, in particular, lived on under the name of 'aether', which for Newton played a quite fundamental part as the medium through which forces could propagate and be transferred from one body to another: for it was obvious that an immediate interaction could only take place by collisions between pieces of gross or subtle matter in direct contact with one another.

However, the transition from Descartes' subtle matter to Newton's aether meant a dramatic turn in the conception of the nature of gross matter. The new knowledge about the laws of motion of material bodies, and especially of the planets, led Newton to realize that the presence of aether itself had so little influence in the motion of gross matter that it could be completely ignored in the investigation of the latter's structure. Thus was this investigation, which built on experience from chemical processes, brought back into atomistic paths: gross matter must consist of separate atoms, and the interspaces between them might simply be reckoned as empty. But was this a return to the atomic theory of the ancients? Not at all, for there was no question of either a really absolute vacuum or of solid atoms – the very nature of atoms was now, as we shall see, a chemical problem accessible to investigation. The motivation that had given rise to these ancient conceptions was then indeed completely unknown. Newton hardly takes account of the old atomists; Boyle says expressly that it is indifferent whether one regards atoms in an Epicurean or a Cartesian sense, for this does not alter the content of the theory with respect to observable phenomena – a very superior attitude.

How correct this point of view was, is amusingly illustrated by the controversy between Cartesians and Epicureans that had just arisen among French scientists. In direct opposition to Descartes, Gassendi, who although a Roman Catholic bishop was a progressive humanist, had tried to revive the Epicurean atomic doctrine – no doubt with some dilution of its materialistic flavour. Among the supporters of Gassendi was the young Pascal, who eagerly advertised the barometric experiment as a decisive argument in favour of the Epicurean claim for the existence of a vacuum: the space above the column of mercury in the barometric tube was undoubtedly empty. This however, involved him in a quarrel with a Jesuit of the name of Noël, and accordingly called in the Catholic style 'le père Noël'. In good Cartesian fashion the latter maintained that the space in question was not empty at all but filled with subtle matter that could penetrate through the glass without difficulty. The ensuing controversy soon showed that the opposition of the two conceptions against each other was a hopeless verbal quarrel. It ended by Pascal silencing the "bon père" with his mordant wit, but this was an admission that he was not able to refute his assertions with genuine arguments. In this matter the behaviour of the English scientists proves much less tradition-bound. They could allow themselves to break with the old tradition because – more quickly than the French – they had acquired a new way of investigating nature and had put its power to the test.

We find a fine example of the range of the Newtonian method precisely in the chemical investigations in which he immersed himself throughout his life. This side of Newton's activity has not gained the same renown as his optical and astronomical investigations; it was, of course, difficult to reach assured results in chemistry before the concept of the pure element had been formulated. Yet with great perspicacity and critical imagination Newton was able to develop a highly detailed conception of the very structure of atoms to account for all the experience he had gathered. The word atom had by then already lost its literal significance: it merely designated the various building blocks that went to make up the structure of bodies, but for Newton there was no obstacle to assuming that these atoms might themselves be built up of yet smaller particles. Thus, from the varying strengths of acids he concluded that there must be various groups of atomic constituents that were held together by binding forces of varying magnitude. He even imagined that the binding forces must proceed from the centre of the atom, so that the various groups of atomic constituents were distributed around this centre at distances the greater as their binding was weaker: strong acids would thus penetrate more deeply into the atomic structure than weak ones. How far this atomic model resembles that we nowadays regard as correct is unimportant in this context; it testifies above all to the superiority of the Newtonian way of thinking. By continually seeking support in experience he had liberated himself from the artificial limitations that abstract speculations had created; he was able to recognize new problems as they arose and to take steady steps towards their solution. As soon as he realized that chemical experience indicated that atoms must be composite systems, he did not hesitate a moment to take up this strange idea and to elaborate it in the direction suggested by experience.

As the last period-picture from the history of atomism, I shall now consider some aspects of nineteenth-century development that add new variations to our theme. In spite of the promising beginning it had been given by Newton's efforts, the investigation of the structure of the atoms had soon reached an impasse, and at the beginning of the nineteenth century this question was in the midst of a serious crisis. The difficulty was that there seemed to be no possibility of determining atomic structure empirically in any unambiguous way, and arbitrary assumptions were useless. An echo of the general disappointment in this field is to be found in Laplace's famous work *Exposition du système du monde*, in which he writes that consideration of the properties of atoms could not further the development of industry. In this statement he also revealed the new motivation that was to dominate the whole further advance of science.

This situation contributed to persuade physicists to being satisfied with an atomic concept from which all determination of structure had been eliminated: atoms were reduced to points, to centres of force, and the forces that proceeded from them were merely presented, in analogy to gravitation, as diminishing with distance from the centre according to some – otherwise unknown – law. This retreat proved enormously fruitful: what was lost in the description of the more specific qualities of atoms was gained in mastering the more general features of their behaviour. On this basis the grand theoretical syntheses could be erected that are known as statistical mechanics and electromagnetism. Although the former of these great theories might well call for many comments, we must be content to linger a little on the latter; and here we must first of all remember that at that time all physical processes were thought of as in essence mechanical: explaining electromagnetic phenomena was necessarily a question of reducing them to certain forms of motion of the constituents of matter.

Now, the introduction of the concept of the atom without structure had given aether a more prominent place in the atomistic description of the world: atoms seemed only to be singularities in this all-filling medium, in which therefore most physical processes had to occur. Thus the mechanical explanation of the polarization of a dielectric body, which Maxwell first attempted to set up, rested on an assumption that certain hypothetical aether particles could be set in motion by an electric force; the presence of inhomogeneities in the aether, corresponding to the presence of the atoms of the body, would then lead to a finite displacement of these aether particles proportional to the electric force, and this would correspond to what we call polarization; in the free aether, on the other hand, no finite displacement would occur. But the tendency to the displacement of the aether particles, the 'displacement current', is present in the whole aether, regardless of whether there are atoms there or not: it was therefore quite natural for Maxwell to reckon with the possibility of displacement currents arising in free aether and to treat these as ordinary conduction currents. Since he had already worked out a mechanical account of the magnetic effects of ordinary conducting currents, he could see at once that a mechanism, completely unintended, now existed for the propagation of electric oscillations in the aether; the recognition of the identity between these and light waves was straightforward, and Maxwell's theory, in its first form, made its entry into science as an account of the mechanical properties of the aether.

Whereas physics could thrive without entering into the problems of the structure of atoms, the renunciation to take up this question proved actually paralysing to theoretical thinking in chemistry. For far too long the attitude of the chemists was characterized by a distrust of all that appeared to them 'hypothetical', to which they added the superstition that they 'exclusively' relied upon experience. This attitude could not but lead them to equally preposterous errors of judgement as Parmenides' blind faith in pure reason. Admittedly they can scarcely be blamed for not appreciating Prout's assertion that there must be a building-block common to the atoms of all elements, for this assertion was not particularly well founded. But the great virtuoso of chemical operations, Stas, committed a logical error when he believed that his extremely accurate determinations of atomic weights provided a decisive rebuttal of Prout's hypothesis: as Kekulé immediately observed, Stas had overlooked the logical (and as we now know actual) possibility that not all atoms of the same chemical element have the same weight. Worse was the failure of the chemists to realize the rational nature of Avogradro's hypothesis: they cut themselves off from the great advantages that the atomic-weight system based on this hypothesis offered – until they were simply forced to recognize it as the only way of creating order among the many new combination possibilities in the organic domain.

It was also the growth of organic chemistry that brought the chemists – although grudgingly – to recognize once more the indispensability of the atomic concept. The occurrence of isomeric combinations showed clearly that the knowledge of the composition of a product was insufficient for characterizing its behaviour in chemical reactions, and purely descriptive classifications had only limited validity. The great pioneer was Kekulé, who by his suggestion of the ring structure of benzene demonstrated that one could come a good step forward merely by assuming that molecules composed of the same atoms but in which the atoms form different geometrical structures may have quite different properties determined by their structure. It is true that Kekulé did add cautiously that these structures ought only to be conceived as symbolic aids, but van 't Hoff, when he took the next step and introduced three-dimensional molecular structures, no longer had any doubts about the physical reality of these structures.

That the idea of atom, through all the perils of its history, was able to remain alive and to adapt itself to the widely varying demands made through the ages upon the description of the world, is one of the most encouraging lessons to be drawn from the history of science. Not without good reason it might be said that the vitality of the atomic concept was due to its being rooted in nature itself; but this is only one side of the matter. Until the close of the nineteenth century the atomic theory was, after all, only a hypothesis, and as such its success can only be explained by pointing out that at any given time it provided the most rational account of experience. The triumph of atomic theory is first and foremost a victory for the scientific method and way of thinking.

Transl. by David Stoner, revised by L. R.

BIBLIOGRAPHY

Boas, M., *Robert Boyle and XVIIth Century Chemistry*, Cambridge University Press, 1958.

Boas, M., 'Newton's Chemical Papers' in *Isaac Newton's Papers and Letters on Natural Philosophy* (ed. by I. B. Cohen), Cambridge University Press, 1958.

G. Bugge (ed.), *Das Buch der grossen Chemiker*, vol. II, Verlag Chemie, Berlin, 1930: See the essay on Canizzaro by B. L. Vanzetti and M. Speler, on Kekulé by R. Winderlich, and on Van 't Hoff by E. Cohen.

Enriques, F. and de Santillana, G.: 'Histoire de la pensée scientifique', *Actualités scientifiques et industrielles*, Nos. 384, 385, 386, Hermann, Paris, 1936.

Farrington, B., *Greek Science*, Pelican Books, London, 1953.

Gillis, J., 'Auguste Kekulé et son oeuvre, réalisée à Gand de 1858 à 1867', *Mémoires in 8°, Cl. des Sciences, Acad. Roy. Belg.* **37** (1966), No. 1.

Levey, M., 'Studies in the Development of Atomic Theory', *Chymia: Annual Studies in the History of Chemistry*, No. 7 (University of Pennsylvania Press, Philadelphia, 1961) pp. 40-56.

van Melsen, A. G. M., *Het wijsgeerig verleden der atoomtheorie*, Amsterdam 1941. [See the later work of van Melsen, *From Atomos to Atom: The History of the Concept 'Atom'* (Harper, New York, 1960) – Ed.].

Pedersen, O. and Pihl, M., *Historisk indledning til den klassiske Fysik. 1. De eksakte naturvidenskaber i oldtid og middelaider*, Munksgaard, Copenhagen, 1963.

Rosenfeld, L., 'The Velocity of Light and the Evolution of Electrodynamics', *Supplemento al Nuovo Cimento* **4** (1956), 1630. [This volume, p. 134].

Rosenfeld, L., 'Newton and the Law of Gravitation', *Archive for History of Exact Sciences* **2** (1965), 365. [This volume, p. 58].

CHAPTER 6

FLICKER IN THE DARKNESS

(Review of *Nicole Oresme and the Medieval Geometry of Qualities and Motions* (ed. M. Clagett) [University of Wisconsin Press, Madison and London, 1968. xiii + 713 pp.])

[1969e]

This volume is the latest instalment of the corpus of mediaeval writing on mathematics and mechanics to the publication of which Professor Clagett, with great competence and unflagging industry, has devoted his department in the University of Wisconsin for many years. It is no mean task to decipher the scribblings of ignorant scribes and transform them into polished critical editions, accompanied by English translations and abundant commentary; it must also be a work of infinite patience to labour through the unspeakably tiresome arrays of pedantic syllogisms into which the schoolmen diluted their primitive arguments. Clagett's own commentary, although making use of mathematical symbols invented to avoid prolixity, is itself so punctilious that it more often than not carries the dilution one step further: in the volume under review, the Oresme text, collated from some fourteen manuscripts, occupies, with its translation, only about one third of the book; the rest consists of an informative introduction, an explanatory commentary, the reproduction and translation of a number of other relevant mediaeval texts, and several detailed indexes. This lavishness – no doubt of great service to the scholar – goes a long way to account for the enormous bulk of the Wisconsin medieval thesaurus.

Confronted with such a display of scholarly zeal, the modern scientist may well wonder whether it could not find more rewarding objects than the barren wastes of long forgotten mediaeval lore. He assumes, reasonably enough, that the founders of the powerful methods of investigation he is still applying today were the best judges of the worthlessness of the scholastic tradition they scornfully rejected, and he is inclined to accept their verdict without taking much trouble to verify it. This summary judgment over the 'dark ages' of Western culture, however, is just what the mediaevalists contest: if they so intently peer into the darkness, it is with the conviction that they are bringing hidden treasures to light.

Although their own exacting studies of the sources have by now exposed the extravagance of Duhem's early claims about mediaeval anticipations of modern scientific conceptions, they are still reluctant to give up the belief that scholastic ideas could have exerted a significant 'influence' on Galileo and other pioneers of modern science. This, then, is the historical problem which at any rate gives the mediaevalists' efforts a valid motivation. Even if it turns out, on critical assessment, that all the documents they are so neatly dusting fail to evidence the existence of any such 'influences', their labour will not have been in vain: their contribution will certainly not have less importance if it leads us definitely to disprove the improbable thesis of a continuity between mediaeval and modern thinking, and documents instead the occurrence of a complete break between them.

Oresme's treatise on the "configurations of qualities and motions", of which we are now given a definitive edition, is a key document in this respect, in view of the popularity it enjoyed in the schools during the fourteenth and all through the fifteenth century. It deals with a favourite exercise of the schoolmen's: the analysis of qualities according to intension and extension – that is to say, the distribution of their intensity over a range determined by the nature of the subject: space, time or any other character. For representing a magnitude of any kind by a linear segment the schoolmen had Aristotle's authority; the innovation Oresme introduced consisted in disposing the segments which represent the intensions perpendicularly to the line along which the degrees of extension were marked: the extremities of the intension segments thus formed a curve, which together with the extension line determined the 'configuration' of the quality. Oresme even contemplated qualities extended over a surface or a volume, whose configurations formed, respectively, a volume or (because there is no fourth dimension) an aggregate of volumes interpenetrating each other. He was quite incapable of handling effectively, however, any other case than what we would call a linear variation of the intension, and there is in his 'imagination' of qualities no shadow of anticipation of analytic geometry. Still, he applies the scheme to a bewildering variety of examples; it was supposed to work equally well for studying a motion of variable velocity, for relating the medicinal virtues of plants to the form of their parts, or for explaining why almond-shaped eyes are more beautiful than round ones.

It is Oresme's application of his method to uniformly accelerated motion that has given rise to speculation about the help Galileo could have derived from it in his own investigation of the law of fall; this issue is indeed raised again by the editor, although it would seem that it was settled long ago in the negative by Dijksterhuis's painstaking discussion of the case in his book *Val en Worp*. Oresme's solution proceeds smoothly enough: the extension being the time and the intension the linearly increasing velocity, he realizes that the space is given by the area of the resulting triangular configuration and he thus finds for it the correct quadratic law. It does not occur to him, however, that this type of motion could have any relation with the fall of material bodies. On the other hand, if Galileo, when he had found the law of spaces by experiment, had known, or remembered, Oresme's formal result, he would not, in his attempt to derive by mathematical reasoning the corresponding law of velocities, have made the false start of assuming the velocity to be proportional to space instead of time. Too much is made of the similarity between Galileo's ultimate formulation of the law of accelerated motion in the *Discorsi* and that given by Oresme: both state that the accelerated body will travel through the same space in a given time as it would in a uniform motion with half its final velocity. This just corresponds to the statement that the area of a right triangle is the same as that of a rectangle of the same base and half the height – a simple thumb-rule familiar, no doubt, to every surveyor, but which our mediaevalists pompously call the 'Merton theorem' because it happened to be a popular subject of verbal exercise in Merton College. Pages teeming with erudition are spent by the editor conjecturing how Galileo could have gained access to Oresme's ideas: why does he neglect the clear evidence we have of the actual course of Galileo's studies, of his early initiation to the works of Archimedes and the resulting break with the scholastic tradition?

It is more to the point to enquire about Oresme's position within the development of mediaeval thought; on this problem much light is thrown by Clagett's sensitive appraisal of his hero's work and personality. To form a fair judgment of him, we must not be put off by the barbarous traits he shares with his time; belief in the conformity of shapes and therapeutic properties of plants, or in the reality of phenomena associated with magic and sorcery. Disconcerting as this may seem, if he includes these things in his treatise on qualities, it is with a rationalistic intention: he fights superstition by rationalizing it; he exorcises the demons by reducing their workings to natural causes. As to astrological predictions, he deprives them of any real foundation by an ingenious argument: because the periods of the heavenly motions are incommensurable, no celestial configuration will ever repeat itself, and the alleged regularities on which the astrologers base their horoscopes are therefore spurious.

Oresme's rationalistic turn of mind applied itself, of course, also to more concrete objects. His familiarity with the optical writings of Robert Grosseteste and Witelo, to which he owed the suggestion of his graphical method (they represented beams of light by cones whose opening indicated the intensity of illumination), reveals an early interest for the study of natural phenomena, which especially Grosseteste and his disciple Roger Bacon endeavoured to foster among the clerics. Oresme's later participation in the affairs of the realm appears indeed as a fulfilment (however modest) of Roger Bacon's vision of a philosophy serviceable to the useful arts: we find an echo of this aspiration in the prologues of Oresme's French writings, which were addressed directly to the merchants, the 'mechanics', the King's civil servants. Yet it could not escape such clear-sighted men as Oresme and his best disciple Henry of Hesse how insufficient the scholastic way of thinking was to cope with real problems, how powerless its formal approach leaves us in front of practical tasks: one perceives in the parts of their writings dealing with natural phenomena an undertone of scepticism and disillusion.

We are here touching on one of the most challenging problems of the history of science. Why did the rationalistic tendencies within the scholastic philosophy ultimately come to nothing? Why did the fire not catch, but remain a mere flicker in the darkness? Clagett is certainly right to point to the intrinsic weakness of the scholastic method as a cause of the decay which set in so soon after Oresme's promising start; but it is not the only one. The schoolmen's pathetic failure must be seen in the perspective of the general crisis which swept away the whole social structure of the middle ages; both the intellectual and the economic aspects of the case must be taken into account. For such a study abundant material is at hand, but historians of science do not seem too keen to analyse it. It is of course more polite, and safer, to bestow praise on the thinkers of the past than to expose their failings; and polite reviewers encourage historians to believe that nothing else is expected of them. Thus the legend of the continuous progress of science is kept alive, and the history of the most spectacular defeat of human reason remains unwritten.

CHAPTER 7

MARCUS MARCI'S INVESTIGATIONS OF THE PRISM AND THEIR RELATION TO NEWTON'S THEORY OF COLOR

[1932c]

The Prague doctor and naturalist Marcus Marci[1] has been frequently cited as a predecessor of Newton in establishing the physical theory of color; he recognized namely that (1) different refractions uniquely correspond to the different colored rays produced by a prism; (2) the monochromatic rays so defined maintain their color unchanged under further refraction: two propositions which in fact form an essential part of the Newtonian theory of color. Now, naturally, this last assertion is correct as far as it goes; however I would not take it as proper to hold Marci up as Newton's predecessor merely on the basis of this fact. For, completely apart from the question of whether or not Newton was acquainted with Marci's arguments, the history of physics should not be comprehended as a mere chronological stringing together of isolated discoveries, but rather the main emphasis should be placed on the logical connections between the individual steps which contribute to the development of a physical theory. In order to judge whether, from this standpoint, a given theory can be designated as a predecessor of another, more complete, theory, the following question must above all be decided: Does an extension and modification, in keeping with their meaning, of the essential categories suffice for arriving at the second theory from the first; or is a complete rejection of the basic concept of the first theory required? Only in the first case could one speak of an historical connection. In this sense, for example, Galileian mechanics would be a predecessor of Newtonian mechanics and this latter in turn a predecessor of Einsteinian mechanics; or the older quantum theory would also be a forerunner of modern quantum mechanics. In no way, however, could one – in spite of the superficial similarity – discern an historical connection between deBroglie's concept of phase waves and the peculiar wave concept that Newton tried to introduce into his corpuscular theory of light in order to explain the 'fits of easy reflexion and transmission'.[2] How do things stand in this respect with Marci's theory of color? In what follows I shall try to answer this question, and, indeed it will turn out that Marci cannot be considered a predecessor of Newton in the sense just explained.

1. The Newtonian theory of color

First we shall recall the well-known train of thought of the Newtonian theory.[3] Then we shall elucidate Marci's investigations – so far as they are of interest for our purposes – in order finally to make a comparison between both theories.

The main feature of the Newtonian arguments is their strictly inductive character, their complete independence of any *a priori* conception of the nature of light. He starts from consideration of the ordinary solar spectrum. According to the law of refraction the observed extension of the spectrum perpendicular to the prism axis stands in contradiction to the assumption of a unique index of refraction identical for all rays. Consequently it is a matter of separating the different kinds of rays and investigating them individually: this was achieved by the *Experimentum crucis* which showed that (1) the ordinary law of refraction holds for each kind of ray, only with different values of the refractive index, (2) a unique constant correspondence exists between refractive index and color. From this the concepts of 'simple' and 'composed' colors are then rigorously derived; in this, the role of the prism as a 'color analyzer' stands out clearly.

2. Marci's investigations of the prism

Let us now turn to an analysis of the portions of interest to us in Marci's optical works, *Thaumantias, Liber de Arcu coelesti.*[4] First it must be emphasized that Marci was still completely under the spell of scholasticism: his method of treatment consists in subtle arguments supported by authorities; if experiments are cited, it is always only as a last resort in order to confirm conclusions arrived at scholastically, and never as the starting point of inductive considerations: facts of experience are treated approximately on the same footing as written authorities. Furthermore, Marci contented himself mostly with superficial, often incorrectly interpreted, observations; he seldom went as far as the quantitative; in no place did he systematically carry out a discussion or analysis of his experimental results.

From the outset he assumes a definite conception of the generation of colors from white light by means of 'condensation'; this 'condensation', moreover, has nothing to do with the intensity of light. Because it creates changes in condensation, the refraction of a beam of white light produces different colored beams (Theorema X, p. 69). But the colors are not caused by the deflection of the beams taking place in refraction (Th. VIII, p. 68). Nevertheless a correspondence may be established between the different colors and the different refractions: "Th. XII (p. 83) Lux non nisi refractione certa in medio denso mutatur in colores: diversaeque colorum species sunt partus refractionum". [Light is changed into colors by means of a certain refraction in an opaque (lit. 'dense') medium: the different appearances of color are the results of refractions.]

Here, by the way, it must be noted that the law of refraction was unknown to Marci.

On p. 94 a thorough investigation of the prism now begins: "De Iride trigonia deque illius proprietatibus et earundem causis". [On the prism spectrum (lit. 'prism rainbow') and its properties and the causes of same.]

After a description of trivial properties of the spectrum, the law of correspondence between spectral color and refraction is easily 'proved' in view of Theorem XII cited above:

"Th. XVII (p. 99) Neque idem color a diversa refractione, neque ab eadem plures colores esse possunt". [It is not possible to have either the same color from different refraction(s) or several colors from the same refraction.]

The following two theorems now state that spectral colors remain unchanged under further reflexion or refraction:

"Th. XIX (p. 100) Reflexio superveniens radio colorato non mutat rationem coloris". [The reflection resulting from a beam of colored light does not change the nature of the color.]

"Th. XX (p. 100) Refraction superveniens radio colorato non mutat speciem coloris". [The refraction resulting from a beam of colored light does not change the appearance of the color.]

This last theorem is just the chief result of the famous *Experimentum crucis* of Newton. But what does Marci say about it? To give, at the same time, an example of his style, I will cite verbatim the beginning of the 'proof', in which the experiments drawn upon are described:

"Constitue pilam crystallinam medio inter visum et trigonum loco: lumenque candelae, una cum coloribus, eversum spectabis: nulla in his facta mutatione. . . . Simili modo in pictura a trigono proxime et mediante pila crystallina, aut per trigonum geminatum transmissa, eosdem prorsus colores videbis. Quin etiam horum singulos per se, nova refractione, in iridem circularem atque unicolorem transformare postea docebo. Sicuti vero refractio superveniens nihil officit radio colorato, ita praevia nihil eidem opitulatur. Quod si enim sphaera pellucida media inter trigonum et luminosum radios praeoccupet, duo[5] simulachra ejusdem flammae recipies, iisdem prorsus coloribus, eodemque situ dispositis: licet hujus simulachrum sit eversum . . .". [Set up a crystal ball half way between an observer and a prism: you will see the candle light disappear (lit. 'swept away'), together with the colors: no change in these matters having taken place. . . . Likewise you will see the very same colors in the image transmitted from the prism near to and through the crystal ball or double prism. I shall also show that each of these (spectra), by means of a new refraction, changes one by one into a circular and monochromatic spectrum (lit. 'rainbow'). As a refraction resulting from a beam of colored light obstructs nothing, so no previous (refraction) helps the same (beam of colored light). But if a clear ball is placed midway between a prism and light beams, you will receive two[5] images of the same flame with the very same colors situated in the same place: although the image of this has disappeared (lit. 'swept away') . . .]

Of these experiments, the one which comes closest to the *Experimentum crucis* is only mentioned in this passage and first described later (Porisma III, p. 254) in another context: the production of a monochromatic circular beam of light. It is a matter, namely, of the image produced by a ring-shaped prism when any monochromatic ray from the spectrum formed by means of an ordinary prism is used as a light source. The monochromatic ray is not picked out, say, by means of a diaphragm; rather the ring-shaped prism is placed at such a great distance from the first prism that it can be completely illuminated by a single, 'monochromatic' spectral beam!

On pages 105–106 the assertion is then made that neither the first refraction alone (i.e. the refraction at the incident surface of the prism) nor the second (at the exit surface) alone can produce the spectrum of colors. For were the spectrum produced e.g. by the first refraction alone, then on the basis of Theorem XX it could not be affected by the second refraction and would consequently occur independently of the angle between both surfaces: this, however, contradicts the experiment with plane-parallel plates.

There now follows on p. 107 a rather unexpected theorem:

"Th. XXXI. Radii defernetes simulachra colorum sunt diversi a radiis deferentibus speciem objecti." [Beams of light showing the images of colors are different from beams of light outlining an object.]

If, namely, say in the experiments used for the proof of Theorem XX, an image of an object with a colored edge is projected by means of a prism, the image will be inverted by a lens placed in between, while the colors along the edge remain unchanged: these colors are thus produced by special rays (radii colorigeni [beams of colored light]) which are not influenced by the lens in the same way as the rays which produce the image. But how is it to be explained that these colors always appear at the edge of the images? The answer goes simply: Occasionally color rays accidentally coincide with image rays ("Per accidens autem fit, quod radii colorati quandoque coincidunt cum specie objecti" [However, it happens by chance that beams of colored light sometimes coincide with the image of the object.]); however, due to the greater intensity of the latter, nothing of the colors is to be noted in the image itself; only along the edge can one observe the irisation (p. 177).

3. CRITIQUE OF THE MARCIAN THEORY

This sketch of the Marcian conception of the origin of the colors actually speaks for itself and we can now settle the question posed at the beginning in a few words.

First, as far as concerns the unique correspondence between colors and refractions, Marci, in spite of the apparently sharp formulation of his Theorem XVIII, overlooked all essential points: he did not believe that the dispersion of light took place through the deflection of rays at a plane of separation; furthermore, he had no concept of the constancy of dispersion, as his use of 'monochromatic' wide spectral beams clearly shows.[6] Consequently he lacked a clear distinction between 'simple' and 'composite' colors (although naturally he had a vague idea of this distinction), which led him to his fantastic account of image production and irisation of the edges.

As concerns the indecomposability of spectral colors, the significance of Theorem XX should also not be over-estimated. He had the *Experimentum crucis* only in primitive and inexact form; but what is more important: he never considered it as a crucial experiment, rather more as an incidental trifle. The other experiments were even completely falsely interpreted: e.g. he asserted that lenses or transparent spheres which were inserted into the path of the light rays produced *no color at all*[7] but only modified the image. Finally, this Theorem XX assumed no central position in Marci's book: he did not use it, say, for a definition of 'simple' color, but only for dialectical purposes (an example of which was given above).

In summary we must thus state that Marci comprehended neither the precise content nor the significance of both his theorems, and that his arguments would have been impossible to use as a starting point for a theory of color.

NOTES

[1] See E. Hoppe, *Archiv für Geschichte der Mathematik* **10** (1927), 282. A portrait of Marci is found in E. Mach's *Mechanik*, 8th edition (Leipzig 1921), p. 311. [See English transl. of the 9th German edition by T. J. McCormack, *The Science of Mechanics* (Open Court, La Salle, Ill., 1960), p. 396 – Ed.]

[2] See e.g. L. Rosenfeld, *Isis* **11** (1928), 116.

[3] See L. Rosenfeld, *Isis* **9** (1927), **44**. [This volume, p. 16.]

[4] *Thaumantias,* Liber de arcu coelesti deque colorum apparentium natura, ortu et causis . . . [Book on the rainbow and the nature, origin and causes of its apparent colors . . .] Authore Joanne Marco Marci (Prague, Typis Academicis, 1648).

[5] He always places the prism in relation to the broad light beam in such a way that two (mirror-image) spectra are cast.

[6] Moreover he explicitly distinguishes only four kinds of color. But Newton, too, indulged in speculations about a finite color scale; although, on the other hand, he rigorously formulated the physical continuity of the spectrum.

[7] What is striking about Marci is the absence of any ability to abstract or generalize: here the essential identity of the lens and prism escape him. In his investigations of collisions, he correctly discussed many special cases but did not arrive at the general law of collisions.

DESCARTES AT UPPSALA

(Review of Rolf Lindborg's *Descartes i Uppsala*. Striderna om "nya filosofien" 1633–1689.
[Almqvist & Wiksell, Stockholm, 1965. xvi + 367 pp.])
[1967a]

Cartesianism had a much more profound influence upon the formation of the modern scientific method and outlook than the merely episodic part too often ascribed to it by superficial historians. Not only did it inspire the greatest creative minds of the time – Huygens, Boyle, Hooke, and even Newton himself – but it had to bear the brunt of the inevitable reaction from the traditional clerical philosophy and thus helped to shield the most critical stage of the scientific expansion from further hindrances. The fight between the Cartesians and the theologians is a long and complicated process, the successive phases of which are most revealing for the mechanisms, of both intellectual and social origin, that govern the progress of scientific epistemology. Although abundant material, especially from the events in the Low Countries, has long been available for the study of this process, its full analysis is still awaiting the historian imaginative enough to attempt it. The book under review does not yet give us this analysis (this was not the author's aim); but besides extremely valuable further material concerning little-known Swedish controversies, it contains a penetrating background study of the Dutch developments which goes a long way toward the elucidation of the major issues.

In a too modest preface, the author disclaims any synthetic intention. He wants to confine himself to the discussion of the purely philosophical issues, while at the same time he declares himself prepared to believe that the key to a deeper understanding of the conflicts would rather be found in the study of the underlying social tensions. I would seriously question the soundness of such an attitude if the author had rigidly adopted it. Fortunately, he does in fact all through his conscientious and objective account of the events duly bring out, with sufficient emphasis, the relevant circumstances throwing light on the social aspects. These aspects force themselves upon him, so to speak, and such a painstaking historian as he is cannot fail to report them in spite of all self-imposed restrictions on the scope of his inquiry.

The most striking outcome of the author's thorough and infinitely patient examination of the long-forgotten struggles of the Swedish churchmen against the introduction of Cartesian ideas in the teaching at the University of Uppsala is to show that they proceeded (discounting differences in the local conditions) along much the same pattern as the similar struggles which had been raging in the Netherlands some twenty years earlier. Even the significant fact that the battle was waged in two stages, by two successive generations (separated by an interval of about twenty-five years), is repeated on the two theaters.

The first round ended in each case with a compromise, which in effect prepared the ultimate defeat of the theologians: they only retained control over questions of religious orthodoxy, but had to concede the *libertas philosophandi* in all problems of natural science. This decision was by no means the result of any serious philosophical discussion of the issues involved, although the Cartesian method found enlightened and active supporters in such strong personalities as the physiologists Regius in Holland and Rudbeck in Sweden. What clinched the issue was on the one hand the wish of the socially powerful medical faculty to assert its independence, and on the other the anxiety of the responsible authorities to prevent disputes which might – considering the rough manners of the time – damage the academic discipline and decorum.

Saving, not the phenomena, but the appearances, was indeed a major pre-occupation in those squabbles. The second clash only occurred, after a period of uneasy coexistence, when the new scientific generation brought up in the Cartesian spirit felt itself strong enough to break the *modus vivendi* that had been agreed upon with the theologians. Meanwhile the real scientific movement went on outside the schools in which the theologians fought their battles and took hardly any notice of their opposition – except as a target for ridicule. Nevertheless, the indirect influence of the schoolmen must not be underrated: after all, they molded the minds not only of the few youngsters that would in spite of all fall under the spell of the new philosophy, but also of the crowd of young noblemen and clerics who were destined to wield the instruments of power on which the existence and activity of the scientists depended.

That the schoolmen were very conscious of this role is shown by the very nature of the leitmotives constantly recurring in their tenacious opposition. They were of course averse to novelties by sheer intellectual inertia: one of their main arguments for upholding the peripatetic jargon was its use in disputes with heretics. Their hostility to Cartesianism, however, had more specific reasons, which are very significant. Above all, they abhorred the methodical doubt advocated by Descartes, that would make the students unruly by depriving them of the firmness and security of mind which the dogmatic philosophy provided. Moreover, the way in which Cartesians proposed 'natural' interpretations of Biblical stories – an imprudence that had been fatal to Galilei – was just as much a thorn in the eye of the protestant theologians, no different in this respect from the Roman inquisitors. Finally, the Copernican cosmology, accepted on both sides as an inherent part of the Cartesian system (Descartes' subterfuge to circumvent the question of the earth's motion deceived nobody), was dreaded for that very consequence, the infinite extension of the world, whose formulation had led Giordano Bruno to the stake. It is not surprising that the stronghold of the Cartesians was the medical faculty, but it is remarkable that in the defense put up by the physicians, the specific Cartesian features and arguments often glided into the background and that the issues of real concern were the experimental method of investigation and the primacy of human reason as a source of knowledge. On the former issue the theologians were ready to make concessions, but they were adamant in their stand regarding the latter.

The above reflections inspired by the reading of the book under review are only intended to illustrate the richness of its content and its importance for the understanding of one of the most decisive spiritual revolutions of human history. They do not by any means exhaust the wealth of out-of-the way information accumulated in its pages. To mention only one detail taken at random, the author, in order to document the intellectual background of one of the protagonists, enumerates most interestingly the main books in his library, as they appeared in the catalogue of the auction sale of his goods after his death. Scholarly care and accuracy are everywhere conspicuous, and the often intricate arguments about points nowadays utterly forgotten are expounded with true Cartesian clarity and distinctness. The book is a worthy addition to the *Lychnos-bibliotek*, edited by the Swedish History of Science Society. It does honor to the standards of quality of Swedish book production as well as to those of scholarship of this active society.

CHAPTER 9

NEWTON AND THE LAW OF GRAVITATION

[1965d]

There has been recently a welcome revival of Newtonian studies. The Royal Society has at long last made a start with the publication of the correspondence[1]. Various scholars have brought to light much interesting new evidence from the too long neglected Newton manuscripts[2]. For the first time a thorough and competent study[3] has been made of a side of his activity which had hitherto remained obscure: his historical researches, coloured by theological considerations, to which he himself attached great importance. By piecing together all this new evidence with long known, but little understood facts, one arrives, as I intend to show, at a view of Newton's personality rather different from the traditional one. The latter is very much influenced by hero worship[4], but it does not help to react to this – as recent biographers are inclined to do – by hero debunking. Newton's personality is not easy to understand: secretive and suspicious as he was, one has to catch him, so to speak, in unguarded moments to get a glimpse of his thoughts and of his passions. To reconstruct a coherent portrait from the scraps of evidence gleaned from his papers, his letters and his actions is a hard detective work, but a rewarding one.

In Newton's thought, the problem of understanding the construction of the universe, which led him to the discovery of the law of gravitation, doubtless occupied a central position, although it was by no means the problem which he regarded as the most decisive and to which he devoted the greatest effort. Indeed, it is strange how casually he dealt with it before Halley with much difficulty managed to wring out of him the work which we esteem his greatest. There is certainly something in this attitude which demands explanation. The circumstances of the discovery of the law of gravitation present a further puzzling feature, which early attracted the attention of historians: after he had the first idea of the identity of the force of gravity on the earth with that governing the planetary motions, why did Newton not follow up the clue at once? Why did he allow twenty years to elapse before enunciating the law of universal gravitation?

To answer this question, a tale was put up, on slender evidence [5], to the effect that Newton's first estimate did not exhibit the expected identity between the two forces, because he used a wrong value of the earth's radius (on which, as we shall see presently, the result sensitively depends); he had to wait until a more accurate value of this radius became available, as a result of Picard's triangulation, before he could repeat the calculation, which then allegedly confirmed him in his previous surmise: in order that the dramatic touch should not fail, it is even added that when he made this last calculation, his hand trembled so much that he had to ask a friend to finish it for him. How unlikely this whole story was did not escape the perspicacity of such competent scientists as Adams (the co-discoverer of the planet Neptune) and Glaisher, a distinguished mathematician: they suggested a seemingly more plausible explanation, which could be supported by Newton's own declaration [6]. In a famous letter to Halley [7] of June 20, 1686, Newton alludes to "a certain demonstration I found the last year" which first gave him full assurance that the inverse square law was accurately valid down to the surface of the earth: obviously, what is here meant is the theorem on the attraction of a spherical shell, without which the argument leading to the identity of the force of gravity and the attraction on the moon has indeed no firm foundation. However, the question of this identity of the two forces cannot be the whole story: Newton had to struggle with many more issues, and the only way to elucidate the matter is to retrace all the stages of his long quest. In doing so, we shall at the same time gain insight into Newton's deeper motivations and the workings of his powerful mind.

Let us start at the beginning, in the autumn of 1665, when the young Cambridge scholar, having sought refuge from the plague in the family mansion of Woolsthorpe, passed the time in studious meditation [8]. For the authenticity of the story of the falling apple starting the decisive train of thought, we have the guarantee of Newton's own testimony, reliably reported by his friend Stukeley [17]. Let us first try to reconstruct the argument: this is inevitably conjectural to some extent, since unfortunately no trace of it has been found so far in Newton's papers. However, there is indirect evidence to tell us that we are on safe ground: thus we know [9] that Hooke, just about this time, was entertaining speculations about a power emanating from the celestial bodies, by which they would attract other bodies and influence their motion; Hooke conceived that when acted upon by such a force a body would deviate from its inertial motion and be constrained to revolve in a closed orbit in a similar way as a conical pendulum or a body attached to a rotating wheel; and in the latter types of motion, he was aware of the interplay of the force deflecting the body from its inertial motion and an "endeavour of recess" or centrifugal force, the two balancing each other along the actual path of the body. Assuming, then, that the young student (whose extraordinary gifts had already impressed his teacher Isaac Barrow[10]) started from a similar conception of the nature of the moon's motion, we see that the main problem for him was to evaluate the centrifugal acceleration of this body on its very nearly circular orbit: for it would give him directly the acceleration due to the attraction from the earth. We shall therefore have to enquire how Newton came to know the expression for the centrifugal acceleration

$$(1) \qquad A=\left(\frac{2\pi}{T}\right)^2 R$$

(or some equivalent one) in terms of the period of revolution T and the radius R of the orbit.

The next step is to assume for the attraction the inverse square law: one then finds its value at the surface of the earth from the proportion $a: A=R^2:r^2$, where the small letters refer to the earth, the capitals to the moon; and one expects this acceleration a to be the same as that of a freely falling body. This step was not difficult to make for Newton if he knew the formula (1): for the inverse square law is an immediate consequence of it, when it is combined with Kepler's third law of planetary motion $T \propto R^{3/2}$. We have direct evidence[11] that Newton did draw this conclusion at the time; indeed, as we shall see, it was also drawn at a later date, independently, by Hooke and other *virtuosi*. It needed Newton's critical acumen, however, to be aware of its possible limitations: but of this later. For a numerical estimate of the acceleration a on the basis of the preceding argument, Newton needed, besides the well-known values of the moon's period of revolution T and the ratio $R:r$ between the radius of the moon's orbit and the earth's radius, also the absolute value of the latter: in fact, the value found for a depends linearly on that adopted for the earth's radius. Here we meet the question alluded to above: which value did

Newton use? Let us now look into his early investigations of the law of centrifugal force; these will also suggest to us a plausible answer to the last question.

The analysis of circular motion in terms of a centrifugal tendency goes back to Galilei. In the second day of his *Dialogo* [19] of 1632, devoted to the examination of the objections raised against the earth's motion, the question comes up whether all bodies, however heavy, would not be hurled into space from the surface of the moving earth as a stone is hurled from a sling. With refined irony Galilei contrives to put the refutation in the very mouth of Simplicio, who is driven by clever questioning to recognize that a body on the earth is not carried away along a tangent to the surface with the full velocity of the earth's motion, but only lifted along the radius towards the tangent at the immediately preceding position. Galilei is not able, however, to give a quantitative analysis of the effect: he just has the correct feeling that it must be much smaller than the force of gravity.

Huygens [20] was the first to bring Galilei's argument to completion. This work dates from 1659, when Huygens, who was then thirty, had reached the maturity of his genius. His treatment of the problem is masterly – better in fact than many a modern textbook exposition. Galilei's point is made most elegantly by considering the motion from the point of view of an observer participating in the rotation: for such an observer the deviation from the inertial motion during a very short time interval may indeed be approximated by a motion towards the centre. This is shown by a minute kinematical analysis, exhibiting Huygens' skill in dealing with problems of continuity by the methods of the ancient geometers. If (Figure 1) the arc *PB* is very small, the radial distance *AB* is easily seen to be approximately equal to the square of the arc *PB*, divided by the diameter, and thus proportional to the square of the time. This is precisely the same law as that of the free fall, so that the centrifugal acceleration is immediately seen to be given by the product of the radius and the square of the angular velocity, as expressed by the formula (1).

However, Huygens did not publish the law of centrifugal acceleration until 1673, when it appeared as an appendix to the *Horologium oscillatorium* [21]. Newton knew it in 1665 because he had discovered it the year before (at the age of twenty-two) by his own exertions. Early papers recently brought to light disclose the devious path by which he arrived at the goal: there is no trace here of the scholarly elegance of the Dutch physicist; Newton's approach appears by contrast curiously simple-minded and uncouth. It bears the mark of his teacher Barrow, whose didactic works herald the final break with the ancient geometrical

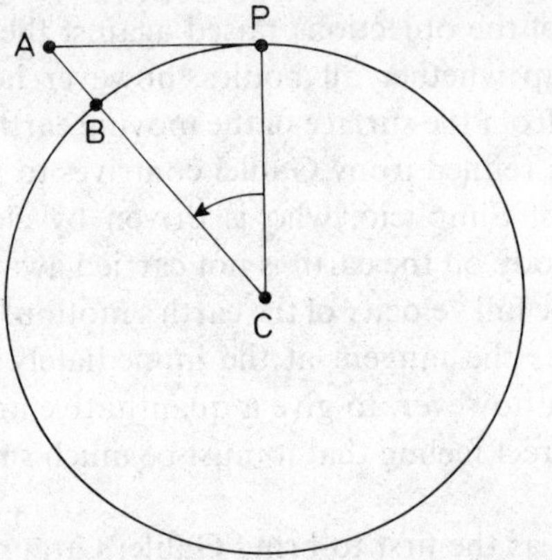

Fig. 1. Analysis of circular motion by Huygens.

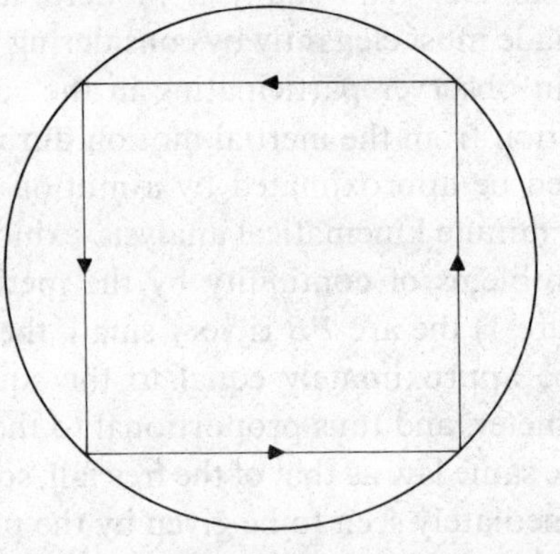

Fig. 2. Analysis of circular motion by Newton.

tradition in favour of the modern analytical methods; a tendency still more evident, of course, in Newton's first attempts, from the same period, at a systematic representation of geometrically defined functions by infinite series.

In his analysis of circular motion[12], Newton discusses the case of a globe moving along a great circle inside a hollow sphere: the inertial motion of the globe is continuously impeded by the spherical surface, which experiences a pressure from the globe; this centrifugal pressure is measured by the change of momentum corresponding to the deviation of the globe's inertial motion. Newton first obtains a lower limit for the effect by observing that in a half-turn the velocity of the globe is just reversed: its total 'endeavour from the center' in a half-turn is thus at least twice its momentum. A better estimate follows from the case (Figure 2) in which the globe, with the same velocity as in the actual circular motion, just bounces four times in a complete turn against the sphere, thus describing the four sides of an inscribed square: this gives the proportion

$$\frac{\text{endeavour from centre at each reflexion}}{\text{momentum}} = \frac{\text{side of inscribed square}}{\text{radius of sphere}},$$

and further

$$\frac{\text{total endeavour from centre in one turn}}{\text{momentum}} = \frac{\text{perimeter of trajectory}}{\text{radius of sphere}}.$$

It is easily seen that this argument holds for any regular inscribed polygon as well as for the square, so that by a passage to the limit familiar from Archimedes, Newton could conclude that in the actual circular motion, the total endeavour from the centre in one revolution is to the momentum as the circumference to the radius: this gave him the value π instead of the lower limit 2 for the same ratio for a half revolution. Also, in view of the perfect uniformity of the process, we may say that the total endeavour from the centre during the time that the body describes an arc equal to the radius is just equal to the momentum. How can one pass from this 'integral' law to an expression for the instantaneous force? For Newton's powerful intuition, the continuous change of direction of the centrifugal force is no embarrassment: its effect must be the same as if the motion, instead of being constrained to the spherical surface, were allowed to proceed on a plane: but then we have the problem solved by Galilei of the effect of a constant force, like gravity, acting perpendicularly to the trajectory of a uniformly moving body. The total effect of the constant acceleration A during a time t is to produce a velocity At: now, we have found that if the time t is R/V (R denoting the radius of the sphere, V the velocity of the circular motion), the acquired velocity is just V. Therefore, $A = V^2/R$, an expression for the centrifugal acceleration equivalent to the formula (1) above.

It is not sure what incited Newton to this highly original study of the circular motion: it may have been, as certain of his notes suggest[13], a reading of Descartes' *Principia philosophiae.* However, while the latter's influence on Huygens as well as Newton remained paramount for their general conception of the transmission of force by contact, Cartesian dynamics was too crude and erroneous to be in the taste of such acute and independent minds. For both of them, the true source of inspiration in their dynamical thinking was Galilei. Other manuscripts[14] from the same time give evidence of Newton's reading the *Dialogo*, an English translation of which by Salusbury had been available since 1661. In one of these[15], one finds a new derivation of the law of centrifugal force very similar to Huygens'; but it was clearly an afterthought. Arguments based on the principle of relativity, of which Galilei and Huygens made such brilliant use, never appealed to Newton; what he took from Galilei was rather, as exemplified by his analysis of circular motion, the more dynamical aspect of the law of inertia, the idea of investigating the forces determining the motion of bodies by the changes of momenta they produce.

One has also retrieved[16], and Herivel has very skilfully deciphered, the scrap of paper on which Newton jotted down the numerical computations leading to the quantitative estimates which were still wanting, as I pointed out above, in Galilei's argument. We see Newton applying his newly acquired knowledge of the law of centrifugal force to calculate this force at the surface of the earth and compare it with the force of gravity. This precious document answers rather definitely the question which value Newton did adopt for the earth's radius in his famous meditation under the apple tree. A learned paper has been written on this question by Cajori [15]: he painstakingly collected all the data he could find in the books on navigation of the time and went on to speculate to what extent Newton could be expected to have cognisance of this lore. He overlooked, however, one item, the one that was perhaps the most obvious and that at any rate has now turned out to give a decisive clue, to wit Galilei's *Dialogo.* In the calculation just mentioned, Newton naturally enough borrows from the book he was studying the value he needed; it is unmistakably quoted as 3500 Italian miles. Admittedly, we cannot be absolutely sure that Newton used the same value in the other calculation, but the proximity of dates and circumstances makes it overwhelmingly probable. Now, this value is very rough indeed, about 16% too small: obviously, Galilei regards it just as a round number easy to memorize and sufficient for rapid estimates. It is noteworthy that the value quoted by Galilei for the acceleration of gravity is much worse: it is about half the true value. This Newton finds unacceptable, and he determines himself a more correct value by means of experiments on the times of oscillation of simple and conical pendulums; by itself another remarkable achievement. He is apparently unconcerned, however, about taking over Galilei's value of the earth's radius.

If he used this value, he found a discrepancy of the order of 16% between the calculated attraction at the surface of the earth and the force of gravity [17]. What did he think of such a result? We have the testimonies of Pemberton and Whiston, who knew Newton well in later years [18]. They give the impression that he regarded the outcome as condemning the idea he wanted to test; they mention the wrong value of the earth's radius as the cause of the failure, but this is a retrospective consideration which throws no light on Newton's possible motive for being so casual about this constant. On the other hand, Newton himself, in a memorandum [19] of 1714, writes that he found his calculations "answer pretty nearly", which would suggest that he was not so dissatisfied with the result. After all, none of the commentators, prone to dramatization, has ever raised the simple-minded question: how nearly did Newton actually expect that the two accelerations would agree? Whiston gives a further piece of information [20] which appears highly relevant in this respect: he states that Newton inclined to the conclusion that besides gravitation, some other cause, such as a Cartesian vortex, might contribute to determine the moon's motion. This is not as unplausible as it might seem: more direct evidence of the influence of Cartesian cosmology on Newton's early views of the moon's motion has indeed recently been brought to light [21]. Only by pursuing the story can we hope to find further clues. So far, however, it will be clear to every scientist that Newton at this stage had opened up for himself an exciting prospect, but had nothing fit to be published.

During the decade following his return to Cambridge, in 1667, we find

Newton engrossed in his optical investigations, as well as busily engaged in mathematical correspondence with Collins. There is important evidence from these years, however, that far from losing sight of the problem of gravitation, he developed about the nature of this force speculations of extraordinary depth and boldness. The first piece of evidence, although somewhat indirect, is significant enough to be quoted. When in 1673 he received from Huygens a copy of the *Horologium oscillatorium* Newton did not fail, in his message of thanks, to intimate in a covert way (as was the custom among the *virtuosi*) that he also had long known all about the centrifugal force: this he contrived to do by giving as an example of its usefulness the application he had made of it to the comparison of the respective attractions exerted by the earth on the moon and by the sun on the earth. Alluding much later[22] to this message, Newton believed that he had even mentioned explicitly in it that such a comparison would lead (in connection with Kepler's third law) to the inverse square law for the attraction; in fact, this is not in the letter, as he soon could ascertain when he found a copy of it; the omission surprised him: obviously, he remembered his original intention of communicating to Huygens a fuller account of his old investigations. This shows at any rate that he had not given them up.

Another curious fact emerges from an exchange of letters[23] he had in January 1680/1 with Thomas Burnet at the occasion of the publication of the latter's book *Telluris theoria sacra*, which represented one of the earliest attempts at a scientific theory of the formation of the earth. Burnet having asked, among other things, for Newton's opinion about the shape of the earth, the latter replies that to the best of his judgement, based on the analogy with the other planets, it is spherical; he intimates that the effect of rotation must be negligible. Besides, he cannot tell what to make of the evidence of geodetic measurements, "not knowing how exactly those measures were made or the latitudes of places taken". This statement helps us at least to understand why he did not show even at this later date more eagerness to look for the best available value of the earth's radius.

The discussion with Burnet is interesting from another, more general, point of view. It gives us a glimpse of an aspect of Newton's thinking so uncongenial to us that it has been mostly neglected or misunderstood by biographers: I mean his attitude to theological problems. In the present instance, while Burnet dismisses the first account of creation in *Genesis* as purely 'ideal', without relation to 'physical reality', Newton defends it, most ingeniously, as a consistent description of physical phenomena they would have appeared to a human observer if any such could have witnessed them from a terrestrial vantage point. As to the phenomena themselves, he imagines the gradual formation of the sun and planets as local condensations of the primeval chaos, perhaps by an entirely natural cause, such as the action of gravitation. The length of the successive 'days' he conceives as gradually decreasing as the earth acquires its present state of rotation: but here he allows for an immediate divine intervention, estimating natural causes insufficient to produce such a rotating motion. This mixture of rationalism and theology is hard for us to appreciate; but I hope to show in the last part of this essay that it can be made more accessible by taking due account of all circumstances. At any event, it would be quite wrong methodically to disregard a side of Newton's activity to which he himself attached perhaps more importance than to his scientific work.

Of still greater interest for our enquiry is another document which reveals to us Newton the scientist engaged in considerations of a character quite different from the penetrating inductive enquiries into the laws of nature for which he is commonly celebrated: it is a long paper[24], written some time about 1675, and in which Newton develops the hypothesis of a universal aether, as the agent by which not only the various forces acting on matter are propagated, but even muscular motion is initiated by the 'soul'. There are several other manuscripts[25], as well as material published in the form of 'queries' at the end of the *Opticks*, which extend and modify the views contained in the paper we are considering; but it is the latter that plays a decisive part in the course of events we are trying to analyse. It was communicated to Oldenburg, the secretary of the Royal Society, on December 7, 1675, the immediate occasion being Newton's controversy with Hooke on the nature of light. Newton had vainly tried to avoid this controversy by arguing that his theory of colours was independent of any assumption about the physical constitution of the 'rays' carrying these colours. Now, he wanted to show Hooke that he had nothing to learn from him on this last question either, and that indeed he had thought more thoroughly about the constitution of the universe than any of those *virtuosi* whose heads 'run much upon hypotheses'.

The deeper motivation, of course, was to acquaint at least the fellows of the Royal Society with a grand conception of the workings of the universe on purely mechanical principles, in the best Cartesian spirit. Newton's attitude about it is curiously ambiguous: he feels that it is not mature and is accordingly reluctant to commit himself; but he clearly regards it as so fundamental that he is impatient to present it to the judgement of his peers. He does not let Oldenburg publish it[26], but he allows it to be 'registered', i.e. copied in the register of the Society, which is only accessible to the fellows. Newton's fondness for these speculations, which so sharply belie his "*hypotheses non fingo*", is a challenge to the historian. I shall take it up in a little while, and have occasion then to stress again the intimate relationship between Newton's scientific speculations and his theologically coloured metaphysics.

For the moment, let us only retain the remarkable interpretation of the force of gravitation proposed by Newton in the framework of his aether hypothesis. In the paper itself, there is only a brief and not too clear indication of it; when, however, in 1686, Newton has occasion, as we shall see, to refer Halley to it, he adds some comments which greatly clarify the idea without modifying it in any way[27]. The idea is that a body like the earth, or the sun, is the seat of a cyclic process of transformation of the aether: a stream of aether falls continually upon the earth and pervades all its parts, its density increasing as it loses momentum in interaction with the gross matter of the earth; this condensed aether would then continually escape from the earth to form the atmosphere and further to disperse into the 'aethereal spaces' where it would resume its original form, thus completing the cycle. In modern terms, a constant inward stream of S aether particles per unit time moving with radial velocity v, has at a distance R from the centre, a density $S/4\pi R^2 v$, increasing in the inverse ratio of the velocity. Now, such a stream will exert on gross matter a pressure $Smv/4\pi R^2$ directed towards the centre and, so long as the velocity does not change appreciably, inversely proportional to the square of the distance, just as is required for the attraction of the sun on the planets by Kepler's third law. Thus we see that in his bold speculations about the constitution of the material world, Newton thought of gravitation as a universal force having all the appearance of an attraction obeying the inverse square law, although it actually proceeded from a contact interaction between aether and matter. Again we understand why he did not publish his ideas at such an immature stage, although his allowing them to be 'registered' is an indication of how confident he was to be on the track of essential truths.

On the other hand, as we have noted, he fully upheld the conclusion of his reflexions in the Woolsthorpe garden, but apparently did not think it worth his while to try and improve the result, for example by resorting to a better value of the earth's radius. I think this behaviour is not so puzzling as it looks. Scientists do not relish the strenuous work which any detailed investigation demands; they only engage in such work under compulsion. Newton was not different in this respect from the rank and file of his craft. Now, it so happened that in 1679 and the following years the moment of compulsion arose from an extraordinary succession of casual events, and literally drove Newton to the completion of his great work.

In December 1679 Hooke, who had succeeded Oldenburg as secretary of the Royal Society, took over the care of the scientific correspondence, previously assumed by a second secretary. He used the opportunity to make another effort to restore his personal relations with Newton, which had been badly strained by the light controversy. A previous attempt, immediately after the dispute, in 1675/6, had been received by Newton (who was Hooke's junior by seven years) with haughty condescension[28], and the kind man fared no better this time (*tantaene animis caelistibus irae*!). In inviting[29] Newton to communicate to the Society "what shall occur to you that is philosophical", he made the diplomatic blunder of soliciting also Newton's "objections against any hypothesis or opinion of mine", and of mentioning particularly his ideas about "compounding the celestial motions of the planets of a direct motion by the tangent and an attractive motion towards the central body". Newton excused himself[30] by the pretext, outrageously improbable, that he had been lately so much out of touch with 'philosophy' as to be ignorant of recent productions: thus, he did not remember having ever heard of Hooke's hypothesis about celestial motions.

Nevertheless, Newton – to 'sweeten' his answer, as he later wrote[31] to Halley – did insert in it a philosophical communication. He pointed out that a proof of the earth's rotation could be obtained by observing the deviation of a falling body from the vertical towards the east. In the figure (Figure 3) illustrating the argument, he prolonged the path *AD* of the

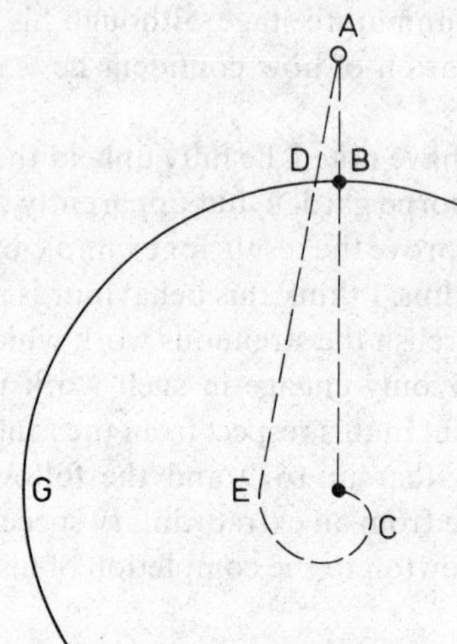

Fig. 3. Path of falling body prolonged to the centre of the earth, according to Newton.

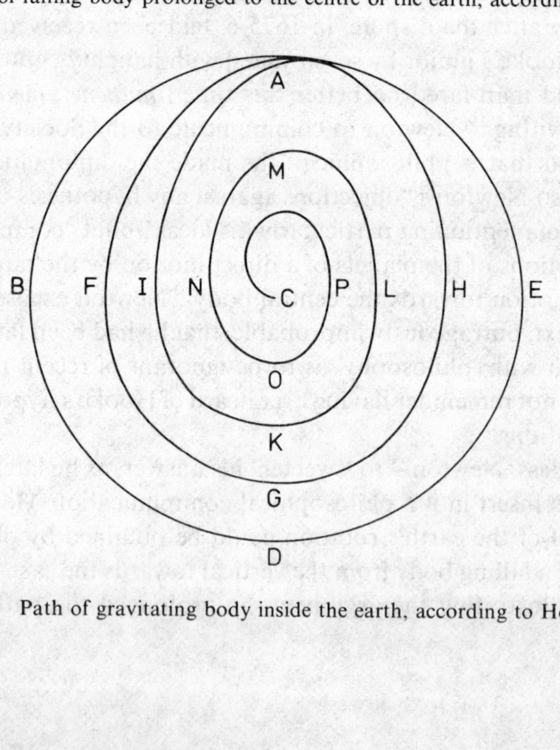

Fig. 4. Path of gravitating body inside the earth, according to Hooke.

falling body to the centre *C* of the earth as a smooth curve *DEC*. This remark elicited a prompt reply[32] from Hooke: if the body could proceed inside the earth, he said, it would not describe such a curve as Newton had sketched, but rather an ellipse-like trajectory around the centre (the curve *AFGH* on Figure 4) bringing it back to its starting point, if there was no resistance to the motion; allowing for such resistance, the path would be a kind of spiral like *AIKLMNOP* ultimately reaching the centre *C*. Moreover, the deviation from the vertical would only be exactly to the east at the equator; in our northern latitudes, however, it would rather be to the south-east, and in London even rather more south than east.

Very much against his wish[33], Newton was thus forced to continue the correspondence, and, worse still, to admit that he had been in error in the shape of the inside part of the trajectory. However, he does not deign to comment on Hooke's solution, but proceeds to discuss the shape that the trajectory would have if one assumed the force of gravity to be constant over the whole inside of the earth. To this Hooke retorts in the next letter[34], which ends the debate, that he had in mind the case in which the attraction is assumed to obey the inverse square law down to the centre. The problem of gravitation was one on which Hooke, in the midst of his busy life, never ceased to meditate in his own acute and imaginative fashion[35]; he has also stated quite plainly[36] the motive for this sustained interest: the utility of a good theory of planetary motions for solving the great problem of navigation, to determine the longitude of a ship's position. He must have realized that the attraction of the sun on the planets must obey the inverse square law as soon as he learned from Huygens' book the law of centrifugal force; we know from entries in his diary that he read the book in November 1675 and at once started thinking about the motion of the planets. Now, we find from this memorable exchange with Newton that by 1679 he had made the correct guess about the form of the orbit resulting from this law.

More than a guess it could not be, for Hooke totally lacked the mathematical powers needed for tackling such a problem; one may even say (as Newton later actually did, in a fit of anger[37]) that it was an easy guess, since after all it was known since Kepler's time that elliptic orbits fitted the observations satisfactorily. But to have clearly recognized that the shape of the orbit is determined by an attractive force from the sun obeying a given law of dependence on the distance, and that the fundamental problem of astronomy was the mathematical derivation of the orbit from the law of attraction, represented a remarkable intellectual feat which is entirely to Hooke's credit. Even Newton, however guardedly he behaved towards him, is candid enough to acknowledge[38] the stimulation he received from Hooke: "Your acute letter having put me upon considering thus far the species of this curve. I might add something about its description by points *quam proxime.*"

This last sentence reveals at the same time that Newton was so well prepared to tackle the problem that in response to Hooke's challenge he could rapidly master it, at least in its essential features. As he explains elsewhere in the same letter, he considers motion "according to the method of indivisibles", as a succession of "innumerable and infinitely little motions...continually generated by gravity"; and this leads him to a construction of the trajectory point by point. If Newton here refers to the method of indivisibles, it is no doubt because he did not think that Hooke could be acquainted with his own method of fluxions, of which nothing was then published. To arrive at a global characterization of the trajectory is of course a much tougher problem, whose general solution was reserved to the Leibnizian school. However, when Newton himself tells us in later memoranda[39] that he did deduce the elliptic form of the orbit from the inverse square law towards the end of the year 1679, we may well trust him to have solved this particular case "by the inverse method of fluxions" even in its rudimentary form.

A touch of comedy is added to this dramatic development by Newton's acquiescence to Hooke's erroneous statement about the southern deviation of the falling body[40]. It is curious that Newton did not notice the error and persisted in it until he took up the whole problem of the form of the earth when writing the *Principia.*

During the month of November 1680, a spectacular comet was observed moving towards the sun, and another one was seen in the latter part of December and in the following January moving away from the sun in the opposite direction with about the same velocity. This prompted Flamsteed, the recently appointed Astronomer Royal, to put forward the view[41] that the two comets were actually one and the same, whose path had undergone a complete reversal in the vicinity of the sun. He endeavoured to account for this effect by some magnetic action of the sun, changing from attraction to repulsion as the comet entered the sun's vortex and was thus deflected from its direct path towards the sun; he naively imagined that the magnetic axis of the comet would follow the deflexion of the trajectory, so that it would eventually present its other pole to the sun: just as a bullet, as gunners apparently believed, always kept the same side forward along its path. Solicited for his opinion, Newton gave a civil, but devastating reply[42]. It would be tempting to dwell on its refutation of Flamsteed's magnetic hypothesis, since it shows us Newton at his best as a natural philosopher; but it is a side issue. I shall only stress one point, because of its interest for the social aspect of the evolution of science: on the opinion just recorded about the behaviour of a bullet, Newton commented that it "may be a tradition of the gunners, but I do not see how it can consist with the laws of motion, and therefore dare venture to say that upon a fair trial it will not succeed excepting sometimes by accident". Such an utterance marks the point at which the science of mechanics emancipates itself from the empiricism of craftsmen and enough confidence is felt in its laws to make predictions about the outcome of yet untried experiments. More to our point, however, is Newton's extraordinary attitude to Flamsteed's identification of the two comets: he is most reluctant to accept it[43], because it would make this comet 'paradoxical'; all comets known so far have been observed to move in the same sense on both sides of the perihelion "in a line almost straight". Not until 1685, when he was working out the theory of cometary orbits for the *Principia*, did he recognize the correctness of Flamsteed's skillful interpretation of the observations[44].

Nevertheless, he is willing to discuss the possibility of a curved orbit and explains[45] to Flamsteed that no repulsive force is needed to bring it about: an attraction from the sun, such as that "whereby the planets are kept in their courses about him", could produce such an orbit. Only the comet would then turn around the sun, and not (as Flamsteed supposed) be deflected before reaching it; and by a direct method he has of computing an orbit (of whatever shape) from exact observations, he finds that the last such observations (including some of his own in February and March 1681), when extrapolated backwards, indicate December positions well beyond the sun. Here then we witness Newton in possession of all the theoretical tools enabling him to assimilate cometary motion to that of the planets, but severely refraining from taking the step because he is not satisfied that it is granted by the data of observation. However convinced he may have been of the universality of the force of gravitation, the strict rules of his natural philosophy forbade him to draw rash conclusions.

Newton and Flamsteed held each other in high esteem; in spite of all disagreement, the tone of their correspondence is a model of courtesy and serenity. When he started work on the *Principia*, at the end of 1684, Newton consulted Flamsteed on various points of astronomical observation[46]. Among other things, he was worried about the fact that Kepler's determination of Saturn's orbit was not in agreement with the third law; he suspected that the discrepancy could be due to the perturbation of this orbit at conjunction with Jupiter, and asked Flamsteed whether he had observed deviations from Kepler's tables at such conjunctions, corresponding to the estimated perturbation. In his reply Flamsteed expressed surprise that the two planets could be thought to influence each other to any appreciable amount, and gave Newton indications about his own observations of their orbits. Thereupon Newton explained that in computing the perturbation of Saturn's orbit, he had assumed an inverse square law of interaction between the two planets; but he could see from the new data that he had overestimated the 'virtue' of Jupiter. "Your information," he adds, "about the error of Kepler's tables for Jupiter and Saturn has eased me of several scruples. I was apt to suspect there might be some cause or other unknown to me, which might disturb the sesquialterate proportion" (by which is meant Kepler's third law). At this late date, as we learn from this incident, he is not yet sure whether the third law "fills the heavens", he still keeps an open mind with regard to its universal validity and still contemplates the possibility of some other 'cause' besides gravitation influencing the planetary motions.

We are now coming to the last act of the drama. In spite of his renewed interest in the problem of gravitation, there is no saying how long Newton would have gone on withholding publication of his powerful methods and the momentous conclusions he had already reached with their help, had it not been for another fortunate circumstance which brought into the picture one of the most gifted men of the new generation, Edmond Halley. Grown up in the midst of those who had first shaken off the shackles of tradition and shaped the modern spirit of enterprise and enquiry, he was following their lead with all the eagerness of youth and the vigour of an uncommon intelligence. Admitted to the Royal Society in 1678, at the age of twenty-two, he was alert to the questions of the day and discussed them with the most distinguished *virtuosi.* Thus it happened[47] that on one Wednesday in January 1683/4, Halley having met Sir Christopher Wren and Hooke in town, the conversation turned as so often before on the great problem of the planetary motions. How the motion had to be analysed as inertial displacement modified by attraction, and how one could conclude that the attraction obeyed the inverse square law, was common knowledge to the three of them, but the *crux* of the matter was how from this knowledge to derive the form of the orbit. Hooke boasted that he could do it, but the others were not to be contented with mere assertions. Sir Christopher offered a prize of a book worth 40 shillings to the one who would produce the solution within two months: needless to say he had not to incur the expense.

Halley, however, did not let the matter rest. In the following May[48], he visited Newton in Cambridge and put the question to him. Newton replied at once that he had shown the orbit to be an ellipse, but he could not immediately reproduce the argument. He promised Halley to send it, and indeed he was hard at work during June and July drafting a treatise[49] *De Motu* in which he enunciated in the traditional style, as a succession of definitions, axioms and propositions, the laws of motion and their application to the case of the inverse square law of attraction: this was the kernel of the future *Principia,* and also, more immediately, the text of his Lucasian lectures for the following Michaelmas term. Halley probably visited him again in August[50], when he "learnt the good news" that Newton "had brought this demonstration to perfection", and in November received at last the promised paper. Halley thereupon paid another visit to Newton and persuaded him that he ought to write up a full-scale book on the whole subject. In order to allow him the required leisure while securing his priority, he proposed to have the tract *De motu* presented to the Royal Society and entered upon their register. With Newton's consent, he could make this announcement at the meeting of the 10th December, and about the middle of the following February, the copy destined to the Society was received and duly registered[51].

Halley had thus not only procured the fellows of the Society the possibility of acquainting themselves without delay with the gist of Newton's ideas, but he had even managed to launch the latter on the elaboration of a complete exposition of them. Newton's interest and energy were now thoroughly aroused, and the work progressed with remarkable speed: on April 28, 1686 the manuscript of the first book of the *Principia* was presented to the Society [52]; the second book followed in the autumn of that year and the third in April 1687; the finished work appeared about midsummer 1687: it was Halley again who had assumed all the chores, as well as the financial burden, of seeing it through the press [53]. How could Halley succeed so easily in overcoming Newton's bent to procrastination and bringing him to such momentous decisions? I think the answer is simple: Newton was a man of very sensitive disposition, who must have felt very lonely among the Boeotian crowd of fellows and students; no wonder that he at once fell under the charm of a brilliant and enterprising young man, to whom he could explain his thoughts with the assurance of an intelligent response. Later, he behaved towards Fatio de Duillier with fatherly kindness and solicitude, no doubt for the same reason [54].

Before he could help the undertaking to its happy completion, Halley had still to weather a storm which put to a severe test his diplomatic talent [55]. At the memorable meeting of April 28, 1686 when much praise was being lavished on Newton's discoveries, Hooke was piqued not to hear any mention of his own contributions to the problem, which indeed, as we have seen, were entirely in the same line as Newton's. He may well be forgiven the human weakness of giving his own ideas a higher estimation than they deserved and even of fancying that Newton had borrowed from him the inverse square law. At the coffee-house where the society adjourned after the meeting, he voiced his claims without success, the others being of opinion that he had only himself to blame "for having taken no more care to secure a discovery which he puts so much value on". The incident was reported to Newton, very tactfully by Halley, but less so by others. It incensed him so much that he threatened to suppress the third book of his work, that which had to treat of the astronomical applications of the theory. He soon relented, however, after another epistolary masterpiece [56] of Halley's, which soothed him to the extent that he expressed regret for his outburst of anger, and touchingly declared [57] that he would acknowledge having learned from Hooke that the deviation of a falling body would be south-east in our latitudes.

Newton's exasperation was indeed out of proportion with the objective facts of the case: Hooke's boastfulness and reckless priority claims, arising from a fiery and uncritical imagination, were well-known, and nobody took them more seriously than could be helped. Newton had never quite forgiven Hooke's questioning of his optical discoveries, and one understands that he might feel irritation at being once more the object of a futile accusation; but in his present position he could well have afforded to ignore it. Instead, he fills passionate pages in defence of the originality of his conceptions, retracing their origin to his student days: precious pages indeed that disclose to us, besides details not otherwise known, how he himself judged the successive stages of his investigations.

It is here that we learn, in the first place, the circumstance mentioned at the beginning of this essay, that Newton had never thought the inverse square law to remain valid down to the surface of the earth, until he had found the theorem on the attraction of spherical shells some time in 1685, while he was writing up the first book of the *Principia.* This would seem to settle the question of the famous discrepancy in the early comparison of gravity and attraction: he could not ascribe to this estimate more than an indicative value, and there was thus little point, he thought, in improving it by using a better value of the earth's radius; he would not have expected the error in the value he adopted to be so large, and was rather inclined to regard the discrepancy in question as the measure of a real physical effect. Only after discovering the theorem on the attraction of the spheres would he suspect the discrepancy to be spurious and verify that it actually disappears if one adopts Picard's value for the earth's radius [58]. By that time however, he had started writing the *Principia,* and the correction of the error had therefore no influence on his decision to publish the results of his studies. In fact, he had by then so many proofs that the law of gravitation "fills the heavens" that the particular argument which had first oriented his thoughts to this enquiry receded into comparative insignificance. In the third book of the *Principia,* it is merely used to show that no other attraction than the universal gravitation acts at the surface of the earth; and another source of inaccuracy appears: the non-sphericity of the earth, itself a consequence of the combined action of gravitation and rotation [59]. We have here an example of a frequent occurrence in the history of science: the degradation, as I would call it, of heuristic arguments which, once they have played their role in guiding the discoverer to some truth of wider scope, appear in a dimmer light when they are contemplated from the higher viewpoint they have helped to reach.

There is, however, in Newton's review of the circumstances manifesting his early understanding of the inverse square law, a remarkable and, I think, very revealing feature: it is the way in which he adduces[60] as evidence the paper of 1675 on the aether hypothesis, for which he refers Halley to the register of the Royal Society. He is a bit embarrassed by the fact that in the short passage on gravitation inserted there as an afterthought (it was 'interlined' at the last moment in the original manuscript) no mention is made of the inverse square law; he is so uneasy about this that he returns to the matter in a further letter[61] to Halley, in which he gives the explicit derivation of the law from the hypothesis of an aether stream which I tried to formulate in modern language earlier in this essay. But his reason for invoking this hypothesis is interesting: he points out that it leads to the inverse square law only 'upward' from the surface of a planet, i.e. in free space, where the velocity of the aether stream remains constant, but not 'downwards', inside the body, where the aether loses momentum. This is why in the correspondence with Hooke about the falling body he was careful not to assume the validity of the inverse square law inside the body – in contrast to Hooke, the 'bungler', who was not aware of this limitation.

Although he describes[62] the hypothesis, rather misleadingly, as "one of my guesses which I did not rely on", he obviously regards it as sufficient to establish the limitation in question – otherwise there would be no point in invoking it to expose Hooke's 'error'. People who want to look upon Newton as the great master of the inductive method (which he is) may be surprised, but there it is: *habemus confitentum reum.* The half-hearted admission that he should not rely on a mere hypothesis while he is in fact relying on it reveals a tension between two tendencies equally powerful in Newton's mind, of which many instances can be found in his writings. There was a contradiction between the rigorous requirements of rational analysis and the urge for a comprehensive, intuitive synthesis, that Newton could never overcome, and whose roots must be sought much deeper than in any question of scientific method. Indeed, the profound meaning of Newton's conception of the aether, and the explanation of the extraordinary value he attached to it in all his life can only be appreciated against the general philosophical and religious background of his activity.

Of paramount influence on Newton's philosophical outlook was Henry More[63], who with Barrow represented at Cambridge, in Newton's student days, the progressive current, open to the modern spirit in science and philosophy. Both men came from the same district of Lincolnshire, a Puritan stronghold, and it was a pupil of More who taught Newton mathematics at Grantham school. At Cambridge they remained close friends until More's death in 1687. More brought to England the ideas of the Italian humanists of the XVIth century, who were fighting the scholastics by playing up Plato against Aristotle: hence he became known as the "Cambridge platonist". This is a misnomer, however, for little of Plato's ideal world remains visible under the strong Puritan colouring More applied to it. Indeed, of the two key concepts of his system, space and God, the first is derived from the Italian school, the second decidedly British. The Italians under the influence of the Copernican view of the world made space infinite and conceived the stars as so many systems similar to ours freely roaming through it; thus space as the substratum of moving things was thought of as existing independently of them and itself eternally at rest. All things were conceived as animated, whether material or immaterial, moving in this substantial space, in which God also was everywhere present. All this cosmology was taken over by More, and through him by Newton as well.

Characteristic for the English aspect of More's system is his conception of God as the absolute master of the world, creating things at will and capable of acting upon them, or even destroying them, at any time according to his own designs. In particular, man's relation to God is that of a servant to his master: it is governed by God's absolute power, not by any of its other attributes: as Newton remarks[64], we say "my God", "my master", but not "my infinite", "my eternal". Yet, God's omnipotence is compatible with man's free will, as well as with the regulation of all natural phenomena by the laws he has designed. This insistence on the personal aspect of God as the ruler of the world is a striking feature of religious thought in England in the latter part of the XVIIth century, and its historical origin is not far to seek: it is indeed a faithful transposition onto the theological plane of the political ideology developed by the bourgeoisie after the failure of its first experiment in self-government and the recall of the king. Incapable of finding a source of sufficient authority among themselves, they put all authority in the hands of a ruler not belonging to their own class, but hedged the king's authority by law in order to preserve their individual liberty. Thus the subject submitted to the king's will, but expected the latter to act according to laws designed for the common good and guaranteeing the freedom of the individual.

This is not a far-fetched theoretical interpretation. We have direct evidence that Newton's political philosophy was precisely what I have just outlined. When King James in February 1686/7 tried to force the University to admit an unqualified monk to the degree of master of arts, Newton advised resistance[65], not in a rebellious spirit, but because it was a clear legalistic issue: "all honest men are obliged by the laws of God and man to obey the king's lawful commands. But if his Majesty be advised to require a matter which cannot be done by law, no man can suffer for neglect of it." And he confidently concludes: "An honest courage in these matters will secure all, having law on our sides."

If the balance of political relations between the king and his subjects was precarious, dealings with the divine power could also be troublesome. Especially the pioneers of modern science, the natural philosophers, were dangerously exposed to the allurements of materialism, and had a difficult course to steer between this Charybdis and the Scylla of pantheism or deism. Against the latter, Newton is careful to warn us: God is not duration or space, he is everlasting and everywhere present; he is not the world-soul, but the world-ruler. The main argument for upholding the conception of a personal God was the old one of design: the regularities of the natural phenomena could not have been produced by chance, they betrayed the existence of a supremely wise and intelligent being who had designed everything according to the function it had to fulfil in the grand harmony of the whole creation. At a time when the wonders of nature were just beginning to reveal themselves to critical scrutiny, this argument carried a great weight: for Huygens, it was indeed the only reason to retain a belief in a deity[66]. To the generation following that of Newton, it was already losing its glamour, and atheism became the fashion. Boyle was so alarmed at this deterioration of morality that he instituted by bequest annual lectures on the evidences of Christianity. The first Boyle lectures were delivered in 1692 by Bentley, who in particular adduced as evidence of design the constitution of the solar system, newly reduced to law by Newton: he obtained the latter's active support in getting his arguments straight, and the Bentley letters[67] illustrate the curious mixture of caution and assurance with which Newton handled finalistic causality.

God has not laid bare his design to man, but he has endowed him with reason so as to enable him to discover it. The great goal of Newton's life was to discover God's design[68], by studying his works and following the clues he had given mankind through his prophets. This motivation throws light on Newton's whole activity and gives it unity and consistency. It is at the root of his choice of method: only by rational analysis of the natural phenomena and rational interpretation of the scriptures can we hope to read God's message, since reason is the tool he has given us for this purpose. It explains his life-long endeavours in search of the meaning hidden in the sacred books (which he treated as a rational problem of de-coding) and his erudite and painstaking historical investigations, which aimed at establishing the great antiquity of the Hebrew people and the authenticity of the prophecies[69]. For him there was no essential difference, either in purpose or in method, between the derivation of the laws of nature from the analysis of the phenomena and the ascertaining of God's intentions about man's fate by a reconstruction of the history of mankind.

We are now better prepared to judge the real place that the aether hypothesis occupied in Newton's mind. In the philosophy of More and Newton, space was occupied by God as well as by the created things: but how did God perceive the things and how could he act upon them? A direct interaction of God and gross matter was out of the question: God has no sense organs and is not affected by the motions of the bodies; the latter do not experience any resistance from God's omnipresence[70]. But could there not be a finer kind of substance providing the missing link? More, as we know from his famous controversy with Descartes, had contemplated such a solution, but shrunk from it because of its materialistic flavour: he was in consequence driven to a mystical conception of 'spirits' emanating from God and animating the created things. Mysticism, however, was wholly averse to Newton's rationalistic turn of mind, Cartesianism much less so. An aethereal fluid filling all space could, as Descartes wanted it, transmit various forces between the bodies by appropriate cyclic motions: "for nature", says Newton[71], "is a perpetual circulatory worker". It could also be the agent of transmission of the sensations from the sense organs to the 'sensorium', in which – according to the rudimentary physiology of the time – these sense impressions were directly perceived by the 'sensible substance' of the animal. Likewise, by a bold analogy[72], space would be as God's sensorium, in which he would perceive the aethereal motions and through them (if such was his will) influence those of the material bodies.

If the aether hypothesis played such a central part in Newton's view of the world, why did he not publish it in the first edition of the *Principia*? Why at least did he not present his tentative theory of the cause of gravitation, which would have spared him the imputation of introducing action at a distance as a primary quality of matter? By a cruel irony of fate, the proof of the universal validity of the inverse square law of gravitation, his greatest triumph, had dealt a severe blow to his whole aethereal construction. He had all the time imagined the celestial bodies moving in the aethereal medium would encounter some resistance from the latter, and accordingly expected the inverse square law to be only approximate. Now, the planets and the comets were found to move through the heavens in all directions without revealing the least presence of any resisting medium. From his own studies of the motion of bodies through fluids, which form the second book of the *Principia*, Newton was forced to conclude that the density of the aether ought to be extremely small: and how could it then fulfil the role for which it was primarily conceived? Confronted with this difficulty, Newton did not relinquish entirely his conception of the aether, but he had to restrict its scope considerably, confining its actions essentially to the inside and immediate vicinity of the material bodies; and the confident assurance of the early days was gone[73]. When, many years later, as an old man, he decided at last to publish his speculations about the aether, together with his unfinished optical studies, the queries in which he propounded them had the pathetic ring of renouncement. There is an undertone of renouncement also in the last page of the *scholium generale* which concludes the second edition of the *Principia*. The arrogance of the sentence "And it is enough that gravitation actually exists and acts according to the laws we have exposed" does not conceal the fact that this 'experimental philosophy' is a position of retreat.

Newton failed in his double quest. His historical constructions fared no better than his aether hypothesis: they were based on conjectural identifications which eluded any possibility of conclusive examination. The common error of his time was to underestimate the magnitude of these problems. The scale of time was hopelessly distorted by the unquestioned acceptance of the biblical tradition; and even though the narrow frame of the medieval world had been shattered by Copernicus, the true dimensions of the universe were still far from being realized. Newton believed that the history of mankind held in the four 'ancient kingdoms' of the classical authors, and that the celestial motions were confined to the solar system. And after a long life of unceasing toil, when he had unravelled the laws of these motions, he found himself on the shore of an "ocean of truth undiscovered" and wrote, regretfully, "*hypotheses non fingo*".

NOTES

[1] Three volumes have so far appeared, covering the period up to 1694 (refs. [1–3]). In the following, documents published in the *Correspondence* will be quoted by their number, followed, for greater convenience, by the reference to the volume; thus no. 288 [2] means the letter no. 288, published in volume 2.

[2] Especially A. R. Hall [4, 5], J. W. Herivel [6–9], A. Koyré [10], D. T. Whiteside [29, 30].

[3] See Manuel's book [11] and my review of it [12].

[4] This is very conspicuous in Brewster's biography [13], but in fairness to this author, it should be stressed that he is remarkably accurate and reliable in his account of the facts. Much worse is the case of Thomson and Tait, whose judgement of the formulation of the fundamental laws of mechanics in the *Principia* (ref. [14], chapter II, especially § 263, 264) is very uncritical and vitiated by a quite unhistorical interpretation of 'actio' as mechanical work.

[5] See the account in Brewster's biography [13], vol. 1, pp. 290–292; Brewster, however, is not entirely uncritical. A full survey of the problem has been given by Cajori [15].

[6] See Cajori [15].

[7] This is letter no. 288 [2]. The theorem in question is proposition 71 of book 1 of the *Principia* [16].

[8] See Brewster [13], vol. 1, p. 25–26.

[9] On Hooke's views about gravitation, see especially Brewster [13], vol. 1, pp. 283–288, and Lohne [18]; in the latter's paper, the relevant texts are reproduced.

[10] To be quite precise, Barrow was not too pleased with Newton when he examined him in geometry in 1664 (Brewster [13], vol. 1, p. 24), but he must have soon formed a better opinion of him, since Newton's discovery of the method of fluxions is recorded as early as May 20, 1665 (*ibid.*, p. 25). The first written expressions of Barrow's esteem for Newton date from 1669 (letters nos. 5–7 [1]). More evidence on Newton's early intellectual development, as well as indirect information on the precise nature of his relations with Barrow in his student years, is found in a well-documented article by D.T. Whiteside [29].

[11] Letter no. 288 [2].

[12] My account is based on the documents published by Herivel [6], but differs in some particulars from his own interpretation of them [6, 8].

[13] See on this point Herivel's [6, 8] remarks. An early essay by Newton pertaining to Descartes' *Principia philosophiae* is published in ref. [5], p. 89–156.

[14] These are the manuscripts first published by Hall [4] and reproduced as documents no. 117 [1] and 347 [3]. A comparison of the latter with Salusbury's translation of Galilei's

Dialogo was performed by the late Prof. H. W. Turnbull, first editor of the *Correspondence*, and is reported by Herivel [7]; it reveals convincing analogies.

[15] This is no. 117 [1].

[16] This is no. 347 [3], which is interpreted by the editor of the third volume of the *Correspondence*, Dr. J. F. Scott, but more fully by Herivel [7].

[17] This is in agreement with the figures quoted by Brewster [13], vol. 1, p. 26, in his account of Newton's work at Woolsthorpe: according to Brewster, he found for the deflexion in one second due to the attraction at the surface of the earth a value of 13.9 feet, whereas that due to gravity is 16.1 feet.

[18] See Brewster [13], vol. 1, pp. 290–292 and Cajori [15].

[19] Quoted e.g. by Cajori [15], p. 160.

[20] This is quoted by Brewster [13], vol. 1, p. 290.

[21] See an important paper by D. T. Whiteside [30], who analyses early astronomical manuscripts of Newton and annotations found in books he read and traces their relation to the contemporary background, of which he makes an extensive study. It is noteworthy that Newton showed as little appreciation as the practical astronomers of his time for the significance of Kepler's two first laws. In particular, he ignored the second law (the law of areas) until he found that it was a consequence of the inverse square law of attraction.

[22] See the beginning of letter no. 116 [1] to Oldenburg, with the editors' comment. Newton refers to this letter in his correspondence with Halley in 1686 (nos. 288 and 291 [2]).

[23] Letters nos. 244, 246, 247 [1].

[24] Letter no. 146 [1].

[25] See especially ref. [5], part III.

[26] Newton's cautious directions to Oldenburg are found in his letters nos. 147, 151 (postscript), 153 [1].

[27] See the end of the letters 288 and 291 [2].

[28] This alludes to the oft-quoted exchange of letters nos. 152 and 154 [1].

[29] Hooke's letter to Newton no. 235 [2].

[30] Newton's reply no. 236 [2].

[31] In his letter no. 288 [2].

[32] Letter no. 237 [2].

[33] Newton's annoyance is vividly conveyed by his later account of this correspondence with Hooke to Halley (letters nos. 286 and 288 [2]). His second reply to Hooke is the letter no. 238 [2]. It must be observed that the figure belonging to this letter is incorrectly reproduced in the *Correspondence;* a photograph of the original figure is given in the paper [18] by Lohne, in which the present controversy between Hooke and Newton is discussed in great detail.

[34] Letter no. 239 [2].

[35] All relevant documents about Hooke's studies on gravitation may be found in Lohne's excellent paper [18]. In this paper, Lohne takes Hooke's side with as much ardour as if the two contending worthies were still alive! Even though I am not able to accept all his arguments or to subscribe to all his conclusions, I derived much help from his careful analysis.

[36] In letter no. 239 [2].

[37] In the passionate postscript of the letter to Halley no. 288 [2].

[38] Letter no. 238 [2] (last paragraph).

[39] These are quoted in Lohne's paper [18]. Nevertheless the latter contends, in my opinion on insufficient grounds, that Newton was not able to solve the problem; he repeats this assertion in another interesting paper [22] on Newton's theory of colours.

40 The nature of the error is very well explained by Lohne [18].
41 Flamsteed's views may be inferred from the surviving part of his correspondence on this topic with Halley (letter no. 250 [2]) and Newton (letters nos. 251 and 254 [2]).
42 Letters nos. 251 and 254 [2].
43 Letter no. 255 [2].
44 See Newton's letter to Flamsteed of 19 September 1685 (no. 281 [2]).
45 Letters nos. 254 and 255 [2].
46 Especially letters nos. 274, 275, 276 [2].
47 The story is told by Halley in his letter to Newton no. 289 [2]. Newton himself (letter no. 286 [2] to Halley) mentions a conversation he had with Wren about the problem of the planetary orbits as early as 1677, when it seems that Wren was already acquainted with the inverse square law.
48 About the dates of Halley's visits to Newton, there is some uncertainty due to discordant testimonies. The question has been ingeniously discussed by Herivel [9], whose conclusions I adopt.
49 One of the four extant versions of this treatise is published in ref. [5], pp. 237–292. The moot problem of ascertaining the relation of these manuscripts with the text communicated to the Royal Society is thoroughly treated by Herivel [9].
50 Letter no. 289 [2].
51 Newton expresses thanks for the registering in his letter no. 278 [2] of 23 February 1684/5, in which he declares his intention "to finish it" (i.e. the proposed book), after he has returned from a journey to Lincolnshire, "as soon as I can conveniently".
52 Letters no. 285 [2].
53 Letters nos. 300, 303, 304, 306, 309 [2].
54 The attitude of Newton towards Fatio de Duillier is revealed by their letters newly published in the third volume of Newton's *Correspondence*. See my review of this volume [23].
55 The acts of this memorable incident are the oft-quoted letters nos. 285, 286, 288, 289, 290, 291 [2].
56 This is letter no. 289 [2].
57 Letter no. 290 [2].
58 In his letter no. 288 [2] to Halley, Newton mentions, in connection with his message to Huygens alluded to above, the early paper (document no. 117 [1]) containing the computations of centrifugal accelerations on which this message was based. However, he also states that in this paper the ratio of the acceleration of gravity to the centrifugal acceleration of the moon "is calculated" (a calculation unfortunately not found in the document as we have it) and then adds, casually, "thought not accurately enough".
59 It is with this complication in mind that Newton, in his letter no. 290 [2] to Halley, returning to the early paper mentioned in the preceding note, says that the calculation of the centrifugal force arising from the earth's rotation "is a thing of far greater difficulty than I was aware of". See further *Principia* [16], book III, proposition IV.
60 In the letter no. 288 [2], both in the body of the letter and in the postscript.
61 In the letter no. 290 [2].
62 Letter no. 288 [2], at the end of the postscript.
63 This point is most forcefully made in an important paper by M. Fierz [24], which also contains a detailed exposition of More's ideas and of their origin in Italian philosophy.
64 In the *scholium generale* at the end of the second edition of the *Principia*. This edition appeared in 1713, but Newton was working on it long before; there is a draft of the *scholium* dating probably from before 1697 (about which see ref. [25] and Brewster [13], vol. 2, p. 154).

65 Letter no. 301 [2]. Similar views were again expressed by Newton on another occasion two years later: see the letter no. 328 [3].
66 Ref. [26], especially pp. 524–528 and more particularly p. 363 (§ 42).
67 Letters nos. 398, 399, 403, 406 [3].
68 Newton's preoccupation with the relation of nature and God is already apparent in the student's essay already quoted about the Cartesian system (published in ref. [5], pp. 89–156). More's influence is very noticeable in this essay.
69 See especially refs. [11] and [12].
70 See the *scholium generale*.
71 Letter no. 146 [1].
72 About the analogy of space as God's sensorium, see especially the query 31 at the end of the *Opticks* [27], and the controversy between Leibniz and Clarke (the latter acting as Newton's spokesman); the documents relating to this controversy have been edited with outstanding accuracy by A. Robinet [28].
73 Lohne [22] exaggerates the difficulty arising for Newton's aether conception from the absence of resistance to planetary motions when he states that it also overthrew his "alchemical cosmogony and essential parts of his theory of light and colours": Newton did uphold the functions of the aether in chemical and optical processes, which only involve short-range interactions; but he had to abandon the hope of a comprehensive synthesis.

BIBLIOGRAPHY

[1] *The Correspondence of Sir Isaac Newton*, vol. 1 (1661–1675). Cambridge University Press, Cambridge, 1959.
[2] *The Correspondence of Sir Isaac Newton*, vol. 2 (1676–1687). Cambridge University Press, Cambridge, 1960.
[3] *The Correspondence of Sir Isaac Newton*, vol. 3 (1688–1694). Cambridge University Press, Cambridge, 1961.
[4] Hall, A. R., *Annals of Science* **13**, (1957), 62.
[5] Hall, A. R. & M. Boas Hall, *Unpublished Scientific Papers of Isaac Newton*. Cambridge University Press, Cambridge, 1962.
[6] Herivel, J. W., *Isis* **51** (1960), 546.
[7] Herivel, J. W., *Isis* **52** (1961), 410.
[8] Herivel, J. W., *Revue d'histoire des sciences* **15** (1962), 105.
[9] Herivel, J. W., *Archives Intern. d'hist. des sciences* **13** (1960), 63, 67, 71.
[10] Koyré, A., *Archives Intern. d'hist. des sciences* **13** (1960), 3.
[11] Manuel, F. E., *Isaac Newton Historian*, Cambridge University Press, Cambridge, 1963.
[12] Rosenfeld, L., *Nature* **202** (1964), 43.
[13] Brewster, D., *Memoirs of the Life, Writings and Discoveries of Sir Isaac Newton*, Thomas Constable, Edinburgh, 1855.
[14] Thomson, W. and P. G. Tait, *Treatise on Natural Philosophy*, Cambridge University Press, Cambridge, 1879.
[15] Cajori, F., in: *Sir Isaac Newton 1727–1927*, pp. 127–188. Williams & Wilkins, Baltimore, 1928.
[16] Newton, I., *Philosophiae naturalis principia mathematica*, Roy. Soc. London, 1687.
[17] Stukeley, W., *Memoirs of Sir Isaac Newton's life* [1752], Taylor & Francis, London, 1936.

[18] Lohne, J., *Centaurus* **7** (1960), 6.
[19] Galilei, G., *Dialogo dei due massimi sistemi del mondo*, Landini, Firenze, 1632. Giornata seconda (from: 'La vertigine veloce ha facultà di estrudere e dissipare").
[20] Huygens, C., *Oeuvres complètes*, Vol. 16, Nijhoff, La Haye, 1929: De vi centrifuga (1659), pp. 237–328.
[21] Huygens, C.: *Horologium oscillatorium*, Muguet, Parisiis, 1673. (*Oeuvres complètes*, Vol. 18. Nijhoff, La Haye, 1934.)
[22] Lohne, J., *Archive for the History of Exact Sciences* **1** (1961), 389.
[23] Rosenfeld, L., *Nature* **195** (1962), 414.
[24] Fierz, M., *Gesnerus* **11** (1954), 62.
[25] Gregory, J. C., *Transactions of the Royal Soc. of Edinburgh* **12** (1829), 64.
[26] Huygens, C., *Oeuvres complètes*, Vol. 21, Nijhoff, La Haye, 1944.
[27] Newton, I., *Opticks*, 2nd edition, W. Innys, London, 1717.
[28] Robinet, A., *Correspondance Leibniz-Clarke*, Presses Universitaires de France, Paris, 1957. [See H. G. Alexander, ed., *The Leibniz-Clarke Correspondence* (Manchester, England, Manchester University Press, 1956) – Ed.]
[29] Whiteside, D. T., *Notes and Records of the Royal Society* **19** (1964), 53.
[30] Whiteside, D. T., *British Journal for Hist. of Science* **2** (1964), 117.

CHAPTER 10

NEWTON'S VIEWS ON AETHER AND GRAVITATION

[1969c]

In the *Principia* [1] Newton presented the theory of gravitation from the point of view which we nowadays call 'phenomenological': he based the whole description of planetary motions on an assumed law of dynamical interaction, and deliberately renounced any attempt at accounting for this particular form of interaction between distant bodies by some mechanism of propagation through an intervening medium. This approach was such an innovation that it was thoroughly misunderstood even by the greatest of the contemporary physicists; Huygens in particular objected to it without apparently realizing that he was himself currently using the same logical procedure in his mechanical investigations, when he justified his postulates by pointing to their empirical origin.

Newton was at great pains to dispel the mistaken suspicion that he regarded gravitation as an intrinsic quality of matter. His first opportunity to do so was his correspondence with Bentley [2], but in the final scholium of the second edition of the *Principia* [3] he is more explicit about the novelty of his method – and more outspoken about his failure to find in the known properties of gravitation any sure clue to its 'cause'. He makes quite clear the deep-lying reason for his dissatisfaction about mechanical explanations of gravitation: to be acceptable according to the exacting criteria of his experimental philosophy, such an explanation ought to be related in a *unique way* to the phenomena – otherwise it is an 'hypothesis' of the kind he does not want to 'feign'. The last paragraph of the scholium, then, in which he alludes to his conception of a universal aether whose multiple functions as the agency of the most varied phenomena he carefully enumerates, can only be intended to intimate that he regards the mechanical causes of these phenomena as more securely founded, or at least less hypothetical, than that of gravitation.

Indeed, the problem of the cause of gravity was of quite a different order of difficulty from the other processes he considered: in the latter, such as the action of gross matter upon the properties of light, one is concerned with immediate, short-range, interactions which easily fit into the generally accepted Cartesian framework of mechanical explanation; the phenomena of gravitation, by contrast, had been extended by Newton himself to the grand scale of the universe: there was here not only a peculiar interaction between aether and matter at play, but a dynamical process within the aether itself had to account for the transmission of a force of unlimited range. This singular position of the gravitation problem is strikingly reflected in the evidence we have of Newton's imaginative attempts at an all-embracing description of the aether's workings: while in these successive formulations he hardly varied in the conception of the short-range interactions between aether and matter responsible for all other effects, the mechanisms he proposed for explaining gravitation exhibit considerable and significant differences.

In the extensive memorandum sent to the Royal Society [4] in 1675, he outlined a remarkable mechanism of kinetic character, according to which material bodies set up around them a cyclical circulation of aether which carries towards their surface the smaller material bodies it 'pervades'. In a letter to Boyle [5] of 1679, however, purporting to acquaint the latter with the content of the memorandum, he surprisingly ends up with an entirely new 'conjecture' respecting gravitation, without even alluding to his previous theory. As if carried away by his eagerness to unfold the whole vision of the world that is disclosing itself to his mind, he tells his friend that the new idea "came into my mind now as I was writing this letter". Although such spontaneity is not a usual mood with Newton, it is in harmony with the tone of confident friendship which marks his relations with Boyle, and there is no reason to doubt his word; what this extraordinary circumstance reveals seems obvious: Newton was so little convinced of the adequacy of his circulatory mechanism that when he was about to put it down in writing, he started thinking again about the problem and readily yielded to the alternative possibility that offered itself to his powerful imagination. In fact, when he later returns to the circulation theory in his correspondence with Halley [6], he calls it in so many words "one of my guesses which I did not rely on" [7]. Nevertheless, this very correspondence shows that he not only did not definitely discard the circulation theory, but on the contrary kept it in mind and even persisted in attaching great importance to it. The new theory summarily formulated in the letter to Boyle, while still in the Cartesian spirit, differed essentially from the previous one by being static rather than kinetic: gravitation was still conceived as the result of a pressure exerted by the aether on material bodies and tending to push them towards each other, but this pressure was now thought to result from the inhomogeneity of the aether, assumed to consist of two kinds of particles, finer and coarser ones, interacting differently with matter. It thus looks as if Newton was contemplating as alternative possibilities the opposite extremes of stationary circulation and static density distribution of aether particles to account for the maintenance of the force of gravity.

If we now turn to his last pronouncement on the subject, which he only put forward in the form of additional Queries at the end of the second English edition of the *Opticks* [8], as late as 1717[1], we see him inclining to the static alternative, which he develops in an apparently simpler form than in the letter to Boyle: he now operates with an aether of homogeneous composition, but whose density is assumed to be influenced by the presence of material bodies in such a way that it is lowest inside and around these bodies and increases continually at larger distances. This superficial resemblance should not mislead us to any facile conclusion, for the aether of Newton's later years, such as he describes it in the Queries, is radically different from the medium bearing the same name, to which, in his youthful speculations, he assigned such a prominent role in the economy of the universe. Originally, Newton took for granted the existence of a medium filling all space, like Descartes' 'matière subtile', not for the metaphysical reasons adduced by the French philosopher, but simply as part of the general mechanical view of the universe, which was then accepted without question. The constitution of the aether discussed in the Queries, on the other hand, is that of an elastic medium of such extremely low density that its single constituent particles must be assumed to be separated by considerable spatial extensions void of every kind of matter: the origin of this aether's elasticity then raised a problem whose solution lay beyond Newton's conceptual horizon.

In order to understand why the development of the aether problem took this dramatic turn, let us now look more closely at its successive phases. The memorandum of 1675 gives us a curious picture of the young Newton's fiery imagination, taking at times a poetical turn, as when he evokes his vision of Nature as a great 'circulatory worker' [4]; but we also observe how he holds his fantasy in check by his other *leitmotiv*, the 'analogy' of Nature: in his boldest speculations he always seeks guidance in comparisons with known phenomena, arguing that Nature proceeds in all her operations according to the same few principles. The mechanism of gravitation he proposes in the memorandum was at first meant to account for the force of gravity experienced on the surface of the earth; as an afterthought [7], he pointed out that it could also be applied to the attraction of the planets by the sun: at that time he was not at all persuaded that this attraction was the only force regulating planetary motions [11], and the problem of gravitation as a universal agency had not yet presented itself to his mind.

He imagines that there is in the aether a component of a 'sticky' nature, which is continually streaming towards the surface of the earth, where it is partly absorbed [2], partly re-emitted into the surrounding space, and is thus involved in a steady circulation. The downward aether stream exerts upon the material bodies it traverses a force which Newton wishes to identify with gravity. To this end, he must specify the way in which the aether particles transfer momentum to those of gross matter: he assumes [12] that the material bodies are acted upon by the aether in proportion to the "superficies of their parts", i.e. by the pressure of the aether stream, so that the radial force upon a unit volume of matter is given by the corresponding pressure gradient in the aether stream. The latter quantity is equal to the product of the local aether density and the amount of momentum which an aether particle transfers on the average per unit time to the gross matter. Newton takes this average force to be proportional to the radial velocity of the aether particle; in this way – as he explains to Halley [12] – since the aether density is inversely proportional to the same velocity, the product is independent of the variation that this velocity undergoes as a result of the interaction of the aether stream with the material bodies. In fact, on Newton's assumption, every material body at the same distance from the centre of the earth experiences the same centripetal acceleration, proportional to the aether flux, and accordingly varying inversely as the square of the distance. The inverse square law is thus only established, however, in the space outside the earth, in which the total aether flux around the earth is not sensibly affected by the bodies it traverses; but it is not expected to hold inside the earth [7], since in this region the 'sticky' aether particles will be gradually absorbed by the matter and the total aether flux will decrease appreciably, in some unspecified way[3].

The interest of this hypothesis lies in Newton's endeavour to give it quantitative precision, aiming at an understanding of the inverse square law – albeit this result could only be achieved at the cost of much arbitrariness. This was at any rate Newton's own motivation for exhuming it from the register of the Royal Society in 1686, when he felt compelled to evidence his early knowledge of the law in question against Hooke's priority contention. By that time, however, his view of the constitution of the aether had already undergone a radical transformation. The exact fulfilment of Kepler's laws throughout the whole solar system, while vindicating the universality of the simple inverse square law of gravitation, had thereby forced Newton to the unwelcome conclusion that since the aether did not oppose any appreciable resistance to the passage of the celestial bodies, it must be a medium of such extremely low density that its role as the agent of gravitation was in jeopardy: thus arose, paradoxically, from the very triumph of the universal law of gravitation, the acute dilemma regarding the origin of this force, which weighed so heavily on Newton's mind during the rest of his life [11].

It was at this critical juncture, in the spring of 1687, that a gifted and enterprising young man, Fatio de Duillier, arrived in London [14]. He came from Holland, where he had been in relation with Huygens, and was well acquainted with the mechanical explanation of gravity that the latter had developed a long time before [15], but had left unpublished. Fatio must have discussed it soon after his arrival at the Royal Society (whose meetings he attended), for on June 14, 1687 he reports to Huygens [16] that some of the members told him he was too much of a Cartesian, and that the ideas set forth in Newton's forthcoming book had changed the whole of physics. Still, the following year, as appears from entries in the journal of the Society [17] for June 27 and July 4, 1688, Fatio was asked, on his accession to membership [14], to give a fuller account of Huygens' theory before the Society. This theory was indeed as Cartesian in spirit as Newton's circulatory hypothesis, inasmuch as it treated the aether as a continuous fluid; in fact, it was a modernization of the Cartesian vortex idea. Huygens, like Newton, had realized the inadequacy of Descartes' plane vortices, and tried to replace them by spherical ones. Thus, he interpreted the force of gravity as the centripetal force which a material body immersed in such an aether vortex around the earth would experience; the quantitative identification requires an angular velocity of the aether particles about 17 times that of the earth's rotation, corresponding to a period of 1 h 25 min (as stated in the report [17] of Fatio's account).

Clearly, Huygens' proposal – which he eventually published in 1690, with critical comments on the *Principia* [18] – was a thing of the past, which could not impress Newton's followers. Fatio seems to have rapidly assimilated the Newtonian principles and nothing was more natural for him than to attempt a new mechanical explanation of gravitation, in harmony with the inferences about the aether's properties that Newton had drawn from his analysis of the planetary motions. His starting point is a kinetic conception of the aether's constitution and action upon matter – as radical indeed as the condition of extremely low density calls for: he visualized the aether as a medium consisting of particles in swift random motion, and so rare that the mutual collisions of these particles may be ignored. Could the impacts of these aether particles on a pair of material bodies immersed in them give rise to a net motion of these bodies towards each other? If so, this apparent attraction would indeed obey the inverse square law. Fatio realized that no such motion could be produced by perfectly elastic collisions between the particles of aether and matter. On the other hand, inelasticity introduces secular alterations of the motions of the system, and it is a moot point whether these can be kept down sufficiently not to conflict with the observed stability of the planetary motions. Fatio's progress was held up by this difficulty for more than two years, until he thought he could prove that the secular effect of inelastic collisions could be reduced as much as desired by decreasing the density of the aether particles and, if need be, increasing their velocity. On February 24, 1689/90, he announces briefly to Newton [19] that his theory is now "clear of objections": from which one may infer that he had previously discussed it verbally with him. On the same day, he writes to Huygens [20] a letter containing a detailed account of his thoughts, and two days later, he reads this letter at a meeting of the Royal Society [21]. Moreover, he composes a prolix essay [14] after the fashion of the time, in which he sets down once more the whole argument at great length; on March 19, 1690, this essay is in the hands of Halley and Newton [22].

Huygens saw at once that Fatio's way of dealing with the consequences of inelastic collisions was a paralogism, and in his reply [23] of March 21, 1690 (n.s.), he pointed out that nothing short of actual absorption of aether particles by the material bodies could produce the expected attraction between them. Fatio presented [24] an ineffectual defense, but not receiving any further reply from Huygens, he fondly imagined that the latter had yielded[4], and his own belief in the soundness of his theory remained unshaken. In fact, as appears from his marginal annotations to Fatio's two letters, as well as from later occasional utterances[5] Huygens thought that the necessity of assuming a continual absorption of aether by the material bodies reduced the whole theory to an absurdity.

As to Newton, we gather from a manuscript [26], as well as from Fatio's reports [25, 22], that he looked more favourably on the latter's theory. He seemed to regard it as a possible mechanical explanation of gravity – indeed he even says: the only possible one, but not with great conviction, for we know from Fatio's own testimony that he often seemed to incline, as an alternative, to the opinion "that Gravity had its Foundation only in the arbitrary Will of God [22]". It is understandable that Newton would not feel as strongly as Huygens the inconvenience of secular perturbations in the solar system; in contrast to Huygens and Leibniz, he never thought that the solar system could subsist indefinitely by itself. His own circulatory mechanism was also essentially dependent on some absorption of aether by matter, and (as we may infer from Flamsteed's testimony [30]) as late as 1685 he was still contemplating the possibility of a slowing down of planetary motions as the result of an increase in mass due to accretion of aetherial matter; he was then in doubt about a suspected slowing down of Saturn's motion, which seemed to follow from tables drawn up by Halley [30] and was too large to be explained as a perturbation due to the attraction of Jupiter [31]. Even after he had convinced himself, in the course of writing the *Principia* [11], that gravitation was the only force governing the solar system, he remained persuaded that in any case the perturbations arising from the mutual attraction of the planets would in the long run upset its structure [32].

The Fatio episode is instructive inasmuch as it evidences the persistence of kinetic conceptions in Newton's mind as offering at least a possible basis for a mechanical theory of gravitation. There is no telling when or why he gave up this prospect and turned again to the static type of mechanism of which he had sketched out an improvised variant in his letter to Boyle. It is first in the new English edition of the *Opticks* [8] (of which two issues were printed in 1717 and 1718) that he resolved to make public, in some additional Queries[6], his last views about the aether and its possible function as the agent of gravitation; these views, however, harmonize with those expressed in the other Queries, which date from the first Latin edition [32] of 1706. We may therefore surmise that Newton's renewed interest in optics, which led him to the elaboration of his curious theory of the fits of easy reflexion and easy transmission, can have been instrumental in orienting his thoughts towards the manifestations of the elasticity of the aether, which he was now considering to be one of its essential attributes.

In order to account for the phenomena of thin plates, Newton had indeed seen the necessity of introducing an element of periodicity in the propagation of light: he imagined that the light corpuscles could excite in the aether waves of definite periods, which would overtake them and put them in fits of easy reflexion or easy transmission. This mechanism fixes a lower limit to the value of the ratio of elasticity to density of the aetherial medium, which is equal to the square of the velocity of propagation of waves in it; this ratio must be at least of the order of 10^{12} times the corresponding ratio for air. On the other hand, the density of the aether must be small enough to make its resistance to the motion of matter through it negligible; if we take it 10^6 times rarer than air, its resistance will be about 10^9 times less than that of water. Its elasticity would then have to be at least 10^6 times larger than that of air. Newton applies here with wonted virtuosity his familiar method of extrapolating from known properties of matter to a plausible construction of a hypothetical medium satisfying the imposed requirements. If he now wanted to fit gravitation into the picture, he had to introduce some inhomogeneity in the distribution of the aether with respect to the material bodies; and he saw that he would achieve his purpose by assuming that the presence of matter produces in the aether a gradient of density, which in its turn maintains the pressure gradient required to bring about the phenomena of gravitation.

It could not escape him, however, that his new aether model, in spite of its apparent simplicity, was much nearer "the arbitrary will of God" than those he had previously devised on the analogy of continuous fluids – an analogy he now firmly rejected (Query 28). For he was now faced with the question of how to account in mechanical terms for the elasticity of the aether. In the *Principia* [1], he had discussed – as a 'mathematical' possibility – a static model of fluids in which a pressure is maintained by a short-range repulsion between the constituent particles of the fluid. This is proposition XXIII of Book II, in which he shows, by a simple argument of proportionality, that a repulsive force between nearest neighbours varying as the $-n$-th power of their distance gives rise to a pressure proportional to the power $\frac{1}{3}(n+2)$ of the density. He does suggest (Query 21) that such a mutual repulsion between the particles of the aether may be the cause of its elasticity; but this cannot be a mechanical cause, for the analogy with ordinary fluids here breaks down on an essential difficulty: the necessity of assuming that the aether particles are on the average separated from each other, without any possibility of material communication between them.

In the famous Query 31 (which was published in 1706, antedating by 7 years the publication of the final scholium added to the *Principia*) Newton raises the problem of action at a distance in full generality: gravitation is only one of the interactions between material particles which can be described in this way; electricity and magnetism offer other examples, and chemical processes give evidence of shorter-range forces of a similar type. Likewise, optical phenomena point to an action of matter upon light corpuscles which is not one of immediate contact. How such actions are transmitted is a question which Newton professes not to be concerned with; they "may be perform'd by impulse, or by some other means unknown to me". Just as in the *Principia*, he is anxious sharply to dissociate his phenomenological approach from the vexed problem of the causes, fully conscious as he is of the immense progress this method represents over the Cartesian philosophy. Nevertheless, he never gave up his dream of a universal aether providing the medium through which an omnipresent God was continually acting upon his creation; but he had to confess that the only workings of the aether he could understand were the transmission of short-range interactions inside the material bodies; these he felt confident could be attributed to "a very subtle spirit pervading the gross bodies and lying hidden in them" ("spiritu quodam subtilissimo corpora crassa pervadente, & in iisdem latente" [3]): for that part of the aether was acting upon sufficiently condensed matter to make its behaviour conceivable in mechanical terms, – in contrast to that of the rare medium of interplanetary space. The noblest function of the aether, the regulation of the heavenly motions, eluded him to the last.

NOTES

[1] Essentially similar ideas occur already in manuscript drafts which can be dated to the period of writing of the *Principia* (1687), and recur in drafts of the final scholium of the second edition (1713) [10]. A general discussion of Newton's thoughts on the constitution of matter has been given by A. R and M. B. Hall [10].

[2] The summary account of Newton's circulatory mechanism given in my paper on Newton's theory of gravitation [11] misses the essential part played by this absorption, and is accordingly quite inadequate. This was noticed by Dr. E. J. Aiton in the course of an investigation of the controversy between Hooke and Newton about the priority of the discovery of the inverse square law. I am most thankful to Dr. Aiton for calling my attention to this point and for sending me the manuscript of his forthcoming paper, which contains in particular the correct interpretation of Newton's argument [13].

[3] A total number S of aether particles per unit time steadily streaming towards the centre of the earth produce at the distance R from the centre an aether flux $F = S/4\pi R^2$, and a local aether density F/v, if v is the radial velocity at this distance. With the assumption that the average force exerted by an aether particle on the gross matter is proportional to v, the total force on a unit volume of matter will be independent of v and proportional to the aether flux F. This again is inversely proportional to R^2 so long as the total aether flux S is sensibly constant.

[4] He says so in a letter of March 30, 1694 written to the Braunschweig resident in London and intended for Leibniz [25], and also in a later memorandum [22].

[5] In a letter of December 29, 1692 (n.s.) to the Marquis de l'Hospital [27] and twice [28] in reply to enquiries which Leibniz directed to him [29] after receiving Fatio's letter mentioned in the preceding footnote.

[6] The first edition of the *Opticks* (1704) contained only Queries 1–16, which were kept (with some additions) in all the following editions. In the Latin edition [32] of 1706, a number of new Queries appear, there numbered 17–23: they coincide in all essentials with the Queries numbered 25–31 in all later editions, and they contain a detailed discussion of the aether's role in optical and chemical phenomena, as well as (in the last Query) physical and theological considerations about the constitution of the universe. New Queries, numbered 17–24, were finally inserted in the second English edition of 1717–18 and the later ones.

[7] All dates are quoted in the Julian calendar, except those marked '(n.s.)' (new style).

BIBLIOGRAPHY[7]

[1] Newton, I., *Philosophiae naturalis principia mathematica*, first edition 1687.

[2] *The Correspondence of Isaac Newton*, (further quoted as *Correspondence*), vol. 3: letter 399 of January 17, 1692/93, esp. p. 240. [Cambridge University Press, 1959 *et Seq*. – Ed.].

[3] Newton, I., *Philosophiae naturalis principia mathematica*, second edition 1713: Scholium generale.

[4] *Correspondence*, vol. 1: letter 146 to Oldenburg of December 7, 1675, esp. p. 366.

[5] *Correspondence*, vol. 2: letter 233 of February 28, 1678/79, esp. p. 295.
[6] *Correspondence*, vol. 2: letter 288 of June 20, 1686 and letter 291 of July 27, 1686.
[7] *Correspondence*, vol. 2: letter 288, p. 440.
[8] Newton, I., *Opticks*, second edition 1717–18; new Queries numbered 17–24.
[9] Hall, A. R. and M. B., *Unpublished Scientific Papers of Isaac Newton* (Cambridge University Press, 1962), pp. 321–333, 349–359.
[10] *Ibid.*, pp. 183–213.
[11] Rosenfeld, L., 'Newton and the Law of Gravitation', *Archive for History of Exact Sciences* **2** (1965), 365. [This volume, p. 58].
[12] *Correspondence*, vol. 2: letter 291, p. 447.
[13] Aiton, E. J., 'Newton's Aether Stream Hypothesis and the Inverse Square Law of Gravitation', *Annals of Science* **25** (1969), 255–260.
[14] Gagnebin, B., 'De la cause de la pesanteur, mémoire de Nicolas Fatio de Duillier', *Notes and Records of the Royal Society of London* **6** (1949), 105 (further quoted as 'Gagnebin').
[15] *Œuvres complètes* de C. Huygens (further quoted as '*Œuvres*'), vol. 19, p. 619; vol. 21, p. 379. [Nijhoff, The Hague.]
[16] *Œuvres*, vol. 9, p. 167 (also *Correspondence*, vol. 2: letter 307, and Gagnebin, p. 109).
[17] Gagnebin, pp. 114–115 (also *Correspondence*, vol. 3: letter 352, note (1), p. 69, where the second entry, however, is wrongly interpreted as referring to the exposition of Fatio's own theory).
[18] *Œuvres*, vol. 21, p. 429.
[19] *Correspondence*, vol. 3: letter 463, p. 390.
[20] *Œuvres*, vol. 9, p. 381.
[21] Gagnebin, pp. 115–116.
[22] Gagnebin, pp. 116–117.
[23] Gagnebin, pp. 154–158 (also, but incomplete, *Œuvres*, vol. 9, p. 391).
[24] *Œuvres*, vol. 9, p. 407 (letter to Huygens of April 11, 1690).
[25] *Œuvres*, vol. 10, p. 608 (also *Correspondence*, vol. 3: letter 440).
[26] *Loc. cit.* in note 9, p. 313.
[27] *Œuvres*, vol. 10, p. 354.
[28] *Œuvres*, vol. 10, p. 613 (letter to Leibniz of May 29, 1694 (n.s.)) and p. 669 (letter to Leibniz of August 24, 1694 (n.s.)).
[29] *Œuvres*, vol. 10, p. 603 (letter to Huygens of April 26, 1694 (n.s.)) and p. 644 (letter to Huygens of June 22, 1694 (n.s.)).
[30] *Correspondence*, vol. 2, note (4) to letter 275, p. 412.
[31] *Correspondence*, vol. 2: letter 276 of January 12, 1684/85 to Flamsteed.
[32] Newton, I., *Optiks* (translated by S. Clarke) 1706, quaestio 23 (= Query 31 in later editions); see also Robinet, A., (ed.), *Correspondance Leibniz-Clarke* (Presses Universitaires de France, Paris 1957). [See H. G. Alexander, ed., *The Leibniz-Clarke Correspondence* (Manchester University Press, Manchester, England, 1956) – Ed.].

CHAPTER 11

THE GENESIS OF THE LAWS OF THERMODYNAMICS

[1941a]

The history of the establishment of the two great laws of thermodynamics is instructive on more than one score. Considered, as it usually is, from its purely logical aspect, it raises varied and important questions of scientific method, which occur here with particular simplicity and clarity. In fact, the logical structure of thermodynamics exhibits very pure lines: It reduces to two laws of universal significance expressing two independent but closely linked aspects of the most general impossibility of "perpetual motion". The manner in which these two laws were successively disentangled, in which their independence was recognized, and in which the problem of their compatibility was then raised, offers for analysis a clearly delimited group of facts without complicated ramifications or cumbersome accessory details. Three more or less successive phases may be distinguished: The first, dominated by the great solitary figure of Carnot, includes the statement of the second law (1824) in a still imperfect form and its slow diffusion throughout the scientific public (from 1834 to 1847); the second phase occupies the period from 1839 to 1847 in the course of which several, as yet isolated pioneers, notably R. Mayer and Joule, arrived at the formulation of the first law, or law of conservation of natural forces; the third phase (1848–1851) culminated in the logical synthesis of these two laws, thanks to the efforts of W. Thomson and Clausius. The depth and originality of the researchers who constructed this great theory make the study of their guiding viewpoints and the analysis of the development of their ideas singularly attractive. This analysis, as we know,[1] has notably afforded us precious lessons on the respective roles of experimental induction and theoretical or philosophical speculation in discovery, and on the logical structure of natural laws.

But the progress of science does not take place by a concatenation of observations and arguments determined solely by an intrinsic logical coherence: In its concrete reality it appears with all the complexity of a social phenomenon, simultaneously a consequence of the development of the organization of society and a powerful tool for perfecting that organization. From this point of view as well, the history of thermodynamics is of unique interest, for it is intimately linked to that 'industrial revolution' which is one of the most decisive phases in the progress of civilization. It is rare to find in the history of a single well-delimited problem such a striking illustration of the various aspects of this fruitful interaction between scientific thought and social evolution. It is above all this side of the question that will concern us in the following pages.

If we first examine the sources from which the discovery of these two laws derive, we see that they belong, with equal significance, to two great areas of concern which have always determined the progress of science and which assert themselves particularly sharply during the great epochs of social transformation, such as the one which concerns us here: They are, on one hand, the technical problems which the economic organization of production and transportation poses at each new stage of its development, and, on the other hand, the efforts which tend to ameliorate the conditions of life and health of human society. In the course of the first two phases of the evolution sketched above, the pioneers of thermodynamics were exclusively either engineers, like Carnot and Joule, or doctors, like Mayer: Carnot undertook to unravel the relationships between heat and mechanical work in the action of the steam engine; Joule did likewise in the case of the electric motor; and Mayer in the case of the human body.

Even though it is social necessities which have determined the nature of new scientific problems, these necessities are far from being generally understood at the outset, nor are these problems immediately perceived: On the contrary, their solution is always due to isolated pioneers who are, it is said, 'ahead of their time'. And a very lively struggle is always required before a fundamentally new idea manages to be assimilated, to be incorporated into the common thinking of the epoch and to play the role in science and its technical and biological applications which is due it. For the intellectual life of a period rests on a foundation of concepts and theories which has itself been laboriously constructed in an attempt to embrace the knowledge already acquired; so only as a last resort is a decision made to alter it profoundly.[2] We find, in the history of thermodynamics, a very accentuated divorce between the young self-taught pioneers, who were inspired directly by the 'needs of the day', and the contemporary schools of physicists, no doubt very brilliant, but continuing the traditions of another age. Only after a long period of a quarter of a century, during which the great new laws, so clearly recognized by isolated researchers, encountered general incomprehension, do we see for the first time, during the last phase of synthesis of the two laws, 'professional' physicists taking an active part in constructing a branch of the science that had until then painfully grown up without them, and even to a certain degree, despite them. But the reason for this is precisely that at this moment, toward 1850, the first generation educated entirely in the spirit of the new industrial civilization reached maturity and its most brilliant young members began to occupy university chairs! Hence the logical completion of the new doctrine coincided with the definitive triumph on the social and cultural level of the new men who had helped to create it.

If we consider the discovery of the laws of thermodynamics from the point of view rapidly outlined in the preceding sections, a drama then unfolds before us; the three phases mentioned earlier are its three acts, each inspired by this double *leitmotiv*: social needs inspiring discovery, the struggle necessary to bring about the adaptation of society to the new perspectives offered by the discovery.

It was, as we have said, within the framework of the 'industrial revolution' that thermodynamics was born. What inspires Carnot's work[3] is the desire to analyze the functioning of steam engines in order to discover the principles of their better utilization. Of course, ancient science already possessed the prototype for these engines: such was the aeolipile [primitive steam engine] described by Hero of Alexandria. But to see its application to technology, it was necessary to wait until the moment when, in the XVIIIth century, the economic expansion of the western European states, enriched by overseas commerce and the resultant growth of industrial production, gave rise to the need for seeking new sources of motive power in hitherto unexplored natural phenomena. Only then was the aeolipile – like the magnet and the piece of rubbed amber – taken out of the collections of curiosities, and only then were the natural laws of which they are manifestations systematically studied. While Joseph Black, in particular, established the principal concepts which describe thermal phenomena, his student James Watt invented the first practical models of steam engines. Naturally, this industrial application did not wait for the completion of the scientific analysis of the problem. In Carnot's time, the already flourishing art of constructing the new engines still rested on very rudimentary empirical gropings.

But sooner or later the need to perfect newly introduced machines, to monitor their functioning and their output, forces a detailed analysis of the phenomena involved, and this analysis quite often reveals the existence of hitherto unknown natural laws whose significance often goes far beyond the technical problem which served as starting point for the scientific investigation. Hence, in his masterful analysis of the 'motive power of fire', Carnot reduced the complicated action of a heat engine to a very simple, very general scheme in which the mechanical work produced is equated to the reversible transfer of an invariable quantity of caloric between two bodies at different temperatures. To this schema, Carnot applied the principle of the impossibility of perpetual motion, well established for all purely mechanical engines, but whose validity he assumed extended to heat engines as well. Thus he managed to bring out the essential content of what we now call the "second law" of thermodynamics. In truth, the real position of this law in the structure of thermodynamics could only be established later; but Carnot recognized and even then clearly stressed its universal significance: He demonstrated its fruitfulness in a series of remarkable applications to the study of the properties of gases. In this sense, Carnot's work is one of the most magnificent examples of scientific discovery inspired by a technical problem. Moreover, the complete obscurity in which his contemporaries left this work gives us a no less striking first impression of that force of inertia, which we shall encounter again confronting other pioneers in thermodynamics, and whose causes we shall later attempt to analyze. Carnot knew, nonetheless, that his work responded to an immediate and serious concern of the builders of steam engines: Aware of the usefulness and importance of his discovery, he endeavored, as his brother testifies,[4] to put his explanation into a clear form accessible to everyone; finally, it would seem that the very simplicity of his reasoning and conclusions should have made them easily acceptable to physicists as well as engineers. But ten years passed without Carnot's book evoking the slightest response from one group or the other. Not only this silence, but equally the very fact that a contribution of such genius arose in such isolation, without any precursors other than one or two insignificant attempts,[5] demonstrates how few minds were prepared to receive it.

Only in 1834 did the engineer Clapeyron[6] take up and paraphrase Carnot's ideas, without adding any essentially new contribution. And Clapeyron's paper itself did not begin to attract attention – and thus indirectly to make Carnot's point of view known – until 1843 when Poggendorff published a translation of it in his *Annalen der Physik.*[7] Moreover, at this moment, completely independently of Carnot's principle and its consequences, the second phase of our story was beginning, a phase which ended in the establishment of the general law of the equivalence of the various forces of nature.

This law of equivalence presents a more immediate character than Carnot's principle. In the form of the extension to all natural agents of the impossibility of perpetual motion, it is, in short, simply a concrete formulation of the principle of causality. So although the two laws are obviously equally universal, the second, less direct, was still confined to the domain of thermal phenomena and was only later extended to chemical and electromagnetic phenomena, while the first revealed itself more easily in all its generality from the outset. From this universal perspective, moreover, it was in harmony with the great idea of the unity of the forces of nature which synthesized, after a fashion, a preoccupation and primordial aspiration of the period of the 'industrial revolution': to view these forces as various manifestations, transforming one into the other, of a single source of motive power. During the period extending from 1839 to 1847, it was assuredly this common aspiration which conditioned the remarkable convergence of the very different paths which led a series of pioneers, isolated and unaware of each other, to the same conclusion.

Among these pioneers, let us pause only at the physician Mayer[8] in whom we see at work – besides the technical problems which serve as the starting point for other investigators: Seguin, Colding, Joule – a different motif, constantly at work in scientific progress, the one to which Descartes in his *Discours de la Méthode* even gives first place: the quest for the rational foundations of medicine. The study of the relationships between the laws of living organisms and those of inanimate nature, dominated directly or indirectly by this motif, has always had a significance which can hardly be exaggerated, not only for the development of biology, but even for that of the physico-chemical sciences themselves. Indeed, it can be said that the evolution of the latter starts out precisely with biological or animistic analogies, which are gradually refined precisely by reason of the progress of our analysis of the conditions which distinguish properly vital activities from the general properties of inanimate matter. One need only think, for example, of the pre-Lavoisian history of chemistry:[9] This science has its origin in metallurgy as well as in medicine and pharmacology, but at the start it operated above all on principles of animistic inspiration such as 'like attracts like': Soon, however, it felt the repercussions of the philosophical systems emanating from the more advanced development of mechanics: Cartesianism, then Newtonianism; we then see physical analogies replace animistic analogies; the principle just cited is notably replaced by the concept of affinity, whose definition is directly inspired by the Newtonian notion of attraction. At a more advanced stage, there are many fruitful connections between biology and physics; besides the example which concerns us here, the question of so-called 'animal electricity' debated by Galvani and Volta may also be recalled. The starting point for Mayer's profound reflections is the theory of the origin of animal heat proposed by Lavoisier. According to the latter, it is a combustion phenomenon in the body tissues that gives rise, not only to the heat itself, but also to the motile activity of the animal. Mayer observes that this motile activity may be used exclusively to engender another quantity of heat (by rubbing) in an external body. From this, a strict interpretation of the principle of causality (*causa aequat effectum*) leads him to postulate a quantitative equivalence between the initial chemical phenomenon and the *total* quantity of heat produced, directly in the body of the animal as well as indirectly in the external bodies, and consequently also an equivalence between the latter quantity of heat and the mechanical work which served to produce it. It is once again the same question of the source of animal heat which inspired Helmholtz,[10] also a physician, to state on his part the general conservation law and to review its application to different types of phenomena – though his remarkable effort of synthesis cannot compare in depth or extent with Mayer's brilliant creation.

Having recalled the work of the pioneers who responded so brilliantly to the exigencies of the new times, the moment has come for us to turn to the contemporaries to whom they addressed themselves and to seek the motives for the latter's incomprehension of such essential problems. At first glance, this lack of comprehension appears, indeed, quite surprising, if one considers the renown of the scientific movement during the period which interests us, not only in France and England, where the new industrial civilization was in full expansion, but in central Europe where it was rapidly being introduced.

But on looking more closely, it appears that the great scientific tradition which inspired this movement, far from being linked to the contemporary upheavals in the European economic structure, rather prolonged and in fact marked the end of the currents of thought born in the preceding period and dominated by the concerns of large-scale maritime commerce and colonial expansion. In the domain of the physical sciences, this was the Newtonian tradition, that 'physics of central forces', of which Laplace's *Exposition du Système du Monde* gives a magnificent synthesis. Concentrating on the difficult problems of astronomy which required the creation of refined mathematical tools, mechanics – particularly in the classic work of Lagrange which represents the most extreme expression of this tendency – cut itself off from the concrete questions of industrial technology and cultivated methods which were inaccessible to artisans, iron-masters, small industrialists who worked by feeling their way in the development of this technology. Thus a deep gap opens between those who directly serve the pressure of the new economic needs, who come to grips with the problems of the nascent science without being equipped to resolve them, and those who know and unceasingly perfect the most powerful scientific methods of their times, but who do not look toward the future.

Very significant in this respect is the attitude of Lagrange and his school toward the notion of mechanical work and the related conservation law, a law whose generalization constitutes the first law of thermodynamics. We see this law playing an essential role at all the decisive stages of development of mechanics. Stevin based statics on the impossibility of a *perpetuum mobile*; Galileo started from other forms of the same principle, as did Huygens, in order to treat the first typical problems of the dynamics of a material point and to derive from it the law of kinetic energy; it was a judicious application of the law in this last form which gave Bernoulli a fundamental law of hydrodynamics; finally, d'Alembert reduced the general equations of dynamics to the same source. Lagrange showed this with admirable sagacity in the masterful historical sketches in his *Mécanique analytique.*[11] But, Newton's system, in which motion is analyzed by means of the concept of an accelerating force, had meanwhile relegated these fertile energetic considerations to the background: Even Stevin's brilliant idea is described by Lagrange as "little known". And the ascendency of the Newtonian tradition was so strong that the reasoning of the founders of mechanics is termed 'indirect' or 'precarious' in this account which nonetheless stresses their full significance.

The divorce between the dynamic point of view of the mathematician-astronomers and the energetic point of view adopted, so to speak, instinctively by the artisan-engineers was so marked that both groups used the same word "force (*Kraft*)" to designate the concept they considered primordial, – the former: accelerating force; the latter: work, – though the distinction between these two concepts was not always recognized from the start. The very loose usage that Lagrange himself[12] makes of the word 'force' already reveals how little awareness there was of the possible danger of a confusion of ideas due to this imprecise terminology. Indeed, on the subject which concerns us, the greatest scientists – Mayer, Oersted, Faraday – fell victim to such a confusion; and there is the further example of Mohr[13] who tries to establish an equivalence between 'quantity of heat' and 'force' in the Newtonian sense of the word. In other areas of the sciences, too, errors having the same origin long persisted: Vital force, contact electromotive force, even gravitational force were then commonly interpreted as principles that were unfailing in action, in flagrant contradiction with the strict causality of science.

However, it would be excessive to try to set up an absolute opposition between the, in a sense, 'official' scholars attached to the ideas of the eighteenth century and the technicians preoccupied with the utilization of new sources of motive power. The economic and social upheavals of the period put their mark on the philosophical systems with which the young generation attempted to formulate simultaneously a synthesis of the teachings of the masters, seen from a new angle, and a program of work intended to complete this synthesis. The conception of the unity of the forces of nature, in particular, reflects (as we observed earlier) the aspirations of an epoch of prodigious expansion of industrial production. In truth, in France, for the young Polytechnic students grouped around August Comte, this idea was obscured by a dogmatic affirmation of the irreducibility of the different types of phenomena – a point of view which was assuredly fatal to a correct appreciation of the new laws; on the other hand, the unity of natural forces plays an extremely important role in the less systematic philosophy which inspired the English physicists; in central Europe, it received a romantic expression within the framework of a 'Naturphilosophie' corresponding to the less advanced state of these countries.

But the idea that close relationships must exist between the different forces of nature, which proved itself so fertile in the hands of a Faraday or an Oersted, at the time of the discovery of the two laws of thermodynamics (that is, during the first two phases of the history of this discovery), was far from helping to orient the ideas of the majority of physicists in a direction favorable to this discovery. On the contrary, seen from an idealistic point of view in this initial period, the conception of the unity of natural forces did not in the least imply the intertransformability of these forces, but only a common origin. '*Naturphilosophie*' – in conformity with its romantic character – considered this common source as inexhaustible. The more flexible English philosophy was closer to the idea of the transmutation of forces and to the problem, directly derived from this, of the quantitative comparison of their effects; but without – even in the case of its most illustrious representative, Faraday – quite reaching it. It is only by putting ourselves back into this hostile or hesitant state of mind that we can appreciate the degree of boldness involved in the extension to thermal phenomena of the impossibility of perpetual motion, and that we can understand the cold reception given, first to Carnot, then – fifteen years later – to the other founders of thermodynamics. Even more, these latter themselves reveal the novelty of their point of view by their errors. Thus, we see Mayer hesitate a moment to extend the validity of his law to solar radiation,[14] while Joule, who arrived at the same law only step by step, had directed his first investigations toward the construction of an electromagnetic motor which would have been a true *perpetuum mobile*![15]

Thus, if the formulation of the laws of thermodynamics gave the concept of the unity of the forces of nature a definitive content, it was not without a radical modification of the meaning heretofore attached to this concept. Those who, in varying ways, first felt the need of this modification were above all the young people whose names fill what we call the second phase of our story; unlike Carnot who still remained an entirely isolated precursor, they appear, though unaware of each other's work, animated more or less consciously by a common inspiration, which is just the idea of a unity in the economy of nature; and their reflections, as we have already emphasized, converge toward the same result. This striking coincidence, followed by the rapid completion of the final phase and general adoption of the new ideas, indicates that we are dealing here with a change in attitude which profoundly affected society during this period. In fact, in the domain of physics the issue was that passage from idealism to materialism which – accompanying the definitive establishment of industrial civilization – so strongly marked the development of ideas towards the middle of the nineteenth century. So, from this perspective, we see that far from being reinforced by the philosophical trends of their times, the pioneers of thermodynamics had, on the contrary, to anticipate them and engage in the same struggle on their ground that the artisans of the new civilization were fighting on all sides.

The foregoing general considerations do not exhaust the analysis of the difficulties which so long delayed acceptance of the laws of thermodynamics. Even the state and orientation of experimental physics during this period offered no corrective for the unfavorable atmosphere of the dominant theories and philosophies. First of all, physics did not yet have an independent position in the organization of teaching and scientific research; it was treated as a simple auxiliary of chemistry – *ancilla chemiae*. It was the problems and trends of the latter science that captured the attention and efforts of the most able experimenters, and work in physics was subordinated to them. This was the case particularly with that 'pneumatic physics' or study of the properties of gases which began a period of vigorous development at the start of the century and whose results were so fortunately exploited by a Carnot or a Mayer. For the experimenters themselves, however, it was only a question of collecting data, the analysis of which still seemed premature. Similarly, the electric and magnetic phenomena which notably played an essential role in the evolution of Joule's ideas, belonged to a domain whose surface had barely been scratched and hence little known to the scientific public.

On the other hand, in the area most directly related to the question, that of caloric phenomena, the long hesitation between the substantialist theory and the kinetic theory of the nature of heat helped to obscure the significance of conclusions which necessarily depended on a choice between these two theories. This last circumstance was all the more serious since it introduced an element of incertitude into the very camp of the innovators. Indeed, Mayer was the only one of them to put the logical aspect of the problem in relief by identifying the law of equivalence with the principle of causality, thus revealing its independence from any special hypothesis on the nature of heat. Almost all the others, and especially Joule, rallied more or less explicitly to the kinetic point of view, which effectively implies an equivalency between mechanical work and quantity of heat. But in so doing, they confused the basis of the law of equivalence with the no less new and more speculative basis of the kinetic theory.

On the contrary, Carnot's considerations, in their original form, were essentially linked to the hypothesis of conservation of caloric and consequently lead to a radically different 'equivalence' between work and heat. This fact forced the appearance of a critical problem of first importance, when Carnot's law was seriously confronted with the law of equivalence: that of the compatibility of the two principles. Helmholtz[16] was the first to narrow the gap between the two, but he still did not clearly reveal the logical problem posed by their synthesis. Nevertheless, the solution was not long in coming. The discussion which led to it and constitutes the final phase of the establishment of the foundations of thermodynamics, brought a resolution which was all the more dramatic in that it bore on the very solidity of the new theoretical structure. By its relative suddenness, coming after its sharpening to the point of crisis, of the very difficulties which we have tried to analyze above, this resolution had the attractive aspect of a 'dialectical' reversal so dear to Hegel.

It is worthy of note that Carnot himself had already clearly posed the logical problem in question. In fact, between 1824 and 1832, the year of his premature death, he had recognized the necessity of abandoning the caloric theory which he had used in his book and of adopting, on the contrary, the kinetic theory, which led him immediately to the law of equivalence. But in abandoning the caloric theory, was it also necessary to abandon the very fertile law whose deduction depended on it? In Carnot's papers, the question is clearly posed, but no answer is given.[17]

This fact is of more than personal interest: It shows just how far a person (in truth, exceptionally keen and independent) could move forward by 1832, by making use of the latest attainments of science.

The course of Carnot's solitary reflections is precisely the one followed between 1847 and 1850 by the young William Thomson, who learned of Carnot's theory through Clapeyron's paper and of the law of equivalence through Joule's work. But this time, Thomson's brilliant analysis was published immediately and the apparent paradox in which it results was forcefully stressed. Even before Thomson himself found the solution, it was found by another young professor, Clausius, whose masterful paper of 1850[18] established the definitive foundations of the new doctrine.

Thus thermodynamics made its entry into the universities with the first generation of scientists who had brought there the new spirit of industrial civilization. From that point on a more regular tradition and development were assured: The 'heroic' period is over – indeed, for this industrial civilization itself, as well as for the great scientific doctrine to which it had given birth.

NOTES

[1] Cf. the works of E. Mach: *History and Root of the Principle of the Conservation of Energy*, transl. by Philip E. B. Jourdain (Open Court Publ. Co., La Salle, Ill., 1911); *The Science of Mechanics*, transl. by Thomas J. McCormack (Open Court Publ. Co., La Salle, Ill., 1960); *Principien der Wärmelehre* (Leipzig 1896) [translation in preparation for the *Vienna Circle Collection*, Reidel Publ. Co., Dordrecht, Holland and Boston – Ed.]

[2] Cf. L. Rosenfeld, *Archeion* **21** (1938), 74.

[3] S. Carnot, *Réflexions sur la puissance motrice du feu* (1824) [*Reflections on the Motive Power of Fire*, ed. E. Mendoza (Dover, New York, 1960) – Ed.]. The 1878 reprinting contains a biography of Sadi Carnot by his brother Hippolyte as well as unpublished notes selected from the papers of Sadi Carnot (who died in 1832). Subsequent references refer to this reprinting.

[4] *Loc. cit.*, note (3), p. 76: "Preoccupied with a desire to be clear, Sadi had me read passages of his manuscript to reassure himself that it would be understood by people devoted to other studies" (H. Carnot).

[5] Carnot scrupulously cites an erroneous calculation by A. Petit, *Ann. de Chimie et de Phys.* **294** (1816), and a paper by Clement which has remained unpublished.

[6] E. Clapeyron, *Journal de l'École Polytechnique* **14** (1834), 153.

[7] *Ann. der Phys. u. Chem.* **59** (1843), 446. Poggendorff accompanies the translation with this remark: "This article which up to now has received very little attention, nevertheless seems to deserve mention here even now because of its importance."

[8] R. Mayer, *Die Mechanik der Wärme*, 3rd edition with biographical notes by J. Weyrauch, and a second volume containing his correspondence and various documents (1893).

[9] On this subject see the remarkable works of Hélène Metzger: *La genèse de la science des cristaux* (1918); *Les doctrines chimiques en France* (1923); *Les concepts scientifiques* (1926); *Newton, Stahl, Boerhaave et la doctrine chimique* (1930); *La philosophie de la matière chez Lavoisier* (1935).

[10] H. Helmholtz, *Über die Erhaltung der Kraft*, 1847 (*Wissensch. Abhandl.*, 1882, vol. I, p. 12). [There are two translations of this work: (1) 'On the Conservation of Force' in *Popular Lectures on Scientific Subjects*, transl. by E. Atkinson, ed. J. Tyndall (London 1884); (2) 'The Conservation of Force: A Physical Memoir' in *Selected Writings of Hermann von Helmholtz*, ed. R. Kahl (Wesleyan University Press, Middletown, 1971) – Ed.]

[11] J. Lagrange, *Mécanique analytique*, 3rd edition (revised) by J. Bertrand, 1853. See for Stevin, Part 1, Section 1, Nos. 5 and 9 (vol. I, pp. 7 and 11); for Galileo, Part 2, Section 1, No. 1 (vol. I, p. 209); for Huygens, *ibid.*, Nos. 6 and 7 (pp. 217 and 219–220); for Bernoulli, Part 2, Section 10 (vol. II, p. 247).

[12] Cf. *loc. cit.*, note (11), vol. I, J. Bertrand's notes at the bottom of pp. 33, 39, 126, 140, 153, 178, 187, and 213.

[13] See Mayer, *op. cit.*, vol. II, p. 420.

[14] *Ibid.*, vol. II, p. 139.

[15] Cf. O. Reynolds, 'Memoir of J. P. Joule' (*Proc. Manchester Lit. and Phil. Soc.*, 4th series, vol. 6 (1892)), p. 38.

[16] Helmholtz, *op. cit.*

[17] Carnot, *op. cit.*

[18] R. Clausius, *Ann. d. Phys.* **79** (1850), 368, 500.

CHAPTER 12

JOULE'S SCIENTIFIC OUTLOOK

(Paper Read on 3rd July, 1951, on the Occasion of the Opening of the Joule Museum at Salford)

[1952a]

In commemorating Joule's pioneering work to-day, I should like to discuss it from two different points of view: the one more general and also more historical, the other more specifically concerned with its scientific aspect.

The first point I want to make is that we cannot really understand how a career like Joule's was possible if we do not put it in its proper social background. If we try to consider Joule as an isolated personality, without regard to the environment in which he worked and lived, we get a very distorted view of his significance in the development of physics.

It is a striking fact that the science of thermodynamics (the science of energy in its various forms and transformations) is hardly more than a century old; yet, if we go back in history we find that the Greeks had already some knowledge of the motive power of heat. However, so far as we know, they only used this knowledge to make a sort of toy, whose description we find in Heron's work on Pneumatics. Quite another use was made of steam power in the 18th century in the Cornish mines, where steam engines were employed to pump water out of the mines. From these beginnings steam engines gradually developed, in the course of the 18th and beginning of the 19th century, into a major source of motive power for large industrial machinery. Why this difference in outlook between Hellenistic society and the kind of society that existed in this country in the 18th century? I cannot attempt to tackle this problem to-day; but the contrast that I have just pointed out brings to mind very forcibly the importance of the economic and social background for the development of science.

Scientific ideas and principles are in a sense eternal and universal: they are lying ready to be discovered and contemplated at all times and in all circumstances. But historically things do not happen that way. The laws of Nature are only discovered when there is an incentive to search for them, most often in order to put them to some kind of social use. Science is not at all what it is sometimes represented to be – some contemplation of Nature which could go on in temples erected for the purpose. It is essentially a social activity and one of the most important of all the activities that go to build human society.

The birth and growth of thermodynamics offers a classic illustration of this social aspect of science, because in this case the link between social and economic motives on the one hand and scientific ideas on the other appears in a particularly clear and simple way. Joule's case, moreover, shows the rather subtle nature of this link. It would be quite wrong to say that Joule started his investigation of the connection between heat and mechanical power with a view to perfecting the engines providing such power. That could have been so, but as a historical fact Joule was primarily animated by pure scientific curiosity. His interest was first aroused by the production of heat by the passage of an electric current through a conductor. This phenomenon had been somewhat neglected by Davy and Faraday, who had concentrated their attention on other perhaps more immediate aspects of voltaic electricity. There was thus hope for an alert young man of independent mind to reap some new results by following up this line of research. It was only at the conclusion of the long series of experiments which led him to the formulation of the law of conservation of energy that he gave consideration to the implications of this law for the utilization of natural forces as the source of mechanical power.

The socially significant circumstance in Joule's attitude to a purely scientific problem is, however, the direction that his curiosity took. The attention that he paid to the relations of equivalence between heat and other physical phenomena was an entirely novel outlook at the time. It was a kind of curiosity quite different from that which inspired Davy and Faraday in looking at the same phenomenon. In fact, we have here one of the most striking examples of a physical phenomenon looked upon by great physicists from two quite different points of view. Joule's way of looking at it was so strange, so novel, that it was not immediately understood, even by the most prominent scientists of the time. This is convincingly demonstrated by the fact that the first paper submitted by Joule to the Royal Society was not judged of sufficient interest to be published in full.

It is said that Joule, when asked many years later what he thought of this rejection of his first paper by the Royal Society, answered: "I was not surprised. I could imagine those gentlemen in London sitting round a table and saying to each other, 'what good could come out of a town where they dine in the middle of the day?'" This remark shows, at any rate, that Joule was keenly aware of the real motive underlying this lack of understanding between the gentlemen in London and the lonely young man in Manchester. It is clear, in the first place, that this misunderstanding is not of a scientific nature, since, of course, the phenomenon is there for anybody to see and study. It is a misunderstanding based on a difference of social outlook.

The scientists who then composed the membership of the Royal Society had received the inspiration of their work in the 18th century, and the scientific effort of the 18th century was directed towards quite different problems from those that were in the centre of interest in a growing manufacturing centre such as Manchester in the early 19th century. The scientific tradition of the 18th century derived from Newton's work, and, like the latter, was determined by problems concerning not an industrial community but a commercial one. The problems of navigation focussed attention especially upon astronomy and optics. Considerations of mechanical power and of the utilization of the forces of Nature as sources of such power were not foremost in people's minds in the 18th century because production then was of a very limited scope and was mainly on an artisanal level.

It was only when the accumulation of wealth in the closing years of the 18th century created a demand for increased production that the need was felt of replacing hand tools by machine tools, and then of supplying power to drive the machines. These changes in the methods of production brought about a revolution in the social set-up of society. A new class of people arose whose activity centred on manufacturing rather than on commerce. The interest of those people in science was naturally directed towards the utilization of the forces of Nature in obtaining mechanical power. It is in just such a community that Joule grew up. The son of a wealthy brewer, he could look forward to a life of leisure. Thus, on the one hand, he was able to indulge in investigations inspired by pure scientific curiosity; but, on the other, the background of this curiosity was nevertheless determined by the interests of the people in whose environment he was brought up. He was led to study those aspects of the phenomena which might throw light on the possibilities of transformation of various natural agencies into each other and into mechanical power.

Joule is not, of course, the only example of the influence exerted by the new industrial class. At about the same time as he engaged in his researches, Grove, a distinguished amateur of science, gave his famous lecture 'On the correlation of the forces of Nature' at the London Institution. In this lecture the various aspects of the transformation of natural agencies into each other are treated at length, though only in a qualitative manner, and the novelty of this way of looking at natural phenomena is emphasized. It is significant that Grove's audience was not the Royal Society but the members of an Institution newly founded by the industrialists for the purpose of discussing scientific problems of special interest to them. It took quite a long time before the new way of looking at Nature advocated by Grove and Joule was incorporated into the scientific tradition. This was not achieved before the middle of the century. There is a lapse of about eight years between the time when Joule formulated his main conclusion regarding the conservation of energy and his election in 1850 as a Fellow of the Royal Society, which indicates the complete change of outlook of the leading scientific circles.

This, then, is the first lesson that we may derive from an effort to understand Joule's work and personality. Let us now turn to a more technical analysis of this work and consider the methods which he used with such mastery in his investigations. It is often said – and unfortunately Osborne Reynolds' biography of Joule, so excellent in other respects, has added authority to that view – that Joule is typical of the experimental physicist who bases his conclusions only on the results of experiments without allowing himself to be biased by preconceived theoretical views. He is further contrasted with Robert Mayer, the German physicist who at the same time as Joule – and, of course, unknown to him – arrived at exactly the same discovery of the general law of the conservation of energy, without making a single experiment but, it is stated, by 'pure' reasoning, or even by 'metaphysical' reasoning, which is, I imagine, the worst that can be said of anybody!

Now, I should like to try and convince you that this view is quite superficial and mistaken. If we look more closely at the actual arguments developed by both Joule and Mayer, we find that these two lines of argument are essentially identical. As a theoretical physicist, it gives me pleasure to say that the argument in both cases is mainly theoretical, with, of course, as in all good theory, some reference to experiment.

Let us first examine Mayer's work. It is true that Mayer did not make any experiment himself, but he was quite clear from the start that the decisive conclusion must be based on facts that could only be learned from experiment. His brilliant achievement was just to make an extremely ingenious use of experiments carried out by others, in order to extract from them the data which he needed. In 1806 Gay-Lussac, the great French chemist, had investigated the heating effects that occur when air compressed in a container is allowed to escape into another container which had been evacuated. He was trying to find out whether the specific heat of a gas depends on its density. This was a question that had arisen in connection with the discussions about the nature of heat. He found that no heat is evolved in the phenomenon and left it at that. Gay-Lussac's paper appeared in a very out-of-the-way publication, the *Journal de la Société d'Arcueil*, in which were published the proceedings of informal meetings of a little group of French physicists at the summer house of one of them at Arcueil. It therefore remained generally unknown. Joule, for instance, made exactly the same experiment in 1845 without ever having heard of Gay-Lussac's previous attempt. Mayer, a young physician, was led to reflect on the connection between heat and mechanical work by his interest in Lavoisier's theory of respiration. It was in the course of this study that he had a first intuition of a general relation of equivalence between heat and mechanical power. But he saw (and here his genius appears) that he could only arrive at a clear understanding of such a relation by tracing it to phenomena much simpler than the physiological processes which he was studying; he must go down to a quantitative analysis of elementary physical phenomena. When he accordingly started to learn physics with this aim in view, he came across a description of Gay-Lussac's experiment, probably in a French text book which he was using extensively at that time.

Now, the remarkable event occurred that Mayer, because his mind was oriented in the right direction, was able to read the hidden message contained in that experiment – a message which Gay-Lussac himself failed to understand. He knew, of course, that a gas takes up more heat when it is allowed to expand against a fixed pressure than when it is kept in the same volume. He suspected that the additional heat in the first case was related to the mechanical work accompanying the expansion; but in order to be sure that the relation between heat and work in this case was one of equivalence, one had to enquire whether the expansion by itself, when it was not accompanied by any mechanical work, did not involve any exchange of heat. The answer to this question, as Mayer instantly recognized, was given by Gay-Lussac's experiment. In its cogency and simplicity, Mayer's reasoning is one of the most beautiful examples of the kind of argument which leads to great discoveries in physics. As you see, there is no metaphysics about it! Mayer picked out the significant facts revealed by the experiment, put them together, and the conclusion fell, like ripe fruit, into his hands. Just to collect the right facts and to squeeze the conclusion out of them required, however, not only the keenest insight, but above all the focusing of the attention in the right direction. All the facts had been in the possession of Gay-Lussac, but he failed to draw from them the momentous conclusion that they implied – not because his insight was less penetrating than Mayer's, but because his interest was not fixed on the decisive aspect of the phenomenon.

Now, what about Joule? Joule's train of thought is much more intricate than Mayer's, and looking at it superficially it seems that the way by which he came to his discovery was extremely devious – that he, so to speak, stumbled on it: but a more careful study leads to a very different conclusion. At the very beginning of the paper in which he describes his first experiments about the heat produced by the passage of current through a conductor, he shows that he is already concerned with the idea of equivalence. He explains that in checking by new measurements the generally accepted assertion that the heat produced by the passage of the current is exactly proportional to the resistance, his concern is to ascertain "whether resistance to conduction is the *sole* cause of the heating effects", The only possible meaning of that sentence, with the word 'sole' underlined, is to indicate that the idea at the back of his mind was that of equivalence between the phenomena of heat production and of the passage of the current through the resistance. If he were not looking for an exact balance, there would have been no point in asking whether there were one or more causes of heat production. Here the 'book-keeping' mentality of Joule's environment is conspicuous, albeit hardly conscious to Joule himself.

The next step was to compare the heat produced by the passage of the current with the chemical changes occurring in the battery which produced this current. Again, it is clear that Joule could only have been led to such a comparison by some suspicion of a possible equivalence between the chemical changes on the one hand and the heat produced on the other. However, he is admirably cautious in the analysis of his results. He does not rashly jump to the conclusion that the chemical change in the battery is the cause of the heat produced. He keeps in mind the two possible theories about the nature of heat: the theory that heat is a substance which would simply be transported by the current but not created, and the rival theory according to which heat is a form of motion, in which case it could be generated by suitable agencies. At this stage he appears to be non-committal about the theory of heat, but in a later paper he declares that he was actually convinced that the dynamical theory was the true one. Here, then, we have his own word for stating that he was guided from the start, in the orientation which he gave to his researches, by a definite theoretical view about the nature of heat. But he rightly kept his theoretical convictions in reserve and did not venture to express them before he had collected empirical evidence which he judged sufficiently convincing.

From such clear and unambiguous hints we are, therefore, justified in concluding that Joule was inspired from the very beginning by theoretical ideas: the idea of the dynamical nature of heat and a perhaps not quite articulate idea that there might be an equivalence between heat and other physical agencies. The experiments that he undertook on the connection between heat and electricity were made with a view to testing those ideas.

A striking confirmation of this inference is offered by Joule's next experiment, which is probably the most extraordinary in the whole history of physics. He encloses the revolving armature of an electromagnetic engine in a cylindrical container, carefully protected against heat losses, and filled with a known amount of water. After ascertaining the temperature of the water, he rotates the coil and water-filled container during a certain time between the poles of the fixed electromagnet, and measures the current induced. He then again takes the temperature of the water, and derives the quantity of heat evolved in the process. The qualitative fact that heat appears in this case in the revolving coil was the decisive proof he was looking for that heat is not simply transferred by the current from one place on the circuit to the other, but is actually generated. This was therefore a crucial experiment which disposed of the view of the caloric theory of heat and definitely confirmed the dynamical theory.

The quantitative implications of the experiment, however, were of still more decisive importance in guiding Joule to the recognition of the universal law of equivalence between various physical agencies. The way in which he proceeded further with the analysis of electromagnetic heat production bears testimony, not only to Joule's ingenuity as a natural philosopher, but also to his intellectual integrity, all the more remarkable when one sees him as a self-taught youth grappling single-handed with one of the most fundamental laws of Nature. He did not – as one might, after the event, imagine he could have done – immediately link the heat produced by the rotation of the coil with the mechanical work spent in this operation. He first proved by a striking argument that the heat production must be quantitatively related to the work done in rotating the coil. For this purpose he superposed upon the induced current an additive or subtractive current supplied by a battery, and he showed that in every case the heat produced was proportional to the square of the total current. Now, he argued, the chemical changes in the battery are proportional to the current intensity itself; therefore they cannot account for all the heat produced, and the balance can only be related to the mechanical work spent in producing the induced current. This piece of reasoning is assuredly one of the most beautiful examples of a penetrating and rigorous analysis of a physical phenomenon. Only after he had thus established the necessity of assuming a relation between heat production and mechanical work did he take the last and comparatively easy step in the investigation, viz. to measure the amount of work needed to produce in the apparatus a known quantity of heat.

It was necessary to remind you of the details of this beautiful work of Joule in order to make clear how the view that Joule was 'a pure experimenter' without any 'theoretical bias' could gain ground. Certainly, the way in which he carried out the series of experiments which I have just tried to sketch very briefly and incompletely is, as I have emphasized, a model of rigorous experimental induction; but it would be quite wrong, and lead us to a quite distorted picture of Joule's achievement, to ignore the fact that the conception and planning of these experiments was essentially directed by the wish to test definite theoretical speculations. If we disregard this fact, the complication and strangeness of Joule's experimental arrangements become utterly incomprehensible. If, on the contrary, simply listening to what Joule himself tells us,[1] we judge the adequacy of his apparatus for the purpose of putting to test the dynamical theory of heat and the general idea of equivalence between the physical agencies, everything becomes clear and the admirable cogency of Joule's procedure comes into full light. If you ask me how it was ever possible to overlook Joule's own unambiguous statements about his being guided by theoretical ideas and create the impossible legend of his merely experimenting, I can only remark that the history of science seems to be the last refuge of some people of otherwise sober judgment for satisfying a primeval urge for mythical creation.

When Joule had thus established on a firm empirical foundation his conceptions of the nature of heat and the conservation of energy, he set up a new series of painstaking experiments with a view to a more direct, and accordingly more accurate, determination of the mechanical equivalent of heat. From the point of view of experimental technique these famous experiments are again masterpieces which set us wondering about his independence of judgment and self-confidence. Even now it is hard to imagine how he could have the boldness to devise apparatus for such delicate measurements and trusted their reliability as he did. This, however, is another fascinating aspect of Joule's work upon which I have no time to enter. The study of Joule as an experimenter is actually still to be made. Perhaps the facilities for such a study provided by the Joule Museum may incite somebody to tackle it. I will just mention a small but characteristic detail. Several pages of Joule's laboratory books are filled with notes on the painstaking comparison of the readings of his different thermometers. It is really touching to see that he gave those thermometers individual names – the "Old Standard", the "Old Sensible", and "Fastré Sensible". The thermometers lived for him: they were, so to speak, his faithful helpers.

There is now a last remark about Joule's achievement which I should like to make very briefly. It concerns Joule's position with respect to the scientific philosophy of his time. Joule, as we have seen, pursued concurrently two great ideas: the dynamical theory of heat and the equivalence between heat and mechanical work. In his mind, these two ideas were united by a quite essential connection: the picture of heat as a mode of molecular motion was the justification of the law of equivalence, the atomic and mechanistic model of the universe was the *raison d'être* of the conservation of energy. This conviction is strongly expressed in the remarkable lecture which Joule delivered in Manchester in 1847. It was shared by all the leading physicists – Thomson, Clausius, Maxwell, Boltzmann – who brought to completion the new science significantly called "thermo*dynamics*" or "mechanical theory of heat". Strictly speaking, however, the principle of conservation of energy is not logically tied to any particular view on the nature of heat. This was always emphasized by Robert Mayer, who insisted on the fact that the establishment of a quantitative equivalence between different classes of phenomena does not at all mean that those phenomena are in any sense identical. Mayer stressed the qualitative differences between various phenomena and their continual transformation into each other.

In this attitude of Mayer we recognize, of course, the influence of the German Romantic philosophy which, in spite of all its shortcomings, had at any rate the merit of reminding physicists of the importance of qualitative changes, which they are sometimes prone to forget. It is quite typical of the 'book-keeping' mentality of Joule and the British physicists that they never understood Mayer's point of view. In one of Joule's notebooks there is a translation of Mayer's first paper which is very revealing. It is, I must say, a rather bad translation; the more abstruse sentences especially have proved to be quite unintelligible to Joule. And he was obviously so impatient of Mayer's views that he could not help inserting in the course of his translation various disparaging remarks. This curious irritation shows that Joule's own mechanistic philosophy was not the result of cool, logical thinking but was rooted in deeper emotions.

I think that the reason why Joule and his followers believed so strongly in a completely mechanical picture of the material world must again be sought in their social environment. The second half of the 19th century witnessed a rapid growth and concentration of the production process, calling for an enormous increase in the development of the sources of motive power, and for a bolder use of materials in building lofty bridges and powerful machinery. Scientists interested in such developments were influenced by their intimate contact with matter in its more solid and bulky aspects. This unwittingly reflected itself, so to speak, in their philosophical views. An extreme example of this kind of mechanical philosophy is afforded by Joule's friend, William Thomson, later Lord Kelvin, who did not shrink from imagining intricate models of the ether involving gyroscopes and strings, and who even identified the understanding of a phenomenon with the construction of a mechanical model for it.

Yet those people would have been grievously shocked if anybody had described them as 'materialists'. Joule himself was deeply religious. We have, for instance, the draft of his presidential address to the British Association in 1873, which he was prevented by illness from delivering, in which he makes a case in favour of scientific education. The chief purpose and justification of the study of science, according to him, is that it is a way of adoring God by the contemplation of His works. Of course, he also deals at great length with the practical uses of science; but he presents them as secondary advantages. Nevertheless, the essential component of his scientific attitude must not be sought in such utterances – too obviously conditioned by uncritical acceptance of social conventions – but in his spontaneous scientific thinking. There he behaved in exactly the same way as the free-thinker Tyndall, that outspoken exponent of "scientific materialism".

Towards the end of the century, there occurred a curious reaction. Ostwald took up Mayer's views again, or rather misrepresented them, in the form of what he called "Energetik"; he endeavoured to do away with atoms and all materialistic interpretations of Nature and to reduce the interpretation of all physical phenomena to the two fundamental laws of thermodynamics. This attempt, which was indeed quite unsound, failed completely. It must be said, incidentally, that as soon as Ostwald saw that he was on the wrong track he honestly acknowledged it and proclaimed his allegiance to the atomic view of Nature. This turn in the development of natural philosophy at the end of the 19th century is not an accident originating in Ostwald's personality: it is again, a sort of echo in the realm of physics of the turmoil that was going on in the social sphere at that time and which gave rise to a general reaction against the so-called materialistic philosophy. It is a very significant fact that apart from the short-lived and rather trivial episode of the Energetik, social reaction could never get any hold on scientific thought. The tremendous advances of science in our time have been achieved in the same spirit as Joule's pioneering work.

NOTE

[1] The especially significant passages in Joule's papers revealing his theoretical preoccupations are the introduction to the fundamental paper analysed above (*Joule's Scientific Papers*, Vol. I, p. 123) and also the fifth observation at the end of the preceding paper (*ibid.* p. 120) where he boldly predicts the quantitative relation between heat and work in the electromagnetic machine and announces in a footnote that he is "preparing for experiments to test the accuracy of this proposition".

CHAPTER 13

AN ANALYSIS OF JOULE'S EXPERIMENTS ON THE EXPANSION OF AIR*

(Including Original Results not Previously Published)

[with A. P. Hatton**]

[1956e]

In 1845 Joule published one of his most famous papers entitled 'On the Changes of Temperature produced by the Rarefaction and Condensation of Air'.[1] A number of experiments were described in this paper, but it would be generally agreed that the most important is that which was used to show "that no change of temperature occurs when air is allowed to expand in such a manner as not to develop mechanical power".[2]

The experiment is shown diagrammatically in Figure 1. *X* and *Y* are two copper vessels connected by a pipe with a valve, the whole being completely immersed in water. *X* was charged with air at high pressure and *Y* was evacuated. When conditions were steady the valve was opened and air flowed from *X* to *Y* until the pressures equalised. No change of temperature of the water was observed by Joule which implies that there was no net heat flow to the air or water and hence no net change of air temperature.

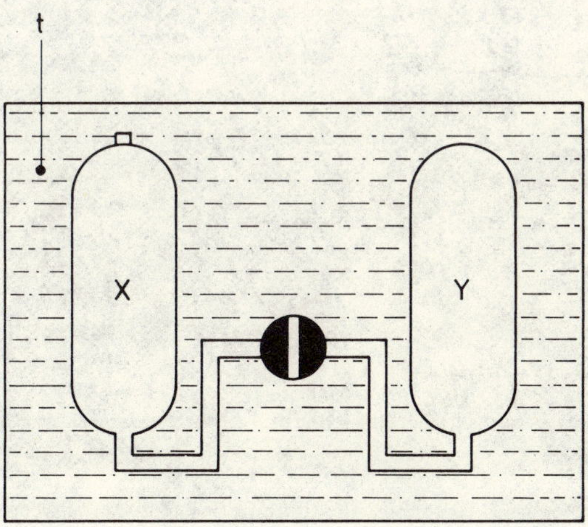

Fig. 1.

With the same apparatus Joule performed further experiments[3] with the vessels immersed in separate water containers with the object of analysing this result. In his paper, however, Joule does not describe these experiments in any detail, although in view of the first experiment they are of great importance. The results are given in Joule's original note-

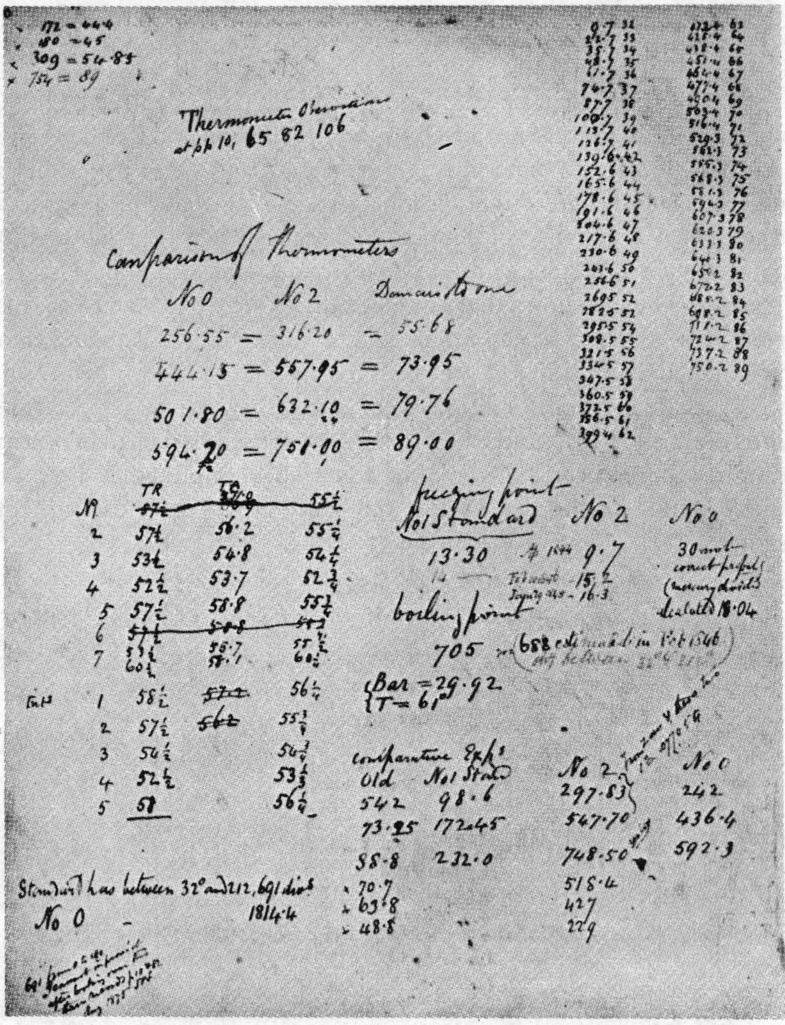
Comparison of Thermometers

No 0	No 2	
256·55 =	316·20	= 55·68
444·15 =	557·95	= 73·95
501·80 =	632·10	= 79·76
594·70 =	751·00	= 89·00

freezing point
No 1 Standard No 2 No 0
13·30 9·7
boiling point
705
Bar = 29·92
T = 61°

Fig. 2. Joule's Laboratory Book, p. 10 (half original size).

book[4] and it is the object of this paper to present and discuss these figures.

The important figures have been photographed from the notebook and are shown in Figures 2–5. The apparatus for these experiments is shown diagrammatically in Figure 6.

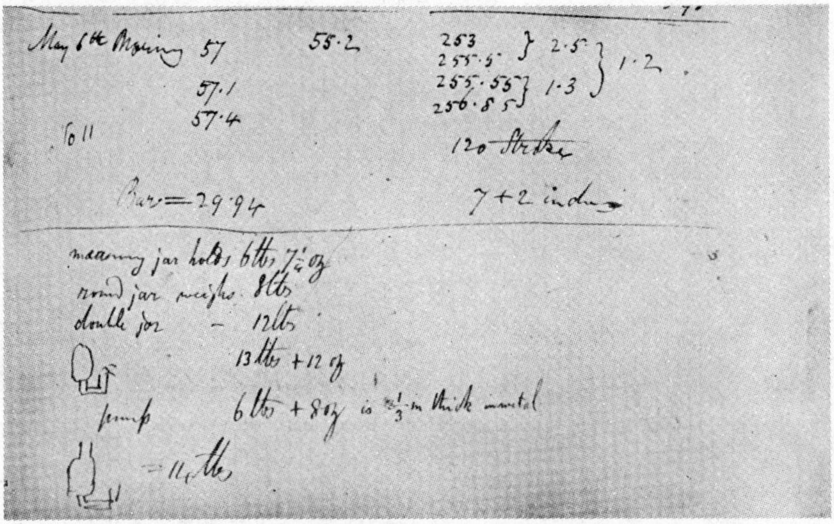

Fig. 3. Joule's Laboratory Book, lower half of p. 14.

1. Joule's experimental results

The weights of the various parts of the apparatus are shown at the bottom entry of Figure 3 and on Figures 4 and 5 are the results of three experiments. It is not clear whether three separate thermometers were used, but at the start of the whole series of experiments Joule calibrated his thermometers and it is clear from the figures (Figure 2) that the relevant readings refer to the thermometer No. 0, for which the calibration factor is approximately 10 divisions equal to 1 °F.

In the original paper Joule gives the additional information that the vessels had a capacity of 136.5 and 134 cu. in. and that vessel X was initially charged to 22 atmospheres, Y being evacuated. This pressure is due to the compression of 2828 cu. in. of air into the vessel which also

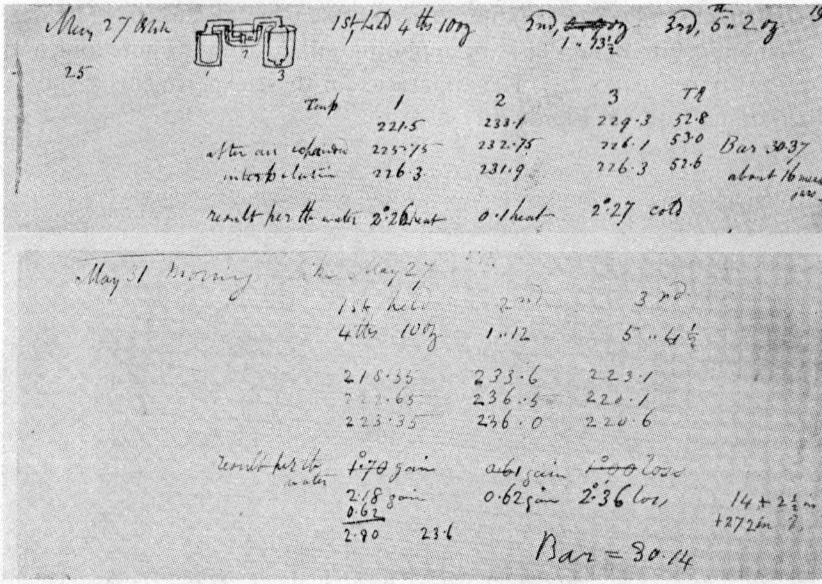
May 27 ... 1st held 4 lbs 10 oz 2nd, 1 .. 13½ 3rd, 5 .. 2 oz 19

	1	2	3	Th
Temp	221.5	233.1	229.3	52.8
After air expansion	225.75	232.75	226.1	53.0
interpolation	226.3	231.9	226.3	52.6
result per lb water	2°.26 heat	0.1 heat	2°.27 cold	

Bar 30.37 about 16 ...

May 31 Morning like May 27

1st held	2nd	3rd
4 lbs 10 oz	1 .. 12	5 .. 4½
218.35	233.6	223.1
222.65	236.5	220.1
223.35	236.0	220.6

result per lb water 1°.70 gain ... gain ... loss
2.18 gain 0.62 gain 2°.36 loss
0.62
2.80 23.6
14 + 2½ in +272 in ...
Bar = 30.14

Fig. 4. Joule's Laboratory Book, parts of p. 19.

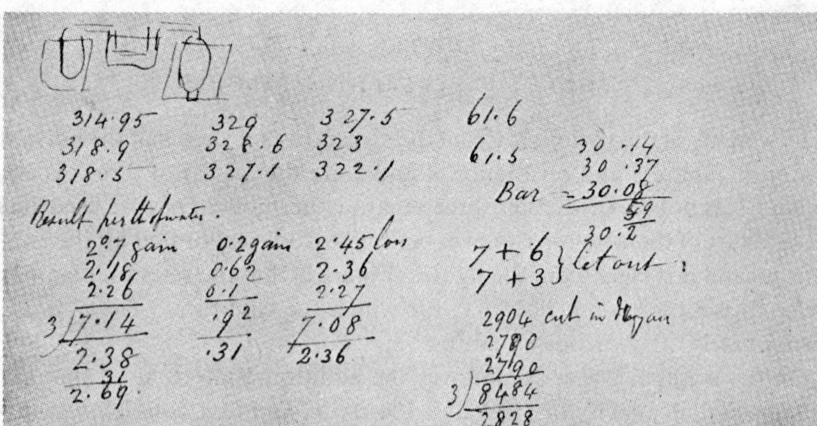
314.95	329	327.5	61.6
318.9	328.6	323	61.5
318.5	327.1	322.1	

30.14
30.37
Bar = 30.08
30.2

Result per lb of water.

2°.7 gain	0.2 gain	2.45 loss
2.18	0.62	2.36
2.26	0.1	2.27
3) 7.14	.92	7.08
2.38	.31	2.36
.31		
2.69		

7 + 6 } let out
7 + 3 }

2904 cut in ...
2790
2790
3) 8484
2828

Fig. 5. Joule's Laboratory Book, lower half of p. 20 (entry dated June 1, 1844).

contained atmospheric air initially. This figure is an average for the three tests as can be seen from Figure 5.

Joule states his conclusion in the original paper as follows: "I found that 2.36° of cold per pound of water had been produced in the receiver from which the air had been expanded, while 2.38° of heat had been produced in the other receiver, and 0.31° of heat also in the can in which the connecting piece was immersed".

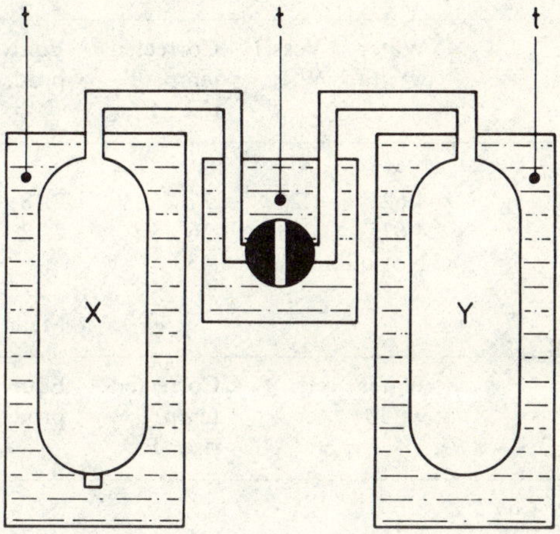

Fig. 6.

By "degree of heat per pound of water" Joule implies the pound – °F unit of heat, or British Thermal Unit.

It is not at all easy to see how Joule arrived at the above conclusion from his experimental readings. The main difficulty is in the understanding of the temperature readings. Joule took three temperature values for each test and for each vessel, namely: – before expansion, after expansion, and "interpolation". Joule does not explain how he took this final observation, but it is clear that it corresponds to what he describes in his paper as "alternation", "to eliminate the effects of stirring, evaporation, &c." It is assumed that after the air expansion was completed the apparatus was left for a time equal to the time between the first two temperature readings and then the "interpolation" reading was taken. In other experiments Joule has "weighted" his correction to allow for

Test	Water wt lb	Vessel W.E.	Corrected temp. rise °F	Equiv. cold prod.	Joule's Value
Vessel X					
1	5.125	1.28	.34	2.17	2.27
2	5.281	1.28	.35	2.29	2.36
3	Not quoted		.36		2.45
				Mean...........	2.36

Test	Water wt lb	Vessel W.E.	Corrected temp. rise °F	Equiv. heat prod.	Joule's Value
Vessel Y					
1	4.625	1.26	.37	2.18	2.26
2	4.625	1.26	.36	2.12	2.18
3	Not quoted		.435		2.7
				Mean...........	2.38

Test	Water wt lb	Corrected temp. rise °F	Equiv. heat prod.	Joule's Value
Valve Can				
1	1.844	.04	.074	0.1
2	1.75	.34	.584	0.62
3	Not quoted	.11		0.2
			Mean...........	0.31

Water equiv. of vessel $X = 14. \times .092 = 1.28$ lb.
Water equiv. of vessel $Y = 13.75 \times .092 = 1.26$ lb.

changes of the temperature difference between the water and the air of the room, but since in these experiments such changes were very small, we shall assume that the corrections were applied directly. For example, the first readings of the first test are 221.5, 225.75 and 226.3. The first difference is 4.25, the second difference is 0.55 and the difference between these values is 3.7 equivalent to 0.36°.

From such corrected temperature changes and using the weight of water and water equivalent of his apparatus, Joule worked out the amount of heat or "cold" produced on each side. Joule gives no details of the outside containers apart from stating in his paper that they were double walled and made of thin material. Two such containers are kept at the Joule Museum. It has not been thought worth while, however, to determine their water equivalents. The containers have thus been ignored in our calculations, which means that the estimated values of heat and cold produced would be expected to be less than Joule's values. The table shows the values derived in this manner and they can indeed be seen to be all slightly less than Joule's values.

2. Theoretical analysis of Joule's experiment

Having interpreted Joule's experimental readings, we can now consider a theoretical analysis of the process.

Consider the arrangement of two vessels X and Y shown in Figure 7. X is charged with gas at a higher pressure than that in Y.

It is necessary to decide the type of process which occurs in Joule's experiment and this is influenced by whether the expansion from X to Y occurs rapidly or slowly.

(a) If the expansion occurs slowly then it will tend to be isothermal and the temperature of the whole apparatus can be assumed constant throughout the process.

(b) The expansion may be rapid and tend to the adiabatic whilst the pressure is falling. After the pressure equilibrium has been reached heat flow will take place until all parts of the apparatus are restored to room temperature; also further mass exchange between the vessels will occur.

Let suffix x denote conditions inside X and suffix y conditions inside Y, suffix 1 denote conditions before valve is opened and suffix 2 conditions when pressure equilibrium is attained.

Assuming that there is no net change of internal energy, we have

$$w_{x1}C_vT_{x1}-w_{x2}C_vT_{x2}=w_{y2}C_vT_{y2}-w_{y1}C_vT_{y1},$$

where w refers to weight, and C_v is the specific heat. Therefore,

$$w_{x1}T_{x1}-w_{x2}T_{x2}=w_{y2}T_{y2}-w_{y1}T_{y1}.$$

From the gas law $pv=wRT$ this gives

$$p_{x1}v_x-p_{x2}v_x=p_{y2}v_y-p_{y1}v_y.$$

Let $v_x/v_y = r$; then

(1) $$r(p_{x1} - p_{x2}) = p_{y2} - p_{y1}.$$

For conservation of mass

$$w_{x1} - w_{x2} = w_{y2} - w_{y1},$$

$$\frac{p_{x2}v_x}{T_{x1}} - \frac{p_{x2}v_x}{T_{x2}} = \frac{p_{y2}v_y}{T_{y2}} - \frac{p_{y1}v_y}{T_{y1}},$$

i.e.

(2) $$r\left\{\frac{p_{x1}}{T_{x1}} - \frac{p_{x2}}{T_{x2}}\right\} = \frac{p_{y2}}{T_{y2}} - \frac{p_{y1}}{T_{y1}}.$$

Equations (1) and (2) apply irrespective of the type of process occurring. For Joule's experiment, $r = 1$, $p_{y1} = 0$ and $p_{x2} = p_{y2}$; hence we get from (1) and (2)

$$p_{x2} = \tfrac{1}{2}\, p_{x1}, \quad \frac{1}{T_{x1}} = \tfrac{1}{2}\left(\frac{1}{T_{x2}} + \frac{1}{T_{y2}}\right).$$

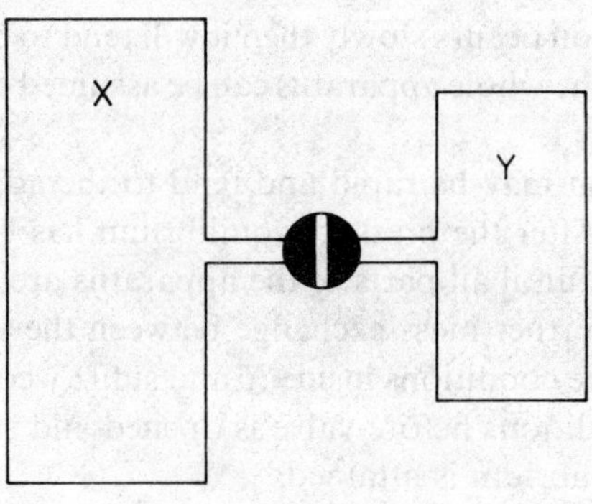

Fig. 7.

However, these two equations will not give information about the changes taking place on one side only of the apparatus.

It must be observed that the thermodynamic processes on each side involve changing mass. We use the fundamental equation of unsteady flow. Consider a tank containing gas under pressure. If a valve is opened and gas flows out the following relation is obeyed:

$$(3) \qquad Q=\int h\cdot dw+\Delta E_v,$$

where

h = enthalpy of fluid/1b inside tank,
dw = increment of mass outflow,
ΔE_v = total internal gain of internal energy,
Q = heat flow in through the walls of the tank.

2.1. *Isothermal Process*

Consider vessel X. We have in this case

$$(4) \qquad Q=C_pT\int dw+C_vT(w_2-w_1)=$$

$$\begin{aligned}&=C_pT(w_1-w_2)+C_vT(w_2-w_1)=\\&=(C_p-C_v)\,T(w_1-w_2)=\\&=RT(w_1-w_2)=\\&=v_x(p_{x1}-p_{x2}).\end{aligned}$$

With

$p_{x1}=22$ atm,
$p_{x2}=11$ atm,
$v_x=136.5$ cu. in.,

we get

$Q=2.36$ B.t.u.,

in remarkable agreement with Joule's result.

2.2. *Adiabatic Process*

The adiabatic process in X will occur so that

$$\frac{T_{x2}}{T_{x1}}=\left(\frac{p_{x1}}{p_{x2}}\right)^{(\gamma-1)/\gamma}=2^{(\gamma-1)/\gamma}.$$

This gives $T_{x2}=425\,°$F abs. if T_{x1} is taken at $520\,°$F abs.

The bottle will now absorb heat *at constant pressure*, certain outflow taking place until the temperature returns to 520°. For this process,

$$\Delta E_v = C_v \Delta(w_x T_x) = \frac{C_v}{R}\Delta(p_x v_x) = 0.$$

Hence

$$Q = \int C_p T dw$$
$$= C\frac{p_{x2} v_x}{R}\int \frac{dw}{w},$$

and denoting the final state by the suffix 3,

$$Q = C_p \frac{p_{x2} v_x}{R} \log \frac{w_{x2}}{w_{x3}} =$$
$$= C_p \frac{p_{x2} v_x}{R} \log \frac{T_{x3}}{T_{x2}} =$$
$$= \frac{\gamma}{\gamma - 1} p_{x2} v_x \log \frac{T_{x3}}{T_{x2}}$$
$$= p_{x2} v_x \log 2,$$

since $T_{x3} = T_{x1}$. This gives a value for Q too low by a factor 0.69. It thus appears that in Joule's experiment the process tended to the isothermal. This would be expected from the design of the valve between the two vessels which was a needle type with a fine thread. This presumably allowed the air to pass only slowly and made the process closely isothermal.

The experiment is a remarkable tribute to Joule's care and skill. The temperature changes involved are very small; yet he has obtained remarkable precision in his results. The only point which remains unexplained is the heat produced in the valve can and indeed there seems to be no simple reason for this. The air temperature will be low when passing the valve at high speed and it is not easy to understand why temperatures higher than the dynamic or total value should be produced.

Joule's first experiment is well known and it is hoped that this article will bring to the second experiment an equal amount of attention, together with credit to the great physicist.

NOTES

* Joule Museum Publication No. 2.

** M. Sc. Tech., A. M. I. Mech. E., College of Technology, Manchester.

[1] *Phil. Mag.*, May 1845 (*Scientific Papers*, vol. I, p. 172).

[2] *Ibid.* p. 182.

[3] *Ibid.* pp. 182–183.

[4] This is item 5 (*b*) of the catalogue of the Joule Collection given by H. Lowery, *Journal of Scientific Instruments* **8**, (1931), 3. The collection now forms the Joule Museum at Salford.

CHAPTER 14

THE VELOCITY OF LIGHT AND THE EVOLUTION OF ELECTRODYNAMICS

[1956a]

1. Introduction

The rise of the velocity *c* from the humble status of a mere conversion factor for two systems of units to the dignity of a universal constant of nature is a most instructive story, illustrating as it does not only the main stages in the evolution of electrodynamics, but a variety of epistemological points in the development of scientific thought.

The first decisive step was the recognition of the identity of light vibrations and electromagnetic waves. Although this step is rightly associated in the annals of science with the name of Maxwell, it is interesting to reflect that it was also taken, quite independently and from a different approach, by his Danish contemporary Lorenz. In some way, Lorenz' argument was more direct than Maxwell's: it stemmed in straight filiation from Weber's and Kirchhoff's work; but it was of a more limited scope. Both Lorenz and Maxwell had, of course, to break away from the conception of action at a distance; Lorenz did it by the formal introduction of retarded potentials, Maxwell by a deeper analysis of the mechanism of propagation of electromagnetic action. Lorenz still proceeded in the spirit of the mathematical school of Ampère, Weber and Kirchhoff; Maxwell took his inspiration from the very different, more descriptive views of Faraday. From this first momentous advance the constant *c* emerges with a clear, but still limited, physical meaning: it was the velocity of propagation of disturbances in a hypothetical material medium, the luminiferous aether, whose elastic properties had somehow to be described in terms of electromagnetic field intensities.

The next period, during which attention was centred on the properties of fast moving electrified bodies, culminated in a crisis which shook to its very foundations the science of mechanics, hitherto considered as the basis of all physical thinking. In the establishment of the theory of relativity, by which this crisis was solved, the velocity of light played an essential part: it was now regarded as a truly universal constant of nature, representing the greatest possible velocity of any physical process, be it the propagation of an electromagnetic wave or the motion of a material body. It is the latter conception of a maximum velocity of motion of all matter which is the great novelty of Einstein's theory: it entails a radical change in the definition of the fundamental kinematic and dynamic concepts, beginning with the concept of simultaneity. But even though the relativistic account of electromagnetic phenomena does not differ in form from Maxwell's theory, the whole position of electromagnetism in the general conception of the physical world has also undergone a complete transformation. The idea of the aether as a material medium has become inconsistent and the electromagnetic field must be treated as a primary constituent of nature on the same footing as material particles. By giving the principle of relativity its utmost extension Einstein was further able to incorporate gravitation into the picture, and thus to round off the comprehensive description of physical phenomena which is now denoted as 'classical'.

It is noteworthy that the conceptual frame of classical physics is essentially dualistic. In the course of time the dualism of matter and force underwent curious vicissitudes, but in spite of repeated efforts, it could never be reduced to unity. During the mechanistic period the aim was, naturally, to eliminate the concept of force. In Hertz' mechanics, for instance, this is done by the expedient of 'hidden masses', not more illuminating than Descartes' 'matière subtile'. The failure to account for the electromagnetic phenomena in mechanical terms precipitated a dialectic reversal of the dilemma: the field concept was given the pre-eminence and the corpuscular appearances of matter were variously, and never convincingly, imagined as condensations or singularities in the field distribution.

The dilemma became quite acute after the discovery of the quantum of action [1] by Planck and the introduction of the concept of light quantum by Einstein. The question was no longer to decide in a general way whether particles could be explained in terms of fields, or vice-versa: in the case of electromagnetic radiation, one had to face the queer situation of the same physical agency behaving like a field or like a particle according to the conditions of observation. The very seriousness of the problem (which, at that early stage, was especially stressed by Planck) helped to bring it back from the somewhat hazy realm of inconclusive speculation to the foreground of theoretical and experimental investigation. The crisis experienced a further sharpening when Niels Bohr, by linking the quantal properties of radiation with the problem of atomic structure, recognised in the existence of the quantum of action the keystone, not only of electrodynamics, but of the whole stability and constitution of matter. However, as Bohr aptly remarked [21], "it is just by bringing out the contrast that it may be possible in due course to give the new ideas a certain coherence". In fact, the solution of the dilemma which eventually emerged from the further elaboration of quantum theory presents the character of the synthesis in which a dialectic process culminates: the dualism of particle and field is now effectively overcome by a synthetic conception in which they both occur as complementary aspects. The scope of this synthesis is universal: it has been found to apply not only to electromagnetic radiation, but just as well to old-fashioned 'matter' and also, as the triumph of Yukawa's prediction of the meson shows, to the newfangled agencies responsible for nuclear interactions and transmutations. It achieves, in fact, a tremendous extension and unification of the concept of matter and puts the problem of the structure of the physical world in an entirely new light.

The relationship of complementarity between physical concepts completely elucidates the apparent paradoxes arising from an uncritical use of such concepts. In particular, it imposes reciprocal limitations upon their quantitative determination which render meaningless any attempt at a deterministic analysis of the classical type. In this connexion, the quantisation of the electromagnetic field raises interesting problems, into the solution of which the universal constant c enters in a new role, viz. as representing the dimension of velocity. For the consistency of the quantum theory of pure radiation fields, it is essential that this theory does not involve any scale of space-time dimensions, and therefore does not impose any restriction upon the construction of the idealised test-bodies serving to explore the field. Now, this property, which is reflected in the conformal invariance (or even the gauge-invariance in the sense of Weyl [22]) of the field equations and of the commutation relations (cf. ref. [23]), is an immediate consequence of the fact that the theory contains only two universal constants, c and h, which are clearly insufficient to fix units of space and time dimensions. In this case the particle, or photon, aspect does not introduce any non-vanishing mass constant, because the complementary field aspect deals with forces of infinite range.

After this preliminary survey of the winding path of history, it would be worth while dwelling a little longer at each of the three main turnings which opened up such unexpected vistas. In this essay, however, we shall not discuss the last two phases any further, but concentrate exclusively on the first. The conceptions and speculations which guided the pioneers have already receded into the haze of the past; it is by no means easy to think about these problems in terms now so unfamiliar to us: yet, the effort is amply repaid by the fascination of the great minds who struggled to solve them and the light their example throws upon essential features of the development of scientific ideas.

2. From Weber to Lorenz

The introduction of two distinct units of electric charge is a direct consequence of the conception of the current as a flow of electricity; for one has then the possibility of basing the measurement of charge either on the ponderomotoric action of charges at rest or on that of moving charges, expressed by Ampère's law. The systematic representation of currents by the motion of charges was part of the atomistic theory of electricity initiated by Fechner's somewhat erratic genius and brought to fruition by Weber's unrelenting effort. In this theory, an electric current is imagined to result from the superposition of two flows of equal and opposite charge density with equal and opposite velocities. By means of this queer conception, Weber [5*a*] succeeded in reducing both the laws of electrodynamics and those of electromagnetic induction to a single formula giving the force between two elements of charge e, e' whose distance r varies with time in consequence of the motion of the charges. If the Coulomb force between the charges is added, the formula becomes

$$(1) \qquad \frac{ee'}{r^2}\left[1-\frac{1}{c_{\mathrm{w}}^2}\left(\frac{\mathrm{d}r}{\mathrm{d}t}\right)^2+\frac{2}{c_{\mathrm{w}}^2}\,r\,\frac{\mathrm{d}^2 r}{\mathrm{d}t^2}\right],$$

Weber's famous law of electrodynamic action. Since it embodies both electrostatic and electrodynamic forces, it contains a constant c_{w}, proportional to the convention factor between the two units of charge.

Although the constant c_{w} was thus introduced by Weber as early as 1846, its actual evaluation [5*b*] was not carried out before the latter part of 1855. The main reason why this task was postponed to so late a stage in Weber's systematic programme of basic measurements of electromagnetic quantities is the technical difficulty of the necessary experiments, especially the static charge determination, for which Kohlrausch's expert assistance was enlisted. Weber had, of course, a rough idea of the expected order of magnitude of the unit ratio he had to measure, which has the dimension of a velocity; but its relationship with the velocity of light was by no means obvious. In fact, Weber's constant c_w, referring to electrodynamic rather than electromagnetic units, is not our c, but $\sqrt{2}$ times this quantity; therefore the numerical result he obtains, when he compares it with the value of the velocity of light, does not fire his imagination. He merely mentions the purely formal interpretation of the velocity c_w which is suggested by the electrodynamical formula (1) (or rather, as we know, by an illicit extrapolation of this formula), viz. the relative velocity which two charges must have so as not to exert any force upon each other, and he observes that the velocity of propagation of a disturbance in an elastic medium is something quite different, anyhow, from an actual velocity of motion of particles.

Nor is Weber more impressed when he discovers, soon afterwards [5*c*], that his law of electrodynamic action predicts the existence, in a *perfectly conducting* circuit, of periodic modes of oscillation of the electric current whose velocity of propagation has indeed the same value $c_w/\sqrt{2}$ as the velocity of light. He cautiously points out that for a circuit of finite resistance these waves of current have neither a constant amplitude nor a unique velocity of propagation; in fact, for a very thin wire, the latter quantity has the form $c[1-(k\lambda\!\!\!^{-}/c)^2]^{1/2}$: it depends on the wave-length $2\pi\lambda\!\!\!^{-}$ and on the damping constant k, i.e. on the resistance and dimensions of the circuit. There is so little similarity between such waves and aether vibrations of the optical type that he not unreasonably doubts whether the near-numerical coincidence of their velocities of propagation has any significance: "it does not seem to be such", he declares, "as to raise great expectations" ("welche nicht derart zu sein scheint, daß sich große Erwartungen daran knüpfen lassen" [5*d*]).

The derivation of the waves of electric current in conducting circuits had been carried out simultaneously by Kirchhoff [6*a*], who forestalled Weber in the publication of his less thorough, but essentially equivalent analysis. Kirchhoff lays somewhat more stress on the limiting case of perfect conductors and on the fact that the velocity of propagation in this case is independent of the state of the conductor; he also mentions its numerical equality with the velocity of light. However, he does not risk any comment on these points, and his lukewarm attitude later on towards Maxwell's theory suggests that he must not have rated their significance much higher than Weber did.

How refreshing, by contrast, is Riemann's youthful reaction! "My discovery of the connexion between electricity and light": by these proud words he describes, in a letter to his sister [7*a*], the contents of a short paper [7*b*] which he submitted to the Göttingen Royal Society in February 1858. I understand, he goes on, that "Gauss has set up another theory different from mine, which he has communicated to his nearest friends. But I am quite convinced that mine is the right one and that in a year or two it will be generally acknowledged as such". His idea was to derive the law of electrodynamic action from the conception of a finite velocity of propagation of this action, and to show that agreement with Weber's results was obtained if this velocity was identified with that of light. With admirable directness he writes down the generalised Poisson equation, involving the operator now called 'd'Alembertian', and the expression for its solution in the form of a retarded potential [2]. He then founders, however, in the mistaken attempt to prove that the retarded *scalar* potential of the charge elements in two conductors is equivalent to the electrodynamic potential derived from Weber's law. It is a strange freak of fate that Riemann, of all people, should stumble into one of those traps which Mathesis usually reserves for her less favoured servants: the unallowed permutation of two integrations! The discovery of this error, and the withdrawal of the paper, must have been a bitter disappointment for him; although he later repeatedly lectured on the theory of electricity and magnetism, he did not venture again into untrodden paths. Riemann's intervention thus remained an obscure incident [3] which did not influence the development of the subject; nevertheless it is well worth being put on record, as a fitting background against which to appreciate the extraordinary failure of Kirchhoff and Lorenz' splendid achievement.

Shortly after his work on the movement of electricity in wires, Kirchhoff published, in the same year 1857, another paper [6*b*] in which the fundamental equations were written down in the local form appropriate to a continuous conducting medium. To this end, he now uses the concept of current density $\mathbf{j}$ connected with the density ϱ of free electricity by the continuity equation[4]

$$(2) \qquad \operatorname{div}\mathbf{j}+\frac{\partial\varrho}{\partial t}=0.$$

Ohm's law is assumed to hold locally even for a non-uniform distribution of electricity. Calling σ the coefficient of conductivity, it may be written

$$(3) \qquad \mathbf{j}=\sigma(\mathbf{E}_{\mathrm{el}}+\mathbf{E}_{\mathrm{ind}}),$$

where $\mathbf{E}_{\mathrm{ind}}$ represents the induced electromotive force, derived from Weber's law, and $\mathbf{E}_{\mathrm{el}}$ the e.m.f. arising from the distribution of free electricity, for which Kirchhoff upholds the expression valid for a static distribution:

$$(4) \qquad \mathbf{E}_{\mathrm{el}}=-\operatorname{grad} V, \quad \text{with} \quad V(\mathrm{P})=\int\frac{\varrho(\mathrm{P}')\,\mathrm{d}v_{\mathrm{P}'}}{r_{\mathrm{PP}'}}$$

($r_{\mathrm{PP}'}$ denotes the distance of the points P, P′ of the medium). For the induced e.m.f. one may write[5]

$$(5) \qquad \mathbf{E}_{\mathrm{ind}}=-\frac{1}{c^2}\frac{\partial\mathbf{A}}{\partial t}, \quad \text{with} \quad \mathbf{A}(\mathrm{P})=\int\frac{\mathbf{j}(\mathrm{P}')\,\mathrm{d}v_{\mathrm{P}'}}{r_{\mathrm{PP}'}}.$$

The potentials V, $\mathbf{A}$ satisfy the Poisson equations

$$(6) \qquad \Delta V=-4\pi\varrho, \quad \Delta\mathbf{A}=-4\pi\mathbf{j}$$

and the auxiliary condition

$$(7) \qquad \operatorname{div}\mathbf{A}+\frac{\partial V}{\partial t}=0.$$

From this set of local equations Kirchhoff derives again those he used in his first paper for thin conducting wires. If i denotes the current intensity and e the charge contained in an element of unit length of the wire, the scalar and vector potentials at a point of the wire (the latter projected along the tangent to the wire) are readily seen to be approximately proportional to e and i respectively, with the same factor of proportionality γ, which depends only on the dimensions of the circuit. Thus,

Equation (3), with (4) and (5), leads to a relation of the form

$$(8) \qquad i = -R^{-1}\gamma\left(\frac{\partial e}{\partial s} + \frac{1}{c^2}\frac{\partial i}{\partial t}\right),$$

(s is the length of arc of the wire, R its resistance per unit length), while Equation (2) yields

$$(9) \qquad \frac{\partial e}{\partial t} + \frac{\partial i}{\partial s} = 0.$$

Elimination of e leads to the wave equation

$$(10) \qquad \frac{\partial^2 i}{\partial s^2} - \frac{1}{c^2}\frac{\partial^2 i}{\partial t^2} - \frac{R}{\gamma}\frac{\partial i}{\partial t} = 0,$$

on which Kirchhoff's and Weber's previous discussion of current waves was based. Now, however, Kirchhoff was more anxious to treat the opposite extreme case of long circuits, in which no periodic currents are possible; this problem, first tackled two years before by W. Thomson, was then of immediate practical interest in connexion with the laying of the first transatlantic telegraph cable.

While the integrated Equation (10) has a wide domain of approximate validity, Kirchhoff's local equations are incomplete and lead to wrong consequences; for the distribution of electricity, e.g., one finds that the expression $(\partial\varrho/\partial t) + 4\pi\sigma\varrho$ would be $= -(\sigma/c^2)(\partial^2 V/\partial t^2)$ instead of $=0$. The effect of the displacement current, which is, of course, the lacking element, could be introduced – as Maxwell [15] does – simply by defining the vector potential in terms of the *total* current rather than the conduction current; more elegantly, however, by taking for *both* V and $\mathbf{A}$ retarded potentials (i.e. replacing the Poisson Equations (6) by the corresponding "d'Alembertian" equations). This is just the decisive step which was taken by Lorenz [8*a*], when in 1866, he takes up the thread of the argument left by Kirchhoff in such an inconclusive state.

Inspired by the great idea of the unity of the forces of nature, Lorenz was consciously looking for an analogy between light and electricity. Just like Riemann, he realises that the first feature which such an analogy must embody is the finite propagation of electrodynamic actions; and he knows from previous investigations [8*b*] how to express this property by means of retarded potentials. He has the enormous advantage over

Riemann, however, of being able to start from the much better basis afforded by Kirchhoff's paper. The local equations established there, representing electrodynamic phenomena in an indefinite medium, were indeed, in contrast to the law of interaction of elementary charges considered by Riemann, directly suited to a comparison with the phenomenon of light propagation. The introduction of retarded potentials into Kirchhoff's equations does not appreciably modify their consequences within the domain of quasi-stationary electrodynamics, which, as Lorenz pointedly emphasises, is their only empirical basis. The local distribution of charge and current, however, as well as their variation in time, are radically altered. From Equation (3), taking Equation (4) into account, one derives successively

$$\frac{\partial \mathbf{j}}{\partial t} = \sigma \operatorname{grad} \operatorname{div} \mathbf{A} - \frac{\sigma}{c^2} \frac{\partial^2 \mathbf{A}}{\partial t^2},$$

$$\operatorname{rot} \operatorname{rot} \mathbf{j} = \frac{\sigma}{c^2} \operatorname{rot} \operatorname{rot} \frac{\partial \mathbf{A}}{\partial t},$$

whence

$$-\operatorname{rot} \operatorname{rot} \mathbf{j} - \frac{1}{c^2} \frac{\partial^2 \mathbf{j}}{\partial t^2} = -\frac{\sigma}{c^2} \frac{\partial}{\partial t} \left(\Delta \mathbf{A} - \frac{1}{c^2} \frac{\partial^2 \mathbf{A}}{\partial t^2} \right).$$

If, therefore, the velocity of propagation entering into the d'Alembertian equation for the vector potential is identified with the constant c of electrodynamics, the last equation takes the simple form

$$(11) \qquad -\operatorname{rot} \operatorname{rot} \mathbf{j} - \frac{1}{c^2} \frac{\partial^2 \mathbf{j}}{\partial t^2} = \frac{4\pi\sigma}{c^2} \frac{\partial \mathbf{j}}{\partial t}.$$

Now, Lorenz recognises in the left-hand side of this equation the expression which, equated to zero, represents the propagation *in vacuo* of the light vector in the elastic theory of light which he had developed some time before. The essential point, which he immediately stresses, is that Equation (11) describes current waves which are purely transversal: this clinches the argument for the identification. A further confirmation lies in the relation, exhibited by Equation (11), between absorptive power and conductivity; a perfectly transparent medium, in which the velocity of propagation has the value c, is one of *vanishing* conductivity.

Lorenz is quite clear about the far-reaching consequences of his work.

Above all, he perfectly realises that the local equations are the fundamental expression for the laws of electromagnetic interactions, and that therefore the whole conception of action at a distance has to be abandoned in favour of that of interactions between neighbouring elements, advocated by Faraday. More remarkable still is his concluding remark about the further problem of reducing the electrical currents to some sort of motion of particles, which is of course the ideal aim of mechanistic philosophy: after some suggestions of striking originality, he emphasises that in any case the main result of his investigation, the identity of light vibrations and current waves, 'does not rest, and therefore does not depend either, upon any physical hypothesis'. Nothing could better illustrate how far the dialectical process which was going to transform the common view of the physical world by the end of the century had already accomplished itself in the mind of the lonely Danish thinker.

3. The Philosophical Background

> La science découvrira (ce que beaucoup de philosophes ont vu) qu'ici encore, la vérité est dans l'action, dans le mouvement de la pensée, et non pas dans telle ou telle position déterminée.
>
> H. Høffding [24]

3.1. *Mechanicism and Phenomenology*

It is indeed time to pause at this stage of the story and try to appraise its significance for the evolution of physical thinking as a whole. The three protagonists, Weber, Kirchhoff and Lorenz, are representatives of the three main currents of ideas which, by the middle of the XIXth century, were putting up rival claims to leadership in science and philosophy. The overruling influence was still that of the mechanistic ideal originating in Newton's revolutionary conceptions and elaborated by Laplace to an all-embracing system of natural philosophy. A first crack in this proud edifice had been the failure of the corpuscular theory of light; a less apparent, but in fact more serious undermining of its foundations resulted from its inability to respond with sufficient flexibility to the question of the mutual relationships between the various natural agencies, to which the requirements of a new phase of economic development had given an overwhelming prominence. As a result, two tendencies had arisen among the new generation of physicists: one of retreat, one of challenge; both threatening the positions of mechanistic philosophy.

The programme of reducing the world to matter and motion was common to all the pioneers of science in the XVIIth century: it stemmed from a natural urge to explain the phenomena in terms of familiar experience. This programme only took shape, however, after the prolonged struggle between the adepts of Descartes and those of Newton had ended by the latter's victory; this was not before the middle of the XVIIIth century. There then emerged a somewhat streamlined Newtonianism, from which all the deeply significant hesitations of the creator's thought, arising from his intimate insight into the difficulties of the subject, had been ironed out by less subtle epigons. This system (in contrast to the Cartesian one) was essentially atomistic, its specific feature being the conception that the atoms are centres of forces acting at a distance, on the model of gravitation. Laplace gave this "physique des forces centrales" its consecration, not so much by formulating its principles in a sharp and lucid way, and showing the universality of its scope, as rather by demonstrating the real power of the seemingly vague and hazardous approach to the analysis of the properties of matter which the atomic hypothesis offered. In his theory of capillarity he showed how the phenomena could be derived in every detail from the mere assumption that the molecular interactions are of short range [1]. It was permissible, therefore, to treat the atoms as structureless sources of central forces; in fact, speculations about their structure are deprecated as "vagues et inutiles à l'avancement des sciences" [1]. The true significance of the atomic theory (at this early stage) is quite rightly seen by Laplace in its suggesting relationships between phenomena which would otherwise appear disconnected; thus (he adds with characteristic emphasis on the pratical role of science) giving rise to chemical theories with increased possibilities of "applications aux arts" [1].

The ban on arbitrary speculations about the structure of atoms, thus imposed by Laplace, was undoubtedly a source of strength for mechanistic physics; but it also contained a seed of discord, which, as soon as partial failure bred disillusion, led to a weakening of the mechanistic positions. It invited cautious minds to distrust atomistic considerations to such an extent that they regarded it as desirable or even imperative to eliminate them altogether from the formulations of physical laws: the aim of physical theories was simply to give an accurate 'description' of the phenomena in terms of concepts which could be unambiguously referred to observation. This "phenomenological"[6] school, of which Kirchhoff was one of the most authoritative exponents, did not explicitly reject the tenets of mechanistic philosophy, but was content to leave the attainment of its ideals to a dim and distant future, it did not deny the existence of the atoms, but pretended to ignore them.

Originally, this attitude arose as a simple matter of practical policy in scientific method; in this sense it is lucidly set out by Ampère [4*a*]: first ascertain the laws which give an accurate account of the observations; then proceed to seek their relation to the assumed structure of the physical world. So long as the criticism of mechanistic speculations remained at this methodological level, reflecting real uncertainties in our knowledge, it could only be stimulating. Inherent in the method itself, however, was the danger of emphasising formal relationships at the cost of concrete physical reality: the unity which it was the role of the atomistic picture to provide came to be sought in mere formal consistency. This danger was very soon illustrated by the vulgarised version of phenomenology given in Comte's system of "philosophie positive": by representing the qualitative differences between the manifestations of the various natural forces as irreducible, just when the question of their mutual transformations and deeper-lying connexions was entering a critical stage, it had, in the French school, a stifling influence upon the progress of this fundamental inquiry.

3.2. *Weber and Kirchhoff*

The impact of the two currents of ideas so far considered upon the development of electrodynamics is very conspicuous. Right from the beginning, through Ampère's dominating intervention, the fundamental laws of interaction of currents and magnets were cast in the mould of central force physics. Weber went on in the same spirit; in fact, he set himself the task of developing the mechanistic conception of electricity and magnetism to its utmost consequences. In particular, he endowed the carriers of electricity with inertia, thus originating the concept of *electron*. Although in a sense premature, since it was as yet incapable of experimental test, this idea gave Weber unfailing guidance in the elaboration of his theories; in fact, as we know, only the electron concept could lead to an essentially correct conception of such phenomena as diamagnetism.

A good example of Weber's method, which at the same time illustrates its difference from Kirchhoff's, is afforded by the question of finding the relation between current intensity and e.m.f. in the case of variable currents (which, as we saw, played an important part in the theory of current oscillations). Kirchhoff takes the purely formal course of extending Ohm's law to this case, and hoping for the best; Weber, consistent with his mechanistic picture, starts from the equations of motion of the electrons and appeals to Ohm's law only to fix the value of the force exerted by the 'ponderable matter' on the electrons in terms of the empirical conductivity coefficient. He accordingly obtains an equation differing from Kirchhoff's by an inertial term involving the time-derivative of the current intensity and depending on the unknown charge-to-mass ratio η of the electrons. The effect of this term is merely to multiply the coefficient of $\partial^2 i/\partial t^2$ in eq. (10) by the constant factor $[1+(c^2/\eta\gamma e)]$. Thus, Weber's treatment does not modify the main argument; but, as Weber points out in some detail, it opens the way, at least in principle, to a determination of the product ηe (involving the charge, mass and number per unit volume of the carriers of electrical current), i.e. to the investigation of a quite new aspect of electrical phenomena.

3.3. *Dynamical Philosophy*

However, in spite of its suggestiveness and its powerful formal resources, central force physics was doomed to ultimate failure in the domain of electromagnetic phenomena, where the basic idea of instantaneous action at a distance meets with an essential limitation of its validity. To break the spell of action at a distance required fresh inspiration. It came from a third current of ideas, to which we must now turn our attention. Phenomenology was not the only reaction to the shortcomings of the mechanistic views in the first half of the century. Especially those physicists who gave prominence to the idea of force, urged upon them by the practical problems raised by the growing utilisation of new sources of power, found little help in the current ideas, and many came to reject atomism altogether. We may conveniently call them 'dynamical' physicists, although there was among them no doctrinal unity at all. The two protagonists of dynamical philosophy with whom we are here concerned are Ørsted and Faraday; another great pioneer of the same tendency is Robert Mayer. These three thinkers developed their ideas in quite different environments without any mutual influence (except for the trivial fact that

Faraday built upon the foundation laid by Ørsted's discovery), and apart from a general similarity of outlook, their views took rather different and highly individual shapes. We will now analyse Ørsted's ideas and trace their influence on Lorenz.

3.4. *Ørsted*

Ørsted was consciously metaphysical in his approach to physics; he upheld more explicitly and consistently than most scientists of his time the XVIIIth century belief, equally shared by the French mechanicists like d'Alembert and their German opponents of the Leibnizian school, that science could be established upon *a priori* principles regarded as necessary elements of our thinking. As a young man he eagerly followed the trend of philosophical speculation initiated by Kant in Germany, even when it dissolved itself into the mists of Schelling's *Naturphilosophie.* He stopped short, however, of the last, synthetic, phase of this evolution, when the subsiding mists brought to view the weird shape of Hegel's system. By that time, Ørsted's sound physical intuition had got the better of his romantic propensities, and he settled down to a view of the physical universe which embodied in sober scientific form valuable conceptions due on the one hand to Kant's emphasis of the dynamical aspect and on the other to Schelling's deep sense of the unity of nature [7].

Ørsted adopted, with some variations, the Kantian view that the essential element underlying the phenomena which we describe as "matter" is some distribution of fundamental forces in space. The necessity of providing an explanation for permanence and stability as well as for change and motion naturally led, in this view, to imagine in every aspect of nature a "conflict" of opposing forces, resulting in rest or motion according as attraction and repulsion did or did not balance each other. This "dynamical" picture of the phenomena harmonised with Schelling's conception of the universe as a process of continual change, in which qualitatively different aspects appeared as various manifestations of the interplay of the same fundamental forces: hence the conclusion that all the agencies of nature are interrelated and can be transformed into each other. Thus, Ørsted pictures the electric current as the result of a conflict of electric forces, whose balance was upset and restored in continuous succession; and this representation led him to suspect that the repercussions of the electric conflict in the space around the conductor might give rise not only to heat and light phenomena, but also to those effects which we describe as magnetic. Experiment, confirming this last suspicion, at once made the conception of the electrical conflict more precise, by suggesting that this conflict must be accompanied by some kind of rotational motion around the axis of the current [3*a*].

Tenuous as such ideas may appear, it is a remarkable fact that their suggestive power was strong enough to put men like Ørsted and Robert Mayer on the track of great discoveries: of course, they could not, and did not, provide more than an inspiration, inclining the mind in the right direction; the actual discovery could only be the result of the usual kind of rigorous logical thinking about skilfully selected observations. This essential identity of method, however, obscured by the unfamiliar twist given to familiar concepts, was not properly recognised by the contemporary mechanistic school: the perfect rigour of Mayer's argument was not appreciated (not even by his generous champion Tyndall), and Ørsted's careful electrodynamical experiments remained almost completely ignored. To be accepted, the new principles had first to be formulated into the current language: when the French physicists, for instance, received the announcement of Ørsted's discovery, they did not stop a moment to consider what he meant by his "electrical conflict"; with utter contempt for an idea so much lacking in Cartesian clarity, Ampère attacked the problem afresh with the refined methods of central force physics. Thus, the reaction of the mechanicists to the dynamical opposition was simply a passive one: lack of understanding for its aims and methods, lack of interest for the vast, if somewhat hazy, prospects it opened. In fact, the mechanistic philosophy, as early as 1820, was so rapidly losing its flexibility that it was becoming unaware of the hypothetical character of its own postulates: thus, no less a thinker than Ampère could describe his formulation of electrodynamics as "uniquement déduite de l'expérience" [4*a*]. Above all, the 'dialectical' way of looking at things, which with Ørsted makes its first entry into physics, was quite foreign to mechanistic thinking; indeed, it has remained so until our own days, when it has been finally forced upon physicists by the advent of quantum theory.

3.5. *Lorenz*

Ørsted's influence on scientific thought was naturally stronger in his own country, where he enjoyed immense prestige. There is no evidence that he had any personal contact with Lorenz (he died in 1851, when Lorenz was a student at the Technical High School at Copenhagen); but we find distinct echoes of his teachings in the latter's original and vigorously expressed ideas. In the first place, Lorenz, like Ørsted, acknowledged as his guiding principle the idea of the unity of natural forces. In the second, he shares Ørsted's dislike of mechanistic hypotheses, but not for metaphysical reasons: it is to the arbitrariness of these pictures that he rightly objects. Accordingly, he regards as the foremost task of physics the precise formulation of the laws of the phenomena in terms independent of any such hypotheses. In this, he does not differ from the phenomenologists, but his belief in an underlying unity prevents him from stopping at the formal, descriptive stage, and urges him onwards in a search for analogies which might reveal an essential identity of natural agencies whose manifestations appear qualitatively different. He has the great advantage over Ørsted (like Maxwell over Faraday) of being a skilled mathematician, and he thus avoids the vagueness which so much weakened the case of dynamic philosophy.

Lorenz' great paper on the identification of light with oscillating electrical currents clearly illustrates these points. This paper is the culmination of a persevering inquiry into the nature of light, which he started in 1860; he soon abandoned the beaten track as he recognised the lack of cogency of any mechanical interpretation in terms of some aether model; he deliberately aimed at a purely abstract description of the laws of light propagation, and he succeeded in showing that under very general conditions these laws could be concisely expressed by a wave equation for transversal vibrations

$$\text{rot rot}\,\mathbf{u}+\frac{1}{a^2}\frac{\partial^2\mathbf{u}}{\partial t^2}=0.$$

Moreover, he found how to connect the light vibrations with their sources by means of retarded potentials. He was thus well prepared, when he saw Kirchhoff's paper, to take up the clue and make the decisive step. The essential point, however, is that, in contrast to Weber and Kirchhoff, he was all the time looking for some analogy between light and other phenomena.

The case we are discussing shows the atomistic point of view (here represented by Weber) decidedly at a disadvantage in comparison with Lorenz' 'dynamical' position. In order to get a balanced assessment of the respective merits of the two opposing conceptions, however, we must not lose sight of the fact that in the other great creation of XIXth century physics, the mechanical theory of heat, they drew level: while Robert Mayer was of the 'dynamical' type, definitely hostile to atomism, his rival Joule found a sure guidance in the kinetic conception of heat. Soon afterwards (1857–60), Clausius' and Maxwell's brilliant elaboration of the kinetic theory of gases heralded a momentous advance in the evolution of scientific ideas; for this theory provided the starting point for the first fruitful synthesis of the dynamical and atomistic viewpoints: statistical thermodynamics, a synthesis which will doubtless remain for ever one of the main pillars of the temple of Nature.

It is interesting to watch Lorenz' changing attitude to this problem: about 1850, in his earliest (never published) investigations [8*c*], he attempted to establish a relation between heat and light, on the basis of a complicated aether model; but he was quick, following the lead of the pioneers, to realise the heuristic value of the kinetic theory, and he adopted it in his later work, albeit with a keen eye for its difficulties and limitations. To appreciate Lorenz' foresight and shrewd judgment in this matter, one has to reflect that he, who had so forcefully emphasised the ambiguities of mechanistic ideas, made this choice at a time when the new advance of atomistic theory was challenged by a sharpening of the phenomenological criticism – encouraged, it must be said, more by a reaction accompanying the economic and social evolution [8] than by any inherent necessity in the development of physics.

While the idea of the unity of force, which had by then found its quantitative expression in the law of conservation of energy, was travestied in the ill-founded phenomenological doctrine of 'energetics', it continued to inspire in a most fruitful way Lorenz' profound investigations. After his recognition of light as an electromagnetic phenomenon, he engaged upon a protracted study of optical refraction and dispersion (1869–83), which he undertook with the view of elucidating the molecular constitution of material media from the way in which the molecules of matter interact with light. Concurrently, he discussed the relationships between heat and electricity (1870–72), again on an atomistic basis, and derived (by arguments lacking in rigour, it is true) such striking results as the temperature-dependence of the Wiedemann-Franz constant and the number of molecules in an equivalent-gram of an electrolyte.

Thus, the identification of light vibrations with electric currents is by no means an isolated incident, a stroke of luck in Lorenz' career: it is the first, and most successful, stage in a quest of very wide scope pursued with remarkable singleness of purpose. The conclusions of Lorenz' paper are brief, but they go to the root of the matter. Pointing out that the differential equation (10) allows of a simpler extension to heterogeneous media than eq. (3), he regards it as the more fundamental one, and stresses the fact that it corresponds to a conception of transmission of electrical action from element to element of the medium, in conformity with Faraday's views. He also mentions Ampère as having suggested such a view; this refers to an interesting discussion by Ampère [4*b*] of a possible mechanism of transmission of the attraction between two conductors carrying currents by a 'neutral fluid' surrounding the conductors. Ampère, like all clear-sighted adherents of the concept of central force, – in fact, like Newton himself – was perfectly aware of the provisional character of this concept and of the desirability of underpinning it by some transmission mechanism; but he rightly thought that this last problem would not for a long time be capable of unambiguous solution.

The detailed picture of the transmission of electrical action outlined by Lorenz retains the rotational character which Ørsted had already recognised as a feature of his electrical conflict; it is so striking, in the light of our present knowledge, that it deserves to be quoted *verbatim*:

"If we imagine light as a *rotational* vibration in the interior of the bodies around axes whose direction is that which in the elastic theory is regarded as the direction of vibration, the electric current will not be any translatory motion, but just a continued rotation in a definite direction, and the direction of the rotation axis is then the direction of the current. This rotation can only persist in good conductors, and the motion is then propagated along the direction of the axis; while in bad conductors it becomes periodic and is propagated, by what, in the theory of electricity, is called induction, in a direction perpendicular to the axis of rotation. On this view, there is hardly any reason to uphold the hypothesis of an aether, since one may very well assume that in all the so-called empty space there is enough material substance left to form a sufficient substratum for the motion."

This may perhaps appear a mere adumbration of Maxwell's powerful vision; at any rate the sketch is aptly drawn, and shows all the essential features of the accomplished masterpiece. The last sentence of the passage quoted betrays the particular aversion to the aether concept which

Lorenz had formed in the course of his study of the mechanical theory of light, when he saw how little helpful and unreliable it was. The immediately following paragraph, on which the paper ends, is more clearly revealing of his detached attitude with regard to all mechanistic constructions: it is the remark, already alluded to above, that his main conclusion is in any case independent of any 'physical' i.e. mechanical, hypothesis. Here Lorenz' thought is decidedly more advanced, thanks to its non-mechanistic component, than Maxwell's: the latter will only reach a similar point of view, as we shall see, at a late stage in the elaboration of his theory. Taking leave with due respect of a work which so worthily crowns the endeavour of the Continental school of electrodynamics, let us now cross the Channel and watch the developments which brought electromagnetic theory to full completion.

4. Faraday and Thomson

4.1. *The British School*

Maxwell's twin sources of inspiration, like those of Lorenz, were on the one hand a powerful dynamical conception of nature, that of Faraday; on the other the great mathematical tradition of the French school. But the emphasis in this case was much more on the dynamical side: the French ideas came to him through the channel of William Thomson, who (himself under Faraday's influence) had been more struck by Fourier's mathematical treatment of the process of heat conduction as one of continuous transmission than by the atomistic trend of the Laplacian methods. Yet neither Thomson nor Faraday stood in any radical opposition to the atom concept: but they insisted on the primary importance of the dynamical aspect of the phenomena and tended to regard atoms as singular occurrences in an essentially continuous medium transmitting force through space. Faraday was content to think of the atoms as structureless centres of force; Thomson wanted to visualize them as definite structures arising from the continuous aether by some mechanical process: thus, as soon as he heard of Helmholtz' beautiful theory of vortices (1867), he started imagining the atoms as vortices in the aether and developed this picture in great detail, and with evident relish. Anyhow, the mainspring of the phenomena was for both of them some distribution of mechanical stresses in the continuous aether.

This blending of atomistic and dynamical conceptions cannot be described as a synthesis in the dialectical sense. It certainly paved the way for a true synthesis in the dynamical theory of heat, as Maxwell's later contribution to the foundation of statistical mechanics clearly shows; but in the domain of electromagnetism, it initiated a very different dialectic process, which brought to the open a deep-seated contradiction between the atomistic and dynamical aspects of electromagnetic phenomena. In the hands of Thomson and Maxwell, the dynamical treatment of these phenomena took a peculiar turn away from concrete representation and towards mere formalism: the mechanical constructions ceased to aim at picturing real processes, but became 'models' with no other purpose than that of illustrating the compatiblity of a set of phenomenological equations with the laws of mechanics. This retreat into abstraction was at first meant to be no more than strategic: it soon became clear that it heralded the end of dynamical philosophy. The electromagnetic field came to be regarded as a primary constituent of the physical world on the same footing as ponderable matter; then, with the discovery of the electron and of the essentially electrical structure of matter, the reversal of the respective positions of mechanics and electromagnetics was completed: the question was raised whether electric constituents of atoms could not be accounted for in terms of the electromagnetic field alone. The impossibility of fulfilling this programme opened the protracted crisis out of which, through the incorporation of the fundamental quantal features, the modern particle-field synthesis eventually emerged.

4.2. *Faraday*

At the origin, there was not a shadow of doubt about the concrete meaning of mechanical pictures. Faraday's astounding power of creative imagination was of the same kind as that of the artist whose searching look discloses the revealing features out of which he composes a portrait presenting a true likeness of his model. The directness of his method has been well characterised by Hertz [17*a*]: "a mind of great freshness, who approached the phenomena anew, uncommitted, who started again from what he saw, not from what he had heard, learned or read." Hertz then aptly describes Faraday's conceptions and contrasts them with those of central force physics:

Faraday heard that when a body was electrified, something is put into it, but he saw that the modifications produced manifested themselves only outside and not at all inside it. Faraday was taught that the forces simply passed over space, but he saw that it makes all the difference for the force with what matter the allegedly passed-over space is filled. Faraday read that the existence of electric fluids was certain but that there was some debate about electric forces; and yet he saw how these forces palpably developed their effects, while he could not perceive anything of the electric fluids themselves. Thus the situation was reversed in his mind. The electric and magnetic forces became for him the existing, the real, the palpable things; electricity and magnetism became the things whose existence was debatable. The lines of force, as he called the forces conceived as subsisting by themselves, appeared to his mind's eye in space, as states of space, as tensions, as vortices, as currents, or whatever they might be – this he could not tell – but there they were, influencing each other, pushing and pulling bodies to and fro, and spreading out from point to point, transmitting the excitation.

With this vivid picture of the forces was coupled in Faraday's mind the conviction of their fundamental identity and convertibility. It is this belief which decided the orientation of the research which led to his most beautiful discoveries [10]. A striking example is the discovery in 1845 of the rotation of the plane of polarisation of light by a magnetic field [9*a*]: this phenomenon was the first hint of a deep-lying relationship between optics and electromagnetism, and made a profound impression on both Thomson and Maxwell. For Faraday himself, it was the occasion of remarkable speculations, his "Thoughts on ray-vibrations" (1846) [9*b*], which reveal us the uncanny precision of his thinking by means of mental pictures of 'lines of force'. Of his general views about the structure of matter we are informed by another short paper [9*c*], written some time previously (1844): there he conceives ponderable matter as consisting of structureless, punctiform atoms, surrounded by a continuous distribution of their various 'powers', which extends from each atom to all the others, not necessarily uniformly in all directions. Now, he considers, more precisely, the atoms as linked together by lines of force along which these powers are transmitted, and he clearly visualises the propagation along such a line of a slight displacement of an atom as a 'shake or lateral vibration', i.e. a transversal wave-pulse. The lines of force seem to him more apt to transmit transversal vibrations than a uniform medium like the aether, which he accordingly proposes to 'dismiss'.

After this sample of Faraday's method, it would be very superficial indeed to represent him as a pure 'experimental philosopher', in opposition with the mathematically minded physicists of the French type. His strength lies indeed in his perfect awareness of the value and limitation of theoretical thinking. He always readily expressed the theoretical ideas in which he constantly sought guidance in his experimenting. As to speculations which could not directly be founded on experiments, he was naturally more reticent; but after all, he *did* publish the two papers just mentioned, and even though the occasion which prompted the second one was accidental [11], the writing-up of the papers, at any rate, was deliberate: of the first one he explicitly declares [9*c*] that it would "serve as the record of my opinions and views, as far as they are at present formed". Moreover, he created his own set of theoretical concepts of a topological and dynamical nature, such as that of 'line of force', which fulfilled in his reasoning the same function as more conventional mathematics. In fact, as Maxwell says, the virtuosity with which he operated with these representations stamps him as "a mathematician of a very high order" [14*a*].

Maxwell's relationship to Faraday is similar to that of Lorenz to Ørsted; when in 1854, at the age of 23, Maxwell turned his attention to the subject of electricity, Faraday, 40 years his senior, had accomplished his life's work: his ideas, embodied in the monumental series of his *Experimental researches in Electricity*, had received their final formulation. Maxwell set himself the task [15*a*] to translate them into the usual mathematical language in order to investigate their precise connexion with the electrodynamical theories based on action at a distance, and to give them, if possible, a firm foundation upon mechanical principles. Some letters he exchanged with Faraday after the publication of his first paper (1857) [16*a*, *b*, *c*] show how eager the old master was to respond to the attempt of a fellow-physicist to enter into his modes of thinking, and how touchingly he gave expression to a satisfaction mingled with awe that his conceptions should have stood so well the "mathematical force made to bear upon" them [19*a*].

4.3. *William Thomson*

Thomson's influence upon Maxwell's thought, hardly less profound than that of Faraday, was of a different character. The two men, similarly distinguished by the precocity and originality of their talent, equally moulded by Cantabrigian education, with a difference of age of only seven years, entertained personal relations of easy cordiality, which, however, never went very deep. Thomson, appointed to the Glasgow chair at the age of 22, had soon acquired so much authority in the field of electricity and magnetism that it was to him that young Maxwell turned for advice [13*a*] as soon as he had passed the ordeal of the Tripos examination. A year later (1855), we see Maxwell seeking [13*b*] and obtaining [16*d*] Thomson's permission "to poach on his electrical preserves".

Thomson's first paper on the subject [12*a*] is indeed one of the most extraordinary productions which have left their mark in the history of science. He wrote it at the age of 17, in the summer of 1841, just before he took residence in Cambridge[9]. In the immature style of a schoolboy's exercise, it presents in full generality the fundamental theorems of the 'theory of attraction' (now called potential theory) with their application to the case of ellipsoids. It is true that the schoolboy had been forestalled by Chasles and by Gauss, and these in turn, as was later discovered, by George Green; but more astonishing than the results was the method by which they were obtained – a method which is entirely Thomson's. It is based on the formal analogy between potential theory and that of stationary heat flow. Although the name 'potential' was not yet in use, the concept itself and the differential equation which characterises this quantity were well-known since Laplace; and Fourier had shown that the theory of heat, in the stationary case, was governed by the same equation. Thomson now makes use of the correspondence between equipotential and isothermal surfaces, and between the derived quantities, electric intensity and heat flow, in order to deduce the general properties of electrostatic forces and charge distributions from those of heat flows and distributions of temperature. The latter he regards as 'obvious' to physical intuition; about two years later, in a letter to his father [13*d*][10], he describes them, with due emphasis, as "propositions, deduced *entirely from physical considerations*": only then was he able to give analytical proofs of them [12*b*].

In order to appreciate the novelty of this juvenile onslaught upon the problems which had occupied the deepest minds of the French school – Poisson, Gauss and Green, to mention only the greatest, – it must be realised that the two monumental works of this school, Laplace's *Mécanique Céleste* and Fourier's *Théorie analytique de la Chaleur*, had peacefully co-existed for twenty years without anyone attempting to follow up the mathematical analogy in question. We witness here an effect of that tendency to metaphysical rigidity which was beginning to hamper the development of mechanistic thought: Fourier, stressing the essential difference between the analytical character of the law of heat conduction (in the general case) and those of mechanics, peremptorily represented this difference as for ever irreducible, independently of any hypothesis about the nature of heat [2].

One may imagine, therefore, the sensation which the young Cambridge graduate created among the Paris physicists when he arrived, in 1845, bringing with him on the one hand Green's *Essay* (which he had just then been able to see for the first time [11]), on the other his own productions. Green's work was the logical completion of the analytical treatment of central forces in the pure Laplacian tradition. But Thomson's derivation of the same results from a formal analogy with the theory of heat conduction had much farther-reaching implications: it opened the prospect of a possible physical analogy on the lines suggested by Faraday's discoveries. In a paper [12*c*] written in Paris, at Liouville's request [13*f*], Thomson developed with great vigour this aspect of the problem, emphasising the heuristic superiority of Faraday's views and the strong corroboration they had received from the discovery of the influence of the dielectric medium on the electrostatic forces. "It is, no doubt, possible" he wrote "that such forces at a distance may be discovered to be produced entirely by the action of contiguous particles of some intervening medium, and we have an analogy for this in the case of heat, where certain effects which follow the same laws are undoubtedly propagated from particle to particle".

The following year marked a further significant progress in Thomson's approach to a dynamical theory of electro-magnetism. Deeply impressed by Faraday's discovery of the action of a magnetic field upon light traversing a transparent solid, he now looked for an analogy between electromagnetic situations and states of strain in an elastic body. His friend Stokes had just given (1845) the equilibrium conditions for such a body (assumed isotropic) in the form of a differential equation for the displacement vector $\mathbf{u}$:

$$\Delta\mathbf{u} = \operatorname{grad}\theta, \quad \text{with} \quad \theta = -k \operatorname{div}\mathbf{u};$$

the latter equation, in which k is proportional to the compression modulus, reducing for an incompressible solid to $\operatorname{div}\mathbf{u} = 0$. In a laconic note [11*d*], published in 1847, Thomson points out that these equations have indeed particular solutions which exhibit the characteristic properties of electrostatic, magnetostatic and electrodynamic quantities. If **r** denotes the vector from some fixed point O to the point of space considered, r the corresponding distance and **m**, **j** constant vectors, these solutions and their analogies are as follows:

$\mathbf{u} = -\text{grad}\,(1/r)$,	electrostatic field of a unit point charge at O;
$\mathbf{u} = \text{rot}\,(\mathbf{m}/r)$,	vector potential of a magnet of moment **m** at O;
$\mathbf{u} = \mathbf{j}/r - \frac{1}{2}\,\text{grad}\,(\mathbf{j}\cdot\mathbf{r}/r)$,	vector potential of an element of current of density **j** at O.

There is nothing in Thomson's brief paper to disclose the depth of his searching or the hopes he entertained. His notebooks [13*g*], however, are extremely revealing in these respects: We see him grappling for nearly a month (November 1846) with the then quite new subject of elasticity before he could express in formulae his intuitive guesses. One must not forget that the concepts of vector analysis, so familiar to us, were then utterly unknown and had to be constructed in the very process of development of the new physical ideas. The final entry leaves no doubt about his expectations: "I have at last succeeded in working out the *mechanico-cinematical* (!) representation of electric, magnetic and galvanic f^{ces}...". Later (1847), in a letter to Faraday [13*h*], commenting upon his paper, he shows himself as clearly aware of the preliminary character of his results as of the ultimate aim to be achieved: "a physical theory of the propagation of electric and magnetic forces".

5. Maxwell

> I have also a paper afloat, with an electromagnetic theory of light, which, till I am convinced to the contrary, I hold to be great guns.
>
> J. C. Maxwell (letter of 5 Jan. 1865) [16g]

5.1. *Hydrodynamical Analogies*

It is at this stage that Maxwell takes up the problem, in 1855. Not that Thomson had been idle in the intervening years; he had written voluminous treatises in which he attempted a more systematic treatment of electricity and magnetism, embodying Faraday's discoveries of the influence of the material media, dielectric, diamagnetic or paramagnetic:

these are very heavy and, although of great value as pioneering work, lack the inspiring suggestiveness of the preceding short notes; but it was from them that Maxwell acquired the mathematical preparation for his own investigations. Moreover, in those critical years, during which the whole science of thermodynamics was created, Thomson had devoted a still larger part of his amazing activity to this new and exciting task. His eager curiosity and quick intelligence made him a sower of ideas in many fields; his inconstancy too often prevented him from reaping the harvest. So it was with thermodynamics, where he was forestalled by Clausius; so it was to be with electromagnetism, where he abandoned to Maxwell the promising quest.

Maxwell's first effort [14*b*] was still of limited scope, though no longer, like Thomson's work, confined to static or stationary phenomena; he also discussed electromagnetic induction, but his treatment of electric currents could only claim validity for the case of closed circuits. In the first part of the paper, Maxwell goes again over the same ground as Thomson: electrostatics, magnetostatics and stationary currents, but he follows a new analogy, that of a flow of incompressible fluid in a resisting medium. Maxwell's treatment of the dynamical analogy is characteristically different from Thomson's: to the latter's analytical power he adds a fine geometrical intuition, which allows him to present the whole discussion in a simple and straightforward language, from which all but a few algebraic formulae are banished. This is indeed one of the great papers of Maxwell's, which remain as unsurpassed models of elegant exposition and of beautifully simple and clear scientific English[12]. The hydrodynamical analogy is the natural one to illustrate Faraday's concept of 'line of force', and this was Maxwell's main purpose in this work. If $\mathbf{v}$ is the velocity of flow of an incompressible fluid of unit density, and if the resisting force per unit volume is assumed to be of the form $R\mathbf{v}$, the equations describing the stationary situation are

$$\operatorname{div}\mathbf{v}=0, \quad R\mathbf{v}=\operatorname{grad}p,$$

the latter expressing the fact the resistance of the medium is balanced by the pressure gradient. Maxwell follows Faraday in treating the three main groups of electric, magnetic and 'galvanic' phenomena in a completely parallel way: he applies to all of them the same hydrodynamical picture. In each case, he distinguishes two characteristics, a 'quantity' and an 'intensity', linked by a relation of proportionality, whose factor corresponds to a 'resistance' of the material medium; the 'quantity' obeys the solenoidal condition (except at sources or sinks), while the 'intensity' appears as the gradient of a 'pressure' or 'tension'. The triple analogy is summarised in the following table:

	Quantity	Intensity	Inverse of resistance
Electrostatics	'inductive action' **D**	electric intensity **E**	'specific inductive capacity' ε
Magnetostatics	'induction' **B**	magnetic intensity **H**	'magnetic conductive power' μ
Electrodynamics	current density **j**	e.m.f. $\mathbf{E}_{el}$	electric conductivity σ

The distinctive success of Maxwell's approach, however, is the interpretation of Faraday's 'electro-tonic state', which he gives in the second part of the paper. According to Faraday, the induction of a current in a closed conductor suggests that this conductor, in the presence of a magnetic field, is in a peculiar state, which does not give rise to observable effects so long as conditions are stationary, but whose changes are revealed by the occurrence of an electromotive force in the conductor. In order to translate this conception in mathematical terms, Maxwell has recourse to an argument of a type quite novel at that time: Helmholtz had just shown (in his famous essay of 1847) how the law of electromagnetic induction could be derived from those of stationary currents by considering the transformations of energy involved in the process and applying the conservation law. Maxwell repeats the argument, but on the basis of new expressions for the magnetic energy and the Joule heat, consonant with the view of a continuous distribution of this energy in space. In modern notation, the equation he sets up for the energy changes during a time δt when the magnetic induction is varied by $\delta\mathbf{B}$ may be written

$$\delta t \int \mathbf{j}\cdot\mathbf{E}\,\mathrm{d}v + \frac{1}{4\pi}\int \mathbf{H}\cdot\delta\mathbf{B}\,\mathrm{d}v = 0.$$

Now, expressing the induction as $\mathbf{B} = \mathrm{rot}\,\mathbf{A}$ by means of the vector potential **A**, and using Ampère's law written in its local form[13]

$$\text{(12)} \qquad \operatorname{rot}\mathbf{H}=4\pi\mathbf{j},$$

the condition of energy conservation is transformed into

$$\text{(13)} \qquad \int \mathbf{j}\cdot\left(\mathbf{E}+\frac{\delta\mathbf{A}}{\delta t}\right)\mathrm{d}v=0:$$

the variation of the induction flux in the space embraced by the conductor has been replaced by the variation of the quantity **A** *inside* the conductor; and from the validity of the equation (13) for arbitrary dimensions of the conductor and magnitude of the current density, one derives the local equation

$$\text{(14)} \qquad \mathbf{E}=-\frac{\partial\mathbf{A}}{\partial t},$$

immediately suggesting the identification of the vector potential with the electro-tonic state visualised by Faraday.

5.2. *Vortices and Idle Wheels*

Maxwell does not at all regard this conclusion as a satisfying answer to the question; in fact, he is quite apologetic about its abstract character: there is as yet no clear dynamical picture of the link which must exist between electric and magnetic lines of force. How deeply he had assimilated the conception of physical line of force is shown by an interesting letter to Faraday [16*c*] of 1857, in which he applies it to the interpretation of gravitation: in this case, the lines must "have a *pushing* force instead of a *pulling* one", a feature which greatly impressed Faraday [26*b*], and baffled Maxwell completely when he returned to the subject later [14*e*]. As to the problem of electromagnetism, he seems to have expected further light from the analogy, already noticed by Thomson, with elastic solids or (what formally amounts to the same) viscous fluids. Very soon, however, another brilliant remark of Thomson led his thoughts into rather different channels. In 1856, Thomson [12*d*, 15*b*] returned to the effect of a magnetic field on the velocity of propagation of circularly polarised light and discussed it from the then unfamiliar point of view of its mirroring transformation properties: he came to the conclusion that the circular 'luminiferous' motion must be compounded with another circular motion linked in a definite way with the magnetic field.

Thomson's paper, as usual, is verbose and far from clear, but the hint it contained was enough to give Maxwell the right start. The decisive point was that one had for the first time an indication from experimental fact sufficiently definite to narrow down the range of possible mechanical analogies to a considerable extent. Working progressively from the representation of magnetic tubes of force as vortices to its successive implications about the nature of electromagnetic induction, electric current and electric polarisation, Maxwell built up piece by piece this startling machinery of the electromagnetic aether, which eventually disclosed to his wondering gaze the longsought mechanism of propagation of a purely transversal disturbance like that of light. No trace has been left, unfortunately, of the progress of Maxwell's thought in these years: we have only the final description of the perfected work in his second great paper [14*c*] of 1861–62. This period was one of the most active in Maxwell's life: we see him carrying out his monumental investigation of Saturn's rings, which fully occupied the years 1857–59, starting his researches in the kinetic theory of gases (1860) and completing his theory of colours (1860). In the published correspondence, the only allusions to electromagnetism date from 1861 and 1862: it is conceivable that the whole picture, intricate as it is, may have crystallised in his mind in the course of a few months.

A vortex-tube of axial symmetry gives rise to a system of stresses which can be resolved into a lateral pressure p and a longitudinal pressure or tension of the form $C\varrho v^2 - p$, where ϱ and v are the mean density of the fluid and its velocity at the boundary of the tube, and the numerical constant C depends on the distribution of density and angular velocity, taking the value $\frac{1}{4}$ if this distribution is uniform. It is readily seen that this tensor reproduces the known forces of magnetic origin, provided that the magnetic intensity $\mathbf{H}$ is interpreted as a vector of magnitude v directed along the axis of the tube (i.e. proportional to the angular velocity of the vortex) and the magnetic permeability μ is identified with the quantity $4\pi C\varrho$. The density of magnetic poles appears as $(1/4\pi)\,\mathrm{div}\,\mathbf{B}$ and the density of (stationary) current as $(1/4\pi)\,\mathrm{rot}\,\mathbf{H}$. As to the pressure p, Maxwell leaves it unspecified; to fix it, it would be necessary to consider in more detail the magnetisation of the medium.

After this merely descriptive picture of the magnetic field as a distribution of vortices, Maxwell takes up the dynamical problem of how these vortices are set in motion and arrange themselves according to the laws of electromagnetism. This is the first time in the history of the search for mechanical analogies that the crucial part of such an analogy, its dynamical aspect, is developed. The very strangeness of Maxwell's solution illustrates the difficulty of the problem. To account for the transmission of rotation *in the same sense* from vortex to vortex, Maxwell conceived that neighbouring vortices are separated by a layer of particles capable of rolling upon the surfaces of the vortices, like 'idle wheels': the translation of such a layer of 'idle wheel' particles is the image of the electric current. In fact, one finds that, under stationary conditions, the flow of particles is proportional to rot **H**, while the interplay of vortices and particles imitates all the features of electromagnetic induction. From energy considerations, the electromotive force **E** acting upon the particles is found to be connected to the change of the vortex motion by the equation

$$\text{(15)} \qquad \text{rot}\,\mathbf{E} = -\frac{\partial \mathbf{B}}{\partial t};$$

i.e. it is expressed in terms of the vector potential by Equation (14): the e.m.f. appears as a mechanical force resulting from a gradual change in the rotation of the vortices; the electrotonic state corresponding to the impulse which would arise from a sudden stoppage of this rotation.

We thus picture the medium transmitting electromagnetic actions as an agglomeration of 'cells' capable of rotation, separated by layers of 'idle wheel' particles, which form the substance of electricity. The last step is now to visualise more concretely the mechanism of interaction between the cells and the particles. Guidance is here afforded by the behaviour of material bodies under a gradient of electric potential, which simply corresponds to a pressure gradient in the layer of electric particles. If we imagine the structure of cells and intervening layer to be on a finer scale than the molecular structure of matter, we may conceive electric conduction, in such a pressure gradient, as the passage of particles from molecule to molecule, only hampered by some resistance; but in a dielectric, the displacement of the particles appears to remain confined within each molecule, which thus becomes polarised. This strongly suggests that some elastic mechanism comes into play: Maxwell assumes that the substance of the cells yields to some extent to the tangential action exerted by the particles when they are displaced; this distortion of the cells calls forth an elastic reaction, which, if the initial difference of potential is removed, acts as an equal and opposite electromotive force upon the particles. The relation between the displacement **D** and the electromotive force **E** which produces it is accordingly of the form

$$(16) \qquad \mathbf{E} = 4\pi c_1^2 \mathbf{D};$$

the constant c_1 is a function of the elastic parameters of the cellular substance, depending on the structure of the cells.

The consideration of the electric displacement at once brought Maxwell to the threshold of his greatest discovery. "This displacement" he observes "does not amount to a current, because when it has attained a certain value it remains constant, but it is the commencement of a current, and its variations constitute currents." The discharge of a condenser is a typical example of the production of such a displacement current: as the strain on the cells is released, their elastic reaction moves the particles back to their original positions. The displacement current $\partial\mathbf{D}/\partial t$ is of the same kind as the conduction current and has to be introduced into the equation connecting currents and magnetic vortices; one has accordingly, instead of (12)[14],

$$(17) \qquad \mathbf{j} + \frac{\partial \mathbf{D}}{\partial t} = \frac{1}{4\pi}\,\mathrm{rot}\,\mathbf{H}.$$

Now, an argument similar to that underlying Weber's comparison of the electrostatic and electromagnetic units of charge leads to a very simple interpretation of the constant c_1. The equation of continuity, combined with Equation (17), shows that the charge density, in electromagnetic units, is just given by $\mathrm{div}\mathbf{D}$. In the static case the corresponding energy, computed as the work done by the e.m.f. **E** in the displacement **D**, is $\frac{1}{2}\int \mathbf{E}\cdot\mathbf{D}\,\mathrm{d}v$, i.e. on account of Equation (16), c_1^2 times the familiar expression for the electrostatic energy of a charge distribution of density ϱ, in air, $\frac{1}{2}\int \varrho_{\mathrm{P}}\,\mathrm{d}v_{\mathrm{P}}\,\varrho_{\mathrm{p}'}\,\mathrm{d}v_{\mathrm{p}'}/r_{\mathrm{PP}'}$; this shows that the value c of the constant c_1 in air precisely represents the ratio of the two units of charge.

This ratio is thus connected in a definite way with the elastic properties of the cellular substance, and more particularly, since the mechanism devised by Maxwell involves tangential stresses, with its torsion modulus: it is thus closely related to the velocity of propagation of transverse vibrations in this substance. Maxwell now proceeds to determine this relation by specifying in an arbitrary way the properties of the cell substance:

he assumes that the cells are of spherical shape and have the elastic properties of a solid of the Navier type; this gives him a definite value for the torsion modulus in terms of the constant c^2. Moreover, he needs the density of the cells, to the inverse of which the square of the propagation velocity is proportional; to this end, he fixes the constant C connecting the density with the magnetic permeability of the material medium at the value $\frac{1}{4}$ corresponding to uniform conditions. Then, in working out the velocity of propagation, he inadvertently omits a factor $1/\sqrt{2}$ and by sheer luck obtains for this velocity the correct value $c/\sqrt{\mu}$!

We have unambiguous evidence that this is indeed the extraordinary way in which he arrived at the momentous conclusion that "light consists in the transverse undulations of the same medium which is the cause of electric and magnetic phenomena." A letter of October 1861 has fortunately been preserved [16*e*], in which he announces to Faraday the results he has just obtained: and we find there the whole above argument plainly stated in non-mathematical language. After giving in miles per second the calculated value of the velocity of propagation and quoting for comparison Fizeau's determination of the velocity of light, he adds the interesting comment: "This coincidence is not merely numerical. I worked out the formulae in the country before seeing Weber's number, which is in millimeters, and I think we have now strong reason to believe, whether my theory is a fact or not, that the luminiferous and the electromagnetic medium are one."

Maxwell's train of thought at this stage would seem to have been about the following: to visualize the transmission of electromagnetic effects we need a substance possessing transverse rigidity, like the luminiferous aether; if we could show that by assigning to it reasonable elastic properties (such as those suggested by Navier on an atomistic basis) we could make it transmit transverse disturbances with a velocity not merely proportional but equal to that of light, we should get a very strong presumption of the identity of the two media. The conviction that this must certainly be possible, even in more than one way, may account for the apparently negligent way in which he carried out the calculation.

Anyhow, Maxwell had then not yet explored to any great extent the consequences of the identification of the electromagnetic and luminiferous media. The only point he discusses both in the paper and in the letter to Faraday is the influence of dielectric matter on the velocity of propagation; this is another consequence of the fundamental Equation (16) and of the relation $\varrho = c$ div**D** derived, as we have seen, from the idea of displacement current. From these equations follows the relation $-4\pi\varrho = (c^2/c_1^2)\, \Delta V$ between charge density and electrostatic potential, which shows that the specific inductive power of the dielectric ε is given by the ratio c^2/c_1^2 of the squares of the velocities of propagation in air and in the medium, respectively; or if n denotes the refractive index, one has between the electromagnetic and optical parameters of the medium the relation $\varepsilon = n^2/\mu$: a formula of great practical importance, as it opens the possibility of an extensive experimental test of the theory.

Maxwell was guided to his conception of electric displacement and displacement current by such concrete phenomena as the polarisation of a dielectric and the discharge of a condenser; but it is interesting to observe how readily he took the decisive step of making such displacements independent of the presence of ponderable matter. It would not be right to say that he treated the aether as any ordinary material medium; but, imbued as he was with Faraday's ideas, he regarded all material bodies as essentially composed of the same medium, modified only by the molecules of ponderable matter somehow embedded in it. More and more distinctly he perceived that the whole variety of electromagnetic and optical phenomena were but manifestations of a complicated interplay of elastic stresses and strains in a gigantic, all-pervading mechanism.

5.3. *Dynamical Models*

Maxwell's second paper marks the climax of the dynamical conception of nature: in it this conception reached its highest achievements; in fact, as soon became apparent, it overreached itself. In 1864, Maxwell published a third paper [14*d*], entitled *A dynamical theory of the electromagnetic field*; beautiful as it is, it comes as an anticlimax: after the soaring flight of creative imagination, the sober and sobering searchings of criticism. There is no longer any question of indulging in details of ingenious transmission mechanisms; all that is given is a careful enumeration, in logical order, of the essential concepts and equations of the theory, interspersed (this is typical of Maxwell's turn of mind) by practical applications worked out in every detail. As an exposition of the abstract logical framework of the theory, this paper is of supreme elegance; but the contrast with the concrete character of the preceding one is all the more striking. The programme of giving a dynamical account of electromagnetic phenomena is by no means abandoned (as indeed the title indicates), and the whole first part of the paper is devoted to an admirable summary in ordinary language of the mechanical ideas of the previous paper; what is left out is the concrete image of cells and particles with their peculiar elastic linkage. Certainly, Maxwell had already in his second paper explicitly disclaimed for this "somewhat awkward" conception any intention of describing "a mode of connexion existing in nature"; but he had firmly asserted its value as a "conceivable, and easily investigated" one, serving "to bring out the actual mechanical connexions between the known electro-magnetic phenomena [14*f*]." Its abandonment is not merely the discarding of a scaffolding after the completion of the building; it marks a decisive turning of the road leading away from dynamical philosophy.

It is not difficult to conjecture what happened. Maxwell, as we would expect, has by now attained full clarity about the foundation of the electromagnetic theory of light. As he expresses it in a letter of September 1864, he has "cleared [it] from all unwarrantable assumption [16*f*]": he has seen how to base it on the field equations alone, without any appeal to the theory of elasticity. The derivation he gives of the wave-equation is rather clumsy, but the essential points emerge: transversality of the vibrations, unique value of the velocity of propagation. Some time later, when he wrote his last great electrical paper (1868) [14*h*], Maxwell's grasp of the theory was so thorough that he was able to crystallise its whole content in just a few pages of admirable lucidity; it is there, in particular, that its characteristic feature is for the first time explicitly expressed in the statement that all electrical circuits are closed. It was especially important to recognise the fact that the velocity of propagation comes out as $c_1/\sqrt{\mu}$, independent of the elastic structure of the aether: this structure is only represented by the constant $4\pi c_1^2$ which connects electromotive force and displacement, and which can directly be related to the ratio of the units of charge, determined "by measuring the attraction between bodies kept at a given difference of potential, the value of which is known in electromagnetic measure [16*f*]". Just at that time, new determinations of the unit ratio were being carried out on this very principle (suggested by Thomson [13*i*] as part of the programme of the British Association Committee on Electrical Standards, and they provided (cf. the adjoining table) a welcome confirmation of the result of Weber and Kohlrausch: Maxwell himself performed one of these new measurements, which he described in his 1868 paper [14*h*].

Determinations of *c*	($\cdot 10^{10}$ cm s^{-1})
Ratio of electrical units	
Weber and Kohlrausch (1856)	3.11
Maxwell (1868)	2.88
Thomson (1869)	2.93
Velocity of light	
Astronomical (from aberration)	3.08
Fizeau (1849)	3.15
Foucault (1862)	2.98

But just as the theory gained a firmer empirical foundation and a neater formal structure, the very definiteness and universality of its enunciations brought sharply to the fore the difficulty of reconciling it with any elastic theory of the propagation of light. Thus, the fact that only transversal vibrations are propagated by the electromagnetic medium at once conjured up the unsolved riddle of how to dispose of the longitudinal vibrations inherent in elastic disturbances of the ordinary type. It is significant, in this respect, that Maxwell did not succeed in giving a derivation of the laws of reflexion and refraction – the *crux* of all elastic theories: a letter to Stokes of 1864 reveals his unsuccessful grappling with this problem [19*a*]. He thus came to recognise that the difficulties besetting theoretical optics had now to be faced just as well in electromagnetics; in the paper of 1864 he admits as much in the telling sentence: "both sciences are at a loss when called on to affirm or deny the existence of normal [i.e. *longitudinal*] vibrations [14*g*]".

It is clear, therefore, that Maxwell felt himself in the grip of an acute dilemma. The conviction of the real existence, in a concrete physical sense, of an underlying mechanism of transmission of electromagnetic actions had led him to the discovery of the identity of this mechanism with that of light propagation; and now, the very unfolding of the logical consequences of this discovery, far from helping him to elucidate the mechanism, was throwing him back against the same obstacles which had so long hampered progress in optics. The course he took in this quandary was undoubtedly a wise one: he stressed those features of the theory which were independent of all uncertainties about the underlying mechanism, and renounced all explicit consideration of the latter. Like Newton or Ampère before him, Maxwell did not see in this kind of renunciation anything but the temporary laying aside of a difficult problem, unripe for fruitful discussion. Yet we can understand Thomson, the staunch believer in the dynamical nature of ultimate reality, shaking his head and regarding the electromagnetic theory of light as a backward step [13*j*]. He made his meaning quite clear in one of the Baltimore lectures [13*k*]; "I firmly believe in an electromagnetic theory of light, and that when we understand electricity and magnetism and light we shall see them all together as parts of a whole. But I want to understand light as well as I can, without introducing things that we understand even less of. That is why I take plain dynamics. I can get a model in plain dynamics; I cannot in electromagnetics." Thomson was in a way of reactionary temperament, clinging to the last to the ideas he had conceived in his youth; but he cannot be denied clearsightedness in his judgment of the trend of ideas in physics. In this case, if he opposed the change of emphasis from the equations of dynamics to those of electromagnetism, introduced in Maxwell's third paper, it was because he foresaw more clearly than anybody else where it would ultimately lead.

The word 'model' in Thomson's remarks just quoted concentrates in itself, so to speak, the whole problem of dynamical philosophy. Mechanical analogies of the kind considered by Thomson and Maxwell are indeed in the first instance *models* intended to imitate certain physical effects whose dynamical nature is not immediately apparent, and thereby reduce these effects to a more familiar form. In this function, they are an aid to thinking, they have heuristic value. But this is by no means their only, or even their essential epistemological significance: in the search for dynamical models is implied the postulate that such models can always be found, because all phenomena are fundamentally dynamical in nature. In other words, dynamical models are not imitations of things non-dynamical, but attempts to reproduce, from the observation of their effects, the features of real dynamical systems. Dynamical models are thus meant to be images of reality, the increasing quality of which will be judged by their ability to account for a larger and larger number of the observed phenomena. Let us not scorn this ideology: should we be tempted to find it crude or naive, let us remember what Maxwell achieved with his vortices and idle wheels – and bow with reverence.

Yet, Maxwell is perhaps greater still in our eyes because in this dynamical symphony his critical ear perceived a jarring note. His efforts to fix the really essential features of the elusive dynamical picture he was trying to outline led him eventually to realise that such a picture was in fact left largely undetermined by the conditions of the problem. The derivation of the field equations given in the paper of 1864 starts from the consideration of a system of closed circuits, each characterised by a single variable (the current intensity). The electrodynamical laws are first derived in integrated form, and then reduced to the local form by the now familiar procedures; the idea that all currents are closed when variations of displacement are taken into account makes the argument perfectly general. Now, in this reasoning, the analysis of the system of coupled circuits was still compared to a special 'dynamical illustration', referring to the reduced momentum of a fly-wheel. But in the *Treatise* of 1873, which embodies the latest state of Maxwell's electromagnetic meditations, the argument is presented [15*c*] in a purely abstract manner, based on the Lagrangian formalism of point dynamics. What is here shown is thus that the laws of electromagnetism are compatible with those of dynamics, and consequently allow of an infinity of mechanical models, all fulfilling their function equally well. Any particular one will play the part of an 'existence proof' just as well as any other. This remarkable conclusion is formulated with all possible clarity in a later part of the *Treatise* [15*b*] (every feeling ranging from reproach to regret has often enough been expressed regarding the bad composition of this wonderful book!); here it is in full:

> The attempt which I then [in 1862] made to imagine a working model of this mechanism must be taken for no more than it really is, a demonstration that mechanism may be imagined capable of producing a connexion mechanically equivalent to the actual connexion of the parts of the electromagnetic field. The problem of determining the mechanism required to establish a given species of connexion between the motions of the parts of a system always admits of an infinite number of solutions. Of these, some may be more clumsy or more complex than others, but all must satisfy the conditions of mechanism in general.

The recognition of the existence of an infinity of possible mechanical models of electromagnetic phenomena and of their essential equivalence was a death blow to any pretension of deriving from mechanical considerations a faithful picture of physical reality. At the end point of the eventful evolution of Maxwell's thought, we see fully vindicated the view of the essential ambiguity of the aether conception, held long before by Lorenz. No doubt, the ideal of an ultimate mechanical explanation of all natural phenomena will still linger on for some time. Boltzmann, who was as unflinching an exponent of this philosophy as Thomson, will still ask, in the peroration of his magnificent speech *On the development of the methods of modern theoretical physics* [18] at Munich in 1899: "Will the conviction ever arise that certain pictures are such that they cannot be superseded by any simpler, more comprehensive ones, that they are 'true'...?" Hertz himself begins his book on the principles of mechanics, in 1894, with the sentence [17*b*]: "All physicists agree that it is the task of physics to reduce the natural phenomena to the simple laws of mechanics." But while paying lip-service to a scientific philosophy which had accomplished its historical part, Hertz was undermining its last positions in electromagnetic theory, by carrying to its ultimate consequences Maxwell's criticism of the mechanical models of this theory.

6. Hertz

Hertz cannot properly be described as a disciple of Maxwell: if he appears to pursue a line of development initiated by the latter, it is by no means as a result of a direct filiation of ideas. Hertz was introduced to Maxwell's equations by his teacher Helmholtz, a natural philosopher steeped in the modes of thinking of central force physics. Among the Continental physicists, Helmholtz had the great merit of recognising at once the importance of Maxwell's conceptions; but in laying on them a rather heavy hand he completely spoiled their subtle harmony. What he did was to construct a general scheme of electrodynamics, from the point of view of action at a distance, which would embody all possible laws of electrodynamic action leading to the same result for closed circuits, and clearly show wherein they could differ in their predictions regarding open circuits. In order to compare Maxwell's formulation with the other versions of electrodynamics, he tried to cast it into the same mould: not only was such an approach to Maxwell's theory entirely alien to its spirit, but it tended to obscure its characteristic features[15] and to make the theory appear as a somewhat singular limiting case of the scheme.

In his first contribution to electromagnetic theory (1884) [17*d*], we see Hertz attacking the problem from the same angle, but with a strong bias in favour of Maxwell's theory. He gives an ingenious derivation of Maxwell's equations from "premises which are accepted also by the rival electrodynamics, and using a chain of reasoning familiar to this electrodynamics"; and he presents it as a compelling argument for the uniqueness of the Maxwell scheme, vindicating Riemann's and Lorenz' pioneering efforts. Nevertheless, when he discovered the electromagnetic waves, he discussed his experimental results in terms of Helmholtz' ideas, and only afterwards, in his famous paper [17*e*] of 1888, did he show how they could be interpreted on the basis of Maxwell's theory. Even so, he still betrays in this last paper a total ignorance of Maxwell's dynamical ideas: he asserts that "Maxwell arrived at his theory by going out from action at a distance and attributing to the aether the properties of a highly polarizable dielectric medium". This is in fact an apt description of the point of view adopted by Helmholtz, and only shows that Hertz had not taken the trouble to read the papers which preceded Maxwell's *Treatise.* It is not surprising, therefore, that in the systematic exposition of the electromagnetic theory which he gave in 1890 [17*f*], and which is still the one currently adopted in Germany, only the formal skeleton of Maxwell's theory subsists: Hertz himself declares [17*c*]: "To the qucstion 'What is Maxwell's theory?' I would not know any shorter and more definite answer than this: Maxwell's theory is.the system of Maxwell's equations".

This oft-quoted statement must be understood in the spirit of phenomenology in its constructive, scientific aspect, for which Kirchhoff had set an example among the German physicists. Hertz has by no means forsaken the aether: on the contrary, he visualises it as the primordial substance, whose properties would not only provide the explanation of the electromagnetic and optical phenomena, but would also give the key to an understanding of the essential characteristics of ordinary matter itself: gravity and inertia. This is the programme which he outlines with glowing optimism in his celebrated Heidelberg speech [17*a*] of 1889. It was not given him to proceed very far with, it, but the work to which he devoted the last years of his brief career (he died in 1894, at the age of 37) was certainly, in his mind, the first step towards an all-embracing theory of aether and matter.

Hertz' effort in attempting a reformulation of the principles of mechanics [17*g*] was in fact directed against a logical flaw which had all the time affected the conception of force transmission by a medium itself dynamically described. As Maxwell, with his usual acuteness, expresses it [15*d*]: "The observed action at a considerable distance is... explained by means of a great number of forces acting between bodies at very small distances, for which we are as little able to account as for the action at any distance however great". Obviously, there is no escape from this difficulty except in the complete elimination of the concept of force, and its reduction to that of motion, in accordance with the old Cartesian ideal. This is exactly what Hertz attempts to do: he shows that the way in which dynamical constraints applied to a system of particles affect its motion can always be described as the effect of adding to the system 'hidden' particles in suitably chosen states of motion. As an illustration of what he has in mind, he expressly mentions Maxwell's mechanical conception of electrodynamic action. Hertz' thinking was thus here moving along the lines of the British dynamical school. But his profound work remained a torso: Hertz could not carry it beyond the foundations, and nobody after him ever took it up in the spirit in which it was conceived.

In fact, the spirit of the dynamical tradition, which had achieved so much, was on the wane. The discovery of the electron and the study of fast electrified particles raised more pressing problems, and in this new context, the emphasis put by Hertz on the phenomenological value of Maxwell's equations acquired its full significance. For practical purposes at least, the question of the structure of the aether appeared quite superfluous: the electromagnetic concepts and equations formed a self-contained, consistent system, offering a firm starting point for the analysis of the new phenomena. There was, however, in this phenomenological point of view a quite novel element: as a result of Hertz' discovery, the electromagnetic field had now become as concrete a reality as any dynamical system, and could serve as a complete substitute for the aether; Maxwell's equations could be regarded as the expression for laws of nature as fundamental as those of dynamics. Thus, the old issue of the mechanical description of nature was not just pushed once more into the background: it was actually eliminated.

The way was then prepared for the emergence of the field concept as denoting a fundamental aspect of all natural phenomena. At the same time, the direct observation of atomic processes removed all doubts raised, at a different level of knowledge, about the existence of atoms. With the recognition of these two general features of the physical world, the fruitfulness of the great guiding ideas we saw at work in the course of this development exhausted itself. Both the dynamical philosophy, which had played such a decisive part in the elaboration of the field concept, and the mechanistic conception, achieving its greatest triumph in the atomistic interpretation of thermodynamics, became obsolete as a result of the broadening of outlook they had themselves brought about. As they vanished from the scene, they left unresolved the question whether the ultimate structure of matter and electricity was continuous or discontinuous. We know how deeply hidden the solution of this conflict was: a synthesis of the opposite aspects exhibited by all natural agencies could only be obtained from the recognition of the novel relationship of complementarity in which they stand to each other. Such a synthetical view (quite apart from the factual knowledge of the existence of the quantum of action, on which it is based) was undreamt of in the scientific philosophy of the XIXth century. Physics then was not ripe for any truly synthetic process: this was the pioneering epoch which prepared the elements for future syntheses. Hence this tendency we noticed to one-sided generalisations and sharp oppositions, which gave the development of theoretical conceptions such a dramatic character. There is no more striking illustration of the tenseness of physical thought in that heroic period than Lorenz' and Maxwell's struggle to establish the electromagnetic theory of light.

NOTES

[1] For the history of the first phase of development of quantum theory, see ref. [20], in which references to the original papers may be found.

[2] The first *published* occurrence of retarded potentials is in a paper of 1861 on the theory of elasticity by L. Lorentz [8b].

[3] The withdrawn paper was published posthumously in 1876, and immediately criticised by Clausius.

[4] I slightly modernise the argument by using the vector notation and the electromagnetic units rather than the electrodynamic ones, i.e. directly the factor c rather than c_W. The current density is expressed in electrostatic units.

[5] This is actually the formula proposed by F. Neumann in 1845 (see ref. [19], chapt. VII). Weber's law leads for the vector potential to the equivalent expression

$$\int (\mathbf{j}(P')\cdot\mathbf{r}_0)\,\mathbf{r}_0\,\mathrm{d}v/r_{PP'}.$$

($\mathbf{r}_0$ is the unit vector along PP′), which is also adopted by Kirchhoff.

[6] This apt, but ugly denomination was only introduced towards the end of the XIXth century; I apologise for the anachronism and the cacophony.

[7] It is significant that, although his general philosophy of life was a form of spiritualism, Ørsted rejected Hegel's system because of its idealistic cosmology, whose topsy-turviness he

exposes, in a most revealing short note [3b], with a pungency which would have done honour to Marx. His main point is that the evolution of the universe had been going on for a long time before reasonable beings appeared on the scene; the universe is therefore not produced by thought, but "thought is nature in us, awakening to our consciousness". He is so clear about the implications of his criticism that he takes the trouble to warn us that these "indubitable truths do not throw us into materialism" because "the essence of all nature... is spiritual"!

This is only one of many examples of the cleavage which took place in the XIXth century in the minds of most scientists between their class-determined spiritualistic beliefs and their scientific philosophy, which could not help being materialistic. For this reason I have deliberately refrained from referring the various tendencies discussed in the text either to materialism or to idealism: in my view, such labelling would be futile and even misleading; the ideological conflict symbolised by these two words was then fought in the social sphere, not in that of science. Marxist doctrinaires may remind me that one should not separate science from its social background: but what can I do if the scientists themselves chose to make this very separation in their own minds? *Why* they behaved like this is of course a most interesting sociological problem, which I have, unfortunately, no space to tackle here.

[8] The transition of capitalism to the imperialist stage brought with it that predominance of idealistic and spiritualistic conceptions which is still paralysing the philosophical thinking of the bourgeoisie to-day.

[9] The otherwise reliable biography by S. Thompson [13c] is in error on this point: it mentions the date of August 1842 instead of 1841. The date of publication in the *Cambridge Mathematical Journal* is February 1842.

[10] Lest anybody should think that he received any inspiration from the Cambridge environment, I may just mention that in the same letter he quotes his tutor as "hoping I do not now lose any time with the *Math*[*ematical*] *Jour*[*nal*]"! These were the dons who subjected the self-taught genius George Green, at the age of 44, to the torture of the Tripos, and kept his great *Essay* unread on their shelves. Equally edifying is the story [25] of how, some years later, the (figuratively) deaf and blind astronomer Challis managed to miss the planet Neptune at the place correctly predicted by the young graduate John Couch Adams.

[11] See the story of the 'discovery' of Green's *Essay* in S. Thompson's biography [13e].

[12] Every young physicist with any ambition to culture ought to have read this paper, as well as the second one, described below. This can be done in an hour or two; thus no one may excuse himself by pleading that it would take up too much of his valuable time.

[13] In electromagnetic units.

[14] In Maxwell's paper, the displacement occurs with the opposite sign in the equations equivalent to (16) and (17): this is because he analyses the 'discharge' process, but counts the displacement from the position of the particles in the undisturbed state.

[15] This is clearly set out by Hertz himself in the introduction to his collected papers (1891) [17c], in which he compares Helmholtz' and Maxwell's treatment of dielectric media. Helmholtz represents the polarisation of the dielectric in a condenser as the result of the action at a distance of charges coating the plates of the condenser: each 'molecule' is thus charged positively on the side turning towards the condenser plate at the *lower* potential. For Maxwell, the electric displacement removes all charge from the condenser plates and brings it into the dielectric: the net result is thus that each 'molecule' appears positively charged on the side turning towards the condenser plate at the *higher* potential.

BIBLIOGRAPHY

[1] Laplace, S., *Exposition du système du monde* (Paris 1796), 4th ed. 1813; (livre IV, chap. XVIII: *De l'attraction moléculaire*).

[2] Fourier, J., *Théorie analytique de la chaleur* (Paris 1822). [English translation by A. Freeman: *The Analytical Theory of Heat* (Cambridge University Press, 1878; Dover reprint, 1955) – Ed.].

[3] (*a*) Ørsted, H. C., *Experimenta circa effectum conflictus electrici in acum magneticum.* (Hafniae 1820) (facsimile edition with translations by A. Larsen, Copenhagen 1920).

(*b*) *Samlede og efterladte Skrifter* (Kjøbenhavn 1852), vol. 9, p. 75.

[4] Ampère, A. M., in *Collection de mémoires relatifs à la physique* (Paris 1887).

(*a*) (vol. III, p. 1) *Mémoire sur la théorie mathématique des phénomènes électrodynamiques, uniquement déduite de l'expérience* (1820–25).

(*b*) (vol. II, pp. 249–251) (1822).

[5] Weber, W., *Werke* (Berlin; vol. III, 1893; vol. IV, 1894).

(*a*) (vol. III, p. 25) *Elektrodynamische Massbestimmungen über ein allgemeines Grundgesetz der elektrischen Wirkung* (1846).

(*b*) (vol. III, p. 609) (mit R. Kohlrausch) *Elektrodynamische Massbestimmungen insbesondere Zurückführung der Stromintensitäts-Messungen auf mechanisches Mass* (1957) [Announcement (vol. III, p. 591) on 20 October 1855].

(*e*) (vol. IV, p. 105) *Elektrodynamische Massbestimmungen, insbesondere über elektrische Schwingungen* (1864).

(*d*) (vol. IV, p. 157) *ibid.*

[6] Kirchhoff, G., *Gesammelte Abhandlungen* (Leipzig 1882).

(*a*) (p. 131) *Über die Bewegung der Electricität in Drähten* (1857).

(*b*) (p. 154) *Über die Bewegung der Electricität in Leitern* (1857).

[7] Riemann, B., *Gesammelte mathematische Werke*, 2nd Ed. (Leipzig 1892).

(*a*) (p. 553); (*b*) (p. 288) *Ein Beitrag zur Elektrodynamik* (1858, published 1867).

[8] Lorenz, L., *Oeuvres scientifiques* (revues et annotées par H. Valentiner) (Copenhagen 1896).

(*a*) (vol. I, p. 173) *Om Identiteten af Lyssvingninger og elektriske Strømme* (1867).

(*b*) (vol. II, p. 3) *Mémoires sur la théorie de l'élasticité des corps homogènes à élasticité constante* (1861).

(*c*) Pihl, M., *Der Physiker L. V. Lorenz* (Copenhagen 1939), p. 101.

[9] Faraday, M., *Experimental Researches in Electricity* (London; vol. I, 1839; vol. II, 1844; vol. III, 1855). [Dover reprint, New York, 1965 – Ed.].

(*a*) (vol. III, p. 1); (*b*) (vol. III, p. 447) (1846); (*c*) (vol. II, p. 284) (1844).

[10] Jones, B., *The Life and Letters of Faraday* (London1869), vol. II, p. 378.

[11] Thompson, S., *Michael Faraday – His Life and Work* (Cassell, London, 1898), p. 197.

[12] Thomson, Sir W., *Reprint of Papers on Electrostatics and Magnetism* (London 1872).

(*a*) (p. 1); (*b*) (p. 126); (*c*) (p. 15); (*d*) (p. 419).

(*e*) *Mathematical and Physical Papers* (London 1882), vol. I, p. 76.

[13] Thompson, S., *The Life of William Thomson* (London 1910).

(*a*) (vol. I, pp. 304–5); (*b*) (vol. I, p. 311); (*c*) (vol. I, p. 42); (*d*) (vol. I, p. 56); (*e*) (vol. I, p. 113); (*f*) (vol. I, pp. 127, 128, 131); (*g*) (vol. I, p. 197); (*h*) (vol. I, p. 203); (*i*) (vol. I, p. 524); (*j*) (vol. II, p. 1024); (*k*) (vol. II, pp. 835–6).

[14] Maxwell, J. C., *Scientific Papers* (Cambridge 1890). [Dover reprint, New York, 1965 – Ed.].

(*a*) (vol. II, p. 360).
(*b*) (vol. I, p. 155) *On Faraday's Lines of Force* (1855–6).
(*c*) (vol. I, p. 451) *On Physical Lines of Force* (1861–2).
(*d*) (vol. I, p. 526) *A Dynamical Theory of the Electromagnetic Field* (1864).
(*e*) (vol. I, pp. 570–1); (*f*) (vol. I, p. 486); (*g*) (vol. I, p. 582).
(*h*) (vol. II, p. 125) *On a Method of Making a Direct Comparison of Electrostatic with Electromagnetic Force; with a Note on the Electromagnetic Theory of Light* (1868).

[15] Maxwell, J. C., *A Treatise on Electricity and Magnetism* (Oxford 1873). [Dover reprint of the 3rd ed. of 1891, New York, 1954 – Ed.].
(*a*) (Preface); (*b*) (vol. II, §831); (*c*) (vol. II, chap. VII–VIII); (*d*) (vol. I, §105).

[16] Campbell, L. and Garnett, V., *The Life of James Clark Maxwell* (London 1882; 2nd ed., 1884).
(*a*) (pp. 519–520); (*b*) (pp. 288–90); (*c*) (2nd ed., p. 202); (*d*) (p. 216); (*e*) (2nd ed., p. 243); (*f*) (p. 340); (*g*) (p. 342).

[17] Hertz, H., *Gesammelte Werke* (Barth, Leipzig 1895).
(*a*) (vol. I, p. 339) *Über die Beziehungen zwischen Licht und Elektrizität* (1889).
(*b*) (vol. III, p. xxiii); (*c*) (vol. II, pp. 23–31).
(*d*) (vol. I, p. 295) *Über die Beziehungen zwischen den Maxwellschen elektrodynamischen Grundgleichungen und den Grundgleichungen der gegnerischen Elektrodynamik* (1884).
(*e*) (vol. II, p. 147) *Die Kräfte elektrischer Schwingungen, behandelt nach der Maxwellschen Theorie* (1888).
(*f*) (vol. II, p. 208) *Über die Grundgleichungen der Elektrodynamik für ruhende Körper* (1890).
(*g*) (vol. III) *Die Prinzipien der Mechanik in neuem Zusammenhange dargestellt* (1894). [English translation: *The Principles of Mechanics*, tr. D. E. Jones and J. T. Walley, ed. with an introductory essay by R. S. Cohen, Dover, New York, 1956 – Ed.].

[18] Boltzmann, L., *Populäre Schriften* (Leipzig 1905), p. 227.

[19] Whittaker, E. T., *A History of the Theories of Aether and Electricity* (London 1910).
(*a*) (p. 296); (*b*) (p. 353 sqq.).

[20] Rosenfeld, L., *La Première phase de l'évolution de la théorie des quanta*, in *Osiris*, **2** (1936), 149. [This volume, p. 193].

[21] Bohr, N., *Drei Aufsätze über Spektren und Atombau* (Braunschweig 1922): Erster Aufsatz (1913), p. 19.

[22] Weyl, H., *Raum-Zeit-Materie* (Berlin 1919), 3rd ed. [English translation: *Space-Time-Matter* by H. L. Brose, from the 4th ed. of 1922, Dover reprint, New York, 1950 – Ed.].

[23] Dantzig, D. van, *Proc. Cambr. Phil. Soc.* **30** (1934), 421; *Proc. Amst.* **37** (1934), 521, 526, 643, 825 and especially **39** (1936), 3.

[24] *Correspondance entre Harald Høffding et Emile Meyerson* (Munksgaard, Copenhagen, 1939), p. 43.

[25] Smart, W. M., *John Couch Adams and the Discovery of Neptune* (London 1947).

Note. – Duhem, P.: *Les théories électriques de J. Clerk Maxwell* (Paris, 1902). This book is entirely devoted to a disparaging criticism of Maxwell's work. Duhem discovers in particular a large number of "errors of sign" in Maxwell's papers, which according to him reveal serious inconsistencies in his ideas. It must be admitted that Maxwell's shifting choices of sign are rather confusing; there is even a definite inconsistency in the sign conventions of his 1864 paper, which gives rise to some obvious mistakes: but such formal blemishes do not affect the soundness of the physical argument.

CHAPTER 15

THE EVOLUTION OF OERSTED'S SCIENTIFIC CONCEPTS

[1970b]

Lecture to the Royal Danish Academy of Science and Letters on October 30th, 1970

Mind in Nature expresses the view of the world which Oersted reached in the full maturity of his thought. The romantic appearance he gives to it is undoubtedly an echo of his first love for '*Naturphilosophie*', but in the dominant role he assigns to reason we can clearly discern not only the fruit of his own experience as a scientist, but above all the traces of an inspiration coming from the French school, which set the tone for scientific thought during the period in which he accomplished his creative work. Nevertheless, it would be unjust to call him an eclectic: he never sought a compromise between the opposing forces by which he was influenced; he strove rather to draw forth from them a synthesis which, while keeping what was most valid in each tendency, could offer a harmonious view of nature and man's place in it. In this respect, his attitude recalls that of Niels Bohr. They were both open to all the currents of thought of their day, but did not accept them without subjecting them to a penetrating critique; moreover, they were able to transform them and blend them into a harmonious unity. Bohr often said that he was happy to have been born in a small country where no tradition limited one's thinking but where, on the contrary, there was free access to all the external cultural impulses; it is precisely this fusion of different traditions, he observed, which has given rise in the course of history to ever richer new civilizations, and this same law governs the formation of new guiding ideas in science.

Retracing the long path which led Oersted to philosophical maturity is not a simple task, because in the way of direct eyewitness accounts we have only incidental information contained either in his own statements or in those of his contemporaries. We must, then, depend above all on the conclusions that may be drawn from the analysis of his writings and various events in his scientific career. The ensemble of all these elements does, however, allow us to develop a fairly clear idea of the evolution of his attitude with regard to the philosophical currents of his time. In this essay we will not undertake to follow this evolution through the full rich range of his thought to which no area of human activity was foreign; we will limit ourselves to studying it in relation to one of the great scientific ideas which dominated the nineteenth century: the concept of mutual transformations of natural forces. This example will suffice to elucidate the themes which guided the progress of Oersted's scientific thought and to illustrate its strengths and limitations.

The idea of the transmutability of natural forces has distant origins in the common heritage of prescientific traditions of the middle ages and the Renaissance, from which modern science took form in the seventeenth century. A first attempt at systematization may be found in Aristotle's hylomorphism, whose guiding principle is the variability of the qualities, the forms, which can place their imprint successively on the same matter. The artisans who adopted this learned concept transformed it into an esoteric doctrine, alchemy, where the primacy of form over matter implied mystical ties, established by similarities of form and gradations of quality, among the various 'kingdoms' of nature; and notably between man, the microcosm, and the universe, the macrocosm. At the beginning of the Renaissance, these speculations were given the beginnings of rationalization, a fact which led to all sorts of extravagances such as the 'bishop fish', described and illustrated among other marine monsters by Rondelet in his book *De piscibus marinis* (1554). These aquatic dignitaries, undoubtedly walruses, were seen along the coasts of Nordic countries, sometimes accompanied by 'monks' which were perhaps seals. Guillaume du Bartas' very popular contemporary poem on the creation of the world alludes to them in speaking of the sea.[1]

> From her profound depths sometimes she draws forth
> Her monk, and her prelate, and casting them on shore,
> Makes them seem monstrous to humans who live in the North.

The doctrine of the 'signatures' of plants derives from the same vein due to Paracelsus and is developed by Porta in his *Phytognomonica* (1588); certain plant organs – leaves, fruits, roots, etc. – show a resemblance to some part of the human body on which the plant in question may have a therapeutic effect. This notion, borrowed from ancient folklore where, indeed, it still flourishes, could not withstand the criticism of the Age of Reason, and by the time of the Victorian era it inspired nothing more than the fantastic images of Edward Lear in his *Nonsense Botany*.[2]

In the seventeenth century, on the contrary, the process of rationalization of the relationships between forms and matter was still very active and soon took a new turn; the distribution of qualities which, according to the alchemists, characterized different species of bodies, gave way to the more general idea of a ladder or chain of beings, extending from the microcosm to the macrocosm by increasingly perfected levels of organization, but deriving from the same plan, and because of this presenting structural and functional analogies. In Newton's thought, the concept of the 'analogy of nature' plays a heuristic role of primary importance: in particular, this concept allows him to attribute to constitutive elements of matter, which cannot be observed, properties analogous to those of directly observable material bodies. Newton was so conscious of the importance of this argument from analogy for his atomistic concept of the material world that he explicitly states it as one of the 'rules' on which he bases his method of 'natural philosophy'.[3]

It would seem that the hypothesis of the atomic constitution of matter must certainly have favored the idea of the transmutability of natural forces, since its aim is to reduce all the qualitatively different manifestations of these forces to a single cause, the behavior of atoms. Nonetheless, although this is in effect the current epistemological position of atomism, it is important to realize that this is the result of a long dialectical process whose first impulse, on the contrary, tended to mask the whole problem of the nature or even the existence of relationships between qualitatively distinct phenomena. In the first place, the Cartesian ideal of a purely mechanical description of the universe, which also dominated Newton's atomism, inclined Newton to eliminate the consideration of qualitative appearances from natural philosophy, in order to situate the problem in the realm of 'ideas', or as we would say, of psychology. It is in this context that the work of Locke, of directly Newtonian inspiration, is situated; it was so understood by the followers of Newton in the eighteenth century. Hence, in this initial phase, atomism led not to an explanation of qualities, but rather to their 'reduction' to a vast dynamics of atoms. The final step taken along this path was undoubtedly Laplace's[4] attempt to reduce psychology itself to a dynamics of the 'sensorium'.

Moreover, the atomism of the eighteenth century presented another peculiarity whose effect was to isolate the natural forces from each other, instead of suggesting possible relationships among them: it associated with each of them specific material agents. Despite the rule of the 'analogy of nature', which might have led him to compare the propagation of light to that of sound, Newton opted in the end for the attribution of luminous phenomena to specific corpuscles. Later Lavoisier and Laplace, although they realized that heat might be only a manifestation of the motion of material atoms, still preferred the hypothesis of a specific weightless element, caloric.[5] It was still during Newton's time that Becher and Stahl introduced phlogiston as the universal agent of the chemical reactions of combustion and reduction; and when Lavoisier replaced it with an acidification agent, oxygen, he was still following the same 'substantialist'[6] tendency. Thus a concept of natural forces was established in the French school, sanctioned by the somewhat tyrannical authority of Laplace, and soon incorporated by August Comte into his system of 'positive philosophy'. This concept resulted in the belief that these forces were attached to specific material agents and consequently not reducible one to the other.

Faced with the vigorous growth of French science, continuing the Newtonian tradition in mechanics and astronomy, but extending throughout the domains of chemistry and physics investigations clearly oriented toward industrial applications – 'applications to the arts', as Laplace said,[7] contemporary German philosophy took the role of reaction against the rationalism of the scientific method and the materialism which – openly or in the guise of deism – expressed its tendency and its conclusions. This ideology of frustration and evasion reflected the resentment of a bourgeoisie as yet too weak to take part in the economic expansion of its rivals and which, turning away from the well-springs of this expansion, sought as it were in contemplative thought, a short-cut which would lead it directly to the results of active research. Kant was the first to cast doubt on the efficacy of rational thought, followed by Schelling and Oken who rejected all appeal to rational thought and claimed to derive a complete description of the world from pure *a priori* 'ideas'. It is hardly surprising to rediscover in these 'ideas' the archetypes whose presence we noted earlier in the doctrines of the alchemists and humanists of the Renaissance. In particular, the primacy of form over matter and the idea of the fundamental identity of the forms in which natural forces manifest themselves are prominent. In attacking the analytical method of science, *Naturphilosophie*, in effect, accuses it of breaking the unity of phenomena, by arbitrarily cutting them apart; *Naturphilosophie* replaces the analytical method with intuition [Anschauung], the synthetic view, which reveals this unity. "The triple character of the universal, the particular, and the lack of difference between the two", Schelling assures us,[8] "expressed by their identity is magnetism, expressed by their difference is electricity, and expressed by the totality is the chemical process. Thus, these forms are all one, and the chemical process is a simple transfer of the three points of magnetism to the triangle of chemistry."

The fact that Oersted drew from these tainted waters the idea which eventually led to his great discovery is not sufficient reason to attribute to the *Naturphilosophie* a decisive role in the development of electromagnetism as has recently been done[9] – not only in the case of Oersted, but even, against all evidence, in that of Faraday. Certainly Schelling, in a famous speech,[10] did comment on the experiments of these two physicists which cstablished the mutual relationship of electric current and magnetism: but for what purpose? He states it with perfect candor: "The intention . . . was to demonstrate and analyze the necessary connection among the cited discoveries . . . as the discoveries . . . were more or less anticipated by thoughtful investigators of nature". Thus Schelling claimed priority for these discoveries for *Naturphilosophie* and not at all an inspirational role, which he would have considered inconsequential. He even went so far as to say that Oersted made his discovery by chance. One could more properly wonder how Oersted succeeded in making productive an idea presented in such a nebulous form. The answer, as we shall see, is quite simple; it was by abandoning the speculative path which had initially attracted him and by adopting the scientific method whose model he found in Paris.

But first, it is important to dismiss the impression – which would be totally false – that the school of *Naturphilosophie* had an appreciable influence on the scientific movement in Germany itself. This school, faced with a generation which felt a profound contradiction between its aspirations for political and economic emancipation and its inability to realize them, could only keep its hold on those who abandoned the struggle and took refuge in dreams. But the most clear-sighted, on the contrary, quickly turned away from it, and endeavored rather to assimilate the methods the French and English were finding so successful. Such was the thankless task of the German chemists and physicists of the early nineteenth century; and as the results show, they accomplished it well. Among these honest and conscientious workers, *Naturphilosophie* met only with undisguised hostility. Moreover, the situation was no different in the life sciences and in medicine where Schelling's 'daring flights' might have been expected to strike a more sympathetic chord. Let us listen, for example, to the testimony of the embryologist K. von Baer[11]. "Schelling's philosophy, I thought, cannot be as empty as some of its calumniators would have it, since many scientific men still feel warmed by its rays." Put off by the foggy uncertainty [nebelhafte Unbestimmtheit] of Schelling, he studied Oken's work, whose precision and logic he appreciated, but with which he often disagreed. But, he concluded, this disagreement was more profitable to him than "keen flights undertaken on the wings of yearning and aesthetic fantasy through mist and dawn, to find the source of light. Though such flights attracted me, because originally I myself had this yearning within me, I soon had to recognize that mist and dawn soon vanish when the light source is fully visible, and where they are apparent they just offer proof that the sun does not yet fully shine." Even more interesting is what he learned from his teacher, the zoologist Döllinger:[12] "Previously he eagerly studied Kant's philosophy, but was then swept along by Schelling to whom he had become personally closer. But with his critical intelligence and ordered imagination, he must soon have recognized that Schelling had made the most difficult tasks of philosophy into the starting point, even into the pedestal, of his doctrinal edifice of natural philosophy. Later he did not like to speak about this period, and expected physiology to be erected on specific observations which would then have to be grasped in a philosophic spirit." A striking comparison can be made between this conclusion and the one Oersted reached independently by 1809 or 1810, as reported by C. Hauch who attended his courses during those years:[13] "Having noticed perhaps that I had more taste for general considerations about nature than for detailed research, he immediately put me on guard against following this penchant too freely. "I am similarly persuaded", he said, "as my lessons demonstrate, that a great fundamental unity reigns throughout all of nature; but it is precisely when one has acquired this conviction that it is proper to turn one's attention to the manifold world where alone this truth can be confirmed. Otherwise, the unity itself becomes a sterile and empty idea, incapable of leading to any true understanding."'

It was during his first trip abroad, from 1801 to 1804, that Oersted was confronted on a much vaster stage than the provincial milieu of Copenhagen with the contradictions of contemporary epistemology: speculative philosophy versus empirical science, form versus matter, or in more precise terms: dynamical physics, relating phenomena to 'conflicts' between forces of opposing polarities, versus atomic physics stressing material corpuscles from which forces emanate and on which they act. The journal he kept during those years and the letters he wrote to his mentor, Professor Manthey, and to his brother abound in precious information on his state of mind during these formative years.[14] It is important to note that he was carrying these contradictions within himself even prior to his departure: though imbued with speculative philosophy, he had through his practice of experimentation already acquired great respect for the facts it reveals. To a *Naturphilosoph* ignorant of facts, he ironically presented himself as an 'empiricist'[15] but on meeting a passionate defender of pure empiricism, he cited the intervention of the mind in the form of mathematics.[16] He recognized the danger of illusory explanations which the 'dynamical system' offered but he criticized the 'atomic system', not without reason, for being satisfied with attributing to the atoms the very properties of matter whose explanation was being sought.[17] In this regard, it is amusing to note his indignation at "the contempt for his country and the blind admiration for the foreign" of the German physicist Pfaff, who was disseminating the French ideas.

The great event of this trip, in Jena, in September of 1802, was Oersted's meeting with Ritter, a bizarre man, in whom the contradictions of the age were exasperated without the control of any criticism.[18] He pushed speculation to extravagance and indulged in qualitative experiments on which he hastily constructed new speculations. But it happened that in this manner he was the first to observe certain important properties of bodies, notably the parallel variation of the contact potentials of the metals and of their affinity for oxygen; and then the demonstration of the existence of ultraviolet radiation by its chemical action; properties that had to be 'rediscovered' by a more reliable method before they could be incorporated into science. Indeed, one can hardly blame the physicists of the period for being put off by the confusion of Ritter's writings. However, Oersted immediately proved receptive to a mode of thought which, in fact – all critical judgment aside – was similar to his own. Through the obscurity of expression, he perceived in Ritter's work on galvanism an argument whose rigor so fascinated him that he compared it to that of Euclid![19] A close friendship soon sprang up between the two young men (they were both 26 years old). Consequently, while staying in Paris the following year, Oersted became Ritter's interpreter among the French physicists presenting the latest results of the latter's research on voltaic electricity. In particular, Ritter had just constructed a 'Ladungssäule' (charge column) which was, in fact, a precursor of the storage battery. Biot was sufficiently impressed to suggest that Oersted invite his friend to compete for the grand prize of the Institute.[20] Oersted gave such clear explanation in French of the formless memorandum that Ritter had sent him that, the author himself declared that he "had learned a great deal from it".[21] Oersted subsequently assisted the commission of the Institute responsible for judging the work to duplicate all the experiments described. Unfortunately Ritter had decided to stress, as the most important phenomenon that his storage battery had revealed, the existence of an electrical polarization of the earth similar to its magnetic polarization: Oersted and the commissioners had no alternative but to conclude that this 'discovery' was illusory, and the imprudent candidate lost the prize. The latter was hardly discouraged by such a trifle,[21] but for Oersted, who had unhesitatingly supported the assertions of his friend,[22] the adventure was a healthy reminder to double his critical spirit, both as regards speculation and in the execution of experiments.

A short while after his return to Copenhagen, another lesson of the same sort awaited him, far more cruel this time as it involved him personally.[23] By the end of 1801, he had become an admirer of one of the most obscure and fantastic works inspired by *Naturphilosophie*, a system of speculative chemistry developed by a certain Winterl and modestly entitled *Prolusiones ad Chemiam Seculi Decimi Noni.* Encouraged by Ritter, Oersted prepared a paraphrase of it in German which was published in 1803. Chemists gave this unfortunate publication the reception it deserved: the reviews which appeared in 1804, in England and France as well as in

Germany, were merciless. "For the glory of the nineteenth century," we read in the *Annales de Chemie et de Physique* "it must be hoped that it will make haste to reject Mr. Oersted's offering and Mr. Winterl's chemistry". This 'thrashing' had no immediate effect on Oersted; his philosophical convictions apparently remained unshaken and the popularity of his courses was not affected; but his reputation did not fail to suffer somewhat. I will mention only one rather amusing repercussion: P. A. Heilberg, who was living in exile in Paris, had met Oersted there and had not taken kindly to him; after all, had this young man not dared one day[24] to argue against him that Danish laws were not iniquitous! Thus, having seen the review in the *Annales de Chimie*, the mordant satirist lost no time in communicating it to Rasmus Nyerup,[25] with most uncharitable comments: "Assuredly if these two gentlemen (Oersted and Winterl) said the idiotic things attributed to them, they each deserve a place in the mad houses of their countries. I will only mention two of these idiocies.... First, they say that the most correct way to define architecture is to call it 'musique congelée' [frozen music],[26] then they call the gods of mythology 'dcs cristallisations intellectuelles' [intellectual crystallizations]. I suppose that this Oersted must be the brother of the one who wrote in praise of the decree of 1799 which was intended to freeze reason in Denmark and turn it into an intellectual crystallization...."[27]

Oersted took the necessary time to convince himself of the inanity of Winterl's fantasies, using the method learned from the French scientists, that is – by careful experimentation. It is evident that by 1806, without abandoning the dynamic point of view he would never depart from, he had formed a more balanced idea of the nature of scientific theories; any theory implies a certain understanding of the laws of nature, that is, of that reason which rules nature; that the expression of this understanding is imperfect and cannot be handed down without correction from one generation to another is the fault of neither individuals nor of generations, but must rather be explained by the laws which govern the development of mental life on earth. He now spoke of speculative systems with detachment, as a stage passed on his own road toward knowledge; "At the time when we first met," he wrote to Oehlenschläger (who had just noisily renounced Steffens and the romantic school),[28] "you remember that my brother and I, through our own studies, had already become friends of the new philosophy and new poetry while you were still calling the

Schlegels impudent urchins. I remember very well that on my return from my trip you wanted to prove to me that the Schlegels were excellent, and that I didn't adore Schelling or Steffens fervently enough and considered them inferior to the great Fichte and the tremendously rich and fertile mind of Ritter. And now it is you who want to open my eyes to the weak sides of these men whom you admired so much: ... I am happy, as a man of science, to have progressed continuously, step by step, toward my goal...." What he says of Steffens a little further on shows that he had learned to relegate philosophical speculation to a more modest and proper role: "He combines many fortunate insights with an unusual range of knowledge, but without depth. If his insights are not taken for incontestable results of the most profound philosophy, but for what they are: flashes of light whose brilliance can guide us, one may learn much from him. This is what I have done and what I can continue to do without danger."

When Oersted wrote those lines, he was still thirteen years away from the discovery of electromagnetism: the truth of the last remark could not be better illustrated than by the history of that discovery. The idea of a relationship between electric current and magnetism, a corollary of the general concept of the unity of natural forces, was a commonplace of *Naturphilosophie*; but it had remained sterile until the moment when Oersted was put on the path to the decisive experiment. The experiment appears simple *post eventum*; but its success depended on a fundamental characteristic of the phenomenon – its symmetry in relation to the direction of the current – which went against generally accepted ideas: for, reasoning in terms of electrical and magnetic polarities, one would be led to expect the most marked effect in a perpendicular, not a parallel, position of the magnet and current. If Oersted was able to surmount the inhibiting influence of this inference, it is because he had been prepared for it by a long sequence of investigations embracing all the previously known phenomena which accompany an electric current; chemical reactions, emission of light and heat. Oersted's method was far from being comparable to the one that Faraday would soon demonstrate so brilliantly: he was still satisfied with qualitative experiments, and the theoretical constructions derived from them were burdened with imprecise hypotheses; but what must be stressed is that, at this time, he was subordinating theoretical reflection to experimentation. Only the general concept of the unity underlying the multiple manifestations of natural forces continued to determine the orientation of his thinking and gave his theoretical representations their dynamical character.

This dynamical point of view[29] implied, above all, a continuous distribution of forces in space. Given electrical forces of opposing polarity, the state of electrical neutrality results from a compensation of these opposing forces at each point. From this, Oersted formulated an original concept of electric current: if the equilibrium of electrical forces is broken, their 'conflict' causes a momentary separation of the opposing electricities; but as these tend to recombine, a local oscillation of charges is generated, and it is the succession of these local oscillations along the conductor which constitutes the current. As, on a large scale, a discharge is accompanied by a luminous spark, the tiny local discharges which succeed each other along the conductor are the origin of the vibratory movements, similar to sparks, which we perceive, according to their frequency, in the form of light or heat radiation. One of the numerous ideas tossed out by Ritter – that oscillatory movement is a universal characteristic of nature – may have influenced Oersted; but in its details, the development of his theory derives, not from *Naturphilosophie*, but from the sort of imagination that we observe in Newton with his appeal to the 'analogy of nature'. It is the same constructive imagination that would soon reappear in Faraday, but with infinitely more precision and rigor. The concepts Faraday developed belong to the dynamic tradition, like those of Oersted, but it would be difficult to establish a thread between the views of these two pioneers of electromagnetism. The concept of field of force dominates any dynamic theory, but it was Faraday who clearly distinguished it and made it the guiding concept that it still is for us. The creative power of Faraday had no need of models and, moreover, could only have achieved such a faithful representation of phenomena through direct contact with observation.

But what is the role of magnetism in Oersted's theory? By his own account,[30] he began to suspect, in the course of his 1819–20 winter lectures, that the magnetic effect of current would be emitted radially around the conductor, as heat and light are. It would then manifest itself most strongly on a magnetized needle parallel to the axis of the conductor – and not perpendicular to the axis, as had so often been tried in vain.

But even this liberating idea did not immediately produce the desired result; believing that the intensity of the magnetic effect would be greater if the caloric and luminous effects were also intense, he first performed the experiment with a thin platinum wire and obtained only a weak, irregular effect. Three more months passed before he had time to return to his experiments and then he soon recognized that large diameter conductors produced the best results. Thus the axial symmetry of the magnetic force of current, which provoked general surprise when it was announced, was for Oersted a property he had foreseen (or at least conjectured) and establishing it experimentally was, in his eyes, a confirmation of his theoretical views and even contributed to making them more precise.[31] He did not limit himself to demonstrating the phenomenon, but he varied his experiments to explore all its particulars. Like Ampère, he established that a closed circuit is set in motion by a magnet and itself behaves like a magnet with its two faces acting as poles. But all this work, like the dynamic theory which inspired it, evoked no response: Ampère's arrival on the scene gave them the *coup de grâce*.

Ampère appropriated the phenomenon discovered by Oersted and, with unheard-of energy and virtuosity, brought the formidable analytical apparatus of French physics to bear on it. He paid no attention to Oersted's dynamic concepts; indeed he would have been unable to enter into a mode of thought so foreign to his own. For him the problem was to reduce electromagnetic actions to the universal scheme of central forces; and the difficulty they presented was precisely to integrate the axial symmetry of these actions. He immediately circumscribed the problem by showing the equivalence of magnet and closed circuit, thus 'reducing' electromagnetic actions to forces between currents. He then needed sources and points of application for his forces; audaciously, he atomized the current, cutting it up (in his mind) into infinitesimal 'elements', and constructed a mathematical expression of the force between two elements of current, whose global effect corresponds in all cases to that indicated by experiment. In fact, simple experimentation, stripping the phenomenon of anything which might obscure it, hand in hand with mathematical analysis, one reinforcing the other with perfect rigor. In truth, when Ampère presented his result – the law of force between the elements of a current – as deduced 'uniquely' from experiment, he brushed aside the dose of abstraction involved in the very notion of the element of current. Later, when Kirchhoff wanted to extend Ampère's theory to conductors that are not wire shaped, he would have to introduce a continuous distribution of current density, preparing the way for a return to a dynamic description of electromagnetism.

Oersted could not compete with Ampère. When he met him for the first time during a trip through Paris, conversation proved difficult: "He is very awkward in a discussion", Oersted wrote to his wife;[32] "he can't seem to understand the arguments of others very well, nor present his own clearly; nevertheless, he is a profound thinker". Another interview, attended by several physicists, two of whom were Ampère's students, gave Oersted the opportunity to present his theory to them; according to him, they later agreed, in the end, that Oersted's theory like that of Ampère, accounted for all the phenomena. The difference, he reported, was that Ampère, "transferred the circulation I discovered from the conductor to the magnet."[33] The weakness of Oersted's position, which his hosts must surely have perceived but had the courtesy not to insist upon, is obviously his inability to leave behind the imprecision of the qualitative. In this he is inferior not only to Ampère, but also to Faraday, because the latter, as ignorant as Oersted of mathematical language, nevertheless managed to define and manipulate concepts with the precision of a mathematician. This is an essential aspect of French scientific thought which always remained foreign to Oersted, not that he did not appreciate its value, but because at the time of his first stay in Paris, he was too poorly prepared to assimilate its principles. The masters of mathematical physics are absent from the lengthy list of scientists with whom he associated for various reasons; at the most, he mentioned unenthusiastically one of Monge's lessons which he attended.[34] "One cannot say that he is extraordinarily talented. He spoke very mathematically about physics on the general properties of bodies. They do enormous amounts of mathematics at this school (the École Polytechnique), and chemistry plays only a secondary role, though they do it very well." "If I could wish something else for myself that would not be a complete transformation," he confided to Manthey,[35] "I would take Berthollet as my model because he has, *as a theoretician,*[36] enriched his country by discoveries which were important for factories and mills". With great insight, he thus defined himself.

Finally, what should we think of the religious feeling which so strongly and profoundly marks Oersted's attitude toward nature? It is clear that in this area, too, he drew away from the romanticism of *Naturphilosophie* very early to adopt a strictly rationalist attitude which brought him (perhaps without his knowing it) closer to Newton. In the construction of a rational science of nature, God plays the epistemological role of guarantor of the necessity of the laws of nature; for human reason which formulates these laws is an emanation of the divine reason which has established them. As long as the ideal of science was to arrive at a statement of the laws of phenomena which was independent of the manner in which the latter were observed, it was indeed necessary to have recourse to a system of reference transcending the set of observers: for Oersted as for Newton, this could only be God, the Creator, whose existence was, moreover, equally necessary to provide an incontestable foundation for the rules of morality and society.

Laplace, who judged the 'hypothesis' of God the Creator useless, but who wanted to keep the Newtonian ideal, also felt the need to refer to an 'intelligence' to whom he attributed more or less the same epistemological function as Newton's God, while depriving him of the attribute of spatio-temporal omnipresence. But he did not see the inconsistency of this secularization; to carry out its function, the intelligence in question would have to remain outside of the description of the universe of which it was supposed to be a part. At present we have taken another step in the elimination of useless hypotheses: we recognize that we have no other system of reference than this very set of observers, and that scientific language has no other guarantee of truth than the agreement among them. Oersted's hypothesis of a divine creator does not seem simply useless to us as it did to Laplace; it is an epistemological mistake.

NOTES

[1] du Bartas, G., *La Sepmaine ou Création du Monde* (1578), Cinquiesme Jour, v. 45–47.

[2] Lear, E., *Nonsense Songs, Stories, Botany and Alphabets* (1871); *More Nonsense* (1872); *Laughable Lyrics* (1877).

[3] This question is the object of a very thorough recent study by J. E. McGuire, 'Atoms and the "Analogy of Nature": Newton's Third Rule of Philosophizing', *Studies in the History and Philosophy of Science* **1** (1970), 3–58.

[4] Laplace, P. S., *Essai philosophique sur les probabilités* (4th ed., 1819), pp. 241–242.

[English translation by F. W. Truscott and F. L. Emory: *A Philosophical Essay on Probabilities* (New York, Dover, 1951) – Ed.]

[5] Lavoisier, A. L. and Laplace, P. S., *Mémoire sur la chaleur* (*Mém. Acad. Sc.*, 1780).

[6] On this point see H. Metzger, *La philosophie de la matière chez Lavoisier* (Paris 1935).

[7] Laplace, P. S., *Exposition du système du monde* (4th ed., 1813), p. 356.

[8] Schelling, F., *Ideen zu einer Philosophie der Natur* (1797).

[9] Williams, L. Pearce, *Michael Faraday, a Biography* (Chapman and Hall, London, and Basic Books, New York, 1965); Hermann, A., 'Der Kraftbegriff bei Michael Faraday und seine historische Wurzel', *Physik. Blätter* **26** (1970), 247.

[10] Schelling, F., *Über Faradays neueste Entdeckung* (1832).

[11] *Nachrichten über Leben und Schriften des Herrn Geheimraths Dr. Karl Ernst von Baer, mitgetheilt von ihm selbst* (St. Petersburg 1866), pp. 289–290.

[12] *Ibid.*, p. 184.

[13] Hauch, C., 'Hans Christian Ørsteds Levnet', *Samlede og efterladte Skrifter af H. C. Ørsted* **9** (1852), 109–194, cf. p. 120.

[14] *Breve fra og til Hans Christian Ørsted*, Første Samling (Copenhagen 1870). This collection will be cited as *Breve I*.

[15] *Breve I*, p. 82.

[16] *Breve I*, pp. 58–59.

[17] *Breve I*, pp. 54–55.

[18] Ritter is the subject of a very complete study by A. Hermann, accompanied by an anthology: Johann Wilhelm Ritter, *Die Begründung der Elektrochemie* (Ostwalds Klassiker der exakten Wissenschaften, neue Folge, Vol. 2, Frankfurt a. Main, 1968).

[19] *Breve I*, pp. 77–80.

[20] *Breve I*, p. 137.

[21] *Correspondance de H. C. Ørsted avec divers savants*, Vol. II (Copenhagen 1920), p. 55; see also p. 46.

[22] *Breve I*, pp. 155–156.

[23] A more detailed account of this incident is given by K. Meyer, 'Scientific Life and Works of H. C. Oersted' in H. C. Oersted, *Scientific Papers*, Vol. 1 (Copenhagen 1920), pp. xxvi–xxix, xxxiv–xxxvi.

[24] *Breve I*, p. 127.

[25] Smith, S. Birket, *Til Belysning af literaere Personer og Forhold i Slutningen af det 18de og Begyndelsen af det 19de Århundrede*, in Samling Breve (Copenhagen 1884), pp. 52–54.

[26] In French in the original text.

[27] Allusion to an article by A. S. Oersted defending the decree on the limitations of freedom of the press.

[28] *Breve I*, pp. 226–227.

[29] K. Meyer's memoir [note 23] contains a detailed description of Oersted's research; see particularly pp. xlv–lx and lxvii–ci.

[30] *Ibid.*, pp. lxix–lxx.

[31] This is what Oersted states at the end of the note announcing his discovery; see *ibid.*, p. xciii.

[32] Collection cited in note 14, Anden Samling, p. 50.

[33] *Ibid.*, pp. 65–66.

[34] *Breve I*, p. 126.

[35] *Breve I*, p. 161.

[36] Underlined in the original text.

CHAPTER 16

THE FIRST PHASE IN THE EVOLUTION OF THE QUANTUM THEORY

[1936b]

In its universality and continuity the development of science appears to contrast sharply with the apparently chaotic fluctuations of other human activities; but this is indeed only a superficial contrast because the relative simplicity of scientific problems is produced by defining the idealizations which form the proper domain of science through abstraction from an infinity of circumstances which are, on the contrary, essential to other points of view. An historical perspective shows us clearly that, indeed, the progress of science is an integral part of man's constant work of adaptation to constantly changing conditions of existence. Scientific theories are systems of concepts and representations which are utilized at each stage of development to classify and describe the totality of our experience as economically as possible.[1] Each new acquisition of science appears historically as the end result of a crisis provoked by evidence of an essential discordance between a given theory and a newly discovered group of phenomena; it consists precisely, in the final analysis, in a modification of our mode of description of phenomena, designed to adapt the latter to the new situation. Hence, the meaning and significance of a theory can only be fully understood on condition that it be considered not only in its final structure, as a method of describing certain phenomena, but as well in its entire evolution as an intellectual endeavor toward greater harmony. Thus, the history of science has an important function to fulfill, as a source of information for the theory of knowledge. Taking for our example the introduction of the idea of the quantum of action into science, it is from this point of view that we will try, in the following pages, to analyze in detail the birth of a theory from the moment when the crisis breaks out until the moment when the road to its solution is finally opened.

I

From 1895 to 1899 Max Planck, in a series of papers which offers a "beautiful example of scientific concentration",[2] systematically developed the thermodynamics of radiation on the basis of the electromagnetic theory of light. He thus solved a problem which was the natural culmination of the great stream of ideas arising from the enunciation, toward the middle of the nineteenth century, of the two general laws of thermodynamics. It will be useful to glance first at the development of these ideas.

The thermodynamics of purely material systems, that is, those for which one can abstract from energy exchanges by means of radiation, from its inception was to be fatally bound up with the 'mechanistic' and atomistic trend then flourishing. The question of the mechanistic interpretation of the properties of matter, which had thus reached maturity, was attacked with great energy by Maxwell and Boltzmann.[3] The first phase of their work culminated with the announcement of the 'H-theorem' in the following representation of the thermodynamic law of the increase of entropy: a system of molecules undergoing mutual collisions tend constantly, by virtue of the laws of motion, toward a well-determined equilibrium distribution, as long as these collisions always satisfy a certain condition interpreted by Boltzmann as a reflection of the 'molecular disorder'. But since the hypothesis of the permanence of molecular disorder was in contradiction with the reversibility of the equations of dynamics, the interpretation of the H-theorem had to be modified in a statistical sense: The monotonic increase of entropy only represented the probable behavior of its variation for a determined system; the apparent irreversibility being explained, for systems with an immense number of degrees of freedom, by the rarity of notable fluctuations from this probable behavior. These profound considerations were based on the study of mechanical systems of a so-called 'ergodic' character, whose compatibility with the equations of motion could, in truth, not be demonstrated;[4] a fundamental property of these systems is expressed by the 'energy equipartition theorem', by virtue of which each translational or rotational degree of freedom of the system makes an equal contribution to the specific heat, independently of the temperature, and each vibrational degree of freedom makes a contribution double that of the others. However, the experimental evidence bearing on this theoretical prediction, which went beyond phenomenological thermodynamics, was divided. In the area of high temperatures, one could obtain remarkable agreement with experiment by applying the theorem to a model with a solid body consisting of atoms executing simple vibrations around equilibrium positions or to a gas formed of molecules with translatory motion and, possibly, axial rotation. But far from being independent of the temperature, the specific heats of solids diminished rapidly at low temperatures, as if a part of the atomic vibrations "became ankylosed", that is, ceased to exchange energy with the others. Concerning gases, the most direct interpretation of their characteristic line spectrum consisted of regarding them as the result of some vibratory motion of the molecules, necessarily with a considerable number of degrees of freedom; but this interpretation was impossible to reconcile with the equipartition theorem unless one also admitted that these vibrations, corresponding to frequencies in the visible or ultraviolet ranges, became ankylosed even at high temperatures. Faced with this situation, most physicists (Lord Kelvin was a notable exception) saw in the partial confirmation of the equipartition theorem a sufficient justification for the statistical viewpoint, that is, they were disposed to attribute the difficulties of the theory of specific heats not to an incompatibility of the equipartition theorem with the principles of classical mechanics, but rather to the inadequacy of our representation of the atomic structure of bodies. This attitude is indicative of the value which the physicist, consciously or not, attributes to a theory. Knowing that it concerns extremely schematic idealizations, he would not think of requiring it to be universally valid; he modestly declares himself satisfied if he finds that the theory manages to explain a certain number of empirical relationships other than those which serve directly as its point of departure; thus a fruitful theory appears to be simply a mode of condensed description of a more or less limited area of experimentation; and the fewer its postulates and the broader its domain of applicability, the more fruitful and harmonious it is.

Parallel to the development of statistical mechanics, the thermodynamics of radiant energy, inaugurated by the work of Kirchhoff, was carried to a high degree of perfection by Boltzmann and Wien.[5] Boltzmann was one of the rare continental scientists who recognized immediately the import of Maxwell's views on the electromagnetic nature of radiation, and applied them for the first time to a thermodynamic problem; indeed, he based his deduction of Stefan's law on the notion of radiation pressure taken from electromagnetic theory. Subsequently, on the basis of the laws of reflection of light by a moving mirror as given by Maxwell's theory, Wien studied in detail the modifications undergone by a monochromatic beam of radiation contained in an enclosure with reflecting walls during a variation in the volume of this enclosure. He was thus able to extend the definition of the concept of temperature to such a ray, the equilibrium of the radiation in an isothermal enclosure being characterized by the equality of the temperatures of all the monochromatic beams of which it is composed and of the material bodies with which it interacts. By considering an adiabatic compression of the radiation, he obtained a fundamental property of the universal function of frequency and temperature giving the distribution of the energy density of black-body radiation per unit interval of wavelength: when the temperature varies, the quotient of this distribution function divided by the fifth power of the temperature depends only on the product of the wave length and the temperature. This 'displacement law', as it was later called, thus reduced the quest for the equilibrium distribution to the determination of a function of a single argument; one could, for example, limit himself to the study of monochromatic beams of a given frequency and study the way in which their energy depends on thc temperature.

Thermodynamic methods combined with the optical properties of beams of radiation could, however, lead no further than the displacement law. To complete the solution of the problem, it was necessary to analyze the energy exchanges between the beams of radiation and material bodies, in order to relate the desired distribution function to the known thermal properties of these bodies. With regard to the structure of the latter, it could be chosen arbitrarily, as Kirchhoff had already shown, the only condition being that the absorption coefficient must be different from zero for all frequencies. The first rational attempt in the direction just indicated was undertaken by Wien[6] in 1896 as an immediate continuation of his thermodynamic studies. However, Wien did not attempt to treat the mechanism of absorption and emission of radiation in detail; he limited himself to making the least arbitrary possible hypothesis for this mechanism which seemed to him likely to lead to an acceptable distribution function. Using an earlier idea of Michelson's, he tried to formulate this hypothesis in order to bring the sought-for distribution function closer to Maxwell's formula for the distribution of molecular speeds of gas in thermal equilibrium. He thus took as the material body a gas at temperature T and assumed that the frequency and intensity of the radiation emitted by a molecule of speed v depends only on this speed. It followed immediately that the density of radiant energy[7] $u(\nu, T)\,d\nu$ in the frequency interval $(\nu, \nu+d\nu)$, being proportional to the number of molecules with a certain speed v determined by the frequency ν, necessarily had the form:

$$u(\nu, T) = F(\nu)\, e^{-f(\nu)/T}.$$

On the basis of this relationship, the complete determination of $u(\nu, T)$ was achieved by simple application of the displacement law, taking additionally into account the condition imposed by Stefan's law on the way in which the total energy density $\int_0^\infty u(\nu, T)\,d\nu$ must depend on the temperature. Thus Wien obtained the law:

$$(1)\qquad u(\nu, T) = b\nu^3 e^{-a\nu/T},$$

depending on two universal constants a and b. Paschen's provisional measurements appeared in acceptable agreement with the form of this law, which accounted, in particular, for the existence of a maximum in the spectral distribution. But as simple as it was, Wien's hypothesis could only be considered a provisional construction supporting a happy intuitive guess; for it would hardly have been possible to relate it to the laws of radiation of ordinary molecular models. The value of Wien's work was rather of a methodological order and, very importantly, it revealed for the first time the power of the displacement law.

II

This is the stage at which Planck found the problem when, pursuing the general line of his research, he was quite naturally led to study it.[8] From the start of his career Planck had, in effect, devoted himself, with the methodical application which is the dominant feature of his scientific activity, to the study of the foundations and consequences of the second law of thermodynamics. After having made important contributions in the applications of this law to physico-chemical questions, he turned to a still incomplete part of the vast theoretical edifice, which he rightly judged ripe for more profound examination. The electromagnetic theory of light was then solidly established as a consequence of Hertz's experiments and the rapid development of the new technique of wireless telegraphy, and moreover, the experimental determination of the spectral distribution of black-body radiation was being actively pursued and could provide both a check and a guide for theory. In all his earlier work Planck had followed a rigorously thermodynamic method, carefully dividing his results into those which could be deduced solely from the two laws and those requiring supplementary hypotheses. But this very rigor permitted him to realize the limitations as well as the value of this method. Hence, he vigorously fought[9] the exaggerated reaction of the energetics school, a reaction against the mechanistic tendency which had dominated scientific thought in the nineteenth century, and fully recognized the importance of the statistical viewpoint in thermodynamics. He saw a close analogy between the problem of the spectral distribution of radiation, which exceeds the scope of purely thermodynamic reasoning, and that of the thermal equilibrium distribution of material bodies, which was resolved by statistical mechanics. His systematic mind immediately went on to a precise statement of the general program whose execution should, according to this analogy, give him, in particular, the form of the function $u(\nu, T)$. It was a matter "of understanding, and if possible demonstrating the second law of thermodynamics in its application to radiation from a purely electromagnetic point of view": in summary, to establish, on the basis of Maxwell's equations which govern the propagation of electromagnetic energy and are thus a counterpart of the laws of molecular motion, the analogue for radiation of the H Theorem in statistical mechanics; as the latter gave an unambiguous expression for the equilibrium distribution of mechanical systems, a similar result could be hoped for in the case of radiation.

Nevertheless, in the beginning, Planck[9a] misunderstood the true character of this analogy. He believed for some time that the very mechanism of emission and absorption of radiation by a source of electromagnetic oscillations could, because of the decay of the vibrations of the source, present an asymmetry implying the irreversibility characteristic of the principle of increase of entropy. He thus undertook a meticulous analysis of this mechanism; moreover, profiting from Kirchhoff's law, he took as his starting point a particularly simple model of a radiation source, a resonator of the type studied by Hertz, which is formed by a dipole oscillating harmonically at a given frequency. But, aided by Boltzmann's criticism, he soon realized his error: like the laws of mechanics, Maxwell's equations are invariant with respect to the sign of the time, and the source of the second law must be sought outside of them. For a mechanical system it is the hypothesis of molecular disorder which gives the H Theorem its statistically irreversible character; the problem which then arose was that of finding a statistical element for radiation comparable to the molecular disorder of material gases. Wien, as we have seen, placed this statistical element in the complexity of the material system in equilibrium with the radiation, a system which he supposed to be composed of a great number of molecules in thermal agitation. But, of course, this could certainly not be the essential point, since it was obviously possible to imagine logically admissible models of material bodies that were not composed of a large number of interacting particles; for example, it would suffice to take for each frequency a resonator oscillating at this frequency, and to suppose the centers of gravity of all these resonators immovable, the exchanges of energy between the resonators and a reservoir at a fixed temperature taking place in any manner independent of the radiation, for example by the mediation of mobile particles transmitting energy by collisions. Hence, it was necessary to attribute a disorganized character to the very structure of the thermal radiation. It is one of Planck's most remarkable scientific contributions that he clearly posed this problem, and that he resolved it by showing that the ordinary concept of the imperfectly monochromatic beam of radiation involved sufficient indeterminacy to allow room for statistical considerations, when it was envisaged as the superposition of an infinite number of electromagnetic waves with well-determined frequencies and different amplitudes and phases. From this point on, the disorder of thermal radiation could be characterized by the hypothesis of 'natural radiation', in complete analogy with the hypothesis of molecular disorder.

Having formulated the mathematical conditions which express the character of natural radiation, Planck applied the results of his preceding study to the exchanges of energy between an oscillator of fixed frequency ν and a beam of this natural radiation in resonance with it, that is, as the analysis itself shows, a beam formed of waves in a frequency interval $(\nu, \nu+d\nu)$. In the case where this radiation is non-polarized, isotropic and of constant intensity, the desired fundamental relationship is obtained between the energy density $u(\nu)$ of the radiation and the average energy U

of the resonator at frequency ν in equilibrium with it:

(2) $$u(\nu) = R(\nu)\, U,$$

with

(3) $$R(\nu) = \frac{8\pi\nu^2}{c^3}.$$

Having reached this point, in order to complete his reasoning, i.e. to find, in the case of thermal equilibrium, how $u(\nu)$ depends on the temperature T, Planck had only to combine formula (2) with the relationship which expresses the energy of the resonator as a function of T. The way which seems quite naturally to suggest itself here would consist of applying the equipartition theorem to the system of resonators maintained at temperature T. But Planck, instead of having recourse to this theorem, proceeds by a detour which, as we shall see, deviates from it radically. Drawing on the methods of pure thermodynamics, of which he had perfect mastery, he initially sought to define the entropy density $s(\nu)$ of the radiation as a function of the energy density $u(\nu)$ as well as the entropy S of a resonator of frequency ν and energy U, in order to deduce from the variations in the total entropy of the radiation and of the system of resonators the existence and form of the equilibrium distribution of radiant energy. It is clear that in choosing the functions $s(\nu)$ and S, Planck must have been guided by Wien's law (1) which he knew suitably represented the experimental results. If this expression was theoretically correct, the sought for entropy density had to satisfy the equation

$$\frac{\partial s}{\partial u} = \frac{1}{T} = -\frac{1}{a\nu} \log \frac{u}{b\nu^3},$$

from which

(4) $$s = -\frac{u}{a\nu} \log \frac{u}{eb\nu^3},$$

where e designates the base of Naperian logarithms. Similarly from relationships (2), (3) and

$$\frac{\partial S}{\partial U} = \frac{\partial s}{\partial u},$$

which are valid for a resonator in equilibrium with radiation, there resulted for S an equation of the form

$$(5) \qquad S = -\frac{U}{a\nu} \log \frac{U}{eh\nu},$$

where the universal constant h is related to the constant b by the relationship

$$(6) \qquad b = \frac{8\pi}{c^3} h.$$

It was natural to investigate, first, whether expressions of this form were capable of playing the role of entropy or whether they had to be modified, which would have implied the need to correct Wien's law. Thus Planck started from expressions (4) and (5) postulated *a priori* as definitions and proved that by virtue of the laws of energy exchanges between resonators and natural radiation, the variation of the total entropy in the course of time indeed cannot be negative and that the maximum of this entropy corresponds exactly to Wien's law for radiation as well as to the relation

$$(7) \qquad U = h\nu e^{-a\nu/T},$$

giving the average energy of a resonator of frequency ν as a function of temperature. Hence definitions (4) and (5) of the entropies $s(\nu)$ and S were acceptable. But were they the only ones possible? Planck gave an affirmative answer to this essential question, but he could not (and for good reason) back it up with a strict proof. He simply said that he had tried in vain to modify the definition of the entropy of the resonator so that it would still satisfy all the conditions imposed by thermodynamics and electromagnetic theory. One could, for example, think of a generalization of the type

$$(8) \qquad S = -\frac{U}{G(\nu)} \log \frac{U}{g(\nu)}.$$

But Stefan's law and Wien's displacement law impose restrictions on the functions $G(\nu)$ and $g(\nu)$ which specifically limit the new expression to the particular form (5). Thus Wien's law seemed to be exactly like that of Maxwell-Boltzmann for material systems, a necessary consequence of the general law of the increase of entropy.

At the very moment when Planck was arriving at this conclusion, new measurements by O. Lummer and E. Pringsheim, going to longer wavelengths than the earlier observations of Paschen which corroborated Wien's law, on the contrary, showed serious divergences between this law and experiment. Planck's reaction to this new element of the problem is interesting because it is typical of the way in which the scientist uses theory as a guide to determine his attitude toward experiment. Planck immediately recognized the extremely grave nature a divergence of the type announced by Lummer and Pringsheim would have if it were indeed established; indeed, it was a question not of some phenomenon whose interpretation could depend on particular models offering a certain latitude, but of a fundamental law which, in his view, was connected directly and apparently without any arbitrariness to the very foundations of thermodynamics and electromagnetism. Rather than admitting that the results of new measurements, still contested by Paschen, really revealed the existence of a weakness in these theories, Planck was initially inclined to attribute the divergence rather to some source of experimental error. He would no doubt simply have continued to hold this reserved attitude if he had been completely sure of the rigor of his theory. But he knew his reasoning had a weak point and so, before making a decision, he conscientiously returned to the problem of determining whether his definition of the entropy of a resonator was really unique.[10] To decide this point, he first investigated what restrictions the law of its increase imposed on the way in which the entropy depends on the energy. To do this, he started with a state of the system in which the radiation is stationary, but where the energy of the resonator deviates from its average value U by a small quantity ΔU. With the help of the laws of natural radiation, it is not difficult to put the variation $\mathrm{d}\Sigma$ of the total entropy during the time $\mathrm{d}t$ in the form:

$$\text{(9)} \qquad \frac{\mathrm{d}\Sigma}{\mathrm{d}t} = \frac{3}{5}\frac{\mathrm{d}^2 S}{\mathrm{d}U^2}\frac{\mathrm{d}U}{\mathrm{d}t}\Delta U.$$

As ΔU and $\mathrm{d}U/\mathrm{d}t$ always have opposite signs, it follows from the law of increase of entropy that necessarily

$$\text{(10)} \qquad \frac{\mathrm{d}^2 S}{\mathrm{d}U^2} = -f(U),$$

the function $f(U)$ always being positive. But it is impossible to go further by utilizing only the law of the increase of entropy; in fact, by repeating the calculations in inverse order, one can even verify that taking for $f(U)$ any essentially positive function, this law is always satisfied. Such a conclusion directly contradicted Planck's prior opinion according to which Wien's law had to be a direct consequence of the second law. Nevertheless, Planck still believed it possible to demonstrate Wien's law by another consideration deriving from formula (9). He took it as evident that, for a system of N resonators, the variation of the total entropy $d\Sigma/dt$ can only depend on the quantities dU/dt, ΔU and $f(U)$ for this system and that this functional dependence is then given as well by formula (9). However, as this variation of entropy is necessarily equal to N times that of a single resonator, a functional relationship is obtained for $f(U)$ whose solution is:

$$(11) \qquad f(U) = \frac{\alpha(\nu)}{U}.$$

The value of the entropy corresponding to this determination is again of the form (8), which leads to Wien's law. As can be seen, this new demonstration made the relationship of Wien's law to thermodynamics seem less close than Planck first believed and even left it open to some doubt. But if Planck at this time still exaggerated the theoretical significance of Wien's law, he at least had the great merit of insisting strongly on the fundamental nature of the distribution law for black-body radiation in relation to the general laws of thermodynamics and electromagnetic theory.

III

While the German physicists attacked the experimental and theoretical study of the problem with much energy and meticulous care, the great artist of science who continued the tradition of Maxwell in England was dealing with the question on his part by an extremely original method inspired by his fundamental work on wave motions. In the two-page note[11] in which he sketched the application of this method to radiation, Lord Rayleigh did not solve the problem, but he mastered it superlatively and straight off completely uncovered the nature and depth of the difficulties that the validity, even approximately, of Wien's law implied for statistical and electromagnetic theories. In effect, he showed that these theories necessarily led to an entirely different law and indicated that this inconsistency represented a new case where the equipartition theorem did not apply to a system of oscillators: "Speculation upon this subject," he said, "is hampered by the difficulties which attend the Boltzmann-Maxwell doctrine of the partition of energy. According to this doctrine every mode of vibration should be alike favoured; and although for some reason not yet explained the doctrine fails in general, it seems possible that it may apply to the graver modes."

The method, at once simple and rigorous, which permitted him thus to reach the heart of the question rests on the following considerations. An enclosure with reflecting walls which, for reasons of simplicity, we will assume to be cubic, admits standing electromagnetic waves with only a denumerable number of modes of proper vibration, each with a well-determined frequency. Any superposition of such proper vibrations will maintain itself without change in the enclosure, so that if by means of a material body capable of absorbing and emitting all the proper frequencies, the total energy initially present is redistributed among these modes in conformity with the law of black-body radiation; this equilibrium distribution will then subsist indefinitely, even after removal of the material body. To calculate the energy density $u(\nu, T)\,d\nu$ contained in frequency interval $(\nu, \nu+d\nu)$, it thus suffices to know the quantity of energy $U(\nu_i)$ which, in the case of thermal equilibrium, is possessed by each proper vibration of frequency ν_i and, to sum per unit volume all the $U(\nu_i)$ for $\nu<\nu_i<\nu+d\nu$. Hence the equipartition theorem permits $U(\nu_i)$ to be determined precisely without it being necessary to enter into the details of the energy exchange mechanism; in fact, it can be applied equally well to the proper vibrations of radiation as to those of any mechanical system, and it may be deduced from this that each proper vibration must have an average energy $U(\nu_i)$ proportional to the temperature and independent of the frequency. Moreover, a very simple enumeration of the number of proper vibrations per unit volume in the frequency interval $(\nu, \nu+d\nu)$ leads to an asymptotic value

$$(12) \qquad N(\nu)=\frac{8\pi\nu^2}{c^3};$$

the result for the distribution being sought is:

$$(13) \qquad u(\nu, T) = N(\nu)\, kT.$$

The meaning of the universal constant k may be taken from the statistical theory of mechanical systems; it is, as we know, the perfect gas constant per molecule. Its appearance in the theory of radiation thus offers a means of obtaining the absolute values of molecular quantities by measuring the black-body spectrum. Nevertheless, it is clear that 'Rayleigh's law' expressed by formula (13) could not be valid for all frequencies since, by reason of the infinite number of radiative degrees of freedom, it would imply an infinite total density for radiation. In addition, it does not take into account the experimentally observed maximum of the function $u(\nu)$. We can see that in this new example, as in the other cases described above, the failure of the equipartition theorem is due to the fact that in the actually observed equilibrium distribution, the high frequency vibrations receive less energy than required by equipartition. The close relationship between those different problems was well recognized by Lord Rayleigh; on another occasion[11a] he wrote concerning this problem: "A full comprehension here would probably carry with it a solution of the specific heat difficulty". In the case of radiation, the discordance is particularly serious because it affects, directly and independently of all hypotheses on the structure of matter, the theory of the pure electromagnetic field, combined with a completely general consequence of the statistical theory of systems governed by Hamiltonian-type equations.

But Rayleigh was not satisfied with establishing this catastrophic contradiction. He presented a positive suggestion which is not only highly valuable from the point of view of interpreting the experiments, but which also touches a fundamental aspect of the theoretical difficulties. We have seen that in the theory of specific heats the failure of the equipartition theorem was only partial and that its consequences seemed to be well verified as long as the frequencies of vibration were not too high or the temperature too low. Now, Rayleigh's formula, which satisfies the displacement law, depends specifically on the ration ν/T. Would it not then be possible that it applied to that part of the spectrum corresponding to sufficiently small values of the parameter ν/T? In this case it would produce a spectral distribution going like ν^2, essentially different from Wien's law which in this case goes like $b\nu^3$.

The critical part of Rayleigh's note could not have an immediate influence on the development of the question; indeed, it brought out the difficulties of the problem rather than indicating a way toward the solution. Thus several years had to pass before the importance of the point of view indicated therein was fully recognized. As for his suggestion regarding the form of the long wavelength part of the spectrum, this, on the contrary, received striking confirmation a short time after the article was published. New measurements[12] by Lummer and Pringsheim, Paschen and notably by Rubens and Kurlbaum revealed, in this region of the spectrum, systematic departures from Wien's law, exactly in the direction foreseen by Rayleigh.[13]

IV

Once the inaccuracy of Wien's law had been proven by these careful and consistent measurements, Planck[14] immediately returned to the investigation of his last proof, but this time with the intention of determining how the premises could be modified to obtain a result which agreed better with the experiments. He continued to find solid ground in formula (10), but was forced to renounce the misleading obviousness of the considerations which led him to the specialization (11) of function $f(U)$, which gives with changed sign, the second derivative of the entropy S of a resonator as a function of its energy U. Then he simply began to try arbitrary (positive) expressions for $f(U)$ and to study their consequences. Among these expressions, the following held his attention:

$$(14) \qquad f(U)=\frac{\alpha(\nu)}{U[\beta(\nu)+U]};$$

"It is," he said, "by far the simplest of all the expressions which give a logarithmic function of U for S (a form suggested by the calculus of probabilities) and which, in addition, for small values of U reduces to the expression corresponding to Wien's law." From formula (14) combined with formula (2) and the displacement law, one obtains by a first integration the expression

$$(15) \qquad U=h\nu(e^{a\nu/T}-1)^{-1},$$

and by means of (6) the distribution law

$$(16) \qquad u(\nu, T) = b\nu^3(e^{a\nu/T} - 1)^{-1},$$

which indeed reduces to Wien's law for large values of $a\nu/T$ and which accounts perfectly for all the observations. In this connection we may note that the essential correction this formula brings to Wien's law is precisely the one suggested by Lord Rayleigh since for small values of $a\nu/T$, function (16) takes the form $(b/a)\, \nu^2 T$ of Lord Rayleigh's law (13). But what attracted Planck even more than agreement with experiment was the elegant form of expression (16) and especially of the one for the entropy of S of the resonator which is obtained by another integration:

$$(17) \qquad \frac{a}{h}S = \left(1 + \frac{U}{h\nu}\right) \log \left(1 + \frac{U}{h\nu}\right) - \frac{U}{h\nu} \log \frac{U}{h\nu};$$

indeed he hoped that in allowing himself to be guided by this form, he would be led to probability arguments destined in his mind to connect the definition of the entropy of the resonator thus empirically discovered to the general principles of statistical mechanics.

In fact, less than two years later – "nach einigen Wochen der angespanntesten Arbeit meines Lebens", as he himself said[14a] – Planck[15] presented to the German Physical Society a statistical derivation of formula (17) in which there appeared, in an as yet rather obscure manner, the new notion of an "element of energy". In order to apply the statistical method, Planck considered a system composed of N identical resonators, interacting through natural radiation of the same frequency, the element of statistical disorder being introduced just by the 'natural' character of radiation, which implies a random distribution of phase and amplitude for each resonator. Consequently, to a state of the system of total energy U_N, equal to N times the average energy U of one resonator, there corresponds an entropy $S_N = NS$ which Planck, following Boltzmann's example, set equal to k times the logarithm of the 'probability' W of the state under consideration. Boltzmann had, indeed, shown that for a mechanical system the definition of entropy

$$(18) \qquad S_N = k \log W$$

satisfied all the conditions imposed by thermodynamics, if the probability W was defined as the number of 'complexions' or *a priori* equiprobable microscopic states giving rise to the same macroscopic state of the system. Consequently, for Planck, the whole problem was reduced to defining *a priori* equiprobable complexions for his resonators, and calculating the number W by means of combinational analysis. Naturally,[16] if the resonators were to be treated as systems governed by the laws of classical mechanics, the definition of the complexions could not be chosen arbitrarily, and for an ergodic system one was led directly to an expression for W implying the equipartition theorem. For a resonator, one would then have

$$U=kT, \tag{19}$$

which, according to (2), as the comparison of expressions (3) and (12) shows, would lead directly to Rayleigh's law (13). It was precisely at this point that Planck took a decisive step which was irreconcilable with classical theory; he introduced a new definition of the complexion for a system of N resonators and postulated that it was the complexions thus defined which were *a priori* equiprobable and which consequently determine W and S_N. He saw in this postulate another more precise way of expressing his hypothesis of natural radiation. As for the choice of the complexions which he had to adopt to recover formula (17), it was suggested to him directly by the latter, or rather by the one for S_N which can be deduced from it:

$$S_N=\frac{h}{a}\left\{\left(N+\frac{U_N}{h\nu}\right)\log\left(N+\frac{U_N}{h\nu}\right)-\frac{U_N}{h\nu}\log\frac{U_N}{h\nu}-N\log N\right\}.$$

Indeed, if we put

$$\frac{h}{a}=k \tag{20}$$

and

$$U_N=P\cdot h\nu, \tag{21}$$

we can put this expression into the form (18) with

$$W=\frac{(N+P)^{N+P}}{P^P N^N}\approx\frac{(N+P)!}{P!N!}\approx\frac{(N+P-1)!}{P!(N-1)!};$$

Now, this latter value of W is none other than the formula of combinatorial analysis giving the number of distinct distributions of P objects into N separate groups.[17] Taking into account the meaning of P given by (21), one is led to the following definition: A complexion of the system of N resonators is defined by a distribution among these N resonators of a certain number P of elements with energy ε, determined by the ratio of the total energy of the system divided by this elementary quantity ε. Formula (18) then gives for S the value:

$$S=k\left\{\left(1+\frac{U}{\varepsilon}\right)\log\left(1+\frac{U}{\varepsilon}\right)-\frac{U}{\varepsilon}\log\frac{U}{\varepsilon}\right\},$$

and the displacement law then imposes on ε the form $h\nu$. Thus formula (17) is obtained, and the distribution law (16), with relationship (20) among the constants. As Planck noted, this relationship allows molecular quantities to be determined on the basis of the measurements of black-body radiation. Moreover, from (20) and (6) one deduces $b/a=(8\pi/c^3)k$, which shows that Planck's formula reduces precisely, and not just up to a factor, to that of Rayleigh, as $\nu/T\to 0$.

The immense value of Planck's analysis lies in his having thus established, by methodically and tenaciously exploring the theoretical possibilities, the absolute necessity of introducing this strange division of energy into elements of finite size in order to describe the experimental observations. It is remarkable to note that he was led to this conclusion virtually in spite of himself: when he began his work, he believed he was putting the final touches on the construction of a theory uniting all the laws of radiant energy on the solid foundation of classical theories; and what should happen but that on reaching the end of his task, he saw himself finally constrained to abandon this very foundation which he recognized as too narrow. Though he had not yet assimilated the profound views of Lord Rayleigh, Planck fully recognized the considerable import of this renunciation; he knew by virtue of Kirchhoff's law that the fragmentation of energy into finite elements could not be a peculiarity of his resonators, but had to correspond to a completely universal characteristic of the mechanism of interaction between material systems and the electromagnetic radiation field. He also noted that his way of defining the complexions still did not reach the core of the problem; it had to be possible to relate it to a property of the statistical weights of the states of vibration of a resonator. But, he added, it would be premature to push the investigation in that direction immediately. In fact, – and this is an oft-noted phenomenon[18] – Planck did not follow the new path he had opened to the end. He stopped first, seeking a diversion in work of a totally different nature; when he returned to the question five years later, it was still not to go forward, but rather to retrace, in a masterful volume,[19] the whole of the road already traveled. And in the impassioned discussions of the following years, we do not find him in the *avant-garde*, but, indeed, among those who were struggling to attenuate the radical character of the new ideas and to propose compromises.

V

It was not until the year 1905 that the energy elements introduced into science by Planck became an object of general attention among physicists. At first sight, this long delay may seem surprising; but it must not be forgotten that the problem of black-body radiation, despite its great theoretical and practical importance, really belongs to a somewhat remote area of physics, cultivated at that period only by a limited group of specialists. At the same time, progress in experimental technique was beginning to open up to research the inexhaustible field of atomic and radioactive phenomena; the abundance of newly obtained results and the need to put them in order provisionally demanded a concentration of efforts in that area. Moreover, it was difficult to judge the true significance of Planck's memoir only on the basis of his success in the one case of black-body radiation; and even the manner in which Planck expressed himself, stressing the parallel between his work and that of Boltzmann without sufficiently emphasizing the essential point where they differed, did not fail to somewhat mislead those who already realized, better than Planck, the flagrant contradiction produced by strict application of the equipartition theorem. This discomfiture was expressed by Lord Rayleigh in particular[20] and we will soon see another noteworthy example. In reality, the problem could not advance beyond the point to which Planck had so brilliantly brought it without a new impetus. This impetus was given in this memorable year 1905 by a brain in full creative activity which, in the space of a few months, conceived the concept of light quanta, predicted the existence of Brownian movement, and established the foundations of the theory of relativity.[21]

In his 1905 memoir, Einstein described, as he stated explicitly in the introduction, the train of thought which led him to adopt an absolutely revolutionary point of view on the nature of light, to which he was content to attribute an 'heuristic' character. His starting point was precisely the theory of black-body radiation. He began by proving the incompatibility of classical theory with experiment; by combining formula (2) established by Planck, for the equilibrium between radiation and resonator, with formula (19), expressing a consequence of classical statistical mechanics applied to the equilibrium between a resonator and a thermal reservoir. One directly obtains the law (13) which Lord Rayleigh had deduced in a different way. Einstein next noted that Planck's formula (16) indeed reduced to this law for long wavelengths, from which it follows that the determination of molecular quantities, through the constant k, on the basis of the black-body radiation spectrum is, in reality, independent of the divergence found between experiment and classical theory at the violet end of the spectrum; and that the satisfactory agreement found between the numerical values thus obtained for these quantities and other determinations could simply be interpreted as indicating the applicability of the classical theory in the limiting case $\nu/T \to 0$. In the other extreme case, $\nu/T \to \infty$, where this theory is, on the contrary, absolutely false, it is Wien's law (1) which gives a representation sufficiently close to experiment. Without worrying about Planck's formula, whose purpose was to reproduce the shape of the entire spectrum, Einstein consequently asked whether examination of Wien's simple limiting law would not allow him to obtain information on the character of the essentially non-classical properties of radiant energy which the existence of this law made it necessary to admit.

To this end, in the spirit of statistical theory, he considered instead of energy density, the equivalent expression (4) for the entropy density. From this formula one obtains for the total entropy S_ν of radiation of frequency ν and energy E_ν in an enclosure of volume V, the value

$$S_\nu = -\frac{E_\nu}{a\nu} \log \frac{E_\nu}{eb\nu^3 V},$$

so that the change of entropy when the volume of the enclosure goes *ceteris paribus* from V to V_0 is

$$(22) \qquad S_\nu(V) - S_\nu(V_0) = \frac{E_\nu}{a\nu} \log \frac{V}{V_0}.$$

This formula presents a remarkable analogy to that which expresses the corresponding change of entropy

$$S-S_0=kP\log\frac{V}{V_0}$$

for a perfect gas made up of P molecules. The statistical interpretation of this relationship, according to Boltzmann's formula, lies in the fact that the probability of finding these P independent molecules, at a given moment in a part V of the total volume V_0 is equal to $(V/V_0)^P$. The linking of these two formulas in the light of this statistical interpretation thus permits the following conclusion: "Monochromatic radiation of low density (in the domain of validity of Wien's formula) behaves from a thermodynamic point of view as if it consisted of independent energy quanta of magnitude *kay*," which equals, according to (20), $h\nu$. It is interesting to observe that it was a formal analogy with the properties of perfect gases which had already suggested the form of his law to Wien; but the analogy Einstein established touches another more profound aspect of the problem. This imaginative mind *par excellence* was guided not by the form but by the physical meaning of 'Boltzmann's principle'; thus, not only did he immediately reach the conclusion already recognized independently by Lord Rayleigh, that Wien's law, considered as an empirically given fact, implied the existence of a characteristic of radiation incompatible with classical theories, but he even contrived to put this characteristic, in the extreme case he was considering, into very striking intuitive form. It is a precious talent for the researcher who is exploring a new domain to be able to call to his aid images which, even if they be only approximate, permit him to profit in the study of new objects from his intimate knowledge acquired in more familiar terrain of the multiple relationships among phenomena and from modes of reasoning capable of leading to new relationships. It is this sort of linkage between a still unorganized group of phenomena and a better known domain which is characteristic of a discovery made from a fertile viewpoint. The end of Einstein's memoir offers a classical example of the way in which such a discovery suddenly illuminates a host of phenomena previously considered disparate but which now seem to come of their own accord to take their place in the newly enlarged framework.

Having established that the behavior of radiation in empty space could, to a certain extent, be compared to that of a discontinuous ensemble of quanta of energy, Einstein wondered whether the laws of the emission, absorption and diffusion of light did not also reflect the same discontinuous character. He discussed Stokes' rule, the photoelectric effect, and the ionization of gases by ultraviolet light from this viewpoint, three groups of laws particularly well adapted to bring out the fertility of the corpuscular image of radiation, while at the same time making it more precise. Indeed, one is led to picture light energy as emitted in a certain number of spatially localized quanta which propagate without dividing and can only be absorbed as a unit. As for the question of discovering what relationship this image might have with the wave representation of optical phenomena given by Maxwell's theory, Einstein hardly concerned himself with it. He was satisfied with suggesting that the waves could adequately describe the average behavior of a great number of quanta, but that the latter should play an essential role in the elementary phenomena of interaction of light with material particles. In this attitude there was not only the intellectual boldness of genius, but above all a mental step which necessarily accompanies all great progress in science. When experiment reveals a group of phenomena which do not fit into the framework of a theory, it is necessary, in order to describe them, to introduce concepts and laws contradictory to this theory; a radical rupture must then occur in the continuity of development, a rupture which no compromise can repair: in exploring the domain where the crisis arose, one is then forced to renounce old concepts resolutely and to pursue the new in all their consequences. Naturally this does not mean that the theory which has been found inadequate is entirely abandoned; the task of the period of systematization which follows that of exploration consists of precisely drawing the limits of its area of validity. The renunciation implied by this limitation is the price paid to acquire the freedom and the enrichment of thought given by the use of the new theory.[22]

VI

Einstein drew his first conclusions on the existence of quanta of radiant energy uniquely from Wien's law and independently of Planck's theory. He even believed at first that the foundations of Planck's theory were incompatible with this idea and it was only a year later[23] that he established the link between quanta of light $h\nu$ and the elements of energy used by Planck in his calculation of the number of complexions W in a system of N resonators. To obtain Planck's formula, starting from the relationship (2) between the average energies of the radiation and of the resonator in equilibrium, it was necessary to abandon the value kT for the latter, which was the result of the calculation of W carried out by the classical methods of statistical mechanics; it is in this way that Planck's way of determining W introduced an essentially non-classical modification to these methods. Einstein then recognized that this modification lay in attributing to the different values of the energy of a resonator statistical weights radically different from those demanded by classical theory. While, according to classical theory, all the continuous multiplicity of the possible values of energy were equiprobable *a priori*, the introduction of the energy elements $h\nu$ necessarily imposed a considerable restriction on the choice of these values, because for sufficiently high frequencies these elements became much greater than the corresponding average energies of the resonator, given by formula (15). Einstein first examined the question by considering the microcanonical ensemble of N resonators, and in a later work[24] he returned to this problem by the method of canonical ensembles. He showed that to get the value of W given by Planck, it was necessary to agree that only the denumerable series of values $nh\nu$ ($n=0$, 1, 2, . . .) should be counted with a non-zero weight, the same for each of them.[24a] That was precisely the answer to the question Planck had raised without finding a solution, as we saw at the end of Part IV, in reference to the deeper significance of the definition of complexion which he had proposed. In providing a precise answer to this question, Einstein on the one hand discovered the close relationship between Planck's theory and the hypothesis of the light quanta, thus providing the latter with new support and, on the other hand, he was clearing the ground for the difficult study of the logical coherence of the new postulates which would have to be undertaken. Apart from the non-classical hypothesis of the quanta of energy, it was in fact necessary to keep relationship (2) in all rigor, that is, to suppose that, at least for the average values of the energy, classical electromagnetic theory gave a correct result and this despite the fact that for high frequencies the average energy of a resonator became less than a quantum.

But in showing that the quanta of energy played such a fundamental role in the statistical mechanics of systems of resonators, Einstein saw the significance of his hypothesis broaden considerably.[25] Was it not, in fact, necessary to introduce the same modification for all the systems of oscillators which, for various reasons, entered into the interpretation of the thermal and optical properties of matter? Should one not, consequently, expect to encounter facts in this domain as well which contradicted classical statistical mechanics and which might find an explanation within the framework of the new theory? It is clear that, in pursuing this idea, Einstein must have been struck very early by the fact that the new expression (15) of the average energy of an oscillator of frequency ν precisely translates into mathematical language the anomalous behavior which we called earlier the 'ankylosis' of the oscillator, a behavior which moreover occurs at a lower temperature the lower the frequency. Quantitative analysis then showed him that thanks to the quantum hypothesis, the whole theory of specific heats was put right, at least to good approximation. Even more, an important theoretical relationship appeared between the behaviour of the specific heat of a solid and the proper frequencies of vibration given by its infra-red spectrum. Such success, confirming the profound views of Lord Rayleigh, definitively established the universal import of the new concept. This universality could doubtless have been predicted from the beginning because of the fundamental nature of the problem of radiation, which we have stressed repeatedly. But Planck's idea still had to be put in a form which was general and flexible enough to become a powerful theoretical instrument. Moreover, even after Einstein's contribution, it was still far from possible to speak of a solution which resolved the difficulties; from this moment, however, the quantum theory was able to develop vigorously. As always, this development took place in two opposite and equally necessary directions. On one side, the foundations of the theory underwent critical study from the point of view of their logical coherence and their degree of generality; at the same time, its sphere of applicability was explored by seeking to extend it further and to make it quantitatively more precise. Thus, after being convinced that the new element for describing and predicting phenomena was indispensable, one sought to understand it better, test its strength, and modify it further to increase its solidity and its scope.

Before entering into the study of these new efforts, let us stop for a moment to glance backward. It is perhaps at first astonishing that this omnipresent quantum was 'discovered' just through its statistical aspect where it operates in a seemingly more hidden manner than in other phenomena directly reflecting the individual quantum processes. But it must not be forgotten that the theory of radiant energy was distinguished by the universality of its laws, which offered precisely the possibility of certain conclusions regarding the inadequacy of classical theories. While the quantitative importance of the divergences from classical predictions made it impossible to doubt this inadequacy, at the same time the simplicity of the quantization of the harmonic oscillator, leading to quanta of equal energy, facilitated the isolation of the principal features of the new element which had to be introduced in the description of the phenomena. After the fact it is evident that, basically, the theory of specific heats of solids, for example, presents a similar simplicity; but here the simplicity is masked by the need to resort to a particular model of the structure of the solid which manifests itself especially in the occurrence of proper frequencies of vibration differing from one substance to another. Consequently, it is understandable that numerous phenomena which are immediately dependent on the universal constant *h* could remain unexplained for so many years that it was the study of pure radiation which provoked the decisive crisis.

The role of chance is no less hidden in the history of a scientific discovery than in that of other social phenomena; it may be reduced simply to the more or less accentuated variations which the individual qualities and human limitations of the prophets can give to the course of events. It is difficult to imagine two more different temperaments than those of the two creators of the quantum theory, and the contrast is all the more striking in this case since the point of departure for their research was basically the same: to compare the approximate empirical law of Wien to the consequences of Maxwell's theory and statistical mechanics. Planck is a product of the German scientific school of the second half of the nineteenth century: systematic, meticulous, he patiently tried to deduce the law from the theory; he recognized only gradually that the problem hid a fundamental difficulty; prudently he ventured to seek the solution and it was by pure power of abstraction that he found it. Einstein, who had a more profoundly assimilated knowledge of statistical mechanics, saw immediately that Wien's law was incompatible with the latter; by a wonderful leap of imagination, he went from there directly to a concept which was radically opposed to the classical view; without allowing himself to be stopped by this opposition, he actively investigated the consequences of his hypothesis, and it was only after assuring himself of its productivity that he became concerned with re-establishing the necessary liaison with the base he had temporarily abandoned. However, when it was a question of finding the essential relationship between the hypothesis of light quanta and the foundations of statistical mechanics, the two lines of attack, so different in appearance, complement each other well in the conquest of their common objective.

VII

This picture of the acquisition of a great advance in our knowledge of nature would lack a very human feature if, alongside of the enthusiastic efforts of pioneers like Planck and Einstein, the crisis generated by the problem of black-body radiation had not also provoked a reactionary attitude in other minds. Such an attitude consists in refusing to see that the solution to a crisis requires a renunciation of previously accepted ideas; consequently, the reactionary tries, in order to deny the difficulties, to disguise their nature with the help of some distinctions which can only be purely verbal. In the example which concerns us, it is J. H. Jeans, the currently fashionable author of several popular works, whom we find as a representative of this tendency. In the course of his studies on the kinetic theory of gases,[26] Jeans believed he had found a way of escaping the disastrous consequences of the equipartition theorem by introducing the time interval needed to achieve this equipartition. In particular, through the use of simple models, he had estimated the order of magnitude of the speed with which the translational energy of the molecules in a gas is transferred by collisions to the oscillatory degrees of freedom of these molecules; he had found that this speed varied as e^{-n} as a function of the ratio n between the collision time of two molecules (binary molecular collision) and the period of the oscillation considered. For oscillation of optical frequencies, for example, the exchange of energy in question would be so slow, according to this formula, that in the states of the bodies observed, in which the latter dissipate a part of their energy by radiation, these oscillations could never reach equipartition and would make no noticeable contribution to the specific heat.

Lord Rayleigh,[26a] taking up a comment already made by Maxwell, raised the objection to this idea that the difficulty which it attempted to avoid would reappear, but infinitely aggravated, when considering a body sealed in an enclosure with reflecting walls. To this Jeans responded[27] that the same sort of considerations might extend to the vibrations of the ether: the equilibrium state of the radiation could very well (at least as long as the spectrum is limited on the high frequency side) be defined by Rayleigh's law, but it might not be realized in a finite time because the passage of energy toward the high frequencies were so slow. Hence the experimentally tested black bodies would have their property only at the longer wavelengths and the spectral distribution of their emission would not correspond to a true equilibrium but only to a state of steady energy flow towards the high frequencies, depending on the nature of the material bodies in contact with the radiation; in the course of time, this distribution would change extremely slowly; and more particularly, the observed maximum, which had forced Wien and Planck to introduce into the theory, in addition to k, a second constant h, would not have universal significance. In truth, this ingenious construction made short shrift of the excellent agreement observed between the spectra coming from black bodies of different constructions and the direct verifications of the perfect absorbing power of these bodies even for short wavelengths. Moreover, setting aside all contradictions of this sort, could Jeans' proposal be granted the character of a scientific theory when it allowed as many arbitrary parameters as desired to reproduce any experimental result whatsoever?

Nevertheless, it was indeed the only imaginable loophole if one wished to remain on the ground of classical mechanics, and it was in this sense that Lorentz[28] drew attention to it, with perhaps too much goodwill, in the famous lecture he gave in 1908 at the 4th International Congress of Mathematicians. In the course of this lecture, Lorentz discussed the problem of black-body radiation in depth by Lord Rayleigh's method and firmly established that the law stated by Rayleigh was indeed an unavoidable consequence of theories based on Hamiltonian-type equations. His indecisive attitude regarding Jeans' suggestion provoked an energetic protest on the part of the experimentalists. Lummer and Pringsheim[29] objected very strongly – a remark in the same vein had already been made by Lord Rayleigh[30] in the course of his discussion with Jeans – that as soon as one accepts Rayleigh's formula for the emissive power of a black body, one is led by simple application of Kirchhoff's law to flagrant contradictions with numerous well-known facts; indeed, according to the empirically known values of the absorbing powers, it turns out that, at ordinary temperatures and for visible light, the emissive power of a surface covered with soot or even with metal would be such that this surface should shine brilliantly in the dark.

While this criticism seemed decisive to Lorentz,[31] Jeans[32] still believed he could escape it and even attempted to give a less vague look to his views by discussing an important example of a radiation source for which calculations could be made explicitly by means of a few simple hypotheses. The problem in question, already discussed in part by Lorentz, was that of the emission and absorption of radiation by free electrons moving in the interior of a metal.[33] But this attempt, besides being tainted with numerous errors, only served to bring out, as shown in the critiques independently presented by McLaren[33a] and in a clearer form by N. Bohr, the complete confusion engendered by this fatal distinction between the unrealizable 'normal' equilibrium state and the 'final' state of stationary energy flow. The electrons under consideration are affected not only by the forces from the atoms of the metal, but also by the radiation field; according to classical theory, the amplitude of this field should become larger and larger for short wavelengths. However Jeans completely ignored this since, as he said, in the stationary state actually observed, the energy would not have the time to accumulate in this region of the spectrum. He believed that a calculation developed on this basis could lead, by means of a proper choice of the force law between atoms and electrons, to a stationary distribution agreeing with experiments. But this, Bohr observed, was manifestly impossible: for if a completely classical calculation, based on the existence of a spectral distribution of the type observed, yielded the conclusion that this distribution was stationary, this distribution must represent a true case of 'normal' equilibrium and not simply 'final'; while the only state of this sort predicted by classical theory is one obeying Rayleigh's law.

VIII

While Jeans vainly denied the need for an essential modification of classical theories, those who were convinced of this necessity did not show themselves, however, to be disposed to go equally far along the road of these reforms. Beginning in 1909, a deep split appeared between the radical point of view of Einstein and the moderate tendency of Planck. The contentious question raised by Einstein's first work was indeed one of the most fundamental: this problem of the structure of light was to continue to dominate the entire later development of the quantum theory and was finally resolved only thanks to Bohr's[34] introduction of the idea of 'complementarity' – one of the greatest modifications that our mode of describing natural phenomena has ever had to undergo.

It was Einstein who opened the debate with his classic paper 'On the Present State of the Problem of Radiation'.[35] In addition to the already known arguments in favor of the hypothesis of light quanta, this paper presents a new one drawn from quite remarkable considerations on fluctuations of the energy and momentum of black-body radiation. It will suffice for us to analyze the case of the energy fluctuations. To calculate them, Einstein used an ingenious 'inversion' of Boltzmann's principle which yielded for the mean square of the energy fluctuation at frequency $(\nu, \nu+d\nu)$ contained in the partial volume V, the general expression

$$(23) \qquad \overline{\Delta E_\nu^2} = -k\left\{\frac{d^2 S_\nu}{dE_\nu^2}\right\}^{-1},$$

with S_ν and E_ν designating respectively the average values of the entropy and the energy contained in this volume. Basically, this formula does not differ from expression (10) which Planck had also found by considering, from another point of view, the deviations from the state of equilibrium, and which had, as we have seen, played a decisive role in the discovery of the law of black-body radiation. In fact, it allowed this law to be calculated by simple integration as soon as one had the expression for the fluctuation $\overline{\Delta E_\nu^2}$ as a function of E_ν. In the classical theory of radiation, the cause of these fluctuations lies in the interference of rays with neighboring directions and frequencies which are superposed with irregularly varying phase differences. The calculation of $\overline{\Delta E_\nu^2}$ thus presents no difficulties of principle; the result, whose form, moreover, Einstein predicted without calculation with the help of some very simple reasoning, may be written using relationship (12):

$$(24) \qquad \left(\overline{\Delta E_\nu^2}\right)_{\text{class}} = \frac{E_\nu^2}{N(\nu)\, V d\nu};$$

introduced into equation (23), it immediately produces Rayleigh's law. The interest of this derivation lies in the fact that it is independent of the equipartition theorem. Similarly, as Einstein and Hopf demonstrated in a later work,[36] complementing the one we are analyzing, the classical expression for the fluctuations of the radiation's momentum implies Rayleigh's law so long as one makes use solely of the equipartition of the translational energy of a material point – a special case well verified by the kinetic theory of monatomic gases. It may already be concluded from this that the cause of the failure of Rayleigh's law must be sought in a more profound property of radiation than the 'ankylosis' of high frequency oscillations. But more precise conclusions can be reached by using equation (23) in another fashion. Following a procedure analogous to the one which had previously led him to formula (22) and to its statistical interpretation, Einstein considered Planck's formula as an experimental fact and calculated the corresponding value of the second term of equation (23); this gave him an expression for the energy fluctuations which must be posited to obtain Planck's law and which, in point of fact, he had guessed simply by writing formula (14). This fluctuation appeared to be due to the superposition of two independent causes:

$$(25)\qquad (\overline{\Delta E_\nu^{\,2}})_{\text{Planck}} = (\overline{\Delta E_\nu^{\,2}})_{\text{class}} + (\overline{\Delta E_\nu^{\,2}})_{\text{quant}},$$

the first of which gives just the classical fluctuation (24) while the other:

$$(26)\qquad (\overline{\Delta E_\nu^{\,2}})_{\text{quant}} = E_\nu \,.\, h_\nu$$

depends on the constant *h*. As can be seen immediately, this 'quantal' term exactly represents the fluctuations that would be obtained if the energy E_ν were composed of a number $E_\nu/h\nu$ of independent quanta $h\nu$. The corpuscular structure of radiation was revealed here in a striking manner, not in opposition to its wave properties, but rather as a complement to them.

It was at the Congress of German Scientists and Physicians held in Salzburg in September 1909 that the thesis of the corpuscular nature of light, brilliantly expounded by Einstein[37], first encountered Plank's decided opposition. The point of view defended by Planck in the course of the discussion which followed Einstein's lecture, and taken up again the following year in a detailed paper,[38] tended in effect to restrict the intervention of quanta to the mechanism of the interaction between matter and light, while the properties of radiation in empty space remained correctly described by classical electromagnetic theory. The argument furnished by the form of the fluctuations of energy and momentum did not seem convincing to Planck because he thought (wrongly) that these fluctuations had to depend essentially, even in the wave theory, on the quantal character of the interaction between matter and radiation. The other experimental data cited by Einstein in support of his thesis relate directly to this interaction, and Planck vividly opposed to them the totality of interference and diffraction phenomena so harmoniously included in wave optics. Indeed formula (25) clearly showed the inadequacy of the hypothesis of entirely independent light quanta, and the major problem was to determine the relationship between the corpuscular and wave properties of radiation. Einstein could, at most, suggest the possibility of a reconciliation of the two groups of phenomena modeled on the theory of electrons, which introduced an atomicity of electrical charge that was just as foreign to Maxwell's theory as the atomicity of light. The light quanta might be singular points of concentration of energy, generating electromagnetic fields whose superposition could give, on the average, the appearance of a classical wave field.

But if such a suggestion could in no way pretend to offer a solution to the dilemma, Planck, for his part, was not less embarrassed in trying to circumscribe clearly the domain within which he believed it possible to confine the influence of the quantum laws. As we know, his theory rested on the quantum formula (15) and the classical formula (2). But since the latter was deduced from the classical laws of absorption and emission, it could no longer be considered as established if these laws were abandoned. There was, in fact, a way, indicated by Debye,[39] of avoiding this defect in logical coherence. By considering, using Rayleigh's method, the proper vibrations of the black-body radiation in an enclosure with reflecting walls, it was possible to calculate by statistical methods the average energy $U(\nu_i)$ of the vibration of frequency ν_i, using only the fact that material resonators of this frequency emit and absorb energy in quanta $h\nu_i$; the result, analogous to (15) is:

$$U(\nu_i) = h\nu_i (e^{h\nu_i/kT} - 1)^{-1}. \tag{27}$$

The energy density in the interval $(\nu, \nu + d\nu)$ is then, by means of (12),

$$u(\nu, T) = N(\nu)\ U(\nu), \tag{28}$$

and it is this formula which, as comparison of (12) and (3) indicates, can now play the role of formula (2). Debye's reasoning would moreover apply just as well in the event that the proper vibrations of radiation were directly supposed to consist of light quanta. Its principal interest is in showing that in any case a detailed analysis of energy exchanges between resonator and radiation, which served to establish formula (2) on the basis of classical theory, can be dispensed with.

But Planck specifically did not wish to renounce this detailed analysis for he hoped that by investigating how it must be modified to rigorously derive the law of black-body radiation, he would obtain more precise information about the nature of the quantum laws of interaction between matter and radiation. To express the fact that a resonator can exchange energy with radiation only by quanta, it seemed necessary to introduce the idea of an 'excitation threshold' for the resonator, or, as H. Poincaré said, an 'antechamber' where energy would accumulate until the quantum was reached. Nevertheless, Planck judged that it would be difficult to admit such a concept in the case of absorption because of the length of time for accumulation which would have to be envisaged. He then proposed[40] to keep the classical theory of absorption and to relegate the intervention of quanta to the emission mechanism. To rigorously develop this theory of 'quanta of emission', Planck considered individual acts of emission as obeying a simple probability law, exactly like the individual transformation of radioactive atoms; moreover, he proposed this hypothesis only reluctantly because it seemed to him an expression of ignorance of a more fundamental causal relationship. The new theory differed from the preceding one by the appearance of a limiting value $h\nu/2$ of average energy of a resonator at the absolute zero of temperature, but this difference is without influence on the law of radiation and the specific heats of solids. Without going beyond the framework of the present study, we cannot retrace the avatars which this remarkable attempt to limit as much as possible the role of the quantum subsequently underwent. It was not entirely unproductive, since it stimulated a number of interesting investigations: however, nothing remains of it today. History once again decided against the thesis that Planck had made his own: "... it is essential to proceed in as conservative a manner as possible, that is, only the changes which have proved to be absolutely necessary should be introduced into the existing theory."[41] The wave-corpuscle dilemma was resolved neither in the sense of Planck nor in the sense of Einstein. But if the concept of light quanta as spatial localizations of energy was insufficient to provide the solution, it is not because the idea was too radical, but rather because it was not radical enough.

IX

At the end of October 1911, an important private congress met in Brussels under the auspices of Ernest Solvay and at the initiative of Nernst, assembling the principal scientists working on the quantum theory.[42] The papers presented at this congress offered eloquent testimony to the fertility of the idea Planck had introduced: both in the original sphere of black-body radiation as well as in that of specific heats and infra-red spectra of solids, new measurements entirely confirmed the theory[43] and Nernst's law in thermodynamics found its place automatically. But it is especially the transcript of the discussions following the reading of the papers which gives a vivid impression of the ferment of ideas stimulated by the 'quantum' crisis. Above all, attention was focussed on the still very shaky foundations of the theory and – aside from Lord Rayleigh who preferred to withhold comment – it was the critically minded such as Lorentz and Poincaré who made new contributions to this question. In addition to the question of the structure of light, they now examined more closely, following the path opened by Einstein, the way in which quantized resonators affect the behavior of purely mechanical systems: as Lorentz said in his opening address, "We cannot simply stipulate that a molecular oscillator exposed to bombardment by the atoms of a gas can only gain energy in finite portions of a given size; we have the right to demand the invention of a mode of action between the gaseous molecules and the oscillator which produces this result". However, another trend also became evident which, though perhaps premature, was symptomatic of the new orientation which would become more definite in the course of the following years: by feeling their way more or less successfully, they were attempting to extend the application of the quantum to phenomena of other kinds than the simple harmonic oscillator. The most notable effort in this direction was that of Sommerfeld, who addressed himself to the discussion of non-periodic atomic phenomena, especially the emission of X-rays, previously studied by Stark.

In his paper, Planck particularly emphasized the question of the basic mechanism of his resonators – a question which was, moreover, the object of general concern, as the preceding quotation from Lorentz's address indicates. But he began by an important explanation of the statistical foundations of his theory in a new form, which he strove to render as general as possible. Despite its imperfection, this attempt, which he related to certain of Larmor's considerations,[44] was to exert a profound influence on the subsequent development of research through the problems it raised and the direction it indicated. As Einstein had formulated it, the quantum hypothesis consisted essentially of imposing a considerable restriction on the manifold of allowed states of the harmonic oscillator. From the statistical point of view, Planck observed, this was equivalent to decreasing the number of *a priori* equiprobable elementary domains of phase space that were considered in defining the states of the system that could be distinguished. While in classical statistical mechanics, the 'volume' of these elementary domains may be chosen arbitrarily small, the decrease in their number imposed by the quantum theory could obviously be obtained by attributing to them, on the contrary, a finite volume. To determine the size of this volume, which for a system with one degree of freedom has the dimension of action, Planck naturally had recourse to his resonators; for these, the domain included between two paths of constant energy has for its volume $\Delta U/\nu$, where ΔU is the difference between the two energies. In order to make the smallest discernible energy difference be $h\nu$, the volume of the elementary domains of phase space, bounded by paths of constant energy, must then be taken exactly equal to the universal constant h. This principle Planck took as basic; the constant h appeared in it primarily as the quantum of action, while the notion of a quantum of energy was relegated to a secondary role. If the additional stipulation is made to take as allowed energy values those which correspond to the limits of the successive elementary domains (which amounts, in the case of the harmonic oscillator, to fixing the zero of energy in accord with the primitive form of the theory), it can be seen that the quantum action principle completely determines the manifold of allowed states of the system: it appears as a 'quantization condition'. In truth, the justification of the new statement, even restricted to one degree of freedom, still depends exclusively on the properties of the resonators; as for extending it to systems with many degrees of freedom, this presents considerable difficulties which H. Poincaré emphasized particularly in the course of the discussion of Planck's paper, difficulties which were not resolved until several years later. But the essential progress inaugurated by Planck's new mode of presentation was to dissociate, at least in principle, the constant *h* from any specific model of a resonator in order to give it a significance which hopefully could be made applicable to any mechanical system. Here we meet one of those bold examples of induction, where an apparently minimal modification in the way of looking at an algebraic relationship suddenly broadened the scope of the concepts involved in it. But then at the end of his paper, as if he feared to have gone too far, Planck once again seemed to want to lessen the importance of the step forward which he had just taken, by suggesting that the domain of application of quanta could very well be limited not only quantitatively, but in principle, to phenomena taking place within atoms and to chemical forces, while classical theories would apply in an absolute way to the motion of atoms and molecules and to the physical forces acting between them!

X

The quantization of the harmonic oscillator, in the new form given by Planck in his report to the Solvay Congress, raised a delicate question to which Lorentz and Poincaré[45] drew attention during the discussion of this paper. At issue was the relationship of the quantum hypothesis to the second law of thermodynamics. When Planck attributed to his resonators the discontinuous character implied by the idea of elements of energy ε, he had been led, by considering their equilibrium with radiation, to impose on them the additional relation

$$\varepsilon = h\nu, \tag{29}$$

so that the equilibrium distribution of radiation would satisfy the displacement law and therewith the second law. But relationship (29) between the energy element of a resonator and its frequency must then hold for other types of energy exchanges between resonators which could, like exchanges by collision for example, be purely mechanical and completely independent of the laws of optics used in deducing the displacement law. This could not be simple coincidence and the problem which arises is then to establish on a sufficiently general basis, as a consequence of the second law, the necessity of this remarkable property of the quantized harmonic oscillator. The principle of finite elementary domains of phase space, which, in short, takes relationship (29) as the starting point for the whole theory, clearly indicates the need for such a demonstration. It was not long in coming: indeed it had already been found in July 1911 by P. Ehrenfest[46] in the course of a beautiful study in which he attempted to bring out the features of the light quantum hypothesis which were really essential for the theory of black-body radiation.

In this paper, Ehrenfest considered, in conformity with Rayleigh's method, the proper vibrations of radiation so that, as he himself noted, his conclusions are in reality applicable, by means of a few simple modifications, to any system of harmonic oscillators. For each frequency interval $(\nu, \nu+d\nu)$ we take a total number of oscillators [in the case of radiation contained in an enclosure of unit volume, this number would be $N(\nu)\, d\nu$, $N(\nu)$ being given by formula (12)], whose phases, in short, are imagined to be uniformly distributed, while the energies are distributed according to any law, defined by the numbers $n(\nu, U)\, dU d\nu$ of oscillators whose energy is contained in the interval $(U, U+dU)$. The probability of such a distribution is not defined unless the statistical weight $\gamma(\nu, U)\, dU$ of the energy range $(U, U+dU)$ is fixed; as we have seen (cf. Part VI), the fundamental divergence between the quantum theory and classical statistical mechanics arises precisely in this definition of weights. However, if one wishes to retain the thermodynamic interpretation of the probability of a distribution, given by Boltzmann's formula, the choice of the statistical weight $\gamma(\nu, U)$ cannot be entirely arbitrary; indeed, Ehrenfest devoted the first part of his work precisely to the search for the restriction the second law thus imposed on $\gamma(\nu, U)$. In the second part he obtained the result that, if to this restriction is added the condition of discontinuity, which implies the very form of Planck's law for the function $\gamma(\nu, U)$, this leads unambiguously to relationship (29). Ehrenfest's great merit lies in the fact that he was the first to clearly pose the problem of the choice of *a priori* probabilities – a task for which he had been well prepared by his profound critical research on the foundations of statistical mechanics.[47] As is often the case, once the problem has been identified, the road to the solution, however arduous it may be, is at least clearly marked.

In the theory of radiation, the second law led to the displacement law when it was applied to an adiabatic change of the volume of the enclosure, that is of the proper vibrations frequencies. Ehrenfest was thus quite naturally led to seek first what condition was imposed on the weights $\gamma(\nu, U)$ by the condition expressing that the logarithm of the probability of any given distribution does not change in the course of a transformation of this sort. His starting point is an important theorem which he deduced from an earlier study by Lord Rayleigh[48] on the pressure of vibrations: if a harmonic oscillator is subjected to an adiabatic transformation consisting of an infinitely slow change of its frequency, the ratio of its energy divided by this frequency remains invariant. Thanks to this property, it is possible to express in terms of a single parameter the changes in all the quantities which enter into the probability calculation and to arrive at the desired fundamental condition by simple reasoning: except for an unimportant factor, the statistical weight $\gamma(\nu, U)$ can only depend on the ration U/ν. Putting

$$(30) \qquad q=\frac{U}{h\nu},$$

the constant h being introduced simply so that q will be dimensionless, one must have

$$(31) \qquad \gamma(\nu, U)=G(q).$$

The energy density $u(\nu, T)$ corresponding to thermal equilibrium, determined by the ordinary statistical method:

$$u(\nu, T)=-N(\nu)\frac{\mathrm{d}}{\mathrm{d}\mu}\log\int_0^\infty \mathrm{d}U\gamma(\nu, U)\, e^{-\mu U}, \left(\mu=\frac{1}{kT}\right),$$

may then be written in the form

$$(32) \qquad u(\nu, T)=-N(\nu)\, h\nu\frac{\mathrm{d}}{\mathrm{d}\sigma}\log\int_0^\infty \mathrm{d}qG(q)\, e^{-\sigma q}, \left(\sigma=\frac{h\nu}{kT}\right),$$

which brings the displacement law to the forefront. Thus this law is obtained by a method which is completely independent of optics and generally valid for any sort of harmonic oscillators. Moreover, the adiabatic invariance of U/ν also allows the displacement law to be demonstrated by purely thermodynamic means, and thus emerges as the true basis of this law; which was already observed incidentally by Ehrenfest in the paper we are analyzing. He would repeat it more emphatically in a later work[49] where he drew attention to a more general theorem of Boltzmann and Clausius concerning arbitrary periodic systems.

Ehrenfest subsequently showed in detail how discontinuities are already necessarily introduced in the choice of the function $G(q)$ if one wishes formula (32) simply to reproduce the asymptotic character, as $\sigma \to \infty$, of the experimental curve of the spectral distribution of black-body radiation. Finally, he noted that if a definite form is imposed on the function $u(\nu, T)$ for all values of σ,

$$u(\nu, T) = N(\nu)\, h\nu f(\sigma),$$

relationship (32) furnishes a functional equation

$$(33) \qquad \int_0^\infty \mathrm{d}q G(q)\, e^{-\sigma q} = e^{-\int^\sigma f(\sigma)\, \mathrm{d}\sigma},$$

which in principle allows the complete determination of the function $G(q)$ (up to a factor); of course, in the event that the function $G(q)$ should show discontinuities, the integral of the first term must be generalized in the sense of Stieltjes, that is, instead of (33), it must be written

$$(34) \qquad \int_0^\infty \mathrm{d}q G(q)\, e^{-\sigma q} + \sum_n G_n e^{-\sigma q_n} \equiv \int_0^\infty e^{-\sigma q} \mathrm{d}F(q) = e^{-\int^\sigma f(\sigma)\, \mathrm{d}\sigma}.$$

Equation (34) expresses, in fact, what is now called the 'spectral decomposition' of the function figuring in the second term. In particular, if for $f(\sigma)$ one takes the form $e^{-\sigma}$ corresponding to Wien's law, or $(e^\sigma - \mathrm{I})^{-1}$ given by Planck's law, this function in the second term of (34) has period $2\pi i$, and the modern theory of equations of this type[50] immediately leads to the result that the corresponding 'spectrum' is entirely discontinuous and is composed of the eigenvalues $q_n = n (n = 1, 2, 3, \ldots)$; in other words, the weight G is null everywhere except for values of the energy of the form $nh\nu$, – a result equivalent to the relation (29). In addition, Planck's law requires the equality of all the G_n, while Wien's law is obtained by taking G_n proportional to $(n!)^{-1}$. It is instructive to compare this second part of Ehrenfest's paper to the brilliant analysis of energy exchange by collisions between resonators given some time later by Poincaré.[51] Using the general methods of statistical mechanics, he treats the equilibrium of a microcanonical ensemble composed of N resonators of the same frequency and N' atoms (or resonators with infinitely long periods) capable of interacting by collisions; by considerations similar to those of Ehrenfest and also based on an equation of type (34), he established the necessity of taking the energy of the resonators to be of the form $n\varepsilon$, in order to recover an expression for the average energy of the type corresponding to Planck's law. But he naturally did not arrive at any relationship between the quantum of energy ε and the frequency ν; Ehrenfest's theorem (31) is essential to the understanding of such a relationship.

We know, moreover, that the historical importance of Ehrenfest's paper considerably surpassed its immediate purpose. Ehrenfest was struck by the fact that the remarkable persistence of the classical displacement law in the quantum theory was due specifically to the adiabatic character of the transformation which lies at the origin of this law. It is this observation which put him on the road to the productive 'adiabatic hypothesis' or 'principle of mechanical transformability',[52] which was to allow a statement of the quantization conditions for a very general class of mechanical systems and the determination of the statistical weights of their stationary states.

XI

Thus we see the theory of the quantum of action gradually consolidating its logical structure and simultaneously, through various fertile ramifications, extending its realm of applicability beyond the statistical phenomena which it had originally helped to interpret, toward the field of individual atomic processes. However, when attacking the latter problem, one was confronted with a difficulty which could largely be avoided in the study of statistical equilibrium: one needed to give a detailed representation of the structure of the atom and the forces at work in atomic processes. To avoid arbitrary constructions, it was necessary to wait for experiment to provide some decisive particulars on this subject. It was in 1911 that Rutherford succeeded in unequivocally deducing his model of a 'nuclear' atom from his studies of the scattering of α particles, characterized essentially by the concentration of the atom's positive electricity in a 'nucleus' of very small diameter compared to the dimensions of the atom. An immediate consequence of this model was the appearance of an irreconcilable opposition between the laws of classical electrodynamics applied to the motion of the electrons around the nucleus and the stability of the atomic structure which was revealed in so many different phenomena. It is needless to recall how Niels Bohr[53] had the courage to squarely confront this contradiction, and with what sure genius he was able to make use of the lessons of the theory of the quantum of action to analyze the non-classical character of the stability of the atom and of its radiation. On the one hand, he introduced the concept of the stationary state, a state in which the atom, contrary to the laws of electromagnetism, does not radiate energy, and he interpreted the monochromatic emissions of atomic systems as being due to non-classical transitions between these stationary states accompanied by the radiation of a quantum of energy proportional to the frequency. On the other hand, he systematically used a master concept which would subsequently continue to play an absolutely essential role, the famous 'correspondence principle' whose source may already be found in Lord Rayleigh's remark (cf. Part III) that the results foreseen by the classical theory should continue to be valid in the limiting case of sufficiently small frequencies. It is from this moment when, thanks to Bohr's work, the quantum of action was solidly grafted onto the theory of atomic constitution, that, stripped of any remaining artificiality, the theory truly acquires universal significance; Bohr's 'quantum postulates' could serve as the foundation of a first attempt to synthesize all atomic phenomena. A close contact was thus established between the quantum theory and a rich body of experimental material of great precision in the area of spectroscopy: in this material the theoretical tools found an opportunity for their sharpening and perfection which had previously been lacking. Consequently, we see a period of exuberant growth of the theory in which problems of a statistical order were relegated to the background while, nevertheless, continuing to make important contributions to the progress of research.

In Bohr's scientific attitude we find to a particularly accentuated degree that intellectual audacity we noted in Einstein. Like the latter, when he introduced into science the relativity of simultaneity or the light quanta hypothesis, Bohr, in his study of the nuclear atom model, understood the need to radically renounce the application of classical laws, and it is this renunciation which gave him the needed freedom of spirit to find the form which had to be given to the new laws. But we have seen that on the question of quanta, Einstein stopped half-way: his renunciation had been only a sort of provisional 'heuristic' exaggeration, he called it, and was soon followed by an effort to reconcile the newly acquired ideas with suitably modified older theories. It was Bohr who took the decisive step; from the beginning of his research, the rejection of the classical theories was for him absolutely definitive; and while skillfully using the lessons provided by these theories through the correspondence principle, it was "precisely by accentuating the opposition"[54] between classical ideas and quantum concepts that he hoped to succeed in organizing the latter into a harmonious whole. We know that the further development of atomic theory has strikingly confirmed this viewpoint and that, with Bohr's constant inspiration, a marvellous enrichment of thought has replaced that irrationality of the quantum action which so tormented those who initiated the new theory.

NOTES

[1] Cf. N. Bohr, *Atomteori og Naturbeskrivelse* (1929). [English edition: *Atomic Theory and the Description of Nature* (Cambridge University Press, Cambridge, England, 1934) – Ed.]

[2] A. Sommerfeld in 'Planck-Heft' of *Naturwissenschaften* **6** (1918), 195–263. The issue contains the bibliography of Planck's works until 1917.

[3] See P. and T. Ehrenfest, *Enzyklop. der math. Wissensch.*, Bd. IV, 2ii, Artikel 32, 1911. [See English translation: *The Conceptual Foundations of the Statistical Approach in Mechanics*, transl. by M. J. Moravcsik (Cornell University Press, Ithaca, N.Y., 1959).]

[4] The impossibility of the ergodic hypothesis was, as we know, demonstrated in 1913 by A. Rosenthal, *Ann. d. Phys.* **42** (1913), 796 and M. Plancherel (*Ibid.*, p. 1081) when the necessary mathematical instrument (theory of sets and Lebesgue integrals) was available. The rigorous foundation of the theory of equipartition has been established recently (J. v. Neumann, T. Carleman, G. Birkhoff, E. Hopf, etc., 1932) thanks to the application of other mathematical theories which were newly studied on account of their utilization in quantum mechanics (theory of groups of linear operators) – a remarkable example of interaction between physics and mathematics! Cf. E. Hopf, *J. of Math. and Phys.* **13** (1934), 51.

[5] See M. Planck, *Theorie der Wärmestrahlung* (1906); on the subject of Stewart's priority over Kirchhoff, cf. Lord Rayleigh, *Phil. Mag.* **I** (1901), 98 [*Scientific Papers*, vol. IV, p. 494].

[6] W. Wien, *Ann. d. Phys.* **58** (1896), 662.

[7] Wien, like all the physicists who were then studying the spectral distribution of black-body radiation, defined this distribution as a function of wavelength. The practice of taking the frequency instead of the wavelength as a variable was only introduced later by Planck; but there is clearly no harm in conforming to this uniformly throughout the article.

[8] Planck's research which we are going to discuss is summarized in his memoir in *Ann. d. Phys.* **I** (1900), 69.

[9] Cf. M. Planck, *Ann. d. Phys.* **57** (1896), 72; and also *Phys. Zs.* **10** (1909), 62.

[9a] It is essential to read the interesting details on the development of his ideas which Planck himself gives with so much modesty and sincerity in his Nobel prize address. See *Les prix Nobel en 1919–1920* (Stockholm 1922).
[10] M. Planck, *Ann. d. Phys.* **I** (1900), 719.
[11] Lord Rayleigh, *Phil. Mag.* **49** (1900), 539 (*Scientific Papers*, vol. IV, p. 483).
[11a] Lord Rayleigh, *Nature* **71** (1905), 559.
[12] O. Lummer and E. Pringsheim, *Verh. d. deutschen physik. Gesellsch.* **2** (1900), 163; F. Paschen, *Ann. d. Phys.* **4** (1901), 277; H. Rubens and F. Kurlbaum, *ibid.*, 649. The results of these measurements came to Planck's attention towards the middle of 1900; see M. Planck, *Verh. d. deutschen physik. Gesellsch.* **2** (1900), 202
[13] At the end of his note, Rayleigh proposed representing the empirical law of black-body radiation by a formula obtained simply by multiplying expression (13) by the exponential factor of Wien's law. The German experimenters pointed out the incorrectness of this formula, but did not stress the correctness of the essential point of Rayleigh's work which concerns the replacement of the factor ν^3 of Wien's law by the factor $\nu^2 T$ in the limit $\nu/T \rightarrow 0$. Cf. the remark added by Lord Rayleigh in his *Scientific Papers, loc. cit.*
[14] See the work cited in note 12.
[14a] See his Nobel address cited in note 9a.
[15] M. Planck, *Verh. d. deutschen physik. Gesellsch.* **2** (1900), 237; *Ann. d. Phys.* **4** (1901), 553.
[16] Cf. A. Einstein, *Phys. Zs.* **10** (1909), 187.
[17] Such a distribution is defined when for each of the P objects, one identifies the group within this distribution to which it belongs.
[18] Cf. W. Ostwald, *L'évolution d'une science, la chimie*, transl. by M. Dufour (Flammarion, Paris, 1909), pp. 313 ff. (Appendix, Chap. I).
[19] M. Planck, *Theorie der Wärmestrahlung*, first edition 1906.
[20] Lord Rayleigh, *Nature* **72** (1905), 54.
[21] The three memoirs of A. Einstein alluded to are published in the *Ann. d. Phys.* **17** (1905), respectively on pp. 132, 549, and 891. [See the English translations: (a) for the relativity paper in the collection *The Principle of Relativity* (Methuen, London, 1923; New York, Dover reprint, many editions): (b) for the Brownian motion paper in *Investigations on the Theory of the Brownian Movement* (Methuen, London, 1926); (c) for the paper on quantum theory discussed here, 'On a Heuristic Point of View about the Creation and Conversion of Light', in D. ter Haar (ed.), *The Old Quantum Theory* (Pergamon, Oxford, 1967), p. 91.–Ed.]
[22] Cf. N. Bohr, 'Lys og Liv', lecture published in *Naturens Verden* **17** (1933), 49. [See English translation: 'Light and life' in *Nature* **131** (1933), 421, 457, especially paragraph 8.]
[23] A. Einstein, *Ann. d. Phys.* **20** (1906), 199.
[24] A. Einstein, *Ann. d. Phys.* **22** (1907), 180.
[24a] In 1906, P. Ehrenfest [*Phys. Zs.* **7** (1906), 528], had independently reached the same conclusions by Boltzmann's complexions method.
[25] Cf. *loc. cit.*, note 24.
[26] J. H. Jeans, *Dynamical Theory of Gases* (Cambridge University Press, Cambridge, England, 1904).
[26a] Cf. *loc. cit.*, note 11a.
[27] J. H. Jeans, *Nature* **71** (1905), 607; **72** (1905), 101.
[28] H. A. Lorentz, *Nuovo Cimento* **16** (1908), 5; cf. also the article cited in note 31.
[29] O. Lummer and E. Pringsheim, *Phys. Zs.* **9** (1908), 562.
[30] Lord Rayleigh, *Nature* **72** (1905), 54.

[31] H. A. Lorentz, *Phys. Zs.* **9** (1908), 562.
[32] J. H. Jeans, *Phys. Zs.* **9** (1908), 853; *Phil. Mag.* **17** (1909), 229.
[33] See the explanation of this question in N. Bohr's thesis *Studier over Metallernes Elektronteori* (Copenhagen 1911); Kap. III, especially p. 103, note 3. [The original, together with an English transl., may be found in: J. Rud. Nielsen (ed.), *Niels Bohr: Collected Works, Vol. I, Early Work (1905–1911)* (North-Holland Publ. Co., Amsterdam, 1972). The note mentioned here will be found as note 1 on p. 379. – Ed.]
[33a] S. McLaren, *Phil. Mag.* **22** (1911), 66; **23** (1912), 513; – on the very original personality of McLaren, see *Scientific Papers* by the late S. B. McLaren (Cambridge University Press, Cambridge, England, 1925).
[34] See works cited in notes 1 and 22; as well as N. Bohr's Faraday Lecture, *Journ. Chem. Soc.* 349 (1932).
[35] A. Einstein, *Phys. Zs.* **10** (1909), 185. Cf. also H. A. Lorentz, *Les théories statistiques en thermodynamique* (lectures given at the Collège de France in 1912), Leipzig 1916.
[36] A. Einstein and L. Hopf, *Ann. d. Phys.* **33** (1910), 1105.
[37] A. Einstein, *Phys. Zs.* **10** (1909), 817; Planck's remarks, *Ibid.*, p. 825.
[38] M. Planck, *Ann. d. Phys.* **31** (1910), 758.
[39] P. Debye, *Ann. d. Phys.* **33** (1910), 1427.
[40] M. Planck, *Verh. d. deutschen phys. Gesellsch.* **13** (1911), 138; *Ann. d. Phys.* **37** (1912), 642.
[41] *Loc. cit.*, note 38, p. 768.
[42] *La théorie du rayonnement et les quanta* (Paris 1912). Cf. also H. Poincaré, *Dernières pensées*, chap. VI and VII (Flammarion, Paris, 1913; reproduction of articles published in 1912). [English translation *Mathematics and Science: Last Essays*, trans. by J. W. Bolduc (New York, Dover reprint, 1963) – Ed.]
[43] There was an obscure point in the theory of specific heats of metals: how is it that the free electrons, whose existence allows the interpretation of many properties of the metallic state, do not make any appreciable contribution to the specific heat? We know that the answer to this enigma depends essentially on Pauli's exclusion principle.
[44] J. Larmour, *Proc. Roy. Soc.* **A83** (1909), 82.
[45] Cf. the works quoted in note 42.
[46] P. Ehrenfest, *Ann. d. Phys.* **36** (1911), 91; the part containing this memoir appeared October, 1911.
[47] Cf. note 3.
[48] Lord Rayleigh, *Phil. Mag.* **3** (1902), 338 [*Scientific Papers*, vol. V, p. 41].
[49] P. Ehrenfest, *Verslag. Akad. Wetensch. Amsterdam* **22** (1913), 586.
[50] See for example J. v. Neumann, *Annals of Math.* **39** (1932), 587. Ehrenfest limits himself to saying that a friend communicated the result to him.
[51] H. Poincaré, *Journ. de Phys.* **2** (1912), 1.
[52] See the discussion of the question and the bibliography in N. Bohr, 'On the Quantum Theory of Line Spectra', *Danske Vid. Selsk. Skrifter Naturvid. Mat. afd.* **4**, I (1918–1922), especially pp. 8–9.
[53] Cf. N. Bohr, *Abhandlungen über Atombau aus den Jahren, 1913–1916* (Vieweg, Braunschweig, 1921). [See the forthcoming Vol. 2 of Bohr's *Collected Works* for the English originals. – Ed.]
[54] Cf. N. Bohr, *Drei Aufsätze über Spektren und Atombau*, 1922; especially the end of the first article dated December, 1913. The French and English [*The Theory of Spectra and Atomic Constitution* (Cambridge Univ. Press, Cambridge, England, 1922) – Ed.] translations of this book are both defective due to a misunderstanding at this important place.

CHAPTER 17

MAX PLANCK AND THE STATISTICAL DEFINITION OF ENTROPY*

[1959b]

ABSTRACT. The statistical interpretation of entropy given by Boltzmann played a decisive role in the thinking of Planck in more than one respect. The study of this aspect of his work – apart from the pleasure derived from reading papers written with such elegance and lucidity – is nowadays still full of lessons. In particular the famous controversy occasioned by the concept of absolute entropy proposed by Planck, touches on very deep epistemological problems which are still current. In the following pages we will review the problems in whose analysis Planck had occasion to develop and apply his outlook on statistical thermodynamics.

1. THE 'ENERGY ELEMENTS'

We know by what strange detours Planck came to establish the spectral distribution law of thermal radiation.[1] Initially he believed [2] that he had proven the formula for a system of oscillators of given frequency in equilibrium with radiation:

$$(1)\qquad \frac{d^2S}{dU^2}=-\frac{K}{U},$$

expressing (with the aid of the constant K, to be determined) how the second derivative of the entropy of an oscillator, with respect to its mean energy U, depends on this energy. In reality, this formula, when combined with the relation defining temperature T,

$$(2)\qquad \frac{dS}{dU}=\frac{1}{T},$$

leads to Wien's distribution law, which is only valid for high frequencies. Planck was well aware that formula (1) was beyond the scope of phenomenological thermodynamics, but he was too unfamiliar with statistical methods to clearly establish its connection with energy fluctuations, which was only highlighted much later by Einstein [3]. It must be stressed, furthermore, that the very notion of statistical fluctuation of macroscopic quantities was only fully understood thanks to Gibbs' treatise [4] published in 1902, where it is systematically discussed under the name of 'Anomaly'.

It was not until [5] that he ascertained that new data at low frequencies were incompatible with Wien's law and (without being aware of the short note in which Rayleigh [6] stated the theoretical law which was valid in this domain) that he attempted to modify formula (1) by trial and error, coming up with the following form:

$$(3) \qquad \frac{d^2S}{dU^2} = -\frac{k}{U(U+\varepsilon)},$$

which depends on two constants, k and ε. This formula gives the correct distribution law for all frequencies if we note that to satisfy Wien's displacement law the constant ε must be taken proportional to the frequency ν:

$$(4) \qquad \varepsilon = h\nu.$$

For the entropy itself we then find

$$(5) \qquad S = k\left\{\left(1+\frac{U}{\varepsilon}\right)\ln\left(1+\frac{U}{\varepsilon}\right) - \frac{U}{\varepsilon}\ln\frac{U}{\varepsilon}\right\}.$$

The constants k and h are universal, because of the universal character of the radiation law. The role of the constant k is purely to relate the units of entropy and of temperature. But the physical meaning of the new constant h, and of the relation (4) in which it appears, required a deeper understanding.

It was then [7], that Planck turned resolutely to the ideas of statistical thermodynamics, which until then he had tried to avoid. He started from Boltzmann's idea according to which entropy (apart from an additive constant) is proportional to the logarithm of the number of complexions that constitute the equilibrium state under consideration; and he undertook an enumeration of these complexions for the case of his oscillator system. Clearly if – like Rayleigh – he had better understood the implications of the classical theory of radiation, he would have known that Boltzmann's formula could only lead to equipartition, and to the form of the distribution law valid at the low frequency limit. Fortunately, his inexperience gave him the audacity to attempt a computation of the number of complexions which he thought in the spirit of classical theory, when in fact from this point of view it is pure heresy.

He imagined that these complexions represent the partition of the total energy NU, of N oscillators, into 'energy elements' ε, whose magnitude is first left arbitrary. If P is the number of these elements $(=NU/\varepsilon)$, then the number of complexions is, to a good approximation, given by

$$(5.1) \qquad C=\frac{(N+P)!}{N!P!},$$

which does indeed lead to the form (5) of the entropy, if the proportionality constant in Boltzmann's formula

$$(6) \qquad S=k \ln C+\text{const}$$

is taken as the same constant k which appears in the heuristic hypothesis (3). In this way, Planck thought he had found a statistical derivation of the radiation law, based solely on Boltzmann's formula (6) for entropy. As to the existence of elements of energy, defined by equation (4), he regards them as a property of the radiation, the radical character of which he does not seem to immediately realize.

2. The Atomic Constants of Matter and Electricity

A consequence that Planck immediately drew from his considerations [8], and to which he justly attached great importance, is the quantitative determination of fundamental constants which characterize the atomic structure of matter and electricity. Boltzmann's formula again plays the major role in the reasoning by which he attains this result – this time because of its universal character, that is because of its applicability both to material systems and to radiation, with the *same* constant of proportionality. Now, for an ideal gas, Boltzmann had obtained for this constant the value R/A, where R denotes the gas constant per mole and A is Avogadro's number. Planck thus wrote

$$(7) \qquad \frac{R}{A}=k,$$

from which he got A and its inverse – the mass in grams of an atom of hydrogen – as a function of the constant k. But the latter is numerically determined simultaneously with h from the experimental values of the constants appearing in Stefan's law and Wien's displacement law. Moreover, in dividing the value of the Faraday by A, we get the value of the elementary electrical charge. Planck insisted on the fact that the system of constants he obtained in this way was much more precise than what direct determinations could give at that time. But even more than this practical advantage, what could thus be gained was a new insight into the mutual dependence of the characteristic constants of matter, electricity, and radiation, guided by the statistical concepts of Boltzmann.

3. Nernst's theorem and the quantum of action

Planck's attention was thus turned to the statistical aspect of the new discontinuity which he had demonstrated in his study of radiation, and he was now to attach more and more importance to this aspect. He was especially encouraged in this by the formulation, in 1906, of Nernst's theorem [9], of which he proposed a more radical interpretation [10], [11] than that advocated by Nernst himself. In the spirit of classical thermodynamics, Nernst restricted himself to an essentially relative formulation of his theorem, expressing the tendency of the entropy of a system to remain constant after any isothermal process when the temperature tends to zero; but Planck concluded that it is possible to define the entropy of the system in an absolute sense, without explicit reference to a state taken as a point of comparison. As stated by Planck, Nernst's theorem fixes the zero of entropy in a universal manner; in particular it excludes the possibility that the entropy takes negative values of arbitrarily large magnitude as the temperature approaches absolute zero.

Passing to the statistical interpretation of this conclusion, Planck recognized that it amounts to universally fixing, in the expression for the probability of a state of the system, the factor expressing the statistical weight of the equal cells, into which one imagines phase space to be divided to calculate the number of complexions. In Boltzmann's view, this division necessarily remained completely arbitrary, the statistical weight of each cell being simply proportional to its volume by virtue of Liouville's theorem. Imposing a finite value on the entropy at absolute zero amounts to imposing a limit on the indefinite divisibility of phase space, by attributing to the elementary cells a universal finite magnitude. Since the invariant element of phase space for a system of f degrees of freedom has the dimensions of action to the power f, it is clear that the magnitude of an elementary cell can only be h^f. It was by means of this hypothesis that Planck related Nernst's theorem to the quantum theory; and it seemed to him so fundamental that at the first Solvay Congress in 1911 he presented it as the expression of deepest significance of the new discontinuity that he had discovered, and that he henceforth called the *quantum of action.*

By the hypothesis of the quantum of action, Planck laid the foundations for a quantum statistics of material systems, whose importance was highlighted by the possibility Nernst's theorem offered of calculating the 'chemical constant' of reactions in the gaseous state. Nevertheless, in the absence of safe rules by which to 'quantize' a mechanical system other than an harmonic oscillator, the calculation of the entropy or the free energy of a gas remained precarious. At the same Solvay Congress, Henri Poincaré immediately raised the seemingly decisive question: What structure should be given to these elementary cells of phase space? It was known that the quantization of the harmonic oscillator defined a set of similar ellipses effectively delineating cells of magnitude h. But did this property extend to more complicated systems? Planck devoted much effort to exploring this possibility: In 1916 he published [12] an attempt to calculate the free energy of a gas, and the chemical constant, based on a particular 'cutting-up' of phase space into elementary cells.

But he was no doubt discouraged by the arbitrariness of this procedure; he returned to the problem again [13] in 1925, this time in a way that emphasized the desire to avoid as much as possible the details of quantization. Unfortunately, his reasoning contained a serious inconsistency, soon pointed out by Schrödinger [14]: For the number of distinct stationary states of energy less than E accessible to a gas of N atoms, Planck wrote the formula

$$(8) \qquad C_E = \frac{1}{h^{3N}} \frac{1}{N!} V(E),$$

in which $V(E)$ represents the volume of the region of the phase space of a gas which corresponds to an energy less than E. This formula expresses the hypothesis that each elementary cell of magnitude h^{3N} corresponds to a unique stationary state; and the factor $(N!)^{-1}$ (to which we will return in a moment) correctly takes into account the reduction of the number of distinct complexions due to the indistinguishability of the atoms. On the other hand, in estimating the number of ways in which these stationary states of the gas can be formed from combinations of stationary states of the atoms, he treats the latter as if they were individualized. The difficulty on which Planck now stumbled was thus related to the problem, raised shortly before by Einstein [15], of the evaluation of the statistical weight of the stationary states of a system of identical elements.

Rather than face this new problem, Planck tried to avoid it even at the price of abandoning Boltzmann's concept of thermodynamic probability: To this end, he proposed [16] to take as the expression for the entropy of an isolated system of energy E

(9) $$S=k \ln C_E,$$

where C_E denotes the number of stationary states of the system in which the energy is less than E; for a gas C_E is given by formula (8). Planck affirmed that this new statistical definition of entropy does not appeal to any probability; in particular, it is compatible with any specification of the weights of the stationary states of a gas. He does not seem to realize the ambiguity remaining here in the notion of "number of stationary states in a given energy interval". As one can see, Planck in fact adopts one of the definitions of entropy given by Gibbs (apart from the modification due to the quantum hypothesis) – without making this connection and perhaps without being aware of it, because he does not appear to have examined the ideas of Gibbs very deeply. To justify the new formula (9), he merely showed that it reproduces well-known results; he did not refer, as he could have done, to the insensitivity of the statistical definition of entropy highlighted by Lorentz [17]. At any rate, it is clear that this method could not lead him, any more than its predecessors, to a completely correct expression for the free energy of gas.

Hence, despite the characteristic perseverance with which he pursued them, Planck's efforts to establish the statistical thermodynamics of a quantized gas ended in failure. It could not have been otherwise, since the symmetry conditions that essentially determine the weight of the stationary states were still unknown. Once these conditions were clearly stated, all the preceding efforts were condemned to an oblivion, from which there is hardly any interest in rescuing them. On the other hand a discussion of the general principle of an absolute definition of entropy, placed by Planck at the foundation of a quantized thermodynamics, remains instructive for several reasons: It can serve to illustrate a delicate problem that even now is still often badly understood; and from the historical point of view it will reveal to us an important characteristic of the scientific attitude of this great physicist.

4. The extensive nature of entropy and the identity of atoms

Let us first pause for a rather minor problem made famous, however, by the surprising differences of opinion which arose from it: Namely, the 'division by $N!$', of which we already encountered an example in equation (8). The point is to know how the characteristic functions – entropy or free energy – of a system composed of identical elements (atoms) will depend on the number of these elements. If we place two systems of this type side by side, composed respectively of N_1 and N_2 atoms, and both in a state defined by the same values of the intensive variables (temperature and for example pressure), the entropy of the total system should, by virtue of its extensive character, be the sum of the entropies of the partial systems. Now, if we consider that a complexion of a system of N atoms is defined by a corresponding cell of phase space, the number of complexions of the total system is not the product of those of the partial systems, but rather this product multiplied by the factor $(N_1+N_2)!/N_1!N_2$. As a consequence, the entropy of a system of N atoms cannot be represented by the logarithm of the number of complexions computed in this way: It is clear that we must divide this number by $N!$, in order to obtain an extensive function.

Planck had some difficulty, it seems, in seeing the reason for this division by $N!$. He first [12] thought he should attribute the appearance of the factor $N!$ to a property of the cutting-up of phase space into cells. It is only in the fourth edition of *Theorie der Wärmestrahlung* (1921) [18] that he finally gives the correct justification for this factor. It is true that for him the question was complicated by the fact that he obtained the distribution law for thermal radiation by the detour of considering a system of identical oscillators in equilibrium with radiation; in this case, the factor in question need not be introduced, and it is therefore still necessary to find a reason for this behavior different from that of the system of oscillators.

The explanation proposed by Planck rests on the observation that, due to the identity of the atoms, two complexions differing only by a permutation of the coordinates of the latter really represent the same physical state. The number of distinct physical states of the system is obtained by dividing the number of complexions by the number of permutations of the N atoms – which is precisely $N!$. It is, of course, upon this number of distinct states that a physical quantity like entropy must depend. As for the imaginary system of oscillators in equilibrium with radiation, the elements (although they have identical structure) are presumed to have no other interaction except that through the intermediary of the radiation. They occupy fixed positions which they cannot exchange, and which consequently completely individualize them.

The distinction between 'specific phase' and 'general phase' can be recognized here, which was introduced by Gibbs in the chapter of his book [4] devoted to 'Systems composed of molecules' – a chapter rich in content, but whose importance was only later recognized. Beyond any doubt, Planck was in no way influenced by these ideas of Gibbs; he only refers to them in connection with the general concept of the ensemble, but without reference to the chapter in question. The opposition of Ehrenfest [19], who declared the simple and clear interpretation given by Planck of the factor $(N!)^{-1}$ to be 'incomprehensible', would surely have been less strong had he better appreciated the methods by which Gibbs established the manner in which the characteristic functions of systems of molecules depend on the number of the latter. But he made no distinction between this profound and original aspect of Gibbs' work, and the general trend of the work, against which he argued as a true disciple of Boltzmann.[2] Nevertheless, on this particular detail, Planck – correctly – adamantly maintained his point of view, notably when Schrödinger [22] addressed the same objection to him.

5. Critique of the Concept of Absolute Entropy

However, the controversy over the factor $(N!)^{-1}$ concerned only a rather superficial aspect of the problem. What the critique, so vigorously formulated by Ehrenfest and Trkal in a justly famous paper [19], was really directed to was the very principle of fixing the absolute value for the entropy of an isolated system – a principle in which, as we have seen,

Planck tended to see the essential content of the concept of the quantum of action. It was really to the domain of logic and epistemology that Ehrenfest and Trkal immediately brought the debate. Rigorously speaking, it is impossible to determine how the entropy of a system varies with the number of atoms without studying a physical process in which this number can effectively change. To do this, we evidently must put the system being considered in contact with another system, in such a way that their interaction causes a variation in the number of atoms they contain.

The authors develop this program by directly studying the case of systems capable of chemical combination. An alternate method, which permits consideration of only one homogeneous substance, and thus nearer to Planck's point of view, is based on the concept of the 'grand ensemble'.[3] It was exactly with this intention that Gibbs introduced the consideration of grand ensembles in the chapter of his book cited above. But again, neither Planck nor Ehrenfest seemed to have been aware of the true scope of Gibbs' methods.

The argument of Ehrenfest and Trkal is clearly inspired by the materialist (others prefer to say 'realist') attitude characteristic of Boltzmann's thought: the problem of statistical thermodynamics is to give a complete and logically coherent deduction of macroscopic laws of thermodynamics from those of mechanics, applied to the atomic model of matter. Considered from this point of view, it is evidently decisive; and it is thus particularly significant that it does not at all impress Planck. No doubt he had qualms about having his ideas fought by 'eminent physicists' [23]; but the manner in which he makes efforts to reply to them shows a very profound difference in the very conception of scientific explanation and the role of theories. As a matter of fact, Planck does not seek at all that harmonious connection leading from the mechanical model to macroscopic laws. No doubt he also starts from the atomistic view, which implies a purely mechanical description of the system being considered. But he is completely satisfied if he has succeeded in constructing, with the help of mechanical variables, an expression which he can show to possess the required properties of the characteristic macroscopic function he was looking for – the question of knowing how this function is logically connected to the purely mechanical properties of the atomic system hardly bothers him.

It is understandable that such a pragmatic attitude was not going to please Ehrenfest, who found in it the tendency he was criticizing in Gibbs. But if Gibbs had turned to a formalistic position quite similar to the one adopted by Planck, it was because of the serious contradictions which prevented the effective realization of Boltzmann's program, and surely not for lack of understanding of or even sympathy for that program. With Planck, the impression rather is that, having come late to atomism, he is far from understanding all the importance it has for a rational explanation of the universe; for him it is above all the basis for a process of calculation justified *a posteriori* by its success. In this sense, it can be said that Planck carries to an extreme degree the fundamentally idealist conception of statistical mechanics first put forward by Gibbs.

However, the sure instinct which characterizes the great physicist, and links him to the reality of things in a better way than conscious reflection does, protected Planck from the worst traps of positivism. From a point of view warped no doubt by a very strict spiritualism, he vigorously defended against the neo-positivists the necessity to admit the reality of the outside world as a foundation of knowledge [24]. And his talk in memory of Maxwell [25] reveals to us that, if he appeared in his work a little estranged from the mechanistic view of the universe, he was at least sensitive to its beauty and grandeur.

In his contribution to the development of statistical thermodynamics, Planck assuredly does not reach the depth of Boltzmann or Gibbs; but despite a few weaknesses, the work he accomplished in this field is substantial enough to deserve the place that he himself assigned to it: it gives us the basis for a rational extension of classical ideas, destined to incorporate the fundamental element of discontinuity expressed by the great law of the quantum of action.

NOTES

* Lecture given April 24, 1958, at the University of Coimbra (Portugal), during the celebration of the 100th anniversary of the birth of Max Planck, organized by the School of Science of that University.

[1] For the history of the first phase of the development of quantum theory see L. Rosenfeld [1].

[2] I tried, elsewhere, to analyze the opposing views of Boltzmann and Gibbs [20]. I take advantage of this opportunity to correct an inaccuracy in the report that I gave on this work, concerning an anecdote about the opinion expressed by Boltzmann on the work of Gibbs.

I attributed to Boltzmann the remark – directed to Gibbs' efforts to extend his methods to systems with any number of degrees of freedom – that there is no interest in applying statistics to a sewing machine. Professor O. Klein, who also heard this anecdote from Ehrenfest, but who remembered it better than I, reminded me that the mentioned flash of wit was, as a matter of fact, by Ehrenfest, whom Boltzmann had asked for his views on Gibbs' book. Naturally, this amused Boltzmann very much; but, Klein added, later on Boltzmann nevertheless talked with admiration about Gibbs. I am not sure, however, that I should agree with Klein on this last point. If we refer to the passage [21] where Boltzmann publicly gives his appreciation of Gibbs, we see that the praise is rather cold; we notice that he takes care to characterize Gibbs' work as a systematization rather than as an innovation.

3 An account of this method, completing the one which was first published by Tolman, can be found in my article cited above [20].

BIBLIOGRAPHY

[1] Rosenfeld, L., *Osiris* **2** (1936), 149. [See this volume, p. 193 – Ed.]
[2] Planck, M., *Ann. d. Phys.* **1** (1900), 719.
[3] Einstein, A., *Phys. Zs.* **10** (1909), 185.
[4] Gibbs, W., *Elementary Principles in Statistical Mechanics* (Yale University Press, New Haven, 1902).
[5] Planck, M., *Verh. d. dtsch. phys. Ges.* **2** (1900), 202.
[6] Lord Rayleigh, *Phil. Mag.* **49** (1900), 539.
[7] Planck, M., *Verh. d. dtsch. phys. Ges.* **2** (1900), 237; *Ann. d. Phys.* **4** (1901), 553.
[8] Planck, M., *Ann. d. Phys.* **4** (1901), 564.
[9] Nernst, W., *Gött. Ber.*, No. I (1906).
[10] Planck, M., in *La théorie du rayonnement et les quanta* (Gauthier-Villars, Paris, 1912), p. 93.
[11] Planck, M. *Phys. Zs.* **13** (1912), 165.
[12] Planck, M., *Berl. Ber.* (1916), 633.
[13] Planck, M., *Berl. Ber.* (1925), 49.
[14] Schrödinger, E., *Berl. Ber.* (1925), 434.
[15] Einstein, A., *Berl. Ber.* (1924), 261; (1925), 3, 18.
[16] Planck, M., *Zs. f. Phys.* **35** (1925), 155.
[17] Lorentz, H. A., *Les théories statistiques en thermodynamique* (Teubner, Leipzig–Berlin, 1916).
[18] Planck, M., *Theorie der Wärmestrahlung*, 4th edition (Barth, Leipzig, 1921).
[19] Ehrenfest, P. and Trkal, V., *Ann. d. Phys.* **65** (1921), 609.
[20] Rosenfeld, L., *Acta Phys. Pol.* **14** (1955), 3. [See this volume, p. 762. – Ed.]
[21] Boltzmann, L., *Populäre Schriften* (Barth, Leipzig, 1905), pp. 360–361. [The relevant passage is translated in Ludwig Boltzmann, *Theoretical Physics and Philosophical Problems*, pp. 171–172 (*Vienna Circle Collection*, vol. 5, D. Reidel, Dordrecht-Holland, and Boston, U.S.A., 1974) – Ed.]
[22] Schrödinger, E., *Phys. Zs.* **25** (1924), 41.
[23] Planck, M., *Ann. d. Phys.* **66** (1925), 365; *Berl. Ber.* (1925), 442.
[24] Planck, M., 'Die Einheit des physikalischen Weltbildes' (1909), 'Positivismus und reale Aussenwelt' (1930) in *Vorträge und Erinnerungen* (Hirzel, Stuttgart, 1949). [The first

paper appeared in English translation by Ann Toulmin: 'The Unity of the Physical World Picture' in *Physical Reality*, S. Toulmin, ed. (Harper & Row, New York, 1970) – Ed.]

[25] Planck, M., in *James Clerk Maxwell: a Commemoration Volume* (Cambridge University Press, Cambridge, England, 1931).

CHAPTER 18

MATTER AND FORCE AFTER FIFTY YEARS OF QUANTUM THEORY

[1963d]

1. Matter and force

If one had asked any physicist about fifty or sixty years ago what was his view of the structure of the physical world, he would not have hesitated to say that it was based on an essential duality, on the interplay of two fundamental entities, which he would have called matter and force. This philosophy was propagated on the Continent in an immensely popular book by Büchner entitled *Stoff und Kraft*; it also gained favour in this country, under the less solemn guise of the character of the ideal physicist endowed by Tait with the name of Hermann Stoffkraft.

There was a general consensus that the structure of matter was atomistic: matter must somehow consist of elements all identical to each other, to which the old traditional name of atoms was given, although it was clear that they were not at all indivisible as the name implies, but must themselves have some kind of structure. However, our friend the physicist would have warned us that this view about the constitution of matter was entirely hypothetical: there was no direct evidence for the existence of the atoms, but only inferences from various chemical and physical phenomena.

About the structure of the atoms he would have been still more reticent. Opinions on this point had varied widely in the course of the century. At the beginning, Laplace had said that enquiry into the structure of atoms was hazardous speculation, useless to the progress of the arts. In fact, Laplace achieved the masterpiece of developing an atomic theory of matter without making any assumption about the actual structure of atoms, just by skilfully developing the consequences of the short-range character of the atomic forces. However, it was soon apparent that the atomic structure must be essentially governed by electromagnetic forces: atoms must consist of parts carrying electric charges of both signs. And by the end of the century, direct evidence was even forthcoming on the nature of the carriers of negative charge, the electrons, which had been separated in the form of cathode rays in electrical discharges in rarefied gases. One learned at that time how to ionise the atoms, how to split them into their electrical constituents, positive ions and negative electrons. One found that the electron has a mass extremely small compared with the total mass of the atom, so that practically all the mass must be attributed to the carrier of the positive charge. This was about all that one could then say definitely about the structure of the atoms.

As regards the nature of force, ideas had also fluctuated greatly. The original conception was that the atoms, in which was condensed, so to speak, the properly material constituent of the world, were held together by another entity, not material (though emanating from matter), which had the generic name of force, and could manifest itself under many different forms. The great achievement of the nineteenth century was to reduce this variety of specific forces, those governing the elastic properties of bodies, including the optical phenomena, those producing the chemical bonds, those exerted by electrical and magnetic bodies, to an essential simplicity: with the exception of gravitation, which remained something of a mystery and was cautiously kept aside, all forces then known were eventually traced to an electromagnetic origin. This result was only reached, however, after a long process of development, implying a radical change in the conception of force itself.

The starting point had been Newton's formulation of the law of gravitation as an immediate interaction between material elements, varying as a certain function of the distance between them. This description of gravitation as a "central force" between structureless elements of matter was in Newton's mind only a provisional one, inasmuch as it was absurd, from the physical point of view, to imagine that material bodies, like the planets of the solar system, could act instantaneously on each other at any distance without intermediary. In spite of Newton's warning, the epigones soon indulged in the facile expedient of regarding the concept of central force as fundamental, and inventing all kinds of specific central forces, with different laws of variation with distance, to account for the various kinds of phenomena.

As soon, however, as one started speculating on the electrical structure of the atoms, the way opened to an electrical interpretation of the chemical affinities and the cohesive forces determining the mechanical behaviour of matter. Still, this first unified account within the framework of central force physics left out one important class of phenomena: as the elastic properties of matter became better known, and the transverse nature of the vibrations of light was recognized, it appeared extremely difficult, if not impossible, to account for the optical phenomena by means of any imaginable elastic medium, obeying the same laws as the known material substances. Thus arose the vexed problem of the optical "aether", to the elucidation of which all the resources of central force physics were applied in vain.

Decisive progress was made by Faraday and Maxwell when they developed, in the realm of the electromagnetic phenomena, the more physical idea of a field of force, involving the intervention of a medium to transmit the force between the interacting bodies by successive and continuous propagation. By exhibiting the possibility of production of electromagnetic waves identical in character with those of light, the introduction of the field concept brilliantly completed the unification of the physical forces; but though in a sense it simplified the aether problem, it did not solve it. The problem was simpler inasmuch as one had to conceive only a single, universal medium with the function of transmitting all forces between the material constituents; but the question of the nature of this medium remained as acute as ever.

In fact, when considered from the point of view of the field concept, it even raised, as Maxwell shrewdly perceived, a peculiar dilemma. It was natural, according to the atomistic conception of matter, to regard the aether itself as a material medium of definite atomic structure; but then, its atoms, which would be separated by some distance, would have also to interact between themselves through this distance in order to transmit forces between the atoms of ordinary matter, and the account of such interaction would necessitate the introduction of another transmitting medium: there was clearly no means to stop the argument from proceeding *ad infinitum*. This difficulty was shunned by hopefully assuming that it would be sufficient to treat the aether as a continuous medium, endowed with suitable material characteristics, without enquiring into its possible atomistic structure. Faraday and Maxwell held the view that the atoms of gross matter were somehow floating in the continuous aether, or were perhaps just peculiar condensations of the substance of the aether moving through it. This conception had the happy consequence of leading Maxwell, in the analysis of the properties of dielectrics, to ascribe to the aether a more essential role than to the atoms of matter accidentally spread through it, and thus to conceive the production of a displacement current in the free aether by the same mechanism which gave rise to a polarisation current in dielectric media.

This is just one example of the remarkable fecundity of the seemingly so vague and hypothetical ideas on which theoretical speculation in the physics of the nineteenth century was based. At the same time, it also helps us to understand why the conclusions reached, even by great physicists like Maxwell and Boltzmann, about the structure of matter could never carry full conviction. Thus, the success of Maxwell's theory was no argument in favour of Maxwell's special conception of the aether and its relation to matter. Scepticism in such respect was most strongly voiced by Hertz, whose discovery of the electromagnetic waves gave Maxwell's theory its physical basis: Maxwell's theory, he said, is just Maxwell's equations. This statement gives pregnant expression indeed to the feeling of resignation produced by the persistent failure to construct any acceptable mechanical model of the aether.

2. Statistical mechanics

The aether problem, however, was not the only one confronting the physicists in their attempt to draw a dynamical picture of the physical universe. The task of the atomistic theory of matter, of accounting for the properties of gross bodies from the view that they are systems of atoms, interacting with given forces, had been accomplished by Boltzmann and Maxwell through the introduction of special methods to cope with such mechanical systems consisting of an immense number of constituents. As Maxwell said, these methods were borrowed from the Department of Statistics: when you have to cope with assemblies of a very large number of individuals, the fate of each individual becomes irrelevant; you are only concerned with the average fate of such assemblies, and you are led quite naturally to concepts of probability and statistics.

Now, the application of statistical methods to large mechanical systems yielded in particular a very simple and striking result about the way in which the energy of such a system is distributed on the average between its elements: each degree of freedom receives the same average amount of kinetic energy. It was therefore a matter of serious concern that such an immediate consequence of the general laws of mechanics sharply conflicted with an inference, apparently just as direct and convincing, about the structure of atoms, drawn from the evidence of the optical emission spectra. The emission of light was naturally interpreted as resulting from vibrations of the charged particles of the atoms around their equilibrium positions. On this view, the immense number of lines of a single emission spectrum, corresponding to different frequencies of oscillation, seemed to indicate that each atom would have an extremely large number of degrees of freedom. On the other hand, the specific heats of gases at ordinary temperatures, interpreted on the basis of the above-mentioned equipartition theorem, showed that the atomic interactions at such temperatures involved only a few degrees of freedom of each atom.

This riddle of the specific heats was regarded by the more critical minds of the time as so disturbing as to cast doubt on the very idea of the atomic constitution of matter. Boltzmann, of course, waved aside the difficulty: he was a man of strong faith. This one failure, which one could blame on our ignorance of the inner structure of atoms, could not outweigh the abundant evidence in favour of the atomistic conception furnished by the successful derivation of the laws of thermodynamics and of many thermal and mechanical properties of bodies. It is understandable, however, as we shall see presently, that other people took a different view.

The use of statistical methods in accounting for the thermal properties of matter raised another issue of a more philosophical character. The structure of mechanics and electro-magnetic theory was essentially deterministic: the physicist could pride himself that if you gave him the initial conditions required by the type of differential equations he had to solve, he could predict in a unique way the behaviour of the system for all future times, or (if there was any curiosity for this) uniquely reconstruct its past behaviour. From this point of view, the introduction of probabilities in thermodynamics was naturally regarded as an imperfection: statistics was introduced because there were no means of ascertaining the state at any instant of systems consisting of immense numbers of atoms; probability, as Poincaré put it, was a measure for our ignorance.

The contrast between deterministic and statistical description was not confined to the academic realm of methodology, however: it had immediate physical consequences of the greatest importance. It was the statistical element in the account of the time evolution of a large system that reflected itself in the irreversibility of this evolution, contrasting with the reversible character of its purely dynamical behaviour. Thus, the application of statistical considerations in this case could not be excused as a mere matter of practical convenience, or practical necessity: it appeared as a logical requirement for the interpretation of the second law of thermodynamics on an atomistic basis.

3. Tensions and contradictions

Let us stop at this stage, and, on the basis of the survey we have made so far, try to visualize the state of mind of physicists at the close of the last century. One sometimes hears the view that they were then in a sort of euphoria: they had mechanics to describe the motions of the atoms and electromagnetic theory to account for the forces causing these motions and their propagation through the aether; thus, in principle at least, everything was explained by these great "classical" theories (the word "classical" suggesting perfection), and there was nothing more to do than to make precise measurements in order to put the last touches to the picture. It is clear from what we have seen that nothing is farther from the truth: what a physicist would then have felt, would be not harmony, but tension: his conscience would have been troubled by the realization that the whole conception of the physical universe was jeopardized by unresolved contradictions.

The endeavour to analyse the mechanism of force transmission between material bodies in terms of the field concept, while highly successful so far as it was confined to a formal description of the electromagnetic actions, seemed doomed to failure when concerned with finding a physical basis for this description in the form of a mechanical model of the transmitting medium. The hypothesis of the atomic constitution of matter was found at the same time to give convincing account of many properties of the bodies and to lead to contradictory inferences from thermal and optical evidence about the number of degrees of freedom of the atomic structures. Moreover, the necessity of using statistical methods in the treatment of the large systems of atoms constituting the material bodies of ordinary experience precluded any explicit fulfilment of the deterministic account of the evolution of the world, which was regarded as the ideal of physical description.

These deep-seated difficulties, it will be noticed, had a common origin: they all arose from the unquestioned requirement that the structure of the material world, including that of the universal aether, should be describable in mechanical terms. Even the most imaginative pioneers, men like Maxwell and Boltzmann, did not manage to free themselves from the mechanistic tradition which had dominated physical thinking since the days of Huygens and Newton. It is true that in many respects the ideal of a mechanical explanation of the world had been an extremely fruitful source of inspiration; but by the end of the nineteenth century, the intractable character of the accumulating contradictions led to a growing suspicion that the mechanistic philosophy had somehow (but nobody could tell how) overreached itself.

Among those who drew up in open revolt against mechanicism, there are two whose heretical views are especially instructive. One of them is Ostwald, who thought not only that there were no atoms, but that the whole of physics could be built up entirely on the sole principle of conservation of energy: all processes of the physical world could be adequately described and analysed just as transformations of energy. Now, this was going too far, and it was easy enough to expose the insufficiency of Ostwald's arguments. Ostwald's was a romantic nature; he was carried away by his enthusiastic advocacy of monism in physics: he wanted to found the whole of physics on a single principle, and he was too hasty in deciding that the universality of the energy concept was sufficient justification for adopting the principle of conservatism of energy as the only basic one. Still, there was a sound intuition in his stressing the significance of conservation laws as the expression for a fundamental aspect of the physical phenomena. He had the merit of bringing forward this view at a time when the prevalence of mechanistic considerations tended to focus attention on the detailed analysis of structure and change and to obscure the energetic aspect.

The other heretic whose independent views deserve more sympathetic consideration than they usually find is Mach. It is very unfair to remember Mach only as "the man who did not believe in atoms", without enquiring why he didn't, although he was careful to tell us, and his reasons were very interesting. Mach could not imagine that such a naïve picture of matter as a mechanical system of atoms could account for the complication of the processes that we actually observe; in other words he cast doubt on the possibility of localising the atoms and describing their motions in a simple-minded mechanical way. He emphasised the possibility that in order to account in detail for the behaviour of matter, we would need not ordinary dynamical co-ordinates, but quite different parameters, corresponding to a mode of description which would not be mechanical.

4. Field theory

The reaction against mechanicism did not stop at the critical attitude evidenced by Ostwald or Mach: it soon took a more constructive turn, aiming at a new conceptual framework from which the contradictions inherent in the mechanistic one would be eliminated. The direction in which to look for a suitable starting point was actually forced upon the physicists by the desperate situation of the aether problem. In view of the deadlock in the search for a mechanical model for the propagation of electromagnetic fields through space, it was tempting indeed to treat these fields as primary entities, sufficiently defined by the laws of their interdependence and ponderomotoric action. At first, in the minds of Hertz or Lorentz, for instance, this tactical decision did not imply any deliberate abandonment of the mechanistic ideal: it was simply recognized that Maxwell's theory furnished a self-sufficient tool for the analysis of the phenomena, thus allowing the troublesome problems of the mechanical substructure to be shelved. It was tempting, however, to swing to the opposite extreme, and to start speculating whether what we call gross matter could not just be the manifestation of some peculiar distribution of electromagnetic fields.

At this juncture, the advocates of the supremacy of the field idea received encouragement from the brilliant completion of electromagnetism achieved by Einstein, and the ulterior extension of the principle of relativity leading to a field theory of gravitation. Einstein's ideas, for all their greatness, did not contribute to resolving the dilemmas of the dualistic view of the universe. What Einstein did was rather to round off the picture of classical physics by disclosing the true invariance properties of spatio-temporal description and dynamical laws, and incorporating the force of gravitation in the same general concept as the electromagnetic forces. It is true that in exhibiting the relativity of simultaneity, and more generally the role of the gravitation field as a metrical field, Einstein, developing to its last consequences the view initiated by Hertz, definitively removed from physics the aether problem: the primary constituents of the universe were now the fields continuously distributed in space and time, and the structure of which was sufficiently defined by their differential equations. However, any such universal field theory faced the task of fitting into the field description the existence and properties of the atomic constituents of matter: the problem of the material structure of aether was just replaced by the no less formidable and no less elusive one of the field structure of matter. Indeed, the problem of matter, in this formulation, proved as much of a blind alley as the aether problem, and the idea of reducing all phenomena to effects of space-time distributions of fields, however attractive it might seem, could achieve nothing in the way of relieving the tensions in the matter-force dualism.

In one important respect, however, the establishment of the theory of relativity exerted a powerful influence on the later developments leading to the solution of the contradictions in the world picture: for the first time since Newton's days, one met here a physical problem whose solution depended on epistemological considerations. Neither the physical insight of Lorentz nor the mathematical power of Poincaré could untangle the contradictions of the electrodynamics of moving bodies; Einstein in one stroke solved the riddle by recognising that the problem was essentially related to the definition of the very concepts of space and time underlying all physical and mathematical analysis of the phenomena. Only on rare occasions are physicists thus forced to reflect on the deeper meaning of the concepts they use without analysing them because they belong to common human experience; but each of these occasions heralds the beginning of a new era in the history of science, marked by an essential improvement of the adaptation of our mental representations to the reality of the external world. Such an occasion presented itself again, as we shall see, with respect to the description of atomic processes; and the great example of Einstein, as well as his direct intervention in the critical discussion of the new conceptions, contributed in no small measure to their elucidation.

The method that Einstein used in his epistemological analysis of the relativity of time has been popularised by the well-known fictions of observers, one at the station, others on passing trains, exchanging light signals, an operation which no traveller or station master would ever dream of performing, but whose consideration nevertheless is a great help to imagination in visualising the implications of the logical argument and convincing us that, however unfamiliar they may be, they are at least not in contradiction with common experience. This is the famous method of *Gedankenexperimente*, of ideal experiments, whose considerable role in the history of scientific discovery had been forcefully pointed out by Mach.

According to Mach, the function of such ideal experiments is to exhibit the relationship between the abstract concepts (in Einstein's case the concepts of space and time) used in physical description and the data of immediate experience to which they originally refer. It is only by making this relationship explicit that the significance of the concepts – as codewords, so to speak, for definite manipulations of specified apparatus – is brought to light, thus allowing us to eliminate the paradoxes which result from the indiscriminate application of insufficiently analysed concepts outside the range of their validity, and to arrive at a more accurate formulation of physical laws. Einstein was very much influenced by Mach's ideas, and his ingenious use of ideal experiments in all his work served in turn as a model for the founders of quantum mechanics.

5. Quantisation and Structure of Radiation

The decisive new element which was brought to bear on the problems of matter and force was the discovery of the quantum of action by Planck. Planck was led to this discovery in studying the thermodynamics of electromagnetic radiation, in which one was confronted with a difficulty of the same kind as that of the specific heat problem, only still more glaringly apparent. The distribution of energy among the modes of oscillation of the radiation field is not at all in conformity with the equipartition theorem, but deviates more and more from the distribution expected on this basis as the frequency increases or the temperature decreases. The occurrence of a maximum in the observed distribution for a certain ratio of frequency and temperature points to the existence of a new fundamental constant of the dimension of a mechanical quantity of action. Planck recognised that there exists in fact an element or quantum of action, such that in any radiative process the action involved can only change by one such indivisible quantum: for radiation in thermal equilibrium, this leads to deviations from equipartition of the kind just needed to account for the energy distribution found experimentally. In the following years this quantisation, this introduction of an atomicity of action, was extended to mechanical systems performing simple oscillatory motions, such as the atoms in a crystal, and it was shown that also for such material systems, the problem of the specific heats could thus be solved by appealing to the quantal character of the process of energy exchange: a striking indication that the scope of Planck's law was truly universal.

Without leaving the domain of radiation, however, another momentous consequence of the existence of the quantum of action was brought to the fore by Einstein, who applied the quantisation idea to the constitution of radiation itself. He gave strong arguments to show that the structure of the radiation field had a peculiar corpuscular character, in the sense that the energy contained in it was not distributed continuously all over space as assumed in the classical Maxwell theory, but was somehow concentrated in the form of photons or quanta of energy and momentum: such photons could be absorbed or emitted by charged material systems in elementary processes obeying the mechanical conservation laws.

This bold conception of Einstein raised, however, a problem of quite a novel character: how had one to understand this dual aspect of a definite physical entity? For there was no question, of course, of just abandoning Maxwell's conception of the electromagnetic field and returning to a purely corpuscular theory of light. Einstein himself knew perfectly well that in doing so he would not get, for instance, Planck's formula but only an approximation to it: the photon concept was an idealisation, valid only for very high frequencies. Moreover, Planck, who spent the rest of his life in efforts to minimise the importance of his own discovery, was stressing the necessity of upholding the classical theory of radiation as the indispensable basis for the explanation of the optical phenomena of interference and diffraction.

Yet, all efforts to ascribe to the electromagnetic field properties which could have simulated its corpuscular features were of no avail: it was impossible to arrive at any reasonable picture of the elementary processes of interaction between matter and radiation. Now, we know that the duality of behaviour of light is irreducible: its paradoxical character, from the classical point of view, has forced us to a radical revision of the most fundamental conceptions of physics, and brought about a revolution in our view of the universe comparable to that which Galileo and Newton initiated when they created modern science.

6. Structure and Quantisation of Matter

The crisis did not only arise from the duality in the structure of light, but also from the equally surprising discovery that a similar duality was exhibited by the elementary constituents of matter. This discovery came at the end of a long development, in the course of which our ideas of the structure of matter took definite shape. It started with Rutherford's fundamental discovery that the positive charge of an atom is concentrated in a nucleus of much smaller dimensions than the atom, so that the negative electrons had to be imagined moving around this nucleus at distances very large compared with its dimensions. Why was the establishing of this picture of the atom of such momentous importance? Because it made it absolutely clear that atomic stability, revealed by all chemical experience, was outside the competence of classical physics. In fact, according to classical electrodynamics, such a planetary system of charged particles as the Rutherford atom would radiate its energy and collapse in a relatively short time.

We were now confronted in the description of material systems with a failure of classical description as irremediable as that of the classical theory of the radiation field. In the latter case, the quantisation of action had just restored stability by preventing the accumulation of the radiation in the lowest modes to be expected on classical theory. Bohr saw that in the case of atomic systems, nothing short of the same radical departure from classical ideas could save the situation. He introduced quantisation here in the form of his famous postulates, according to which a mechanical system can only exist in certain stationary states, whose energies are determined by appropriate conditions of quantisation of action, and radiative transitions can only take place between such stationary states with emission or absorption of a quantum of radiation of the appropriate frequency.

Bohr's postulates not only gave the clue to an understanding, in principle at least, of all chemical and optical properties of matter, but above all exhibited the universal function of the quantum of action as the stabilising element in the structure and interaction of matter and radiation. However, much work remained to be done in order to incorporate these postulates in a logical way in the framework of a general theory. This task was strenuously pursued during several years from the point of view of the quantisation of the mechanical behaviour of systems of particles. It might have been accomplished in this way, but historically its completion was notably accelerated by an unexpected new development.

The quantisation of the dynamical equations carried out by Heisenberg was a rational algorism for determining the stationary states of atomic systems and the probabilities of radiative transitions between such states: Heisenberg claimed that by thus providing a direct approach to these characteristics of the atomic processes, he had eliminated from the theory all features (such as the orbits of the atomic electrons) which were inaccessible to observation. However, as Pauli was quick to remark, he had overlooked situations in which trajectories did fall within the range of observation; after all, we do observe the orbit of the moon, although there is no reason not to apply to the motion of this body around the earth the same rules of quantisation as those of the electrons around the nucleus of an atom. This objection clearly pointed to an essential incompleteness of Heisenberg's quantum mechanics.

It was at this critical stage that help came from a line of thought entirely foreign, and even in a sense antagonistic, to the dynamical considerations pursued by the school of Bohr. The starting point of this attack on the quantum problem had been Louis de Broglie's bold speculation about a fundamental analogy between matter and radiation, manifesting itself, among other things, by the existence of the same duality of aspects for matter as for light: matter was known so far only under its particle aspect, but we ought to expect it also to exhibit wave properties similar to those of light, and connected with its dynamical properties by similar quantum relations. The question of the physical nature of the matter waves was left by de Broglie very much in the dark, but the striking elegance of his formal results stimulated Schrödinger to further speculation in the same vein. The similarity of structure between particle mechanics and geometrical optics, pointed out long ago by Hamilton, suggested that the wave aspect be the fundamental one, whereas the particle trajectories would correspond to an approximate description of the same type as the rays of light. The potential functions of mechanics would then enter into a generalized wave-equation for matter in the role of parameters determining a variable refractive index for the waves. The quantised stationary states of atomic systems found a natural interpretation as the proper modes of wave-motion defined by Schrödinger's equation.

Schrödinger himself was sanguine about the prospects thus opened for a pure field theory of matter, from which all features of discontinuity implied by the concept of particle and the quantum of action would be eliminated: thus, quantum transitions would just be resonance phenomena in the interference of the waves corresponding to the two stationary states involved, and the particles themselves would be identified with 'wavepackets' formed by the superposition of the proper modes with appropriate amplitudes and phases. Closer examination, however, soon disclosed fatal weaknesses in this view: the proposed interpretation of the quantum transitions did not allow, at least not naturally, for the induced emission and could therefore not lead to Planck's formula for the equilibrium distribution of radiation; the wave-packets associated with systems of particles could not at all be pictured in ordinary space, since the wave-functions depended on thc coordinates of all the particles, and moreover, even a wave-packet in ordinary space was generally too unstable to represent a single particle permanently confined to a small spatial domain.

What sealed the fate of Schrödinger's spirited attempt at tilting the balance towards the field aspect of matter, however, was the recognition that the mathematical scheme of his wave theory was actually equivalent to that of Heisenberg's quantum mechanics: the wave-function appeared as an operator effecting the transformation between a mode of description in terms of position co-ordinates and another in terms of quantum numbers of stationary states. Since both modes of description were equally abstract, equally remote from immediate physical intuition, there was no reason to favour either of them as in any sense more fundamental than the other. We were faced with regard to matter with the same duality of structure as for light, and it thus seemed at first sight that the duality of matter and force had now been doubled by a further duality of wave and particle aspects for matter as well as force.

The true implications of the situation, however, unfolded themselves with dramatic swiftness. Through their fusion, the two modes of description of matter clarified and strengthened each other. It became clear that the part played by the matter waves in the framework of the theory is an essentially statistical one: they cannot, except in limiting cases, give a concrete image of particles, but represent varying statistical distributions of particles in space and time. Such distributions, however, can be very unlike those governed by the laws of classical dynamics: for the superposition principle of wave mechanics allows for the occurrence of diffraction patterns analogous to those exhibited by Roentgen rays – a theoretical inference soon impressively verified by Davisson and Germer's fortuitous discovery of such a diffraction effect in the reflection of electrons from crystalline surfaces.

Besides the dynamical laws, modified by quantisation, the account of the behaviour of material particles requires another principle, specifying the symmetry properties of the distribution of any system of such particles between their individual states: for the electrons, in particular, this specification takes the form of an exclusion principle, prescribing that no two electrons can be in the same state. The exclusion principle, like the superposition principle of matter waves, is not amenable to classical pictures, and when combined with the latter leads to still more startling consequences. When applied to systems of identical particles interacting according to a given law, it gives rise to an additional interaction of a novel kind, called exchange interaction to remind us that it has its origin in the identity of the particles, and not at all because it could be attributed to any actual permutation of their locations. These exchange forces are of tremendous practical importance: they account for the homopolar chemical bonds, the internal field responsible for the magnetic behaviour of metals, and other essential features of atomic interactions. For our theme, however, the occurrence of such forces immediately related to the wave properties of matter is of decisive significance: it illustrates the fact that the matter waves can transmit force just as well as the electromagnetic waves.

The further development of quantum theory, involving the incorporation of relativistic features, such as the spin of the electron and the occurrence of positively charged anti-electrons, confirmed this inference and made it more precise: just as the electromagnetic field transmits interactions between the electrons, the field consisting of a pair of electrons of both signs transmits an interaction between photons; just as the electron is the source of an electromagnetic radiation and is acted upon by this field, a photon, in the presence of some system capable of exchanging momentum with it, can create a field consisting of an electron pair and experience a force from such a field. As more and more constituent entities of the universe were discovered, forming the present array of so-called elementary particles, the same mutual relationship was found to hold between any couple of such entities: under its field aspect, the one entity transmits some specific interaction between particles of the other kind; as a particle, it acts as a source of the field of the other kind and is acted upon by this field. Thus, the antagonism of matter and force vanishes from the scene, to be replaced by a universal particle-field duality affecting equally each of the constituent entities.

7. Complementarity

It now remains to reconcile ourselves with this peculiar duality of aspect, which had been such a puzzle since Einstein had found it in electromagnetic radiation, and which is not easier to grasp for being recognised as the common and universal basis of the conceptions of matter and force. Clearly, we are here confronted with an epistemological problem of the same fundamental nature as those of relativity theory. What do we mean by applying such words as 'particle' and 'field' to the same entity although they denote plainly contradictory conceptions? This problem was solved by Niels Bohr through the introduction of a new logical relation which he called complementarity.

We have here to do with a truly revolutionary advance in the theory of knowledge, with a widening of the frame of ordinary logic allowing us to make use of contradictory concepts in such a way as to avoid any mistake which a trivial violation of the principle of contradiction would entail. The possibility of achieving this follows from the simple remark that every one of our concepts has a limited scope of validity, and that you only get into difficulties when you try to give a concept an absolute meaning: for then you risk coming into conflict with the contradictory concept, which may also have its own range of validity. In such cases, the two concepts are said to be complementary: you must then enquire into the limits of their legitimate application, and if you are always careful of keeping within these bounds you do not run any risk of error. Thus, any apparent contradiction implied by a duality of aspects resolves itself and the duality appears as a true synthesis of the two aspects in a complementary relation to each other.

The concrete case of the complementarity of the particle and field aspects will accordingly require a detailed analysis of all the concepts used in connection with each of these aspects, in order to bring out the mutual limitations imposed on their applicability by the physical situation. To this end, it is essential to return to the very definition of such physical concepts. In order to realise what words like position and momentum actually imply, the only method is to go back to the imaginary experiments, the idealised manipulations of measuring apparatus, which allow us to assign to a physical object a definite position or momentum. This is the *raison d'être* of Bohr's famous discussion of measuring processes, by which he disclosed the origin of the reciprocal limitations, expressed by 'uncertainty relations', to the determination of quantities pertaining to the complementary modes of description. He could trace all such limitations to the law of the quantum of action, applied to the interaction between the system investigated and the measuring apparatus: it is the quantal nature of this interaction which prevents us from controlling with unlimited accuracy the two complementary types of behaviour of the system.

The kind of limitation we here meet is by no means an imperfection in our account of the phenomena: ideally, there is no limit in the accuracy with which we can ascribe any single characteristic to the system under investigation. What is implied by the complementarity relations, and so beautifully elucidated by Bohr's argument, is a much more radical transformation of our whole conception of scientific description. The object of such description is to communicate results of our experience in a language so chosen as to convey to fellow investigators a unique meaning, allowing them to reproduce, if they wish, these results or to draw safe conclusions from them. To be intelligible in this very practical sense, physical experience must therefore be formulated in terms of the classical modes of description, which have been elaborated precisely for this purpose. Experience about atoms, however, is necessarily of an indirect character, and the complementarity we here find between the classical modes of description is not of our choosing: it is forced upon us by the laws of nature and the peculiar relationship in which we, compelled as we are to use large amplifying devices to get down to processes on the atomic scale, stand to the world of atoms.

It should be clear that the explicit reference to ourselves (or at least to beings of comparable complication of organisation) in the last statement does not entail any restriction of the objective character of our account of atomic phenomena. This objectivity is fully guaranteed by the very requirement that this account should be reduced, in the last resort, to statements referring to immediate experience, common to all observers. In this connexion, the introduction of complementary relations appears, not as a hindrance, but as an indispensable logical tool to secure the correct use of an objective terminology.

In such a perspective the problem of determinism or statistical causality appears in a new light. Here again, the decision is not one of free choice or of appeal to philosophical considerations, which are nothing else than philosophical prejudices. That the fundamental laws of atomic physics are essentially and irreducibly statistical is a clear consequence of the whole situation just outlined: statistical causality is just the type of causality adapted to the formulation of these laws in a language objectively intelligible.

8. Retrospect and prospect

In the view of the universe to which the exploration of the atomic world has thus led us, hardly anything remains of our friend Stoffkraft's ideal. His picture of a universe governed by the rigidly determined interplay of matter and force was a *fata morgana.* Instead, we have learned to look upon atomic processes as individual responses of atomic systems to our experimental probings, following definite statistical expectations, and presenting us, according to our approach, one or the other of their complementary aspects. Yet, Herr Stoffkraft would be pleased to see that the classical conceptions, to whose elaboration the great masters of his time had devoted so much effort, still play a most essential part in the modern theories as providing the only rational language in which these theories can be formulated.

The criticism of nineteenth century trends voiced by Mach and Ostwald played no part in the modern developments. The two rebels lived long enough to witness the first experiments evidencing individual atomic processes, and to acknowledge very candidly that their scepticism on this score had gone too far; and nobody paid any attention to the other points they had made. Now, we may well recognise, retrospectively, that Mach's rejection of mechanical pictures of atomic phenomena and plea for a more abstract mode of description (although put in too general terms to have really constructive value) was at any rate not so far off the mark. One would fancy that our Schrödinger equations in multidimensional configuration space, involving further strange variables like spin, would have been much to his liking. As to Ostwald, he would hail the fundamental role played in quantum theory by the consideration of groups of transformations of variables and the resulting conservation laws; indeed, he would be entitled to point out that out of the wreckage of the edifice of nineteenth century physics the only construction emerging intact is the principle of conservation of energy.

Thus reminded that scepticism is sometimes not out of place, let us then ask what the future has in stock for us. Well, probably new revolutions in our views which may be just as sweeping and momentous as the one I have tried to describe. In the new domain of the so-called elementary particles which we are now beginning to explore it seems that the principles of quantum theory will meet limits of their validity. It is true that the fundamental postulates of the existence of stationary states and quantal transitions between them give us here continued and unfailing guidance; but with regard to the complementary aspect, the possibility of localising the processes in space and time, there are indications that we may expect some fundamental limitation. Which form it will take is impossible to say, but we may be sure that within this still wider framework the insight in the laws of atomic phenomena gained in these fifty years of struggle will retain its validity. What this long struggle, not only for the unravelling of the phenomena, but also for the elucidation of the nature of our own thinking has achieved, is greater harmony between the two sides of our existence: our multiple interactions with the external world and our endeavour to give a rational and objective account of this experience.

CHAPTER 19

MEN AND IDEAS IN THE HISTORY OF ATOMIC THEORY

[1971d]

The elucidation of the atomic constitution of matter and of the constitution of the atoms themselves has been one of the most significant advances in knowledge in our time; not only because it has placed on a sure foundation a universal conception of the innermost workings of all material systems – from galaxies to living organisms, – but also because, unexpectedly, it has brought about deepened understanding of the structure and meaning of scientific theory. All this progress has been the work of twentieth century physics: at the end of the preceding century, atoms were still hypothetical concepts; nowadays we know every detail of their constitution and properties, and are far advanced in the exploration of a still deeper structural layer – the elements out of which the atomic constituents are built, and the forces ensuring their stability. Every scientific enquiry involves a parallel development, in continual interaction, of experimental techniques and theoretical conceptions; when we lay the emphasis on theoretical issues, we should not forget that these issues have their origin in actual experimental situations, for it is only from this experimental origin that theoretical conclusions derive the peculiar character of inevitability that we express by the phrase "law of nature".

In the present instance, we are not only faced with new laws of nature, but even with a novel form of logical relationship between physical concepts: this is one of the rare cases in which one had to scrutinize the presuppositions for the use of concepts which had so far been uncritically applied without restriction. 'Classical' physics, as it was shaped by the great masters of the XIXth century and brought to final completion by Einstein's genius, proposes us an idealized mode of description of the physical phenomena as they appear to our direct observation – we call it "classical" not only to pay homage to its creators, but also to express the fact that it is so exquisitely adapted to its aim that it will adequately fulfil it for ever. It describes two fundamental aspects under which we perceive the causal relations between physical processes: on the one hand the sequence in time of processes localized in space, on the other the timeless and non-local connexion between them expressed by the laws of conservation of energy and momentum. Now, we have learned that the conditions under which we can observe individual atomic processes are such that the two aspects just mentioned are no longer strictly compatible, but subject to reciprocal limitations: they become 'complementary', and as a result, the causal connexion our conditions of observation enables us to establish between atomic processes is of a statistical type. Obviously, the recognition of such a deep-seated feature in the logical structure of a physical theory can only be the outcome of a long process of gradual clarification, the history of which would by far exceed the limits allotted to this essay[1]. I have accordingly selected for detailed treatment three steps among the most influential in the whole development, and also well-suited to illustrate the human aspects of a creative effort which is not the work of cold intellect alone, but engages the whole man.

The first episode is Niels Bohr's interpretation of the hydrogen spectrum, leading to the enunciation of the quantum postulates; the comparison of Bohr's argumentation with A. Haas' earlier, abortive attempt at the same problem will throw light on the circumstances which were decisive in giving atomic theory its initial impulse and orientation. The next cross-section brings us *in medias res*: it shows us physicists grappling with the dilemma of the nature of electromagnetic radiation – one of the earliest manifestations of the complementarity of space-time description and conservation laws; we shall see how a reluctant young American was caught in the radical mood prevailing in Copenhagen, and how Bohr, Kramers and Slater proposed to solve the dilemma by giving up every causal connexion between individual processes, soon to learn that this was too cheap a way out, that they had not fathomed yet the full depth of the difficulty. Finally, we shall witness the last stage which brought Heisenberg and Bohr, after weeks of tense discussion, to the insight that the newly evolved mathematical formalism of quantum mechanics was wide enough both to allow full scope to each of the two classical modes of description, and to provide rules imposing reciprocal limitations to their application; thus, logical consistency between mutually exclusive aspects of the phenomena could be upheld without sacrificing either of them.

To reconstruct the probable course of such events is not so simple a task as it may seem. The knowledge of the logical and physical issues involved only affords limited guidance, when our concern is not to produce some streamlined account, but to revive those hesitations, gropings, false steps, that the protagonists, in their reminiscences, unconsciously tend to gloss over or mythify. Private documents contemporary with the events are more telling: for example, Slater's hard time in Copenhagen is vividly evoked in an extensive collection of letters and memoranda written on the spur of the moment[2]. Moreover, much of my information comes from conversations with some of the main actors or witnesses; with one exception (A. Haas), it has been my privilege to be acquainted with all the characters of the following stories, with some intimately, with every one at least sufficiently to have observed how he approached scientific work and what science meant to him: this personal experience helped me a great deal in my endeavour to keep down the part of fiction and recapture as much as possible of fugitive truth.

1. The quantum postulates

For the formulation of the "quantum postulates", Niels Bohr's only guidance was a formula expressing the optical properties of the simplest possible atomic system: the hydrogen atom, consisting of a single electron bound to a nucleus. Yet (as he was fond of recalling later), he was convinced that these postulates would form the whole foundation for a rational incorporation of the quantum of action into the dynamics of the most general atomic systems and their electromagnetic interactions. This conviction was not a mere intuition, supported by the general breakdown of classical physics in dealing with atomic properties; it was the result of rigorous inference from a penetrating study of the nature and extent of this breakdown, which had shown him that it was beyond redemption: thus, he had proved in his doctoral thesis that the principles of classical physics left no room for the magnetic properties of metals. The hydrogen atom was by its very simplicity such a transparent case that the hint provided by the optical formula pointed unambiguously to the features in which atomic behaviour radically departed from the classical expectations.

It would be a great mistake, however, to imagine that the road to the theory of atomic constitution was clearly marked for every one; quite on the contrary, the physicists who then sensed the depth of the crisis brought about by the discovery of the quantum of action were no more than a handful, and among them the young Bohr was unique in his keen realization of the epistemological issues involved and his unflinching reliance on their logical implications. We are in the fortunate position of being able to illustrate the peculiarity of Bohr's line of thought by comparing it with that of a fellow-scientist of the same generation, A. Haas, who like Bohr was deeply interested in the philosophical aspects of physics, and who may be taken as a distinguished representative of the then prevailing attitude of mind, which we now call "classical". Outwardly, Haas' and Bohr's investigations look very similar: both bring the radiation emitted by a hydrogen atom in relation with the quantum theory, and both arrive at the conclusion that the parameter occurring in the optical formula is a 'universal constant', which can be expressed in terms of fundamental atomic quantities: the charge and mass of the electron and the quantum of action; there, however, the resemblance stops, and on closer examination, the two lines of argument are as different as they could be.

It will help first to outline the formal background in terms sufficiently general to apply to both calculations. Following J. J. Thomson's example, both Haas and Bohr, in their description of the hydrogen atom, limit themselves to the circular orbits of a negatively charged electron in the electrostatic field of an equal positive charge uniformly distributed over a sphere of finite radius; Bohr, of course, adopts Rutherford's nuclear model, which amounts to assuming the positive charge to be concentrated in a sphere of vanishingly small radius. In each orbit, the electron revolves with a definite frequency, but classical mechanics would allow of any value for this frequency. In order, therefore, to account for the fact that the actually emitted optical frequencies form a discrete sequence, some additional condition is needed to restrict the manifold of possible orbits. The decisive step now is to invoke for this purpose a quantum condition, analogous to that introduced by Planck in his theory of radiation: we may choose it to specify that the kinetic energy of the electron on an allowed orbit must be proportional to the energy quantum $h\nu$ – product of the frequency ν of the electron on this orbit by Planck's constant h; the factor of proportionality, which may be written $\frac{1}{2}k$, has yet to be determined. By combining the quantum condition with the classical Keplerian relation between the frequency and the radius of the orbit, one obtains for the frequencies of the allowed motions the sequence of values

$$(1) \qquad \nu = 2\frac{R_0}{k^3},$$

where R_0 represents a certain combination of the quantities defining the electron's behaviour (its mass and charge) and Planck's constant: thus, R_0 appears as a universal constant, whose numerical value can be calculated in terms of these fundamental atomic quantities. For the kinetic energies K of the electron on the allowed orbits, one has the equivalent sequence of values

$$(2) \qquad K = \frac{hR_0}{k^2}.$$

These were, reduced to the essential, the two relations from which the young men both took their starting point. Let us now examine what use they make of them, and in particular how they confront them with the empirical formulae giving the frequencies of the series in which the lines of the hydrogen spectrum had been arranged. Balmer had been the first to recognize the prominent series

$$(3) \qquad \nu_n = \nu_\infty \left(1 - \frac{4}{n^2}\right), \quad n = 3, 4, \ldots$$

where ν_∞ represents an accurately known limiting frequency; Rydberg had later shown the Balmer series to be a special case (for $m=2$) of a more general class of the type

$$(4) \qquad \nu_{nm} = R\left(\frac{1}{m^2} - \frac{1}{n^2}\right), \quad m, n = 1, 2, \ldots$$

with a constant $R = 4\nu_\infty$.

What strikes us in the first place is the difference in motivation. Bohr regards the quantum of action as a new element essentially foreign to classical physics and tries to incorporate it into the description of atomic structures; Haas on the contrary sees as his task the incorporation of the quantum of action into classical physics, – a programme that Planck himself would not have disavowed. However, when he hopes to achieve this result by quantizing a 'real' atom rather than a Planck 'resonator', he seems to disregard the logical difficulty, duly stressed by Planck, of making a universal feature of the radiation field appear to depend on the peculiarities of an atomic model. He does not attempt to interpret the whole spectral series represented by the Balmer formula (3), but he only tries to identify the state of the atom from which the limiting frequency ν_∞ is emitted, and he naturally assumes that this optical frequency is equal to the frequency of revolution of the electron in the state in question. By a reasoning so weak that it amounts to an arbitrary selection, he picks out as the required state that whose orbit is on the surface of the positive sphere, and which can thus be regarded as a state of harmonic motion; accordingly, he assumes that it fulfils the same quantum relation as the corresponding state of a harmonic oscillator, which means choosing $k=1$ in the relation (1): this leads him then to the identification $\nu_\infty = 2R_0$. It should be noted that this relation is wrong by a factor 8, – a discrepancy which was masked at the time by the uncertainty in the value of the electron charge.

Bohr starts from Rutherford's nuclear atom because he judges that Rutherford's conclusion about the concentration of the positive charge in a nucleus of vanishing dimensions is a clear induction from experiment, which leaves us no alternative. He realizes that on this model the stability of atomic and molecular systems is beyond the reach of classical mechanics: the stable states of these systems must be singled out by some non-mechanical condition. This is precisely the place where one may expect an intervention of the quantum of action. Thus, he arrives at a frequency condition of the form (1), but lacks a further clue to fix the coefficient k. In a first attempt, before he has become aware of Rydberg's formula, Bohr proceeds very much like Haas: limiting himself to the consideration of the stablest orbit, he tries to find the frequency of the electron in it by an appeal to the classical theory of optical dispersion; according to this theory radiation impinging on an electric system in periodic motion enters into resonance with it, giving rise to an enhancement of the scattering of radiation of the same frequency as the periodic motion. Actually, the optical resonance frequency is only known from dispersion measurements for the hydrogen molecule; but he has also worked out a simple mechanical model for this molecule, and uses it for the identification he needs of a mechanical and an optical frequency; he seems to expect that the value of k resulting from this particular identification might have universal validity.

When he sees the Rydberg formula, however, he at once reads from it the true answer to the problem and marvels at the much deeper harmony it reveals between the seemingly irreconcilable features of classical physics and quantum theory. If one takes the natural step from the frequencies ν_{nm} of a spectral series to the corresponding quanta of radiated energy $h\nu_{nm}$, Rydberg's formula (4) shows that these energy quanta are differences of similarly built energy 'terms' hR/n^2. Then, a glance at the expression (2) for the kinetic energy of a stable state (which, except for the sign, is the same as its binding energy) strongly suggests that these terms indeed represent the binding energies of a sequence of such stable or stationary states, characterized by the successive values of an integer n; this interpretation simply requires that the factor k be taken proportional (or equal) to n. The consideration of states containing increasing numbers of energy quanta was, of course, not unfamiliar: the assumption that the modes of harmonic vibration of crystal lattices were somehow restricted to a sequence of energies $nh\nu$ had actually been used in previous calculations of specific heats of solids.

What we now learn from Rydberg's formula is that a system quantized in this way can only emit or absorb radiation from one of its stationary states, and that the process consists in a transition of the system to another stationary state, accompanied by the emission or absorption of a single radiation quantum of definite frequency. A startling aspect of this radiation mechanism is that the frequency of the emitted or absorbed quantum is in general different from those of either the initial or the final state of the atom. Now, no feature of classical radiation theory would seem more natural, more obvious even, than that a vibrating source of definite frequency produces a field oscillating with the same frequency or its higher harmonics (i.e. integral multiples). To abandon this representation was a step requiring uncommon independence of judgment, but for Bohr it was not a foolhardy step: prepared as he was for the necessity of radical change, he found the message of Rydberg's formula so clear as to be compelling, and he did not hesitate to give his conclusions the form of 'postulates' of general scope and fundamental significance. The first postulate enunciates, for every atomic system, the existence of stationary states, in which the system has no interaction with the radiation field; the second postulate describes the transition mechanism by which such an interaction can take place.

On the face of it, the quantum postulates seem to imply a complete denial of the very foundation of classical mechanics and electromagnetism. Yet, the break cannot be complete: it should be possible to throw a bridge between the atomic realm, governed by quantum rules, and the domain of large scale phenomena, so accurately accounted for by the laws of classical physics. The clear recognition of this requirement is conspicuous from the very beginning in Bohr's cautious argumentation: every intervention of the quantum of action must be such as to secure a merging of the quantum relation into a 'corresponding' classical one when the number of quanta involved becomes very large. In the case of the hydrogen atom, it is readily verified that the two postulates can be made to satisfy this essential condition, by observing that the interval between the energies of consecutive stationary states, for large values of the quantum number n, becomes approximately $2hR/n^3$: this shows, in the first place, how the discrete sequence of these stationary states becomes less and less different from the continuous manifold of classical states of binding of the electron; in the second place, comparison with the formula (1) shows that, for large values of n, the frequency $\nu_{n+1,n}$ of the radiation quantum emitted in the transition between consecutive states may indeed become nearly equal to the classical frequency of revolution of the electron on its classical orbit, provided that the Rydberg constant R is identified with R_0 and k taken equal (and not merely proportional) to n. Thus, the requirement that the quantum transition mechanism should go over into the classical one completes the unambiguous determination of the quantum condition, and leads to a very stringent test of Bohr's interpretation of Rydberg's formula: the equality $R=R_0$ unambiguously relates Rydberg's empirical constant to the fundamental atomic quantities. There is of course no comparison between the crucial character of this test in Bohr's case and the loose order-of-magnitude relation set up by Haas: from the latter one could hardly draw any conclusion, whereas the former is the key-stone to a whole theoretical structure combining the nuclear atomic model with a precise formulation of the quantum laws governing its stability and radiative behaviour.

In the preceding analysis the accent has been put on the epistemological and methodical aspects of the question, because these were the decisive ones for its elucidation. Haas' failure is partly due to epistemological misconceptions, partly to methodical shortcomings: he did not realize that the existence of the quantum of action was a feature of the radiation field independent of the structure of matter and incompatible with classical electromagnetic theory, and he accordingly could not appreciate the depth and difficulty of the problem; moreover, he did not set out to scrutinize Balmer's formula in order to discover all the information hidden in it, but was too easily satisfied with a partial interpretation, which he did not try very hard to put on a firm logical basis. In Bohr's approach, we notice, fully developed already, the combination of qualities which will characterize all his future activity: uncompromising reliance on the consequences of rational analysis, and yet instant readiness to revise the basis of such an analysis according to the 'lesson' learned from nature. There was no conflict, in his thinking, between these seemingly opposite tendencies: both were part of his conscious striving for harmony between the observed phenomena and our rational account of them. This comparison, obviously, leads us beyond differences in epistemology or method, or rather to the root of such differences: to differences of temperament, themselves largely conditioned by social environment. Haas regarded science as an object worthy of the refined curiosity of a Viennese intellectual; Bohr conceived it as a human endeavour to achieve harmony with nature, a common task to which it was the earnest duty of each individual scientist to contribute his share. For Haas scientific work was a *geistreich* occupation, for Bohr it was (to use a modern concept) a total *engagement.*

2. The Dilemma of the Nature of Light

At the end of 1923, a young American physicist, John Slater, arrived at Copenhagen to work in Bohr's institute. He did not come empty-handed. During the preceding months, spent at the Cavendish laboratory, he had struggled with one of the most puzzling consequences of the quantum theory of radiation – the particle-like behaviour of radiation quanta in their interaction with atomic systems, and worked out a carefully reasoned proposal for a treatment of these interaction processes incorporating the quantum features without sacrificing the wave-like properties of the radiation field. This was a problem that had remained unsolved since Einstein had raised it in 1905 when he advocated, in the most extreme form, the hypothesis of a particle structure of electromagnetic radiation: radiation of definite frequency ν would consist of particles, the light-quanta (later called photons), of energy $h\nu$ and momentum $h\nu/c$, moving with the velocity of light c. He was prompted to this radical conception by the evidence of processes in which individual atomic systems interact with a radiation field: the course of such processes is indeed quite at variance with the classical picture of an extended distribution of radiation energy, and rather points to the localization of this energy implied by the photon concept. Einstein was content to put off such obvious objections as the incompatibility between the localization of the photon and the strict periodicity ascribed to it, which would demand a process of indefinite extension in space and time, or the difficulty of accounting for interference and diffraction phenomena without appealing to wave-propagation; with most physicists, Planck in particular, such objections weighed very heavily against Einstein's standpoint.

In any case, as Einstein himself noted, the photon concept could not claim more than approximate validity as an extreme idealization. Thus, it would not do to compare radiation in thermal equilibrium within an enclosure to a gas of photons; this simple-minded picture would not yield Planck's law for the equilibrium distribution, but only the limiting form known as Wien's law, which applies to photon energies much larger than the average thermal energy of each atom interacting with the radiation. In order to obtain the correct result, one must add to the entropy of the radiation precisely the contribution which would follow from the classical wave-theory. The origin of this failure of the extreme particle picture was greatly clarified when Einstein resumed the discussion of the problem in 1916. He analysed directly the processes of energy and momentum exchange maintaining thermal equilibrium at a given temperature in an assembly of photons and atoms. In the spirit of Bohr's quantum postulates, these processes could be described as transitions of the atoms between their stationary states, accompanied by absorption or emission of photons of the appropriate frequency. Einstein's method consisted in drawing up the balance of all the processes by which the average population of the atoms in any given stationary state is increased or decreased during a given time. To this end, a definite probability of occurrence per unit time is assigned to each type of process. The probability of absorption of a photon of given momentum will be proportional to the average intensity of radiation in the direction of the photon; the emission of photons may occur spontaneously in any direction, independently of the radiation intensity. These two types of process, however, lead together to Wien's law; in order to obtain Planck's law, it is necessary to introduce a further type of transition: the 'induced' emission of photons, depending – like the inverse absorption process – on the average radiation intensity present in each direction. This, then, is the element foreign to the simple picture of particle interaction that has to qualify the extreme photon idea. The parallelism between absorption and induced emission is immediately suggested by the classical theory of radiation: the interaction of a classical radiation field with a resonator freely oscillating with its proper frequency will increase or decrease the energy of the resonator according to the phase difference between the vibrations of the resonator and those of the impinging field. Thus, impressive as it was by its beautiful simplicity and elegance, Einstein's analysis did not in the least resolve the dilemma of the nature of light: it only presented it under a new light. Nevertheless, it contributed (much after Einstein's intention) to make the photon idea popular: one had only to remember induced emission, and one could interpret and visualize all radiative processes as interactions between photons and atomic particles.

In 1923, the fruitfulness of the corpuscular point of view had been illustrated in a spectacular way by the interpretation of Compton's discovery of the frequency shift accompanying the scattering of high frequency radiation by atomic electrons (which in this reaction behave approximately as free particles): the process could be described as an absorption of the incident photon, followed by the emission of a photon of different frequency, and a straightforward application of the laws of conservation of energy and momentum sufficed to obtain the exact expression for the frequency shift. Thoughtful people pointed out that (as Compton himself had originally noticed) this formula could also be interpreted as the Doppler shift affecting a classical radiation wave scattered by a moving charge (the velocity of this charge would be exactly that of the electron in the frame of reference in which the total momentum of the electron and the photon vanishes). This, however, did not prevent the less cautious from talking about the 'physical reality' of the photons, without giving sufficient thought to the meaning of such a treacherous phrase.

On a physicist educated in the British and American tradition, with its strong liking for simple, concrete representations of the phenomena, offering a workable line of approach to quantitative analysis, the photon conception could not fail to exert its seduction. Slater, however, was critical enough to notice that, for all its apparent clarity, it left a lot of things in the shadow: in fact, he realized that the classical conception of the wave-like distribution of the radiation field in space and time could not be dispensed with, and that it had somehow to be related in a consistent way with the photon conception. In particular, he looked critically upon the quantum postulates themselves: he was concerned about the incompatibility, from the point of view of classical radiation theory, between ascribing a quantum transition a definite frequency and speaking of it as 'instantaneous'. To get a sharply defined frequency we need a long vibration time, and the only long time at our disposal is the life-time of the stationary state from which the transition takes place. Hence, he was led to imagine that an atom, in each of its excited stationary states, generates a radiation field containing components of all the characteristic quantum frequencies; this classical field would not be observable as such – it would be 'virtual' – but it would nevertheless be the agent regulating all observable phenomena: it would react upon the atom creating it, and also act upon other atoms, in such a way as to give rise to the actual quantum transitions of the atoms between all their stationary states, each type of transition being governed by the virtual field component of the appropriate frequency. When a transition occurs in any atom, it would cease to produce the virtual radiation field associated with its initial state, and start generating that corresponding to its new stationary state. Moreover, each transition would be accompanied by the emission or the absorption of a photon, and a further essential function of the virtual radiation field would be to 'guide' the photons along definite paths (determined at each point by the Poynting vector of the field) so as to distribute in space and time the actual electromagnetic energy and momentum carried by the photons in accordance with the observed phenomena of absorption and dispersion, as well as interference and diffraction.

A critical feature of this remarkable synthetic effort was the specification of the connexion between the virtual radiation field and the motions of the atomic electrons. Here, however, Slater had not to innovate: he could incorporate into his theory the views developed by Bohr in the preceding years about the general relationship between the classical and quantal properties of atomic systems and their radiative interactions. Bohr had sought guidance in the 'correspondence' argument, whose significance he had recognized since his earliest work on the hydrogen spectrum; he had now turned this argument from a straightforward limiting condition into a heuristic principle of great power. The motions of complex atomic systems can be represented as superpositions of periodic motions, the frequencies of which are sums of a number of fundamental frequencies and their harmonics; there are simple relations between these frequencies and the quantum conditions defining the stationary states, which make it possible, in the limit of large quantum numbers, to establish a correspondence between the various harmonic components of the multiperiodic motion and the actual quantum transitions between neighbouring stationary states. This correspondence is then postulated to subsist for all transitions, even in the proper domain of quantum theory. In the hands of Bohr and his first disciple and faithful collaborator Kramers, the 'correspondence principle' had proved especially fruitful for studying the optical properties of atoms: thus, the intensities of spectral lines were assumed to have the same mutual relations, if not exactly, at least approximately, as the squares of the amplitudes of the corresponding harmonic components of the classical motion.

This method fitted naturally in Slater's scheme: the sources of the virtual radiation field would not be the moving atomic electrons, but a set of 'virtual oscillators' corresponding to the harmonic components of their motions; according to this correspondence, the proper frequencies of the virtual oscillators could be identified with the transition frequencies, and their amplitudes related to the transition probabilities. Clearly, the recourse to the correspondence principle destroys any direct physical connexion between the radiation field and the classical motions of the atomic electrons; this is particularly striking in the case of aperiodic processes, as we have seen by the example of the connexion between the Compton shift and the classical Doppler effect. Indeed, Slater found himself operating with a disembodied radiation field (he actually spoke of a 'ghost' field), which could only provide a formal link between what were, in his view, the only substantial elements of the phenomena: the atoms and the photons. Nevertheless, he conceived this link as a strictly causal one; admittedly, from correspondence considerations he could only estimate the probabilities with which the virtual field could give rise to quantum transitions, but he expected this statistical stage to be merely provisional. At any rate, he regarded the whole behaviour of the photons, and in particular the exchange of energy and momentum between distant atoms by emission and consecutive absorption of photons, as causally regulated by the virtual field.

No wonder that when, on his arrival at Copenhagen, he presented his ideas to Bohr and Kramers, he was greatly disturbed by their reaction. Long and intensive study of the significance and implications of the correspondence principle had convinced Bohr, and with him Kramers, that any possibility of analysing atomic processes in space and time on the classical deterministic pattern was ruled out, and that a purely statistical account of these processes, such as Einstein's characterization of radiative interactions by specific transition probabilities, was the adequate mode of description one should strive for. Bohr would point out that statistical causality was rooted in the quantum postulates themselves, since there was nothing in the definition of a stationary state that could lead to a unique determination of the quantum transitions it would undergo. He was therefore decidedly averse to Slater's suggestion of establishing a causal connexion between distant atoms by his virtual radiation field. The device itself of this virtual field, which extended considerably the scope of the correspondence principle, appealed to him very much, but the communication between distant atoms it would set up could only be of a statistical character: there could be no question of imagining photons transporting energy and momentum along definite space-time trajectories; indeed, there was no use at all for the photon concept. Bohr's customary way of presenting the problems in their most general context could not make much impression on Slater's pragmatic turn of mind, and the eagerness with which Bohr tried to persuade him to forsake space-time description simply hurt his strong sense of independence: "By God", he thought, "I am determined to get a space-time description!" and he resented paternal admonition. However, he felt the force of Kramers' more specific argumentation: the correlation we can expect between a process of emission by an atom and a process of absorption by another atom owing to the action of the virtual field upon these atoms is only governed, so far at least as the correspondence method goes, by the respective transition probabilities: it would therefore be inconsistent to assume this same field to prescribe uniquely the succession and separation in time of these events by acting in a strictly determined way upon an hypothetical photon involved in them. Clearly, the only course open to Slater was to yield to the others' proposal to drop the photons entirely and only retain the idea of the virtual radiation field establishing a purely statistical correlation between quantum transitions. Still, he could not defend himself from the impression that such an amputation of his theory left little more in it of his original intention than the grin of the Cheshire cat.

The step taken by Bohr, Kramers and Slater was indeed a radical one. The only observable features of radiative processes left over after the elimination of the photons were the quantum transitions of the atoms themselves, and these were assumed to occur at random, without any definite correlation in space and time, and consequently without any strict balancing of the energy and momentum changes accompanying these transitions. The role of the photons, as Slater understood it, in conformity with Einstein's own conception, was precisely to supply a physical basis for the conservation of the energy and momentum exchanged between atom and radiation in each individual process. In the new theory one could only expect the conservation laws to be satisfied on the average for a large number of such processes. This was at any rate a theoretical prediction that could be tested, and it was too challenging not to stimulate the ingenuity of the experimentalists. The Compton effect suggested itself as the simplest phenomenon on which a complete test of the conservation laws could be performed, and within a short time observations of single Compton processes led to unambiguous decision: counter experiments verified the exact correlation in time between the recoil of the electron and the detection of the scattered radiation, and cloud chamber photographs exhibited the precise correlation between the directions of recoil and scattering which the conservation laws led one to expect according to the simple-minded photon picture. These results removed every doubt about the status of the photon concept: it had definitely more than formal significance, and represented an essential quantum property of electromagnetic radiation. On the other hand, they did not affect the equally significant conclusion reached by Bohr about the role of the classical wave-aspect of the radiation in the correspondence considerations yielding statistical indications about the occurrence of the quantum transitions. Thus, the new experiments, far from vindicating Slater's original proposal, reinstated the dilemma of the nature of light on a wider experimental foundation, which only enhanced its acuity.

It may seem strange, at first sight, that Bohr did not show more appreciation for Slater's attempt to deal with this dualism. The fact is that Bohr had always looked with suspicion upon Einstein's photon idea. Certainly, it was clear to him that the dualism of the wave-properties of radiation and the quantal aspect of its interaction with matter was irreducible – this at a time when both Planck and Einstein were hoping against hope that one of the contrasting aspects would in the end be subordinated to the other, Planck defending the supremacy of the classical theory, Einstein holding the corpuscular view for the deeper one. In this debate, however, while he differed from both contestants on the main issue, Bohr's sympathies were definitely on Planck's side: in this view, the correspondence argument allowed us to ascribe 'physical reality' to the classical radiation field in the well-defined sense that this field is – under appropriate conditions – accessible to direct observation. By this criterion, the photon concept would of course appear to be just a *façon de parler*, not necessarily a useful one. Bohr was thus inclined to connect the quantal side of the dualism, not with the radiation, but rather with the atomic systems, since the constituent particles of the latter could also be conferred physical reality in the sense of the correspondence argument. Against this background, it is easier to understand Bohr's reaction to Slater's standpoint: on the one hand, he could not condone Slater's sin against correspondence in applying classical mechanics to photons; on the other, he saw his way to exploiting Slater's idea of the virtual radiation field in order to attempt a synthesis of the dual aspects of radiation phenomena as he conceived them. No one who remembers the bewilderment he experienced on his first contact with Bohr will blame the young American physicist freshly landed in Copenhagen for finding Bohr's views 'hazy'. As it appears, it turned out that, in this case, they were too definite.

Nevertheless, the photon experiments had remarkably little immediate influence on the orientation of Bohr's thinking and the course he and Kramers were keeping in the investigation of radiative processes; the guiding methodical principle remained the same: giving up any attempt at visualizing the behaviour of atomic systems in space and time and adopting instead a purely statistical point of view. Kramers' work on the correspondence theory of optical dispersion was well under way before the Slater intermezzo. Indeed, the discussion of the points raised by Slater offered Bohr and Kramers an eagerly seized opportunity of making their own position clearer to themselves, of stating explicitly in which direction they thought the search for a consistent formulation of the quantum theory of radiation phenomena had to proceed; and the new insight about the detailed balance of energy and momentum in these phenomena, while revealing a deep-seated limitation of the purely statistical approach, did not shake confidence in its extensive validity. It even proved possible to check this validity in the case of scattering of light by atoms by extending to this case the comparison between the correspondence approach and the photon picture which had been successful for the Compton effect: the general dispersion formula derived by Kramers and Heisenberg was found to be in complete harmony with the conclusions of Smekal's treatment of the problem on the basis of the photon conception. A real understanding of this harmony, however, had to wait for the completion of a rational quantum theory, which was itself the outcome of a struggle with conceptual contradictions ranging much farther than the wave-corpuscle dualism of light – contradictions whose existence was then still unsuspected.

We have here an example of the familiar dialectics of scientific thinking: the solution of a crisis does not come about until the underlying contradictions have reached the most extreme form to which they are susceptible. Bohr (in blissful ignorance of such philosophical lore) had a fine intuitive feeling for this dialectics, and he realized that the right policy in the circumstances was just to go ahead exploring the consequences of the correspondence principle from the extreme statistical point of view, without heeding too much its limitations. The whole dispersion work, culminating in Heisenberg's quantum mechanics, progressed under this inspiration: Heisenberg's achievement was to derive from the rules establishing the correspondence between the quantal and classical amplitudes and frequencies a method for dealing directly with the former without referring at all to classical theory; he could thus entirely dispense with such inadequate conceptions as the classical orbits of the electrons and operate exclusively, with the help of an autonomous quantal calculus, with the energies of the stationary states and the amplitudes characterizing the transitions between these states. This was quite in the spirit of the theory of Bohr, Kramers and Slater: in retrospect, this theory appears, not as a short-lived episode in the history of atomic physics, but as a manifesto of lasting influence, outlining a programme whose fulfilment disclosed essential aspects of the structure of matter and light.

3. Form and content of quantum mechanics

In the "golden afternoon" of 1922 when Bohr, during a long walk in the woods around Göttingen, initiated Heisenberg to his innermost thoughts on the unsolved problems of quantum theory, he had found in the keen student a responsive, but not uncritical listener. A German mind, brought up in the respect of philosophical tradition, could not help being impressed by the emphasis Bohr put upon the epistemological aspects of the problems; and Heisenberg was fascinated by the indefinite prospects which the subtleties of Bohr's dialectics presented to his vivid imagination. But the sense for the beauty of mathematical expression and the faith in its power, which his teacher Sommerfeld had inculcated in him, could not be reconciled with the cavalier way in which Bohr seemed to treat the formal side of the theory: in contrast with Sommerfeld, who took the elaborate mechanical models of atomic systems quite seriously, Bohr only considered them as a provisional scaffolding whose only role was to support physical imagination in its groping search for the fundamental laws of atomic processes. One would think that such an opposition of points of view could not go very deep: in the analysis of atomic models fine physical intuition constantly inspired mathematical virtuosity in Munich, and mathematical elegance in the formulation of such intuitive ideas as the correspondence principle was not scorned in Copenhagen; about the ultimate aim of the quest: harmony between mathematical form and physical content of the theory, there was of course complete agreement. Yet, when it came to the critical point of deciding which of the two approaches should be chosen for the final assault, the tension between the two protagonists reached a high pitch until the full elucidation of the issue restored peace and harmony.

Before relating this dramatic development, a third character must be introduced, whose intervention was essential for steering it to its happy conclusion. Pauli was Heisenberg's fellow-student and intimate friend; both soon were drawn into the circle of Bohr's nearest disciples and co-workers, among whom uninhibited exchange of ideas and mutual criticism were encouraged by Bohr's own example. In this free and trustful environment Pauli gained fame for the penetration of his critical judgment and the sarcastic form in which he expressed it; the good-natured disposition he revealed on closer acquaintance did much to soften the effect of his sharp wit. Among the younger physicists who gathered around Bohr in those years, Pauli was undoubtedly the one who by temperament came nearest to Bohr's own philosophical attitude, and who most fully understood the latter's striving for harmony between such mutually exclusive aspects of our mental activity as the conscious experience of our environment and its rational analysis. His irony expressed only one side, and not the dominant one, of an extremely sensitive nature, open to all human problems; in fact, he had only come to physics after hesitating between a literary and a scientific career, for either of which he was equally gifted. During all his life he remained as deeply concerned with the obscure, irrational manifestations of the human soul as with the rational form of thinking, and he dreamt of a philosophy wide enough to embrace both. He thus keenly realized the importance and earnestness of Bohr's constant preoccupation with the epistemological background of the enquiry into the laws of atomic physics; at the same time, following his critical bent, he was more exacting still than Heisenberg in the demand for a neat mathematical formulation yielding as complete as possible a description of the observed phenomena. In the odd rivalry that had arisen between intuitive and formal strategy, his role as arbiter was clearly indicated.

Heisenberg had scored a point when he discovered a simple and neat algebraic formulation of the quantum conditions defining the stationary states and of the rules for the computation of transition amplitudes. He was carried too far, however, when he attempted to represent this formalism as the fulfilment of a new 'philosophical' outlook, according to which only 'observable quantities' ought to be retained in the account of the atomic phenomena: in particular, he wanted to eliminate completely the orbits of the electrons because they were not accessible to observation. At this point, Pauli's criticism intervened. He had put the new quantum mechanics to a severe test by applying it to the hydrogen atom – a problem that had baffled Heisenberg; the Rydberg formula was indeed derived, but only by an algebraic *tour de force*. It is true that the classical trajectories of the electron played no part in the interpretation of the result, but the whole algebra leading to it was in fact a transcription of the classical computation of the Keplerian orbits. This, Pauli pointed out, was no mere coincidence: one had to expect, according to the correspondence principle, that in a stationary state of high quantum number, the classical orbit, from just a mathematical fiction, would somehow become a physical feature of the system. After all, the moon's successive positions along its Keplerian orbit are perfectly observable. Clearly, although originating from the correspondence principle, Heisenberg's theory had lost too much of its connexion with it in the process of its formalization.

Moreover, what about the free motion of atomic particles, whose trajectories are traced out by trails of ionization in rarefied gases or cloud chambers? what about collision processes between such particles, or about the Compton effect? Dirac had attempted to treat this last problem by a bold extension of the quantum algebra to relativistic mechanics; but this work, in which we now discern remarkable anticipations, must have then appeared as a failure. Quantum mechanics was too narrowly limited to the description of periodic phenomena, and the vast domain of aperiodic effects was still beyond its reach; there were more observable things in the world of atoms than was dreamt of in Heisenberg's philosophy. Pauli turned his wrath in a playful way against Heisenberg's Göttingen partners, Born and Jordan, whose contribution had been decisive in making of Heisenberg's somewhat unhandy mathematical rules of computation a fully-fledged algebraic formalism, based on the matrix calculus: he rebuked Heisenberg for having allowed himself to be blinded by a "Göttingen display of erudition", and insisted (as Bohr would have done, though in more flowery terms) on the urgency of a more directly physical approach.

As things were reaching this critical climax within the Copenhagen group, the whole fabric of atomic theory, from its latest proud achievements down to the very quantum postulates on which it rested, was suddenly exposed to a full-scale onslaught from a quite unexpected quarter. Schrödinger, one of the most distinguished continuators of the Boltzmann tradition in statistical thermodynamics, entered the field of atomic theory with all the unconventionality of the outsider. His starting point was a speculation put forward by another outsider, Louis de Broglie, about the possible existence of a corpuscle-wave dualism of matter, in analogy with the photon-wave dualism of light. While de Broglie conceived this dualism in the same way as Slater had tried to do in the case of radiation – matter waves guiding particles along definite trajectories, – Schrödinger, inspired by Hamilton's beautiful comparison of classical mechanics and geometrical optics, argued that the classical trajectories of particles were rather the 'rays' associated with matter waves, and therefore, like the rays of light, had only approximate significance: the exact description of matter, like that of radiation, had to be sought in its wave aspect. Therefore, he concluded, there is nothing more mysterious in the discrete sequence of stationary states of an atomic system than in the similar sequence of proper vibrations in a medium whose state is governed by a wave equation.

Hamilton's analogy directly suggests proportionality between the frequency of a matter wave representing a stationary state of a particle and the energy of the particle in this state; Planck's constant is here introduced into the theory as the coefficient of proportionality. Any state of a system may then be represented by a superposition of the waves corresponding to its stationary states, and such a composite wave will exhibit 'beats' or peaks of intensity occurring with frequencies given by the differences of the frequencies of the stationary states, which are just the radiation frequencies to be expected according to the quantum postulates. Accordingly, one has only to identify this intensity (suitably normalized) with the electric charge density of the system in order to account for its quantal radiation properties without any need (or so it seems) for the notion of quantum transition: the quantum postulates could apparently be replaced by the familiar classical concepts of wave theory. The wave picture was a bit blurred, perhaps, for systems of more than one constituent, since it then proved necessary to consider waves in the configuration space of all the constituents, rather than in ordinary space; but this did not disturb Schrödinger, probably because he was accustomed to the still more abstract phase-space representation of systems of molecules, currently used in statistical mechanics.

At any rate, Schrödinger made no secret of his intention to substitute simple classical pictures for the strange conceptions of quantum mechanics, for whose abstract character he expressed deep "aversion"; he was conscious that this last sentiment was shared by all the older generation of physicists, who had not accepted the necessity of giving up their habitual ways of thinking when dealing with phenomena on the atomic scale. Significantly, he turned towards the chevroned peers of the classical era – Lorentz, Planck, Einstein – who did not grudge him praise and encouragement, and shunned the founders of quantum mechanics. The latter, however, who had taken no notice of de Broglie, looked into Schrödinger's ideas most critically; he had indeed enforced the claim to be taken seriously by solving the wave equation for the hydrogen atom and obtaining Rydberg's formula by a calculation much simpler and more elegant than Pauli's quantum algebra. This could hardly be a fortuitous coincidence, and indeed both Pauli and Schrödinger soon discovered that the two theories so different in conception, were mathematically equivalent: for instance, the quantal amplitude governing the transition between two stationary states could be readily computed with the help of the wave amplitudes corresponding to these states. Would this formal equivalence clinch the issue in favour of Schrödinger's contention that the proper quantal concepts can be altogether dispensed with? Far from it, Heisenberg had at once seen that this contention was untenable: Schrödinger's way of treating the charge density as a classical source of radiation would even prevent him from obtaining Planck's law for the distribution of thermal radiation. How could such an elementary, but fatal objection have escaped, not only Schrödinger himself, but above all the creators of the quantum theory of radiation, who gave him such uncritical support? Almost regretting to have ushered in ideas whose revolutionary consequences they had not foreseen, these great masters desperately clung to the 'sound philosophy' of classical physics, without realizing its limitations; an emotional resistance dimmed their judgement. The pioneers of quantum mechanics, on their side, were now confronted with a new challenge: Schrödinger's wave theory of matter clearly provided another formal approach to the consistent description of atomic phenomena they were striving for, but its physical content still eluded them.

It did not last long, however, before they realized that wave mechanics was just the appropriate technique they wanted for dealing with the aperiodic phenomena intractable by the quantum algebra. This was explicitly demonstrated by Born, who showed how to treat atomic collisions by a transposition of the mathematical methods applied to the analogous classical problem of the scattering of light waves by a polarizable medium. Born's argument, moreover, embodied an essentially new feature, of decisive importance, with regard to the physical interpretation of the intensity of the matter waves. The optical analogy suggested a comparison of this intensity with that of a classical light wave, which the quantum theory of radiation interprets as the density of the statistical distribution of the associated photons. Born pointed out that the usual particle description of the atomic collision process could be maintained if one adopted a similar statistical relationship between the matter waves and the associated atomic particles. He accordingly proposed to interpret the wave intensity, not as the density of an actual distribution of matter, as Schrödinger imagined, but as a density of probability for the presence of a particle. Thus, the formal equivalence of wave mechanics and Heisenberg's quantum mechanics became physically meaningful: it established a complete harmony between the statistical meaning of the wave intensity and the statistical character of the rules of quantum algebra for the calculation of transition probabilities.

With respect to the formalism at least, a clear standpoint had been reached; one had now a consistent mathematical scheme, with definite rules for the transformation of variables that rendered it very flexible: it was readily possible to pass from the choice of variables adapted to one type of description – the evolution of the processes in space and time, say – to any other set of variables appropriate to another aspect of the phenomena, such as the exchange of energy and momentum by quantum transitions between stationary states of the interacting systems. However, in contrast to the initial expectations of Heisenberg and Schrödinger, none of the types of description they thought unwanted had been eliminated: wave mechanics had not made the consideration of quantum transitions superfluous, and it had added space-time localization to the store of 'observable' features. A synthesis of these modes of description was achieved, whose logical consistency was guaranteed by the consistency of the formalism; but its physical nature was as much in the dark as ever: there remained here an epistemological problem, on the solution of which the ultimate effort had now to be concentrated. One fundamental point was clear: the type of causality governing the whole account of atomic phenomena had to be essentially statistical. Schrödinger was distressed by the turn events had taken; Bohr vainly tried to gain him over to the hopeful prospects his own endeavour had opened: "Had I known we should not get rid of this jumping about (*Herumspringerei*)", he spitefully declared, "I would not have meddled in this business!" Indeed, his intervention had truly been that of the 'sleepwalker' which Koestler would like us to believe is not exceptional in the history of science.

Bohr was now impatient to come to grips with the outstanding epistemological issue. He was convinced that in the quantum theory of matter just as in that of radiation, one faced an irreducible particle-wave dualism, and it seemed to him that the way was clear to the elucidation of its epistemological significance. The emphasis must be laid here on the role of a scientific theory as a means of unambiguous communication of experience. In atomic theory it is found convenient to use for this purpose, according to the circumstances, either the language appropriate to the description of wave phenomena, or the language of particle mechanics, – both modes of description being subject to essential limitations: it is a necessary task of the theory to formulate these limitations so as to fix the conditions of validity of each type of idealized description. Night after night in the first months of 1927, Bohr and Heisenberg pondered together over these questions; again, they were at loggerheads over the strategy: Bohr expected the answer would follow from a direct analysis of the definition of the idealized concepts, Heisenberg argued that the answer was hidden in the formal structure of the theory and that a closer scrutiny of this structure should bring it to light.

Towards the end of February, they parted, after a last fruitless discussion, Bohr having decided to seek a much needed recreation in the snowy hills of Norway. On the same night, however, Heisenberg had an inspiration that put him on the right track. He vividly remembered a discussion he had with Einstein a year before about the new quantum mechanics; his attempt at arguing that a good theory ought only to operate with observable quantities had elicited from Einstein the pointed retort, which had made a strong impression on him: "Only the theory itself can decide what is and is not observable". This remark, he thought, was the key to the whole problem: all one had to do was to investigate, for each given phenomenon, which conclusions one could draw from its theoretical analysis, according to the principles of quantum mechanics, about possible limitations to its observability. For instance, one could see at once that the localization of a particle at a point of space would be incompatible with the determination of its momentum, – since according to the rules of quantum algebra the corresponding quantum variables cannot both be assigned precise numerical values. The ensuing weeks of separation brought both Heisenberg and Bohr a welcome relief from the preceding strain: Heisenberg was able to follow up his promising conjecture and work out in detail both the underlying formal relations and the physical side of the argument; Bohr in a more restful mood succeeded in clarifying his own thoughts about the function of physical concepts and their idealized character. When they met again, they were both well prepared for a final confrontation.

Heisenberg based his argumentation on very general 'indeterminacy relations' which he had shown to be deeply rooted in the formal structure of quantum mechanics. The mechanical variables are always associated in 'conjugate' pairs – one variable of each pair referring to the spatial localization of the system, the other to its state of motion; in the new formalism, these conjugate variables represent mathematical operations, and their product, representing the successive performance of the two operations in a certain order, does not yield the same result if they are performed in the reverse order: this is a fundamental property, which has its correspondence in classical mechanics, and which is expressed, in the quantum algebra, by 'commutation relations' of a very abstract character, since they involve the imaginary square root of -1. It is remarkable that these purely kinematical commutation relations play the part of the previous quantum conditions which defined the stationary states: it is through them that the quantum of action enters into the formalism. Thus, if q denotes some spatial coordinate and p the conjugate momentum component, the commutation relation between them has the form

$$(1) \qquad qp - pq = \frac{h\sqrt{-1}}{2\pi}.$$

Now, Heisenberg derived from this abstract equation the following physical consequence. Consider the state of a system in which the value of the coordinate q is determined with a latitude Δq, and the value of the momentum component p with a latitude Δp; then, the product of the indeterminacies Δq, Δp cannot be smaller than a quantity of the order of magnitude $h/2\pi$:

$$(2) \qquad \Delta q \cdot \Delta p \gtrsim \frac{h}{2\pi}.$$

The reciprocal form of this relation is especially illuminating: it shows in the first place that there is no absolute limitation to the possibility of determining any single variable with any accuracy – however, the experimental situation leading to such a determination will then be such as to exclude the possibility of determining the conjugate variable; under less stringent conditions, the relation (2) gives a quantitative expression for the inevitable limitations imposed by the existence of the quantum of action upon the classical modes of description based on either class of conjugate variables.

Heisenberg naturally wanted to clinch the argument by analysing particular examples of idealized experimental situations which could illustrate the application of the indeterminacy relations. He was anxious, in particular, to demonstrate the impossibility of assigning a trajectory to a moving electron: let us imagine we should attempt to localize the electron by means of a microscope; in order to achieve sufficient resolution, we should have to illuminate it with radiation of very high frequency, in the range of the gamma-rays emitted by radioactive substances; the scattering of such photons by the electron, according to the theory of the Compton effect, would throw the electron completely out of the path prescribed by its previous velocity. Bohr's reaction to these considerations was very characteristic: he at once realized that Heisenberg again had forged the weapon that was needed to master the problem, but was soon dissatisfied with the smart way in which he was wielding it. The idea of bringing out the origin of the limitations of classical description by analysing imaginary processes of observation was of the type that would appeal to Bohr, and he took it up eagerly, but in his own infinitely patient manner, examining the argument from all sides, leaving no point in the shadow. He was not long to discover that Heisenberg's discussion of the gamma-ray microscope touched only one aspect of the question – the role of the frequency of the radiation, but did not really go to the heart of the matter: the Compton recoil of the electron could only be a source of indeterminacy for its momentum to the extent to which the scattering process itself is not completely determined. Now, he pointed out, the formation of an image in a microscope requires a bundle of rays of finite aperture, and it is this latitude in the direction of the photons impinging on the electron that implies a corresponding latitude in the electron recoil. On the other hand, a large aperture increases the accuracy of the electron's localization; and a simple computation shows that the two contrary effects of the aperture on the indeterminacies of position and momentum are such as to impose on their product the lower limit given by the relation (2), independently of the aperture and of the frequency of the radiation.

In fact, Bohr looked upon such imaginary experiments in a rather different spirit from Heisenberg. The latter was satisfied with the knowledge that he had now established a complete correspondence between the mathematical structure of the theory and the usual physical concepts describing the various aspects of experience; the indeterminacy relations were part of this correspondence, their physical content could be exhibited by the discussion of appropriate experiments. Bohr wanted to pursue the epistemological analysis one step further, and in particular to understand the logical nature of the mutual exclusion of the two aspects opposed in the particle-wave dualism. From this point of view the indeterminacy relations appear in a new light. The conjugate variables for which they hold refer, respectively, to the two mutually exclusive classical pictures centred on the idealized concepts of particle and wave: to the former are attached the momentum and energy variables, to the latter the coordinates of space and time. The particle concept is used to describe the exchanges of momentum and energy between atomic systems, and between such systems and radiation: it is the physical support, so to speak, of the conservation laws governing such exchanges, irrespectively of any space-time localization. The wave concept, or rather the derived density function, describes the localization of atomic systems and the distribution of radiation in space and time; it excludes the specification of momentum and energy, since only a wave of infinite spatial and temporal extension can have a sharply defined periodicity, necessary for a sharp definition of momentum and energy. The odd situation we meet in atomic theory is the necessity of making use of these two conflicting pictures in order to deal with all the conditions under which we can observe the atomic phenomena. The indeterminacy relations are therefore essential to ensure the consistency of the theory, by assigning the limits within which the use of classical concepts belonging to the two extreme pictures may be applied without contradiction. For this novel logical relationship, which called in Bohr's mind echoes of his philosophical meditations over the duality of our mental activity, he proposed the name "complementarity", conscious that he was here breaking new ground in epistemology.

The solemnity of the occasion did not dawn immediately upon Heisenberg. He was not well-disposed toward the idea of a particle-wave dualism, and Bohr was himself still struggling to find the right expression for the new ideas that were taking shape in his mind. However eager he was to let Heisenberg share his elation at the prospects they disclosed, his eagerness was not sufficiently matched by clarity to make impression on a reluctant interlocutor. The memory of the fruitless discussions with Schrödinger still lingered in Heisenberg's mind, and he regarded with deep suspicion any attempt to make more than formal use of the concept of matter wave. He stressed, quite rightly, that according to the correspondence principle, it is the particle concept that has a direct physical meaning with respect to atomic constituents, whereas matter waves are merely mathematical auxiliaries. To this Bohr would oppose the case of radiation, where the correspondence principle, on the contrary, points to the electromagnetic waves as the fundamental classical concept, and leaves only a symbolic part to the photons. Moreover, Oskar Klein, who was then Bohr's assistant, had carefully discussed the relation of matter waves to the correspondence principle, and shown how the latter gave as reliable a guidance for the physical interpretation of wave mechanics as for that of the quantum algebra: but Klein's paper had not found grace with Heisenberg. He stuck obstinately to the view that his own scheme of quantum mechanics formed a sufficient basis for a complete description of the phenomena.

Matters did not improve when Bohr was shown the hastily written paper in which Heisenberg had developed his arguments. Bohr started handling the manuscript as he would have done one of his own, namely as a rough draft which could eventually lead to an acceptable text; and with his usual optimism, he expected Heisenberg to welcome this scrutiny: the latter was simply annoyed to find Bohr raising so many objections to the inaccuracies and careless statements of the paper, and proposing new formulations with which he was in no mood to agree. There is no telling how things would have ended if Pauli's arrival on the scene had not relieved the tension. Heisenberg had informed Pauli of his ideas and received from him approval and encouragement; now, Pauli was in a position both to lend Bohr a dispassionate ear and to gain a hearing from Heisenberg. He was able to explain to the latter that there was no disagreement between his and Bohr's ways of analysing the physical content of atomic theory, but that Bohr had gone further and deeper in the analysis of its logical structure. Even so, all Heisenberg could be persuaded to do about his paper was to append to it a "remark added on the proof", in which he declared in substance that he had missed essential points, whose clarification would be found in a forthcoming paper by Bohr. This addendum must have puzzled many readers: it is not often that the announcement of a decisive progress in our insight into the workings of nature is qualified by such a warning.

As to Bohr's "forthcoming" publication, more than a year elapsed before it appeared in print: it was a much furbished version of a lecture he delivered, shortly after the events just retraced, at a physicists' conference in Como. He had the greatest misgivings about presenting his conception of complementarity to the community of physicists in a state which he judged immature; but he yielded to the advice of his more practically-minded brother Harald. Upon the latter's urging, he even consented to write up a brief account, that could be promptly published in *Nature*, as a letter to the editor: but this letter never reached its destination. With the help of Klein, he actually managed to complete it just on the night of his reluctant departure for Como. However, when Klein came up to the Institute the next morning, and enquired whether the letter had been sent off, he learned that there had been a double hitch, of a kind to delight Freudians. Bohr had missed the night train, because he could not find his passport (which lay on his desk); he had departed by the next train, taking with him the famous letter.

NOTES

[1] This essay was originally prepared as a contribution to a discussion meeting on the general theme "The making of modern science" organized by the American Academy of Arts and Sciences. I am most grateful to Professor S. R. Graubard and Professor G. Holton, on whose initiative the meeting was convened, for inviting me to take part in it. A selection of the submitted essays has now appeared as the Fall 1970 issue of *Daedalus*.

[2] This documentation, generously put at my disposal by Professor Slater, is now deposited in the Niels Bohr Archive in Copenhagen. Professor Slater has kindly expressed approval of the use I made of it and given his assent to the publication of my account.

BIBLIOGRAPHICAL NOTES

For a general background the following authoritative articles are recommended:

[1] Pauli, W., 'Quantentheorie' in *Handbuch der Physik*, vol. 23, Springer, Berlin, 1926;

[2] Pauli, W., 'Die allgemeinen Prinzipien der Wellenmechanik' in *Handbuch der Physik*, vol. 24/1, Springer, Berlin, 1933; both reproduced in *Collected Scientific Papers* of W. Pauli, vol. 1, Interscience, New York, 1964.

The two following works contain much first-hand historical information; in the first are found English translations of fundamental papers:

[3] Van der Waerden, B. L., *Sources of Quantum Mechanics*, North-Holland, Amsterdam, 1967;

[4] Rozental, S. (ed.), *Niels Bohr, His Life and Work as Seen by His Friends and Colleagues*, North-Holland, Amsterdam, 1967.

The source material for Section 1 is contained in the following books, which also include reprints of the original papers:

[5] Haas, A. E., 'Der erste Quantenansatz für das Atom' (mit Biographie von A. Herrmann), *Dokumente der Naturwissenschaft*, vol. 10, Battenberg, Stuttgart, 1965;

[6] Bohr, Niels, *On the Constitution of Atoms and Molecules* (papers of 1913 reprinted with introduction by L. Rosenfeld) Munksgaard, Copenhagen, 1963.

The subject of Section 2 has received a masterly treatment in the following article (which also contains full references):

[7] Klein, M., 'The First Phase of the Bohr-Einstein Dialogue' published in volume 2 of *Historical Studies in the Physical Sciences* [ed. by Russell McCormach], University of Pennsylvania Press, Philadelphia, 1970.

With regard to Section 3, the reader may consult the following easily available reprints of original papers:

[8] De Broglie, L., 'Recherches sur la théorie des quanta' (réédition du texte de 1924), Masson, Paris, 1963;

[9] Schrödinger, E., 'Abhandlungen zur Wellenmechanik', Barth, Leipzig, 1927;

[10] Schrödinger, E., 'Briefe zur Wellenmechanik', Springer, Wien, 1963;

[11] Born, M., 'Zur statistischen Deutung der Quantentheorie', *Dokumente der Naturwissenschaften*, vol. 1, Battenberg, Stuttgart, 1962.

Dirac's first paper on the Compton effect, mentioned in the text, is the following:

[12] Dirac, P. A. M., *Proc. Roy. Soc. A* **111** (1926), 405;

Klein's paper on the correspondence interpretation of wave mechanics is

[13] Klein, O., *Z. Phys.* **41** (1927), 407;

Heisenberg's paper on the indeterminacy relations is

[14] Heisenberg, W., *Z. Phys.* **43** (1927), 172,

and Bohr's Como lecture is reprinted as the second essay in

[15] Bohr, N., *Atomic Theory and the Description of Nature*, Cambridge University Press, Cambridge, 1934.

A document of major historical importance is Heisenberg's recent book:

[16] Heisenberg, W., *Der Teil und das Ganze*, Piper, München, 1969.

CHAPTER 20

JACQUES SOLOMON*

[1959c]

It was at the Collège de France in Langevin's class that I met Jacques Solomon for the first time toward the end of 1926. Fresh from the University of Liège, I was spending some time at the École Normale Supérieure as a foreign student. At that time, when the Institut Poincaré did not yet exist and when Louis de Broglie's course was attended only by a few rare initiates, it was especially the old physics amphitheater of the Collège de France which served as the meeting place for those curious about modern physics. It was reached by an improbable stairway pierced in an obscure passage which opened onto the top level of the amphitheater. On the bottom lower wall above the blackboard a marble plaque was displayed where the names of the great physicists of the First Empire were methodically engraved in parallel columns. As for the professor, we glimpsed him moving back and forth behind an outsized table which was separated from the public by a balustrade. It was in this old-fashioned setting that Langevin initiated us into the latest progress in physics. That year he was dealing with quantum statistics, and it was one of his most important courses, full of original developments. He brought us, in addition to an analysis of the most recent work, the results of his own thinking which we saw developing and taking form from one lecture to the next. Jacques, I do not know exactly why, had 'spotted' me: when the class ended he came up to me and introduced himself with a simplicity and openness which immediately won me over. A few remarks revealed our common interests and that was the beginning of a long friendship.

After the class he returned to his parents' home on the avenue Trudaine in Montmartre; we soon developed the habit of walking part of the way together. Absorbed in our discussions, we went down the rue Saint-Jacques and along the quays to the Louvre. There I generally left him, but more than once we found ourselves beside the Bourse before realizing the passage of time and the length of the way. Bit by bit he described to me the largely self-taught course of his studies: at sixteen, he told me as if it was the most natural thing in the world, he had unraveled general relativity.

"Its principle is so simple", he added, "that I have never understood why so many people resist it." This erudition did not disconcert him: he had none of that lack of equilibrium which is too often the price of intellectual precociousness. On the contrary, open to everything, generous and sensitive, he responded to all the problems of the day. Our conversation jumped from physics to politics, from art to philosophy, and back again. We passed final judgment equally easily on the latest paper in the *Zeitschrift für Physik* and on the most recent exhibition of paintings.

Soon I was received by his family with a cordiality which immediately put me at ease. It was undoubtedly the harmony of this family circle which was the key to Jacques' beautifully balanced personality. The intellectual superiority of his father was revealed in the perfect simplicity of his approach and the lively humanity of his conversation. His mother radiated affability; his younger brother displayed his artistic temperament with fervor, and his young sister brought a note of pure gaiety to the company. They had a small country house at Hénonville in Normandy; it was there especially, spontaneously, without pedantry, that their intelligent curiosity was given free rein. In turn we botanized, puttered about with a wireless set, extracted fossilized shells from some chalky embankment; or we simply walked about the woods and fields abandoning ourselves to the ever new joy of contact with nature. It is not without melancholy that I evoke these happy carefree years when Jacques' talent was able to expand and assert itself. The following period of his brief existence undoubtedly offered him more intense and deeper moments of happiness, but he never again found the beautiful serenity of his adolescence.

I returned to Paris in 1931, this time to give a course at the Institut Henri Poincaré on quantum electrodynamics, then in its beginnings. Jacques had married Hélène Langevin and it was at the home of his parents-in-law that I saw him then. Langevin had kindly offered me the hospitality of his apartment in the rue Vauquelin and that is where I spent all my time in the company of Jacques. In my course I had mentioned in passing the presence of an infinite residual energy of the radiation field as one of the minor difficulties of the theory. Jacques immediately proposed that we attack this clearly delimited and apparently accessible problem. In fact, it seemed to us that this residual energy (the inverse of the self-energy of the electron) had a purely formal character and thus could be eliminated by an appropriate modification of the formalism. We reached a solution after a fashion, though it would no longer be regarded as satisfactory today: for it is far more probable that the residual radiation energy corresponds to a physical reality and no attempt should be made to cause it to disappear. But this has no importance: this common work had, in any case, the effect of further strengthening our friendship, and I believe that for Jacques it was the decisive test which, by showing him that he was capable of flying with his own wings, reaffirmed his self-confidence.

His enthusiasm and persistence were a joy to watch. He devoted himself completely to research, forgetting the course of daily life. At meals it was not easy to prevent him from continuing the interrupted calculations. He tried to continue them mentally, enumerating the rapid series of operations out loud and evidently expecting of me the same virtuosity of algebraic imagination. Langevin smiled at this holy ardor and gently tried to calm him. Soon the conversation took a less esoteric turn. Whatever the subject might be, Langevin took an interest in it with his habitual kindness and often one of his remarks shed light on the discussion or rectified its course. . . .

From this time, Jacques took flight. A visit to Copenhagen (where we met again soon afterwards) served to put into more concrete form the vast knowledge that he had accumulated in isolation: he saw all the questions that he had asked himself during his solitary meditations debated daily, under the paternal aegis of Niels Bohr, by a host of physicists of the most diverse origins and formations. His judgment matured; his critical faculties sharpened. The example of Bohr, who at this time was putting the finishing touches on his vast conception of complementarity, revived the predilection Jacques had always had for great problems, the most fundamental aspects of physical theories. "Er macht letzte Dinge" (He does eschatology), people said of him; and this bizarre remark amusingly concealed a profound esteem.

Along this line, he took up, in particular, a problem which extended the research we had begun in Paris. I had calculated the gravitational self-energy of a photon by using the ordinary expression of the electromagnetic energy-momentum tensor and shown that this self-energy too was infinite. Jacques wanted to see whether our new expression for the electromagnetic tensor (by which the residual energy was eliminated) would modify this conclusion. He successfully completed this work on his return to Paris and presented the results in his doctoral thesis.

Nonetheless, his ever lively curiosity could not be content with a fragmentary consideration of a reality so rich in physical phenomena. Jacques' preoccupations were not only 'eschatological', they were encyclopedic. He not only studied the depths; he also wanted to understand each facet of the atomic world in its least details. And so we see him, in the course of the following years, direct his attention in turn toward abstract questions of field theory and general relativity (which he studied in Zürich with Pauli) and toward a whole series of concrete problems raised by the discoveries which succeeded each other in the domain of nuclear physics and cosmic radiation. He not only scrutinized the framework of the vast theoretical edifice of modern physics, but penetrated with equal ease into its innermost recesses. The great variety of subjects which he treated bears witness simultaneously to the perfect mastery he had attained and to his insatiable need to understand. His spirit did not tolerate ignorance, or obscurity: if a problem presented itself to him, perhaps in a chance conversation, he could not break away from it until he had resolved it or at least determined its scope and difficulty.

Curious about everything, eager to understand everything, Jacques could not be indifferent to the conflicts presented by the economic and social development of the modern world. His free and upright judgment led him directly to Marx's ideas which he recognized as the first attempt to apply a rigorous method of scientific analysis to these difficult problems. But his activity in this line could not remain simply intellectual. His whole soul resonated with the innumerable souls of the oppressed. The same sincerity which engaged him to concentrate his entire energy on the problem of his choice in his scientific work, clearly dictated another step which also would enlist him totally. His joining the Communist Party was no trivial gesture: it was the total, deliberate, gift of his person to the ideal he had established for himself.

In the course of this last, very tense, period of his life, Jacques and I had only brief encounters at long intervals. Perhaps better than a more continuous contact, this allowed me to measure the profound transformation which was occurring in him. I was struck by the change from one meeting to the next. When we spoke of physics or evoked memories, his gaze brightened once again with the naive joy or benevolent irony which were formerly habitual with him. But more and more frequently, he burned with a sombre fire betraying the internal concentration, the fierce resolution of a man engaged in a merciless struggle. Observing the maturity of his talent, the outstanding productivity of his scientific thought, I secretly regretted the overwhelming ascendency of his political activity within his personality. His humanity was undoubtedly still as broad, but it seemed to have lost some of its former gentleness. Even on a scientific or philosophical plane, he was becoming harder, more cutting. Nevertheless, each time, the frank handshake, the clear smile of goodbye, reassured me: it was still the same optimism, the same confidence in human reason, the same goodness, which were guiding him on his perilous road.

He followed this road straight to the ultimate sacrifice. No sacrifice has been more fully, more purely accepted. With full awareness of motive, a powerful scientific talent engaged itself, ready to bring forth similar talents, through its example and its teaching. This is another example he gave to future generations. Should we weigh the loss and the gain? His fate is not one of those which one laments; it merits, like that of the heroes of Thermopylae extolled by Simonides, πρὸ γόων δὲ μνᾶστις, (instead of lamentations, remembrance).

NOTE

* [Jacques Solomon and his wife Hélène were arrested March 1942 for activities in the French Resistance. He was killed by the Nazis on May 23, 1942 and she was deported to Auschwitz but survived. – Ed.]

CHAPTER 21

QUANTUM THEORY IN 1929: RECOLLECTIONS FROM THE FIRST COPENHAGEN CONFERENCE

[1971c]

Although in the historical perspective the first Copenhagen conference in 1929 rather marks the completion of a heroic period in the life of the Institute than the initiation of its activity fifty years ago, the celebration of the jubilee may be a fitting occasion for calling up recollections of an elating week when the Copenhagen spirit hovered over the troubled waters from which quantum theory had just emerged. The circumstances of this gathering loom large in my own memory: I was then a mere tiro, and the picture I have kept of the events has all the vividness of the first glimpse into a wonderful world; I have therefore no scruple to revive it, as a testimony of the mood in which a new generation was then taking the cue from the pioneers to carry on the unending quest. In fact, I have once before written down some of my impressions of this famous week under the title 'My initiation', which conveys this mood. It was a light-hearted, but truthful tale, which had the honour of publication in the 1945 issue of the *Journal of Jocular Physics*; I do not want the present account to be much more solemn, and I shall take the liberty of quoting occasionally from the previous one.

As most good things in the world, the idea of this conference was due to chance. In Bohr's own words:

> Der Plan zu dieser Konferenz ist dadurch entstanden, dass mehrere Physiker, die früher hier gearbeitet haben, Besuche in Kopenhagen in den Osterferien angemeldet haben. Da wir unter anderen auf die Anwesenheit von Kramers und Pauli rechnen, wird es wohl zu lebhaften und lehrreichen Diskussionen Anlass geben können ...[1].

It was quite in line with Bohr's paternal attitude that this fortuitous conjunction of visits should have prompted him to summon a full-scale family re-union. Not all could manage to be there, but a good twenty of them from all over Europe responded to the call: from Cambridge, Bohr's old friend from the Manchester time, Darwin; from Leiden another near friend, one of the great masters from the classical era, Ehrenfest, accompanied by one of his youngest students, Casimir; from Utrecht, Bohr's first collaborator, Kramers, who had left Copenhagen three years before to take up the Utrecht chair; from Zürich came Pauli, who with his acutely critical mind, was already then, as Bohr used to put it, "the conscience of the physicists"; further from Holland, Germany and Scandinavia a number of those who, each in his domain of predilection, had shared in the edification of atomic theory: Goudsmit and Kronig, Hückel, Fues, Jordan, Heitler and Nordheim, Rosseland, Holtsmark and Waller. My own good luck was due to the circumstance that I had some time before written to Bohr to enquire whether I could come to work under his guidance; he generously counted me as a disciple *in spe*. At the time, four foreign visitors were staying in Copenhagen; they added their distinctive features to a variegated assembly: Mott's Cantabrigian elegance, Trumpy's Norse cheerfulness, the quiet composure of the Chinese Chou were overshadowed by the whimsical fantasy that Gamow had brought to the West from the lively group of young Soviet physicists. As to the Institute's staff, it consisted of two experimenters, S. Werner and J. C. Jacobsen, and of Bohr's closest collaborator, Oskar Klein, who had succeeded Kramers and Heisenberg in this position of trust. Among the handful of Danish students who attended the proceedings, we find the familiar names of Christian Møller, Bengt Strömgren, Mogens Pihl and the regretted Ebbe Rasmussen.

The conference was due to start on Monday the 8th of April, and a number of the pilgrims travelled on the preceding Sunday. They all met inevitably on the deck of the ferry from Warnemünde to Gedser. There was much handshaking, exchange of news and shop talk. This was my maiden trip, and a propitious day for a first view of the Danish scenery, in the timid budding of spring, with flags gaily flying in front of every thatchroofed farmhouse (not, as I learned from my more experienced companions, in honour of the conference, but just as a weekly manifestation of homely content). The old-fashioned look of the ferries and railway carriages, the queer funnels of the locomotives, the easy-going demeanour of the railway people and of the local passengers at the cosy red-brick stations where the train unhurriedly lingered, all concurred to build up the impression of a simple-minded, undemanding peasant community, happily confined to its own well-protected little world. Tourist agents had not yet discovered how idyllic a country it was, how wonderful its capital.

The pleasant feeling of old-time hospitality conveyed by the Danish countryside came to a climax on our arrival at Copenhagen. Niels Bohr himself was awaiting us on the platform, together with his brother Harald, his lieutenant Klein and a few boys of various sizes, obviously his sons. I recognized Harald, whose lectures on almost periodic functions I had attended in Göttingen, and I shook hands with Niels for the first time. He received me with a broad, benevolent smile; I was struck by the cordial simplicity with which he greeted old friends and newcomers alike. From the station I strolled to the boarding house I had been assigned, on the Vester Boulevard, then a spacious promenade lined with stately elms, which formed a worthy setting for the little trumpeter's eternally unavailing last call. At dinner, I had occasion to experience how the Danes could sometimes overdo their hospitable attentions. They subjected me to the ordeal of pronouncing "rødgrød med fløde", and adding insult to injury, expected me to find the stuff itself delicious.

On the next morning, we all flocked to the Institute lecture room; old acquaintances were renewed, new ones formed. I happened to be nearby when a beaming Ehrenfest came in and went straight to greet Bohr, followed by a tall, fair-haired, rosy-cheeked youth of rather indolent gait, who did not quite know what to do with his arms. "Ich bringe Dir diesen Knaben", he said to Bohr, while he affectionately put his hand on the boy's shoulder. "Er kann schon was, aber er braucht noch Prügel". Well, in the course of the week, Casimir was going to show us, unobtrusively, that he indeed could do something. Goudsmit was then applying his extraordinary skill to the classification of hyperfine spectra and was naturally anxious to know what the contribution of the electron spin to the hyperfine structure would look like. As soon as he heard Goudsmit mention the problem, Casimir keenly responded: he asked Goudsmit some brief questions and, lost in meditation, withdrew to a quiet corner; but let us hear the story from himself [2]: "At the time of the conference Dirac's theory of the spinning electron was still fairly new ... I had studied the paper and also Weyl's *Gruppentheorie und Quantenmechanik* and had been much impressed by the beautiful simplicity of the current distribution in the fundamental state of a hydrogen atom. So when Goudsmit asked me whether I could calculate the hyperfine interaction in a S state I realized at once that one had only to calculate the magnetic field of the current. I arrived at the formula

$$\Delta E = \frac{8\pi}{3} \frac{e\hbar}{mc} \frac{e\hbar}{2Mc} |\psi(0)|^2 g_i(s \cdot I)$$

where the nuclear moment is given by $(e\hbar/2Mc)\, g_i I$. Goudsmit supplied an estimate for $|\psi(0)|^2$ for 'Tauchbahnen' and we obtained pretty nice results for Na if I remember well. After the conference I wrote a manuscript – a fairly clumsy one I fear – which I sent to Goudsmit. I got it back, much later but soon afterwards Fermi's manuscript 'Ueber die magnetischen Momente der Atomkerne' was received (18. Dec. 1929) by *Zs. für Physik* and appeared (*Z. für Physik* **60** (1930), 320)". It was quite edifying to watch Casimir's eagerness in working out this problem. One evening, some of us went to see the pictures, and persuaded him to join us. In those days, they showed one reel of film at a time, and lit the room every time they changed reels; at each recurrence of the light, we could see our friend bending over odd scraps of paper and hastily filling them with formulae.

Commenting on the decisive part the unravelling of the spectral regularities had played in the development of atomic theory, and mentioning especially Goudsmit's virtuosity in this art, Bohr told us that the latter's talent for tracking hidden order in apparent capriciousness was by no means limited to physics; he also exercised it on classifying the types of representation of sacred scarabs by the ancient Egyptians. On Goudsmit's first visit to Copenhagen in 1926, Bohr had been with him to see the collection of Egyptian sculpture at the Glyptotek; as he started translating the Danish labels for Goudsmit's benefit, the latter quietly told him it was not necessary, as he could read the inscribed hieroglyphs. Bohr had a large stock of such pointed anecdotes, reflecting his warm interest in people, that he was never tired of telling in those conversations with the older and younger physicists clustered around him, during which he freely expressed his thoughts on the prospects of current research as well as the widest implications of science, and displayed the best of his wisdom and deep humanity.

Bohr as a lecturer is a different matter. It is much glossed, but very little written about. Perhaps the only one who has put his view of it in print so far is Larmor; in a speech (later published[3]) at the Maxwell celebrations in Cambridge in 1931, he commented upon Maxwell's reputation of being 'a poor lecturer' and roundly added: 'So perhaps with our friend Bohr: he might want to instruct us about the correlations of too many things at once ..." I was sitting near Bohr when the speech was delivered; as this judgment was expressed, Bohr whispered to me: "Imagine, he thinks I am a poor lecturer!" Bohr's lectures, composed with tremendous labour, were indeed masterpieces of allusive evocation of a subtle dialectic; the trouble was that the audience was usually unprepared to catch subtle allusions to conceptions and arguments which were anyhow unfamiliar and hard to grasp.

I am not sure whether Bohr's introductory talk at the conference was really worse than the average; perhaps he had not prepared it so thoroughly, since the idea was to have quite informal discussions: no programme had been set up in advance – Bohr took in turn each of the participants aside and asked him what topic he wished to bring up. At any rate, here is the impression this talk has left in my memory, as I described it (with some hindsight) in 1945: "He had begun with a few general considerations calculated, no doubt, to convey to the audience that peculiar sensation of having the ground suddenly removed from under their feet, which is so effective in promoting receptiveness for complementary thinking. This preliminary result being readily achieved, he had eagerly hastened to his main subject and stunned us all (except Pauli) with the non-observability of the electron spin. I spent the afternoon with Heitler pondering on the scanty fragments of the hidden wisdom which we had been able to jot down in our note books."

It was comforting to hear from Klein, when I told him some time ago of our failure to understand what Bohr meant by the impossibility of measuring the spin of the electron, that he had had the same difficulty when Bohr first discussed the matter with him in the autumn of 1928. Guided by the general correspondence idea, Bohr argued that such a purely quantal concept as the electron spin, vanishing from the theory in the classical limit, could not possibly be brought in direct relation with classical quantities like angular momentum or magnetic moment. It was not immediately clear to Klein, however, how this correspondence argument could be reconciled with the Stern-Gerlach effect, which clearly exhibited a contribution to the magnetic moment of an atom from an electron bound in a ^2S state; but what Bohr demonstrated was precisely that with a free electron a Stern-Gerlach experiment could not succeed, because the effect of the Lorentz force would inevitably blur any Stern-Gerlach pattern. This is the point he ineffectually tried to make in his talk. Fortunately, Mott, during his stay at the Institute, had been engaged in the problem of electron polarization, and in the paper[4] in which he brilliantly showed how this property could in principle be ascertained by a double scattering experiment, he gave a very clear account of the whole situation. He finished writing this paper shortly after the conference (it was sent off by Bohr on the 25th of April) and we were thus soon able to appreciate at leisure the full force of Bohr's famous argument.

The incorporation of the spin into the relativistic quantum theory of the electron had not removed from this theory the riddle of the negative energy states. Klein had just given a striking illustration of the acuteness of the difficulty by showing that electrons impinging upon a sufficiently high and steep potential wall would not only suffer reflexion, but that a sizable fraction of them would penetrate through the wall and undergo a transition to states of negative kinetic energy[5]. At the conference, he submitted, tentatively, a way to escape from this 'paradox': it amounted to treating the electrostatic potential as an operator whose eigenvalues would have a finite higher limit. Jordan pointed out that in order to eliminate the paradox, this highest potential value should be of an order of magnitude quite within the range of observation; probably, it ought to have been something like 2 mc^2 (m denoting the electron mass). Potential differences of this magnitude are indeed observed between thunderstorm clouds and the earth.

Jordan's remark caught Bohr's imagination: here was a possible test of Klein's assumption. The limiting potential would presumably entail a limit to the stability of matter: if it did occur in a thunderstorm, a bird flying in the region of highest potential would then be killed. The discussion went on, to and fro, as various consequences of Klein's formalism were either put forth in support of it or found to raise difficulties. Bohr kept musing about the fate of the bird, to Ehrenfest's boundless amusement. Whenever the discussion ebbed out for a moment: "Now, Bohr", he would ask with boyish mischief, "is the bird still alive?"

When we gathered again after lunch, Klein declared that he was now convinced that his proposal would not work and that he withdrew it. This was probably the shortest lifespan ever meted out to any theory. For all his apparent playfulness, Ehrenfest took very much to heart the difficulties in which relativistic quantum theory seemed to be bogged. I got an inkling of this the next day, when, much to my surprise, I was told, shortly after the end of the session, that Bohr wanted me to come to his house (which was adjacent to the Institute, and is now part of it).

I was ushered into his study, where I found him and Ehrenfest installed in comfortable armchairs and, judging by their smiles, engaged in pleasant conversation. "Could you tell us", Ehrenfest asked me with his wonted directness, "why the relativistic Klein-Gordon theory is unacceptable?" – "Because", I replied pedantically, "the charge-current density is not of definite sign." – "Yes", said Ehrenfest, "but could you just give us a simple example in which this property leads to unphysical consequences?" I never felt so sheepish in my life. Ehrenfest turned to Bohr: "You see", he said, "just as I told you!" From this remark I surmised with some relief, – and my guess was soon confirmed by the ensuing conversation, – that they did not impute the fiasco to my personal stupidity, but took it as an illustration of a widespread failing of the young generation, prompt to uncritical acceptance of formal dogmatism. Bohr intervened with a soothing pronouncement about how hard it was to imagine how the behaviour of the electron could be described beyond the limit where correspondence with the classical point-charge picture afforded legitimate guidance; we must here be prepared for further renouncement in the application of classical concepts. How often were we to hear this warning in the following years, until the discoveries of the neutron and the positive electron broke the spell!

In retrospect, one cannot help admiring the penetration and independence of judgment revealed by Ehrenfest's iconoclastic questioning. And yet (as I later learned from Bohr), he was at the same time turning his searching criticism against himself: he realized that progress in such untried regions of thought and imagination as quantum theory had opened, demanded adventurous minds, unhampered by the scruples and doubts that assailed him; he imagined he was losing his grip on physics and – worse still for one always ready to pour out a wealth of affection on the youth – getting out of touch with the coming generation. Of this inner tension there was no outward sign, however: to the last we saw him as cheerful, witty and warmhearted as ever.

Pauli, so far as I remember, was rather subdued, except on one spectacular occasion. Heitler, by lecturing on the theory of the homopolar bond, unexpectedly excited his wrath: for, as it turned out, he had a strong dislike to this theory. Hardly had Heitler finished, than Pauli moved to the blackboard in a state of great agitation; pacing to and fro he angrily started to voice his grievance, while Heitler sat down on a chair at the edge of the podium. "At long distances," Pauli explained, "the theory is certainly wrong, since we have there Van der Waals attraction; at short distances, obviously, it is also entirely wrong. "At this point he had reached the end of the podium opposite to that where Heitler was sitting. He turned round and was now walking towards him, threateningly pointing in his direction the piece of chalk he was holding in his hand: "Und nun," he exclaimed, "gibt es eine an den guten Glauben der Physiker appellierende Aussage, die behauptet, dass diese Näherung, die falsch ist in grossen Abständen und falsch in kleinen Abständen, trotzdem in einem Zwischengebiet qualitativ richtig sein soll!" He was now quite near to Heitler. The latter leaned back suddenly, the back of the chair gave way with a great crash, and poor Heitler tumbled backward (luckily without hurting himself too much).

Casimir, who also remembers the incident, notes that Gamow was the first to shout: "Pauli-effect!" And as an afterthought he adds: "Sometimes I wonder whether Gamow had not done something to the chair beforehand."

I found Gamow as jolly as I had always known him in Göttingen since the first day I saw him appear at Born's institute and heard of his surprising intention of applying quantum mechanics to alpha-radioactivity. His direct approach to the problem was, of course, most congenial to Bohr, who had supported him wholeheartedly when he had to counter Laue's doubts about the lawfulness of the new type of solution he had introduced [6]. I was then too much indoctrinated with Göttingen lore not to side with Laue in this debate; the more so as Born had just shown us how he wanted to treat the problem by a somewhat hazardous, but very ingenious, extension of orthodox perturbation technique [7]. When remonstrating with Gamow in the library, I went so far as to express doubt whether Gamow wavefunctions, with their infinite norm, could at all be regarded as solutions of a Schrödinger equation. Pauli happened to overhear this rather wild statement: "They are certainly solutions," he interjected, "but whether they are allowed in quantum mechanics, this is questionable." Gamow looked at us with an expression of wonder in his face: "But my solution", he said, "just represents a damped resonance process; the exponential increase at large distances has a simple interpretation, and after all I get the decay constant and an improved Geiger-Nuttall law. What else do you want?" – "Anyhow", Pauli concluded as he went away, "it was great fun to see the flood of papers your theory let loose."

"I don't understand what all this fuss is about", Gamow continued.

It is just a solution of a partial differential equation, of the kind we use in mechanics and elasticity. When I showed it to Bohr, he was at once enthusiastic. I was so pleased and proud that I wanted to show him something more. I had noticed that the Schrödinger equation could be written as a classical diffusion equation, – only with a purely imaginary diffusion coefficient Now, I had better kept that to myself: Bohr's reaction to it was anything but enthusiastic...

On Gamow's work, which ushered in nuclear physics, there was no debate. Mott, however, presented his analysis of the scattering of alpha-particles by helium[8], which was received with considerable interest. It was a beautifully simple case of an interference effect arising from the identity of the interacting particles, the first concrete example of the intervention of Bose statistics in an individual process. It was also a good lesson in the correct interpretation of a wave-function – an art then so new that Mott's argument was not accepted without some resistance on the part of one or two participants.

Darwin was deeply interested in these problems of interpretation; he did extremely helpful work discussing in great detail, with consummate mathematical skill in the best Cambridge tradition, a variety of idealized experimental devices illustrating typical features of quantum mechanics. The one he submitted at the conference[9] was a rather complicated case of constructive interference of wave-packets producing "spontaneously the results which simple intuition would suggest could only be due to particles". The analysis was as brilliant as ever, but unfortunately carried him away, at the end, onto slippery ground. He argued that the seat of an "observation" could be shifted from the physical receptor to the retina, and thence to "somewhere in the brain", where "we are absolutely compelled to stop." Right up to the brain the process is "non-committal": "it is only after our consciousness has animated the proceedings that it is possible to infer back and describe what actually happened in the familiar language of particles." Thus, we have a "sub-world" described by a wave-function, "a dead world, not involving definite events, but instead the potentiality for all possible events. It becomes animated by our consciousness, which so to speak cuts sections of it when it makes observations."

This remarkable anticipation of von Neumann's "Schnitt" conception struck me as being quite beside the point. It cannot make any difference by how many eyes, or brains, a scintillation on a screen is observed, and there is no reference in the formalism (then my only anchor!) than to the particle producing scintillation: the structure of the eye or the brain, or even that of the screen, is quite irrelevant. These were (as near as I can figure out after so many years) the confused thoughts that crossed my mind as I listened to Darwin. What troubled me mightily, however, was Bohr's failure to react to propositions which seemed to me so alien to the ideas he had expressed in the Como lecture. I was so disturbed by what I mistook for silent acquiescence that, mustering all my courage, I approached Bohr and cautiously started to voice my doubts. "Oh!" he said, interrupting me, "this was all nonsense!"

Then he motioned me to a neighbouring room, which had been Hevesy's laboratory: it was lined with cupboards, behind whose glass doors one could see rows of flasks containing brilliantly coloured solutions of rare-earth compounds. In the middle of the room there was a long table, and I stood at one extremity of it. I cannot find better words to tell what happened next than those I wrote in my previous article: Bohr described around the table,

> at a rather lively pace, a Keplerian ellipse of large excentricity, of which the place where I was standing was a focus. All the time, he was talking in a soft, low voice, explaining to me the broad outlines of his philosophy. He walked with bent head and knit brows; from time to time, he looked up at me and underlined some important point by a sober gesture. As he spoke, the words and sentences which I had read before in his papers suddenly took life and became loaded with meaning. It was one of the few solemn moments that count in an existence, the revelation of a world of dazzling thought, truly an initiation.

Bohr was particularly well prepared to comment on the problems Darwin had raised: he was just writing up (in a short essay[10] in honour of the 50th anniversary of Planck's doctorate) his reflexions on the complementary aspects of psychical phenomena. As an example of the shift that can occur in the separation between observing subject and object of observation, he mentioned the case of a cane held in the hand: if it is held firmly, it is an instrument extending the range of tactile perception of the subject, who feels it as a part of himself; if it is held loosely, it ceases to serve as an instrument and becomes an object of observation. He was quite excited about this effect. He insisted that I should try it with a pencil that was lying on the table, and he watched me intently during the operation, eager to catch on my face a sign of the joyous wonder I could not fail to experience.

I have not much to tell about the obligatory three-castle tour, with a stop at Bohr's country-house at Tisvilde, a long walk through the wood to Brantebjerg and along the strand to the village. The more enterprising displayed their sportive talents; Bohr was outstanding at ricochet throws of pebbles on the sea, but most spectacular was Jordan's acrobatics (it was rumoured he had lately taken to jiu-jitsu). I would rather recall an afternoon outing of a small group of us, among whom Ehrenfest, to Christianshavn, where Bohr wanted to show us Our Savior's church. We took the boat at Nyhavn and landed near the famous church, with its queer outer staircase winding up along the steeple. I was thrilled at the thought of seeing this monument, about which I had read as a boy in Jules Verne's *Voyage au centre de la terre.* From the landing place the steeple looked curiously flat; Ehrenfest had the same impression: "It looks as if it had an elliptic cross-section – if this were possible," he said. One ascends to the steeple from inside the church; one first arrives at a kind of platform, on which lies, idle and rusty, a huge clockwork dating, as we ascertained, from the time of Huygens and Newton. Here was the real goal of the trip. Bohr exulted: "This shows us what classical mechanics is about", he exclaimed. "Nobody that sees this can doubt that our measuring instruments must be big bodies!"

Next time Jauch comes to Copenhagen, I must take him to Our Saviour's church.

REFERENCES

[1] Letter from N. Bohr to P. Jordan (5 March 1929).
[2] Letter from H. Casimir to L. Rosenfeld (7 December 1970).
[3] James Clerk Maxwell: A Commemoration Volume 1831–1931. (Cambridge University Press, Cambridge, 1931) p. 78.
[4] N. Mott, *Proc. Roy. Soc.* **A124** (1929), 425.
[5] O. Klein, *Z. Physik* **53** (1929), 157.
[6] M. v. Laue, *Z. Physik* **52** (1928), 726; G. Gamow, *Z. Physik* **53** (1929), 601.
[7] M. Born, *Z. Physik* **58** (1929), 306.
[8] N. Mott, *Proc. Roy. Soc.* **A125** (1929), 222; **A126** (1930), 259.
[9] C. G. Darwin, *Proc. Roy. Soc.* **A124** (1929), 375.
[10] N. Bohr, *Naturwiss.* **17** (1929), 483.

CHAPTER 22

NIELS BOHR

An Essay Dedicated to Him on the Occasion of His Sixtieth Birthday, October 7, 1945

[Second Edition Corrected]
1961

[1945e]

Tu, pater, es rerum inventor, tu patria nobis suppeditas praecepta...

Lucretius *De rerum natura III 9–10.*

"Thou, father, art Nature's discoverer, from thee fatherly precepts we receive". This invocation to the great atomist of antiquity we will also address to the creator of modern atomic theory! As soon as the young physicist comes under the spell of the bright and serene atmosphere of the institute in the Blegdamsvej, when he partakes of those animated discussions improvised on any subject offered by chance and superbly disdainful of the flight of time, when he joins the group of disciples around the kindly smiling master, truly he feels himself received into a spiritual family, strongly united under Niels Bohr's paternal aegis. And just as, in the latin poet's restless soul, veneration for Epicure derives from a feeling of decisive liberation – liberation from the 'terrors of religion' by a rational understanding of the universe, – the disciple sees in Bohr the thinker who for contemporary science has brought about a similar liberation by elucidating with his idea of complementarity the apparent paradoxes of quantum theory.

But this unity of view on the problems of science – so characteristic of those who have lived around Bohr that the phrase 'Copenhagen spirit' has been coined to denote it – has nothing in common with dogmatic adhesion to rigid doctrine; the Copenhagen spirit, on the contrary, is *par excellence* that of a complete freedom of judgment and discussion. "Every sentence I utter", Bohr often repeats, "must be understood not as an assertion, but as a question". A mind always alert, always open to doubt, to possible improvement, attentive to the least objection, never resting until it has completely enveloped the object of his search and dominates all its aspects, such is the model of a scrupulous, patient, laborious investigator which Bohr offers to his pupils. With absolute candour he interrogates Nature, with infinite caution he makes himself her interpreter. He allies in the most happy balance boldness and vigour of thought with prudence and critical acumen; whenever he ventures on untrodden paths, abandoning without regret conceptions that seemed most strongly established, he never takes such a decision light-heartedly, but on the contrary – similar in that to all great creators – only after he has made sure of its necessity by careful examination, searching analysis of the requirements of experience. This lesson in scientific method, still more perhaps than the light thrown by Bohr's powerful views on the most fundamental questions of atomic and nuclear constitution, is the inspiration found in Copenhagen by physicists of all nations.

They come to the scientist, but they find a man, in the full sense of the word, as the oft-quoted lapidary formula defines it: "Homo sum, humani nihil a me alienum puto". On every occasion, Bohr's mind is untiringly striving to achieve harmony between the multiple solicitations of his ever wakeful sensibility; his attention is equally arrested by the great problems of existence and the small events of everyday life; he always proceeds with the same originality, with that freshness of outlook and that deep sincerity which he exhibits in his scientific work. His whole attitude, in the laboratory as in the family circle, towards the problems of science as towards those of society, always expresses without effort the same serenity, the same tranquil force, the same desire for universal comprehension, and above all the same radiant, unshakable optimism.

Harmonious existence, with its deep and always present accord between a wealth and plenitude of content and a simplicity and spontaneity of form – of attitudes and reactions. Harmony between form and content: that is how Bohr himself summarizes the aim and object of his whole activity. His method of attack on any problem is from the beginning directed towards that aim: it is a stubborn struggle for an adequate formulation of the various features successively brought to light by patient analysis. "Let us try", he begins, "to survey what we know, and to formulate it as best we can." So a first draft is outlined, on which criticism is at once brought to bear; innumerable retouchings and corrections soon cover the pages, fill the margins, insinuate themselves between the lines. But suddenly – generally some morning after a sleepless night (which he describes as "sleeping on the problem") – all this painfully constructed composition is abandoned: another way of looking into the matter has occurred to him and a fresh start is made along this new line. A second text is elaborated, embodying perhaps some remnants of the first, and is immediately subjected to the same criticism, ruthlessly destructive of any obscurity or looseness of thought. When in full action, Bohr's mind, with its restless probing, its adventurous explorations in all directions, reminds one of the ant's feelers, obstinately palpating on all its faces an obstacle suddenly encountered. Bohr likes to claim for the investigator of nature the unrestricted right to let "his thoughts go freely to and fro". These successive attempts slowly cover the problem with an ever tightening net; none of its aspects can ultimately escape this unceasing search. However, though a keen awareness of the imperfection of all human labour never leaves him, Bohr is too earnestly conscious of the scientist's social duty to lose himself in endless speculation. He often repeats Faraday's exhortation "Work, finish, publish". So at length a last version of the paper is reluctantly abandoned to the printer; every sentence in it, rich in nuances and delicate modulations, endeavours to express the whole variety of aspects and points of view successively recognized, with all the maze of mutual connexions and interactions between them. To nobody better than to Bohr can Buffon's words be applied: "Le style est l'homme même"; his style everywhere displays a subtle ponderation striving to harmonize content and form.

This conscious tendency towards a harmonious synthesis of various elements, seemingly discordant, dominates Bohr's entire scientific career. Thinkers of his type (names like Galilei and Goethe suggest themselves) are wide open to all influences, to all contemporary streams of thought, but their strength, their genius, lies in their ability to pick up from the materials offered by ruling traditions those which can be united into combinations transcending such traditions and lifting science onto a new plane. Bohr counts himself lucky to have grown up in a small country – unhampered by any national pride or illusion of self-sufficiency – in which the great traditions of British and German science found a prosperous meeting place; to this circumstance he ascribes his early and thorough acquaintance with the ideas which inspired his atomic theory – those of Planck and Einstein on the one hand, of J. J. Thomson and Rutherford on the other.

The question whether he would confine himself to experiment or theory did simply not arise with him: his tremendous power of intuition into natural phenomena was served from the start by an equal mastery of all the necessary mental tools, the experimenter's turn of mind as well as the mathematician's most refined methods of analysis. To be sure, the very way in which he handles the latter, as well as his general views on mathematics, betray a bias which sharply separates his from the pure type of mathematical mind; for instance, he has never been able to take axiomatics quite seriously: such abstractions are obviously too much for his strong sense of reality, for his instinctive need to connect every human activity with some concrete aspect of nature. His fine intuition of mathematical concepts, disclosed by his peculiar virtuosity in applying them to the analysis of physical problems, is based on an intimate knowledge of their harmonious adaptation to this living reality, rather than on any deeper feeling for their abstract consistency and beauty.

If, therefore, after a first experimental paper in the best Rayleigh style, he was led to concentrate his efforts almost exclusively on theoretical problems of the most fundamental kind, this was not the consequence of any one-sided orientation of his mental outlook. It was not even the result of any deliberate choice: as a matter of fact, when he went to England, after the completion of his thesis, it was with the intention of carrying out experimental work; but in that memorable period, when a more and more detailed picture of atomic constitution was emerging from a quick succession of discoveries, what could he do but throw himself heart and soul into the discussions and speculations to which these discoveries gave rise? The very earnestness and thoroughness with which he tackled the problems arising in this connexion continually distracted him from pursuing any quiet line of experimental investigation of his own. So it came about that, as he was just starting preparations for such a piece of work, the theory of atoms of which he had been all this time 'dreaming' (to use his favourite expression) suddenly crystallized on the Balmer formula; following Rutherford's advice, he then gave up any other endeavour and settled down to his first great work.

The real clue to an understanding of the underlying motives of this work is to be found in a line of development of Bohr's thought entirely different from the more technical one I have just been commenting upon: a trend of thought which, indeed, determined his whole attitude towards scientific problems, especially his predilection for the features of natural phenomena illustrative of general and fundamental laws, and that characteristic endeavour to achieve a rational synthesis between seemingly contradictory elements. It is natural for the youthful mind awakening to the contemplation of the world to face with bold inexperience the eternal problems of nature and man, and young Bohr was no exception to this rule. But he did it in his own way, receptive as he was to all the messages from nature and struggling to harmonize them. His efforts to find his way in the infinite variety and wealth of the universe incited him to lay peculiar stress on the epistemological problem, and he reached in this respect conclusions which were a powerful guidance and inspiration to him. To this source can in fact be traced the spirit of the 'correspondence argument', from the very beginning his leading idea when dealing with problems of atomic constitution. Balmer's formula, then – with which he became acquainted by casual reading – was the touchstone he needed; for, inwardly convinced as he would be of the soundness of his general principles, he could not let the matter rest until he had laid hands on some concrete bit of reality, allowing him to test their truth and to work out at leisure a more accurate formulation of their content. Indeed, while his mind is always striving after universality, he cannot be satisfied but in the closest scrutiny of detail: while his aim is the formulation of some general law of nature, he does not conceive of any other way of achieving it than to concentrate "im kleinsten Punkte die höchste Kraft", as his favourite quotation from Schiller puts it. This imperious need for a solid grasp on reality provides, in the wonderful balance of his mind, an adequate compensation for a certain northern mistiness, a love of half-tints, resulting from the equally strong urge that no shade of this infinitely complex reality be lost in the process of analysis.

The strength of Bohr's philosophy lies in the exquisite suppleness arising from its essentially pragmatic character. Any attempt to make a system of it would in fact be a betrayal of its very spirit. It does not pretend to reduce the laws of nature to a few basic principles; Bohr characteristically avoids such a word as "principle"; he prefers to speak of "point of view" or, better still, "argument", i.e. line of reasoning; likewise, he rarely mentions the "laws of nature", but rather refers to "regularities of the phenomena". For him, indeed, man, as an analysing subject, always occupies the central position, and the task of science can only be one of rational description, of reducing to compendious order the data of observation. Often, when commenting upon some controversial point, he would say, "The question is not whether this view is true, but what *arguments* about it we can honestly draw from the available information". And the most brilliant conception finds no favour in his eyes if it is not clear "by what arguments it can be defended". So his aim is above all one of method: he strives to attain a general point of view from which the results of the analysis of the phenomena can be comprehended in a rational scheme.

The difficulty here is to do justice to the overwhelming complexity with which nature confronts us. Few men know as well as he does the power of analysis, but at the same time he keenly feels the utter insufficiency of any one-sided analytical procedure: the harmony of things is made up of an interplay of apparently conflicting aspects. But where metaphysically minded people would be driven to despair, he finds salvation in his healthy sense of reality, in his humble readiness to 'learn' from nature the right way and his unwavering resolution to follow that way to the end. He points out that the idealized concepts we use in science must ultimately derive from common experiences of daily life which cannot themselves be further analysed; therefore, whenever any two such idealizations turn out to be incompatible, this can only mean that some mutual limitation is imposed upon their validity. Thus recognizing the futility of any attempt to conciliate them, he claims their right to be both upheld in their full significance, as 'complementary' to each other, within sufficiently restricted domains of validity.

Although (and this deserves to be stressed as a most remarkable circumstance in the history of atomism) this conception of complementarity inspired Bohr from the days of his youthful meditations, its final formulation demanded a long elaboration supported and, indeed, only made possible by the development of quantum theory. The relation of complementarity, like the narrower one of causality, is in itself but a formal scheme which must be filled with some contents to take a definite shape: in each special case, the appropriate form has to be worked out in close contact with the relevant experience. And just as the law of causality could only be given a precise enunciation in connexion with the problems of classical physics, it was quantum theory which ultimately provided the basis for a clear formulation of the point of view of complementarity. From the very beginning Bohr's approach was marked by the complementary mode of thought; particularly revealing in this respect is the concluding sentence of a lecture he gave in 1913 on his explanation of the hydrogen spectrum: "Just by accentuating this contrast it will perhaps be possible in the course of time to bring about a certain consistency also in the new ideas". In this truly prophetic statement we find indeed a striking expression of the typical process of complementary thinking: a rejection of any compromise between the conflicting conceptions, and instead a following up of the implications of each of them in its pure form. The same endeavour is again perceptible in the 'correspondence argument', which dominates the evolution of the theory until the great crisis of the years 1926–27. Then the exaltation of the contrast between the conceptions of point mechanics and wave theory reached the climax needed for prompting Bohr to the decisive step, expounded for the first time (not yet in a perfect form) in the Como lecture. From that time on we have witnessed an ever deepening refinement of the formulation of the complementarity idea, as well as an ever widening extension of its scope; a fruitful epistemological point of view has developed, placing its founder among the foremost thinkers who from Descartes on have pondered upon the meaning of human knowledge.

To Bohr himself, the disclosure of the fundamental complementarity relation of quantum theory brought a real liberation, by affording him the possibility at last to give adequate expression to the strongest yearnings of his soul which influenced not only his rational thinking but even his innermost affective life. How happy he is, when an occasion arises in the course of some present work, to tell in simple words about his past hopes and enthusiasms in his strenuous struggle for inner harmony! The period of his adolescence was marked by a reaction of the rising generation against the flat rationalism of the last epigones of XIXth-century thought. No wonder that Bohr eagerly participated in this revolt, he who could have taken as his motto that verse of Schiller's which he so often recites and which ends with the line

Und im Abgrund wohnt die Wahrheit.

The dominant feature of his mental attitude has indeed always been a keen realization of man's middle position between the infinite expansion of the universe and the unfathomable profundity of his own self. When speaking of the novel situation met with in quantum theory as a result of the intricate interaction of the observing and analysing mind with the world of atoms, he likes to emphasize, in his soft, insistent voice, that one cannot peer into such problems without a feeling of dizziness; adding, with a humorous twinkle in his eye, that as long as you don't feel dizzy, you have not grasped the point. To foster such an attitude, Danish literary tradition offered one of its finest jewels: through Poul Möller's charming "Tale of a Danish Student", young Bohr became indirectly acquainted with the trend of thought of Hegelian philosophy, so highly congenial to his own. The dialectical difficulties of the caricatural licentiate strongly appealed to him. He clearly saw how the conscious renunciation of carrying any analysis beyond a certain limit – the only way to escape the danger of drowning, like this poor licentiate, in the bottomless sea of reflexive thinking – implies a recognition of the co-existence of complementary descriptions starting from different sets of unanalysed concepts.

In this connexion he was led to profound views on the use of language, which he tried to clear up for himself by means of a remarkable analogy with the conception of multiform function and Riemann surface. Every word referring to our own mental activity is ambiguous: it belongs to different 'planes of objectivity', just as the values of a multiform function are distributed on different Riemann planes. So the concept 'consciousness' is like a singularity in consciousness. The use of words in everyday life is subject to the condition that they be kept within the same plane of objectivity; but as soon as a word is used which refers to our own thinking, one is exposed to the danger of gliding onto another plane. For instance, in the seemingly innocuous remark "One might imagine beings on some other planet whose logic would be different from ours", a fatal danger is hidden in the word 'imagine'. Also, mathematics is a game in which any self-reference is forbidden; mathematical concepts are only valid in the plane of objectivity for which they have been shaped, and the paradoxes of the theory of sets arise from an insufficiently guarded extension of such concepts into some other domain of objectivity. But just as the gist of Riemann's conception lies in comprising all the branches of a multiform function within a single concept ('the' function), so it is an essential feature of language that there is one word only for each form of our psychical activity: one word "will", for instance, to designate voluntary behaviour as well as conscious feeling of volition. It is therefore an illusion to believe that one could avoid this fundamental ambiguity by any substructure using "new" concepts. The complementary relations of the planes of objectivity to each other cannot be but primitive, irreducible ones, corresponding to an essential property of our very use of language to describe and communicate our experiences.

While young Bohr was following in solitary meditation such highly original lines of thought, his formal philosophical education remained rather cursory: it was practically limited to Höffding's university lectures. These were, it is true, so masterly in their pregnant exposition of the main systems, especially of the flourishing period from the XVIth to the XVIIIth century, that even though he did not acquire any firsthand knowledge of them, he could form a reliable idea of their tendencies. Open-minded without formless eclecticism, of wide outlook without being superficial, Höffding gave his teaching quite an uncommon orientation by particularly insisting on the philosophical implications of scientific questions and, above all, by "striving to give more consideration to the proposition of problems than to their solution". "The solutions", he explained, "may die, but the problems are still living; otherwise, philosophy would not have had so long a life as she has had." Bohr was particularly attracted by the figure of Spinoza, in whose system he again recognized features of complementary thinking. Another master to whom, besides a priceless schooling in the technique of experimental research, he owes much inspiration in philosophical matters, is his father, the physiologist; by him he was acquainted with the great problems of biology, and in particular he learned from him that no exhaustive account of the phenomena of life would ever be possible without recourse to final causes; this conviction was later for him the incentive to look at the opposition between mechanicism and vitalism in the light of complementarity.

Deeply interested in psychological problems, he was much impressed by Kierkegaard's searching and so acutely sensitive essays. William James he came to know only in later years, but then he was at once enthusiastic about the celebrated picture of the 'stream of consciousness', which fits so well into his own view. About pragmatism as a whole, although it expresses a tendency undeniably akin to his, he is, however, non-committal, as he dislikes being classified under any 'ism'. Neither did religious problems leave him indifferent. His independent mind could not be reconciled with the appeals to revelation and stale commentaries on sacred scriptures of the extant churches. But he was fascinated by some of the great religious figures of the past, especially the Jewish prophets and the Buddha; he earnestly endeavoured to penetrate the human side of their teachings and arrived at interpretations of striking originality. Into these sacred texts, coming from the inmost recesses of the human soul, he read an effort to account in a peculiar language for that deepest complementarity between rational knowledge and living experience of the cosmos which in his eyes characterizes man's ambiguous position and rules his whole activity. In that sense he is even prone to surmise that points of view of complementarity may exert a determining influence on religious thinking of the future.

The preceding evocations, however fragmentary, may have conveyed some idea of the exceptionally rich and many-sided early development of Bohr's philosophy. No wonder that such deep-rooted views were to play a leading part in his thoughts of later years and even pervade every shade of his sensitive reactions to the various impressions of everyday life. I cannot refrain from citing here, as a sample of the poetry of complementarity, a delightful recollection of his Japanese journey which he once told me: At sunset the top of Fujiyama disappeared behind a curtain of gold-fringed clouds; the black mass of the mountain, surmounted by this fulgent crown, conveyed an impression of awe and majesty. On the next morning, it offered an entirely different spectacle: the pointed summit alone, covered with shining snow, emerged from the dense mist filling the valley; the landscape was radiating gladness and joy. So, Bohr mused, the two half-mountains together are not simply equal to a mountain: to each belongs a peculiar, individual impression, and the two are complementary.

The intimate communion with nature that marks Bohr's thinking in all its aspects does not result from any facile yielding to impressions or intuitions; it necessitates, on the contrary, the participation of incessant work of analysis. The spontaneity of its manifestations is never an improvisation. It issues forth from deeper sources of powerful vitality, from an organism of uncommon vigour and endurance, without which such a continual strain could not be borne. As an adolescent, Bohr gave vent to his surplus energy in a practice of sport as passionate as the pursuit of his favourite studies; together with his brother Harald, he won glory in a football team. A few years ago, a bank clerk who had been in the same team and was duly proud of it, confided to me that already in those days he had thought, judging by Bohr's play, that there was a boy who would go far. "You know, he played with his brain", he explained, and there may be a deep truth after all in this observation. Even though the torch of football playing has long since been handed on to the next generation, Bohr's passion for sport (he displayed great skill in skiing and sailing) and his love of nature have retained all the freshness of youth. How thrilling were those long wanderings in his company through Tisvilde forest (already some ten years ago), when our fancy would be caught by the contorted shapes of pines grown in the blasts of the nor'-wester, a few humble ruins hidden in the foliage, the quaint behaviour of an antlion discovered on some sandy path, or the sight of a stranded ship in rain and wind!

The spectacle of his fellow-men would not be less alluring to him. I remember how once, in Brussels, a crowded little restaurant in a busy square unexpectedly captivated him by its picturesque medley. He looks on mankind with unbounded sympathy and confidence; in fact, his warm belief in the good will of men is not sufficiently tempered by a realization of the harshness of contemporary social conditions. His subtle intuition into the human soul cannot here, any more than his open-mindedness, counterbalance the lack of direct experience; in a small democratic country, of relatively uniform and rather high social standard, he grew up in a purely intellectual environment, whose complacent liberalism had not yet been shaken by the tumultuous development of industrial civilization, and his naïvely idealistic representation of social relations is still under the spell of Spencerian individualism. But within this atmosphere of somewhat obsolete distinction, he is at his best when he performs, at times with a slight touch of solemnity, the duties of father and head of the family. No more exquisite harmony, no more delicately poised complementarity can be imagined than that which unites him with his wife. He modestly attributes to her the greater part in the education of his sons, while he describes his own role as "adding a remark or two on reading the proofs"; in fact he managed, in playful or earnest intercourse, to give them a firm lead without interfering with the free development of thought and personality. The boys look upon him, with trustful eyes, as an older comrade. On their father's fiftieth birthday, one of them, Erik, aptly expressed *their* opinion on the matter in a charming speech ending thus: "We are glad to have a father such as we have." Bohr's disciples, young and old, likewise come to him in the same spirit of happy confidence and never appeal in vain to his wisdom. He partakes of their troubles with genuine concern, of their joys and hopes with all the kindness of his heart; and if there is some love affair in hand, well, he is all the more fatherly.

In any circumstance one can be sure that he will take a hopeful view of the situation. Optimism is a dominating feature of his character; it is apparent in the very boldness of his scientific thought and above all in the conviction, underlying his whole outlook and his whole endeavour, that any obscurity, any discordance whatever, must in the end resolve itself into harmony. Sometimes this optimism is rather disconcerting, for instance when it concerns the estimate of the rate at which a paper is to be written. But he needs it somehow: it conditions the dynamics of his personality. Satisfying himself about some problem is not enough for him; his task is not completed until he has done his best to win the others over to his point of view. And also in this last respect the outcome does not always answer to his sanguine expectations. His papers require arduous study; the subtle intricacy of the long periods finds no relief in the *epitheta ornantia* which he thinks advisable to insert here and there; their conventionality rather contributes to make the style heavier still and to dissimulate its real beauty. Furthermore, his typical way of proceeding by allusions rather than by plain statements, of suggesting rather than expressing the ideas, does not make the access to his thought any easier. As a lecturer, in spite of the tremendous amount of labour he spends in preparing himself for the task, he is seldom successful: audiences are generally not disposed to make the effort of concentration he demands from them; as Sir Joseph Larmor once observed (comparing him in this respect with Maxwell), "he might want to instruct us about the correlations of too many things at once, out of the repertory floating before his vision". The real way for him to convey his ideas is in conversation; there lies his force as a teacher, there he is able freely to display the resources of his mind and, above all, to open his heart.

A discussion of some question is not for Bohr a mere affair of exchanging views or information; he envisages every problem with such earnestness, and there is within him such a close and harmonious interweaving of all rational and affective impulses, that the discussion stirs his whole being, brings all his faculties into play. Youthfully enthusiastic, lively and spirited, persuasive and tenacious, he holds the audience under the spell of his speech. Then his shy and abstracted countenance, the irregularity of his features – the heavy lower jaw, hanging cheeks and thick-lipped mouth – fade away behind the fascination of his look, always frank and kind and, under the shield of the thick eyebrows, so wonderfully expressive of intense concentration, sympathetic curiosity or gentle irony. He instinctively conceives his work in a combative spirit, as a struggle to conquer new insight or to dispel prejudice, a struggle for the acquisition of knowledge or for its defence against unfounded doubt and criticism. Hence the remarkably stimulating effect of opposition on him: when faced with an objection, his excitement does not subside until he has seen his way to dispose of it. So no more efficacious incentive to a lively conversation can be wished for than some allusion to opinions or conceptions at variance with his own belief. If he, moreover, chooses to treat the difference in a lighter mood, one may witness a delightful debate, sparkling with humour of the most candid and good-natured kind.

Bohr's humour is indeed a side of his personality which reveals itself in his conversation more than in his writings (the style of his papers lacks the necessary freedom). It is a smile which gives his earnestness the complementary touch and prevents it from degenerating into Helmholtzian solemnity. It has a peculiar quality: neither superficial nor brilliant (he dislikes French 'esprit'), it proceeds by means of finely pointed allusions – a delicate play with ideas. The element of play, by the way, the 'homo ludens' aspect (to use Huizinga's expression), is not a negligible one in Bohr's psychology. It has undoubtedly some part, for instance, in his taste for experiment and for all sorts of techniques. When he was working on the surface tension of water, he had to melt glass tubes to make jets: he took such a delight in this operation that, completely forgetting its original purpose, he spent hours passing tube after tube through the flame. There is also the memorable controversy – one of the bright spots in the minor history of Bohr's institute – about the mechanism of a 'thought transmission' trick which was being performed at the Tivoli: with what genuine enjoyment did he advocate his own view of the case! It was an ingenious theory, too, involving ventriloquism as its basic principle. This playfulness of his is easily awakened by certain widespread opinions about which he has been led to entertain a deep-rooted scepticism: to hear him question the reality of ball lightning (to mention only one subject) is an unforgettable experience. School philosophy of any description stimulates his humorous vein. He is unable to feel any respect for metaphysical problems because they appear to him so utterly barren, so devoid of any utility for the great aim of science, that he deems it entirely superfluous to take them into consideration; they belong to those which he dismisses as 'trivial'. Once he attended a lecture by a prominent positivist philosopher, who set himself to disclose the various pitfalls that threaten addicts of metaphysics. Rising to thank the lecturer, Bohr made the following comment:

> It has been very aptly proposed to define the specialist as the man who has worked so long in some branch of learning that he has come to know all the errors which can be made in it. I think we may say that Professor N – has shown us, in his most interesting lecture, that he truly deserves the name of metaphysician.

In the preceding pages I have endeavoured to sketch as true a picture as I was able to of a great thinker and a lovable man, as I have come to know him and as, I trust, others who have even longer enjoyed his warm friendship will recognize him. Like all prophets, Bohr is liable at times to feel lonely. He once confided to me, not without a melancholy note in his voice, that he considered his fight for complementarity a failure, as people – apart from the small group of his closest friends – did not realize the earnestness of his endeavour. On this day, all those to whom he has given guidance in years past are gathered around him in thankfulness, expecting from him further inspiration: let this convince him that he was mistaken in yielding to such an unusual pessimism. May his fatherly smile enlighten our way for long years to come!

CHAPTER 23

THE CONCEPTION OF THE MESON FIELD*

Some Reminiscences and Epistemological Comments

[1968e]

It is difficult to recapture the sensation which the discovery of the neutron produced among the small group of physicists who were then just beginning the investigation of the nuclear structures. The reason for the excitement was that this discovery at once brought some relief to the critical situation that had arisen concerning the role played by electrons in the constitution of nuclei. Squeezed within the volume of a nucleus, these electrons were expected to show extreme relativistic behaviour, and the relativistic quantum theory of the electron, according to which strong electromagnetic fields would lead to transitions to states of negative energy, offered no reliable guidance. Experimentally, the newly measured value of the spin of ^{14}N suggested that the intranuclear electrons somehow lost their spin; the scattering of γ-rays by nuclear electrons proved equally intractable. Such difficulties instantly vanished if one regarded the neutron, instead of the electron, as a nuclear constituent besides the proton. Heisenberg, taking up the new picture of the nucleus with his wonted vigour and directness of approach, was soon able to show that it could form an adequate basis for a comprehensive account of nuclear phenomena. Certainly, it could not remove all difficulties; those connected with β-decay remained as acute as ever: there was apparently a loss of energy and spin in the process, although the initial and final states had both definite energies and spins.

This last problem, however, could now be shifted to the constitution of the neutron itself. It is significant that Heisenberg, while treating the neutron as a distinct nuclear constituent, still conceived it as a composite particle, which could decay into a proton and an electron. This conception was in a way fruitful; it suggested the introduction of an exchange interaction between proton and neutron, ultimately amounting to the exchange of a virtual electron between two protons. In another way, it was misleading, as Majorana was quick to point out: the Heisenberg interaction was not sufficient to explain the extreme stability of the helium nucleus configuration. The great importance of Majorana's contribution lay in its direct appeal to phenomenological considerations in setting up a suitable form of nuclear interaction. The problem of the neutron structure was just a scaffolding which had helped Heisenberg up to a point, but was now rather embarrassing than useful; by removing it from the scene, Majorana cleared the way for a more unbiased approach to the problem of the origin of the nuclear forces.

The solution of this problem was considerably stimulated by the next great advance, the introduction by Pauli of the concept of the neutrino, as the elusive carrier of the energy and spin which seemed to disappear without leaving any trace in the process of β-decay. The idea was so bold that Pauli himself seemed hesitant to follow it up; we used to tease him at the time because he never formally published it, but was content, after putting it forward at a symposium in Pasadena, to let the rumour of it spread over the small world of physicists. Whatever doubt it might at first have aroused was soon dispelled when Fermi gallantly took up the challenge and – besides baptizing the new-born baby – elaborated the theory of β-decay which logically followed from Pauli's proposal. This double stroke of genius had the effect of liberating the quantum theory of fields from the spell that retarded its progress: the cloud of mystery that hung over the relativistic theory of the electron was blown away, and wide prospects came to view. The puzzle of the intranuclear electrons received the simplest possible answer: there were no such electrons; only, we had to accept the idea that an electron field – accompanied by a neutrino field – could be created in a nuclear transition just as an electromagnetic field was produced in an atomic transition. The ground for this novel conception had been prepared by another concomitant development which I have no time to retrace: the idea of anti-particle, initiated by Dirac, put into its final shape by Pauli and soon substantiated by the experimental discovery of the phenomena of pair production and annihilation. Thus, we first gained real confidence in one of the most fundamental features of quantum theory: the elimination of the dualism of matter and force and its replacement by a relation of complementarity of these two aspects of the behaviour of every constituent of the universe. Every such constituent can be the source of other fields and be acted upon by them, and it can itself play the part of a field transmitting interaction between other constituents.

Before the momentous events I have tried to recall, this general conception could hardly be convincingly established. Apart from the electromagnetic field, with its complementary photon aspect, we only knew of the role of the electron field in chemical binding, which was not a very clear case, and of the gravitation field, whose quantization was of course quite conjectural. The background for my own attempt, at that time, to calculate the gravitational energy of a photon on the basis of quantum theory, was just to provide at least a theoretical test of this idea of the interplay of the field and particle aspects. Now, however, we were confronted with two new and very real types of interactions: those that are now called the strong interactions – the nuclear forces, and the weak interactions – the nucleon-lepton coupling governing the β-decay. These were provisionally described in a phenomenological fashion; there was a clear challenge to investigate them from the point of view of the particle-field complementarity.

Let us at this point interrupt the story for a moment, and reflect upon the epistemological implications of these developments in physical theory. The introduction of the neutrino by Pauli in order to 'save' the principles of conservation of energy and angular momentum reminds us of a similar case in the history of chemistry: Lavoisier's introduction as acidifying agent of a new substance, oxygen, which could carry the mass necessary to uphold the conservation principle in chemical reactions. It is true that Lavoisier also introduced massless 'elements', caloric and light; but he was wise enough to point out that these might be essentially different from the 'material' elements, endowed with mass. In fact, heat turned out to demand a new conservation principle, originally independent of mass conservation; and light was eventually recognized to belong to a primary constituent of the universe, the electromagnetic field, emanating from electrically charged bodies and acting upon such bodies as a 'force'.

In this way arose the dualism of matter and force to which I alluded a moment ago. The particles of matter were conceived as carriers of invariant quantities, mass and electric charge, quantities obeying conservation principles. Every motion, every change in the distribution of these invariant quantities, was then produced by some kind of force. It is this dualism that has been overcome in quantum theory by the recognition that every constituent of the universe can play the part, both of carrier of invariant quantities and of conveyer of force. Philosophers will endlessly dispute whether this is a victory for materialism or for idealism [Laughter]. I am afraid I can only say, in the spirit of complementarity, that it is neither the one nor the other [Laughter]. As you know, quantum theory has forced us to revise not only our ideas concerning the fundamental constituents of the physical world, but also our ideas concerning the way in which we have to speak about the physical phenomena. This epistemological analysis of the conditions of rational description of the phenomena has consequences which go far beyond the realm of physics; they affect our whole attitude towards the function of science. We now recognize that the task of science is not to picture the world as a spectacle watched 'from outside', if I may say so, but rather to give us the means of communicating, in a rational and objective way, the experience derived from our interaction with the world around us, of which we ourselves are a part. It is clear that this relationship between the material universe and the human mind transcends the antiquated opposition of materialism and idealism, but could rather be described as a synthesis of these mutually exclusive points of view, in the sense of complementarity. It should be stressed that this conclusion is not the result of any intention to decide or evade the issue. It has been forced upon us. No physicist would willingly embark upon the hard work of an epistemological analysis if it were not forced upon him by some really acute dilemma, – as those that confronted Einstein and Bohr. For this reason, the conclusions we draw from such analyses are not matters of arbitrary choice or endless debate, but have the same character of indubitable certainty as any other acquisition of science.

I wanted to recall these well-known things in order to present a background against which we shall now be able to measure the greatness of the achievement that the introduction of the concept of the meson field by Yukawa represented at the stage we had reached in our recollections. Just before he entered the scene, an interesting question had been asked and answered by Tamm and Ivanenko, quite in the spirit of the general conception just discussed of the double role of the elementary constituents of nature. Since beta-decay suggested the existence of a direct coupling between nucleons and leptons, could not the interaction between the nucleons be transmitted through the intermediary of the lepton field? Treating the problem by a perturbation method, they found that the range of this leptonic interaction is given by the Compton wave-length of the electron, which is the mass-carrying lepton; the distance dependence being expressed by a fifth-power law. However, as we would expect, the order of magnitude of the coupling strength came out all wrong: there was a discrepancy of a factor 10^{-12} between the strength of the calculated potential between two nucleons and the observed strength of this interaction. It may seem surprising for us that one should have deemed it necessary to perform the whole calculation to reach this conclusion, that there should have been so little feeling for the orders of magnitude to be expected. This reveals how new and unfamiliar these ideas were at the time, and helps us to realize the boldness of Yukawa's intervention at this critical moment.

The decisive point stressed by Yukawa was the necessity of introducing a new kind of field to account for the nuclear interactions, just because they are so much stronger than the electromagnetic ones and those responsible for the beta-decay. When I was asked to give this talk, the first thing I did was to read again Yukawa's paper, and I was very much impressed to see, against the background of our present knowledge, how much this short paper contained, how thoroughly Yukawa had thought out all the consequences of the idea he was putting forward. Above all, he pointed out the relationship between the range of the force transmitted by the field and the mass of the associated particles; a feature which may look very simple, but which was then entirely new. I remember how eagerly it was discussed in Copenhagen, where Wick very elegantly exhibited its deep-lying connexion with the complementarity of the field and particle aspects. Just because it was so deeply rooted in the foundations of the quantum theory, this relation offered the possibility of a severe test of the new idea: one had to expect that at sufficiently high energies, particles of the predicted mass would be actually produced, and one had then to look for them in the cosmic radiation. Now, as you know, this part of the story turned out to be a little more complicated than the optimists thought when a heavy component of the cosmic radiation was first discovered. Indeed, it was found that these heavy particles could be captured by atoms like electrons, and did not, therefore, interact strongly with the nuclei. However, the puzzling situation thus created was not of long duration, for the weakly interacting particles were very soon shown to be decay products of the true mesons we were expecting. Yukawa's paper contained another interesting suggestion: if the meson field were coupled not only to the nucleons, but also – much more weakly – to the leptons, it would bring about a nucleon-lepton interaction, which could be that responsible for beta-decay. This possibility is still *sub judice*, in the modernized form of a hypothetical 'intermediate boson'.

I shall not pursue any further these reminiscences of a most exciting chapter in the history of modern physics. From the point we have reached its further development loses the glamour of the pioneering period and proceeds at a more sedate pace along the conceptual lines firmly drawn out in Yukawa's first paper. We are still to-day thinking in terms of the same general ideas, however much we may be hesitating about the analytical expression we ought to give them. It was quite natural for Yukawa and his early followers to formulate meson theory in analogy to the theory of electromagnetic fields. We are now much in doubt whether this formulation has more than a very limited validity, – in spite of the fact that quantum electrodynamics, on the contrary, has stood the most exacting tests and now appears as one of the best established theories of physics. It quite probably owes this singular distinction to the smallness of the electromagnetic coupling constant, allowing us to uphold (at the price of some renormalization conventions) the correspondence between quantal and classical quantities, which guarantees the consistency of the theory. The phenomena resulting from strong interactions, as we observe them at high energies, present a complexity absent from the electromagnetic effects, and rather suggesting the existence of some underlying structure of the 'hadrons' themselves. It would be pointless at present to speculate about the turn the exploration of this deeper structural level will take. We have no cogent argument yet from experiment to orient theoretical thinking in one definite direction, and our conjectures are at the mercy of unexpected discoveries.

There is one aspect of the modern theories which is not likely to disappear and which is sometimes causing concern: it is the abstract character of the concepts with which they operate. Even in the asymptotic descriptions entering the definition of the *S*-matrix or the group classification of states, one has to use, besides directly measurable quantities like energy, momentum or electric charge, a number of parameters, such as spin, isospin, hypercharge, which do not correspond to classical quantities. Equally abstract are the mathematical procedures by which one attempts to proceed beyond such asymptotic descriptions, in order to estimate cross-sections or other parameters derived from the experimental data. Professor Yukawa, in his introductory talk to the symposium preceding to-day's ceremony, complained about this feature, regretting to see physics drift away from visualizable representations and explanations in simple terms. One is certainly inclined to sympathize with him when one hears the jargon in which some of the younger physicists express themselves. Indeed, some have given up speech altogether, they only draw graphs. Quite recently, I heard a talk about scattering of electrons by nuclei; the speaker – a very distinguished experimental physicist – wanted to convey the remark that an electron moving near a nucleus is deflected by the Coulomb field of the latter; well, what he said was that one has to take account of a continuous distribution of nonphoton lines attached to the electron line [Laughter]. On second thoughts, however, one realizes that the question is not so simple as it looks.

What is, after all, an 'anschaulich' picture? A visualizable representation? Something that corresponds more or less directly to our perceptions: but what is a perception? Certainly not – as Newton believed – the projection of naive images onto a 'sensorium', but a highly schematic transmission of signals from the sense organs to the brain, in fact an abstract coding system of information. When one thinks of it, one may well wonder whether there is much to choose between Feynman graphs and pictures like those you see hanging on the wall [Laughter]. I might try to express the situation by suggesting that we are dealing in physics with two code systems. The one we have just alluded to, related to acts of perception, includes the concepts forming the ordinary language and describing our daily experience of the world on the macroscopic scale. This is the basis of what we call 'classical physics'. This code system is quite essential: any communication we wish to make about our experience, any description of a physical experiment, is only intelligible if it is expressed in that language. On the other hand, for the description of atomic phenomena, we need another language, another code system, operating with concepts that we call abstract because they are not directly related to our perceptions, which just means that they are not part of the first code system. Now, this second code cannot, obviously, be self-contained; in order to be intelligible, it must be linked to the first. It is at this point that the new epistemological feature of atomic theory appears: it turns out that this link is subject to mutual restrictions, expressed as relations of complementarity in the use of classical concepts in the account of atomic processes. It is too optimistic to hope that this abstract way of stating the epistemological situation could satisfy everybody? Why should not everybody be satisfied, by the way? This is another question [Laughter].

Some people seem to ask of philosophy more than philosophy can reasonably be expected to give. Philosophers have always tried to crystallize in their vast systems the sum of knowledge acquired at the time; they have looked for elements in this knowledge that could be thought to be invariable, and to the extent that their systems embody really invariant elements, they keep their validity. The human mind, however, has very few invariant features: it is a delicate biological process subject to continual change. This is the rock against which philosophical systems are likely to founder: they too often ascribe eternal validity to conceptions that only reflect a temporary adaptation of the human mind to a given situation. For this reason, when we are faced in physics with an epistemological problem, we must not allow ourselves to be influenced by whatever philosophers may have said on the subject. We have a better guide to trust, the method that the experimentalist trusts in his laboratory, that the theorist trusts when he is solving his equations – or operating a computer. It is the method of open-mindedness, of freedom from preconceptions, of readiness to adopt new modes of thought leading to greater harmony in our view of the universe.

NOTE

* Lecture delivered in September 1965 on the occasion of the 30th anniversary of the Meson Theory; Prof. Rosenfeld and Prof. Shoichi Sakata delivered two lectures at the initiation of an international conference on elementary particles held at Kyoto. – Ed.

NUCLEAR REMINISCENCES*

[in memory of George Gamow]

[1972c]

I have been asked to tell you some of my memories from the prehistory of nuclear physics, and I gladly respond to this suggestion, even though it is always a bit melancholy to be reminded that one has reached the age of reminiscences. There is another circumstance which will tinge my speech with some melancholy. One character that looms large in my memories is George Gamow; I had hoped to describe some of our experiences of those days with a light touch, but his recent death has changed the character of what I will tell you about him: it will be an homage to the memory of one of the pioneers of nuclear physics.

In my experience nuclear physics starts with the sudden appearance, one morning in the library of the Göttingen Institute, of a fair-haired giant, with shortsighted, half-shut eyes behind his spectacles, who introduced himself, with a broad smile, by declaring: "I am Gamow." This announcement, at that time, could not provoke very much excitement. As it turned out that Professor Born would not be in for some time, I proposed to Gamow to go out for a walk. It was during that walk that he told me what he was doing.

He wanted to understand alpha radioactivity. Now, this seemed to me – and I think most physicists then would have had the same reaction – quite a fantastic idea. All we know about nuclei was that they were very small and that they had a charge and a mass. We hardly knew at that time that they had a spin; this had just emerged from Pauli's interpretation of the hyperfine structure which spectroscopists had detected in the spectra of the heaviest atoms.

However, thinking in retrospect, what Gamow then told me was extremely revealing for the approach to a completely new subject, for the gropings and missteps that pioneers are exposed to. His first attempt had been a complete failure. He had considered an alpha-particle in the repulsive Coulomb potential of a point nucleus, and found – this was not then as obvious as it looks today – that all he could get was a perfectly smooth, continuous spectrum, without enhancement of special states corresponding to the energies of the observed alpha rays.

Then, it occurred to him that he must also take into account what happened inside the nucleus before the alpha particle came out, and this was the decisive step. There was an attractive potential inside the nucleus, which combined with the repulsive potential outside, to form a 'potential barrier' for the alpha-particle. He proceeded to construct a solution of the Schrödinger equation for this type of potential, corresponding to an outgoing wave at large distances. He found that he could not get any such solution for honest, real values of the energy, but guided by a powerful intuition he introduced a solution with a complex value of the energy, the imaginary part of which he interpreted as the disintegration constant. This gave him the explanation of the Geiger-Nuttall relation between this constant and the energy of the emitted particle.

He obtained this result soon after his arrival at Göttingen, and when he presented it at the weekly colloquium, he created quite a sensation. Born immediately saw the importance of the discovery, not only as the first application of quantum mechanics to nuclear physics, but as an example of a peculiar quantal effect of very general character and wide scope. He at once pointed to another example of such a 'tunnel effect': the 'cold' emission of electrons from metals to which a constant electric field is applied. Here there is a constant potential keeping the electrons inside the metal, and another linearly varying potential corresponding to the outside electric field tending to pull them out: this defines a barrier through which electrons can escape.

After this brilliant improvisation, however, Born expressed rather sharp criticism of Gamow's method: it was against the rules to introduce such strange solutions with complex eigenvalues; physical quantities were represented by Hermitian operators, which have real eigenvalues. Soon after the publication of Gamow's paper, von Laue raised the same objection: this unorthodox procedure should not be tolerated. Still, it was not so easy to dismiss Gamow's result, since it was confirmed by the Geiger-Nutall law.

This dilemma provoked Born so much that within a few weeks he produced another derivation operating only with Hermitian operators and perfectly regular states of real energy. It is worth recalling his procedure. In order to make the state in the interior of the nucleus perfectly stationary he imagined that the nuclear potential was infinitely high, defining a completely discrete spectrum inside; the Coulomb repulsion outside gives a perfectly continuous spectrum, which overlaps with the discrete spectrum inside. Neglecting a small overlap, the wave functions in the two regions, corresponding to these two spectra, may be treated as a complete orthogonal set, according to which one may expand the solution representing the decay process. One has still to correct for the error introduced by the infinitely high barrier. Although this may look hazardous at first sight, it may be done by a calculation similar to the perturbation technique; for it turns out that all the matrix elements occurring in this calculation are actually finite.

Well, our views of what is proper and improper in quantum mechanics have changed since that time. I was then very pedantic, as a natural result of my ignorance; I was even more shocked than Born by Gamow's procedure, and very happy to see that the results could be derived by regular methods. Nowadays, I have been led to take the opposite point of view and to defend the idea that just such 'Gamow states' with complex eigenvalues provide the most natural approach to a theory of the resonances that occur not only in spontaneous disintegrations, but in all nuclear reactions.

Nevertheless, in order to calculate the resonance parameters and other properties of the compound system, it is still a procedure very similar to Born's method that is the most appropriate, as can be seen from the recent work of the group headed by Weidenmüller and Mahaux in Heidelberg, which is based on a natural generalization of Born's conception.

Still, one of the latest developments of the theory of nuclear reactions has been the direct use of the Gamow states in the analysis of resonance processes. There had been an objection to such a procedure, arising from a seemingly paradoxical property of the Gamow states: while the eigenfunctions corresponding to complex eigenvalues decay in time, they increase exponentially in space, so that they are not normalizable. Recently, however, two physicists working independently, Berggren in Lund and Romo in this country, discovered that it was possible to get around that difficulty by introducing a converging factor in the definition of scalar products involving Gamow states; this makes it possible to use such decaying states for direct calculations of nuclear cross sections and resonance parameters.

As for Gamow himself, his defense was very characteristic of his scientific attitude:

What is all this fuss about? My solution is directly interpretable in terms of physically observable quantities. Thus, it yields a relation between the disintegration constant and the energy of the emitted particle, which are the physical quantities one measures and which characterize the process. The asymptotic condition defining my solution is simply the expression for the conservation of particle current. I cannot see anything unsatisfactory in this.

The work of Gamow made a very great impression on another young physicist who arrived at Göttingen about that time and who was also a very picturesque figure. Houtermans' 'Bohemian' manners were in marked contrast with the somewhat decorous Göttingen environment. He told us, for instance, that he had "hitchhiked" (to use the modern term, then quite unknown) through the whole of Italy, a feat that struck us with a feeling of wonder verging on scandal.

Well, Houtermans was fascinated by Gamow's analysis, and raised the question about the reverse process, more precisely about the possibility of a proton penetrating into a nucleus and either being captured or exciting the nucleus and producing a reaction. Hc estimated that in order to cross the potential barrier with a reasonable probability, the proton would require an energy of a few hundred kilovolts, which then seemed an enormous quantity, that appeared to limit the practical importance of the process to the cosmic scale. Indeed, Houtermans and Gamow developed remarkable speculations about the cosmical consequences of the proton capture process.

For an electrical engineer like Cockcroft, however, Houtermans' considerations presented quite another challenge. Cockcroft realized that it would be technically possible to build a high tension generator capable of accelerating protons to the required energies and thus to produce controlled nuclear reactions. This brilliant inference demanded uncommon ingenuity and vision, but it was nothing in comparison with the feat of persuading Rutherford that it was worthwhile building such a huge machine.

Rutherford was the last representative of the sealing-wax-and-string type of experimenter; however, as soon as he was won over to the 'big engineering' policy, he pushed ahead with such vigour and enthusiasm that Cockcroft was now hard put to temper his impatience. Well, Cockcroft and Walton did not let him wait too long, and their splendid pioneering work soon opened up a vast new field of experimentation in nuclear physics.

Returning once more to Göttingen, I remember spending some time with Gamow discussing the possibility of finding experimental evidence for the excited nuclear states that were indicated by the theory. We took from the shelves the latest volumes of the *Philosophical Magazine* in order to look up what they were doing in Cambridge, where we vaguely knew that they were playing about with nuclei, and we discovered with as much surprise as satisfaction that they had actually observed discrete gamma-ray spectra, immediately suggesting quantum transitions between such nuclear states. Gamow was so enthusiastic about it that he at once decided to go to Cambridge to be on the spot and pursue work in this new direction.

We met again in 1930 in Zurich. Rosenblum in Paris had just announced the discovery of a fine structure of the alpha rays emitted in spontaneous radioactivity: besides the main beam there occurred secondary beams with somewhat lower energies. This was for Gamow the occasion of one of his most brilliant pieces of analysis. By that time he knew so much about nuclear spectroscopy that he immediately saw the quantitative connection between these alpha ray spectra and the gamma ray spectra, observed by Chadwick, which resulted from transitions between excited states and the ground state of the radioactive nuclei; this gave him the explanation of the new phenomenon. The lower energy alpha rays were alpha particles leaving the nucleus in one of its excited states: then, the spectrum of excited states revealed by the gamma rays observed by Chadwick should be the mirror image of the energy spectrum of the low-energy alpha rays; and this relationship worked out beautifully.

Now, we had arranged, Gamow, Peierls and I, to take a trip together to the mountains. When we set out, Gamow had in his pocket the finished manuscript of a note to *Nature* outlining the explanation of the Rosenblum fine structure. As we walked through the Engadine, his fancy was set working by the queer sounding Romanche names of the mountains. We happened to come to the vicinity of a tall peak called Piz da Daint. This somehow appealed to him so much that he exclaimed: "I must date this note to *Nature* from Piz da Daint!"

He then proceeded to add a sentence at the end of the note, thanking us for our help. As we had not contributed in the least to the work, we protested; but he explained: "You are really going to help me. We must get to the top of Piz da Daint, and I will actually date and sign the note there."

This plan was indeed carried out to the last detail – not without difficulty, because the weather deteriorated and we arrived at the top in the middle of a snow storm which, however, soon abated. Gamow then proceeded to the dating ceremony and ordered Peierls to take a photograph of the scene. I told him, "Don't spoil your manuscript; it will be all wet with the snow. Take some odd piece of paper, nobody will notice any difference on the photograph." He was rather shocked by this suggestion, and insisted on a genuine performance; it is the actual manuscript he is holding in his hand on the famous photograph. He glued copies of the photograph on a number of reprints of the note, destined to his particular friends.

Rosenblum, by the way, is one of the most lovable men I have known; he died prematurely in rather tragic circumstances. He was highly cultivated and extremely sensitive; he had a brilliant imagination. The way in which he managed to carry out those experiments, he told me once, was rather adventurous. He was working in Mme Curie's laboratory and wanted to make more accurate determinations of the energies of the alpha particles emitted by radioactive nuclei. He did not suspect to find anything like this fine structure, he just wanted greater accuracy, and proposed to obtain it by magnetic deviation with a sufficiently powerful electromagnet. Mme Curie was not unsympathetic, but she had no money.

Then Rosenblum heard that a huge electromagnet was being built in Bellevue for quite different purposes. This was his opportunity: he persuaded his colleagues to let him use the magnet when nobody else wanted it. Thus he could, at odd times, make his first experiments and obtain his quite unexpected results. Of course, these were sufficiently impressive to help him to secure better conditions of work.

Let me evoke one or two more aspects of the early developments, associated with the personality of Niels Bohr. The latter followed eagerly the progress of our knowledge of nuclear properties, and especially the work at the Cavendish Laboratory, of which he got first hand information from his old teacher and friend Rutherford. He quickly realized that the research in his own institute ought to shift emphasis from atomic to nuclear physics, and with the energy and determination he could display when it came to the point, he built up a comprehensive equipment which would be sufficiently flexible – that was always the policy at Copenhagen – to allow the physicists to make those experiments that they really wanted to have done: experiments suggested by theoretical considerations, for instance, which would have the definite purpose of testing some theoretical idea. He would never think of just acquiring some apparatus or other and becoming a slave to it.

A big item in this rational equipment was the building of a cyclotron. Lawrence sent over one of his best collaborators, Laslett. I chiefly remember poor Laslett sitting in the canteen quite exhausted and dejected, after hours of unsuccessful efforts to spot a leak. Leaks were the curse of experimentation in those days: the apparatus was always leaking somewhere, and to locate a leak was a harassing business which could go on for days. Still, the Copenhagen cyclotron was, I think, the first to work satisfactorily in Europe.

As soon as the first experiments of the Fermi group with slow neutrons came to our knowledge, their theoretical implications were discussed in our colloquium. It was clear that the observed resonances in the radiative capture of these neutrons could not be accounted for by the usual representation of the interaction between the neutron and the target nucleus as a simple potential well. The contradiction was so acute that Bohr remarked: "In order to understand this we need some crazy idea;" and turning to the audience: "Has anyone a crazy idea?"

It was not long before Bohr himself had the "crazy idea" – which was not so crazy after all. To begin with, feeling that the low energy resonance processes were too complicated to give a useful clue, he looked for a simpler situation. He had not to look very far: he had only to take Chadwick's first experiments with fast neutrons, which had been lost sight of by the physicists who were so worried by the slow neutron processes. The cross sections for the capture of those fast neutrons are of the order of the geometrical dimensions of the nuclei, which means that they are captured as soon as they hit the nucleus.

Now, the wave-packets of the fast neutrons are small compared with the dimensions of the nuclei, and this had for Bohr the decisive advantage that in a first approximation he could treat them as classical particles entering the nucleus. This suggested the famous analogy with colliding billiard balls in a trough, which led him to the conception of the compound system and the resulting mechanism of nuclear reactions. He had great fun having a wooden trough made in the workshop, together with an inclined rail along which a steel ball could roll into the trough and collide with a number of steel balls lying in it: this showed beautifully how the kinetic energy of the impinging ball was rapidly shared among all the balls, leading to its capture and the formation of the 'compound system' of the steel balls in random agitation.

The most spectacular application of the compound system theory was the explanation of the fission phenomenon. The discovery of fission came as a great surprise: nobody had imagined that a nucleus could split so easily into large fragments. I need not recall the circumstances of the discovery: how Hahn and Strassmann established that one of the products of the reaction of uranium with neutrons was no 'transuranian' element, but plain barium, and how Frisch and Lise Meitner concluded that this barium fragment had actually split off the uranium nucleus.

It was at that time, at the beginning of 1939, that Bohr had to leave for Princeton, and I accompanied him on this journey. As we were boarding the ship, Bohr told me he had just been handed a note by Frisch, containing his and Lise Meitner's conclusions; we should "try to understand it." We had bad weather through the whole crossing, and Bohr was rather miserable, all the time on the verge of seasickness. Nevertheless, we worked very steadfastly and before the American coast was in sight Bohr had got a full grasp of the new process and its main implications.

The most puzzling point was the large probability of occurrence of the fission process. Of course, Frisch had pointed out that the heavy nuclei become unstable because of the decrease of the surface tension with increasing electric charge. Still, why did they so frequently decay by this splitting mode? The answer turned out to be very simple, on the basis of Bohr's treatment of the excited compound nucleus as a system in thermodynamic equilibrium: it is just a consequence of the equipartition of energy between all the modes of motion of the system. The relative oscillation of two large fragments of the nucleus competes on equal terms with any other mode, such as the relative motion of a single neutron, leading to its emission.

On arrival in New York we were met by Wheeler; Bohr was detained by some business, so Wheeler and I proceeded to Princeton without him. On that night they happened to have a meeting of their 'journal club' at the physics department. I attended, in spite of the fatigue of the voyage, and was politely asked whether I had anything to tell them. Well, I had: I told them all about the problem we had struggled with during the journey. I did not know that Bohr had no intention of giving out the news so quickly, because he was anxious that Frisch's note should first come out in print. I was under the impression that the note was already sent off and would appear in the next number of *Nature*; in fact it only appeared a few weeks later.

The effect of my talk on the American physicists was more spectacular than the fission phenomenon itself. They rushed about spreading the news in all directions, and very soon the fission fragments had been seen in the oscilloscope in several laboratories in the United States, a very striking demonstration that was quite easy to produce.

In the meantime, Bohr was sending telegrams to Copenhagen to press them to proceed with more elaborate experiments, but (as we later learned) the Copenhagen people completely failed to realize why they should hurry so much – and they didn't. Things came to a climax at the Washington meeting of the American Physical Society at the beginning of February 1939. At long last Bohr had got news from Copenhagen that they had observed and measured out fission fragments in the cloud chamber, and he could announce the results at the meeting.

Meanwhile, Bohr had himself made further progress of fundamental importance in the theory of fission. Some time in January, Placzek, who had just come over from Europe, came to see us as we were sitting at breakfast at the Faculty Club. The conversation soon turned upon fission. Bohr casually remarked: "It is a relief that we are now rid of those transuranians." This elicited Placzek's protest: "The situation is more confused than ever," he said, and he explained to us that there was a capture resonance at about 10 volts both in uranium and thorium, showing, apparently, that transuranians were produced concurrently with fission.

Bohr listened carefully; then he suddenly stood up and, without a word, headed towards Fine Hall, where we had our office. Taking a hasty leave of Placzek, I joined Bohr, who was walking silently, lost in a deep meditation, which I was careful not to disturb. As soon as we entered the office he rushed to the blackboard, telling me: "Now listen: I have it all." And he started – again without uttering a word – drawing graphs on the blackboard. The first graph looked like this:

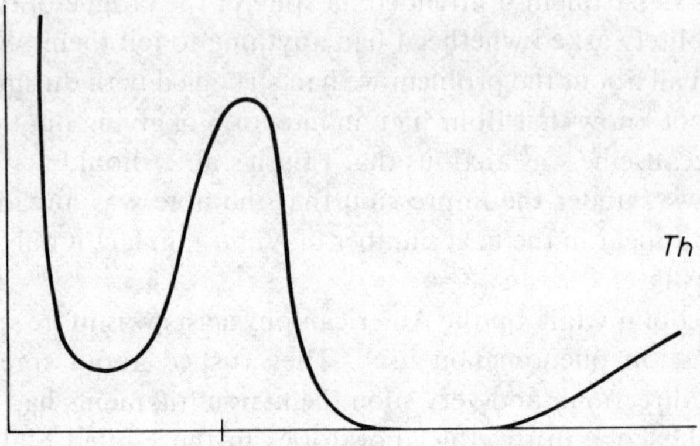

Clearly, the idea was to show, for thorium, the capture cross section, with its resonance at about 10 volts, and the fission cross section starting at a much higher threshold. Then he drew exactly the same graph, with the mention U^{238} instead of Th, and he wrote the mass number 238 with very large figures – he broke several pieces of chalk in the process. Finally, he drew quite a different picture which he labeled U^{235}. This intended to show the fission cross-section, with nonvanishing values over the whole-energy-range:

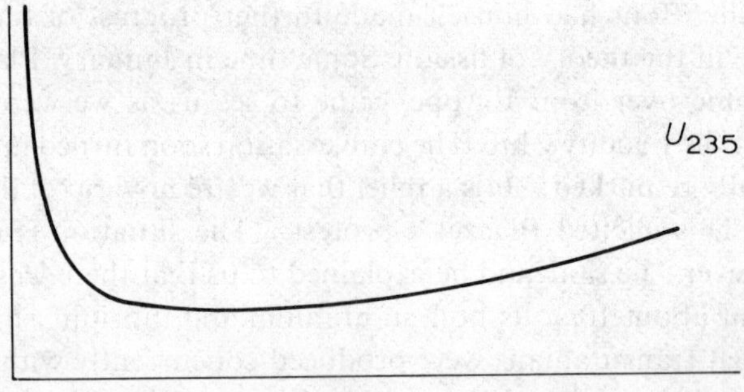

Having drawn the graphs, he started developing his argument: obviously the resonance capture must belong to the abundant uranium isotope, otherwise, its peak value would exceed the limit set by wave theory. For the same reason the fast neutron fission must also be ascribed to the abundant isotope, whose behavior is thus entirely similar to that of thorium. Consequently, the observed slow-neutron fission must be attributed to the rare isotope U^{235}: this is a logical necessity.

The next step was to explain the similarity between the two even-mass nuclei Th and U^{238} and the essential difference respecting fissility between the even-mass and the odd-mass uranium isotope. I need not repeat this explanation, which is now common knowledge.

I shall never forget the thrill of witnessing such a display of logical power in the analysis of experimental data, such a combination of quick thinking and unwavering argumentation. Neither shall I forget the next experience, namely the coolness with which this beautiful argument was received by several eminent physicists when Bohr told them about it.

Fermi was unable to appreciate its compelling logic; it was both the strength and the weakness of Fermi to be so intent on following his own lines of thought that he was impervious to any outside influence: he could not be expected to enter into Bohr's way of thinking. He fancied there could be a different interpretation of the evidence discussed by Bohr, and that only experiment could decide whether Bohr's 'conjecture' about the fissility of the rare uranium isotope was true. Of course, he eventually did make the experiment. Dunning at Columbia was one of the few who immediately accepted Bohr's argument and also started on his side experiments leading to its verification.

Let me conclude here these spread memories. I feel I was very fortunate to live in such an exciting period of physics – as a modest, but not inattentive spectator.

NOTE

* Transcription of a speech given at the State University of New York, Albany, on 5 October 1968.

CHAPTER 25

CELESTIAL AND TERRESTRIAL PHYSICS IN HISTORICAL PERSPECTIVE

[1969b]

Ingenious, but inconclusive, arguments about the hardness of the Moon's surface have recently been settled in a very simple way: by scraping it. It is curious that this experiment has been accepted as the most natural thing to do, not only by the small community of scientists, but by the public at large. It has been hailed as a technological feat, but its whole conceptual background has just been taken for granted. Nobody doubts that the Moon is made of the same material as the Earth and that it is just a question of finding out how soft its soil is, much as one would do on the Earth. This sounds trivial enough, but think how revealing it is for the change of attitude undergone by, at any rate, a large part of mankind in the short span of two centuries under the impact of technological progress. I say 'technological' and not 'scientific', since the true spirit of science, which is of course the prime mover, has not yet taken hold of the minds of men; this will be, perhaps, the next stage on the path to happiness. Even in the late 18th century, while philosophers were discoursing about the 'progress of enlightenment', the peasants of western Europe held beliefs about the celestial phenomena hardly different from those entertained by the peasants of Egypt and Sumer four thousand years earlier.

In those remote times, when nature appeared to men in all its unanalysed complexity, there was no reason to assume any similarity between the celestial and terrestrial phenomena; on the contrary, it was natural to ascribe to them, as to all others, a quality of their own. In fact, they were just regarded as a part of a terrestrial environment that was identified with the Universe. Like all the forces of nature that escaped human control, they were popularly regarded as manifestations of divine powers. On this narrow scene, the first stage of sophistication of which we have documented evidence is a succession of imaginative attempts to reduce the variety of appearances to some underlying simplicity. In such an endeavour it is pretty obvious to associate the Sun, the Moon, and the stars with the fires that burn on the Earth, and even to regard the fire emanating from them as more perfect than the earthly ones, which are always mixed with other elements. But how is it, now, that this celestial fire appears to us in the shape of Sun, Moon, and stars and that these shapes move about as they do?

The answers given to these questions by the ancient cosmogonies illustrate the formidable difficulty of drawing analogies from terrestrial experience. The idea of bodies moving freely through space was the last to occur. Indeed, no arrow or stone hurled by whatever kind of earthly force could be imagined to move on in such regular and never-ending revolutions. It was easier to give a material form to the vault of the heavens and to ascribe to it a continual rotation. In one early cosmogony, the Sun, the Moon, and the stars were interpreted as holes in a revolving spherical screen, through which one had glimpses of the celestial fire burning behind. In others, the stars were thought to be hanging from the heavenly vault like lamps. Eventually, the necessity for giving a more accurate account of the celestial motions – of 'saving the phenomena' – led to a multiplication of the spheres carrying the Sun, Moon, planets, and stars and to an increasing complication of their motions; but the idea that consisted in interpreting these fiery appearances as bodies moving through space without support was truly revolutionary.

Firstly, it was one of those moves towards the unification of apparently heterogeneous phenomena, which are so decisive for the development of a scientific view of the Universe. Moreover, it was a step towards the substitution of a rational explanation of the celestial phenomena for the mystical features attached to them by popular belief. As such, it was a threat to the stability of the social institutions which were sanctified by being related to the heavenly powers. For expressing the view that the Moon and the other celestial bodies were made of the same kind of matter as the Earth, Anaxagoras was accused of impiety by the Athenians: such an offence to the divinities of the heavens could induce them to withdraw the protective influence they exerted on the affairs of the city. This involuntary opposition to popular religion remained for a long time a formidable obstacle to the progress of scientific cosmology – the rational attitude of a handful of philosophers was no match for the social forces bent on maintaining a divine sanction for institutions which it was deemed too hazardous to abandon to the will of men. Whenever these social inhibitions were weakened, as happened in Hellenistic times, and later during the Renaissance in Western Europe, a more tolerant climate prevailed in scientific pursuits. From the former period, we obtain a glimpse of the degree of sophistication attained by the leisured class of society through Plutarch's curious dialogue *On the Face of the Moon*. Here, all the conflicting concepts, including that of the identical nature of the Moon and Earth, are freely discussed, but significantly the question is left undecided: debating it was a pleasant exercise for the dilettanti, but there was no possibility of either testing the ideas or drawing from them consequences of any use.

Likewise, the problem of 'saving the phenomena' by perfecting the machinery of rotating spheres was bogged down in sterile formalism, completely divorced as it was from the mechanics of terrestrial bodies. The latter was itself in too rudimentary a state to be of any help even if this help had been sought, but the rift was deeper: the celestial motions were regarded as radically different from those of terrestrial bodies, and the idea of any relation between the two was beyond the conceptual horizon. This was as true of Copernicus as of the scholastic philosophers. There was indeed a Copernican revolution, but it was not made by Copernicus: it came much later. Copernicus wanted to simplify the Ptolemaic system, and he saw that one did not need so many spheres if they were arranged around the Sun rather than around the Earth, and if the Earth rotated around its axis; as a humanist, he referred these purely kinematic ideas to their source in the Pythagorean tradition, but he could, no more than Plutarch, produce compelling arguments for regarding the Earth's mobility as a dynamical reality. Indeed, by proposing the alternative system in which only the Sun revolved around the Earth, while all the planets revolved in the Copernican way around the Sun, Tycho Brahe made it clear that the Copernican reform did not imply the assimilation of the Earth to a planet. Tycho is a good example of those clever people who apply their ingenuity to preventing revolutions.

The issue had to be faced, however, and it was Tycho's own disciple who, by his remarkably unprejudiced approach to the problem of planetary motions, initiated the radical change of outlook that gave birth to modern science. By putting his trust in the accuracy of Tycho's observations, which showed that the heliocentric motion of the planet Mars could not be circular, Kepler brushed away the whole concept of the celestial spheres and opened up the heavens. The concept of the planets as bodies moving along definite paths around the Sun acquired the character of a law of nature directly derived from observation, and by the same token the whole Copernican system could no longer be denied substantial reality. The ultimate consequences of the Copernican view had received a poetic formulation in Giordano Bruno's vision of an infinite Universe with innumerable suns and planetary systems – a vision so destructive to the very foundations of social order that its prophet had to pay for his imprudence with his life. But how much more forceful were the rigorous arguments by which scientists like Kepler or Otto von Guericke (who allied an equally powerful imagination with a severely critical judgment) established these consequences as facts of observation! Of special importance to our theme is von Guericke's contribution; his concept of the heavenly expanse as an unlimited empty space in which suns and planets were moving was directly inspired by his experiments on the production of such limited empty volumes on Earth. The complete identification of the Earth and the other heavenly bodies is illustrated by Kepler's essay *Somnium* – a 'dream' in which he imagines himself on the Moon and goes on to describe the lunar landscape and the appearance of the heavens from this extra-terrestrial vantage point, exactly as he would have done for some remote region of the surface of the Earth.

In the immensely widened picture of the Universe as it emerged from Kepler's lifework, one essential element was still missing. Kepler realized that the planetary motions must be ascribed to some influence emanating from the Sun, but he did not know how to attack this dynamical problem, because no sound method was yet available for the analysis of the motion of ordinary terrestrial bodies. It is just on this last point that Galileo concentrated his efforts, and, although he did not arrive at a complete solution, he laid the foundations of the dynamics which eventually enabled Newton to find the answer to Kepler's question. It is an irony of history that Galileo himself was unaware of the problem of celestial mechanics to the solution of which his own work contributed. He completely ignored Kepler's laws of planetary motion (if indeed he ever knew of them – even Newton did not appreciate Kepler's law of areas until he found that it was a consequence of the general laws of motion). On the contrary, he clung to the Platonic idea of the perfection of circular motion, and it was circular motion, and not motion in a straight line, that he regarded as inertial (as we would call it); when he applied his principle of inertia to the rectilinear component of the motion of projectiles on Earth, it was as an approximation to the actual circular displacement along the surface of the Earth. Kepler was also engrossed in Platonic philosophy, but the strict demand he imposed on his theories of accurately accounting for the observations led him gradually to abandon it. Galileo had no need of accurate experiments to establish the law of falling bodies on Earth, because he took it for granted that it should be some simple law of proportionality, and rough measurements sufficed to guide him to the right formulation. In his astronomical work, Galileo's intention was quite different from Kepler's: he looked for physical arguments in support of the reality of the Copernican system, and in his quest he proceeded with a surer intuition of physical phenomena (though not without some blundering) than the more mathematically minded Kepler. The latter was a better logician than the sanguine Italian, but it was Galileo's brilliant dialectics that won the field for the Copernican system. Kepler was better than Galileo at understanding optics, but it was Galileo's practical sense that prompted him to take the telescope from the hands of the craftsmen and make of it a working tool for disclosing the physical aspects of the celestial bodies, and thus definitely clinch the argument for the identity of matter in the whole Universe.

For Newton this identity – which he called 'the analogy of nature' – became a heuristic principle of paramount importance. As an instance of the guidance he always sought in this 'analogy', I shall just mention his refutation of a hypothesis proposed by Flamsteed in his correspondence with Newton about the comet of 1680. Flamsteed had correctly interpreted the observations as showing that the comet, first seen moving towards the Sun and later seen receding from it, had been deflected in its neighbourhood (where it was not visible); he went on to suggest that this deflection might be due to magnetic forces exerted by the Sun on the comet, the magnetic axis of which he naïvely imagined would remain parallel to the trajectory, so that the comet, after an initial deflection on entering the vortex of the Sun, would gradually turn towards the Sun the pole opposite to that initially facing it. Newton did not only point out this last mistake in Flamsteed's argument; he observed that the Sun could not possibly exert magnetic forces, because it was too hot: for we know from experience with magnets on Earth that they lose their magnetism on being heated.

Newton's discovery of the law of universal gravitation made such an impression on the scientific community that the celestial mechanics he founded upon this law became the model for a purely dynamical concept of the whole physical Universe: all the phenomena (including those of light and heat) would in principle be explained by the motions of material particles interacting by central forces. This ambitious programme, which received its most comprehensive elaboration in Laplace's great work, had an extraordinarily stimulating influence on the development of our modern theoretical physics. Taking their inspiration from the problems of celestial mechanics, Laplace and his successors – Gauss in Germany, George Green, William Thomson, and Maxwell in Britain – created the refined mathematical methods which found application not only in terrestrial mechanics, but in optics, electricity, and magnetism as well.

A parallel line of development that led to momentous results was the study of radiation from both stellar and terrestrial sources, initiated in the 18th century by Bouguer and Lambert. Its culmination was Kirchhoff's creation of the method of spectral analysis, which provided an additional proof of the universal identity of matter, this time from the point of view of its chemical character. In this domain we have again numerous examples of fruitful influences in both directions, one of the most famous being the spectroscopic discovery, in the solar chromosphere, of a new element, helium, which was then also detected in terrestrial locations.

As the recognition of the uniformity of the material constitution of the Universe thus, step by step, was taking shape, speculation never ceased about the fascinating problems of cosmogony, which likewise passed from the mythical to the scientific level. There is indeed no difference of principle between the study of the constitution of the Universe and that of its formation: in the former we are testing theoretical predictions about the results of experiments; in the latter, we are working backwards from the present evidence, formulating 'retrodictions' instead of predictions. The real difficulty here, apart from the vastness of the object of study, is in fact our impatience to reach definite conclusions: cosmogony may just be retrodiction, but it is mostly retrodiction on insufficient evidence. Hence its chequered history reflects a steady increase in our basic knowledge and a succession of ephemeral inferences rashly drawn from it. Still, from Thomas Wright's first vision of a Universe of separate galaxies – a static vision that inspired Kant's dynamical hypothesis about the formation of galaxies and planetary systems by condensation of nebular matter – to our present sophisticated cosmogonies, a broadening of the World picture and a deepening of its foundations has taken place that completely dwarfs the Copernican revolution.

One might say that this revolution was extended to the time dimension, and the obstacle of the traditional myths with their hopelessly inadequate time scale proved as difficult to overcome as the geocentric prejudice. In this respect, history tells of a continual struggle to push back the origins beyond the artificial barriers raised by timid imagination. In Newton's time, cosmogony was still discussed in terms of the biblical account. Burnet's *Theory of the Earth* is an attempt to interpret this account as a symbolical description of actual events; in his correspondence with Burnet, Newton himself tries to defend a more literal interpretation of it, as giving a description of the process as it would have appeared to a human observer. Only the demand from expanding mining activity for better methods of prospection freed geological science from the strait-jacket of biblical mythology, and the same fight had to be fought again on the question of the antiquity of man and of organic evolution. As regards the time scale of celestial phenomena, it offers a still greater challenge to imagination, and there have been quite a few hurdles raised by the physicists themselves. Let us only recall Lord Kelvin's stubborn opposition to the estimates of the age of the Earth arrived at by the geologists, on the ground that they conflicted with his own estimate, based on the theory that the Sun's heat originated from gravitational contraction – a theory regarded as being beyond doubt.

After putting such emphasis, in the course of this survey of the relations between celestial and terrestrial physics, on the realization of the essential unity of constitution of the Universe, I shall conclude by raising a question which in a sense could suggest a limitation of this general conclusion. It concerns the force of gravitation itself, which was historically the first example of a universal agency, and which indeed dominates all phenomena on the cosmic scale. The temptation is great to treat gravitation on the same basis as the other universal forces of nature – the trinity of strong, electromagnetic, and weak interactions, whose domain of action is primarily that of the phenomena on the atomic and nuclear scale. Is this assimilation justified, however? In particular, is the gravitation field quantized like the others? By its identification with the metrical field, gravitation is ascribed a unique function in the description of nature, a function which it does not seem easy to reconcile with the quantal features of the field. Even in its dynamical action, gravitation differs from other forces in one important respect (Professor Prigogine recently drew my attention to this): a system of particles interacting only by gravitation does not show any tendency towards a state of equilibrium; being of infinite range and always attractive, the force of gravitation does not lead to a statistical distribution of the energy among the particles within any finite average time. Thus, purely gravitating systems are outside the range of application of the second law of thermodynamics – a circumstance well worth remembering in cosmological studies.

PART II

EPISTEMOLOGY

CHAPTER 1

ON THE QUESTION OF THE MEASURABILITY OF ELECTROMAGNETIC FIELD QUANTITIES*

[with Niels Bohr]

[1933b]

1. Introduction

The question of limitations on the measurability of electromagnetic field quantities, rooted in the quantum of action, has acquired particular interest in the course of the discussion of the still unsolved difficulties in relativistic atomic mechanics. On the basis of exploratory considerations, Heisenberg[1] attempted to demonstrate that the connection between the limitation on the measurability of field quantities and the quantum theory of fields is similar to the relationship between the complementary limitations on the measurability of kinematical and dynamical quantities expressed in the indeterminacy principle and the non-relativistic formalism of quantum mechanics. However, in the course of a critical investigation of the foundations of the relativistic generalization of this formalism, Landau and Peierls[2] came to the conclusion that the measurability of field quantities is subjected to further restrictions which go essentially beyond the presuppositions of quantum field theory, and which therefore deprive this theory of any physical basis.

At first glance, one might see in this contradiction a serious dilemma. On the one hand, the quantum theory of fields surely ought to be considered a consistent re-interpretation of classical electromagnetic theory in the sense of the correspondence argument, just as quantum mechanics represents a reformulation of classical mechanics adapted to the existence of the quantum of action. On the other hand, quantum electrodynamics has essentially increased the difficulties, already encountered in classical electron theory, of a harmonious blending of field theory and atomic theory. However, closer consideration shows that the various problems involved here can be to a large extent separated from each other, since the quantum-electromagnetic formalism in itself is independent of all ideas concerning the atomic constitution of matter. This is evident from the fact that in addition to the velocity of light only the quantum of action enters the formalism as a universal constant; for these two constants obviously do not suffice to determine any specific space-time dimensions. In the quantum theory of atomic structure such a determination will be obtained only by the inclusion of the electric charge and the rest masses of the elementary particles.

Just the insufficient distinction between field theory and atomic theory is the principal reason for the conflicting results of previous investigations of the measurability of field quantities in which single electrically charged mass points were used as test bodies. The utilization of classical electron theory, in the sense of the correspondence argument, which underlies current atomic mechanics, rests above all on the smallness of the elementary charge in comparison with the square root of the product of the quantum of action and the velocity of light, which makes it possible to treat all radiation reactions as small compared to the ponderomotive forces exerted on the particles. However, in measurements of field quantities it turns out to be essential to be able to adjust the charge of the test bodies to an extent which would be in conflict with the latter presupposition if one were to consider these bodies as point charges. As we shall see, however, these difficulties disappear if one uses test bodies whose linear extensions are chosen sufficiently large, compared to atomic dimensions, that their charge density can be considered approximately constant over the whole body.

In this connection, it is also of essential importance that the customary description of an electric field in terms of the field components at each space-time point, which characterizes classical field theory and according to which the field should be measurable by means of point charges in the sense of the electron theory, is an idealization which has only a restricted applicability in quantum theory. This circumstance finds its proper expression in the quantum-electromagnetic formalism, in which the field quantities are no longer represented by true point functions but by functions of space-time regions, which formally correspond to the average values of the idealized field components over the regions in question. The formalism only allows the derivation of unambiguous predictions about the measurability of such region-functions, and our task will thus consist in investigating whether the complementary limitations on the measurability of field quantities, defined in this way, are in accordance with the physical possibilities of measurement.

Insofar as we can disregard all restrictions arising from the atomistic structure of the measuring instruments, it is actually possible to demonstrate a complete accord in this respect. Besides a thorough investigation of the construction and handling of the test bodies, this demonstration requires, however, consideration of certain new features of the complementary mode of description, which come to light in the discussion of the measurability question, but which were not included in the customary formulation of the indeterminacy principle in connection with non-relativistic quantum mechanics. Not only is it an essential complication of the problem of field measurements that, when comparing field averages over different space-time regions, we cannot in an unambiguous way speak about a temporal sequence of the measurement processes; but even the interpretation of individual measurement results requires a still greater caution in the case of field measurements than in the usual quantum-mechanical measurement problem.

Characteristic of the latter problem is the possibility of attributing to each individual measurement result a well-defined meaning in the sense of classical mechanics, while the quantum-imposed interaction, uncontrollable in principle, between instrument and object is fully taken into account through the influence of each measuring process on the statistical expectations testable in succeeding measurements. In contrast, in measurements of field quantities, indeed, every measuring result is well defined on the basis of the classical field concept; but, the limited applicability of classical field theory to the description of the unavoidable electromagnetic field effects of the test bodies during the measurement implies, as we shall see, that these field effects to a certain degree influence the measurement result itself in a way which cannot be compensated for. However, a closer investigation of the fundamentally statistical character of the consequences of the quantum-electromagnetic formalism shows that this influence on the object of measurement by the measuring process in no way impairs the possibility of testing such consequences, but rather is to be regarded as an essential feature of the intimate adaptation of quantum field theory to the measurability problem.

Before we turn to a detailed exposition of the considerations indicated above, we want to stress once more that the fundamental difficulties which confront the consistent utilization of field theory in atomic theory remain entirely untouched by the present investigation. Indeed, consideration of the atomistic constitution of all measuring instruments would be essential for an assessment of the connection between these difficulties and the well-known paradoxes of the measurement problem in relativistic quantum mechanics. Also particularly relevant in this context is the limitation which the finite value of the elementary charge compared to the square root of the product of the velocity of light and the quantum of action places on atomic mechanics based on the correspondence argument.[3]

2. Measurability of Fields According to Quantum Theory

The quantum electromagnetic formalism found its starting point in the quantum theory of radiation developed by Dirac, which is characterized by the introduction of a non-commutativity, consistent with the quantum-mechanical commutation relations, of canonically conjugate amplitudes of vibration of a radiation field. On the basis of this theory, Jordan and Pauli set up commutation relations between the electromagnetic field components for the case of charge-free fields, and the formalism was then brought to a certain completion by Heisenberg and Pauli who treated the interaction between field and material charges on correspondence lines. However, the consistent application of the theory to atomic problems is essentially impaired by the occurrence of the well-known paradoxes of the infinite self-energy of the elementary particles, which were not removed by Dirac's proposed modifications in the representation of the formalism.[4] Yet, in our discussion of the limitations of the measurability of field quantities these difficulties play no role, since for this purpose the atomistic structure of matter is not an essential issue. It is true that the measurement of fields requires the use of material charged test bodies, but their unambiguous application as measuring instruments depends exactly on the extent to which we can treat their response to the fields as well as their influence as field sources on the basis of classical electrodynamics.

In these circumstances we may restrict ourselves to the pure field theory, and thus take the commutation relations for charge-free fields as our starting point for the investigation of the consequences of the quantum electromagnetic formalism with respect to the measurability of field quantities. Using the usual notation, $[p, q]=pq-qp$, we thus have the following typical relations between the field components in two space-time points (x_1, y_1, z_1, t_1) and (x_2, y_2, z_2, t_2), from which the remaining commutation relations are obtained by cyclic permutation:[5]

$$
(1)\quad \begin{cases}
[\mathfrak{E}_x^{(1)}, \mathfrak{E}_x^{(2)}] = [\mathfrak{H}_x^{(1)}, \mathfrak{H}_x^{(2)}] = \sqrt{-1}\,\hbar(A_{xx}^{(12)} - A_{xx}^{(21)}) \\
[\mathfrak{E}_x^{(1)}, \mathfrak{E}_y^{(2)}] = [\mathfrak{H}_x^{(1)}, \mathfrak{H}_y^{(2)}] = \sqrt{-1}\,\hbar(A_{xy}^{(12)} - A_{xy}^{(21)}) \\
[\mathfrak{E}_x^{(1)}, \mathfrak{H}_x^{(2)}] = 0 \\
[\mathfrak{E}_x^{(1)}, \mathfrak{H}_y^{(2)}] = -[\mathfrak{H}_x^{(1)}, \mathfrak{E}_y^{(2)}] = \sqrt{-1}\,\hbar(B_{xy}^{(12)} - B_{xy}^{(21)}).
\end{cases}
$$

Here, $\mathfrak{E}_x^{(1)}, \mathfrak{E}_y^{(1)}, \mathfrak{E}_z^{(1)}, \mathfrak{H}_x^{(1)}, \mathfrak{H}_y^{(1)}, \mathfrak{H}_z^{(1)}$ are the electric and magnetic field components at the space-time point (x_1, y_1, z_1, t_1) while the following abbreviations have been used:

$$
(2)\quad \begin{cases}
A_{xx}^{(12)} = -\left(\dfrac{\partial^2}{\partial x_1\,\partial x_2} - \dfrac{1}{c^2}\dfrac{\partial^2}{\partial t_1\,\partial t_2}\right)\left\{\dfrac{1}{r}\,\delta\left(t_2 - t_1 - \dfrac{r}{c}\right)\right\} \\
A_{xy}^{(12)} = -\dfrac{\partial^2}{\partial x_1\,\partial y_2}\left\{\dfrac{1}{r}\,\delta\left(t_2 - t_1 - \dfrac{r}{c}\right)\right\} \\
B_{xy}^{(12)} = -\dfrac{1}{c}\dfrac{\partial}{\partial t_1\,\partial z_2}\left\{\dfrac{1}{r}\,\delta\left(t_2 - t_1 - \dfrac{r}{c}\right)\right\}
\end{cases}
$$

Further, $\hbar$ is Planck's constant divided by 2π, c the velocity of light, and r the spatial separation between the two points. Finally, δ denotes the symbolic function introduced by Dirac which, as is well known, is characterized by the property

$$
(3)\quad \int_{t'}^{t''} \delta(t - t_0)\,\mathrm{d}t = \begin{cases} 1 \text{ for } t' < t_0 < t'' \\ 0 \text{ for } t_0 < t' \text{ or } t_0 > t'' \end{cases}
$$

and which is formally differentiated like an ordinary function.

The occurrence of the δ-function, defined in (3), in the commutation relations (1) brings to the fore the fact mentioned above that the quantum theoretical field quantities are not to be considered as true point functions but that unambiguous meaning can be attached only to space-time integrals of the field components. With a view to the simplest possibility of testing the formalism we shall consider only averages of field components over simply connected space-time regions whose spatial extension remains constant during a given time interval. Thus, for example, if the volume of such a region G is denoted by V and the corresponding time interval is T, we define the average value of $\mathfrak{E}_x$ by the formula

$$\mathfrak{E}_x^{(G)} = \frac{1}{VT} \int_T \mathrm{d}t \int_V \mathfrak{E}_x \, \mathrm{d}v. \tag{4}$$

For the average values of two field components over two given space-time regions I and II there exist commutation relations which follow immediately from (1) by integration over the two regions and division by the product of their four-dimensional extensions. Thus, the value of the bracket symbols $[\bar{\mathfrak{E}}_x^{(\mathrm{I})}, \bar{\mathfrak{E}}_x^{(\mathrm{II})}], \ldots$ are obtained from (1) simply by replacing the quantities $A^{(12)}$, $B^{(12)}$ by their average values over the two regions

$$\left\{\begin{aligned}
\bar{A}_{xx}^{(\mathrm{I,\,II})} &= -\frac{1}{V_{\mathrm{I}} V_{\mathrm{II}} T_{\mathrm{I}} T_{\mathrm{II}}} \int_{T_{\mathrm{I}}} \mathrm{d}t_1 \int_{T_{\mathrm{II}}} \mathrm{d}t_2 \int_{V_{\mathrm{I}}} \mathrm{d}v_1 \int_{V_{\mathrm{II}}} \mathrm{d}v_2 \\
&\qquad \left(\frac{\partial^2}{\partial x_1\, \partial x_2} - \frac{1}{c^2} \frac{\partial^2}{\partial t_1\, \partial t_2}\right) \left\{\frac{1}{r}\, \delta \left(t_2 - t_1 - \frac{r}{c}\right)\right\} \\
\bar{A}_{xy}^{(\mathrm{I,\,II})} &= -\frac{1}{V_{\mathrm{I}} V_{\mathrm{II}} T_{\mathrm{I}} T_{\mathrm{II}}} \int_{T_{\mathrm{I}}} \mathrm{d}t_1 \int_{T_{\mathrm{II}}} \mathrm{d}t_2 \int_{V_{\mathrm{I}}} \mathrm{d}v_1 \int_{V_{\mathrm{II}}} \mathrm{d}v_2 \\
&\qquad \frac{\partial^2}{\partial x_1\, \partial y_2} \left\{\frac{1}{r}\, \delta \left(t_2 - t_1 - \frac{r}{c}\right)\right\} \\
\bar{B}_{xy}^{(\mathrm{I,\,II})} &= -\frac{1}{V_{\mathrm{I}} V_{\mathrm{II}} T_{\mathrm{I}} T_{\mathrm{II}}} \int_{T_{\mathrm{I}}} \mathrm{d}t_1 \int_{T_{\mathrm{II}}} \mathrm{d}t_2 \int_{V_{\mathrm{I}}} \mathrm{d}v_1 \int_{V_{\mathrm{II}}} \mathrm{d}v_2 \\
&\qquad \frac{1}{c} \frac{\partial^2}{\partial t_1\, \partial z_2} \left\{\frac{1}{r}\, \delta \left(t_2 - t_1 - \frac{r}{c}\right)\right\}
\end{aligned}\right. \tag{5}$$

In exactly the same way as the Heisenberg relation for two canonically conjugate mechanical quantities

$$\Delta p\, \Delta q \sim \hbar, \tag{6}$$

which is the basis for the uncertainty principle, can be derived from the general quantum mechanical commutation relation

$$[q, p] = \sqrt{-1}\, \hbar\,, \tag{7}$$

one obtains for the products of the complementary uncertainties of the field averages in question the following typical formulae:

$$\begin{aligned}
\Delta\bar{\mathfrak{E}}_x^{(\mathrm{I})}\,\Delta\bar{\mathfrak{E}}_x^{(\mathrm{II})} &\sim \hbar\,|\bar{A}_{xx}^{(\mathrm{I},\,\mathrm{II})} - \bar{A}_{xx}^{(\mathrm{II},\,\mathrm{I})}|, & \Delta\bar{\mathfrak{E}}_x^{(\mathrm{I})}\,\Delta\bar{\mathfrak{E}}_y^{(\mathrm{II})} &\sim \hbar\,|\bar{A}_{xy}^{(\mathrm{I},\,\mathrm{II})} - \bar{A}_{xy}^{(\mathrm{II},\,\mathrm{I})}| \\
\Delta\bar{\mathfrak{E}}_x^{(\mathrm{I})}\,\Delta\bar{\mathfrak{H}}_x^{(\mathrm{II})} &= 0, & \Delta\bar{\mathfrak{E}}_x^{(\mathrm{I})}\,\Delta\bar{\mathfrak{H}}_y^{(\mathrm{II})} &\sim \hbar\,|\bar{B}_{xy}^{(\mathrm{I},\,\mathrm{II})} - \bar{B}_{xy}^{(\mathrm{II},\,\mathrm{I})}|.
\end{aligned} \tag{8}$$

Several results of importance for our problem follow immediately from the expressions (5) and (8). Above all, we see that, in accord with the property of the δ-function expressed in (3), the quantities $\bar{A}^{(\mathrm{I},\,\mathrm{II})}$, $\bar{B}^{(\mathrm{I},\,\mathrm{II})}$ change continuously as a result of a continuous displacement of the boundaries of regions I and II, as long as the extension of these regions, i.e. the values of V_{I}, T_{I}, V_{II}, T_{II}, remain different from zero. In particular, the differences $\bar{A}^{(\mathrm{I},\,\mathrm{II})} - \bar{A}^{(\mathrm{II},\,\mathrm{I})}$ and $\bar{B}^{(\mathrm{I},\,\mathrm{II})} - \bar{B}^{(\mathrm{II},\,\mathrm{I})}$ vanish without discontinuity when the boundaries of the two regions gradually are made to coincide. From this it follows that the averages of all field components over the same space-time region commute, and thus should be exactly measurable, independently of one another. In fact, this consequence of the theory, which goes essentially beyond the presupposition of unrestricted measurability of each single field quantity, appears as a special case of two more general theorems which follow from the symmetry properties of the quantities $\bar{A}^{(\mathrm{I},\,\mathrm{II})}$ and $\bar{B}^{(\mathrm{I},\,\mathrm{II})}$. For the fact that the expressions $A^{(12)} - A^{(21)}$ change their sign when the times t_1 and t_2 are interchanged, implies that the averages of two components of the same kind (i.e. two electric or two magnetic components) over two arbitrary space-time regions always commute if the associated time intervals coincide. Further, the corresponding anti-symmetry of the expressions $B^{(12)}$ and $B^{(21)}$ with respect to interchange of the spatial points (x_1, y_1, z_1) and (x_2, y_2, z_2) implies that the averages of two components of different kind, e.g. $\mathfrak{E}_x$ and $\mathfrak{H}_y$, over two arbitrary time intervals commute when the corresponding spatial regions coincide.

At first sight these results might seem incompatible with the commutation relations between averages of field quantities at one and the same time and over finite space regions, which can be derived from the Heisenberg–Pauli representation of the formalism and are discussed in the book by Heisenberg previously cited. While it is stated there too that averages of components of the same kind commute, it is deduced that components of different kind over one and the same region of space do not commute. However, this contradiction is solved easily by noting that Heisenberg's treatment involves a limiting process in which two originally non-coinciding space-time regions are brought into coincidence only after their temporal extension has been contracted to one and the same instant of time. For from the symmetry of the expression (2) for $B_{xy}^{(12)}$ with respect to t_1 and t_2, and from the property of the δ-function stated in (3), we find in the case of coinciding time intervals

$$(9) \qquad \bar{B}_{xy}^{(\mathrm{I},\,\mathrm{II})} - \bar{B}_{xy}^{(\mathrm{II},\,\mathrm{I})} = \frac{2}{V_{\mathrm{I}} V_{\mathrm{II}} T^2} \cdot \frac{1}{c} \int_{V_{\mathrm{I}}} \mathrm{d}v_1 \int_{V_{\mathrm{II}}} \mathrm{d}v_2 \frac{\partial}{\partial z_1}\left(\frac{1}{r}\right),$$

where we have put $T_{\mathrm{I}} = T_{\mathrm{II}} = T$ and where the double integration is over all pairs of points, one in each of the spatial regions, whose separation r is smaller than cT. If we now further assume that the two spatial regions have the same volume, $V_{\mathrm{I}} = V_{\mathrm{II}} = V$, and the same shape, but are displaced relative to each other in the z-direction, then in the limit where cT can be considered negligible compared with the linear extensions of the space regions, the space integral in (9) gives by partial integration $\pm 2\pi c^2 T^2 F$, where F is the area enclosed by the projection on the xy-plane of the curve in which the surfaces of V_{I} and V_{II} intersect, and where the sign is + or − depending on whether region II is displaced relative to region I in the positive or the negative z-direction. Thus, if the regions are displaced continuously through each other the difference $\bar{B}_{xy}^{(\mathrm{I},\,\mathrm{II})} - \bar{B}_{xy}^{(\mathrm{II},\,\mathrm{I})}$ undergoes a discontinuous change of $8\pi cF/V^2$, while both expressions $\bar{B}_{xy}^{(\mathrm{I},\,\mathrm{II})}$ and $\bar{B}_{xy}^{(\mathrm{II},\,\mathrm{I})}$ change their sign. Therefore, the commutation relation between the instantaneous spatial averages of $\mathfrak{E}_x$ and $\mathfrak{H}_y$ in the limiting case under discussion displays an essential ambiguity, which is responsible for the apparent contradiction mentioned above.

Furthermore, as we shall see, the supposed demonstration, on the basis of previous investigations of the physical possibilities of measurement, that there is a complementary limitation on the measurability of field components of different kinds in one and same region of space, also depends entirely on the use of point charges as test bodies, which does not permit a sufficiently sharp delimitation of the region of measurement. As we have already emphasized, only measurements employing test bodies with a charge distribution of finite extension should be considered for the testing of the quantum-electromagnetic formalism, since every well-defined statement of this formalism refers to averages of field components over finite space-time regions. Yet this circumstance in no way prevents us from testing by field measurements all unambiguous conclusions that can be drawn from the Heisenberg–Pauli representation concerning the time dependence of spatial field averages. For this purpose we need only introduce averaging regions whose temporal extension T, multiplied by c, is sufficiently small compared to their linear spatial dimensions, the order of magnitude of which we shall in the following always denote by L.

In fact, the case $L > cT$ is particularly suited for a thorough testing of the consequences of the formalism in the properly quantum-theoretical domain. Of course, already in the domain of validity of classical theory the case $L \leqslant cT$ is of little interest, since all the peculiarities of the wave fields inside the volume V are smoothed out to a large extent by the averaging procedure, because of the propagation during the time interval T. In addition to this smoothing out, there are, in the quantum domain, the peculiar fluctuation phenomena which derive from the basically statistical character of the formalism. As we shall see shortly, these fluctuations, while essentially entering the solutions of problems in the case $L \leqslant ct$, play a comparatively small role in the case $L > cT$.

The fluctuations in question are intimately related to the impossibility, which is characteristic of the quantum theory of fields, of visualizing the concept of light quanta in terms of classical concepts. In particular, they give expression to the mutual exclusiveness of an accurate knowledge of the light quantum composition of an electromagnetic field and of knowledge of the average value of any of its components in a well-defined space-time region. Let the density $\omega_i(\kappa_x, \kappa_y, \kappa_z)$ of light quanta with definite polarization parameter i and given momentum and energy, $\hbar\kappa_x$, $\hbar\kappa_y$, $\hbar\kappa_z$ and $\hbar\nu = \hbar c\sqrt{\kappa_x^2 + \kappa_y^2 + \kappa_z^2}$ be known; then the expectation value of all field averages are indeed zero, but the mean square fluctuation of each field quantity, such as $\mathfrak{E}_x^{(G)}$ defined in (4), is given by the easily derivable formula

$$(10)\quad \left\{ \begin{aligned} & S(G) = \frac{1}{V^2T^2}\cdot\frac{\hbar}{3}\int_T \mathrm{d}t_1 \int_T \mathrm{d}t_2 \int_V \mathrm{d}v_1 \int_V \mathrm{d}v_2 \frac{\partial^2}{\partial t_1\, \partial t_2} \int_{-\infty}^{\infty} (\sum_i \omega_i + 1) \\ & \cos\left[\kappa_x(x_1 - x_2) + \kappa_y(y_1 - y_2) + \kappa_z(z_1 - z_2) - \nu(t_1 - t_2)\right] \\ & / \frac{\mathrm{d}\kappa_x\, \mathrm{d}\kappa_y\, \mathrm{d}\kappa_z}{\nu} \end{aligned} \right.$$

From formula (10) we see that for a given light quantum composition the fluctuations in question can never vanish, since even when $\omega_i = 0$, i.e., in the complete absence of light quanta, they assume a finite positive value, which by an easy calculation can be put in the form

$$S_0(G) = \frac{2}{3\pi^2} \frac{\hbar c}{V^2} \int_V dv_1 \int_V dv_2 \frac{1}{r^2[(cT)^2 - r^2]}. \tag{11}$$

For every other light quantum distribution defined by a given density ω_i, the mean square fluctuation of the average of a field component becomes larger than $S_0(G)$. However, the fluctuations of a field average expected according to the formalism can be arbitrarily small when a direct knowledge of field quantities, obtained by measurement for example, is assumed. In such a case the light quantum density ω_i is obviously not well defined, and we must content ourselves with statistical statements about this density.

Furthermore, it is of decisive importance for the discussion of the measurement possibilities that the expression (11) holds not only for the field fluctuations in a region free of light quanta, but also represents the mean square fluctuation of each field average in the more general case where only classically describable current and charge distributions occur as sources of the field. The state of the field is then uniquely defined by the requirement that the expectation value of every field quantity coincide with the classically computed value, and that the numbers of light quanta of given momentum and polarization be distributed around their mean value n_0, estimated by means of the correspondence argument, according to the probability law valid for independent events

$$w(n) = \frac{n_0^n\, e^{-n_0}}{n!} \tag{12}$$

An easy calculation shows that the field fluctuations in this state are given just by the expression (11). Moreover, in correspondence with the characteristics of black-body fluctuations, it follows that also in the general case of a field of given light quantum composition, the inclusion of field effects of any classically describable sources will have no influence on the fluctuation phenomena.

The square root of the expression (11) may be regarded as a critical field strength, $\mathfrak{S}$, in the sense that only when considering field averages essentially larger than $\mathfrak{S}$ are we allowed to neglect the corresponding fluctuations. To assess the possibilities of testing the formalism in the properly quantum domain, still another critical field size, $\mathfrak{U}$, is relevant, which equals the square root of the products, given by (8), of the complementary uncertainties of two field averages over space-time regions that only partially coincide, being displaced relative to each other by spatial and temporal distances of order of magnitude L and T, respectively. For when the field strengths are essentially larger than $\mathfrak{U}$ we obviously enter the domain of validity of classical electromagnetic theory, where all quantum mechanical features of the formalism lose their significance. A simple estimate based on formulae (8) and (11), shows that in the case $L \leqslant cT$ both critical expressions $\mathfrak{U}$ and $\mathfrak{S}$ are of the same order of magnitude

$$\text{(13)} \qquad \mathfrak{U} \sim \mathfrak{S} \sim \frac{\sqrt{\hbar c}}{L \cdot cT}$$

On the other hand, in the case $L > cT$ one has

$$\text{(14)} \qquad \mathfrak{U} \sim \sqrt{\frac{\hbar}{L^3 T}} \text{ and } \mathfrak{S} \sim \frac{\sqrt{\hbar c}}{L^2},$$

so that in the limiting case $L \gg cT$ the critical field strength $\mathfrak{U}$ is much larger than $\mathfrak{S}$ and, therefore, in testing the characteristic consequences of the formalism we can to a large extent disregard the field fluctuations.

Before we turn to the comparison of the consequences of the quantum electromagnetic formalism discussed in this section, with the physical measurement possibilities for field quantities, we want to emphasize here once again that the consistent interpretability of this formalism is in no way endangered by such paradoxical features of its mathematical representation as the infinite zero-point energy. In particular, this latter paradox, which moreover can be removed by a formal change in the representation[6] that does not influence the physical interpretation, has no direct connection with the problem of measurability of field quantities. In fact, a field-theoretic determination of the electromagnetic energy in a given space-time domain would require knowledge of the values of the field components at each space-time point of a region, which are inaccessible to measurement. A physical measurement of the field energy can be carried out only by means of a suitable mechanical device that would make it possible to separate the electromagnetic fields in a given region from the rest of the field, so that the energy contained in the region could be measured subsequently by application of the conservation law. However, because of the interaction with the measuring mechanism, any such separation of the fields would be accompanied by an uncontrollable change in the field energy in the region in question, the consideration of which is essential for clarification of the well-known paradoxes that arise in the discussion of energy fluctuations in black-body radiation.[7]

3. Presuppositions for Physical Field Measurements

The measurement of electromagnetic field quantities rests by definition on the transfer of momentum to suitable electric or magnetic test bodies situated in the field. Quite apart from the caution required by quantum theory in applying the customary idealization of field components defined in each space-time point, we are here always concerned with averages of these components over the finite time intervals necessary for the momentum transfer as well as over the spatial domains in which the electric charges or magnetic pole strengths of the test bodies in question are distributed. Obviously, even the assumption of a uniform charge distribution on a test body is an idealization, subject to a certain restriction because of the atomic constitution of all material bodies, but indispensable for the unambiguous definition of field quantities.

In order to have a definite case in mind we consider the measurement of the average of the electrical field component in the x-direction, $\mathfrak{E}_x$, over a space-time domain of volume V and duration T. For this purpose we therefore use a test body whose electric charge is uniformly distributed over the volume V with a density ρ, and determine the values p'_x and p''_x of this body's momentum components in the x-direction at the beginning t' and at the end t'' of the interval T. The average $\mathfrak{E}_x$ we are looking for is then determined by the equation

$$(15) \qquad p''_x - p'_x = \rho \mathfrak{E}_x VT,$$

where it is assumed that the time intervals required for the momentum measurements, whose order of magnitude we shall denote by Δt, can be regarded as negligibly small compared to T, and that we can disregard the displacements suffered by the test body due to the momentum measurements as well as the acceleration given to it during the time interval T by the field that is being measured, in comparison with the linear dimensions L of the spatial domain V.

By choosing a sufficiently heavy test body we can obviously make its acceleration due to the field arbitrarily small. In the momentum measurements, however, we encounter conditions which are independent of the mass of the test body. As a consequence of the indeterminacy principle, any measurement of the momentum component p_x carried out with the accuracy Δp_x is accompanied by a loss Δx in the knowledge of the position of the body in question, the order of magnitude of which is given by the relation contained in (6)

$$(16) \qquad \Delta p_x \, \Delta x \sim \hbar.$$

Nevertheless, in itself this state of affairs does not imply any restriction on the accuracy to be achieved by the field measurement, because we still have at our disposal the value of the charge density. In fact, if we neglect Δt and Δx in comparison with T and L, we get from (15) and (16) for the order of magnitude of the accuracy $\Delta\bar{\mathfrak{E}}_x$ of the field measurement

$$(17) \qquad \Delta\bar{\mathfrak{E}}_x \sim \frac{\hbar}{\rho \Delta x \cdot VT},$$

which for any value of Δx, however small, can be made arbitrarily small by choosing a sufficiently large value of ρ.

Strictly speaking, the accuracy of the field measurement is also dependent on the absolute magnitude of the value of $\bar{\mathfrak{E}}_x$ itself, for with given ranges of Δt and Δx the value of $\bar{\mathfrak{E}}_x$ ascertained from (15), even if Δp_x were zero, would be affected with an uncertainty arising from the latitude in the delineation of the domain of measurement which would surpass any limit as $\bar{\mathfrak{E}}_x$ increases indefinitely. Yet, the latter circumstance merely reflects the general limitation on all physical measurements, for which a knowledge of the order of magnitude of the expected effects is always required in order to choose the appropriate measuring instruments. In our problem an upper limit to the effects that we are interested in is set by the fact that as the magnitude of the field components increases we gradually reach the domain of validity of classical electromagnetic theory. As mentioned in the previous section, in the case $L > cT$, which is particularly suited for testing the quantum electromagnetic formalism, the expression

(18) $$Q=\sqrt{\frac{\hbar}{VT}},$$

which is equivalent to the right-hand side of the first formula (14), represents a critical field size in this respect. Substituting this into (17), the latter relation assumes the form

(19) $$\Delta\mathfrak{E}_x \sim \lambda Q,$$

where

(20) $$\lambda=\frac{Q}{\rho\Delta x}$$

is a dimensionless factor determining the accuracy of the field measurement.

The requirement that λ be small compared to unity and simultaneously Δx be small compared to L means that the total electric charge of the test body must consist of a large number of elementary charges ε. In fact, according to (20) this number is given by

(21) $$N=\frac{\rho V}{\varepsilon}=\frac{QV}{\lambda\varepsilon\Delta x}=\frac{1}{\lambda}\cdot\frac{L}{\Delta x}\cdot\sqrt{\frac{L}{cT}}\cdot\sqrt{\frac{\hbar c}{\varepsilon^2}}$$

and is very large when the above requirements are fulfilled and when $L > cT$ as assumed. The last factor is of course the reciprocal square root of the fine-structure constant whose smallness, as we already mentioned in the Introduction, is an essential presupposition for the correspondence approach to electron theory. As emphasized there, essential restrictions are imposed on a field measurement with an elementary charge as a test body, a fact which is also directly visible from (21) if one puts $N=1$.[8] Moreover, the assumption of a large value of N is a necessary condition for the physical realization of a uniform distribution of the charge of the test body over the volume V; and as long as the linear dimensions of the test body are large compared to the atomic dimensions, its fulfillment obviously presents no difficulties in principle. It need hardly be mentioned that with this presupposition the assumption used above about the mass of the test body, equivalent to the requirement that this mass be very large compared to that of a light quantum of wave length L, always can be satisfied.

Thus far we have completely disregarded the electromagnetic field effects which accompany the acceleration of any test body during the momentum measurement. These effects superpose themselves on the original field and must be included in the field averages defined by equations of type (15). Hence, the main task of the following investigation will be to find a measuring arrangement in which the field effects of the test bodies can be controlled or compensated to the largest possible extent. Yet here we must first of all discuss the question of whether the reaction of the radiation fields produced by the acceleration of the test body in the momentum measurements could impair even the practicability of measuring the values occurring in (15) of the test body's momentum components at the beginning and end of the measuring interval. It was just this possibility that led Landau and Peierls, in the work cited at the beginning, to doubt the reliability of the indeterminacy relation (16) for charged bodies and to conclude that it should be replaced by another even more restrictive relation in which the charge of the test body enters in an essential way. However, they likened the electromagnetic behavior of such a body to that of a point charge e, and consequently used the following expression for estimating the order of magnitude of the test body's momentum change, brought about by radiation recoil, during the time Δt

$$(22) \qquad \delta_e p_x \sim \frac{e^2}{c^3} \frac{\Delta x}{\Delta t^2}.$$

If, however, $\delta_e p_x$ is considered an additional indeterminacy of the momentum measurement, then if one puts $\rho V = e$ and does not distinguish between $\bar{\mathfrak{E}}_x$ and $\mathfrak{E}_x$, one gets instead of (17)

$$(23) \qquad \Delta_e \mathfrak{E}_x \sim \frac{\hbar}{eT\Delta x} + \frac{e\Delta x}{c^3 T \Delta t^2},$$

whose minimum under variation of e is obviously given by

$$(24) \qquad \Delta_m \mathfrak{E}_x \sim \frac{\sqrt{\hbar c}}{c^2 T \Delta t}.$$

If, still following Landau and Peierls, one does not distinguish between T and Δt, this expression agrees with the absolute limit on the measurability of field components that they gave, on which they based their criticism of the foundations of the quantum electromagnetic formalism.

However, the supposed difficulties of the momentum measurement disappear as soon as sufficient account is taken of the finite extension of the test body's electric charge. Using the idealization of a uniform, rigidly displaceable charge distribution, to be more closely examined below, the electric field strengths in the region V during the acceleration of the test body within the time Δt can at most reach a value of the order of magnitude $\rho\Delta x$; since, according to Maxwell's equations, their time derivatives are at most of the same order of magnitude as that of the current density, given by $\rho(\Delta x/\Delta t)$. Hence, any electromagnetic reaction on the body during the measuring interval Δt can only contribute a momentum transfer of the order of magnitude

$$(25) \qquad \delta_\rho p_x \sim \rho^2 V \Delta x \Delta t.$$

Thus, in view of (18) and (20), we get by comparison of (16) and (25)

$$(26) \qquad \delta_\rho p_x \sim \Delta p_x \cdot \lambda^{-2} \frac{\Delta t}{T},$$

which implies that for any desired accuracy of the field measurement, symbolized by a given value of λ, the influence of the electromagnetic reaction on the momentum measurement of the test body can be neglected if only Δt is chosen sufficiently small in comparison with T. It is precisely this circumstance that is decisive for assessing the accuracy of the field measurements; for it turns out to be impossible to directly take into account the influence of the radiation reaction on the momentum and energy balance in the individual momentum measurements. For example, Pauli's proposal[9] to measure subsequently by means of a special device the momentum and energy contained in the emitted radiation would already be impracticable because of the fact that, at least in the case $L > cT$ which is of particular importance for field measurements, the radiation fields that are produced in the momentum measurements at the beginning and at the end of the interval T cannot be separated from each other to the degree sufficient for this purpose. In fact, in the following sections we shall show quite generally that any attempt at such a control of the test body's field effects would essentially impair the realization of the field measurement in question.

Besides, not only for discussing the behavior of an individual test body during the measurement but also for assessing the mutual influence of several test bodies, it is essential to treat these not as point charges but as continuous charge distributions. This is because the customary identification of the position indeterminacy of a test body, considered as a point charge, with the linear dimensions of the domain of measurement is an arbitrary assumption that is foreign to the measurability problem. For this reason, not only do the estimates of the product of the uncertainties of $\mathfrak{E}_x$ and $\mathfrak{H}_y$ inside the same space-time domain, obtained by Heisenberg and by Landau and Peierls by considering point charges, deviate from the predictions of the quantum-electromagnetic formalism as already mentioned; they are in agreement with each other only in the special case $L \sim cT$. In this case, both estimates give the expression Q^2, which corresponds to the order of magnitude, to be expected from the formalism, of the value of the product of the complementary indeterminacies of two field averages in space-time regions that are displaced relative to each other through space-time distances of the same order of magnitude as L and T. Moreover, it is an essential feature of the formalism that the product in question vanishes identically for coinciding regions. The physical meaning of this result becomes obvious as soon as one takes into account the uniform charge distribution of the test body used to measure $\bar{\mathfrak{E}}_x$; since the magnetic field strength which is produced at a point P_2 of the volume V by the displacement of the charge ρdv contained in a volume element situated at the point P_1 is exactly equal and opposite to the magnetic field strength produced at the point P_1 by the same displacement of the charge ρdv at the point P_2, so that the average over the volume V of every magnetic field component produced by the displacement of the test body disappears.

From the foregoing it emerges that in the investigation of the measureability of field quantities it is decisively important to assume that the test bodies to be used behave like uniformly charged rigid bodies whose momenta can be measured, in any given arbitrarily small time interval, with an accuracy, expressed by (16), complementary to the accompanying uncontrollable displacement. Of course, in view of the finite propagation of all forces, we should not think here of the usual mechanical idealization of rigid bodies, but must think of every test body as a system of individual components of sufficiently small dimensions; and think of the measurement of the total momentum of this system as carried out in such a way that, to a sufficient approximation, all the components undergo the same displacement during the momentum measurement. That this requirement can be fulfilled without difficulties of principle, at least insofar as one can disregard the atomic constitution of the test body, is due to the fact that the required momentum measurements can be fully described on a classical basis, irrespective of whether they depend on looking at a collision process between the test body and a suitable material colliding body; or, say, on the study of the Doppler effect involved in the reflection of radiation from the test body. For, if only the mass of the colliding body is sufficiently large or if the packet of radiation that is used to measure the Doppler effect contains a sufficient number of light quanta, then the interaction between the test body and the colliding body can be described classically to any approximation. In fact, the loss of knowledge of the position of the test body, which accompanies the momentum measurement, is due solely to the impossibility of simultaneously fixing the course of the collision process relative to a well-defined space-time reference frame. Indeed, the peculiar complementarity of the mode of description ultimately derives from the fact that any such fixation is bound up with an unavoidable, and in principle uncontrollable, transfer of energy and momentum to the scales and clocks needed to establish the co-ordinate system.[10]

We recall that, according to the indeterminacy principle, the latitude in the time Δt that is left open in any description and the accuracy with which the energy exchanged in the collision process between colliding body and test body is known are connected by the well-known relation

(27) $\Delta E \cdot \Delta t \sim \hbar.$

Because of the relation between energy and momentum and velocity components

(28) $\mathrm{d}E = v_x \, \mathrm{d}p_x$

which is valid for both bodies, it follows directly that

(29) $\Delta p_x |v''_x - v'_x| \Delta t \sim \hbar.$

Even though, as noted above, the change of velocity $|v''_x - v'_x|$ of the test body in the momentum measurement can be considered as arbitrarily well known for a sufficiently heavy test body, the factor

(30) $|v''_x - v'_x| \Delta t = \Delta x$

obviously implies a completely free latitude in the position of the body relative to the fixed frame of reference, in complete agreement with the indeterminacy relation (16). From (30) the condition

(31) $\Delta x < c\Delta t,$

follows immediately, which, because of (16), imposes an absolute limit on the accuracy Δp_x that can be achieved in a momentum measurement with a given upper limit for the time latitude Δt. However, in view of the relativistic invariance of the relations (16) and (27) and in particular of formula (28), this circumstance implies no restriction in the formulation and applicability of the indeterminacy principle. In our problem it is even permissible to disregard all mechanical relativistic effects, for by using sufficiently heavy test bodies we can always arrange that the velocities of all test bodies during the whole measurement process remain small compared to the velocity of light. Consequently, we can even consider any displacement Δx in the momentum measurements very small compared to the corresponding value of $c\Delta t$ which itself must be chosen arbitrarily small.

It is exactly the possibility of accurately tracing the relative space-time course of the process serving as momentum measurement that enables us to measure the total momentum of an extended body within any given time interval with the required accuracy expressed by (16). Thus, we can determine the total momentum of the system of charged material component bodies serving as test body by a single collision process if we make use of a colliding body of special construction which intervenes everywhere in the test body system and gives every component the same acceleration at the same time. It is true that this device imposes severe demands on the construction of the colliding and test bodies, but these demands do not present any difficulties of principle as long as we can disregard the atomic constitution of the bodies. The measurement of the total momentum of the test body would presumably be performed most simply by optical means, i.e. by determination of the Doppler effect; for this purpose one might proceed as follows: imagine that every component body is equipped with a small mirror at right angles to the x-direction, and that a number of other mirrors are placed in fixed positions in such a way that the light path from the radiation source to each component body is the same. If now by means of a suitable device we produce a packet of radiation of duration Δt and containing a number of light quanta sufficiently large compared to the number of component bodies, then all these bodies will suffer a collision simultaneously and undergo an acceleration which for all component bodies can be made equal with arbitrary accuracy.

In order to show that one can in fact measure the total momentum of the test body with an accuracy satisfying relation (16) by means of such an arrangement, we shall consider somewhat more closely the interaction between the test body system and the packet of radiation. In view of the assumed smallness of the velocity of the test body compared to the velocity of light. we have for each component body

$$\text{(32)} \qquad \begin{cases} m_\tau(v''_{\tau,x} - v'_{\tau,x}) = \dfrac{\hbar}{c} \sum\limits_{n_\tau} (\nu' + \nu''), \\[2ex] \dfrac{1}{2} m_\tau(v''^{\,2}_{\tau,x} - v'^{\,2}_{\tau,x}) = \hbar \sum\limits_{n_\tau} (\nu' - \nu''), \end{cases}$$

where m_τ denotes the mass of a component body, $v'_{\tau,x}$, $v''_{\tau,x}$ are its velocity before and after the reflection, and where the summation extends over the n_τ light quanta reflected from the component body whose frequency (reciprocal period times 2π) before and after the reflection are denoted by ν' and ν'', respectively. It follows from (32) that the momentum of the component body in question before and after the collision is

$$\text{(33)} \qquad \left.\begin{aligned} p'_{\tau,x} &= m_\tau v'_{\tau,x} = \\ p''_{\tau,x} &= m_\tau v''_{\tau,x} = \end{aligned}\right\} m_\tau c \frac{\sum\limits_{n_\tau} (\nu' - \nu'')}{\sum\limits_{n_\tau} (\nu' + \nu'')} \mp \frac{1}{2} \frac{\hbar}{c} \sum_{n_\tau} (\nu' + \nu''),$$

If we now assume that the mean spectral frequency ν_0 of the radiation packet is very large compared to the mean width $(\Delta t)^{-1}$ of its frequency distribution as well as to all frequency changes $\nu' - \nu''$, then we can take the change of velocity of the component body in the collision, to a sufficient approximation, as

$$\text{(34)} \qquad v''_{\tau,x} - v'_{\tau,x} = \frac{\hbar}{m_\tau c} \sum_{n_\tau} (\nu' + \nu'') = \frac{2 n_\tau \hbar \nu_0}{m_\tau c}$$

and we can assume that it is the same for all component bodies. Thus, in the collision all component bodies suffer displacements which, although uncontrollable, are arbitrarily close to being identical and whose order of magnitude Δx satisfies relation (30) where $|v''_x - v'_x|$ is to be identified with the common velocity change of the whole test body system. Consequently, since according to our assumptions Δx can be considered negligibly small compared to $c\Delta t$, we get from (33) and (34), for the product of Δx and the uncertainty of the total momentum of the test body, approximately

$$(35) \qquad \Delta p_x \Delta x \sim \Delta t \cdot \Delta\left(\sum_{\tau}\sum_{n_\tau} \hbar v' - \sum_{\tau}\sum_{n_\tau} \hbar v''\right).$$

The quantities in the bracket in (35) are precisely the total energies of the radiation packets impinging on and reflected from the test body. The energy of the latter packet can be measured with arbitrary accuracy, e.g. by spectral analysis of the reflected radiation. For the incoming radiation packet, however, such an analysis would obviously be incompatible with the experimental conditions. Yet the total energy of this radiation can be measured with an accuracy that is complementary to Δt, as given by (27). To do that it suffices to use a purely mechanical device by means of which the packet in question is separated from a radiation field, whose energy before and after the separation can be determined with arbitrary accuracy, e.g. by spectral measurements. Thus, the relation (35) is identical with the usual indeterminacy relation (16). Note further that the demonstration of this identity is essentially dependent on the fact that in accordance with the described arrangement we obtain no information about the momentum of the single component bodies but only about the total momentum of the test body.

The fact that the test body system suffers a common translation during the required momentum measurements is not only important for the calculation of the field effects of the test body which accompany these measurements, but also gives us the possibility to arrange things in such a way that outside the short time intervals occupied by the momentum measurements all the test bodies employed in the field measurement can be considered as being at rest, which greatly simplifies the calculation. For immediately after each momentum measurement, i.e. practically speaking still inside the interval Δt, we can give the test body system a second push in the opposite direction by means of a suitable device, such that the velocity change which every component body suffered in the first collision is cancelled out; this can be done with an arbitrary accuracy, i.e.

an accuracy that is inversely proportional to the mass of the component body, and without losing the desired knowledge of the total momentum of the test body. However, with this arrangement it is impossible to know the time interval between the two collision processes with a latitude smaller than Δt, and so, as required by the indeterminacy principle, the test body is not returned to its original position by the counter-collision, but rather is brought to rest to the required approximation at an unknown position, displaced by a distance of the order Δx.

To assess the complementary limitations on the measurability of field quantities, which will be more closely investigated in the following sections, we must follow the behavior of the test body as accurately as possible during the whole measuring process. It turns out that for this purpose it is necessary first of all to know accurately the position of each test body at all times before and after its use in the measurement. This is achieved most expediently by having the test body firmly attached to a rigid frame serving as a spatial reference system, except in the time interval during which the momentum transfer to the test body from the field is to be determined. At the beginning of this interval the attachment must be disconnected, and the momentum component of the test body in the direction of the field component that is to be determined must be measured. We always assume that by an immediately following counter-impulse, as discussed above, the body is brought back to rest with an accuracy inversely proportional to its mass, at a position which is not accurately predictable. At the end of the time interval and after renewed measurement of the momentum component in question, the firm attachment is re-established; here it turns out to be not unessential that the test body be brought back into exactly the same position as it had originally. If the space-time averaging domains are to be sufficiently sharply defined, these prescriptions alone impose far-reaching demands on the detailed construction of the test body system. For due to the retardation of all forces it is strictly speaking necessary that the severance as well as the re-establishment of the attachment of the test body system to the fixed frame are performed in such a way that all its independent component bodies, whose linear dimensions must be at least as small as the smallest relevant value of $c\Delta t$, are unfastened and fastened simultaneously, i.e. within the time latitude Δt of the momentum measurement, which itself must be chosen sufficiently small compared to the time interval T.

Still more far-reaching demands on idealization with respect to the construction and handling of the test body system are obviously needed to measure field averages over two partially overlapping space-time regions. For in this case we must have test bodies at our disposal which can be displaced inside each other without mutual mechanical influence. In order that the electromagnetic field to be measured be disturbed as little as possible by the presence of the test body system, we shall imagine, moreover, that every electric or magnetic component body is placed adjacent to a neutralization body with exactly the opposite charge. In the case of a magnetic test body system it is to be noted that a uniform pole strength distribution cannot exist on a strictly delimited body. However, one can imagine, at least in principle, that every component body of such a system is connected with the corresponding neutralization body by magnetizable flexible threads. All these neutralization bodies are to remain connected with the fixed frame during the whole measurement process without mechanically influencing the free mobility of the component bodies belonging to the test body system proper. Of course, the idealizations entailed in such presuppositions, as well as in the still needed compensation mechanisms to be introduced below, are justifiable only as long as we can neglect the atomic constitution of the test body. However, as already mentioned, this neglect does not imply any restrictions in principle on the possibility of testing the quantum electromagnetic formalism, since no universal space-time dimensions appear in this formalism. Accordingly, the purpose of the preceding considerations was above all to show that in the purely mechanical problems which are relevant to the field measurements, it is possible to distinguish strictly between the restrictions on the constitution of the test body stemming from the atomic structure of matter and the restrictions on the handling of these bodies that are due to the quantum of action, formulated in particular in the principle of indeterminacy.

4. Calculation of the Field Effects of the Test Body

After having investigated the physical presuppositions for the constitution of the test body we now turn to a closer consideration of the electromagnetic field effects of the test body which accompany the measurement of field quantities and which are of decisive importance for the measurability question. In accordance with the discussion above we shall treat each test body as a charge distribution which uniformly fills the spatial averaging domain and which undergoes a simple translation during the momentum measurement. We shall first carry out the calculation of the electromagnetic fields thereby produced on the basis of classical electrodynamics, and only afterwards discuss the restriction on the validity of this treatment due to the quantum of action.

Let us consider two space-time regions, I and II, with volumes V_{I} and V_{II} and durations T_{I} and T_{II}, and let us ask for the electromagnetic field 'which is produced at a point (x_2, y_2, z_2, t_2) of region II by a measurement of the average of $\mathfrak{E}_x$ over the region I. Thus, we assume that in volume V_{I} there are originally two electric charge distributions with the constant densities $+\rho_{\mathrm{I}}$ and $-\rho_{\mathrm{I}}$. In the interval from t'_{I} to $t'_{\mathrm{I}}+\Delta t_{\mathrm{I}}$ the first charge distribution experiences a simple non-uniform translation in the x-direction through a distance $D_x^{(\mathrm{I})}$; in the interval from $t'_{\mathrm{I}}+\Delta t_{\mathrm{I}}$ to t''_{I} it remains at rest at the displaced position; finally, in the interval from t''_{I} to $t''_{\mathrm{I}}+\Delta t_{\mathrm{I}}$ it moves non-uniformly parallel to the x-axis back to its original position, which coincides with that of the neutralization distribution. In accordance with the requirement discussed in thc preceding sections, we assume further that Δt_{I} is very small compared to $T_{\mathrm{I}}=t''_{\mathrm{I}}-t'_{\mathrm{I}}$ and that $D_x^{(\mathrm{I})}$ is very small not only compared to the linear dimensions of the spatial averaging region of volume V_{I}, but also small compared with $c\Delta t_{\mathrm{I}}$.

Hence, in the limiting case of vanishingly small Δt_{I}, the sources of the field that we are looking for may be represented as a polarization in the x-direction of constant density, $P_x^{(\mathrm{I})}=\rho_{\mathrm{I}} D_x^{(\mathrm{I})}$, existing in the region I during the time interval from t'_{I} to t''_{I}, as well as a current density present only in the immediate vicinity of the times t'_{I} and t''_{I}, which we can write as

$$(36) \qquad J_x^{(\mathrm{I})}=\rho_{\mathrm{I}} D_x^{(\mathrm{I})}[\delta(t-t'_{\mathrm{I}})-\delta(t-t''_{\mathrm{I}})],$$

using δ symbol defined by formula (3). By means of the same symbol we can similarly express the polarization at an arbitrary time t by the formula

$$(37) \qquad P_x^{(\mathrm{I})}=\rho_{\mathrm{I}} D_x^{(\mathrm{I})}\int_{t'_{\mathrm{I}}}^{t''_{\mathrm{I}}}\delta(t-t_1)\,\mathrm{d}t_1.$$

As is well-known, the components of the fields at the space time point

(x_2, y_2, z_2, t_2) produced by these sources may be calculated from the formulae

$$(38)\quad \begin{cases} E_x^{(\mathrm{I})} = -\dfrac{\partial \varphi^{(\mathrm{I})}}{\partial x_2} - \dfrac{1}{c}\dfrac{\partial \psi_x^{(\mathrm{I})}}{\partial t_2}, & E_y^{(\mathrm{I})} = -\dfrac{\partial \varphi^{(\mathrm{I})}}{\partial y_2}, \quad E_z^{(\mathrm{I})} = -\dfrac{\partial \varphi^{(\mathrm{I})}}{\partial z_2}, \\ H_x^{(\mathrm{I})} = 0, & H_y^{(\mathrm{I})} = \dfrac{\partial \psi_x^{(\mathrm{I})}}{\partial z_2}, \quad H_z^{(\mathrm{I})} = -\dfrac{\partial \psi_x^{(\mathrm{I})}}{\partial y_2}, \end{cases}$$

where we have used latin letters to distinguish these fields from the field components that are to be measured. In (38), $\varphi^{(\mathrm{I})}$ signifies the retarded scalar potential

$$(39)\qquad \varphi^{(\mathrm{I})} = \int_{V_1} \frac{\partial}{\partial x_1}\left[\frac{P_x^{(\mathrm{I})}(t_2 - r/c)}{r}\right] \mathrm{d}v_1$$

and $\psi_x^{(\mathrm{I})}$ the retarded vector potential component

$$(40)\qquad \psi_x^{(\mathrm{I})} = \frac{1}{c}\int_{V_1} \frac{J_x^{(\mathrm{I})}(t_2 - r/c)}{r}\, \mathrm{d}v_1,$$

where r is the distance between the space points (x_1, y_1, z_1) and (x_2, y_2, z_2). Noting that the expression (36) can also be written in the form

$$(41)\qquad J_x^{(\mathrm{I})} = -\rho_1 D_x^{(\mathrm{I})} \int_{t_1'}^{t_1''} \frac{\partial}{\partial t_1}\, \delta(t - t_1)\, \mathrm{d}t_1$$

and taking into account (37) and (41) one sees that the field components given by (38), (39) and (40) can be expressed by the typical formulae

$$(42)\quad \begin{cases} E_x^{(\mathrm{I})} = \rho_1 D_x^{(\mathrm{I})} \displaystyle\int_{V_1} \mathrm{d}v_1 \int_{T_1} \mathrm{d}t_1 A_{xx}^{(12)}, & E_y^{(\mathrm{I})} = \rho_1 D_x^{(\mathrm{I})} \displaystyle\int_{V_1} \mathrm{d}v_1 \int_{T_1} \mathrm{d}t_1 A_{xy}^{(12)}, \\ H_x^{(\mathrm{I})} = 0, & H_y^{(\mathrm{I})} = \rho_1 D_x^{(\mathrm{I})} \displaystyle\int_{V_1} \mathrm{d}v_1 \int_{T_1} \mathrm{d}t_1 B_{xy}^{(12)}, \end{cases}$$

where the abbreviations defined in (2) have been used.

In view of the properties of the symbolic δ-function it is easy to see that the field components given by (42) always remain finite and even cannot surpass a value of the order of magnitude $\rho_{\mathrm{I}} D_x^{(\mathrm{I})}$ at any space-time point (x_2, y_2, z_2, t_2). As already mentioned, the electromagnetic forces which occur during the momentum measurement of the test body in the time interval Δt are just of this order of magnitude (cf. p. 372). The fact that the field intensities do not subsequently increase essentially is solely a consequence of the counter collision taking place right after the momentum measurement which brings the body back to rest, and finds its idealized mathematical expressions in (36) and (37).

The averages of these field components over the region II, which are of particular interest to us, are obtained from (42) by a simple space-time integration and, in accordance with (5), are given by the formulae

$$(43)\qquad \begin{cases} \bar{E}_x^{(\mathrm{I},\,\mathrm{II})} = D_x^{(\mathrm{I})} \rho_{\mathrm{I}} V_{\mathrm{I}} T_{\mathrm{I}} \bar{A}_{xx}^{(\mathrm{I},\,\mathrm{II})}, & \bar{E}_y^{(\mathrm{I},\,\mathrm{II})} = D_x^{(\mathrm{I})} \rho_{\mathrm{I}} V_{\mathrm{I}} T_{\mathrm{I}} \bar{A}_{xy}^{(\mathrm{I},\,\mathrm{II})} \\ \bar{H}_x^{(\mathrm{I},\,\mathrm{II})} = 0, & \bar{H}_y^{(\mathrm{I},\,\mathrm{II})} = D_x^{(\mathrm{I})} \rho_{\mathrm{I}} V_{\mathrm{I}} T_{\mathrm{I}} \bar{B}_{xy}^{(\mathrm{I},\,\mathrm{II})}. \end{cases}$$

As a result of the properties of the expressions $\bar{A}$ and $\bar{B}$, already discussed in Section 2, we see that for a given value of $D_x^{(\mathrm{I})}$ the field averages given by (43) are well-defined continuous functions of the regions I and II. Hence, for decreasing latitudes, Δt and Δx, of the duration of the momentum measurements and of the accompanying unpredictable displacements, these field averages are completely independent of the detailed space-time course of the collision process, and simply proportional to the constant displacement of the test body in the measuring interval T_{I}. As we shall see, this very fact turns out to be decisive for the possibility of an extensive compensation of the uncontrollable field effects of the test bodies.

Thus far, the calculation of the field effects has been carried out on a purely classical basis. Yet for a more detailed comparison of the measurement possibilities and the requirements of the quantum-electromagnetic formalism one must also take into account the limitation imposed on the classical mode of calculation by the quantum-theoretical features of any field effect, symbolized by the concept of light quanta. In order to get a general view of the situation we assume that the averaging regions in question are of the same order of magnitude and spatially displaced relative to each other through distances of the same order of magnitude as their linear dimensions, which we denote by L; and that further the corresponding time intervals of order of magnitude T are smaller than L/c. Under these conditions, the spectral decomposition of the field effects contains essentially only waves of wave length of the same order of magnitude as L. Since, furthermore, in the case under consideration the intensity of the field produced in the momentum measurement is of the order $\rho\Delta x$, and consequently the field energy contained in the volume V is of the order $\rho^2\Delta x^2 V$, then the number of light quanta in question is approximated by the expression

$$n \sim \rho^2 \Delta x^2 V \frac{L}{\hbar c} = \lambda^{-2} \frac{L}{cT}, \tag{44}$$

where λ is the factor, defined in (20), that provides the measure for the accuracy of the measurement. Thus we see that in our case n is always large compared to unity if an accuracy of measurement is required that permits field strengths to be measured which are smaller than the critical field quantity Q.

Evidently, the classically calculated expressions (42) and (43) for the field effects become relatively more exact, the greater the accuracy aimed at for the field measurements. However, it is essential to note that the absolute accuracy of these expressions does not change for increasing values of n. For in our case the statistical range of fluctuation of the field averages is approximately given by

$$\frac{\rho\Delta x}{\sqrt{n}} \sim \sqrt{\frac{\hbar c}{VL}} \sim \frac{\sqrt{\hbar c}}{L^2}.$$

This expression for the range of fluctuations of the field effects of the test body, which depends only on the linear dimensions of the measurement domain and which always remains finite, agrees in fact with the expression (14) for the order of magnitude of the pure black-body fluctuations which was derived from the formalism in the case $L > cT$. Actually, in the above consideration we are dealing merely with an example of the general relation, mentioned in Section 2, between black-body fluctuations and the deviations, only describable statistically, of field averages from field quantities that are calculated according to classical theory from specification of the sources. Furthermore, as was already there pointed out, in the case $L > cT$, especially important for testing the formalism, the black-body fluctuations are always smaller than the field strength Q which is a measure of the complementary measurability of field quantities; and, indeed, so much the smaller, the larger the ratio between L and cT. Thus, in the following comparison between field measurements and formalism we shall always start from the classically calculated expressions (43), and only afterwards discuss the significance of the fluctuation phenomena for the consistency of the formalism.

5. Measurement of single field averages

By definition we base the investigation of the measurement possibilities for field averages on equation (15) which expresses the classically described momentum balance for a test body situated in the field. According to the preceding arguments, each field component, such as $\mathfrak{E}_x$, is in this context to be regarded as the superposition of the fields originating from all field sources, including the test body itself, and the core of the measurement problem is precisely the question of the extent to which these fields can be associated with the various sources. However, we must emphasize here immediately that the strict applicability of the classical field concept in defining field averages is not in itself impaired by the previously discussed limited validity of the classical description of the field effects of the test body. Quite apart from the question of the accuracy attainable in the momentum measurements of the test body at the beginning and end of the measuring interval, which was discussed in Section 3, the unambiguous character of this definition would itself require that the masses of the test bodies be chosen sufficiently large that any modifications of the electromagnetic fields, stemming from their accelerations under the influence of these fields during the measuring interval, may be neglected. If one were inclined to regard this neglect as in contradiction with the atomic character of the momentum transfer between electromagnetic wave fields and material bodies, then it must be recalled that in the measurement problem under consideration there is no question of tracing well-defined elementary processes in the sense of the light quantum concept. In particular, in the measuring arrangement described, an uncontrollable amount of momentum is absorbed by the rigid frame to which every test body is attached before and after the measuring interval. In the limiting case of a classically describable interaction between an electromagnetic wave train and a sufficiently heavy charged body, the momentum transfer just mentioned would obviously exactly compensate the momentum absorbed by the test body in the measuring interval.

As a preparation for the general discussion of the measurement problem, we first consider a single field measurement and, as in Section 3, ask for the average value of $\mathfrak{E}_x$ in a certain space-time domain which we denote, as in the previous sections, by I. Thus, from the fundamental equation (15), we get for the momentum balance of the test body

$$(45) \qquad p_x^{(\mathrm{I})''} - p_x^{(\mathrm{I})'} = \rho_\mathrm{I} V_\mathrm{I} T_\mathrm{I} (\mathfrak{E}_x^{(\mathrm{I})} + \bar{E}_x^{(\mathrm{I},\,\mathrm{I})}),$$

where $\mathfrak{E}_x^{(\mathrm{I})}$ represents the part of the average of $\mathfrak{E}_x$ which would be present in the space-time domain I under consideration if no momentum measurement were made on the test body at time t', while $\bar{E}_x^{(\mathrm{I},\,\mathrm{I})}$ is the part of the field average that arises from this measurement, whose classically estimated expression is given by (43), if regions I and II are set equal.

According to the arguments of Section 3, the sum of the field averages $\mathfrak{E}_x^{(\mathrm{I})}$ and $\bar{E}_x^{(\mathrm{I},\,\mathrm{I})}$ appearing in (45) can be determined with arbitrary accuracy by choosing a sufficiently large value of ρ_I. However, the larger ρ_I is chosen, the larger will be the uncontrollable value of $\bar{E}_x^{(\mathrm{I},\,\mathrm{I})}$; and therefore, the attainable accuracy in the determination of $\mathfrak{E}_\mathrm{I}^{(\mathrm{I})}$ by means of the simple measuring arrangement previously described, which according to (45) is given by

$$(46) \qquad \Delta\mathfrak{E}_x^{(\mathrm{I})} \sim \frac{\Delta p_x^{(\mathrm{I})}}{\rho_\mathrm{I} V_\mathrm{I} T_\mathrm{I}} + \Delta\bar{E}_x^{(\mathrm{I},\,\mathrm{I})},$$

has a limit imposed upon it. Indeed, due to relation (16) and to the fact that the quantity $D_x^{(\mathrm{I})}$ appearing in (43) is predictable only with a latitude Δx_I, we get from (46) the expression

$$(47) \qquad \Delta\mathfrak{E}_x^{(\mathrm{I})} \sim \frac{\hbar}{\rho_\mathrm{I} \Delta x_\mathrm{I} V_\mathrm{I} T_\mathrm{I}} + \rho_\mathrm{I} \Delta x_\mathrm{I} V_\mathrm{I} T_\mathrm{I} |\bar{A}_{xx}^{(\mathrm{I},\,\mathrm{I})}|,$$

for $\Delta\mathfrak{E}_x^{(\mathrm{I})}$, whose minimal value obviously is

$$(48) \qquad \Delta_m \mathfrak{E}_x^{(\mathrm{I})} \sim \sqrt{\hbar |\bar{A}_{xx}^{(\mathrm{I},\,\mathrm{I})}|}$$

and in the case $L_\mathrm{I} > cT_\mathrm{I}$ precisely equals the critical quantity Q_I. It is true that when L_I is large compared to cT_I, (48) is essentially smaller than the expression (24) which was given by Landau and Peierls as the absolute limit on the measurability of field quantities; yet if (48) were to be considered as an unavoidable limit on the accuracy of measurement, we should still arrive at the conclusion, in agreement with the view of these authors, that the quantum-electromagnetic formalism admits of no test in the properly quantum domain, and that therefore physical reality can be ascribed to the entire field theory only in the classical limit.

However, this conclusion cannot be maintained, for the fact that according to (43) the coefficient of the unpredictable displacement $D_x^{(I)}$ in $\bar{E}_x^{(I,I)}$ is a well-defined quantity depending solely on geometrical relations, allows us to so arrange things in the measurements that the effect of the field $E_x^{(I)}$ is completely compensated except for the unavoidable field fluctuations. This is achieved by a measuring arrangement in which the test body is not freely movable, even during the measuring interval T_I, but remains connected with the rigid frame through a spring mechanism whose tension is proportional to $D_x^{(I)}$. If the force in the x-direction exerted by this mechanism on the test body is $-F_x D_x^{(I)}$, then the total momentum transferred from the field $E_x^{(I)}$ to this body will obviously be completely cancelled by the spring if the spring constant is chosen to be

$$(49) \qquad F_I = \rho_I^2 V_I^2 T_I \bar{A}_{xx}^{(I,I)}.$$

At any rate, this holds when the test body is so heavy that its oscillation period under the influence of the spring is large compared to T_I and thus its displacement due to the spring tension during the time T_I is small compared to $D_x^{(I)}$. Furthermore, the action of the spring, which strictly speaking is classically describable only in the asymptotic limiting case, may be calculated on the basis of classical mechanics with an accuracy which is the greater the larger the mass of the test body. Apart from the limitations due to the atomic structure of all bodies, no objection of principle could exist against such a compensation device. In the first place, by using a mechanical spring all electromagnetic fields are avoided, which would be inseparable from the fields to be measured. Secondly, if the length of the spring is sufficiently small, i.e. small compared to cT_I, one may obviously disregard all retardation effects. In doing so, if the test body system is sufficiently heavy, it is clearly immaterial whether the spring acts only on a component body or whether one uses a system of springs that affects each component body uniformly.

Thus we see that the sharpness of a single field measurement is restricted solely by the limit set for the classical description of the field effects of the test body. However, even in the case $L_{\mathrm{I}} \leqslant cT_{\mathrm{I}}$ this limitation, which is the more insignificant the larger L_{I} is compared to cT_{I}, implies no restriction at all on the possibility of testing the consequences of the quantum-electromagnetic formalism. In assessing this question, we must distinguish sharply between the testing of theoretical predictions which presuppose data concerning electric or magnetic forces obtained by field measurements, and of those which depend on knowledge of the state of the field in question obtained on some other basis. As for the former predictions, their testing obviously requires a closer investigation of the mutual relations between several field measurements; thus, to begin with, here it can only be a matter of testing predictions of the latter kind.

Now, as mentioned in Section 2, it is a major result of the quantum theory of fields that all predictions concerning field averages which do not rest on true field measurements, but on the light quantum composition of the field to be investigated or on the knowledge of classically described field sources, must be of an essentially statistical nature. Further, the more detailed argument presented there shows that inclusion of the fluctuations of the test body's field effects around their classically estimated value brings about no change whatsoever in these statistical predictions. Without further correction, the measurement results obtained by means of the experimental arrangement described thus appear as the desired field averages for testing the theoretical statements. Such a view of the measuring results, whose general justification we shall investigate more closely in the following, is also suggested by the fact that all measurements of physical quantities, by definition, must be a matter of the application of classical concepts; and that, therefore, in field measurements any consideration of limitations on the strict applicability of classical electrodynamics would be in contradiction with the measurement concept itself.

Even though, consequently, as already stressed in the Introduction, the measurement concept is to be applied with even greater caution in field measurements than in the usual quantum mechanical measurement problems; nevertheless, as regards the inseparability of phenomenon and measuring process the situation described exhibits a far-reaching analogy to these problems. Indeed, even in a position or momentum measurement on the electron in a hydrogen atom in a given stationary state one can assert with a certain right that the measuring result is produced only by the measurement itself. It is here not a question of a limitation on the sharpness of the measuring result on the basis of classical mechanics, indeed; but only of abandonment of any control over the influence of the measuring process on the state of the atom. In field measurements, this complementary feature of the description, essential for consistency, corresponds to the fact that the knowledge of the light quantum composition of the field is lost through the field effects of the test body; and in fact, according to (44) the more so, the greater the desired accuracy of the measurement. Moreover, it will appear from the following discussion that any attempt to re-establish the knowledge of the light quantum composition of the field through a subsequent measurement by means of any suitable device would at the same time prevent any further utilization of the field measurement in question.

However, the fact that pure black-body fluctuations appear as the common limitation in the demonstration of the correspondence between the testability of the consequences of the quantum-electromagnetic formalism by means of a single field measurement and the interpretability of such a measurement on the basis of classical electrodynamics in no way implies that these fluctuations set an absolute limit for any utilization of field measurements. Indeed, such a general limitation exists neither for the consequences of the formalism regarding relations between averages of a field component over different regions, nor for the testing of such relations through direct field measurements. This will become clear from the considerations in the following sections; and in particular it will be shown that the requirement of repeatability of measurements of kinematical and dynamical quantities, essential for the discussion of the consistency of the usual quantum mechanics, possesses its natural analog in field measurements.

6. Measurability of Two-Fold Averages of a Field Component

In investigating the measurability of two field quantities it is convenient to start with the measurement of the average of one and the same field component over two different regions, I and II. Thus, considering as above the field component $\mathfrak{E}_x$, and disregarding to begin with the limitations of the classical describability of the test bodies' field effects, we have in this case for the momentum balance of the two test bodies, instead of (45):

$$(50)\quad \begin{cases} p_x^{(\mathrm{I})''} - p_x^{(\mathrm{I})'} = \rho_{\mathrm{I}} V_{\mathrm{I}} T_{\mathrm{I}} (\bar{\mathfrak{E}}_x^{(\mathrm{I})} + \bar{E}_x^{(\mathrm{I},\mathrm{I})} + \bar{E}_x^{(\mathrm{II},\mathrm{I})}), \\ p_x^{(\mathrm{II})''} - p_x^{(\mathrm{II})'} = \rho_{\mathrm{II}} V_{\mathrm{II}} T_{\mathrm{II}} (\bar{\mathfrak{E}}_x^{(\mathrm{II})} + \bar{E}_x^{(\mathrm{II},\mathrm{II})} + \bar{E}_x^{(\mathrm{I},\mathrm{II})}), \end{cases}$$

where $\bar{E}_x^{(\mathrm{I},\mathrm{II})}$ is defined by expression (43), and $\bar{E}_x^{(\mathrm{II},\mathrm{I})}$ is obtained from this expression by simple interchange of the indices I and II.

According to the considerations in the previous sections, the appearance of the expressions $\bar{E}_x^{(\mathrm{I},\mathrm{I})}$ and $\bar{E}_x^{(\mathrm{II},\mathrm{II})}$ in equations (50) implies that each of the desired field averages, $\bar{\mathfrak{E}}_x^{(\mathrm{I})}$ and $\bar{\mathfrak{E}}_x^{(\mathrm{II})}$, can only be determined with a limited accuracy, given by (48), by means of a simple measuring arrangement. Thus, it is evident from the beginning that a compensation procedure is unavoidable, and for preliminary orientation about the more complicated measuring problem considered here we therefore first use a measuring arrangement in which the reactions, $\rho_{\mathrm{I}} V_{\mathrm{I}} T_{\mathrm{I}} \bar{E}_x^{(\mathrm{I},\mathrm{I})}$ and $\rho_{\mathrm{II}} V_{\mathrm{II}} T_{\mathrm{II}}$-$\bar{E}_x^{(\mathrm{II},\mathrm{II})}$, are cancelled by means of two springs acting on the test bodies I and II, the spring constants being given by (49) and an analogous expression.

From equations (50), with $\bar{E}_x^{(\mathrm{I},\mathrm{I})}$ and $\bar{E}_x^{(\mathrm{II},\mathrm{II})}$ omitted, it follows, using (16) and (43), that in this arrangement the uncertainties of the two field measurements, taking into account that the displacements of the test bodies, $D_x^{(\mathrm{I})}$ and $D_x^{(\mathrm{II})}$, appearing in $\bar{E}_x^{(\mathrm{I},\mathrm{II})}$ and $\bar{E}_x^{(\mathrm{II},\mathrm{I})}$ are completely independent of each other and known only with the latitudes Δx_{I} and Δx_{II}, are given by

$$(51)\quad \begin{cases} \Delta \bar{\mathfrak{E}}_x^{(\mathrm{I})} \sim \dfrac{\hbar}{\rho_{\mathrm{I}} \Delta x_{\mathrm{I}} V_{\mathrm{I}} T_{\mathrm{I}}} + \rho_{\mathrm{II}} \Delta x_{\mathrm{II}} V_{\mathrm{II}} T_{\mathrm{II}} |\bar{A}_{xx}^{(\mathrm{II},\mathrm{I})}| \\ \Delta \bar{\mathfrak{E}}_x^{(\mathrm{II})} \sim \dfrac{\hbar}{\rho_{\mathrm{II}} \Delta x_{\mathrm{II}} V_{\mathrm{II}} T_{\mathrm{II}}} + \rho_{\mathrm{I}} \Delta x_{\mathrm{I}} V_{\mathrm{I}} T_{\mathrm{I}} |\bar{A}_{xx}^{(\mathrm{I},\mathrm{II})}|. \end{cases}$$

By suitable choice of the values of $\rho_{\mathrm{I}} \Delta x_{\mathrm{I}}$ and $\rho_{\mathrm{II}} \Delta x_{\mathrm{II}}$ either one of the quantities $\Delta \bar{\mathfrak{E}}_x^{(\mathrm{I})}$, $\Delta \bar{\mathfrak{E}}_x^{(\mathrm{II})}$ can obviously be arbitrarily diminished, but only at the expense of an increase of the other. For according to (51) we get for the product of the two quantities the minimum value

$$(52)\quad \Delta \bar{\mathfrak{E}}_x^{(\mathrm{I})} \Delta \bar{\mathfrak{E}}_x^{(\mathrm{II})} \sim \hbar [|\bar{A}_{xx}^{(\mathrm{I},\mathrm{II})}| + |\bar{A}_{xx}^{(\mathrm{II},\mathrm{I})}|].$$

In spite of the great similarity of relation (52) to the uncertainty relations (8) required by the formalism, there is, nevertheless, a fundamental difference in that the latter contains not the sum of the magnitudes of the quantities $\bar{A}_{xx}^{(\mathrm{I,\,II})}$ and $\bar{A}_{xx}^{(\mathrm{II,\,I})}$ but their algebraic difference. It is true that (8) and (52) are in general in agreement as to order of magnitude when the regions I and II are spatially and temporally displaced relative to each other by distances of order of magnitude L and T, in which case they both have the approximate value Q^2. However, as mentioned in Section 2, the difference sign appearing in the indeterminacy relation (8) has the effect that in important cases the product of the complementary uncertainties vanishes identically, even though the quantities $\bar{A}_{xx}^{(\mathrm{I,\,II})}$ and $\bar{A}_{xx}^{(\mathrm{II,\,I})}$ each remain different from zero. This happens, for example, when the temporal averaging intervals T_{I} and T_{II} coincide, and in particular when the two averaging regions I and II completely overlap. In the latter case even the limit on the measurability of two field averages given by (52) would be in glaring contradiction to the result of the previous discussion of measurement of a single field average. In general, the two expressions (52) and (8) agree exactly only when at least one of the quantities $\bar{A}_{xx}^{(\mathrm{I,\,II})}$ or $\bar{A}_{xx}^{(\mathrm{II,\,I})}$ vanishes which in general requires that one of the expressions $t_1 - t_2 - r/c$ or $t_2 - t_1 - r/c$, appearing as arguments of the δ-function in the integrals (5), remain different from zero for every pair of points (x_1, y_1, z_1, t_1) and (x_2, y_2, z_2, t_2) of regions I and II.

Thus, apart from the last mentioned case in which there exists no correlation, or at any rate only a one-way correlation, between the two field averages, the demonstration of the agreement between measurability and quantum electromagnetic formalism requires a more refined measuring arrangement in which the uncontrollable effects can be compensated to a larger extent. It is true that there appears here, in comparison with the compensation procedure needed already for measuring a single field quantity, the further complication that the displacements of the two test bodies not only must remain unknown but are also completely independent of each other. However, this circumstance implies no fundamental difficulty; only a somewhat more complicated procedure is necessary in order to compensate as much as possible the influence of the relative displacement of the test bodies on the field measurements. For this purpose we select two component bodies ε_{I} and $\varepsilon_{\mathrm{II}}$, one from each test body system I and II, for which the expression $r - c(t_1 - t_2)$ vanishes for two times t_{I}^* and t_{II}^* lying in the time intervals T_{I} and T_{II}, respectively. If such a choice were not possible, then as said above the agreement between measurability and formalism would already be attained without further compensation. To establish the necessary correlation between the test bodies one might at first think of a spring which should connect the bodies ε_{I} and $\varepsilon_{\mathrm{II}}$ directly with each other; however, due to the retardation of the forces one would thereby run into difficulties. But one can manage with a short spring, i.e. small compared to cT, if one adds to the second test body system a neutral component body $\varepsilon_{\mathrm{III}}$ which is situated in the immediate neighborhood of component body ε_{I}, belonging to the first system, and connected with it by a spring.

Like all component bodies of the two test body systems, the body $\varepsilon_{\mathrm{III}}$ is initially to be bound to the rigid frame. Then at time t'_{I}, its momentum is to be measured, after severing of this link, with the same accuracy as that of test body system II. It thereby undergoes an unknown displacement $D_x^{(\mathrm{III})}$ in the x-direction which is of the same order of magnitude as Δx_{II}. If now the tension of the spring mounted between $\varepsilon_{\mathrm{III}}$ and ε_{I} is chosen equal to $\frac{1}{2}\rho_{\mathrm{I}}\rho_{\mathrm{II}}V_{\mathrm{I}}V_{\mathrm{II}}T_{\mathrm{II}}(\bar{A}_{xx}^{(\mathrm{I},\,\mathrm{II})}+\bar{A}_{xx}^{(\mathrm{II},\,\mathrm{I})}$, then in the time interval T_{I} the momentum

$$(53) \qquad P=\tfrac{1}{2}\rho_{\mathrm{I}}\rho_{\mathrm{II}}V_{\mathrm{I}}V_{\mathrm{II}}T_{\mathrm{I}}T_{\mathrm{II}}(\bar{A}_{xx}^{(\mathrm{I},\,\mathrm{II})}+\bar{A}_{xx}^{(\mathrm{II},\,\mathrm{I})})(D_x^{(\mathrm{I})}-D_x^{(\mathrm{III})})$$

will be transferred from $\varepsilon_{\mathrm{III}}$ to ε_{I}, while $\varepsilon_{\mathrm{III}}$ undergoes the momentum change $-P$ during the same time interval. At time t'' the momentum of $\varepsilon_{\mathrm{III}}$ is measured again with the same accuracy. However, before this measurement, and in fact at time t^*_{II}, a short light signal is to be sent from $\varepsilon_{\mathrm{II}}$ to $\varepsilon_{\mathrm{III}}$, by which the relative displacement $D_x^{(\mathrm{III})}-D_x^{(\mathrm{II})}$ of these bodies can be measured with arbitrary accuracy by means of a suitable device. At the emission and absorption of the signal the two bodies undergo momentum changes which indeed remain completely unknown, but cancel each other exactly in the sum of the momentum changes measured on the bodies.

Thus, for the momentum balance of the two test body systems during the measurement we have, if we include the body $\varepsilon_{\mathrm{III}}$ in system II,

$$(54) \qquad \begin{cases} p_x^{(\mathrm{I})''}-p_x^{(\mathrm{I})'}=\rho_{\mathrm{I}}V_{\mathrm{I}}T_{\mathrm{I}}(\mathfrak{E}_x^{(\mathrm{I})}+\bar{E}_x^{(\mathrm{II},\,\mathrm{I})})+P \\ p_x^{(\mathrm{II})''}-p_x^{(\mathrm{II})'}+p_x^{(\mathrm{III})''}-p_x^{(\mathrm{III})'}=\rho_{\mathrm{II}}V_{\mathrm{II}}T_{\mathrm{II}}(\bar{\mathfrak{E}}_x^{(\mathrm{II})}+\bar{E}_x^{(\mathrm{I},\,\mathrm{II})})-P. \end{cases}$$

Taking into account (43) and (53), these formulae can be put into the form

$$
(55)\quad \begin{cases}
p_x^{(\mathrm{I})''} - p_x^{(\mathrm{I})'} = \rho_{\mathrm{I}} V_{\mathrm{I}} T_{\mathrm{I}} \mathfrak{E}_x^{(\mathrm{I})} \\
+\tfrac{1}{2}\rho_{\mathrm{I}}\rho_{\mathrm{II}} V_{\mathrm{I}} V_{\mathrm{II}} T_{\mathrm{I}} T_{\mathrm{II}} \{ -D_x^{(\mathrm{II})} (\bar{A}_{xx}^{(\mathrm{I,II})} - \bar{A}_{xx}^{(\mathrm{II,I})}) \\
+ (D_x^{(\mathrm{II})} - D_x^{(\mathrm{III})})(\bar{A}_{xx}^{(\mathrm{I,II})} + \bar{A}_{xx}^{(\mathrm{II,I})}) + D_x^{(\mathrm{I})} (\bar{A}_{xx}^{(\mathrm{I,II})} + \bar{A}_{xx}^{(\mathrm{II,I})}) \} \\
p_x^{(\mathrm{II})''} - p_x^{(\mathrm{II})'} + p_x^{(\mathrm{III})''} - p_x^{(\mathrm{III})'} = \rho_{\mathrm{II}} V_{\mathrm{II}} T_{\mathrm{II}} \mathfrak{E}_x^{(\mathrm{II})} \\
+\tfrac{1}{2}\rho_{\mathrm{I}}\rho_{\mathrm{II}} V_{\mathrm{I}} V_{\mathrm{II}} T_{\mathrm{I}} T_{\mathrm{II}} \{ D_x^{(\mathrm{I})} (\bar{A}_{xx}^{(\mathrm{I,II})} - \bar{A}_{xx}^{(\mathrm{II,I})}) \\
- (D_x^{(\mathrm{II})} - D_x^{(\mathrm{III})})(\bar{A}_{xx}^{(\mathrm{I,II})} + \bar{A}_{xx}^{(\mathrm{II,I})}) + D_x^{(\mathrm{II})} (\bar{A}_{xx}^{(\mathrm{I,II})} + \bar{A}_{xx}^{(\mathrm{II,I})}) \}
\end{cases}
$$

The last terms in the curly brackets in (55) are proportional to the unknown displacements of the test bodies I and II and can therefore, exactly as the simple reactions of each test body on itself, be cancelled by means of suitable spring connections with the rigid frame. This simply amounts to replacing the expression (49) for the tension of the spring acting on body I by

$$
(56)\qquad F_{\mathrm{I,II}} = \rho_{\mathrm{I}}^2 V_{\mathrm{I}}^2 T_{\mathrm{I}} \bar{A}_{xx}^{(\mathrm{I,I})} + \tfrac{1}{2}\rho_{\mathrm{I}}\rho_{\mathrm{II}} V_{\mathrm{I}} V_{\mathrm{II}} T_{\mathrm{II}} (\bar{A}_{xx}^{(\mathrm{I,II})} + \bar{A}_{xx}^{(\mathrm{II,I})})
$$

and similarly changing the tension of the spring between the frame and body II. Furthermore, in the measuring arrangement described, the terms proportional to the relative displacement $D_x^{(\mathrm{III})} - D_x^{(\mathrm{II})}$ are known with arbitrary accuracy and can therefore easily be taken into account in the field measurements. In fact, by means of a somewhat more complicated device one could even obtain the vanishing of the difference $D_x^{(\mathrm{III})} - D_x^{(\mathrm{II})}$ by using $P_x^{(\mathrm{II})} + P_x^{(\mathrm{III})}$ (in analogy to the arrangement for measuring the total momentum of a test body system described in Section 3) to determine one and the same radiation packet and, by means of suitably placed fixed mirrors, by regulating the light path in such a way that in the first momentum measurement the reflections at body $\varepsilon_{\mathrm{III}}$ and at all component bodies of system II occur at the times t_{I}' and t_{II}', respectively, and in the second momentum measurement occur at the times t_{I}'' and t_{II}''.

By means of all these contrivances, whose considerable complexity lies in the nature of the problem, being due solely to the finite propagation of all field effects, we now have actually removed the apparent conflict between the determination of single and two-fold averages of a field component described at the beginning of this section. For from (55) we now obtain for the indeterminacies of $\mathfrak{E}_x^{(\mathrm{I})}$ and $\mathfrak{E}_x^{(\mathrm{II})}$, instead of (51),

$$(57)\quad \begin{cases} \Delta\mathfrak{E}_x^{(\mathrm{I})} \sim \dfrac{\hbar}{\rho_{\mathrm{I}}\Delta x_{\mathrm{I}} V_{\mathrm{I}} T_{\mathrm{I}}} + \dfrac{1}{2}\rho_{\mathrm{II}}\Delta x_{\mathrm{II}} V_{\mathrm{II}} T_{\mathrm{II}} |\bar{A}_{xx}^{(\mathrm{I,II})} - \bar{A}_{xx}^{(\mathrm{II,I})}| \\ \Delta\mathfrak{E}_x^{(\mathrm{II})} \sim \dfrac{\hbar}{\rho_{\mathrm{II}}\Delta x_{\mathrm{II}} V_{\mathrm{II}} T_{\mathrm{II}}} + \dfrac{1}{2}\rho_{\mathrm{I}}\Delta x_{\mathrm{I}} V_{\mathrm{I}} T_{\mathrm{I}} |\bar{A}_{xx}^{(\mathrm{I,II})} - \bar{A}_{xx}^{(\mathrm{II,I})}|, \end{cases}$$

which immediately yields for the minimal value of the product of the uncertainties

$$(58)\qquad \Delta\mathfrak{E}_x^{(\mathrm{I})}\Delta\mathfrak{E}_x^{(\mathrm{II})} \sim \hbar|\bar{A}_{xx}^{(\mathrm{I,II})} - \bar{A}_{xx}^{(\mathrm{II,I})}|$$

in agreement with the consequence of the quantum theory of fields expressed by (8).

However, in order to demonstrate the complete accord between the measurability of the averages of a field component over two space-time regions and the requirements of the quantum electromagnetic formalism, we must go somewhat further into the question of the extent to which the assumption of the classical describability of the test body's field effects impairs the possibilities of testing theoretical predictions. For, as already indicated, exactly in the case of measurement of several field averages one might think beforehand that the neglect of the fluctuations of all field effects of the test body, which cannot be followed classically, and are inseparable from the pure black-body fluctuations, in this respect signifies an essential renunciation. At any rate, as long as we are dealing with averaging regions which are displaced spatially and temporally relative to each other by distances of the same order of magnitude as their linear dimensions L and corresponding time intervals T, this neglect is of little significance in the important case where L is large compared to cT. However, if L is of the same order of magnitude or smaller than cT, then the black-body fluctuations, as mentioned in Section 2, are of just the same order of magnitude as the critical field strength $\mathfrak{U}$, which is defined for such displaced regions by means of the indeterminacy relations and is to be regarded as the limit of the classical field description. The neglect in question might appear even more doubtful and seem to imply a complete renunciation of the repeatability of field measurements in the case of two almost coinciding domains, in which the product of the complementary uncertainties of the field averages, given by (8), tends to zero independently of the ratio between L and cT, and where thus the critical field strength $\mathfrak{U}$ can be arbitrarily small compared to the black-body fluctuations.

Nevertheless, a closer consideration shows that we obtain a consistent interpretation of all consequences of the quantum theory of fields, if, in a necessary generalization of the measurement concept, we interpret the measuring results obtained by the arrangement described as the desired field averages. For the classically undescribable fluctuations included in the field effects of all test bodies cannot be separated in any way from the fundamentally statistical features of every theoretical assertion whose conditions do not refer to actual field measurements. Without in any way limiting the given measurement problem we can therefore always regard the fluctuations in question as an integral part of the field to be measured. The situation in multiple field measurements differs in this respect from that obtaining in measurement of a single field average only insofar as the state of the field, with which we are concerned in every single measurement in the general case, is codetermined by the result of the other field measurements.

However, with regard to this state of affairs it may not be superfluous to point out that in the correlation of several field measurements we have to do with a feature of the general complementarity of description which is alien to the usual measuring problem of non-relativistic quantum mechanics. Indeed, the fundamental simplification which we meet in the latter theory lies precisely in the separation made there between spatial coordination and temporal evolution, which makes it possible to order all measuring processes in a simple temporal sequence. On the other hand, it is possible to speak of such a sequence of measuring processes during the measurement of two field averages only when the corresponding time intervals do not overlap at all. In general, in accordance with the formalism, the correlation of the two measurements is also a reciprocal one; and only when one of the quantities $r-c(t_1-t_2)$ and $r-c(t_2-t_1)$ remains different from zero for all pairs of points of the regions I and II, do we encounter conditions similar to those of the usual measurement problem of atomic mechanics, since the result of the one field measurement may then be simply included in the preconditions used in the predictions to be tested by the other measurement.

We meet an instructive example of an intimate reciprocal correlation in measurements of the averages of a field component over two almost coinciding space-time domains. In conformity with the requirement of the repeatability of measurement results, the theory demands in this case that both measurements yield the same result to an arbitrary degree of approximation quite independently of the statistical assertions about the values of the field quantities to be measured which are implied by the preconditions. That this requirement is actually fulfilled in our experimental arrangement follows from the fact that in this case we have to do with two test body systems which occupy almost the same spatial region and are used during almost the same interval of time. Thus, by definition, they are exposed to almost the same field, quite irrespective of the sources producing this field, and of which contribution comes from one or the other test body.

Actually, it follows from the last remark that in the case of coinciding averaging regions we would get exactly identical results of the two measurements even without any compensation. However, on account of the field effects of the test bodies the measurement results so obtained would differ in an unpredictable way from the theoretical predictions to be tested, the more so, the greater the desired measuring accuracy. It is true that by means of the compensation mechanism suitable for single field measurements, which we had retained unaltered at the beginning of this section, these deviations are in general diminished; but at the same time any strict correlation of the measuring results is prevented by the effects of the springs which are proportional to the independent displacements of the test bodies. In the arrangement for two field measurements finally adopted, in which all well-defined differences between measurement results and theoretical predictions are compensated, such a correlation is also re-established just in the case of coinciding regions. For, as one easily sees, quite independently of the relation between the magnitudes of the uncontrollable displacements of the test bodies, the momenta transferred to each test body through the combined effect of all the springs, divided by the corresponding charge densities, are exactly identical in this case.

As far as the consistency of the description is concerned, we might still remark that any attempt to control the change in the light quantum composition of the field caused by the field measurement by investigation of the test body's radiation, as already mentioned several times, would exclude the possibility of utilizing the measurement result for a comparison with a second field measurement. For in order even to be able to speak of such a utilization, there must exist pairs of points from the regions I and

II, respectively, for which one of the expressions $r-c(t_1-t_2)$ or $r-c(t_2-t_1)$ vanishes. But this implies that the radiation fields produced by the test bodies I and II during the measurement cannot be intercepted and analyzed on their way from one test body to the other without at the same time essentially influencing the fields to be measured by these bodies. Only after the completion of all field measurements, when their direct utilization is no longer of concern, is it possible to perform an arbitrarily accurate analysis of the light quantum composition of the entire field without adversely affecting the given measurement problem.

7. Measurability of Two Averages of Different Field Components

As far as measurements of averages of different field components are concerned, only the case of perpendicular, similar or dissimilar components needs closer investigation; for the complete commutativity and independent measurability of averages of parallel dissimilar components required by the quantum electromagnetic formalism finds its direct interpretation in the identical vanishing of the component $H_x^{(\mathrm{I})}$ of the field produced by the measurement of $\bar{\mathfrak{E}}_x^{(\mathrm{I})}$, as shown by (42). Besides, on the basis of equations (43), the measurement of averages of perpendicular field components allows a method of treatment analogous to the procedure described in the previous section.

Let us consider the measurement of the average of $\mathfrak{E}_x$ over the region I and the average of $\mathfrak{E}_y$ or $\mathfrak{H}_y$ over the region II. If to begin with we use a measuring arrangement in which the field effects of each test body on itself during the measurement are compensated in the manner described in Section 5, we get equations of the following type for the momentum balance of the two test bodies to be used:

$$
(59)\quad \begin{cases} p_x^{(\mathrm{I})''} - p_x^{(\mathrm{I})'} = \rho_{\mathrm{I}} V_{\mathrm{I}} T_{\mathrm{I}}(\bar{\mathfrak{E}}_x^{(\mathrm{I})} + D_y^{(\mathrm{II})} \sigma_{\mathrm{II}} V_{\mathrm{II}} T_{\mathrm{II}} \bar{C}_{xy}^{(\mathrm{II},\,\mathrm{I})}) \\ p_y^{(\mathrm{II})''} - p_y^{(\mathrm{II})'} = \sigma_{\mathrm{II}} V_{\mathrm{II}} T_{\mathrm{II}} \bar{\mathfrak{R}}_y^{(\mathrm{II})} + D_x^{(\mathrm{I})} \rho_{\mathrm{I}} V_{\mathrm{I}} T_{\mathrm{I}} \bar{C}_{xy}^{(\mathrm{I},\,\mathrm{II})}) \end{cases}
$$

Depending on whether we are dealing with a measurement of similar or dissimilar components, the letter $\mathfrak{R}$ here represents the usual designation of the field components $\mathfrak{E}$ or $\mathfrak{H}$, while C is written instead of the symbols A or B appearing in (43); further, the designation σ_{II} represents the charge density or the pole strength distribution of the test body II accordingly.

In a manner similar to the derivation of (52) one obtains from (59) the relation

(60) $$\Delta\bar{\mathfrak{E}}_x^{(\mathrm{I})}\Delta\bar{\mathfrak{H}}_y^{(\mathrm{II})}\sim\hbar[|\bar{C}_{xy}^{(\mathrm{I},\,\mathrm{II})}|+|\bar{C}_{xy}^{(\mathrm{II},\,\mathrm{I})}|],$$

which, like (52), does not generally, but only in certain cases, represent an agreement between measurability and quantum electromagnetic formalism. Of such cases we might mention in particular the measurement of dissimilar perpendicular field components inside the same spatial region, for which, as stressed in Section 2, both expressions $\bar{B}_{xy}^{(\mathrm{I},\,\mathrm{II})}$ and $\bar{B}_{xy}^{(\mathrm{II},\,\mathrm{I})}$ vanish. The correctness of interpreting this fact as an arbitrarily accurate independent measurability of the field quantities in question was suggested already in Section 3 by elementary considerations in the case of coinciding space-time regions.

For the general treatment of the measurability problem of perpendicular field components we choose, as in the previous section, two component bodies ε_{I} and $\varepsilon_{\mathrm{II}}$ of the test body systems I and II, respectively, whose separation is $r=c(t_{\mathrm{I}}^*-t_{\mathrm{II}}^*)$, where t_{I}^* and t_{II}^* are within the time intervals T_{I} and T_{II}, respectively. Furthermore, in the immediate neighborhood of ε_{I} we introduce a third body $\varepsilon_{\mathrm{III}}$, whose momentum in the y-direction is measured at the times t_{I}' and t_{I}''; the relative displacements $D_y^{(\mathrm{III})}-D_y^{(\mathrm{II})}$ of the bodies $\varepsilon_{\mathrm{III}}$ and $\varepsilon_{\mathrm{II}}$ are again determined by means of a light signal, as a result of which both bodies undergo equal and opposite momentum changes. Instead of connecting $\varepsilon_{\mathrm{III}}$ directly with ε_{I} by a spring we must, however, in order to make the force transfer through the spring mechanism proportional to $D_y^{(\mathrm{III})}-D_x^{(\mathrm{I})}$, use a device which consists of two springs and an angular level with two equally long mutually perpendicular arms which can rotate on a hinge mounted on the rigid frame, and the arms of which are initially parallel to the x- and y-directions, respectively. A spring parallel to the y-axis is fastened between the first arm and the body $\varepsilon_{\mathrm{III}}$, and a spring parallel to the x-axis acts between the second arm and the body ε_{I}. Let the tension of the springs be so chosen that the force which acts on the body $\varepsilon_{\mathrm{III}}$ in the y-direction and on the body ε_{I} in the x-direction during the time interval T_{I} is given by

$$\tfrac{1}{2}\rho_{\mathrm{I}}\sigma_{\mathrm{II}}V_{\mathrm{I}}V_{\mathrm{II}}T_{\mathrm{II}}(\bar{C}_{xy}^{(\mathrm{I},\,\mathrm{II})}+\bar{C}_{xy}^{(\mathrm{II},\,\mathrm{I})})(D_x^{(\mathrm{I})}-D_y^{(\mathrm{III})}).$$

Thus, the momentum balance of the two test body systems, after suitable rearrangement, may be expressed in the following form, analogous to (55)

$$(61)\quad \begin{cases} p_x^{(\mathrm{I})''} - p_x^{(\mathrm{I})'} = \rho_{\mathrm{I}} V_{\mathrm{I}} T_{\mathrm{I}} \bar{\mathfrak{E}}_x^{(\mathrm{I})} \\ +\frac{1}{2}\rho_{\mathrm{I}}\sigma_{\mathrm{II}} V_{\mathrm{I}} V_{\mathrm{II}} T_{\mathrm{I}} T_{\mathrm{II}} \{ -D_y^{(\mathrm{II})}(\bar{C}_{xy}^{(\mathrm{I},\,\mathrm{II})} - \bar{C}_{xy}^{(\mathrm{II},\,\mathrm{I})}) \\ +(D_y^{(\mathrm{II})} - D_y^{(\mathrm{III})})(\bar{C}_{xy}^{(\mathrm{I},\,\mathrm{II})} + \bar{C}_{xy}^{(\mathrm{II},\,\mathrm{I})}) + D_x^{(\mathrm{I})}(\bar{C}_{xy}^{(\mathrm{I},\,\mathrm{II})} + \bar{C}_{xy}^{(\mathrm{II},\,\mathrm{I})})\} \\ p_y^{(\mathrm{II})''} - p_y^{(\mathrm{II})'} + p_y^{(\mathrm{III})''} - p_y^{(\mathrm{III})'} = \sigma_{\mathrm{II}} V_{\mathrm{II}} T_{\mathrm{II}} \mathfrak{R}_y^{(\mathrm{II})} \\ +\frac{1}{2}\rho_{\mathrm{I}}\sigma_{\mathrm{II}} V_{\mathrm{I}} V_{\mathrm{II}} T_{\mathrm{I}} T_{\mathrm{II}} \{ D_x^{(\mathrm{I})}(\bar{C}_{xy}^{(\mathrm{I},\,\mathrm{II})} - \bar{C}_{xy}^{(\mathrm{II},\,\mathrm{I})}) \\ +(D_y^{(\mathrm{II})} - D_y^{(\mathrm{III})})(\bar{C}_{xy}^{(\mathrm{I},\,\mathrm{II})} + \bar{C}_{xy}^{(\mathrm{II},\,\mathrm{I})}) - D_y^{(\mathrm{II})}(\bar{C}_{xy}^{(\mathrm{I},\,\mathrm{II})} + \bar{C}_{xy}^{(\mathrm{II},\,\mathrm{I})})\}. \end{cases}$$

After compensation of the last terms we thus obtain for the uncertainties of the field averages:

$$(62)\quad \begin{cases} \Delta\bar{\mathfrak{E}}_x^{(\mathrm{I})} \sim \dfrac{\hbar}{\rho_{\mathrm{I}} \Delta x_{\mathrm{I}} V_{\mathrm{I}} T_{\mathrm{I}}} + \dfrac{1}{2}\sigma_{\mathrm{II}} \Delta y_{\mathrm{II}} V_{\mathrm{II}} T_{\mathrm{II}} |\bar{C}_{xy}^{(\mathrm{I},\,\mathrm{II})} - \bar{C}_{xy}^{(\mathrm{II},\,\mathrm{I})}| \\ \Delta\mathfrak{R}_y^{(\mathrm{II})} \sim \dfrac{\hbar}{\sigma_{\mathrm{II}} \Delta y_{\mathrm{II}} V_{\mathrm{II}} T_{\mathrm{II}}} + \dfrac{1}{2}\rho_{\mathrm{I}} \Delta x_{\mathrm{I}} V_{\mathrm{I}} T_{\mathrm{I}} |\bar{C}_{xy}^{(\mathrm{I},\,\mathrm{II})} - \bar{C}_{xy}^{(\mathrm{II},\,\mathrm{I})}|, \end{cases}$$

from which the minimal value of their product is obtained as

$$(63)\quad \Delta\bar{\mathfrak{E}}_x^{(\mathrm{I})} \Delta\mathfrak{R}_y^{(\mathrm{II})} \sim \hbar |\bar{C}_{xy}^{(\mathrm{I},\,\mathrm{II})} - \bar{C}_{xy}^{(\mathrm{II},\,\mathrm{I})}|,$$

again in complete agreement with the quantum electromagnetic formalism.

Furthermore, from the general arguments at the end of the previous section it follows that also in the case considered here the utilization of field measurements for testing the formalism's assertions in no way is impaired by the classical evaluation of the field effects. Besides, in measurements of averages of dissimilar field components the question of repeatability does not arise at all, and the pure black-body fluctuations are included in all theoretical assertions as an unavoidable statistical feature.

8. Concluding Remarks

We thus arrive at the conclusion already stated at the beginning, that with respect to the measurability question the quantum theory of fields represents a consistent idealization to the extent that we can disregard all limitations due to the atomic structure of the field sources and the measuring instruments. As already emphasized in the Introduction, this result should properly be regarded as an immediate consequence of the fact that both the quantum electromagnetic formalism and the viewpoints on which the possibilities of testing this formalism are to be assessed have as their common foundation the correspondence argument. Nevertheless, it would seem that the somewhat complicated character of the considerations used to demonstrate the agreement between formalism and measurability are hardly avoidable. For in the first place the physical requirements to be imposed on the measuring arrangement are conditioned by the integral form in which the assertions of the quantum-electromagnetic formalism are expressed, whereby the peculiar simplicity of the classical field theory as a purely differential theory is lost. Furthermore, as we have seen, the interpretation of the measuring results and their utilization by means of the formalism require consideration of certain features of the complementary mode of description which do not appear in the measurement problems of non-relativistic quantum mechanics.

At the completion of this work we should not like to leave unmentioned that we have found much stimulation and help in many discussions of the questions considered with past and present colleagues at the Institute, among them Heisenberg and Pauli as well as Landau and Peierls.

Universitetets Institut for teoretisk Fysik
Copenhagen, April 1933

NOTES

* Translated by Prof. Aage Petersen; revised by RSC and JS.

[1] W. Heisenberg, *The Physical Principles of the Quantum Theory*, transl. by C. Eckart and F. Hoyt (Dover, New York, 1930), pp. 42 ff.

[2] L. Landau and R. Peierls, *Zs. f. Phys.* **69** (1931), 56.

[3] Cf. N. Bohr, 'Atomic Stability and Conservation Laws', *Atti del Congresso di Fisica Nucleare* (1932). Added in the proof: A separate publication to appear shortly will contain a discussion of the consequences for the problems discussed in the cited reference implied by the recent discovery of the occurrence, under special circumstances, of so-called 'positive electrons'; and by the recognition of the connection of this discovery with Dirac's relativistic electron theory.

[4] Cf. L. Rosenfeld, *Zs. f. Phys.* **76** (1932), 729.
[5] Cf. P. Jordan and W. Pauli, *Zs. f. Phys.* **47** (1928), 151 and also W. Heisenberg and W. Pauli, *Zs. f. Phys.* **56** (1929), 33. Apart from an unessential difference in sign resulting from a difference in the choice of time direction in the Fourier decomposition of the field strengths, the formulae above are equivalent in content with those derived in the papers quoted. In particular, the notation used here, where all terms appear as retarded, is a purely formal change which aims at an interpretation of the measurement problems that is as intuitive as possible.
[6] Cf. L. Rosenfeld and J. Solomon, *J. de Physique* **2** (1931), 139 and also W. Pauli, *Handbuch d. Physik*, 2nd edition, Vol. 24/1 (1933), p. 255.
[7] Cf. W. Heisenberg, *Leipziger Berichte* **83** (1931), 1.
[8] Cf. V. Fock and P. Jordan, *Zs. f. Phys.* **66** (1930), 206, where reference is made to such restrictions on field measurements, which are unrelated to the quantum theory of fields. Cf. also J. Solomon, *J. de Physique* **4** (1933), 368.
[9] Cf. W. Pauli, *Handbuch d. Physik*, 2nd edition, Vol. 24/1 (1933), p. 257.
[10] See N. Bohr, *Atomic Theory and the Description of Nature* (Cambridge University Press, Cambridge, England, 1934). In the meantime, this question has been treated in more detail by the author in a guest lecture in Vienna, to appear shortly, in which in particular the paradoxes arising in the interpretation of the indeterminacy principle when account is taken of the requirements of relativity are further discussed.

CHAPTER 2

FIELD AND CHARGE MEASUREMENTS IN QUANTUM ELECTRODYNAMICS

[with N. Bohr]

[1950a]

ABSTRACT. A survey is given of the problem of measurability in quantum electrodynamics and it is shown that it is possible in principle, by the use of idealized measuring arrangements, to achieve full conformity with the interpretation of the formalism as regards the determination of field and charge quantities.

INTRODUCTION

Recent important contributions[1] to quantum electrodynamics by Tomonaga. Schwinger and others have shown that the problem of the interaction between charged particles and electromagnetic fields can be treated in a manner satisfying at every step the requirements of relativistic covariance. In this formulation, essential use is made of a representation of the electromagnetic field components on the one hand, and of the quantities specifying the electrified particles on the other, corresponding to a vanishing interaction between field and particles. The account of such interaction is subsequently introduced by an approximation procedure based on an expansion in powers of the nondimensional constant $e^2/\hbar c$. As regards the interpretation of the formalism, this method has the advantage of a clear emphasis on the dualistic aspect of electrodynamics. In fact, an unambiguous definition of the electromagnetic field quantities rests solely on the consideration of the momentum imparted to appropriate test bodies carrying charges or currents, while the charge-current distributions referring to the presence of particles are ultimately defined by the fields to which these distributions give rise.

Just from this point of view the problem of the measurability of field quantities has been discussed by the authors in a previous paper.[2] A similar investigation of the measurability of electric charge density was then also undertaken, but, owing to various circumstances, its publication has been delayed.[3] When recently the work was resumed, it appeared that by making use of the new development as regards the formulation of quantum electrodynamics a more general and exhaustive treatment could be obtained.[4] As these considerations my be helpful in the current discussion of the situation in atomic physics, we shall here give a brief account of the implications of present electron theory for measurements of charge-current densities. For this purpose, it will be convenient to start with a summary of our earlier treatment of the measurability of field quantities.[5]

1. Measurements of electromagnetic fields

Classical electrodynamics operates with the idealization of field components $f_{\mu\nu}(x)$ defined at every point (x) of space-time. Although in the quantum theory of fields these concepts are formally upheld, it is essential to realize that only averages of such field components over finite space-time regions R, like

$$(1) \qquad F_{\mu\nu}(R) = \frac{1}{R} \int_R f_{\mu\nu}(x)\, d^4x$$

have a well-defined meaning (I, §2). In the initial step of approximation, in which all effects involving $e^2/\hbar c$ are disregarded, these averages obey commutation relations of the general form

$$(2) \qquad [F_{\mu\nu}(R), F_{\kappa\lambda}(R')] = i\hbar c [A_{\mu\nu,\kappa\lambda}(R, R') - A_{\kappa\lambda,\mu\nu}(R', R)],$$

where the expressions of the type $A_{\mu\nu,\kappa\lambda}(R, R')$, defined as integrals over the space-time regions R and R' of certain singular functions, have finite values depending on the shapes and relative situation of the regions R and R'.

The measurement of a field average $F_{\mu\nu}(R)$ demands the control of the total momentum transferred within the space-time region R to a system of movable test bodies with an appropriate distribution of charge or current, of density ρ_ν, covering the whole part of space which at any time belongs to the region R. In the case of an electric field component F_{4l}, we shall take a distribution of charge, with constant density ρ_4, and in the case of a magnetic field component F_{mn}, a uniform distribution of current in a perpendicular direction, with density components ρ_m and ρ_n. The field action of such charge-current distribution, so far as it does not originate from the displacements of the test bodies accompanying the momentum control, can in principle be eliminated by the use of fixed auxiliary bodies carrying a charge-current distribution of opposite sign, and constructed in a way which does not hinder the free motion of the test bodies. In the case of a current distribution, such auxiliary bodies are even indispensable in providing closed circuits for the currents by means of some flexible conducting connection with the test bodies. As a result of this compensation, the field sources of the whole measuring arrangement will thus merely be described by a polarization $P_{\mu\nu}$ arising from the uncontrollable displacements of the test bodies in the course of the field measurements.

If the test bodies are chosen sufficiently heavy, we can throughout disregard any latitude in their velocities, but the control of their momentum will of course imply an essential latitude in their position, to the extent demanded by the indeterminacy relation. Still, it is possible, without violating any requirement of quantum mechanics, not only to keep every test body fixed in its original position except during the time interval within the region R corresponding to this position, but also to secure that, during such time intervals, the displacements of all test bodies in the direction of the momentum transfer to be measured, although uncontrollable, are exactly the same. This common displacement D_μ is described, in the case of the measurement of an electric field, by the component D_l parallel to the field component F_{4l}, and when a magnetic field is measured, by the components D_m and D_n perpendicular to F_{mn}. Without imposing any limit on the accuracy of the field measurement, it is, moreover, possible to keep the displacement D_μ arbitrarily small, if only the charge-current density ρ_ν of the test bodies is chosen sufficiently large. By a further refinement of the composite measuring arrangement described in our earlier paper (1, §3), it is even possible to reduce the measurement of any field average to the momentum control of a single supplementary body, and thus to obtain a still more compendious expression for the ultimate consequences of the general indeterminacy relation.

An essential point in field measurements is, however, the necessity of eliminating so far as possible the uncontrollable contribution to the average field present in R, arising from the displacement of the test bodies in the course of the measurement. In fact, the expectation value of this contribution will vary in inverse proportion to the latitude allowed in the field measurement, since it is proportional to the polarization $P_{\mu\nu} = D_\mu\rho_\nu - D_\nu\rho_\mu$ within the region R. Just this circumstance, however, makes it possible, by a suitable mechanical device, by which a force proportional to their displacement is exerted on the test bodies, to compensate the momentum transferred to these bodies by the uncontrollable field, insofar as the relation of this field to its sources is expressed by classical field theory. With the compensation procedure described, the resulting measurement of $F_{\mu\nu}(R)$ actually fulfils all requirements of the quantum theory of fields as regards the definition of field averages (I, §5). In fact, the incompensable part of the field action of the text bodies due to the essentially statistical character of the elementary processes involving photon emission and absorption, corresponds exactly to the characteristic field fluctuations which in quantum electrodynamics are superposed on all expectation values determined by the field sources.

When the measurement of two field averages $F_{\mu\nu}(R)$ and $F_{\kappa\lambda}(R')$ is considered, it appears (I, §4) that the expectation value of the average field component $\Phi_{\mu\nu,\kappa\lambda}(R, R')$ which the displacement of the test bodies operated in the region R produces in the region R' is equal to the product of $\frac{1}{2}RP_{\mu\nu}$ with the quantity $A_{\mu\nu,\kappa\lambda}(R, R')$ occurring in the commutation relation (2). Likewise, the expectation value of the average component $\Phi_{\kappa\lambda,\mu\nu}(R', R)$ of the field in R due to the test bodies in R' is equal to $\frac{1}{2}R'P'_{\kappa\lambda}A_{\kappa\lambda,\mu\nu}(R', R)$. When optimum compensation of the momenta transferred to the test bodies by these fields is established by suitable devices, making use of a correlation by light signals transmitted between points of the two regions R and R', it can be deduced from the reciprocal indeterminacy of position and momentum control that the only limitations of the measurability of the two field averages considered correspond exactly to the consequences of the commutation rule (2) for such averages (I, §6, 7). In this connection, it must be stressed that the field fluctuations which are inseparable form the incompensable parts of the fields created by the operation of the test bodies, do not imply any restriction in the measurability of a field component in two asymptotically coinciding space-time regions. In fact, we have here to do with a complete analog to the reproducibility of the fixation of observables in quantum mechanics by immediately repeated measurements.

2. Charge-Current Measurements in Initial Approximation

In the formalism of quantum electrodynamics, charge-current densities, like field quantities, are introduced by components $j_\nu(x)$ at every space-time point, but, even in the initial approximation in which such symbols are formally commutable, well-defined expressions are only given by integrals of the type

$$(3)\qquad J_\nu(R)=\frac{1}{R}\int\limits_R j_\nu(x)\,\mathrm{d}^4x,$$

representing the average charge-current density within the finite space-time region R. From the fundamental equations of electrodynamics it follows quite generally that

$$(4)\qquad RJ_\nu(R)=\int\limits_R \frac{\partial f_{\nu\mu}}{\partial x_\mu}\,\mathrm{d}^4x=\int\limits_S f_{\nu\mu}\,\mathrm{d}\sigma_\mu,$$

which expresses the definition of the average charge-current density over the region R in terms of the flux of the electromagnetic field through the boundary S of this region. In this four-dimensional representation, such generalized fluxes comprise, of course, besides the ordinary electric field flux defining the average charge density, other expressions pertaining to the average current densities and representing magnetic field circulations and displacement currents.

In the simple special case in which the region R is defined by a fixed spatial extension V and a constant time interval T, the average charge density, in accordance with (4), will be given, in the ordinary vectorial representation, by

$$(5)\qquad J_4(V,T)=\frac{1}{VT}\int\limits_T \mathrm{d}t\int\limits_S \mathbf{E}\mathbf{n}\,\mathrm{d}\sigma,$$

where S is the surface limiting the extension V, and $\mathbf{n}$ the unit vector in the outward normal direction on this surface. In such representation, the average current density will be given by

$$(6)\qquad \mathbf{J}(V,T)=\frac{1}{VT}\int\limits_T \mathrm{d}t\int\limits_S \mathbf{n}\wedge\mathbf{H}\,\mathrm{d}\sigma-\frac{1}{VT}\int\limits_V \mathbf{E}\,\mathrm{d}v\,\Big|_{t_1}^{t_2},$$

where the first term on the right-hand side represents the time integral of the tangential component of the magnetic field integrated over the surface S, while the last term expresses the difference of the volume integrals of the electric field at the beginning and at the end of the time interval T.

The determination of an average charge-current density $J_\nu(R)$ thus demands the measurement of a field flux through the boundary S of the space-time region R. The approach to the problem of such measurement must rationally start from the consideration of the average flux over a thin four-dimensional shell situated at the boundary S, and which for simplicity we shall assume to have a constant thickness in space-time. As in the situation met with in the measurement of an average field component $F_{\mu\nu}(R)$, we shall require for this purpose a system of movable test bodies, filling the space which belongs to the shell at any time with an appropriate uniform charge-current distribution, and whose field actions are ordinarily neutralized by a distribution of opposite sign on fixed, penetrable, auxiliary bodies. For the measurement of an average charge density J_4, it suffices to take a set of test bodies with a uniform charge distribution of density ρ_4, while in the measurement of a current component, J_l, we shall have to use, besides such test bodies, another independent set of freely movable test bodies with a uniform current distribution ρ_l parallel to the current component to be measured.

In the measurement of an average charge density, the estimation of the flux over the shell demands the determination of the algebraic sum of the momenta transferred to the test bodies in the direction of the normal to the instantaneous spatial boundary. The evaluation of this sum, however, does not require independent measurements of the momenta transferred to the individual test bodies within the time intervals during which their positions belong to the space-time shell, but can be obtained by a composite measuring process in which the positions of all test bodies are correlated by suitable devices to secure during these intervals a displacement of every test body in the normal direction by the same amount. By choosing the product of the thickness of the shell and the charge density of the test bodies sufficiently large, it is possible to keep the uncontrollable common displacement D of all the test bodies in the normal direction arbitrarily small, and still to obtain unlimited accuracy for the average flux over the shell. Like in the measurement of a simple field average, it is further possible to achieve an automatic compensation of the uncontrollable contribution to this average flux, due to the fields created by the displacement of the test bodies, and proportional to $D\rho_4$. This compensation will even be complete, in the initial approximation considered, because the field fluctuations, owing to their source-free character, do not give any contribution to the flux. Since these considerations hold for any given thickness of the shell, it is in principle possible, in the asymptotic limit of a sharp boundary, to measure accurately the average charge density within a well-defined space-time region.

In measurements of an average current component J_l, we have to take into account the magnetic circulation as well as the electric field in the space-time shell. Thus, in the special case in which R is defined by a spatial extension V and a time interval T, we have to do, according to (6), not only with a contribution from the time average over T of the magnetic circulation around the direction l within a thin spatial shell on the boundary of V, but also with a contribution representing the difference between the volume integrals over V of the electric field component in the direction l, averaged over two short time invervals at the beginning and at the end of the interval T. The evaluation of these contributions requires measuring procedures of a similar kind as those described above in the case of measurements of simple field averages. While the measurement of the latter contribution demands the control of the momentum in the direction l transferred to a set of test bodies with uniform charge density ρ, the evaluation of the former contribution demands the control of the momentum normal to the spatial boundary transferred to another set of test bodies with uniform current density ρ_l.

Just as in the field or change measurements discussed above, all these operations can be correlated in such a way that the determination of the algebraic sum of the momenta transferred to each test body within the time interval and in the direction required can be reduced to the momentum control of some supplementary body. In such a correlation, all the test bodies of charge density ρ will be subjected during the appropriate time intervals to the same displacement D_l and all the test bodies of current density ρ_l to the same normal displacement D. The interpretation of the current measurement requires further the establishment of a correlation between these two displacement, satisfying the condition $\rho D_l = \rho_l D$. Under such circumstances, it is possible, by choosing ρ and ρ_l sufficiently large, to achieve that the displacements D_l and D be arbitrarily small without imposing any limitation upon the accuracy of the measurement. Moreover, it is possible, by suitable mechanical devices of the kind already mentioned, to obtain a complete automatic elimination of the uncontrollable contributions from the operation of the test bodies to the average current to be measured.

It need hardly be added that the procedure can be extended to quite general space-time regions R, by using an arrangement in which each test body is displaced just in the time interval during which its position belongs to the space-time shell surrounding the region R. In this connection, it may be noted that a compendious four-dimensional description of all the measuring processes pertaining to charge-current components involves a uniform four-vector current distribution in the shell, parallel to the charge-current component to be measured.

Like in charge measurements, all the considerations concerning current measurements are independent of the thickness of the shell, and in principle it is therefore possible, in the initial approximation considered, to determine with unlimited accuracy any average charge-current component $J_\nu(R)$ within a sharply bounded region R. As regards charge-current measurements over two space-time regions, it can easily be seen that, in the limiting case of sharp boundaries, all field actions accompanying the flux measurements will vanish at any point of space-time which does not belong to the boundaries. In conformity with the formalism, there will therefore, to the approximation concerned, be no mutual influence of measurements of average charge-current densities in different space-time regions.

The situation so far described is of course merely an illustration of the compatibility of a consistent mathematical scheme with a strict application of the definition of the physical concepts to which it refers, and is in particular quite independent of the question of the possibility of actually constructing and manipulating test bodies with the required properties. The disregard of all limitations in this respect, which may originate in the atomic constitution of matter, is, however, entirely justified when dealing with quantum electrodynamics in the initial stage of approximation. In fact, at this stage, the formalism is essentially independent of space-time scale, since it contains only the universal constants c and $\hbar$ which alone do not suffice to define any quantity of the dimensions of a length or time interval.

3. Charge-current measurements in pair theory

New aspects of the problem of measurements arise in quantum electrodynamics in the next approximation, in which effects proportional to $e^2/\hbar c$ are taken into consideration, and where we meet with additional features connected with electron pair production induced by the electromagnetic fields. For the commutation rules of the field components, this means in general only a smaller modification expressed by additional terms containing $e^2/\hbar c$. The charge-current quantities, however, will no longer be commutable but will obey commutation relations of the form.

$$[J_\nu(R), J_\mu(R')] = i\hbar c[B_{\nu\mu}(R, R') - B_{\mu\nu}(R', R)], \tag{7}$$

where the expressions $B_{\nu\mu}(R, R')$ are integrals of singular functions over the regions R and R'. In contrast to the quantities $A_{\mu\nu,\kappa\lambda}(R, R')$ occurring in (2), which depend only on simple spatio-temporal characteristics of the problem, the B's will, however, besides such characteristics, also essentially involve the length $\hbar/mc$ and the period $\hbar/mc^2$, related to the electron mass m.

To approach the problem of the measurability of a charge-current quantity $J_\nu(R)$ in this approximation, we must again consider systems of electrified test bodies operated in a space-time shell on the boundary of the region R, but we shall now have to examine the effect of the charge-current density appearing as a consequence of actual or virtual electron pair production by the field action of the displacement of the test bodies during the measuring process. As we shall see, these effects, which are inseparably connected with the measurements, do not in any way limit the possibilities of testing the theory.[6]

In the first place, the average effect of the polarization of the vacuum by virtual and actual pair production in the measuring process can be eliminated by a compensation arrangement like that previously described. It is true that a direct estimate of these polarization effects in quantum electrodynamics involves divergent expressions which can only be given finite values by some renormalization or regularization procedure.[7] By such a procedure the average polarization effects will give rise to a contribution to the charge current density which is proportional to the common displacement of the test bodies. Thus in the limit of a sharp boundary of the region R we get, denoting the surface polarization on the boundary by P_ν, the expression $RP_\nu B_{\mu\nu}(R, R)$, where the last factor represents the value of $B_{\mu\nu}(R, R')$ in (7) for coinciding space-time extensions.

Moreover, the statistical effects caused by actual production of electron pairs in the measurement process are inseparably connected with the interpretation of the fluctuations of average charge-current densities in quantum electrodynamics. While the mean square deviation of the field component $F_{\mu\nu}(R)$ over a sharply bounded space-time region R has a finite value, finite mean-square fluctuations of charge-current quantities can only be obtained, however, by further averaging over an ensemble of regions R whose boundaries are allowed a certain latitude around some given surface.[8]

This feature finds its exact counterpart in the estimate of the statistical effects of the real pairs which are produced in measurements of charge-current quantities by the indicated procedure. In fact, the mean square fluctuation of an average flux will increase indefinitely with decreasing thickness of the shell in which the test bodies are operated, in just the same way as, according to the formalism, the mean-square fluctuation of the corresponding charge-current density will vary with the latitude of the ensemble of space-time extensions over which the averaging is performed. The appearance of an infinite mean-square fluctuation in a sharply limited space-time region is in no way connected with the divergencies which appear in vacuum polarization effects but is a direct consequence of the fundamental assumptions of the theory, according to which the electrons are regarded as point charges.

In the case of measurements of charge-current averages over two space-time regions, it can be shown that the polarization effects of the manipulation of the test bodies used for the measurement of $J_\nu(R)$ will give rise, in the limit of sharp boundaries, to a contribution to the average charge-current density component of index μ in the region R', equal to the product of the quantity $B_{\mu\nu}(R', R)$ occurring in formula (7) with RP_ν, where P_ν is the surface polarization created on the boundary of R during the measuring process. Conversely, the measurement of $J_\mu(R')$ will give a contribution $R'P'_\mu B_{\nu\mu}(R, R')$ to the average charge-current density of index ν in R. By similar compensation devices as required for two field measurements, it is therefore possible, as readily seen, to obtain an accuracy of measurements of average charge-current densities in two space-time regions subject only to the reciprocal limitation expressed by the commutation relation (7).

4. Concluding remarks

The conformity of the formalism of quantum electrodynamics with the interpretation of idealized field and charge measurements has of course no immediate relation to the question of the scope of the theory and of the actual possibility of measuring the physical quantities with which it deals.

In the present state of atomic physics, the problem of an actual limitation of measurements interpreted by means of the concepts of classical electrodynamics can hardly be fully explored. Still, in view of the great success of quantum electrodynamics in accounting for numerous phenomena, the formal interpretation of which involves space-time coordination of electrons within regions of dimensions far smaller than $\hbar/mc$ and $\hbar/mc^2$, it may be reasonable to assume that measurements within such regions are in principle possible Indeed, the comparatively heavy and highly charged test bodies of such small dimensions and operated over such short time intervals, which would be required for these measurements, might be conceived to be built up of nuclear particles.

Yet, an ultimate limitation of the consistent application of the formalism is indicated by the necessity of introducing forces of short range in nuclear theory, with no analog in classical electrodynamics, and by the circumstance that the ratio between the electron mass and the rest mass of the quanta of the nuclear field has the same order of magnitude as the fundamental parameter $e^2/\hbar c$ of quantum electrodynamics.[9] The further exploration of such problems may, however, demand a radical revision of the foundation for the application of the basic dual concepts of fields and particles.

NOTES

[1] S. Tomonaga, *Prog. Theor. Phys.* **1** (1946), 27; *Phys. Rev.* **74** (1948), 224. J. Schwinger, *Phys. Rev.* **74** (1948), 1439; **75** (1949), 651; **75** (1949), 1912; **76** (1949), 790. F. Dyson, *Phys. Rev.* **75** (1949), 486; **75** (1949), 1736. R. Feynman, *Phys. Rev.* **76** (1949), 749; **76** (1949), 769.
[2] N. Bohr and L. Rosenfeld, *Kgl. Danske Vid. Seis., Math.-fys. Medd.* **12** (1933), No. 8. This paper will be referred to in the following as I. [English translation: this volume, p. 357].
[3] An account of the preliminary results of the investigation, which were discussed at several physical conferences in 1938, has recently been included in the monograph by

A. Pais, *Developments in the Theory of the Electron*, Princeton University Press, Princeton, New Jersey, 1948.

[4] The bearing of this development on the elucidation of the problem of measurability was brought to the attention of the writers in a stimulating correspondence with Professor Pauli.

[5] A more detailed account of the subject with fuller references to the literature will appear later in the Communications of the Copenhagen Academy.

[6] In a paper by Halpern and Johnson, *Phys. Rev.* **59** (1941), 896, arguments are brought forward pointing to a far more restrictive limitation of the field and charge measurements. In these arguments, however, no sufficient separation is made between such actions of the charged test bodies as are directly connected with their use in the measuring procedure and those actions which can be eliminated by appropriate neutralization by auxiliary bodies of opposite charge.

[7] Cf. W. Pauli and F. Villars, *Rev. Mod. Phys.* **21** (1949), 434.

[8] Cf. W. Heisenberg, *Leipziger Ber.* **86** (1934), 317. We are indebted to Drs. Jost and Luttinger for information about their more precise evaluation of charge-current fluctuations, showing that the unlimited increase of the charge-current fluctuations in a space-time region with decreasing latitude in the fixation of its boundary involves only the logarithm of the ratio between the linear dimensions of the region and the width of this latitude. Even a latitude very small compared with $\hbar/mc$ will therefore imply no excessive effect of the charge fluctuations. A situation entirely similar in all such respects to that in electron theory is met with in a quantum electrodynamics dealing with electrical particles of spin zero which obey Bose statistics. We are indebted to Dr. Corinaldesi for the communication of his results regarding the charge-current fluctuations and pair production effects in such a theory.

[9] Cf., e.g., N. Bohr, *Report of the Solvay Council* (1948).

CHAPTER 3

ON QUANTUM ELECTRODYNAMICS

[*Among Essays Dedicated to Niels Bohr on the Occasion of his 70th Birthday*]

[1955b]

... et discors concordia fetibus apta est.
OVID, *Metam.* I, 433

When I arrived at the Institute on the last day of February, 1931, for my annual stay, the first person I saw was Gamow. As I asked him about the news, he replied in his own picturesque way by showing me a neat pen drawing he had just made[1]. It represented Landau, tightly bound to a chair and gagged, while Bohr, standing before him with upraised forefinger, was saying: "Bitte, bitte, Landau, muss ich[2] nur ein Wort sagen!" I learned that Landau and Peierls had just come a few days before with some new paper of theirs which they wanted to show Bohr, "but" (Gamow added airily) "he does not seem to agree – and this is the kind of discussion which has been going on all the time." Peierls had left the day before, "in a state of complete exhaustion," Gamow said. Landau stayed for a few weeks longer, and I had the opportunity of ascertaining that Gamow's representation of the situation was only exaggerated to the extent usually conceded to artistic fantasy.

There was indeed reason for excitement, for the point raised by Landau and Peierls [1] was a very fundamental one. They questioned the logical consistency of quantum electrodynamics by contending that the very concept of electromagnetic field is not susceptible, in quantum theory, to any physical determination by means of measurements. The measurement of a field component requires determinations of the momentum of a charged test-body; and the reaction from the field radiated by the test-body in the course of these operations would (except in trivial cases) lead to a limitation of the accuracy of the field measurement, entirely at variance with the premises of the theory. In fact, the quantization of the field only entails reciprocal limitations of the measurements of pairs of components, arising from their noncommutability, but no limitation whatsoever to the definition of any single field component. On the other hand, one had to face another inescapable consequence of the field quantization, the occurrence of irregular fluctuations in the value of any component; the existence of this fluctuating "zero-field" (as it was called because it persists even in a vacuum) was known to be responsible for one of the divergent contributions to the self-energy of charged particles, but its meaning was very obscure. Landau and Peierls, somewhat illogically, tried to bring it in relation with their alleged limitation of measurability of the field. and this only further confused an already tangled issue.

1. Measurability of electromagnetic fields[3]

Bohr's state of mind when he attacked the problem reminded me of an anecdote about Pasteur [3]. When the latter set about investigating the silkworm sickness, he went to Avignon to consult Fabre. "I should like to see cocoons," he said, "I have never seen any, I know them only by name." Fabre gave him a handful: "he took one, turned it between his fingers, examined it curiously as we would some singular object brought from the other end of the world. He shook it near his ear. 'It rattles,' he said, much surprised, 'there is something inside.'" My first task was to lecture Bohr on the fundamentals of field quantization; the mathematical structure of the commutation relations and the underlying physical assumptions of the theory were subjected to unrelenting scrutiny. After a very short time, needless to say, the roles were inverted and he was pointing out to me essential features to which nobody had as yet paid sufficient attention.

His first remark, which threw decisive light on the problem, was that field components taken at definite space-time points are used in the formalism as idealizations without immediate physical meaning; the only meaningful statements of the theory concern averages of such field components over finite space-time regions. This meant that in studying the measurability of field components we must use as test-bodies finite distributions of charge and current, and not point charges as had been loosely done so far. The consideration of finite test-bodies immediately disposed of Landau and Peierls' argument concerning the perturbation of the momentum measurements by the radiation reaction: it is easily seen that this reaction is so much reduced, for finite test-bodies, as to be always negligible.

On the other hand, the construction and manipulation of extended test-bodies proved a most perplexing affair. To get a rough idea of the kind of problems that had to be faced, let us just consider the measurement of the electric field component E_x averaged over a volume V and a time interval T. We take a test-body filling the volume V with uniform density ρ and we measure its momentum p'_x, p''_x at the beginning and the end of the time interval T. By making the test-body sufficiently heavy, we may arbitrarily reduce its displacement during this interval, and we thus get for the average $\bar{E}_x$.

$$\text{(1)} \qquad \bar{E}_x \cdot \rho V T = p''_x - p'_x.$$

The momentum determinations of the test-body, however, entail a sacrifice of the knowledge of its position to a certain extent Δx, and the resulting error $\Delta\bar{E}_x$ on $\bar{E}_x$ will be, on account of (1) and the indeterminacy relation,

$$\text{(2)} \qquad \Delta\bar{E}_x \approx \frac{\hbar}{\rho\Delta x \cdot VT}.$$

This shows that we can reduce the indeterminacy of the field average indefinitely by increasing the charge density of the test-body.

Here we meet with a question of principle directly affecting the logical structure of the quantum theory of fields. So long as we treat all sources of electromagnetic fields as classical distributions of charge and current, and only quantize the field quantities themselves, no universal scale of space-time dimensions is fixed by the formalism. It is then consistent to disregard the atomistic structure of the test-bodies and there is no restriction to the logically admissible values of the charge density. It remains to be seen whether it is legitimate to isolate this theory of quantized fields interacting with classical sources from the more elaborate formalism in which it is attempted to treat the interaction of electromagnetic and material fields, both subjected to the appropriate quantization. Such a question could of course not be decided by formal considerations, since the mathematical consistency of the formalism was then very much in doubt. Bohr, however, went straight to the root of the matter by reminding us that the very formulation of the fundamental quantum postulates implies a neglect of all radiative couplings, which is only justified by the actual smallness of the fine structure constant. The natural approach to field quantization can therefore only be regarded as consisting of a succession of steps, in which the effects of radiative interactions are gradually taken into account: the first step is just the quantization of the pure radiation field, the next step the quantization of the field associated with the charged particles. Such a step-wise analysis of the content of the formalism is thus not merely a matter of practical convenience, but one of logical necessity. It is but one aspect of the situation we meet everywhere in quantum theory: the physical interpretation of the quantal formalism must be based in the last resort upon classical idealizations.

The formula (2) was encouraging insofar as it showed that the average field acting on a finite test-body can be accurately determined; but what is the relation of the field measured in this way to that produced by the given source distribution? The test-body itself is added to these sources, and in how far does the reaction of its own field influence the result of the measurement? This was indeed the central question, which could only be answered by a minute analysis of the behaviour of the test-body in the measuring process. The first thing to do was to devise a mode of measurement in which the perturbing influence of the motion of the test-body would be reduced as much as possible. Outside the time interval T a complete neutralization is achieved by superposing to the test-body another body of opposite charge, and the time-intervals Δt during which the test-body is in motion, at the beginning and end of the interval T, can be made arbitrarily small. What remains is thus an essentially uncontrollable displacement D_x of the test-body, constant during the whole time-interval T.

It was then an easy matter to work out, at any rate to the classical approximation, the field produced by such a distribution of electric polarization of density ρD_x subsisting during a time T, and the momentum imparted by this field to another test-body during a given time-interval. It was highly gratifying to find that the expression for the latter quantity coincides in form with either of the terms, symmetrical with respect to the two space-time regions considered, whose difference forms the commutator of the corresponding pair of average field components.

This formal relationship may be presented in full generality as follows: Let $D_r(x-x')$ be the Green's function expressing the retarded potential of a point source. The commutation relations between field components may be written

$$(3)\qquad [f_{\mu\nu}(x), f_{\kappa\lambda}(x')] = i\hbar c\, a_{\mu\nu,\kappa\lambda}(x, x')\, D(x-x'),$$

with the notation

$$(4)\qquad a_{\mu\nu,\kappa\lambda}(x, x') = a_{\kappa\lambda,\mu\nu}(x', x) \equiv \left(\delta_{\nu\lambda}\frac{\partial}{\partial x'_\kappa} - \delta_{\nu\kappa}\frac{\partial}{\partial x'_\lambda}\right)\frac{\partial}{\partial x_\mu} - \left(\delta_{\mu\lambda}\frac{\partial}{\partial x'_\kappa} - \delta_{\mu\kappa}\frac{\partial}{\partial x'_\lambda}\right)\frac{\partial}{\partial x_\nu}$$

and

$$(5)\qquad D(x-x') = -D_r(x-x') + D_r(x'-x).$$

For the average field components over space-time regions R, R',

$$(6)\qquad F_{\mu y}(R) = \frac{1}{R}\int_R f_{\mu\nu}(x)\, d^4x,$$

this becomes

$$(7)\qquad [F_{\mu\nu}(R), F_{\kappa\lambda}(R')] = i\hbar c\,[A_{\mu\nu,\kappa\lambda}(R, R') - A_{\kappa\lambda,\mu\nu}(R', R)],$$

with

$$(8)\qquad A_{\mu\nu,\kappa\lambda}(R, R') = \frac{1}{RR'}\int_R d^4x \int_{R'} d^4x'\, a_{\mu\nu,\kappa\lambda}(x, x')\, D_r(x'-x).$$

Now, the measurement of $F_{\mu\nu}(R)$ involves test-bodies of charge-current density ρ_ν, whose various elements $\rho_\nu d^4x$ undergo displacements D_μ which, though unknown, may be made equal; there results an electric or magnetic polarization, uniform over the region R, of density

$$P_{\mu\nu}(x'') = P_{\mu\nu}\int_R \delta(x-x'')\, d^4x$$

with

$$P_{\mu\nu} \equiv D_\mu\rho_\nu - D_\nu\rho_\mu.$$

This corresponds to a charge-current density

$$-\frac{\partial P_{\mu\nu}(x'')}{\partial x''_\mu} = P_{\mu\nu}\int_R d^4x \frac{\partial}{\partial x_\mu}\,\delta(x-x'')$$

which gives rise to a potential

$$A_\nu(x') = P_{\mu\nu} \int d^4x'' D_r(x' - x'') \int_R d^4x \frac{\partial}{\partial x_\mu} \delta(x - x'')$$

$$= P_{\mu\nu} \int_R d^4x \frac{\partial}{\partial x_\mu} D_r(x' - x).$$

The corresponding field component $\phi_{\kappa\lambda}(x')$ may be written

$$\phi_{\kappa\lambda}(x') = P_{\mu\nu} \int_R d^4x \left(\delta_{\nu\lambda} \frac{\partial}{\partial x'_\kappa} - \delta_{\nu\kappa} \frac{\partial}{\partial x'_\lambda} \right) \frac{\partial}{\partial x_\mu} D_r(x' - x)$$

$$= \tfrac{1}{2} P_{\mu\nu} \int_R d^4x \, a_{\mu\nu,\kappa\lambda}(x, x') \, D_r(x' - x),$$

and its average over the region R' is accordingly

(9) $$\Phi_{\mu\nu,\kappa\lambda}(R, R') = \tfrac{1}{2} P_{\mu\nu} R A_{\mu\nu,\kappa\lambda}(R, R').$$

It will be noticed that the coincidence between the quantities occurring in the formulae (9) and (7) has its origin in the relation expressed by formula (5), between the commutator and the Green's function. We shall return later to this remarkable relation which we are now able to understand a bit better than was then possible. At the time, the expressions of the type (9) found for the interactions between the test-bodies gave us the hope that we were on the right track; in fact, it seemed as if we had hit upon a mode of measurement exactly suited to give the best combined accuracy, for pairs of field determinations, compatible with the theoretical limitation. We little imagined that it would still take us almost two years to reach that goal.

At first, all went well. We had a lively surprise when we found out that the only case of coupled measurements for which a reciprocal indeterminacy relation had been explicitly written down and discussed in the literature, namely that of an electric field and a perpendicular magnetic field in the same volume element, was one in which unlimited accuracy had to be expected from the correctly integrated commutation law. It was immediately clear that in such a case the mutual perturbations of our extended test-bodies would cancel out and that we could fulfil the theoretical prediction. More exciting was the realization that the reaction of the test-body upon itself in the course of the measurement of any single field component could be automatically compensated, at least so far as it can be classically computed. In fact, this reaction, being proportional to the displacement of the test-body, can be matched exactly by an elastic spring mechanism of known strength.

I must not by-pass the problem of the actual momentum measurement of the extended test-bodies. To take account of relativistic effects, these bodies must be imagined to consist of a large number of independent elements, and it was far from obvious how the total momentum of all these elements could be obtained without multiplying the error far beyond the optimum limit which we required. Moreover, one had to make sure that the relativity requirements could be met without further restriction to the measurability of the momentum. This necessitated a much more detailed analysis of the measuring process than one was wont to in ordinary quantum mechanics. Bohr succeeded in showing that the measurement of the total momentum can even be performed in such a way that the displacements of the elements, though uncontrollable within a finite latitude Δx, are all equal, and that the determination of the total momentum is only limited by the uncertainty of the common displacement Δx to the extent $\hbar/\Delta x$ indicated by the indeterminacy relation. The interest of this result transcends its immediate application to the problem of field measurement: it affords a specially clear example of a measuring process which can be entirely described in a purely classical way, and in which the origin of the reciprocal indeterminacies is thus directly traced to the impossibility of specifying the dynamical characteristics of the system without loosening its connection with the space-time frame of reference[4]. The solution of this problem is one of the most striking products of Bohr's uncanny virtuosity in this subtle kind of analysis. It cost him some hard thinking, but was for him a source of intense enjoyment.

Having satisfied ourselves of the possibility of defining with unlimited accuracy the space-time average of any single field-component, we hopefully proceeded to the discussion of a pair of such measurements. The investigation, however, was very soon brought to an apparent deadlock by a difficulty of a most baffling nature. What precipitated the crisis was the minus sign on the right-hand side of formula (7) for the commutator of the field averages. This harmless-looking feature revealed itself a most crucial and excruciating riddle. It was child's play, with our arsenal of test-bodies and springs, to reduce the product of the indeterminacies $\Delta F_{\mu\nu} \cdot \Delta F_{\kappa\lambda}$ to the *sum*

$$\hbar c\left[|A_{\mu\nu,\kappa\lambda}(R, R')| + |A_{\kappa\lambda,\mu\nu}(R', R)|\right].$$

But how could one ever hope to get down to the *difference*

$$|A_{\mu\nu,\kappa\lambda}(R, R') - A_{\kappa\lambda,\mu\nu}(R', R)|?$$

At this juncture, Bohr did not hesitate to challenge the commutation rules themselves. Repeatedly we went through every step of their derivation, on the look-out for possible ambiguities. I was much impressed by this sacrilegious attitude, revealing as it did the sovereign independence of a powerful mind and its readiness to submit to any well-founded conclusion, however at variance with common expectation. Eventually, we noticed that in the troublesome cases, i.e. those in which none of the two terms of the difference vanishes, so that there is a mutual perturbation of the two measurements, it is also possible to exchange light signals between the two regions R, R'. Closer examination soon showed that the 'messages' which could thus be automatically conveyed by the test-bodies contained just enough information about their respective displacements to conjure up the difference predicted by the theory. It is very striking indeed to see how the greatest accuracy compatible with the commutation laws can only be achieved by exploiting to the utmost the possibilities, afforded by the physical situation, of controlling the course of the measuring process.

The realization of such a complete harmony between the formalism and its physical interpretation was felt as a fitting reward for our tribulations, and with great alacrity Bohr presented our conclusions and handed in the manuscript of our paper at a meeting of the Danish Academy on the 2nd December, 1932. The reading of the fourteen or so successive proofs only took about one more year. One point especially still gave us much trouble to the very last: what part has one to assign to the field fluctuations in the logical structure of the theory? This is a quite fundamental question, which I have kept to the last for some comment.

The self-reactions and mutual perturbations of the test-bodies can only be compensated by spring mechanisms to the extent of their classically evaluated average magnitudes. Their actual values, however, owing to the quantal character of the electromagnetic field, deviate from the classical averages to an extent which necessarily escapes our control: for such a control would involve a determination of the actual numbers of photons emitted in the various modes of oscillation and accordingly entail the loss of our knowledge of the phase relationships between these modes. Such statistical fluctuations accompanying any field defined by classical sources have, however, a universal character: in fact, they are just the zero-field fluctuations. This may be seen in the following way:

The radiation part of any field may be represented, in the usual way, in the form

$$(10) \qquad F = \sum_i (a_i f_i + a_i^\dagger f_i^*),$$

in terms of the annihilation and creation amplitudes a_i, $a_i^\dagger$ belonging to the different modes labelled by the index i. We may take formula (10) to represent the space-time average of some field component over a given space-time region; the f_i, f_i^* will then be the corresponding space-time integrals over the appropriate progressive phase functions. If the field is defined by a classical source distribution, the expectation values $\langle a_i \rangle$, $\langle a_i^\dagger \rangle$ are given in modulus and phase:

$$(11) \qquad \langle a_i \rangle = \bar{N}_i^{1/2} e^{i\chi_i}, \ \langle a_i^\dagger \rangle = \bar{N}_i^{1/2} e^{-i\chi_i};$$

in this notation, $\bar{N}_i$ represents the average number of photons of mode i.

Now, we can readily set up the expression for the state-vector describing this situation. In fact, since the emission processes which build up the radiation field are statistically independent events, the actual numbers N_i of photons in the different modes are distributed around the averages $\bar{N}_i$ according to the Poisson formula

$$(12) \qquad p(N_i) = (N_i!)^{-1} \bar{N}_i^{N_i} e^{-\bar{N}_i}$$

and the state-vector is of the form

$$(13) \qquad \Psi(N_1 N_2 \ldots) = \Pi \phi(N_i)$$

with

$$|\phi(N_i)|^2 = p(N_i).$$

In order to obtain the expressions (11) for the average amplitudes, it suffices to fix the phases of the factors $\phi(N_i)$ by

$$(14) \qquad \phi(N_i) = |\phi(N_i)|\, e^{iN_i\chi_i}.$$

Using the state-vector (13), (14), a straightforward calculation shows that the mean square fluctuation of F,

$$\langle F^2 \rangle - \langle F \rangle^2,$$

reduces *exactly* to the expression $\sum_i |f_i|^2$, quite independent of the $\bar{N}_i$'s and χ_i's.

The Poisson distribution of photons has lately been established by very elegant calculations utilizing the new methods of quantum electrodynamics [4]. In those days, however, people were not so learned. Judging from the surprise which the formula (12) excited in those who saw it, I think Bohr and, no doubt, Pauli must have been the only ones who knew that this property is an inherent part of the photon idea.

The property just discussed of the quantal field fluctuations gives the clue to the assessment of their significance for the consistency of the theory. In fact, it becomes clear that the fluctuations affecting the fields produced by the test-bodies during the measurements cannot be separated from those which equally affect the field which we want to measure. The impossibility of compensating or controlling them in any way, far from being an imperfection of the measuring device, is a property which it must necessarily possess to ensure that all the consequences of the theory are in principle verifiable by measurement. The fact that the zero-field fluctuations are superposed on to the classical field distribution is indeed a well-defined theoretical prediction, and we see that we are able to suppress the perturbations arising from the manipulation of the test-bodies to an extent which just leaves scope for the test of this prediction.

Now the existence of such fluctuations certainly means that the causal connection of classical theory between the fields and their sources is lost in quantum theory: this is just one aspect of the breakdown of determinism and its replacement by a wider form of statistical causality. This issue, however, should not be confused with the question whether the field concept can be consistently upheld in the quantal mode of description. Our analysis of field measurability shows that any measurement of a field average will always give a definite answer, which will be reproducible (this means that measurements of the same field component over two almost coinciding space-time regions will yield the same answer): but the result of any such measurement may differ by any amount from its classical estimate. The breakdown of the classical connection between field and source is a feature of the quantum theory of fields just as fundamental as the reciprocal limitation of measurability of field components; both are in fact direct and formally quite parallel consequences of the commutation laws which express the quantization of the field.

The classical solution of the field equations, expressed in terms of the retarded potentials corresponding to the given charge and current distribution, does not satisfy the commutation rules. Owing to the linear character of the equations, however, it can always be supplemented by a solution F_0 of the type (10) of the homogeneous equations, which represents a pure radiation field due to far-away sources not included in the system. If there is no such external source at infinity, the radiation field F_o would in classical theory reduce to zero, but in quantum theory it does not vanish identically. The condition $a_i^{\dagger} a_i = 0$ expressing that the field does not contain any photon, implies that the expectation value of each component of F_o vanishes everywhere, but any space-time average of the field component will exhibit fluctuations with a finite quadratic mean

$$\langle F_o^2 \rangle = \sum_i |f_i|^2 .$$

This leads us again to our previous conclusion, but the present argument shows how essentially it is linked to the characteristic noncommutability of the quantal variables. Indeed, one may say that the existence of the zero-field immediately follows from the additional requirement imposed on the solutions of the field equations by the commutation laws.

At the end of our laborious inquiry, we had thus completely vindicated the consistency of quantum electrodynamics, at least in its simplest form. Our increased insight invited a re-assessment of the scope of the analysis we had just completed. We had set out on the suspicion of grave defects in the logical structure of the theory, and used the direct method of testing definitions of concepts by investigating the concrete measuring processes they embody. Knowing, however, that the formalism is free from logical flaws and that the physical interpretation of its symbols is in strict accordance with their relation to observable quantities, it becomes tautological to assert the possibility of constructing measuring devices capable of verifying all theoretical predictions. All that the actual design of such instruments can do is to illustrate the logical relationships and, perhaps, elucidate their meaning.

2. Measurability of Charge and Current Densities

This somewhat sobering reflection did not deter us, however, from carrying on the investigation a step further. Heisenberg, who was actively exploring the possible repercussions of the explicit consideration of space-time averages in quantum electrodynamics, came across remarkable properties of charge fluctuations [5], the interpretation of which was a strong incentive for us to extend the analysis of measurability to the charged fields describing the sources of electromagnetic radiation. Apart from this immediate motive, however, such an analysis had an interest of its own as a further illustration of the point of view of complementarity, with its wider implications for general human problems never absent form Bohr's mind even when he seems exclusively engrossed with abstract points of epistemology or the ingenious technicalities of measuring contraptions. More deeply than anyone else Pauli grasped the earnestness of Bohr's endeavour. His criticism – pointed in every sense of the world – had all the time sustained our exertions and he was now egging us on to continue.

The problem now requiring attention was the use of the classical idealization of charge and current density even when this quantity is so large that the quantal properties of the charged fields must be taken into account. To take a simple case, consider the average charge contained within a given volume V during a given time-interval T. It is measured by the flux of electric displacement issuing from the volume V during the time T. We are well-equipped to measure this flux. We surround the volume V with a 'shell' of test-bodies and by appropriate lever transmissions we transfer upon a single auxiliary body the sum of the normal components of momentum of all the testbodies; the measurement of the flux is thus essentially reduced to the same manipulations of the test-bodies as that of a single field component. It is easily seen that in this case all electromagnetic interactions of the test-bodies exactly cancel out, including even the net effect of the zero-field, which obviously has no flux through a closed surface. Moreover, the electromagnetic field produced by the shell of test-bodies is entirely confined within the shell, and measurements of average charge densities over two different space-time regions can therefore be performed without any reciprocal limitation of accuracy on this count. These rather obvious results provide the physical justification for the stagewise treatment we have adopted, in the first approximation of which the charge-current densities are treated as classical quantities; formally, of course, the consistency of this approximation is guaranteed by the fact that the field divergencies $\partial f_{\mu\nu}/\partial x_\nu$ commute with the field components.

The real problem starts at the next stage of approximation, in which effects linear with respect to the fine structure constant are considered. The new fundamental feature which then makes its appearance is the phenomenon of electron pair production, or more generally, the production of a "pair field" by the electromagnetic field [5]. There is also a modification of the expression for the commutators of electromagnetic field components, but this is a generally small correction which does not essentially affect the conclusions reached in the initial stage of approximation.[6]

The part played by the pair field in the problem of the measurability of charge-current densities is an almost identical pendant to that of the photon field in the measurability of electromagnetic field averages. The manipulation of the test-bodies produces inside the shell an electromagnetic field, which is the source of a pair field manifesting itself as a polarization of the surrounding 'vacuum' as well as the appearance of actual pairs of electrons of opposite signs. The expectation value of this pair field can be calculated and since it is proportional to the displacement of the test-bodies, its effect on single density measurements can be completely compensated, while the mutual disturbance of a pair of measurements can be reduced to the optimum limit derived from the commutation relations for charge-current densities. The fluctuating part of the pair field, which cannot be compensated, is again an inherent property of every charge-current density, in quite the same sense as the electromagnetic zero-field.

There is no need to enlarge further upon this parallelism; I could do no more, within the compass of this article, than transcribe the more precise, but still very condensed account which is given in our second paper [7]. However, I might just illustrate it from the more formal side by exposing the skeleton of the argument, in close similarity with the electromagnetic case. The polarization current $I_\nu(x')$ produced by an electromagnetic field of potential $A_\mu(x)$ can be expressed by means of a Green's function of tensor character $b_{\nu\mu}(x', x)$:

$$(15)\qquad I_\nu(x') = \int b_{\nu\mu}(x', x)\, A_\mu(x)\, \mathrm{d}^4 x.$$

If the commutator between charge-current densities $j_\mu(x)$ is written in the form

$$(16)\qquad [j_\mu(x), j_\nu(x')] = i\hbar c K_{\mu\nu}(x, x'),$$

we have the relation

$$(17)\qquad K_{\mu\nu}(x, x') = -b_{\mu\nu}(x, x') + b_{\nu\mu}(x', x),$$

of the same general type as that expressed by formula (5). For spacetime averages

$$(18)\qquad J_\mu(R) = \frac{1}{R} \int_R j_\mu(x)\, \mathrm{d}^4 x$$

we get

$$(19)\qquad [J_\mu(R), J_\nu(R')] = i\hbar c [B_{\mu\nu}(R, R') - B_{\nu\mu}(R', R)],$$

with

$$(20)\qquad B_{\mu\nu}(R, R') = \frac{1}{RR'} \int_R \mathrm{d}^4 x \int_{R'} \mathrm{d}^4 x' b_{\nu\mu}(x', x).$$

Now, we must evaluate the potential $A_\mu(x)$ arising from the manipulation of the test-bodies involved in the measurement of $J_\mu(R)$. If we idealize the shell of test-bodies by a surface distribution of constant charge-current density ρ_μ over the boundary S of the region R, to be subjected to a constant normal displacement of magnitude D, we obtain a polarization density

$$P_{\lambda\mu}(x'') = P_\mu \int_S \delta(x'' - \bar{x})\, \mathrm{d}\sigma_\lambda,$$

with

$$P_\mu \equiv \rho_\mu D.$$

(We denote by $\bar{x}$ the point of integration on the boundary S.) The corresponding vector potential $A_\mu(x''')$ is

$$(21) \qquad A_\mu(x''') = -P_\mu \int d^4x'' D_r(x''' - x'') \frac{\partial}{\partial x''_\lambda} \int_S \delta(x'' - \bar{x})\, d\sigma_\lambda$$

$$= P_\mu \int_S \frac{\partial}{\partial \bar{x}_\lambda} D_r(x''' - \bar{x})\, d\sigma_\lambda = P_\mu \int_R \Box D_r(x''' - x)\, d^4x$$

$$= P_\mu \int_R \delta(x''' - x)\, d^4x,$$

giving rise to an electromagnetic field which vanishes everywhere except on the boundary S. According to (15), the polarization current at x' due to the field (21) is

$$I_\nu(x') = P_\mu \int_R b_{\nu\mu}(x', x)\, d^4x$$

and its average over the region R' is

$$(22) \qquad I_\nu(R, R') = P_\mu R B_{\mu\nu}(R, R').$$

The analysis of the measurability of charge-current densities thus confirms and extends our previous conclusions concerning the logical consistency of quantum electrodynamics. The consideration of the quantal fluctuations of these quantities, however, raises a quite new issue, which has a bearing on the entirely distinct problem of the domain of validity of the theory for the description of the physical phenomena. This issue arises from the fact that with the introduction of particles of finite mass, we must face the existence of an absolute space-time scale, and the restrictions this may impose upon the construction and manipulation of our test-bodies. The behaviour of the characteristic pair-field fluctuations warns us that when we try to test theoretical predictions pertaining to regions of very small dimensions we move on dangerous ground. In fact, Heisenberg's investigation reveals that – in striking contrast with the fluctuations of the electromagnetic field – the charge and current fluctuations within any sharply limited space-time region are infinite. It is only by considering the average fluctuation over an ensemble of space-time regions whose boundaries are contained in a shell of finite thickness that one gets a finite value, which, of course, tends to infinity as the thickness of the shell vanishes.

The physical meaning of this result directly follows from the analysis of charge and current measurements. The fluctuations of the charge or current inside a given space-time region are inseparable from those of the pair-field produced by the shell of test-bodies in the course of the measuring process. We see that if we want to increase the sharpness of definition of the space-time region by making the shell thinner, we pay for this by a corresponding increase in the fluctuation of the charge or current enclosed in the region. The existence of these fluctuations by no means impairs the possibility of definition of the charge-current density; it is a necessary consequence of the quantization of the charged field. From the practical point of view, however, it is clear that if the fluctuation of a quantity becomes of the same order as its expectation value, or larger, the knowledge of the latter, however well defined, becomes less and less significant. In this sense, the study of the increase of the charge and current fluctuations with decreasing dimensions of the regions in which these quantities are ascertained points to a degree of smallness of these dimensions beyond which the information conveyed by the theory ceases to be useful in guiding our expectations.

We are anxious to test those predictions of quantum electrodynamics which exhibit typically relativistic features; the domain in which such effects become prominent has dimensions smaller than the Compton wave-length (divided by 2π) $\lambda\!\!\!^{-}=\hbar/mc$ of the charged particles. The shells of test-bodies must accordingly have a thickness b considerably smaller than $\lambda\!\!\!^{-}$. In that case, it is found [8] that the root mean square fluctuation of the charge contained in a volume of linear dimensions L and averaged over a time of the order L/c is approximately given, in units of the electron charge, by $[\log(L/b)]^{1/2}$. This logarithmic dependence of the charge fluctuation on the linear dimensions b of the test-bodies composing the shell is a very favourable circumstance[7]: it allows us to go down to values of the ratio b/L of the order of a few per cent without unduly increasing the fluctuation. Even so, if L is smaller than the Compton wave-length of the electron, this brings b down to nuclear dimensions. It is still conceivable that our measuring apparatus could be built up of nucleons, or perhaps hyperons (they have to last only for times of the order $\lambda\!\!\!^{-}/c$), and we thus arrive at the satisfactory conclusion that the predictions of relativistic electron theory are in principle susceptible to verification; but we are obviously on the brink of the abyss: it is questionable, for instance, whether the same theory is at all meaningful for nucleons. On general grounds, nuclear dimensions would seem to mark the limit of validity of the kind of idealization which makes up the conceptual framework of electrodynamics. Beyond this limit, strongly interacting fields, with a limited range of force transmission, call for a radically different mode of description, in which the idealization of point charge presumably has no room.

3. The formal structure of quantum electrodynamics

It is necessary to emphasize again, in view of the confusion prevailing in certain circles, that the last remarks, while directly suggested by our discussion, do not at all imply that this discussion leads to the disclosure of the insufficiency of quantum electrodynamics. The limits of validity of the theory can only be derived from physical arguments, quite distinct from the logical analysis of its internal consistency. This analysis shows in fact that, whatever shape any more comprehensive theory will take in future, the present theory does give a coherent description of electromagnetic fields and electrons within a domain amply covering all practical applications.

The steady progress of the last few years in the unravelling of the formal structure of quantum electrodynamics leads to the same conclusions. In particular, it is now widely realized that the mathematical divergencies, which were such a disturbing feature so long as they could not be isolated and identified in a covariant way, are not due to any inherent inconsistency of the formalism, but only arise from a failure to take account of the non-electromagnetic elements which come into play in domains of nuclear dimensions.

In this respect, the situation was in no way different in the classical electron theory. Critical conditions under which the classical picture of the electron as a moving point charge breaks down, are reached when radiative reactions become of the same order as static interactions; the size of the critical domains, given by the 'classical electron radius' e^2/mc^2, coincides with nuclear dimensions – owing, no doubt, to some unknown, deep-lying law of nature. It is impossible, however, as was stressed by Poincaré, to account for the cohesion of an extended electron without appealing to forces of extraneous origin.

One often speaks in this connection of "open" theories, as opposed, I imagine, to some sort of ideal theory which would suffer no restriction of the range of variation of the physical quantities involved, and would be completely self-contained. It seems to me, however, that this opposition is quite artificial; so far as I can see, all physical theories have always been of the so-called 'open' type, and a 'closed' theory is just as much a will o' the wisp as universal determinism.

Bohr certainly never showed any respect for the noble elegance of a Lagrangian principle. In the stalemate of the first period of development of quantum electrodynamics, he repeatedly stressed that our only guidance in judging the adequacy of the formalism remains the correspondence with classical theory, that sesame which in more heroic days had served him so well. One aspect of the 'openness' of the formalism is the divergence [10] of the series expansions in powers of the fine structure constant α; however, owing to the dimensionless character of this constant and its numerical smallness, the series (regarded as functions of α^{-1}) are of the asymptotic type, and the successive terms are decreasing up to powers of α much larger than are ever likely to enter into the most refined experiments. The correspondence argument then assures us that such asymptotic series will not meet with any difficulty of interpretation.

From this correspondence point of view, the condensation of the contents of the theory into a variational principle may in a sense appear deceptive. A remarkable feature of the variational approach, however, is its ability to bring out the fundamental laws in integral form. It is unlikely (to say the least) that this form could help to circumvent the divergence of the usual procedure of piecemeal construction of solutions according to the differential mode of analysis. But it casts interesting light on certain aspects of the harmony between formalism and physical reality.

It is clear that in an integral formulation a central part will be played by the Green's functions describing the propagation of the fields in space-time. These propagators are not limited, of course, to the simple forms $D_r(x'-x)$, $b_{\nu\mu}(x', x)$ already considered, but form an extended class of functions involving connections between increasing numbers of space-time points. Schwinger [11] has shown that the starting point for a theory of propagators is a Lagrangian principle in which the interaction of the quantized fields with external sources, treated classically, is explicitly taken into consideration. It is in fact by subjecting the classical source densities to arbitrary variations that the fundamental integral equations satisfied by the propagators are derived [8]. Moreover, by selecting special variations of the external sources, according to a method independently proposed by Peierls [12] (in a somewhat different context), it is possible to derive general expressions for the commutators or anti-commutators between the field variables in terms of appropriate propagators. The discussion in the preceding sections may have prepared us to appreciate the significance of this introduction of classical sources into the very foundations of the formalism: in fact, this step will appear quite natural when it is realized how essentially all physical interpretation of the formalism is based on reference to classical quantities in the sense of the correspondence argument. There is truly no occasion here to speak (as Schwinger does) of 'fictitious' sources; on the contrary, such an idealistic terminology conveys a topsyturvy idea of the situation and fails to do justice to the practical utility of a classical description of the sources in many problems of radiation theory.

Let L_o be the usual Lagrangian density referring to a system of an electromagnetic field $A_\mu(x)$ and a pair field $\psi(x)$, $\bar{\psi}(x)$, coupled together with a strength of as yet undetermined magnitude $e^2/\hbar c$. Let us now consider a system of external sources for these two fields, defined by a charge-current density $J_\mu(x)$ and a potential $\alpha_\mu(x)$, respectively. If we put

$$(23) \qquad \bar{\eta}(x) = \frac{ie}{2}\,\bar{\psi}(x)\,\gamma_\mu \alpha_\mu(x), \quad \eta(x) = \frac{ie}{2}\,\alpha_\mu(x)\,\gamma_\mu \psi(x),$$

we may write the total Lagrangian density in the form $L = L_0 - H$, with

$$(24) \qquad H = -\theta^{1/2} A_\mu(x)\,J_\mu(x) - \lambda^{1/2}\left[\bar{\eta}(x)\,\psi(x) + \bar{\psi}(x)\,\eta(x)\right];$$

in this formula there appear two undetermined constants, θ, λ, to whose interpretation we shall return in a moment. Let Ψ_n be a system of eigenvectors of the Hamiltonian in the absence of external sources; the 'vacuum' state Ψ_o is defined as that corresponding to the least eigenvalue. We may describe[9] the effect of the external sources by a canonical transformation of these eigenvectors,

$$\Psi(\sigma_1) = U(\sigma_1, \sigma_2 \mid \eta J)\, \Psi(\sigma_2), \tag{25}$$

defined, with respect to an arbitrary family of space-like surfaces $\sigma(x) = \sigma$, by the equation

$$i\hbar c \frac{\delta U(\sigma_1, \sigma_2 \mid \eta J)}{\delta \sigma(x)} = H(x)\, U(\sigma, \sigma_2 \mid \eta J). \tag{26}$$

Every operator $F(x)$ pertaining to the source-free system will thus be transformed, in presence of the external sources, into the operator

$$F^{(s)}(x) = U^{-1} F U. \tag{27}$$

Consider, in particular, the effect of a point source

$$J_\mu(x) = \varepsilon \delta(x \quad x')\, \delta_{\mu\nu}$$

of the electromagnetic field upon the potential $A_\mu(x)$. On the one hand, we have by definition of the Green's function $D_r(x - x')$,

$$A_\mu^{(s)}(x) = A_\mu(x) + \varepsilon \delta_{\mu\nu} D_r(x - x'),$$

on the other, by (27), $A_\mu^{(s)}(x) = U^{-1} A_\mu(x)\, U$. The transformation U is the solution of (26) in which (putting $\theta = 1$) $H = \varepsilon A_\nu(x')\, \delta(x - x')$, i.e.

$$U(\sigma) = 1 \text{ for } \sigma < \sigma',$$
$$U(\sigma) = 1 + (i\hbar c)^{-1} \varepsilon A_\nu(x')\, U(\sigma') \text{ for } \sigma > \sigma'.$$

By assuming ε sufficiently small, we may use the approximate expression

$$U(\sigma) \approx 1 + (i\hbar c)^{-1} \varepsilon A_\nu(x') \text{ for } \sigma > \sigma',$$

which gives

$$A_\mu^{(s)}(x) \approx A_\mu(x) + (i\hbar c)^{-1} \varepsilon [A_\mu(x), A_\nu(x')].$$

The comparison of the two expressions for $A_\mu^{(s)}(x)$ yields

$$[A_\mu(x), A_\nu(x')] = i\hbar c \delta_{\mu\nu} D_r(x - x') \text{ for } \sigma > \sigma'.$$

By interchanging the roles of x and x', one finds for the commutator the value $-i\hbar c\delta_{\mu\nu}D_r(x'-x)$ in the case $\sigma'>\sigma$, and therefore, for arbitrary x and x',

$$(28) \qquad [A_\mu(x), A_\nu(x')] = -i\hbar c\delta_{\mu\nu}D(x-x'),$$

with the definition (5) of $D(x-x')$. This is an example of Peierls' general argument [12] establishing the close relationship between commutator and Green's function, which we found of fundamental significance in testing the consistency of the interpretation of the formalism. The formula (16) for the commutator of current densities can be derived by an entirely similar argument.[10]

We have seen that an equally fundamental part is played by the 'vacuum' fluctuations of the field variables, or more generally by the correlation functions

$$\langle[A_\mu(x), A_\nu(x')]_+\rangle, \quad \langle[j_\mu(x), j_\nu(x')]_+\rangle$$

and their derivatives, integrated with respect to x and x' over the same space-time region R; the notation $\langle F\rangle$ represents the 'vacuum' expectation value of the operator F (in absence of external sources). These correlation functions are connected with the corresponding commutators by a quantitative relation, most simply expressed in terms of Fourier transforms. If, e.g. the Fourier components of wave vector k of

$$\langle[A_\mu(x), A_\nu(x')]_+\rangle \quad \text{and} \quad [A_\mu(x), A_\nu(x')]$$

are denoted by $A^+_{\mu\nu}(k)$, $A^-_{\mu\nu}(k)$, it is easily seen that

$$(29) \qquad A^+_{\mu\nu}(k) = \varepsilon(k)\, A^-_{\mu\nu}(k),$$

where $\varepsilon(k) = \pm 1$ according as the time component k_o is positive or negative; this follows from the definition of the vacuum state as the state of minimum energy. Only a slight extension of this argument is needed to establish a similar relation in the case of the current densities. An interesting consequence of Equation (29) is to show that the order of magnitude of the quadratic field or current fluctuation is a critical one, not only for the deviation of field or current averages from their classical determinations, but also for the reciprocal indeterminacy of pairs of such field or current averages. More precisely, let us consider two congruent space-time regions consisting of a volume and a time interval, displaced with respect to each other by a space-time vector whose components are of the same order as the corresponding linear dimensions of the regions. The absolute value of the commutator of two field or current components averaged over these regions gives an estimate of the largest reciprocal limitation of measurability to be expected. With the help of relation (29), it can be shown [6] that this quantity is always of the same order as the quadratic fluctuation of a field or current component within the same region (assuming a boundary of finite width in the latter case)[11].

Commutators and correlation functions appear in a somewhat different relation as parts of the special propagators introduced by Feynman, which embody past and future in a symmetrical way. The Feynman propagators may be defined by the following formulae:

$$(30)\qquad D^{(F)}_{\mu\nu}(x, x') = \langle P[A_\mu(x)\, A_\nu(x')]\rangle,$$
$$K^{(F)}_{\mu\nu}(x, x') = \langle P[j_\mu(x)\, j_\nu(x')]\rangle,$$

in which the notation of the chronological product

$$(31)\qquad P[F(x)\, G(x')] = \tfrac{1}{2}[F(x), G(x')]_+ + \tfrac{1}{2}\varepsilon(\sigma, \sigma')\,[F(x), G(x')],$$

with

$$\varepsilon(\sigma, \sigma') = \pm 1 \text{ according as } \sigma \text{ is } > \text{ or } < \sigma',$$

has been used. The Feynman propagators, from which the quantities of direct physical interpretation, such as the Green's functions, and transition probabilities, can readily be derived, are conveniently chosen as the fundamental quantities in terms of which the integral field theory is formulated. More precisely, one has of course to consider the whole manifold of propagators of the general form

$$(32)\qquad \langle P[\psi(x_1)\cdots\psi(x_m)\, \bar{\psi}(x'_1)\cdots\bar{\psi}(x'_n)\, A_{\mu_1}(z_1)\cdots A_{\mu l}(z_l)]\rangle.$$

The remarkable result of Umezawa and Visconti's analysis [15] is that all these propagators are contained in the transformation U which expresses, according to the Equation (26), the effect of external sources on the system; the integral equations by which they are mutually connected are accordingly contained in this equation. We saw how the fundamental commutation relations appear as consequences of introducing weak external sources of a special form. To obtain the propagators one has to examine the effect of the most general infinitesimal variation of the external sources. It is easily seen that, e.g.,

$$\frac{\delta^2 U(\sigma_1, \sigma_2 \mid \eta J)}{\delta J_\mu(z)\, \delta J_\nu(z')} = (i\theta^{1/2})^2 \cdot P[U(\sigma_1, \sigma \mid \eta J)\, A_\mu(z) \times$$
$$\times\, U(\sigma, \sigma' \mid \eta J)\, A_\nu(z')\, U(\sigma', \sigma_2 \mid \eta J)],$$

whence

$$D_{\mu\nu}(z, z' \mid \eta J) \equiv \Psi_o^\dagger(\sigma_1)\, \frac{\delta^2 U(\sigma_1, \sigma_2 \mid \eta J)}{\delta J_\mu(z)\, \delta J_\nu(z')}\, \Psi_o(\sigma_2)$$
$$= (i\theta^{1/2})^2 \cdot P[\Psi_o^\dagger(\sigma)\, A_\mu(z)\, U(\sigma, \sigma' \mid \eta J)\, A_\nu(z')\, \Psi_o(\sigma')],$$

and therefore

(33) $$D_{\mu\nu}^{(F)}(z, z') = -D_{\mu\nu}(z, z' \mid 00), \text{ with } \theta = 1.$$

Quite generally, the transformation $U(\sigma_1, \sigma_2 | \eta J)$, regarded as a functional of the external source densities η, J, is a *generating functional* of the propagators; and the equation (26) gives rise, in the same sense, to a simple system of *generating equations* for an infinite set of integral equations linking together propagators of successive orders.

We shall not go further into the details of this beautiful formal scheme, but only briefly point to some of its most salient features. In the first place, it is noteworthy that all problems of 'renormalization' arising from the upholding of the point charge and local interaction idealizations are confined to the first stage of the theory, i.e. the determination of the elementary propagators involving just one photon or one electron [12]. The equations relating higher order propagators to the elementary ones do not involve any further divergence.

To illustrate the renormalization problem, it will suffice to consider the one-electron propagator (whose definition is slightly modified by the adjunction of a normalization factor)

(34) $$G^{(N)}(x, x' \mid 0J) =$$
$$= (i\lambda^{1/2})^2\, \frac{P[\Psi_o^\dagger(\sigma)\, \psi(x)\, U(\sigma, \sigma' \mid 0J)\, \bar{\psi}(x')\, \Psi_o(\sigma')]}{\Psi_o^\dagger(\sigma)\, U(\sigma, \sigma' \mid 0J)\, \Psi_o(\sigma')}.$$

It satisfies the integro-differential equation

(35) $$i\left[\gamma_\mu \frac{\partial}{\partial x_\mu} - \frac{ie_1}{\hbar c}\, \gamma_\mu \mathscr{A}_\mu(x) + \frac{mc}{\hbar}\right] G^{(N)}(x, x' \mid 0J)$$
$$+ \frac{c}{\hbar} \int M(x, \xi \mid 0J)\, G^{(N)}(\xi, x' \mid 0J)\, d^4\xi = \lambda\delta(x - x').$$

In this formula, $\mathscr{A}_\mu(x)$ represents the classical value of the vector potential due to the external current distribution J_μ, while the charge parameter has been replaced by the renormalized value $e_1 = e\theta^{-1/2}$. The constant θ, which thus effects the charge renormalization, represents the undetermined polarizability of the 'vacuum', whose dielectric constant is θ^{-1} and magnetic permeability is $c^2\theta$: its role is merely to make the adjustment of the electromagnetic units rendered necessary by the introduction of the pair field. As regards the 'mass-operator' $M(x, \xi \mid 0J)$, which is defined in terms of the propagators and the 'vertex operator'

$$(36) \qquad \Gamma_\nu(z; x, x' \mid 0J) = \frac{\delta [G^{(N)}(x, x' \mid 0J)]^{-1}}{e_1 \delta \mathscr{A}_\nu(z)},$$

it embodies the modification of the undetermined mass parameter m of the electron due to its coupling with the electromagnetic field: if m_1 denotes the observed mass, one finds that

$$(37) \qquad \gamma_4 \int M(x, \xi \mid 00)\, G^{(N)}(\xi, x' \mid 00)\, d^4\xi = \\ = (m_1 - m)\, G^{(N)}(x, x' \mid 00) \text{ for } \sigma > \sigma'.$$

It is apparent from equation (35) that the propagator for a free electron will be affected by the renormalization factor $\lambda\mu^{-1}$, where μ can be expressed as

$$(38) \qquad \mu = \frac{\partial}{\partial m_1} [m + M(p)] \quad (\text{for } \gamma_\mu p_\mu + m_1 = 0)$$

in terms of the Fourier transform $M(p)$ of the mass operator $M(x, \xi \mid 00)$. The reduction to the usual space-time units is thus effected by taking $\lambda = \mu$. The constant λ, which affects the particle density[13] $\bar{\psi}\psi$, serves to adjust the spatial dimensions to the value of the Compton wave-length $\hbar/m_1 c$ fixed by the mass value m_1.

While it is gratifying to find that the renormalization problem can thus be circumscribed to the definition of the elementary propagators, and does not arise with the higher order ones, there is another feature in the treatment of the latter which is rather disturbing. As we have just seen, the elementary propagators and the renormalization parameters are susceptible to a direct and quite transparent physical interpretation. This is not the case with expressions of the general form

(32); such quantities can only be related to transition probabilities of actual physical processes if the variables x, x', z consist of two sets $(x_1,\ldots, x_m; z_1,\ldots, z_k)$, $(x'_1,\ldots, x'_n; z_{k+1},\ldots z_l)$, belonging to two space-like surfaces σ, σ', respectively. The integral equations for these higher order propagators, however, do not allow of a restriction of this type in the range of variation of the space-time coordinates: they make the determination of the transition probabilities dependent upon that of purely mathematical symbols which cannot correspond to any well-defined physical situation. This would not in itself be objectionable, provided that it could be shown that the equations in question do have solutions of the required character. The mathematical problem thus raised is one of great difficulty, especially in view of the fact that such solutions must not only refer to the continuous manifold of states corresponding to scattering processes, but also include the bound states of many-particle systems (e.g. the 'positonium'). Although the issue is very much in doubt, it is at any rate a merit of the integral formulation of the theory to present it in such a clear-cut way.

From the point of view of the correspondence argument, however, one might well regard any rigoristic conception of this problem as superfluous, or even ill-founded: one would, of course, attack it by introducing series expansions in powers of the coupling constant α; and even though these expansions would not converge, one could attach a well-defined meaning to those few terms which could actually be computed. But the justification for this attitude is limited to the case of electromagnetic interactions, and the issue becomes most acute as soon as we try to apply the methods of quantum electrodynamics to the description of the couplings between other fields.

4. Meson couplings

Heisenberg was the first to point out, as long ago as 1936, that the behaviour of interacting fields should be essentially different according as the coupling strength could be expressed by a dimensionless constant, as in electrodynamics, or had the dimension of some positive power of a length [14]; in the latter case, one expects effects of multiple interaction to become predominant at high energy. The formalism of quantum electrodynamics cannot unambiguously be extended to couplings of this type, since it would involve an infinity of renormalization parameters [16].

The coupling of charged bosons with the electromagnetic field is of the first or second type according as the spin of the boson is zero or larger than zero. The explicit calculation of the charge-current fluctuations for bosons of spins 0 and 1, performed by Corinaldesi [6, 8, 9], strikingly illustrates the radical difference between the two cases: charged bosons of spin 0 behave exactly as charged fermions, whereas the charge fluctuations of spin-one bosons within a space-time region of unsharp boundary exhibit a much stronger divergence than those of fermions as the thickness of the boundary layer tends to zero. This makes it extremely doubtful whether the point charge idealization can be applied at all usefully to such particles.

When we come to consider the coupling between nucleons and π-mesons, however, we are in a quandary. We are in fact faced with two distinct possibilities, the pseudoscalar coupling which is of the first type, and the pseudovector coupling which is of the second, without as yet sufficiently convincing clues from observation as to their relative importance. That the interaction of the nucleons with the pseudoscalar mesons is commonly treated as a pseudoscalar coupling – in what very hurried or very lazy people call the 'ps-ps' theory – is more an illustration of human weakness than of human reason. It is true that the 'ps-ps' theory has booked a certain amount of success, but this is not conclusive, since the two forms of coupling become undistinguishable at low energies and we have no reliable estimate of the extent to which the pseudovector coupling would affect high-energy phenomena.

I shall only call attention to one of the problems recently discussed in this context, because of its relation to electrodynamics: the mass-difference between charged and neutral particles of the nucleon and π-meson species. It is tempting to interpret such mass-differences as purely electromagnetic effects, but the computation of the electromagnetic self-energies, involving the 'regularization' of divergent integrals, introduces an arbitrariness which would seem to frustrate any attempt at substantiating this interpretation. However, Petermann [17] has succeeded to show, by an ingenious argument to a large extent independent of the regularization arbitrariness, that the signs and orders of magnitude of the two mass differences can indeed be accounted for on a purely electromagnetic basis. This interesting conclusion, which, of course, does not preclude the existence of other contributions to the mass-differences, is not likely to be affected by the uncertainty regarding the meson-nucleon coupling.

Even if we adopt the pseudoscalar form of meson-nucleon interaction, we are faced with a situation differing from that of electrodynamics by another important circumstance: far from being a small quantity, like α, the dimensionless coupling constant g of meson theory is certainly larger than unity, probably of the order of 10. Power series in g are no longer asymptotic, but diverge from the start, and another approach is needed to solve the fundamental equations of the theory. A most interesting step in this direction has been taken by Edwards [18]. He sets up for the vertex operator (36) of quantum electrodynamics an approximate integral equation obtained by expanding all other quantities in power series and retaining only the lowest terms, and he proceeds to solve this equation exactly. The corresponding equation in meson theory would be quite similar. It is true that, as the result shows, this procedure is still too rough even for the electromagnetic case, but even so it reveals a remarkable property, which certainly subsists when the approximation is refined.

The problem is eventually reduced to the consideration of a linear differential equation, whose indicial equation at infinity involves the coupling constant. The precise form of this equation is immaterial; the essential point is that the reality character of the roots, and with it the analytical properties of the solution, change when the coupling constnat passes a certain critical value. This feature establishes a sharp distinction between the weak electromagnetic coupling on the one hand and the strong pseudoscalar coupling of meson theory on the other. By approaching the problem from the strong coupling side and suitably extending Edwards' method, Petermann [19] has recently been able to elucidate the essential features of the anomalous magnetic moments of the nucleons, hitherto so obscure: it appears from his analysis that the nucleonic contribution to the magnetic moment is considerably reduced by a damping effect, whereas the mesic contribution, which is not subject to such a reduction, yields equal and opposite moments for the proton and the neutron. It would seem, therefore, that the study of the meson field leads us into quite unfamiliar regions, in which progress is not only hampered by mathematical difficulties, but still more by the absence of any physical principle which could fulfil a part comparable to that played in quantum electrodynamics by Bohr's correspondence argument.

NOTES

[1] I am afraid this work of art has been allowed to disintegrate before its historical value could be realized.

[2] This is a familiar danicism of Bohr's for "darf ich."

[3] This section contains an analysis of the paper on the subject by Bohr and Rosenfeld [2], to which the reader is referred for further details.

[4] It has apparently, like so many things, escaped the attention of those quixotic young physicists who, spell-bound by distorted echoes of a lore unfathomed, try to split the quantum with the rusty sword of mechanistic materialism.

[5] For the sake of clarity I confine myself here to the consideration of electrons (or at any rate fermions) as the carriers of electric charge and current. Later on I shall have occasion to refer briefly to the problems raised by charged bosons.

[6] This point has been investigated by E. Corinaldesi in his (unpublished) Manchester thesis. Cf. his survey article [6], p. 93–95.

[7] The fluctuations of current components defined in a similar way show a much worse behaviour [9]; they actually increase as $(L/b)^{1/2}$. These quantities, however, have no such direct physical meaning as the charge fluctuations.

[8] The line of investigation initiated by Ferretti [13] and pursued by Caianiello [14], in which variations of the coupling parameter are considered, falls under the same scheme.

[9] I adopt the elegant mode of exposition of the theory proposed by Umezawa and Visconti [15].

[10] If we apply the same reasoning directly to the fermion field variable $\psi(x)$, by considering a source density of the form $\eta(x')\,\delta(x-x')$, we get the relation

$$[\psi(x), \bar{\psi}(x')\,\eta(x')] = i\hbar c S(x-x')\,\eta(x'),$$

where $S(x-x')$ is the corresponding propagator. Assuming, in harmony with (23), that the source strength $\eta(x')$ anticommutes with $\psi(x)$, we arrive at the fundamental relation of anticommutation

$$[\psi(x), \bar{\psi}(x')]_{+} = i\hbar c S(x-x').$$

[11] Our paper [2] on field measurability must be amended on this point; the correction does not affect the argument.

[12] The propagator $K^{(F)}_{\mu\nu}$ in (30) can be reduced to the simpler one $\langle P[\psi(x)\bar{\psi}(x')]\rangle$.

[13] The factor λ does not occur with the charge-current density, because of a compensation of divergences, known as 'Ward's identity,' which is a consequence of the gauge-invariance of electromagnetic theory.

[14] The case of a coupling strength with the dimension of a negative power of a length, of which no physical example is known, is similar to that of a dimensionless coupling constant.

BIBLIOGRAPHY

[1] L. Landau and R. Peierls, *Z. Physik* **69** (1931), 56.

[2] N. Bohr and L. Rosenfeld, *Dan. Mat.-fys. Medd.* **12** (1933), No. 8. [English translation: this volume, p. 357].

See also B. Ferretti, *Nuovo Cimento* **12** (1954), 558.
[3] J. H. Fabre, *Souvenirs entomologiques*, **IX**, Chapter XXIII.
[4] W. Thirring and B. Touschek, *Phil. Mag.* **62** (1951), 244.
R. Glauber, *Phys. Rev.* **84** (1951), 395.
H. Umezawa, Y. Takahashi and S. Kamefuchi, *Phys. Rev.* **85** (1952), 505.
J. Schwinger, *Phys. Rev.* **91** (1953), 728.
[5] W. Heisenberg, *Leipziger Ber.* **86** (1934), 317.
[6] E. Corinaldesi, *Supplemento al Nuovo Cimento* **10** (1953), 83.
[7] N. Bohr and L. Rosenfeld, *Phys. Rev.* **78** (1950), 794. [This volume, p. 401].
[8] E. Corinaldesi, *Nuovo Cimento* **8** (1951), 494.
[9] E. Corinaldesi, *Nuovo Cimento* **9** (1952), 194.
[10] C. Hurst, *Proc. Cambr. Phil. Soc.* **48** (1952), 625.
W. Thirring, *Helv. Phys. Acta* **26** (1953), 33.
A. Petermann, *Phys. Rev.* **89** (1953), 1160; *Archives Genève* **6** (1953), 5; *Helv. Phys. Acta* **26** (1953), 291, 731.
[11] J. Schwinger, *Proc. Nat. Acad. Sci.* **37** (1951), 452; *Phys. Rev.* **82** (1951), 914; **91** (1953), 713, 728; **92** (1953), 1283; **93** (1953), 615; **94** (1954), 1362.
[12] R. Peierls, *Proc. Roy. Soc. A* **214** (1952), 143.
[13] B. Ferretti, *Nuovo Cimento* **8** (1952), 108; **10** (1953), 1079; **12** (1954), 457.
[14] E. R. Caianiello, *Nuovo Cimento* **10** (1953), 1634; **11** (1954), 492; **12** (1954), 561.
[15] H. Umezawa and A. Visconti, *Nuovo Cimento* **1** (1955), 1079.
[16] S. Sakata, H. Umezawa and S. Kamefuchi, *Progress in Theoretical Physics* **7** (1952), 377, 551.
[17] A. Petermann, *Helv. Phys. Acta* **27** (1954), 441.
[18] S. F. Edwards, *Phys. Rev.* **90** (1953), 284.
[19] A. Petermann, *Nuclear Physics* **1** (1956), 355, 357.

CHAPTER 4

ON QUANTIZATION OF FIELDS

[1963a]

ABSTRACT. Critical remarks on some formal and physical aspects of field quantization, with special application to the case of gravitation.

The textbook *Elementary quantum field theory* by E. M. Henley and W. Thirring (McGraw-Hill, 1962) opens with general considerations on the quantization of fields which, coming from such distinguished physicists, cannot pass unchallenged, for they give a completely lopsided view of the physical situation.

The authors start with the assertion that elementary quantum mechanics, i.e. the quantization of the kinematical and dynamical variables, is inconsistent with classical electromagnetic theory and therefore implies the necessity of quantizing the electromagnetic field. Certainly, nobody will wish seriously to question the factual truth of this statement: the quantization of the motion of charged particles is inseparably linked with that of the radiation field in the very formulation of the fundamental quantum postulates. However, it is characteristic of the authors' formalistic tendencies that, instead of plainly acknowledging the empirical origin of the connexion between quantized motion and quantized radiation expressed by these postulates, they endeavour to 'prove' it as if it were a logical necessity.

In this undertaking they seek support in the paper by N. Bohr and L. Rosenfeld on the measurability of electromagnetic fields: according to them, this paper would contain proof that the electromagnetic field must necessarily be quantized in order to avoid contradiction to the uncertainty relations between position and momentum of charged particles in interaction with this field. The argument of the paper mentioned, however, clearly shows that so long as the electromagnetic field is treated classically, no limitation arises for the measurability of space-time averages of any one component or any pair of components of the field, even when due account is taken of the uncertainty relations to which the displacements and momenta of the test-bodies are subjected.

This conclusion is indeed necessary to guarantee the *limited* consistency of classical electromagnetic theory with quantum mechanics for the description of those phenomena in which only the average behaviour of the sources of the electromagnetic field is of importance. The ultimate necessity of quantizing the electromagnetic field (or any other field) can only be founded on experience, and all that considerations of measurability of field components can do is to illustrate the consistency of the way in which the mathematical formalism of a theory embodying such quantization is linked with the classical concepts on which its use in analysing the phenomena rests.

It is not superfluous to insist on this obvious point in view of the fact that a similar, and equally fallacious, argument is sometimes adduced in support of the alleged necessity of quantizing the gravitation field[1]. There is no denying that, considering the universality of the quantum of action, it is very tempting to regard any classical theory as a limiting case to some quantal theory. In the absence of empirical evidence, however, this temptation should be resisted. The case for quantizing gravitation, in particular, far from being straightforward, appears very dubious on closer examination.

The essential difference (duly stressed by the experts) between the situation in gravitation theory and that in the theory of the free electromagnetic field is that whereas the latter sets no absolute scale of space-time dimensions or mass, a quantum theory of gravitation involves the existence of a critical interval and a critical mass corresponding to the concentration of a quantum of gravitation energy and momentum within a limited space-time domain; the significance of such a quantized theory will therefore decisively depend on the comparison of these critical values with possible limitations of other origin to the definition of the quantities in question. Thus, to take a simple analogy, the limit of validity of the classical dynamics of the electron, expressed by the classical electron radius, loses any meaning since quantum effects become essential for the motion of electrons within domains of much larger dimensions. Now, in quantum electrodynamics and meson theory, an absolute limit to the possibilities of space-time localization is suggested (although its precise nature is still obscure) by the increase of the fluctuations of the charge and current contained within an unsharply defined space-time domain when the thickness of the shell limiting this domain decreases. It would seem that the critical thickness is of the order of the nuclear length unit (1 fm), which is much larger than the critical length for quantal effects of gravitation to become appreciable. This leads at least to a strong suspicion that quantization of the gravitation field would be meaningless.

The combination of a classical description of the gravitation field with a quantal description of the other fields need not present any difficulty, either of practical formulation or of principle. In formulating such a theory, one would naturally assume that the source of the gravitation field would be determined by the expectation value of the energy-momentum density operator. As to the occurrence of classical field variables in a comprehensive quantal scheme, one must not lose sight of the fact that the formulation of any theory in its application to given physical situations involves the specification of the system under consideration and of the external conditions under which it is investigated: such specifications, which represent the essential link between the theoretical description and the physical observation, are necessarily expressed in terms of classical concepts, and therefore enter into the equations in the form of *c*-number parameters. From this point of view, the metrical tensor, at any rate for all practical purposes, appears as such a *c*-number specification of conditions of observation, and there is no logical imperfection in regarding as fundamental a classical, unquantized, form of the equations expressing the connexion of the metrical or gravitation field with the other fields.

The opinion has been expressed that nothing short of a completely quantized theory is required in order to analyse the process of measurement of field quantities, by which such quantities are defined, entirely in terms of a unified description, in which the experimental arrangement itself would be included. It should be clear from what has just been said, however, that the requirement of such a comprehensive account of the measuring process is actually pointless, since (apart form providing a rather trivial check of the consistency of the theory) it would only obscure the essential function of the experimental arrangement in establishing the connexion between the system and the classical concepts indispensable for its description. Disregard of this simple point, which is an essential feature of any field theory, has led to futile attempts at circumventing complementarity in ordinary quantum theory by arguing that a solution of the wave equation for the total system including the experimental arrangement would yield a uniquely determined account of the whole measuring process and its result; in such an argumentation it is not sufficiently realized that the equation in question would contain c-number parameters which in the last resort would refer to just the conditions of observation directly expressed in the classical description of the experimental arrangement. In the particular case of the measurement of gravitation quantities, it is unavoidable to have some classical metrical substratum for the localization of the test-bodies.

There is a further erroneous view to which misplaced emphasis on formal aspects of quantum theory has led, and which is so wide-spread that it has even (I am sorry to say) found expression in the incriminated introductory section of the textbook quoted at the beginning of this note. It is there asserted that the wave functions associated with material particles represent 'classical matter fields', and from this patently wrong statement it is then argued that these 'matter fields' should also be quantized. Actually, as ought to be well known, the scheme of quantum mechanics (in any representation) constitutes a complete incorporation of the quantum postulates in the description of the behaviour of material particles, and what is misleadingly called 'second quantization' is nothing else than an equivalent formulation of this scheme with the help of convenient operators in a representation characterized by the choice of the numbers of particles in given individual states as variables. In particular, the commutation or anti-commutation relations satisfied by the annihilation and creation operators have nothing to do with the quantization of the field (they do not contain Planck's constant), but are just algebraic properties of these operators.

The preceding hints may suffice to expose the inanity of a reasoning which amounts to putting physics upside down. It is nice to have at one's disposal such exquisite mathematical tools as the present methods of quantum field theory, but one should not forget that these methods have been elaborated in order to describe definite empirical situations, in which they find their only justification. Any question as to their range of application can only be answered by experience, not by formal argumentation. Even the legendary Chicago machine cannot deliver the sausages if it is not supplied with hogs.

NOTE

[1] This question was discussed at the Jablonna Conference on Gravitation in July 1962; it may be convenient to repeat the main points in the present context.

CHAPTER 5

THE EVOLUTION OF THE IDEA OF CAUSALITY

[1942d]

Physicists generally have the reputation of being peaceful people, little inclined to reveal in the forum of the republic of letters the hesitations and doubts which assail them in their patient attempts to analyze the phenomena of inanimate nature. But, from time to time, it does happen that they throw this republic into the greatest ferment. On the basis of some seemingly minute difficulty which they, nevertheless, consider decisive, they raise the most serious problems. Then they do not hesitate to question the unlimited validity of our most fundamental concepts without the slightest regard for the innumerable philosophical dissertations which had made these concepts the unshakeable *a priori* forms or categories of our understanding. Thus, because they cannot succeed in comprehending the stability of atomic structures and analyzing the transformations occurring therein, they propose, scandalizing their good sound colleagues, to declare that stability incomprehensible, those transformations unanalyzable, and our usual concept of causality itself inapplicable to these phenomena. But what brings the scandal to its peak is that they turn out to be right. However disconcerting the thing may seem to metaphysical intellects, this renunciation of classical causality in the domain of atoms, far from denying us access to this domain, in fact permits us to order its laws rationally in a sufficiently enlarged logical framework.

Perhaps this situation will seem less strange to us if we realize that it is not new: Looking back, in fact, we see that each stage in the development of science has required a corresponding adaptation of the notion of causality, and these successive vicissitudes were not without disputes that were just as resounding as those of recent years on the acausal behavior of atoms. We can also satisfy ourselves that in this latest case, as in the preceding ones, it is not lightly, but under the pressure of extreme distress, that physicists decided to admit the necessity for a radical change in viewpoint. We will then, doubtless, understand better that the new attitude which physicists have taken toward causality, the fruit of a lesson drawn directly from the study of nature, is a definitive enrichment of our representation of the world, and we will be more at ease in asking ourselves, finally, what repercussions this change in attitude would have in other spheres of thought.

How, within the framework of science, could the least doubt arise concerning the universal validity of the principle of causality? In its most direct aspect, this principle affirms that the course of a phenomenon can be predicted unambiguously as soon as we know certain other phenomena which we call its 'causes'. Does this not formulate the most immediate, and hence the most general condition for the very possibility of all science? On looking more closely, however, this simplicity is misleading, for it is immediately evident that, at the risk of reducing the principle to a tautology, we must choose among the innumerable circumstances which accompany the phenomenon in question, those on which we will confer a determining influence: And further, if, with our poor means, we wish in practice to ensure the desired prediction of the future course of the phenomenon, this choice must be very strict. Of the exuberant wealth of concrete reality, there will remain only a skeletal scheme: In the last analysis, the causal law expressing the temporal development of the phenomenon will be a mathematical relationship among the numbers representing the measurements of a few physical quantities which we bring into this law. *Causa aequat effectum*, said Robert Mayer; there is a quantitative equivalence between the measurement of the cause and that of the effect; such is, for science, the most general rigorous formulation of the principle of causality.

But the more such a principle tends to be universal, the more it loses its concrete content. Hence, what is more important than the simple affirmation of the existence of quantitative causal relationships is the precise definition, in each case, of what should properly be considered the measurements of the cause and the effect. These are the specifications which provide the whole value, but also the whole difficulty, of the principle of causality: Indeed, the fertility of the principle depends on the judicious choice of these elements, and the extent of its domain of applicability may depend on the very possibility of this choice. Let us now take, as an example, one of the most general forms of physical causality: If the 'effect' on which we focus our attention is the product of mechanical work, the principle of causality implies the impossibility of 'perpetual motion', i.e. of a machine capable of producing (or expending) work without another modification equivalent to this work appearing in some form. But the utility of such a principle cannot become apparent until the quantitative measure of the equivalent of this work has been determined for each sort of phenomenon: thermal, electric, chemical or any other, accompanying the production of mechanical work. These 'mechanical equivalents' can only be determined by direct appeal to experiment. Moreover, in principle, a single individual experiment suffices to obtain the universal expression of the desired equivalent. The genius of Robert Mayer, whom we mentioned earlier, lies precisely in his having clearly recognized this situation and having actually established the mechanical equivalent of heat through a penetrating analysis of the thermal properties of gases. From this point of view, the various manifestations of natural forces appear as so many equivalent aspects of a physical quantity: the 'cause' of mechanical work, or 'energy'; for each of the forms it may take, energy is well defined by the corresponding mechanical equivalent.

The principle of the equivalence of the different forms of energy, or principle of the conservation of energy, has this remarkable characteristic: There remains no trace of the problems of fixing the course of phenomena in time or localizing them in space which characterizes the primitive idea of causality; whatever the order of succession of the transformations under consideration, whatever the place where they occur, the balance sheet of the energies brought into play must be closed with neither gain nor loss. What is left of causality here is simply the idea of well-determined relationships among certain quantities, without the manner in which these quantities vary in space and time necessarily intervening. Certain minds, fascinated by this high degree of abstraction, have not failed to see in this the unique ideal of science, and to try to confine the analysis of all phenomena to a scheme of energy exchanges. The cricitism of the 'energetics school' was aimed especially at atomism; this doctrine, based on the hypothesis that matter consists of atoms in motion, on the contrary, led quite naturally to a more extensive analysis of the course of phenomena in space and time; but this analysis is also much more uncertain, since it is a matter of imagining in greater or lesser detail the modalities of this saraband of atoms of which we perceive only the global effects. Energetics claimed to avoid all useless speculation and to rest content with directly measurable quantities. But one cannot with impunity try to impose limits on our means of action, however precarious they may seem, in the search for the laws of nature: After having imagined atoms, scientists managed to reveal them experimentally, and even to isolate the atoms of electricity, or electrons, which enter into the constitution of material atoms. And the career of energetics was terminated by the conversion of its most fiery protagonist to triumphant atomism. The principle of the conservation of energy naturally continues to express an essential aspect of causality in the physical world, but it is just as indispensable to establish a theory which permits us to situate atoms in space and time.

From its appearance in the history of Greek science with Leucippus dispensable to establish a theory which permits us to situate atoms in space and time.

> *et quo quaeque modo fiant opera sine divom.*
> (Lucretius, *De rerum natura*. I. 158)
> [whence each thing can be created and how can come to be, without the help of gods – A. D. Winspear tr., p. 9, The Harbor Press, New York, 1956]

It is at this summit, where ancient thought maintained itself for just an instant, that we see the first refinement of the notion of causality applied to the atomic picture of the world. Surmounting the last vestiges of animism, which were still apparent in his predecessors, Democritus renounced the sterile search for a first 'cause' of atomic motion in a vacuum; he recognized that this motion must be considered to be a given fact, just as primitive as the very existence of the 'form' of atoms: The question of its 'cause', that is, its production by some external intervention, cannot be raised. Here we can already recognize the point of view which will lead Galileo to formulate the principle of mechanical inertia in opposition to the Aristotelian conceptions still prevalent at his time. No force is needed to keep a body in motion at a uniform speed; any effort applied to the body will have the primary effect of modifying not its position, but its velocity, i.e. of accelerating it. Thus, the quantity which measures the motion from the causal point of view is not, as Aristotle believed, the velocity but rather the change in velocity, or better, the change in 'momentum' of the body (the product of the mass times the velocity): the principle of causality is expressed here by the equality of the force applied to the body – the cause of the motion – with the change in momentum which it produces. To arrive at this law, it was necessary to have the courage to limit the causality of motion to changes in momentum, while accepting the initial momentum as a new given datum of the problem, independent of the initial position. That was surely, in common opinion, an attitude unworthy of a philosopher! But with their subtle distinctions among primary, secondary, efficient, final, instrumental and other causes, the scholastic philosophers were, in the end, prevented from correctly describing the motion of a thrown stone; and in contrast to this age-old sterility, the prodigious development of mechanics during the half-century between Galileo and Newton offers one of the most striking examples of the power of scientific method in developing and modifying the conception of causality in accord with the evidence of experiment.

No one was more daring in this direction than Newton himself. For his grandiose conception of universal gravitation rests once again on a conscious renunciation of causality as it seemed to be required by the mechanical picture of the world, generally considered to be the only rational one in his time. The development of mechanics had, in effect, put its mark on the ideas which could then be conceived to provide the inner constitution of bodies; whether the existence of the vacuum was accepted or denied, atoms were represented by a picture which had become familiar, a picture of solid bodies like very hard corpuscles which transmit their movements only by direct contact. Now such a picture of the world had no place for a force like gravitation, which acts instantaneously at any distance from the body from which it emanates. Newton, more than anyone else, had long sought a mechanical explanation for the law of gravity; it was due to the failure of these efforts that he crossed the Rubicon and stated this law as a given fact, irreducible to mechanical causes. The greatest of his contemporaries refused to follow him. Speaking of universal gravitation, Huygens[1] wrote: "Which I could not accept because I believe that I see clearly that the cause of such an attraction is explicable neither by any principle of Mechanics, nor by the laws of motion." And the father of pre-established harmony declared without flinching:[2] "... I maintain that the attraction of bodies, properly so called, is a miraculous thing, since it cannot be explained by the nature of bodies."

The opposition to Newtonianism lasted well into the eighteenth century, especially with regard to the famous question of the shape of the earth. But gradually, as Newton's theory asserted itself through brilliant verification of all its consequences, scientists were more and more tempted to forget that its character was irrational from the viewpoint of mechanistic atomism. Scientists progressively grew more accustomed to the concept of a 'central force', based on the model of gravitation: a force emanating from a center and acting instantaneously at a distance according to a deterministic law. The introduction of this sort of force gave the greatest of services: In particular, it was possible to incorporate the fundamental laws of electric and magnetic phenomena. Toward the end of the century, the turnabout was complete. Atomism had taken on a Newtonian form: Atoms were no longer compared to solid particles influencing each other by contact; they had been reduced to centers of forces, to geometrical points from which emanated central forces of various types, insuring, among other things, the cohesion of matter; even more significantly, heat and light were atomized in the same fashion, with calorific and luminous atoms also reacting upon material atoms through properly specified central forces. And now it is this 'physics of central forces' which becomes the ideal of scientific causality. We have reached Laplace's well-known fable:[3] "Given for one instant an intelligence which could comprehend all the forces by which nature is animated and the respective situation of the beings who compose it, – an intelligence sufficiently vast to submit these data to analysis – it would embrace in the same formula the movements of the greatest bodies of the universe and those of the lightest atom; for it, nothing would be uncertain, and the future, as the past, would be present to its eyes."

Through the somewhat solemn eloquence of this excerpt, we feel the emotion which grips the thinker at the prospect of a grandiose rational synthesis of the phenomena of inanimate nature. Though the idea of a central force continued to play an essential role in the development of physics in the nineteenth century, this hope for a universal synthesis was soon disappointed. Fresnel's experiments on the phenomena of diffraction and polarization show that light (as Huygens had already claimed) is propagated by waves in the same manner as a shock in an iron bar; thus it was necessary to imagine that matter was plunged in some sort of an elastic medium, the ether, which was capable of propagating light vibrations. Thereby Fresnel not only destroyed the theory of light corpuscles, but – a much more important event – he appealed to a new kind of mechanical analogy, borrowed from the theory of the elastic properties of bodies. This latter theory, like that of the motion of fluids, is assuredly not incompatible with an atomic conception of bodies; but what gives it its peculiar character is that it can, to a very large extent, do without such a representation and treat material bodies as continuous media. The state of a continuous medium is characterized by certain quantities (density, pressure, rate of flow, etc.) which are defined at each point in space occupied by the medium: It is said then that such a quantity constitutes a field. It is this notion of field which is the fundamental concept of the physics of continuous media. It proved particularly essential in the domain of electricity and magnetism. The electromagnetic field, that is, the distribution in space of the electric or magnetic force emanating from an electrified or magnetized body, can, in fact, as Faraday suggests, be interpreted as indicating a modification of the ether surrounding the body in question; from this field concept, and guided by the analogy of the transmission of forces in an elastic bar, a natural interpretation of forces which appear to act at a distance between electrically charged and magnetized bodies can be formulated: The electromagnetic field puts the ether in a particular state of tension, thanks to which it can, from one neighboring point to the next, transmit these forces from one body to another. However attractive this conception may appear, we shall not be astonished to see it become the butt of disdain from the partisans of central forces. The eminent astronomer Airy, for example, could not imagine anyone hesitating an instant between the "simple and precise" notion of action at a distance and "such a vague and variable thing" as the mechanism for transmitting forces proposed by Faraday. Nevertheless, Maxwell, in his approach to the study of time-dependent electromagnetic phenomena, which were the stumbling-block for theories of central forces, showed the superiority of the new concept in this domain: Everything reduced to a very simple law of propagation of an electromagnetic field in the ether, and this law was identical to the one Fresnel had postulated for the propagation of light waves. So, Hertz's experimental confirmation of these predictions of Maxwell's theory was able to sanction not only the synthesis of optics and electromagnetism, but above all the triumph, in this vast domain, of the theory of fields over the theory of central forces.

In the framework of the mechanistic ideas of the seventeenth century, the concept of action at a distance, without being intrinsically incompatible with these ideas, reflected a point of view in a sense opposed to that of transmission of forces by collision, then considered the fundamental concept. We have seen how the force of events accentuated this antagonism until it ended in a complete reversal, implying a radical change in attitude: analyzing the idea of central force was renounced and central forces were placed, such as they were, at the base of the new world system. We shall now witness a new dialectical process of the same sort. With Hertz, the conflict between the two antithetical points of view, the physics of central forces and the physics of fields, was resolved by abandoning any effort to give a mechanical model of the ether: the electromagnetic field was then considered to be a primitive quantity, and its laws, Maxwell's equations, summarized the new synthesis of electromagnetism obtained at the price of this renunciation: a richer and more fertile synthesis since it includes, in addition to the old laws of central forces, a host of more general laws. Maxwell's equations thus became the prototype for a new form of causality: variations in the field quantities in the course of time are determined as soon as the initial values of these quantities throughout space are supposed to be known. When, by a theory of rarely equalled simplicity and beauty, Einstein had incorporated into the physics of fields the central force par excellence, gravity, it was once again possible – as in the time of Laplace – to believe science to be on the threshold of a general synthesis of the physical world; it is a faith which Einstein himself still holds today.[4] Yet, to achieve this synthesis, it would have been necessary to incorporate into it the atoms and electrons themselves, the 'sources' or causes of gravitational and electromagnetic fields. But it is precisely here that the scope of the physics of fields ends; atomism has shown itself to be irreducible and, following a suggestion of Lorentz, we are obliged to accept atomism, along with the fundamental fields, as a primitive element in our picture of the world. At the moment when the development of the idea of the field, initiated by Fresnel, reaches completion, we find ourselves still in the presence of the same dualism, but now in a much more acute form than at the beginning of this development, for we have successively had to abandon all hopes of resolving this dualism, either in the sense of mechanistic atomism or in the sense of the physics of fields. Hence the result of so much effort to formulate a conception of causality capable of embracing all the laws of the physical world is, in the end, a disappointing hybrid: the laws of electromagnetic and gravitational fields, on one hand, and the laws of the motion of atoms and electrons, on the other, are united only by uncertain and fragile connections. Such a critical situation can only end in a radical solution: to arrive at a satisfactory synthesis, this time it was necessary to sacrifice our habitual conception of causality itself.

Before taking this last step, it would perhaps be good to stop for a moment. The road on which we have just traveled so rapidly through the historical development of the idea of causality in physics may well have seemed difficult. It was necessary to touch upon one after another of the most diverse aspects of phenomena and to marshal a whole arsenal of concepts and laws whose beauty, so deeply felt by the initiates, must seem rather harsh at first contact. But they are all elements we must be familiar with before seeing how they will come together again in the theory of atomic phenomena and blend into a harmony hitherto unknown. How might we briefly summarize the general structure of the laws which together constitute what is called 'classical' physics? At the base we have the laws which regulate the development in space and time of the fundamental fields and the movements of material or electric particles: the spatio-temporal variations of the field quantities have as 'causes' the movements of particles, and these in turn are determined by forces wholly defined with the help of field quantities. The principle of the conservation of the total energy of fields and particles, may be derived from this ensemble of laws to which it would be correct to add an analogous principle of the conservation of momentum. This is, as we have seen, another aspect of physical causality, completely independent of any spatio-temporal localization of phenomena. But the particular lesson which emerges from our glance at history is that at no moment did the principle of causality appear to formulate anything but a practical demand of scientific method, a logical schema according to which attempts were made to group phenomena in the most concise manner possible, following their mutual relationships; in this sense causality takes on meaning only through the concrete content given it, a content which varies with the progress of our knowledge of nature. That this has often given rise to illusions, and that a provisional program corresponding to a stage chosen at random along the infinite path of science has been transformed into an eternal truth, nothing could be more human – but that is another story! We have seen, in fact, that the ideal of physical causality in each of its successive forms, however audacious it appeared, always became too narrow in the end. Neither Laplace's ideal nor Einstein's was able to be realized entirely: nor has anyone succeeded in combining them harmoniously. Consequently, if new experiments emerge which reveal to us the existence of natural laws incompatible with the requirements of classical causality, why should we refuse to abandon these requirements? Why not accept once more the decisive step which alone makes a broader synthesis possible?

The new laws we have just alluded to are those of the constitution of atoms and their principal interactions with the electromagnetic field; they are now codified in a vast synthesis, the 'quantum theory', which encompasses and goes beyond the entire classical physics of fields and particles. If we are willing to close our eyes to certain still unresolved problems relative to the interaction of fields and particles, we can say that the quantum theory is in no way inferior to classical physics in beauty and simplicity. But the particular character of quantum laws is profoundly different from classical causality: The atomic processes governed by these laws present a kind of individuality, refractory to any analysis by means of classical concepts. Hence the very stability of an atom is incomprehensible if the electrons constituting it are imagined as moving in conformity with the laws of mechanics and electromagnetism: these electrons are more properly compared to stationary waves, similar to those which can be established and maintained along a vibrating string. Moreover, light energy is not emitted or absorbed by atoms in the same way as an electromagnetic wave would be but, on the contrary, in a discontinuous manner in the form of a 'quantum of light', that is, a well-determined quantity of energy for each kind of radiation: it is the old corpuscular theory of light which seems to have returned to favor. Clearly in quantum theory the antagonism between the conceptions of field physics and particle physics takes on a totally unexpected sharpness. These conceptions are now idealizations applicable in certain circumstances to light as well as to matter. In their spatio-temporal distribution, light and matter give rise to diffraction phenomena whose interpretation requires a wave theory; in their energy and momentum exchanges, they both behave in conformity with the laws of conservation which apply to systems of particles. In the first case, it would even be possible in principle to realize, through an appropriate concentration of waves, a precise spatio-temporal localization of matter or light, reminiscent of the particle of classical physics; then we would not be dealing with a wave possessing a well-determined quantum of energy and momentum, but rather with a conjunction of numerous waves corresponding to different energies and momenta; and such that in this case no use could be made of the concepts of energy and momentum; for example, it would no longer be possible to speak of a well-determined trajectory of a particle. Conversely, if it is the energy and momentum balance of some atomic process that is being considered, we must introduce waves distributed throughout space and continuing indefinitely in time; hence there can be no question of attributing to the process in question any particular position in space and time. Thus, in the world of atoms, two mutually exclusive points of view are found to be equally indispensable to describe these elusive individualities completely:

> I am a bird, see my wings . . .
>
> I am a mouse, long live the rats!

Never in the memory of physicists had there been such a dilemma. Among the great pioneers of the quantum theory, we see Planck and Einstein vainly trying to resolve it within a framework of classical ideas; Bohr, on the contrary, considered from the start that it was "precisely accentuating this opposition"[5] that it would be possible to release a new harmony. It is from the point of view Bohr finally reached that today we can master the unusual situation presented to us in the quantum domain. In the final analysis, in order to describe the results of our observations precisely, we have at our disposition only the concepts of space and time on the one hand, and those of energy and momentum on the other. It is no longer possible, in describing atomic phenomena, to combine these two systems of concepts in the classical manner: they relate to two mutually exclusive, though equally necessary, aspects of the phenomena. This is a totally new type of relationship between concepts, a new logical category, if you wish, which, with Bohr, we call 'complementarity'. Of the two complementary aspects of physical reality, one – which includes the laws of conservation of energy and momentum – is also, as we know, an aspect of classical causality; the other, on the contrary – the spatio-temporal description – no longer satisfies the requirements of that causality – except in an approximate manner in those cases (which are, moreover, those of everyday life) where the quantum discontinuities can be neglected and where the laws of ordinary physics can then be regained as limiting cases. Naturally we will rarely have circumstances which come sufficiently close to the ideal conditions in which one or the other mode of description can be employed alone; most often we will have to resort to both of them at the same time: the spatio-temporal localization will not be ideally precise; the balance of energy and momentum will not be known exactly. We then find – and this is what Heisenberg's famous 'uncertainty relations' express – that the inaccuracies which necessarily affect the two complementary descriptions influence each other reciprocally: each of them can in principle be reduced at will, but only at the expense of a corresponding increase in the other. Under these conditions, the predictions permitted by the quantum theory concerning the values that physical quantities may assume are reduced to probabilities. Of course, the use of the notion of probability in physics is not a novelty; on the contrary, it constitutes an essential element of the classical theory of the atomic constitution of bodies, brilliantly developed by Boltzmann and Gibbs, and of Lorentz's theory of electrons. From the moment that we deal with the properties of bodies composed of trillions of atoms, it is not, practically speaking, the motions of individual atoms which interest us, but only their mean effects, and to calculate these it is enough to know the most probable behavior of the atoms or electrons under the given conditions. It is no less true that this 'statistical' theory of matter and electricity continued to be based, at least in principle, on a causal substratum. In quantum theory, on the contrary, probability assumes the dignity of a primary notion, conditioned, not by our ignorance of causes which are accessible in principle, but by the limitation imposed on our very possibilities of causal analysis, a limitation imposed by the fundamental quantum laws.

Faced with such serious conclusions, it is only right to ask: are they really necessary? Is the quantum theory not rather an incomplete description of physical reality? Are there not, behind the quantum laws, other entirely causal laws which still elude us and whose statement would require concepts other than those we have utilized thus far? Bohr offered a decisive answer to these questions. To reach it, he had to undertake an extremely profound analysis of the process of physical measurement itself, in its relationship to quantum laws. How can we, in effect, define the quantities which we use in physics other than by denoting a concrete means of measuring them? But the description of our measurement procedures will necessarily use only the classical concepts of space and time, energy and momentum, since these concepts – as we have already observed – are precisely adapted to all descriptions of directly observable phenomena. But, as Bohr's detailed analysis shows us, the inevitable interaction between the object of the measurement – an electron whose position we wish to determine, for example – and the appropriate measurement apparatus, partially escapes our control due to the fact that it is subject to quantum laws, and this uncertainty bears on the complementary aspect of what we wish to grasp through measurement. In the example chosen, measuring the position of the electron with some precision will necessarily involve an uncontrollable exchange of momentum between the electron and the measuring apparatus, with the result that the momentum of the electron can be defined only with a certain latitude linked by Heisenberg's relation to the inaccuracy of the measurement of position. The complementarity is of a sort which is directly connected with the very manner in which we intervene in the course of natural phenomena to submit them to our analysis. In the causal conception of classical physics, the disturbance occasioned by this intervention, far from inevitably having a finite size, was, in principle, supposed to be reducible to an arbitrarily small quantity: we observed the spectacle of nature somewhat like the public at a meeting of Parliament; we could prepare an 'objective' representation of it. With the quantum theory, we are, for the first time in the history of physics, touching a limit imposed on such objectification by natural law. In the study of atomic phenomena, we remain free to choose, in each case, the point of view we wish to take; but once the conditions of observation are fixed, the act of observation itself, by virtue of the quantum laws, creates, so to speak, a phenomenon bearing the mark of the chosen viewpoint and excludes all logical possibilities of employing the complementary mode of description. So then, if the perspective of classical causality proves too narrow, this is no illusion due to a gap in the theory or a flaw in the concepts it uses. Far from being incomplete, the methods of quantum mechanics, within their vast domain of application, permit us to control all possibilities of observation that are compatible with the new situation. But the novelty of the situation lies precisely in the decisive influence of the observer on the very definition of the phenomenon observed, and in the resulting separation of his possibilities of analysis into two complementary aspects.

In the development of Bohr's ideas, we easily recognize the same characteristics which had struck us during our discussion of the earlier crises in the notion of causality. The discovery of the quantum laws, which are irrational from the standpoint of classical causality, obliges us to renounce this unlimited causality: but such a renunciation, imposed by the very enrichment of our experience, obviously has no admission of defeat: we are renouncing a prejudice to rise to a new synthesis. But on this score, what must we think of the fate in store for today's triumphant synthesis? Will not the moment arrive when complementarity, still taught to young students as one of the clear foundations of human reason, will be called 'prejudiced' by the theorists of that day, grappling with the great problem of living matter? Assuredly it would deny all the lessons which we have attempted to draw from the past development of science suddenly to claim that complementarity has been established as a dogma never to be surpassed. Quite to the contrary, if there is a sure prognosis which can be drawn from the lessons of history, it is that other renewals of our conception of the relationships among the various aspects of phenomena must be expected, not, certainly, in the sense of a return to classical causality, but rather in establishing new complementary relationships as much in the domain of physics itself as, perhaps, in other areas where the physicist as yet ventures only timidly. But it would be just as dangerous an error to read into the incessant revision of even the most fundamental conceptions of science an invitation to skepticism. Because these revisions, in the end, affect only the boundaries of the domain of validity of these conceptions, while, integrated as they are into a more and more extended body of doctrine and harmoniously linked among themselves, these conceptions keep their full significance unaltered. Newton's law remains the necessary foundation for the theory of the solar system, despite the enlargement of our cosmic horizon and the refinements of Einstein's theory. Maxwell's equations will always be indispensable to those working with electricity, though it is now realized that they are inadequate to describe electromagnetic interactions on the atomic scale. The predominance of the perspective of complementarity does not prevent classical causality from still being the ideal schema for describing and predicting phenomena in which quantum laws play only a negligible role. This is not simply a practical, but also it is a logical, necessity: in their origins, the classical concepts were adapted to a specific aspect of experience; they may subsequently see their domain of applicability extended or reduced but they can never lose their primary significance. In this sense it may be said that complementarity is a definitive acquisition and, by the depth of its penetration into the analysis of man's relationship with the external world, it is one of the most fundamental contributions physics ever made to the general progress of thought.

What remains for us to examine is precisely what vistas the perspective of complementarity can open outside of physics. Here, fortunately, I will be very brief, for the subject itself permits only a few observations of a general nature, but all the more suggestive than lengthy exposition would be. Indeed, the introduction of new concepts in physics very often exerts an influence on the life sciences, less by applying the results directly to biological or psychological questions than by suggesting new ways of viewing these questions: for physics, due to the relative simplicity of its problems, generally arrives at a rigorous formulation of general principles before the other sciences do, and therefore can offer them models of the scientific method. To cite a single example, we have only to think of the expansion of the English School of physiology in the seventeenth century under the inspiration, not of Galileo's and Newton's results, but of their methods of experimental physics. It is above all an inspiration of the same nature which we can draw from the perspective of complementarity.

This situation has frequently been misunderstood in the course of discussions in recent years concerning the possible role of quantum laws in understanding the elementary phenomena of life. For example, an eminent physicist, having noticed a direct intervention – undoubtedly very real and very important – of these laws in the mechanism of biological processes, decided somewhat precipitously that herein lay the sole basis of a theory of organisms. Equally serious, though in another sense, is the error of those who believe that they have found in complementarity the penetration of irrational or extra-rational tendencies into the very citadel of determinism. This is a tenacious illusion with examples at each step in the history of the development of science. Each step forward in scientific causality, we have seen, implies a critical stage where a conflict explodes between accepted conceptions and the new element which does not fit into them. Those who have fixed their eyes on this stage easily imagine that an appeal to 'causes' that are inaccessible to human reason has been imposed on science. This sort of illusion is never shared by intellects far-sighted enough to understand that no compromise is possible between the scientific spirit and mysticism. Those people who are convinced of the preeminence of the latter do not hesitate to adopt the only logical attitude implied by this conviction: the scholar-inquisitors who condemned Galileo because he was trying to establish the Copernican system of the world on a rational basis certainly saw more clearly than the theologians of the school of Newton who naively believed they had found in this system the proof of the active intervention of Providence in the course of material phenomena.

In reality, it is in quite another sense that the point of view of complementarity could throw some light on the great controversies which, on strictly scientific grounds, are linked to the problems of life. Here we encounter another form of causality based on the notion of final cause. This notion was previously considered fundamental to our entire representation of the world by Aristotle and the Scholastics, and then in a renewed form by Leibniz; but it has been definitively banished from physics since Laplace, and has an extremely poor reputation among most biologists today. The evolution it has undergone, parallel to causality in physics, is marked only by defeats. The peripatetic doctrine of the 'natural place' of the bodies was destroyed by Galileo's mechanics, the 'horror of a vacuum' could not withstand the idea of the weight of air, and the 'principle of least action' was soon stripped of the mystical coloration that Maupertuis still wished to give it. Similarly, in the life sciences, the inaccuracy and sterility of finalist considerations provoke their continual regression, as the analysis of phenomena progresses with the aid of physico-chemical methods. Without mentioning the 'dormative power' of opium, all the successive forms of vitalism, for example, have disappeared without making in the least any lasting contribution to science.

However, there is here an essential difference from the case of physics: though the purely finalist theories of life are hopelessly sterile, it is no less true that we have no other purely physico-chemical explanation to put in their place. And it is undeniable (at least in the opinion of eminent physiologists) that in terms, not of the system, but of methodology, consideration of biological reactions according to their end, their role in the vital organization, is, in any case, very useful, perhaps even indispensable. In order for this organization to subsist, would it not require conditions which no longer allow the organism to be defined as a physico-chemical system, strictly delimited in relation to the milieu in which it lives? However that may be, practically speaking we are faced with a relationship of complementarity here. The biological phenomena confront our attempts at analysis with two complementary aspects: a physico-chemical mechanism, on the one hand, and a vital organization on the other; each must be described and studied according to its own methods, which reciprocally limit their areas of competence. Clearly, there is nothing more in this suggestion of complementarity than an opportunity to make use, without contradiction, of two opposing points of view, both useful, by leaving to each its entire range. But would it not already be a great deal to replace a sterile opposition of systems with a conception which allows the relationships among the laws of life and the laws of inanimate matter to be grasped without doing violence to one or the other?

The domain where teleological considerations have an incontestable relevance, the one, in fact, where they originate, is the domain of psychology. There they reflect the essential properties of psychic pehenomena as they are presented to consciousness. Nevertheless, as soon as these questions are approached, we encounter conflicts of perspective which are just as intense as in biology: the problem of instinct, that of free will, and so on, down to the difficulty of establishing a clear distinction between the subject and object of a psychic experience. To the conscious, subjective aspect of phenomena is opposed a purely biological aspect, a 'behavior' whose laws can be formulated in total ignorance of the phenomenon of consciousness. Here, too, the introduction of relations of complementarity seems indicated in order to do justice to these diverse aspects of psychic life. But if we consider the relationships between the experience of conscious life and objective observations of behavior from the angle of complementarity, it is not, as in physics or biology, an impression of novelty that we feel. We perceive rather something of an echo from the most profound meditations of the philosophers. Listen, for example, to Spinoza: "... mind and body are one and the same thing, conceived first under the attribute of Thought, secondly under the attribute of extension".[6] Or again: William James' beautiful description of a 'stream of consciousness' or the Buddhist comparison of existence to a flame fed by sensations and desires; are not they, too, in different languages so many expressions of the same harmony? When physics, itself, for the first time on its own terrain encountered the active intervention of the observer in the analysis of phenomena, it forged a new language, that of complementarity, to account for a situation new to itself but, in fact, as old as human thought. The physicist now meets with the complementarity between the harmony of existence and the effort of analysis in science as vividly as the Vedic poet who cried out:

> "Who then knows, who has declared it here,
> from whence was born this creation? The
> Gods came later than this creation, who
> then knows whence it arose?"
>
> "He from whom this creation arose, whether
> he made it or did not make it, the highest
> seer in the highest heaven, he forsooth
> knows, or does even he not know?"
>
> (*Hymn of the Rigveda*, X, 129.)[7]

In this way today, the science of inanimate nature is joined again with ancient wisdom; less Olympian, but more human, will it not have greater value in our eyes?

NOTES

[1] Huygens, C., *Discours de la cause de la Pesanteur* (1690) [an appendix to his *Traité de la lumière*; cf. Oeuvres XXI, p. 471 – Ed.].

[2] Leibniz, G., 'Third paper against Clarke', §17 (1716) [Eng. edition: H. G. Alexander, ed., *The Leibniz–Clarke Correspondence* (University Press, Manchester, 1956), p. 30 – Ed.].

[3] Laplace, P. S., *Essai philosophique sur les probabilités* (1814) [Truscott and Emory translation (Dover ed., New York, 1951), p. 4 – Ed.].

[4] Cf. especially an essay by Albert Einstein in *James Clerk Maxwell, a Commemoration*

Volume (Cambridge University Press, Cambridge, England, 1931), pp. 66–73.

[5] Bohr, N., *Drei Aufsätze über Spektren und Atombau*, 1. Aufsatz (1913), p. 19. [Eng. edition: *The Theory of Spectra and Atomic Constitution: Three Essays* (Cambridge Univ. Press, Cambridge, England, 1922), p. 19. As Rosenfeld points out in Note 54 to [1936b] the English translation is defective at this point and we have translated from the German text. – Ed.]

[6] [Spinoza, *Ethics*, transl. by R. H. M. Elwes, Part III, Prop. II, Note – Ed.]

[7] [Translated by Max Müller, cited by S. Radhakrishnan in his *Indian Philosophy*, volume 1 (London 1948), pp. 100–101. – Ed.]

CHAPTER 6

STRIFE ABOUT COMPLEMENTARITY*

[1953c]

Property was thus appall'd
That the self was not the same;
Single nature's double name
Neither two nor one was called.

The Phoenix and the Turtle.

If one wanted to condense into a pregnant formula the fundamental achievement of quantum theory, one would say that it has done away with the dualism of matter and force, which had been the *leitmotif* of scientific thinking since the days of Newton. According to quantum theory, every physical agency partakes of the properties expressed by these two concepts, and its material or dynamical aspects can, so to speak, be conjured up at will by suitably selecting the conditions of observation. Thus, the same agency that usually presents itself to us as a field of electromagnetic force, spread out in space and time, can also appear in the form of a photon, *i.e.* a quantum of energy and momentum satisfying the conservation laws in the way characteristic for material particles. Again, the agency familiarly known under its aspect of material particle as the electron has dynamical functions of essential importance, *e.g.* in establishing molecular bonds of the homopolar type, which can only be described in terms of fields of force obeying specific wave-equations.

This synthesis of matter and force is no idle philosophical flourish: its heuristic power has been strikingly illustrated by Yukawa's beautiful analysis of the nuclear forces. Applying the synthetic conception of quantum theory to the still unknown agency responsible for the binding of nucleons. Yukawa pointed out that this agency, producing a force of limited range, must also show up as a massive particle, whose mass would have a definite relation to the range of the force. The empirical verification of Yukawa's predictions has turned out to be more eventful than one had imagined: but the guiding idea has stood the test splendidly.

The integration of the new synthesis into our scientific thinking, however, raised a logical problem of a kind unprecedented in physics.

How were we to apply to the same physical agency two modes of description so utterly contrasting as that of material particle and wave-field of force without running into contradictions? It took years of hard thinking before the logical relationship between the two 'complementary' aspects of physical reality could be completely elucidated. But the resulting doctrine of 'complementarity,' elaborated by Niels Bohr, exhibits such novel features that even today many people are reluctant to accept it. If we try to analyse the motives for this reluctance, we find that they can all be referred to a common failure to realise the dialectic movement of scientific thought. One cannot attempt to arrest this thought at any stage of its unceasing development without closing one's mind to the understanding of its further progress. Bohr often repeats, with his modest smile, that after all the idea of complementarity only expresses perfectly obvious relationships. If, nevertheless, long explanations are needed to make this obvious concept accessible, it is because one has first to clear a metaphysical jungle which obscures its view. I am not going to tackle this thankless job. All I should like to do is to discuss as briefly as possible some of those aspects of complementarity which have been the subject of recent controversy and try to dissipate the confusion which these controversies have again thrown into a problem long since elucidated.

1. Complementarity and Experience

The first point to realise is that the conception of complementarity forces itself upon us with logical necessity. It is not some fanciful speculation which we could at will accept or reject according to whether we find it conformable to some philosophical criterion or other. It arises from an effort to adapt our ideas to a novel experimental situation in the realm of atomic physics. We can therefore only judge whether is adequately fulfils its function by considering it in relation to our experience and its immediate interpretation. If we consider it from the standpoint of epistemology, we may be led to modify the theory of knowledge; we cannot possibly change the lessons of experience.

Now, from the physical point of view the situation is crystal clear. Starting from the wave-corpuscle synthesis, which embodies the law of the existence of the quantum of action, we are first led by purely mathematical deduction to the uncertainty relations. Indeed, these relations follow from two others, one of which expresses a property of wave groups and the other the way in which the parameters characterising a wave are related to those which define corpuscles. On the one hand we have the well-known reciprocity between the spread in the wave numbers and frequencies of a wave group and the dimensions of the space-time region in which this group is confined. For each wave of the group, on the other hand, the wave number and frequency are proportional to the momentum and energy of the associated corpuscle.

But the decisive step is the physical interpretation of the uncertainty relations. It is at this stage that the idea of complementarity is introduced by Bohr's famous analysis. In any account of atomic processes, a restriction is imposed upon the concurrent use of 'conjugate' concepts, such as position and momentum, because the interaction between the system under investigation and the apparatus used to observe it underlies the quantal laws. Owing to the quantal features of this interaction, the measurement of any quantity pertaining to the system modifies the conjugate quantity to an extent which partly escapes our control and consequently limits the possibility of defining the second quantity in the circumstances of that particular measurment. This limitation, arising as it does from the very manipulation of the measuring apparatus, is clearly inevitable.

We meet here with a deep-lying difference between the classical way of envisaging phenomena and the way in which we have to consider them in the atomic domain. In classical physics it is possible to establish a sharp distinction between the system investigated and the means of observation, and therefore to ignore the latter in framing our conception of the phenomenon. The existence of the quantum of action makes such a distinction impossible because it imposes a limit upon the analysis of the interaction between the system and the apparatus which fixes the circumstances in which we observe it. It is therefore the indivisible whole formed by the system and the instruments of observation which now defines the 'phenomenon.' Any atomic scale phenomenon must thus be conceived as a whole; any attempt to apply to it the same kind of analysis as in classical physics would simply make it vanish. The word 'atomic' here resumes its etymological sense with a more subtle connotation.

From this point of view complementarity appears as a relation between phenomena, of a type unknown in classical physics. It occurs whenever the definition of one of the phenomena to a given approximation only allows the definition of the complementary phenomenon with a proportionally greater latitude, and consequently the strict definition of the one entirely excludes the definition of the other. Such is the situation which is formally expressed by an uncertainty relation.

No hypothetical element eneters into Bohr's interpretation. The reader can convince himself of this by studying the admirable account recently given by Bohr [1] of his long controversy with Einstein. He will see there how Bohr, in countering Einstein's tenacious criticism, gradually refined the expression of his thought and eventually arrived at a formulation which, by Einstein's own admission, is unimpeachable within the scope of our present knowledge of physical phenomena.

The fact that a coherent interpretation of quantal phenomena necessitates a thoroughgoing revision of our most fundamental concepts has nothing to surprise us. Not so long ago a similar situation arose when Einstein showed how the theory of relativity forces us to revise the concept of simultaneity. As soon as the quantum of action was discovered, everybody felt that the incorporation of so strange an element into the theory of atomic phenomena would not be achieved without a radical reshaping of this theory. One might say, for that matter, that the whole history of science is made of successive refinements of our concepts aiming at their adaptation to a more and more detailed knowledge of the phenomena.

In the case of quantum theory, however, the conclusion to which we are driven seems more disquieting than on previous occasions because the explicit consideration of the interaction between the system and the means of observation seems to question the determinism and the objectivity of physical theories. This is why the idea of complementarity is still, in spite of all, looked upon with suspicion; these are the problems which we must face in order to elucidate the true significance of the new idea for the philosophy of science. But it is essential to remember that the complementarity idea is first of all the most direct expression of a fact which the theory of knowledge, just as well as physics, has from now on to accept.

2. The Statistical Character of Quantal Laws

Let us start with one of the most immediate consequences of the relation of complementarity. Two phenomena linked by such a relation can only be connected with each other by a statistical law. This fundamental intervention of the idea of probability in quantum theory is a stumbling block for many physicists, even the greatest. A jest of Einstein's is often quoted in this connection: "That the Lord should play with dice, all right; but that He should gamble according to definite rules, that is beyond me." Since the concept of probability is essentially relative to the degree of our knowledge of the phenomena to which we apply it, it is clear that according to the classical conceptions it could not possibly enter in a fundamental way into the formulation of natural laws, purporting to describe the course of the phenomena independently of our observation. But we have just seen that in quantum physics such a purport becomes illusory; the very definition of the phenomena must specify the circumstances of observation. There is then no objection to conceiving that phenomena might have a proper statistical character, inherent in their nature and accordingly irreducible.

As a matter of fact a re-examination of the role of statistical conceptions in physics [2], in the light of the lessons derived from quantum theory, leads to a complete reversal of the situation. In contrast to the traditional view, one finds that the idea of probability occupies in thermodynamics just as fundamental a position as in quantum mechanics. It has in the two cases exactly the same epistemological meaning.

The fundamental problem of statistical thermodynamics is to reconcile the irreversibility of the evolution of a system, expressed by the second law, with the reversibility of the elementary laws which govern the behaviour of this system. The irreversibility arises from a statistical element superimposed upon the elementary laws, namely, the incompleteness of our knowledge of the initial conditions which determine the evolution of the system in virtue of those laws. Within the frame of classical physics recourse to the calculus of probabilities is justified by the very existence of the elementary laws which establish, so to speak, a deterministic substratum for the statistical arguments. In principle, one argues, it would always be possible to fix the initial conditions and to arrive at a perfectly determinate view of the course of the phenomena. Reasoning by analogy, one is then led to require a similar substratum for the statistical laws of quantum mechanics.

On closer reflection, however, this conception of statistical thermodynamics is very artificial. The fact that the initial conditions of the system are known imperfectly is not accidental; it is just this fact which defines the thermodynamical aspect of the behaviour of the system. Certainly, by a flight of imagination, we may look at the system from the point of view of Maxwell's demon, but this point of view has nothing to do with the actual situation in which we find ourselves in thermodynamics. On the contrary, it excludes the thermodynamical aspect, since it brings out the reversible behaviour of the system on the microscopic scale in contrast to its thermodynamical irreversibility. Already Clausius had a very fine intuition of this peculiar situation, although he was not able to express it as clearly as one would wish, when he said of the second law that "it is not concerned with what heat can do with the help of demons, but with what it can do by itself" [3].

We are now in a position to characterise the situation in a deeper and more precise way than Clausius or Boltzmann could do. The relationship between the thermodynamical aspect and the microscopic aspect of the evolution of the system is truly and properly a relationship of complementarity. Of course, this thermodynamical complementarity has nothing whatever to do, from the physical point of view, with the complementarity between atomic phenomena. The use of the same term in the two cases is only justified by the similarity of the epistemological situations. In particular it does not matter at all whether the elementary laws are assumed to be those of classical mechanics or of quantum theory: the statistical aspect of the second law is entirely independent of the deterministic or statistical character of the elementary laws. This circumstance, incidentally, by itself suffices to show how precarious is the view that the second law would need a deterministic substratum. But I do not want to insist on this point. What I want to stress is the logical feature common to the two problems: the occurrence of a relation of complementarity intimately associated with an essential use of the idea of probability.

3. Complementarity and determinism

Probability arguments have thus from the start been introduced in physics in the spirit of complementarity. Of course, in the initial stage of the development, physicists could not possibly be aware of the new logical relationship which they were thus setting up, so to speak, instinctively.

What hindered a clear recognition of the situation was obviously their belief in determinism. Hence the view that probability can only refer to an imperfect and provisional state of knowledge – can only be, as Poincaré said, a measure for the depth of our ignorance; hence this superficial interpretation of the relation between the thermodynamical aspect and the microscopic aspect of the phenomena, according to which the latter is the deterministic substratum of the former. In classical theory this idea of the deterministic substratum could still put us off. In quantum mechanics we have no longer any choice: the existence of the quantum of action is incompatible with the classical laws, and complementarity appears as the only possible rational interpretation.

We have been forced to the conclusion that classical determinism has only a limited validity by arguments so obvious as to preclude any alternative. This limitation of determinism is thus by no means some arbitrary decree contrary to the spirit of science, but a necessary step which opens the way to a generalisation of considerable scope. We are here witnessing with striking clarity the unfolding of a dialectic movement according to the typical scheme. The impossibility of integrating the quantum of action into the body of deterministic laws of classical physics corresponds to the phase of 'negation'; in fact, as Engels [4] observes, the dialectic negation consists in recognising a limitation to the validity of some concept, in the present instance the concept of classical determinism. This phase, dominated by the contradictions of the old quantum theory, ends with the formulation of quantum mechanics and its interpretation in the frame of complementarity. This is the 'synthesis' in which those contradictions vanish to make room for a new harmony.

Determinism, as a general conception of the form of natural laws, is perfectly adapted to the description of phenomena on the macroscopic scale; its validity in that field is of course not in question: there determinism reigns as supreme as ever. But the danger lies in attempting to enforce it universally. When Laplace formulated the ideal of determinism by his famous fiction of an 'intelligence' which would embrace the state of the whole universe at a given instant and for whom therefore the future would no longer have any secret, he thought that the grand outline which he thus drew would be vast enough to encompass all that human science could ever attain to.

But however bold they seem to our imagination, scientific extrapolations are always superseded in the end by reality. Then their philosophical character alters: they cease to point truthfully in the direction of the dialectic movement and they become fossilised in metaphysical dogmas; they cease to be fertile and become obstacles to progress. Determinism has not escaped this fate; the physicist who still clings to it, who shuts his eyes to the evidence of complementarity, exchanges (whether he likes it or not) the rational attitude of the scientist for that of the metaphysician. The latter, as Engels [5] aptly describes him, considers things "in isolation, the one after the other and the one without the other," as if they were "fixed, rigid, given once for all." Those who want to 'save' determinism must indeed leave the synthetic position expressed by the idea of complementarity and try to restore in some absolute meaning the one or the other of the antithetic aspects of matter, waves or corpuscles. They are of course unable to obtain such a result without doing violence to logic and they fatally fall back into the contradictions which only complementarity can eliminate.

This metaphysical attitude is strikingly illustrated by the way in which Schrödinger [6] tries to limit the scope of quantum theory to the sole wave aspect. To him the field governed by the wave equation is the ultimate reality; the particle concept would just be a name for the element of discontinuity introduced by the 'quantisation' of the field. This, however, is too easy an escape from the central problem of interpretation of quantum mechanics, namely, to express in ordinary language the content of the purely formal concept of quantisation. It is hard to understand that anybody could imagine solving this problem by conjuring it away. In fact we may not renounce the use of one or the other classical concept in the description of phenomena. This description may just as well require the establishment of an energy-momentum balance as a space-time localisation. Unless it be grossly inadequate the theory must enable us to answer all such questions, since we are actually asking them when we decide on our experimental arrangement; and the answers must necessarily be formulated with the help of the appropriate classical concepts, for the very simple reason that it is the function of these concepts and of no others to describe all the arrangements and manipulations of apparatus by which we contact the world of atoms.

It is remarkable how light-heartedly Schrödinger disregards these obvious exigencies of physical interpretation. He extols the precedence of the wave 'picture', but he fails to explain what kind of picture a wave in a multi-dimensional space could possibly convey. At the same time, he tries to exploit just this symbolical character of the waves as an argument against the legitimacy of the particle concept. According to quantum theory, he tells us, there are no longer any 'true' particles because they are deprived of the individuality which we are wont to associate with the particle concept. This is, of course, a complete misrepresentation of the actual situation. The idealisation of particle is used in quantum theory in exactly the same sense as in classical physics; but we must exercise more caution in applying it to the constituents of atoms than to the planets of our solar system. There are phenomena to whose description the particle idea is not applicable at all, but whenever this idea is used in accordance with the rules of quantum theory it does mean a fully fledged particle endowed with as much individuality as the planet Venus.

Determinism is not a formal but a real feature of physical phenomena. The fact that the quantum laws can be formally expressed by a wave equation has nothing to do with it. The type of causality – whether deterministic or statistical – defined by such differential equations can only be ascertained by experience. The misplaced emphasis on the formal aspect is typical of the idealists, who imagine that it is our mind which impresses sovereign stamp on the phenomena. It is therefore surprising to find that a formalistic approach very similar to Schrödinger's is also advocated by physicists like Frenkel [7], who profess to derive their inspiration from a 'materialistic' philosophy. Engels [8] has given a remarkably clear characterisation of what essentially constitutes materialism: in contrast to idealism, it regards the formation of our ideas and concepts as the result of a process of gradual adaptation of mind to experience; ideas are not at the origin but at the end of this process. In this sense materialism is nothing else than the position spontaneously adopted, though not always consciously, by those scientists who have reflected most deeply upon the conditions of their creative activity. Clearly Frenkel has not in mind this scientific materialism; what misleads him is a metaphysical form of materialism which was in vogue in the latter half of the nineteenth century but is now utterly antiquated. This 'mechanistic' philosophy was indeed based upon a purely formal conception of determinism. [1]

Another physicist sharing Schrödinger's and Frenkel's bias for the field idea is Blokhintzev [9]. One of his arguments is based on a simple misunderstanding. He seems worried by the fact that the operations necessary for the localisation of a charged particle may give rise to the creation of pairs of electrons; this, he hastily concludes, would make the localisation of a *single* electron impossible and would force us to regard the electron field as the sole reality. It is true that the localisation of electric charge is a rather delicate problem, but its detailed study [10] shows that the difficult raised by Blokhintzev is illusory.

The other point on which he insists is in itself of great interest, but it must be said that it has nothing to do with the problem at issue. Blokhintzev reminds us that when different fields are strongly coupled together there may exist states of the system in which the particles which would be associated with these fields if they were isolated completely lose their identity. The field idea, therefore, has wider potentialities than that of the particle. It is indeed quite possible that this remarkable circumstance may come to play a part in the analysis of phenomena as yet hardly understood, such as the multiple production of mesons, which are in any case outside the scope of the present methods of quantum theory. But is it good logic to cast doubt on the consistency of a theory by invoking a problematic situation to which this theory may certainly not be applied?

It is therefore not surprising that Blokhintzev's argumentation is wide of the mark. He observes quite rightly that the quantal relation between the parameters of a wave representing a stationary state of the field and those of a corpuscle (its momentum and energy) is only rigorously true in the absence of all interaction. In fact these quantities, like all scientific concepts, are abstractions, whose correspondence with reality can never be more than approximate. The usefulness of such idealisations is measured by the extension of the domain in which they can be applied to the analysis of the phenomena with an accuracy corresponding to the experimental requirements. Everybody knows – this is Bohr's idea of 'correspondence' – that the whole edifice of quantum mechanics and electrodynamics, including the complementarity relations, owes its consistency and solidity essentially to the experimental fact that the electromagnetic coupling is weak. The solidity of the edifice is therefore not shaken by the remark that when this fundamental restriction is abandoned the field concept opens prospects which extend beyond the wave-corpuscle synthesis.

Moreover, let us suppose, to take a definite example, that the coupling between nucleons and mesons is sufficiently strong to give rise to excited nuclear states of the type contemplated by Blokhintzev, in which the particles would be, so to speak, melted together. Even so, after the explosive disintegration of such a state, we shall again need the concept of particle to describe the resulting 'star.' At this level therefore, which is the level of direct observation of the process, we would necessarily fall back upon complementarity. Whatever form the future theory will take, it will have to embody complementarity as a limiting case, just as complementarity itself embodies classical determinism.

While Schrödinger and the Russian physicists each in their own way vainly try to eliminate the particle concept from quantum theory, others, and foremost among them Bohm [11], seem on the contrary to have a foible for this concept. According to Bohm, a description of the phenomena entirely in the spirit of corpuscular mechanics would supply a deterministic substratum for quantum mechanics, whereas the de Broglie wave would only occur as an auxiliary concept; its function would be to represent in some fashion an average residual interaction of the atomic system under consideration with the macroscopic bodies which fix the conditions of the phenomenon. This programme is closely similar to de Broglie's own attempt at interpretation of wave mechanics immediately after its momentous discovery; an attempt which he very soon abandoned, for cogent reasons [12]. It is nevertheless, worth while to criticise Bohm's work, which is certainly the ablest possible effort to explore the furthest consequences of de Broglie's old ideas.

Bohm's argument is very cleverly contrived. One would look in vain for any weakness in its formal construction. What a paradox! Here is a faithful translation of all the formulae of quantum mechanics into a language which to all appearances is that of classical mechanics. Certainly the properly quantal element is introduced in a statistical guise, as an average 'quantal potential' defined in terms of the associated wave, but it is permissible to imagine, and our author does not fail to do so, that this statistical feature has itself a deterministic basis.

Yet, all this seductive construction is just a sham. It is Bohm's pleasure to give his 'hidden parameters' such names as coordinate and momentum, but it is a far cry from the name to the thing. In order to be sure that such and such a parameter really represents the position of a particle it is necessary to examine its relation to the spatial system of reference of some observer, in other words to analyse the *measurement* of the position. But then, as one would expect and as Bohm conscientiously proves, one finds that the identification of the parameters with the corresponding physical concepts is only justified within the limits of the uncertainty relations. Thus, in the end, this subtle and laborious circuit leads us back again to complementarity.

I intentionally confine the debate to the field of epistemology, for the crucial issue is one of logic, not of physics. In any case there is no need to dwell upon the medley of so-called deterministic interpretations of quantum theory, which are now shooting up like mushrooms after the rain: they are all doomed to one fate. Vigier [13], for instance, entertains the hope of finding solutions of the wave equation with a moving point singularity, which he imagines could be identified with the particle associated to the wave; in this way the wave would determine a trajectory of the classical type for the associated particle. There is no need to speculate about the more than dubious existence of such singular solutions; without directly contradicting the fundamental predictions of current theory (which nobody wishes), they could not possibly change anything in the uncertainty relations and their physical interpretation.

The idea of the 'double solution,' revived by Vigier, was a prominent feature of Louis de Broglie's early attempt at a consistent formulation of the physical meaning of wave mechanics [14]. It has a chequered history. In 1909, in the course of the famous Salzburg debate on the quantum theory of radiation [15], Einstein had suggested this very idea of the 'double solution' as a possible means of harmonising the wave properties of light with its corpuscular aspect, symbolised by the concept of photon. The same idea naturally occurred to Louis de Broglie when he was confronted with the similar problem of reconciling the wave and corpuscle aspects of matter. Neither Einstein nor de Broglie, however, could make anything from it. There is less reason than ever today to revive it, since in the meantime the crisis of the wave-corpuscle dualism has been resolved on a higher plane of the theory of knowledge.

It is indeed right to say that we rise higher in the realm of rational knowledge when we get rid of some metaphysical tether. People are too much wont to regard determinism as a symbol of scientific 'certainty' to which they metaphysically oppose arbitrariness. Probability, however, does not express chance without rule, but just the contrary: that which in chance is subject to rule. A statistical law is above all a law, that is, the expression for some regularity, and as such an instrument of prevision. The scope of the calculus of probability is not narrower but much wider than that of the classical methods, since this calculus may be applied to an immense variety of situations in which deterministic laws, assuming that they exist, would be of no help whatever. When we substitute for the deterministic conception the wider and more flexible idea of probability, we do not lose by the change. Far from restricting our means of scientific prevision, we increase their power. See how neatly the statistical predictions of quantum mechanics are adapted to the interpretation of the experimental results! They are all susceptible to verification and they embody every single detail that experience reveals to us. Neither too much nor too little: what more do people want? They want to 'justify' the use of the calculus of probability in quantum physics and they believe that they cannot do it except by some underlying determinism. They seem to forget that determinism itself, in its legitimate domain of classical physics, has no other justification than that which it derives from experience. The justification for complementarity is exactly of the same kind.

4. Complementarity and Objectivity

The generalisation of Laplacian determinism is not the only new element with which atomic physics has enriched the theory of knowledge. This generalisation, as we have seen, has forced us to define more precisely what we mean by phenomenon in order to do justice to the 'atomic,' indivisible character of the processes into which the quantum of action enters essentially. The unfamiliar features of this atomicity of quantum processes are of course reflected in the new conception of phenomenon. For the first time in the history of physics we are compelled to abandon the distinction, so convenient, between the observer and the object of observation. We can only describe an atomic system by specifying the conditions under which we observe it, and the various aspects of the system which we discover by varying these conditions can only be combined into a complete description within the frame of the relations of complementarity.

This situation might easily appear to undermine the very objectivity of scientific fact, and many people have actually been misled by the appearances. On the one hand we see the incorrigible idealists again heralding the bankruptcy of science. Edouard Le Roy's [16] old paradox 'le savant crée le fait' is again held in honour, now, so it seems, by the scientists themselves. On the other hand, adherents of the narrow and antiquated brand of materialism alluded to above, in a sincere but misplaced effort to defend the scientific ideal, vituperate Bohr and his school, whom they accuse of disavowing it. Let us examine in all simplicity, *sine ira et studio*, how things really stand.

The theory of relativity is rightly regarded as a paragon of objective analysis. Let us therefore ask the relativist by which signs he recognises this objectivity. "Look," he would reply, "at this set of equations which describes the course of some phenomenon. Each symbol represents the measure of some physical quantity, but the way in which this measure is obtained has no influence upon the description of the phenomenon. It does not enter into the equations." "I notice, though, that those symbols bristle with indices. What do such indices represent?" "They serve to relate the physical quantities to the fixed system of reference with respect to which the observer describes the course of the phenomenon." "Thus the observer does have some influence on the expression of the physical law, for the numerical values of the symbols with indices will vary according to the standpoint from which he determines them. Is not thereby a subjective element surreptitiously introduced into the theory?" "Not in the least. The observer may choose his mode of reference as he pleases. The *form* of the equations will always be the same. It is this *invariance* which confers to the theory its objectivity." "All right. Shall we now change places and resume our dialogue, this time about the equations of quantum mechanics? *Mutatis mutandis* it can be repeated from beginning to end." Now, physical quantities will correspond with operators susceptible to an infinity of numerical representations. Each of these representations refers to particular conditions of observation, but the equations connecting the operators are invariant for the canonical transformations which express the passage from one mode of observation to another. These equations represent the objective content of the theory, the objective expression for the quantal laws of nature.

We thus identify the objectivity of a physical theory with its invariance for an appropriate group of transformations. It is not by chance, of course, that this argument is exactly that already used by Henri Poincaré [16] to refute Edouard Le Roy. Obvious as it is, it suffices to dispel the current misunderstanding that "in atomic physics the observer influences the course of the phenomenon." I know that one can imagine certain experimental situations for which the predictions of quantum mechanics have a rather strange look. According to whether the observer decides to act in one way or another upon some part of the system, he will make different predictions about the behaviour of some other part even when it is no longer coupled with the first. Just in such cases the careful terminology established by Bohr reveals its usefulness: the two possible interventions of the observer define two different phenomena; to each of these phenomena there corresponds a well-defined and perfectly objective set of possible predictions. The freedom of the observer reduces to his deciding upon the particular intervention which defines the phenomenon he wishes to study. This makes it impossible for him to study the complementary phenomenon upon the same system, but he can always do so by experimenting on some other identical system, placing it in the appropriate conditions. Thus the observer's intervention does not introduce any arbitrariness into our knowledge of atomic processes. In contrast to Einstein's [17] contention, our control of these processes is as complete as possible; the reciprocal limitations expressed by the uncertainty relations are objective, inherent in the nature of such processes. But between the objectivity of the relations of complementarity and the objectivity of the laws of classical physics there is a difference in *quality*, arising from the dialectical character of the passage from the one to the other.

5. Complementarity and Materialism

We may thus say that the idea of complementarity succeeds in reconciling a complete objectivity of the description of natural phenomena with the necessity of taking account explicitly, in this description, of the conditions of observation. This conclusion may appear bewildering if it is appraised 'metaphysically,' but not if it is regarded as the outcome of a dialectical movement. From the dialectical point of view it is almost self-evident to observe that the essential part played by the observer in the definition of the phenomena is perfectly consonant with the fundamentally materialistic character of science. For materialism, in the scientific sense, is just (as I pointed out before) the philosophical expression for the process of gradual refinement of our mental representation of the external world. Trouble can only arise if materialism becomes fossilised into some metaphysical system and anything not 'dreamt of in this philosophy' is branded as 'idealistic.'

Certainly our representation of the world is no longer the crude chap-book imagery which contented the naïve mechanistic materialism of the nineteenth century. Quantum theory rather calls to mind Picasso's art. It does in the last resort make use of the concepts which describe everyday experience, but these concepts no longer occur in their familiar interconnections. A whole set of precise and complicated rules is needed to express, with the help of these concepts, the objective relations between phenomena disclosed by experience. But this growing abstraction of our view of the external world should by no means be confused with an abandonment of materialism. On the contrary, it means its consolidation, since it represents a finer adaptation of our thinking to the exigencies of nature.

It is a pity that the creators of the conception of complementarity have not paid sufficient attention to this important aspect of the problem, and have sometimes expressed themselves in ambiguous or even frankly idealistic terms. But this ought not to disconcert us. Are we going to lay complementarity under an interdict because Heisenberg is an idealist? We might just as well condemn the *Principia* because Newton dishes up his dialectics in the guise of Puritan theology. In the theory of knowledge, just as in physics, what counts is concrete reality and not the metaphysical attire which clothes it at the whim of fashion.

An example will show what one ought to think of Heisenberg's idealism. Let us take the recent article [18] in which he analyses the process of successive adaptation of physical theories to our growing knowledge of natural laws. With a sure and profound intuition, he perceives and describes the dialectical character of this process. The whole masterly discussion is conducted in the purest spirit of scientific materialism. Then, suddenly, as a screeching dissonance, comes the idealistic conclusion, which nothing in the preceding arguments prepares or justifies. From the plane of scientific thinking the author somersaults to a mystical plane without any rational connection with the first. The word "geistig," which he insistently repeats in his peroration has no epistemological significance: it is the shibboleth of a class. Heisenberg's idealism concerns sociology, not the theory of knowledge. In this last domain the power of his creative thought belies the poverty of his philosophy.

6. THE LIMITS OF COMPLEMENTARITY

Bohr is too subtle a dialectician to fall into the same inconsistency as Heisenberg. He is hostile to all metaphysics, to all dogmatism, because he experiences the dialectical movement of nature as a living reality with which he has completely identified his thought and his feelings. Naturally enough, dialectical relations take in his mind the shape of complementarity relations. It is in this sense that Bohr's efforts to extend the idea of complementarity to all domains of knowledge should be understood. He insists above all on the sterility of metaphysical oppositions and on the broader and more flexible framework which complementarity offers for possible syntheses. Certainly it would be premature to assert that we shall be able to fit the dialectics of life or consciousness to such a framework. But in any case the relation of complementarity is the first example of a precise dialectical scheme, whose formal structure has been accurately analysed by the logicians [19]. This is a new fact whose importance could not be exaggerated.

While emphasising the universal aspect of complementarity, Bohr is more keenly aware than anyone else of the danger of treating this conception metaphysically. He insistently warns us that in the evolution of physics complementarity is just a stage which we shall soon have to leave behind us. We can already clearly discern the limits of the domain within which one can define complementary phenomena in the sense of the uncertainty relations. The problems raised by the field of nuclear force and the various particles associated with it demand for their elucidation new methods which will very probably mean a qualitative change in our conceptions comparable to that embodied in the passage from determinism to complementarity. But just as the laws of quantal phenomena and their complementarity relationships cannot be formulated without essentially making use of the deterministic laws of classical physics, likewise complementarity will necessarily form the basis of the new conceptions which will transcend it. In generalising determinism, complementarity does not destroy it; it rather makes it more fruitful and firmer by assigning it its proper limits. Likewise the future theory will reinforce complementarity by fixing its place within a still wider synthesis.

I have now reached the end of a long discussion in which I have chiefly tried to say what atomic phenomena are not in the hope of conveying a clearer idea of what they are. This is not without some drawbacks: I seem to be at pains to plead a cause long since decided, and I seem to quarrel with everybody. However, in the words of the candid Machiavelli [20], "my intention being to write something which might be of use to him who understands it, it seemed to me more advisable to consider things as they are in reality rather than in imagination."

NOTES

* The present essay is a revised version of the author's contribution, entitled 'L'évidence de la complémentarité', to the collective volume, *Louis de Broglie, physicien et penseur*, Paris 1952.

[1] The prevailing muddle between mechanistic and scientific materialism may be traced back to Lenin. In his famous politico-philosophic pamphlet *Materialism and Empirio-criticism* (1908), Lenin, while no doubt intending to follow Engels, in effect advocated mechanistic materialism, with its metaphysical conception of determinism. See in this connection the remarkable and too-little-known analysis of Lenin's ideas by A. Pannekoek, *Lenin as Philosopher*, New York 1948.

BIBLIOGRAPHY

[1] Bohr, N., Discussion with Einstein on Epistemological Problems in Atomic Physics, *Albert Einstein: Philosopher-Scientist*, Library of Living Philosophers, La Salle, Ill., Vol. VII, 1949.

[2] Bohr, N., Faraday Lecture, *J. Chem. Soc.*, 1932, pp. 376 *et seq.*

[3] Clausius, R., *Die mechanische Wärmetheorie*, 2nd edn., 1876, p. 316.

[4] Engels, F., *Herrn Eugen Dührings Umwälzung der Wissenschaft ('Anti-Dühring')*, Moscow edn., 1946, pp. 173–4. [See Engels, *Anti-Dühring*, Part I, Chap. 13, first published 1878 – Ed.].

[5] *Ibid.*, p. 24.

[6] Schrödinger, E., 'Are there Quantum Jumps?', *Brit. J. Philos. Sci.*, **3** (1952), 109, 233. See also E. Schrödinger, 'The Meaning of Wave Mechanics', *Louis de Broglie, physicien et penseur*, Albin Michel, Paris 1952.

[7] Frenkel, J., *Uspekhi fizich. nauk* **42** (1950), 69; **44** (1951), 110.

[8] Engels, F., *Ludwig Feuerbach and the End of Classical German Philosophy* [first published 1888], Chap. II.

[9] Blokhintzev, D., *Uspekhi fizich. nauk* **42** (1950), 76; **44** (1951), 104.

[10] Bohr, N. and Rosenfeld, L., *Phys. Rev.* **78** (1950), 794. [English translation; this volume, p. 401].

[11] Bohm, D., *ibid.*, **85** (1952), 166, 180.

[12] *See* de Broglie, L., *C.R. Acad. Sci.* **233** (1951), 641.

[13] Vigier, J., *ibid.*, **233** (1951), 1010.

[14] *See* de Broglie, L., *ibid.*, **233** (1951), 1012; and 'Vue d'ensemble de mes travaux scientifiques', *Louis de Broglie, physicien et penseur*, Albin Michel, Paris, 1952, pp. 465 *et seq.*

[15] *See*, for example, Rosenfeld, L., 'La première phase de l'évolution de la théorie des quanta', *Osiris* **2** (1936), 149. [English translation this volume, p. 193].

[16] *See* Poincaré, H., *La valeur de la science*, 1905, Chap. X. [English translation by G. B. Halsted: *The Value of Science*, Science Press, New York, 1921 and various reprints, Dover, New York – Ed.].

[17] Einstein, A., Rosen, N., and Podolsky, B., *Phys. Rev.* **47** (1935), 777.

[18] Heisenberg, W., *Dialectica* **2** (1948), 331.

[19] *See* in this connection the interesting treatise by Mme P. Destouches-Février, *La structure des théories physiques*, PUF, Paris, 1951.

[20] Machiavelli, N., *Il Principe*, Chap. XV. [*The Prince*].

CHAPTER 7

COMPLEMENTARITY AND STATISTICS I AND II

[1958b]

PART I

There still lingers a popular view of science (which scientists did much to encourage and nothing to dispel), according to which a scientific theory is accepted when it is in accord with *all* the facts it purports to describe: hence the unrockable certainty of scientific statements in contrast to the frailty of human judgments of other origin. Apart from the obvious tautology implied by the consideration of the totality of the evidence, this naive picture of the way in which science is made is as remote from reality as the meticulous preciseness of pre-raphaelite painting. Physicists know very well, when they are at work (although they do not usually put it down in so many words when they write about their work), that their decisions to adopt or reject theories are always based on a *very small* number of facts, which they regard as crucial.

The conviction that quantum mechanics adequately accounts, as Dirac put it, for 'a large part of physics and the whole of chemistry' has not been acquired by an exhaustive theoretical survey of this vast field – a purely mythical proposition – but by the knowledge that in a few selected cases, sufficiently simple but nevertheless involving the essential features of the theory, it was actually possible to carry through the quantal computations to the same exquisite accuracy as the spectroscopic measurements. I still vividly remember the intense satisfaction with which the result of Hylleraas' work on the ground state of Helium was received. Such a computation was viewed at the time as a pioneering enterprise of high virtuosity, and it was virtuosity well-spent. Once quantum mechanics had passed through this severe ordeal, everyone was convinced that with it we held the key to the whole problem of atomic and molecular structure.

This behaviour of the physicists is not as reckless as it seems; not more so, in fact, than that of the engineer wanting to assess the strength of a steel girder: he will not probe every inch of it, but only perform a small number of adequately distributed *sample* measurements. Likewise, the scientist proceeds by *sample tests* of his theories and relies upon this sampling in assessing their adequacy; whether the sampling is judicious, or how extended is its scope, can only be ascertained by trial and error. Thus, in science as in every other walk of life, all decisions, all actions are taken on the strength of truth-judgments which are of a fundamentally statistical character. There is here no sharp distinction between 'practical' and 'philosophical' truth: attempts to draw such a distinction have only led to 'Scheinprobleme', as for instance the very problem of the 'foundation' of statistical causality.

Mathematicians know that the concepts and axioms of the theory of probabilities form a self-contained logical system which does not need any extraneous foundation. Professional statisticians use this theory as a tool perfectly adapted to their needs, and never had any qualms about its 'ultimate' justification. Only the physicists, by their uncritical acceptance of a deterministic philosophy, were long prevented from recognising in statistical causality the actual form of all scientific reasoning. When, however, in order to formulate in rational terms (i.e. without running into *trivial* contradictions) the fundamental laws of atomic processes, dominated by the existence of the quantum of action, it was found necessary to introduce *complementarity* as a new kind of logical relationship between physical concepts, it became clear that the causal links between atomic phenomena are no longer necessarily determinate, like in classical physics, but do indeed belong to the more general, statistical types.

To imagine that the elucidation of the conceptual framework of quantum theory, brought about by the idea of complementarity, must have won the universal adhesion of the physicists, would be to underestimate that tenacity of metaphysical prejudice, which calls forth what might be described as the Palmström attitude. One remembers how Christian Morgenstern's hero reacts to a traffic accident of which he has been the victim:

> Eingehüllt in feuchte Tücher,
> prüft er die Gesetzesbücher
> und ist alsobald im klaren:
> Wagen durften dort nicht fahren!

Und er kommt zu dem Ergebnis:
Nur ein Traum war das Erlebnis.
Weil, so schliesst er messerscharf,
nicht sein *kann*, was nicht sein *darf*.

What appeared, at the beginning of this century, to be just a spark of light fantasy, has, by the middle of it, become a grim reality: the Palmstrøm argument has intruded the domain of physics, that stronghold of rational thinking. Faced with the plainly apparent breakdown of deterministic causality in atomic processes, our Palmströms decide that this must be impossible, since they cannot find anything of the kind in the *Anti-Dühring*. Thus arises the paradoxical situation of a physical theory, well-established for a generation, being suddenly questioned without the slightest indication of any experimental ground for revising it. Quantum theory, we are told, may be allright so far as it goes, as a superficial description of experience, but its 'philosophy', since it is not permissible, must be wrong.

The stumbling block to the understanding of complementarity and the appreciation of its significance is undoubtedly the belief in the absolute validity of the deterministic form of causality: for those who entertain this belief, statistics must appear as an incomplete mode of description, which has to be based upon some deterministic substratum. Thus, by postulating the necessity of such a substratum in the atomic case, they create for themselves the problem of finding it, – without realising, apparently, the deceptive character of such an argument.

The traditional conception of determinism is not so simple as it looks. The idea is that, if you know all the *relevant* circumstances at some instant, you are able to predict in a unique way what is going to happen next. The restriction implied by the word 'relevant' is obviously essential in order to avoid a tautology: but the very necessity of thus selecting from the infinite wealth of events only a limited number of data on which inferences about the future (or the past) are to be based emphasises the highly abstract character of deterministic causality. That such inferences should at all be possible, that it should be possible to set up differential equations involving only a few variables and allowing valid predictions or retrodictions about the states describable in terms of these variables, is indeed very remarkable; but it is a situation of fact, a 'law of nature', and as such it partakes of the limitations inevitably affecting all the concepts and statements by which we try to express in finite terms the salient features of an infinite reality. No doubt, the domain of validity of determinism as the form of causality characteristic of the laws of matter in bulk – the laws of motion of material bodies as well as those of electromagnetic radiation and gravitation – is so vast that it was natural to regard this type of causality as the universal one, so long as there was no suggestion in experience of situations to which it would not apply: but it is fatal to lose sight of the provisional character of conceptual extrapolations of this kind and to proclaim them 'necessary' features of reality.

The *practical* situations in which we find ourselves with respect to causal prediction are at most approximations to the ideal case of determinism; strictly speaking, our predictions, hampered by incomplete knowledge of the initial conditions, are always of the *statistical* type. Certainly, in many cases the statistical element is a secondary feature, which could conceivably be removed, or at least reduced, by increasing the accuracy of the determination of the initial data. This is even true for typically statistical processes, like the casting of dice, whenever the initial conditions are effectively, or ideally, under our control. It is just the consideration of such cases which encouraged the identification of statistical causality with incomplete knowledge of an essentially deterministic substratum. Insistence on this aspect of statistical situations, however, tends to obscure a much more important circumstance: if the specification of the initial conditions would actually reach the accuracy required for quasi-deterministic predictions, the phenomenon to which such predictions would refer would be qualitatively different from that corresponding to the less detailed specification on which statistical predictions are based; if we were in possession of a complete mechanical description of the motion of the dice, there would be no point in betting on the outcome of the throw. This obvious remark warns us that the problem of causality cannot be discussed *in abstracto*, but that the type of causality involved in the description of a phenomenon is essentially determined by the very conditions by which this phenomenon is defined.

The character and implications of this fundamental dependence of causality upon the definition of the phenomena will become clearer if we pass on to less trivial problems. Let us consider the way in which the macroscopic, directly observable properties of a gas, for example, are derived from the picture of this gas as a system of atoms; let us for the moment disregard quantal effects and just assume that the atoms obey the laws of classical mechanics: this is anyhow a good approximation within wide limits of pressure and temperature. The macroscopic parameters, such as temperature and entropy, appear as statistical averages over the variables describing the states of the individual atoms: but in this case, it is obviously futile to pretend that the macroscopic behaviour of the gas is thereby reduced to that of a mechanical system. Not only is it practically impossible to obtain the initial conditions of the atomic system, but even if we did know them, it would take an electronic computer decades to ascertain the steady state of the gas and work out its temperature. One is tempted to say that nevertheless the operation is possible 'in principle', that it can be 'conceived' without conflicting with the basic assumptions of the theory. Such an unrealistic view, however, would be a misunderstanding of the meaning of scientific theories: it would disregard, in this specific case, a fundamental function of the statistical element in the atomic description of the gas, viz. to help us, quite efficiently, over an obstacle which a straightforward application of the methods of mechanics could not possibly surmount.

The essential point to notice here is that the introduction of statistical causality appears as a tool especially well adapted to the conditions under which we actually *observe* the behaviour of systems of atoms. The macroscopic concepts have been defined in immediate relation to these conditions of observation; with their help we can give a consistent account of the properties of the gas on the scale of direct observation. Now, we want to compare this macroscopic picture with an entirely different one, on the atomic scale; and we find that the link between the macroscopic and the atomic parameters is essentially a statistical one. It would be pointless to add that this statistical point of view is 'merely' practical: for what would be the use of the atomic theory if it could not yield practical results? In fact, here is a case in which we are strongly reminded of the impossibility of dissociating theory from practice without foundering into barren idealistic speculations.

PART II

There is still much more to it, however. It is not true that the two modes of description, the macroscopic and the atomic one, are entirely equivalent. Indeed, they are contradictory in one important feature: a mechanical system of atoms does not exhibit the irreversibility in time characteristic of macroscopic behaviour. This is a dialectical contradiction, the solution of which, given by the ergodic theorems of statistical thermodynamics, has resulted in a true synthesis of the atomic and macroscopic viewpoints, – a synthesis most fundamental for our whole conception of the physical world. Now, the decisive step in bringing about this synthesis has precisely been the introduction of statistical causality in the passage from the atomic to the macroscopic description. Moreover, the ergodic theorems, which form the framework of the statistical theory, are so general that they remain valid when we abandon the deterministic laws of classical mechanics for the atoms and replace them by the statistical laws of quantum mechanics; in other words, in its essential application to the atomistic interpretation of thermodynamics, statistical causality does not at all require any deterministic substratum.

From the atomistic point of view, the primordial type of causality is statistical; an apparent determinism results from particular statistical distributions with very sharp and narrow peaks. The very circumstance which makes statistical causality indispensable in the atomistic treatment of matter in bulk – the immense number of atoms in any macroscopic body – is also that which makes possible the occurrence of such nearly deterministic distributions. It is interesting to observe that when dealing with systems small enough to exhibit noticeable effects of atomic interactions – such as Brownian motion – we may apply with some latitude either mode of description: the macroscopic, deterministic one or the atomistic, statistical one. The approximations involved are governed by the square root of the number of atoms in the system, and depend on this quantity in a reciprocal fashion: the relative fluctuations about the macroscopic behaviour are inversely proportional to it, whereas the mean deviations from the statistical equilibrium distribution increase in the same proportion. The product of the fluctuations of any two directly measurable quantities respectively representative of the macroscopic and atomistic aspects, such as the position and the entropy of the system, is therefore independent of its size; the order of magnitude of this product is determined by a universal factor characteristic of atomic dimensions, for which we may take, for instance, Boltzmann's constant, or the inverse of Avogadro's number.

I have now said enough to bring out the striking similarity of this *logical* situation with the one which was later encountered in quantum mechanics and was given the name of complementarity. We are here dealing with an aspect of atomism which has nothing to do with the quantum of action, but essentially only with the large number of atomic elements forming a unit of macroscopic description; and we see how the dialectics of this problem finds its natural expression in a typical relation of complementarity. It need not surprise us that the epistemological implications of Boltzmann's great work were not clearly recognised at a time when even the few scientists who, like Boltzmann himself, resisted the idealistic obscurantism of the official philosophy, remained unaware of the dialectical character of scientific thinking. A much more acute crisis was required to shake physicists from their philosophical numbness.

In dealing with quantal processes, we are indeed faced with such a sharp dilemma that it is impossible to dodge it. The very definition of the quantum of action exhibits a seemingly paradoxical character, since the two quantities – energy-momentum and period-wavelength – whose product is equal to Planck's constant refer to mutually exclusive aspects of the atomic object, respectively idealised as a punctiform particle and an indefinitely extended plane wave. It is therefore clear that the existence of the quantum of action imposes a limitation to any analysis of an atomic process in terms of such classical idealisations, and thereby introduces a statistical element into the description. It is impossible, for instance, to assign the electron in a stationary state of the hydrogen atom a continuous motion along a trajectory in the classical sense, because such a motion would imply a continuous variation of action, in violation of the quantum law. The quantal transitions thus appear as phenomena only definable as a whole, in terms of the initial *and the final* state; and in view of the multiplicity of possible final states, their occurrence can only be characterised by relative probabilities. It is also clear that no introduction of 'new concepts' can have any influence on this situation, since such concepts (in order to be intelligible) must be somehow related to the classical concepts adapted to the account of our direct experience, and this relation must be compatible with the law of the quantum of action: thus, all typically quantal concepts, like spin or parity, are defined in classical terms, but their definition necessarily contains a statistical element.

It is of decisive importance for the consistency of quantum theory that the limitations imposed by the existence of the quantum upon the use of classical ideas should not be absolute, but only relative, and, in fact, reciprocal. Every classical concept corresponds to a situation which can be reproduced, in principle, with any accuracy by an appropriate experimental arrangement: but every such arrangement precludes (up to a reciprocal latitude) the possibility of defining the concept which characterises the complementary situation. This ensures that the statistical causality inherent in the description of quantal processes does not imply any incompleteness of this description. Complementarity provides a logical framework wide enough to comprehend in a rational way those aspects of the atomic processes which, while mutually exclusive, are equally necessary parts of an exhaustive account.

In this connection, it must be stressed that the requirement of completeness cannot be formulated absolutely, but depends on the type of causality of the mode of description envisaged: it is different for a deterministic and for a statistical theory, and it is unreasonable to demand from the latter that it should conform to criteria applicable to the former. Failure to realise this has led to a curious misapprehension of the consequences of the reciprocal limitations of measuring processes for the completeness of the quantal description of the phenomena. When we have ascertained by observation some characteristic of a system, we have irrevocably deprived ourselves of the possibility of ever knowing anything about the complementary feature of this system: how then, it is asked, dare we assert that our description of the system can ever be complete? Here again, it is apparently overlooked that whatever observations are made upon atomic systems are in the nature of a *sampling*, a very minute sampling indeed, of the innumerable *identical* systems which make up the universe. The identity of constitution of all atomic systems of the same species, implied in the formulation of such a fundamental law of nature as the exclusion principle, is quite essential for the atomistic picture of the universe to make any sense at all. Only this identity gives a meaning to probability statements about the behaviour of a system of definite species. The complementarity of different types of behaviour implies that we cannot observe these types of behaviour on the same sample; but we have many samples at our disposal, and the fate of a particular sample is of no interest: we are only concerned with predictions valid for *any* sample.

No less essential than completeness is the requirement of objectivity which every scientific theory must fulfil. On this point there is also a considerable amount of confusion: the terms 'subjective' and 'objective' are used about as ambiguously as 'tant pis' and 'tant mieux' in the *Sentimental Journey*. G. N. Lewis regarded entropy as a 'subjective' concept because of its statistical origin, and seemed to think it was the worse for it. Weizsäcker calls the mode of description of quantum theory 'subjective', but has his own reasons for hailing this as 'tant mieux'. What these eminent physicists intend to express, albeit with conflicting feelings, by the word 'subjective' is the important fact that in thermodynamics as in quantum theory the specification of the conditions of observation enters explicitly into the definition of the basic concepts. This fact, however, is not as characteristic as they think; to give it such prominence rather tends to obscure the deeper epistemological significance of complementarity.

In fact, the widening of our conception of causality implied in complementarity does not in the least conflict with the requirement of objectivity as commonly understood. The argument which led to the recognition of the new form of complementary relation between quantal phenomena is a quite rigorous one: at no stage has any arbitrary postulate been introduced about what 'reality' ought or ought not to be like. It is true that in our account of atomic phenomena explicit reference must be made to the interaction of the atomic objects with the systems of macroscopic bodies constituting the apparatus by means of which these objects are observed. But this circumstance, far from bringing in any subjective element, enables us to put the fundamental laws of quantum theory in a form valid for any observer, and accordingly objective in exactly the same sense as the laws of classical physics. It is clear that, strictly speaking, all physical concepts imply some reference to conditions of observation. In classical physics, we may usually forget about it and indulge in the illusion that we are contemplating the unrolling of the phenomena as a spectacle in which we have no part; when dealing with atoms, we must adapt our epistemology more closely to the real situation we occupy in the universe.

At this stage, it would seem that complementarity, vindicated from the worst misconceptions, appears in its true meaning, as a decisive step in the dialectical unfolding of our knowledge of Nature. In dealing with the atomistic constitution of matter, and again with the laws of atomic systems, this dialectic process has involved an extension of our conception of causality in a statistical sense, presenting at both stages the same synthetic character expressed by relations of complementarity. It is a cruel irony of fate that physicists who believe themselves the guardians of the Marxist tradition should be blind to such a momentous dialectical development and oppose to it the most outworn metaphysical preconceptions. Having failed in the attempt to uphold determinism in quantum theory, they are now trying to argue that one ought to consider a kind of hybrid causality, partaking in some confused way of both determinism and chance: at each 'level' of description, there would be a statistical residue, which would have to be underpinned by the determinism of some hidden processes going on at a 'deeper' level. They are thus driven to a frantic chase of determinism into an infinite sequence of deeper and deeper levels of reality: a purely imaginary construction, as ingenious and as misguided as those of medieval scholasticism. Unsupported as it is by any concrete evidence, it indeed appears to serve no other purpose than to provide our critics with a perpetual line of escape from embarrassing difficulties.

This extraordinary behaviour is prompted to a large extent by another Palmström motive: the wickedness of positivism. There is an undeniable similarity between the epistemological conclusions, drawn from the peculiar character of the quantal laws, about the active part of the observer in defining the phenomena, and the insistence of the early positivists on the essential part played by our sensations in determining our knowledge of the external world. This only means that, to that extent, the early positivist movement was a healthy reaction against the shallow metaphysics of mechanistic materialism. But why should scientists be made responsible for the later positivists' blundering into a metaphysics of their own? No scientist would accept the extreme positivist contention that there is nothing more in statements about phenomena than the conceptual expression of relations between sensations: he would maintain that such statements refer primarily to real processes of the external world; our mental representation of these processes being itself, of course, subject to definite laws depending upon our sensorium.

Clearly, the point of view of complementarity finds its natural place in the line of development of a philosophy rooted in a straightforward common sense approach to the realities of existence, – a philosophy which historically found its most vigorous expression in Josef Dietzgen's early writings. This is the true line of development of the philosophy of science, because on the one hand scientific thought can only thrive if it is in harmony with the social function of scientific activity, and on the other hand all philosophical abstractions can only find in science a sure foundation.

CHAPTER 8

MISUNDERSTANDINGS ABOUT THE FOUNDATIONS OF QUANTUM THEORY

[1957a]

Recent criticism of the foundations of quantum theory originates from a number of physical and epistemological misconceptions. To point out, with painful explicitness, the most serious of these might be helpful to those whom this criticism seems to have caught unprepared.

1. 'Interpretation of the formalism'

The critics have put forward the novel suggestion that an alternative 'interpretation' could be found for the mathematical formalism of quantum theory. Physicists know what it means to interpret physical experience in terms of mathematical concepts; but to imagine that they might have to interpret somehow pre-existing mathematical symbols in physical terms is an idealistic construction alien to the spirit of science. A physical theory is a consistent set of relations between certain concepts; these concepts usually refer to magnitudes and are therefore associated with numbers, but they are primarily the mental representation of concrete physical realities and as such must be defined in words, linking them (either directly or indirectly) to *everyday experience*, i.e. to features of common observation not further defined. The ordinary language (spiced with technical jargon for the sake of conciseness) is thus *inseparably* united, in a good theory, with whatever mathematical apparatus is necessary to deal with the quantitative aspects. It is only too true that, isolated from their physical context, the mathematical equations are meaningless: but if the theory is any good, the physical meaning which can be attached to them is *unique.*

Historically, the false problem ('Scheinproblem') of 'interpreting a formalism' appears as a short-lived decay-product of the mechanistic philosophy of the nineteenth century. So long as the luminiferous aether was firmly believed to be an indispensable constituent of the universe, the search for a dynamical theory of light was a real physical problem, and it was tackled by the great masters in a perfectly realistic way: their papers are in fact wonderful illustrations of this inseparable unity of physical representation and mathematical symbolism which characterizes true theoretical thinking. When it was realized, however, that the aether problem had no *unique* solution, a process of disintegration set in, in which the mathematical set of equations expressing the laws of electro-magnetic phenomena became detached from its 'dynamical interpretations'. The unity was then soon restored by recognizing the electromagnetic field as a primordial element of physical description, irreducible to dynamical concepts.

Now, quantum theory eminently possesses this character of uniqueness; every feature of it has been forced upon us as the only way to avoid the ambiguities which would essentially affect any attempt at an analysis in classical terms of typical quantum phenomena. The suggestion that such ambiguities might be circumvented by the introduction of 'hidden parameters' is empty talk: for such parameters, in order to be of any use, should be linked with observable quantities, i.e. with classical concepts: and this link, whatever it is, could not possibly violate the restrictions in the use of the latter concepts imposed by the existence of the quantum of action. Consider, for instance, essentially quantal concepts, like spin or parity, which cannot be defined in classical terms: the definite prescriptions which bring them into relation with observable phenomena are in full harmony with the law of the quantum of action.

2. 'Deterministic substratum'

The critics are perturbed by the fundamentally statistical character of the causal relationships of quantum theory. There is, however, no logical justification for requiring a 'deterministic substratum' to a given set of statistical laws. There may or may not be such a substratum: this is a question to be decided by experience, not by metaphysics. Now, the peculiar *wholeness* of quantal processes, which implies the reciprocal lack of determination of canonically conjugate quantities, is an immediate consequence of the *physical content* of the law of quantification of action. No logically consistent formal device can therefore produce a deterministic substratum without doing violence to the immense body of experience embodied in the quantum laws.

3. The point of view of Sirius

The wholeness of quantal processes necessitates a revision of the concept of phenomenon. Since the concepts which in classical theory describe the state of a physical system are actually subject to mutual limitations, they can no longer be regarded as denoting *attributes* of the system. Their true logical function is rather to express *relations* between the system and certain apparatus of entirely classical (i.e. directly controllable) character which serve to fix the conditions of observation and register the results. A phenomenon is therefore a process (endowed with the characteristic quantal wholeness) involving a definite type of interaction between the system and the apparatus.

This new definition of the concept of phenomenon contains no arbitrary element: it is part of a terminology inherent in quantum theory and expressing a situation of fact, i.e. a situation which is not of our doing, but which we have learned from experience. It is perfectly true that this novel experience has profound epistemological implications. We must relinquish the illusion of being capable of unrolling a picture of the events on the Earth as they might be contemplated by some sharp-sighted, but not otherwise interfering, dweller of Sirius. But why should such a picture be at all possible? This is one of those metaphysical extrapolations which invariably lead us astray, because we attach absolute validity to presuppositions derived from limited experience. Only experience can teach us the limits of validity of such presuppositions, and in the present case, one of the lessons we have learned from the investigation of atomic phenomena is just that the quantum of action sets a limit to the unambiguous application of the mode of description of classical physics. To object to a lesson of experience by appealing to metaphysical preconceptions is unscientific.

The necessity of including the specification of the conditions of observation into the definition of the phenomenon has nothing to do with the question of the objectivity of the resulting mode of description. With respect to the latter question, the logical situation in quantum theory is exactly the same as in classical physics. If Micromegas paid us another visit, he would perfectly understand our statements about atoms; in fact, being like ourselves a highly organized compound of many atoms, he could not discuss atoms with us in any other terms than those of quantum theory.

4. The curse of positivism

According to our critics, the epistemological point of view of quantum theory undermines the sound belief in the reality of the external world, in which all physical thinking is rooted, and opens the door to the barren doctrine of positivism: we are no longer concerned with things, but only with the way to speak about things; science is degraded from a quest for truth to a verbal exercise. Dogmatic assertions that certain questions are meaningless bar the way to further inquiry; inability to understand the riddles of the quantum is hidden in the clouds of mystical renouncement.

This picture would be alarming if it were true. However, it is just another dream, a nightmare perhaps, of our critics. Obviously, it cannot result from a serious assessment of the immense broadening and enrichment of scientific thought which has resulted from the endeavour to formulate the laws of atomic phenomena. In fact, it is based on the most futile casuistics: the critics diligently excerpt from the writings in which the principles of quantum theory are discussed isolated sentences on which they put arbitrary interpretations. No wonder that they should find (as they freely confess) some difficulty in 'understanding Bohr': which, incidentally, does not prevent them from branding him as a positivist. There is no difficulty, at any rate, in understanding the critics' philosophy and exposing its unscientific character.

The realization of the mutual limitations imposed upon the use of classical concepts by the conditions of observation has forcefully reminded us of our own position in the world, and of the function of science in relation to this position. We are not merely contemplating the world, but acting upon it and thereby modifying its course. Accordingly, the scientific description of the phenomena is fundamentally concerned with the interaction of external agencies with the human observer; or, at least, in the narrower domain of physics, with material systems under the latter's control. The mode of description of classical physics appears, from this point of view, as a special case of wide validity, in which the quantitative effects of the interaction between observational devices and observed systems may be neglected: which does not mean, however, that the presence of these observational devices is not just as essential as in quantum theory for the very definition of the physical concepts.

It must again be stressed that there is in this view of the nature of science no arbitrary element: it is just an explicit statement of a situation which has always existed, even though it was not always so clearly recognized. Certainly, it puts an emphasis unknown to the outdated materialistic metaphysics of the nineteenth century on the active role of the observer in defining the phenomena: but in so doing, it brings the whole structure of science nearer to reality, in closer conformity with our real relationship to the external world.

There is an undeniable similarity between the epistemological conclusion drawn in such a straightforward, unambiguous way from the peculiar character of the quantum laws, and the insistence of the early positivists on the essential part played by our sensations in determining our knowledge of the external world. This only means that, to that extent, the early positivist movement was a healthy reaction against the shallow metaphysics of mechanistic materialism. But why should scientists be made responsible for the later positivists' blundering into a metaphysics of their own? No scientist would accept the extreme positivist contention that there is nothing more in statements about phenomena than the conceptual expression of relations between sensations: he would maintain that such statements refer primarily to real processes of the external world; our mental representation of these processes being itself, of course, subject to definite laws depending upon our sensorium.

To point out that certain relations between classical concepts cease to be meaningful in quantum theory has none of the sinister implications fancied by our critics: it is a plain statement of fact, founded in a law of nature. The words 'renouncement' or 'resignation' often used in this context are ambiguous in their emotional connotation: renouncement may be felt as privation or as liberation. Some critics seem to take the invitation to 'renouncement' as an attempt on their personal freedom: the right to indulge in metaphysical dreams is not disputed; only, this activity is not science.

5. The insufficiency of quantum theory

As any other scientific theory, quantum theory has its limitations. Our methods of field quantization break down when we attempt to apply them to processes confined within space-time regions of linear dimensions much smaller than 10^{-13} cm, or (what amounts to the same) involving momentum and energy exchanges much larger than a GeV. There is evidence that the 'particles' which we treat at present as the elementary constituents of nuclei have some kind of 'structure' extending over such small regions, and are susceptible of metastable excited states of some tenths of a GeV energy. These excitations are connected with several types of meson fields, which are coupled in various ways to each other and to the lepton fields. It is clear that we are here on the threshold of a new world with laws of its own, which we have not yet discovered.

The critics try to exploit this situation to their advantage, but it is just here that the dilettantism of their approach is most conspicuous. In fact, it is idle to 'hope' that the cure of our troubles will come from underpinning quantum theory with some deterministic substratum, unless this surmise could be justified by arguments derived from the analysis of the phenomena which have to be explained. For instance, these phenomena might reveal some limitation of the law of quantization of action: this would indeed give substance to the critics' belief in 'hidden' sub-quantal features. Unfortunately for them, all the evidence points with merciless definiteness in the opposite direction: however strange they may be in other respects, all the processes involving the unknown interactions invariably conform to the fundamental quantum law.

Moreover, recent developments have strikingly demonstrated the fruitfulness, for the interpretation of these processes, of concepts like parity and particle-antiparticle conjugation, which are specifically quantal, i.e. irreducible to classical representations. While our critics, in their numerous speeches and profuse writings, were prophesying the ultimate vindication of their views by future discoveries in the domain of high-energy processes, one major discovery, of truly universal import, about deep-lying symmetry properties of the fundamental laws of nature, did actually take place: and o! its substance and meaning can only be formulated in terms of the essentially quantal concepts just mentioned. The true road to future progress in this domain is clearly marked: the new conceptions which we need will be obtained not by a return to a mode of description already found too narrow, but by a rational extension of quantum theory. This will indeed require a sound criticism of the foundations of the theory.

* * *

[The following passage appeared in the transcript of the discussion of the above paper – Ed.]

Rosenfeld: Relativity is part of classical physics – by classical physics I mean physics in our scale, including relativity – so, therefore, ...as long as we remain within the domain of classical phenomena we attribute the length to the object; that is an approximation we are allowed to make at that level. In connection with this ... Bohr uses classical language for this kind of experience because classical language is our only tool for describing this kind of experience although its application is limited by laws of nature which are independent of us. This brings me to the question of Vigier who absolutely wants to put the accusation of positivism upon me. I think there is a great difference between the early positivists for whom I have much respect – even for Mach, in spite of his aversion against atomism – and the later positivists. I have not the same respect for the later positivists because I think that where positivism went wrong was when it became a system in which those nice points which I think were made already by Mach, were fossilized into principles of absolute validity; it was maintained that science was only a language and that one could decide about statements only by reducing them to sense data and so on. Mach himself was not very consistent and he made many mistakes which encouraged that direction. However, as far as quantum mechanics is concerned, I would say that it is impossible to understand it without assuming that there is an external world which is independent of what we think and which is the ultimate origin of all our ideas. In that sense I absolutely reject the suggestion of present-day positivists about the subjectivity of our statements. As regards the atoms and the atomic theory which has been invoked by Bohm and also by other people as an example of a theory with hidden parameters which has been successful, I have to say this: of course there are atoms and therefore people who spoke of atoms before it was possible for us to observe them were very lucky; Francis Bacon did not know much about science but happened to put out certain ideas which proved later to be right, together with many others which turned out to be wrong. Now as regards the atomic theory in the nineteenth century it is necessary to go a bit closer into the historical development and to follow up the transformation of this idea from a speculation which might be true or not into a scientific theory. I think that the fact that individual atomic processes were only observed at the end of the nineteenth century is, from our present point of view, more or less an accident. The atomists like Boltzmann were convinced of the reality of atoms for very cogent reasons although they had never seen individual atomic processes. For Maxwell the connexion between the various parameters which are undetermined on the phenomenological level, was a very strong argument in favour of the reality of the underlying mechanism. Another very important point was that it was possible to make predictions, and to find relations between phenomenological parameters without going into details about the structure of atoms; in this way the degree of arbitrariness in the atomic theory was very much reduced. It is true that Maxwell invented his law of distribution but that was not essential. The most beautiful application of atomic theory was the earliest one, namely, Laplace's theory of capillarity. In this theory he only assumed that the forces between molecules are forces of short range and he could deduce all the macroscopic laws of capillarity from that assumption alone. This shows that there was very little arbitrariness in that conception and it is only that feature of early atomic theory which in my view conferred upon it the quality of a scientific theory as opposed to a speculation. When Mach criticized the atomic theory he of course bet upon the wrong horse and he has to bear the blame for it. But apart from that his paper on the *Erhaltung der Arbeit* was just a very sound warning against the danger of introducing arbitrary elements into atomic theory. His whole point was that one has no right to introduce and to apply to atoms the mechanical concepts which have been derived from experiences about macroscopic bodies, unless one has cogent experimental reasons for doing so. I therefore think that it is unfair to criticize Mach just because he happened to draw the wrong conclusion, as his criticisms were completely sound. Now Vigier has presented us with an example where the same physical laws are interpreted in two different ways. I would not describe this example as he does. I would rather say that we have two different theories which are both selfcontained – a microscopic description and a phenomenological description: the latter is a definite system of concepts, connected with mathematical symbols; and the former is another system of concepts, namely the properties of systems of a large number of particles assumed to move according to the laws of mechanics; and we define also quantities in that theory which are uniquely connected with the other mode of description and which, of course, satisfy the same relationships.

CHAPTER 9

FOUNDATIONS OF QUANTUM THEORY AND COMPLEMENTARITY*

[1961e]

The foundations of quantum mechanics were laid some thirty years ago. Thus, a whole generation has passed since these foundations were established, but for the new generation which is now coming up the situation is perhaps stranger still than it was for the physicists who had to struggle with those problems for the first time. The people who face a problem for the first time and have to think very hard to try to solve it have a more direct and lively impression of what is involved in the new ideas which bring about its solution. The next generation is presented with the finished work and often does not appreciate properly the difficulties that were involved in the solution of the problem. So it happened with relativity theory one generation after Einstein; some people started then to raise doubts about the very principles underlying the general theory of relativity. This episode is now forgotten; but we see the same phenomenon happening with quantum mechanics. Young physicists are raising doubts about the correctness of the basic ideas of quantum mechanics, and try to do it better. These efforts are, I am afraid, rather futile, because they rest on a complete misunderstanding of the problem, on a lack of appreciation of the really very difficult conceptual situation which quantum mechanics presents to us. This situation is not something that depends on any free choice, about which one might have this or that opinion. It is a problem which is imposed on us by Nature. After all, the existence of the quantum of action is a law of Nature: it is not something we have invented, but an aspect of Nature we have discovered. This discovery created, when it was made, a very embarrassing situation, and quantum mechanics is, in fact, the answer which physicists after many years of effort have been able to give to this challenge of Nature.

The discovery of the quantum of action was made in connexion with properties of electromagnetic radiation. In order to have something definite before us, we can therefore speak about the situation in the theory of light or electromagnetic radiation, although exactly the same can be said of electrons or any other constituent of matter. According to the classical theory of electromagnetism, light or any electromagnetic radiation can be described as a field, in the simplest case a field endowed with periodicity in space and time; the element of radiation is thus represented by a plane monochromatic wave, that is, a field distribution which has strict periodicity in space, with a certain wave-length, and in time a certain period connected with the wave-length and velocity of propagation by the wave equation. In order to have this strict periodicity, the field must be spread out over the whole of space, and it must persist during unlimited time: a pulse which only exists during a finite time does not have a sharply defined frequency, but a whole spectrum of frequencies; likewise, a limitation in space of the 'wave packet' implies a distribution of wave numbers (or inverse wave-lengths).

Now, the discovery of the quantum implied that when radiation interacts with matter, as in the photoeffect or in the Compton effect, the amounts of energy and momentum exchanged in this interaction are not at all related to the distribution of energy and momentum in the radiation field, but are completely determined by the mechanical laws of conservation governing collisions between particles of definite energy and momentum; in this picture, a monochromatic radiation is represented by a light quantum or photon, behaving like a particle the energy and momentum of which are proportional to the frequency and wave number of the radiation. Thus, one was here led to describe radiation by means of another classical concept, which had hitherto been applied only to matter: the concept of the particle, which represents a concentration of energy and momentum at a single point of space and at any single instant of time. This is also, like the monochromatic wave, an ideal conception, and it is the extreme opposite of the latter; in fact, the two are mutually exclusive. Yet, they are intimately related by the equations:

(1) $$E = h\nu, \quad p = h\sigma$$

which express the proportionality between the energy and momentum E, p, concentrated in the particle, and the frequency and wave number ν, σ, defined by the plane wave.

These are at first sight very paradoxical equations: they relate together characteristics of the radiation which are contradictory to each other, and yet both are necessary to account for all phenomena (for the interference of waves is just as fundamental a phenomenon of radiation as the scattering of photons). The coefficient of proportionality, Planck's universal constant *h*, appears in the equations (1) as the product of two physical quantities – the energy of the photon and the period of the wave – which relate to two descriptions of radiation that are mutually exclusive. Thus such a situation demands more than just a mathematical analysis. The mathematics are extremely simple, but the question is to understand what these simple equations mean, as they seem to combine things that are mutually exclusive. What we meet with here is actually a problem of logic: What are we to do when we are confronted with such a situation, in which we have to use two concepts that are mutually exclusive, and yet both of them necessary for a complete description of the phenomena?

It is this logical problem which confers a very peculiar character on quantum mechanics. The situation was not quite the same with the relativity theory. Of course, in creating this theory, Einstein had also to make a very deep analysis of the fundamental concepts of space and time and of the relation between them, but although the result was quite revolutionary, he could perform this analysis within the framework of classical physics. Here we see that it is this framework itself that is too narrow logically: the task is to find a consistent way of using classical concepts so as to give well-defined meaning to mathematical relations like the equations (1) in spite of the fact that the parameters involved refer to two contradictory images. The solution of this problem, to which a fundamental contribution was given especially by Heisenberg, received its final formulation from Niels Bohr; the new logical instrument which was created by Bohr is called complementarity. Complementarity denotes the logical relation, of quite a new type, between concepts which are mutually exclusive, and which therefore cannot be considered at the same time because that would lead to logical mistakes, but which nevertheless must both be used in order to give a complete description of the situation.

The idea underlying complementarity and making it possible to surmount the logical contradiction is, in fact, very simple. Contradiction arises when one tries to give a universal validity to the concepts that are mutually exclusive. If one tries to apply both of them to any situation irrespective of the circumstances of the situation, then, of course, one gets into contradictions. However, if one reflects on the use of all physical concepts, one soon realizes that any such concept can only be used within a limited domain of validity. These limitations may be disregarded so long as one is dealing with simple situations, as in classical physics: the constant h, which marks the limit of applicability of classical concepts, is so small compared with the actions involved in macroscopic experiments that we can ignore the limitation. Therefore the physicists of the nineteenth century had no need to think very deeply about this question. But after the discovery of the quantum of action has confronted us with such a situation, we are forced to analyse the logical relations between concepts, and to recognize their limitations. But once one has realized the simple fact that every physical concept is only defined with respect to a certain class of phenomena, to certain well-defined physical situations, it becomes easier to understand that different sets of concepts, although incompatible, may nevertheless be applicable to different physical situations.

Domains of validity of complementary concepts are not, in practice, rigidly exclusive of each other; they may overlap to some extent. Where they overlap, the concepts in question can both be applied, but of course not rigorously. A well-known example is that of a 'wave packet' of radiation. There can be no rigorously monochromatic wave: a real wave is always limited to a finite time-interval, Δt, and a finite region of space, the dimension in one direction of which we shall denote by Δx; such a wave can be analysed into monochromatic components with frequencies and wave numbers (in the direction x) distributed over intervals $\Delta \nu$, $\Delta \sigma_x$, the order of magnitude of which, according to the mathematics of Fourier expansions, is just $(\Delta t)^{-1}$, $(\Delta x)^{-1}$. We may regard such a wave packets as approximating the localization of the phenomenon in time and space with latitudes Δt, Δx, and at the same time, because of the relations (1), fixing the energy and momentum of the radiation within an accuracy $\Delta E = h\Delta \nu$, $\Delta p_x = h\Delta \sigma_x$. Thus, between the latitudes of the complementary pictures there is a reciprocity:

(2) $$\Delta E \cdot \Delta t \approx h, \quad \Delta p_x \cdot \Delta x \approx h$$

which is an example of the famous 'uncertainty relations' holding for any pair of physical quantities complementary to each other.

This reciprocity illustrates a fundamental feature of the complementarity relationship. On one hand, there is no absolute limitation to the possibility of defining either of the complementary quantities: by making the wave packet infinite we obtain the idealization of a photon of well-defined energy and momentum; by making it infinitely small we have the ideal localization of the radiation at a single point of space-time. On the other hand, when we approach either of these extreme idealizations, we see that the complementary one becomes less and less defined, and is finally excluded: the fixation of the energy and momentum of the photon precludes its localization, and conversely, if we localize the radiation field, we lose the definition of the energy and momentum contained in it. Such reciprocal limitations correspond, of course, to real physical effects, and the complementarity relationship just expresses the fact that the experimental set-ups required for the determination and recording of each of the quantities in question are such as to impose on the complementary quantity the lack of definition expressed by a relation of the type (2). To realize quite clearly what is involved here, we must sharpen our language and ask ourselves, more searchingly than was necessary in classical physics, what we understand by a physical phenomenon. The precise definition of a phenomenon must include the description of the apparatus which defines the conditions of the experiment, and also some recording process (an index reading, a spot on a photographic plate, etc.) by which alone the phenomenon can be regarded as terminated: the physical quantities we are dealing with in our mathematical theories are in fact just such records of phenomena. Phenomena so defined can be in a complementarity relationship, and the accuracies of their recordings then stand in a reciprocal uncertainty relation.

Reciprocal uncertainty relations are in fact precisely what we must expect, from a purely logical point of view, if we are, as in quantum theory, in a situation in which contradictory pictures of the phenomena have to be resorted to for a complete account. In order to avoid trivial mistakes which would arise from an indiscriminate use of such contradictory concepts, we must have some way of delimiting their domains of validity, some prescription enabling us to decide in each case how far we can go with help of each of them. The uncertainty relations tell us, so to speak, the price we have to pay for applying to the description of a phenomenon two sets of concepts which are strictly contradictory. It will be realized, without going into further detail, that the conception of complementarity means a very considerable extension of our way of speaking about Nature and about natural phenomena. It involves a new logical framework within which to arrange our concepts, compared with which the classical way of speaking appears very narrow because it is confined to the simple classical logic, governed by the principle of contradiction. Of course the classical logic remains the foundation of any argument we use, and certainly of our mathematics. But we find that when we deal with quantum phenomena it is too narrow: we must have a wider way of connecting together the various concepts, and this is provided by complementarity.

The most considerable change in our view of natural phenomena which is implied by this widening of our outlook is related to our conception of causality. In classical physics the expression of causality is the result of the mathematical properties of the equations expressing the laws of motion of particles and of propagation of the electromagnetic field. When we are given the initial conditions, involving the initial values of the quantities and perhaps some of their derivatives, the solution of the equations is completely determined; and accordingly, the type of causality of classical physics is determinism. This is no longer the case when we have a complementary relationship, as is already suggested by the fact that we cannot then speak of all physical quantities with unlimited accuracy. One of the most obvious examples illustrating the new situation is the emission of the spectral lines: according to the quantum postulate, an atom in a high state of excitation can effect transitions to any one of the lower excited states; but there is nothing in the conditions of the phenomenon to tell us which of these possibilities will be realized in a definite case. Therefore, all we can do is to try to find the relative probabilities of their occurrence; it is just to cope with such situations that the concept of probability has been invented. The argument applies quite generally to all complementary situations; if we consider one set-up, in which, let us say, the field is localized within a certain space-time region, we have a probability distribution for all possible values of the energy and momentum, and conversely. These distributions are such that the corresponding average fluctuations are connected with the complementary time and space latitudes by reciprocity relations: if the field distribution is very narrow in time, it is very wide in energy, etc. In other words, the kind of causality that is implied by the existence of complementary relations between concepts is statistical causality.

As is known, statistical causality was introduced, historically, in dealing with the affairs of human societies or states, which are notoriously undetermined: that is the origin of the word 'statistics'. It was for the first time applied in physics in connexion with problems of atomic theory, though not those concerned with the individual properties of atoms; it was introduced in thermodynamics, that is, in the study of the properties of very large systems of atoms, and later also of systems composed of a very large number of monochromatic waves of radiation. This occurred in the second half of the nineteenth century, when, of course, nobody dreamt of the possibility of such queer things as complementarity. At that time the justification for introducing statistical causality was simply this. We aim at a complete description of a gas as a mechanical system of molecules; but this could only be obtained if we could know at a certain time the exact positions and velocities of all the molecules of the gas; then we could solve, at least in principle, the mechanical equations for the system, and deduce in a unique, deterministic way the state of the gas at any future, or past, time. However, this ideal aim is obviously unattainable, and therefore the best we can do is to speak of the states of the system in terms of probability, just as, in view of the impossibility of predicting the exact time of a man's death, we speak of the probability of his living for so many years.

The concept of probability was thus used with a sort of apology, in default of something better. This attitude was of course quite natural for people believing in determinism, believing that the Creator had imposed on Nature, once for all, rigid laws that had to be obeyed and were so contrived that they could only yield a single, predetermined, result. Unprejudiced observation of the world around us, however, will soon reveal that this idea of rigid determinism is a fiction, an unwarranted extrapolation from a very limited experience, and that there is no real justification for it. On the contrary, if we want to describe immediate experience, surely we must say that all our judgments about Nature or about ourselves are judgments of probability: we are never in a situation in which we know all the conditions for applying laws which would give us a unique answer. Thus, statistical causality appears in fact as more natural, more directly adapted to our actual needs than determinism; and it is much better to look at determinism as a limiting case of statistical causality, corresponding to statistical distributions very sharply peaked around the average values. From this point of view we are better prepared to accept the fact that the type of causality of quantum mechanics is fundamentally statistical; and we are even in a position to give exactly the limit of validity of classical determinism: the statistical deviations from classical behaviour become negligible in phenomena involving a large number of quanta of action.

Now, returning to the case of statistical thermodynamics, it is instructive to consider an objection to the point of view just outlined which is still often heard. It is all right to say that statistical mechanics is a self-contained theory, as I have tried to suggest; nevertheless it is conceivable that, if not human beings, at least imaginary beings, which Maxwell and his friends called demons, could be so small (and yet endowed with sense organs similar to ours) that they could observe the motions of the individual molecules of a gas; such beings could then give a deterministic description of it. I think the answer to this was already given very pointedly by Clausius, when Tait tried to use this fiction of the Maxwell demon to question the universal validity of the second law of thermodynamics by pointing out that those demons could bring about situations contradictory to the second law. Clausius's answer was: "We are not concerned with what demons can do, but with what heat can do by itself", which clearly means: with the phenomenon of heat which we can observe. I think there is a very deep truth in that answer, deeper perhaps than Clausius himself could realize at the time.

In fact, what the situation in quantum mechanics has helped us to realize is that science is primarily concerned with the communication of experience about phenomena, in other words, with statements made by human beings, and which other human beings are able to understand. For this very reason all such statements must make use of the concepts of classical physics, because classical physics is just the language describing the results of direct observation, the phenomena which we can observe with our senses, either directly or with instruments to increase the sensitivity, in any event, with recording systems composed of many atoms. If, then, we talk in this language about atoms which do not follow the laws of classical physics, we meet with the limitations of its use expressed by relations of complementarity. We thus see that this complementarity has its source just in the peculiar situation in which we find ourselves, as macroscopic systems, trying to describe the possible macroscopic interactions of our apparatus with the smallest elements of which Nature consists.

Now, if we look back at thermodynamics from this point of view, we see that it presents a logical problem which is just the same as that of quantum mechanics. Here also we are trying to describe atomic structures in terms of macroscopic concepts, and we may do this in two different ways; we may be able to give a more or less detailed dynamical description of the atomic system, in terms of the motions of its parts; or we may describe its state in terms of the macroscopic quantities, such as volume, pressure, temperature, entropy, which are related by the laws of thermodynamics. Between these two modes of description there is a relation of complementarity; this is very clear if we consider the temperature and the energy, representing the thermodynamical and the dynamical points of view respectively. These two quantities are complementary to each other in the sense that if a system has to have a well-defined energy, it must be thermally isolated, and we cannot then ascribe to it any temperature, because in order to define the temperature we must establish thermal contact between the system and another, very large system, serving as a thermostat; conversely, if the temperature of the system is defined in this way, it exchanges energy with the thermostat, which means that its energy is not determined; in fact, we know that in this case this energy has a definite statistical distribution, the canonical one, the modulus of which is proportional to the temperature of the system.

This complementarity we can formulate in a more general and precise way, in wide-going analogy with the case of quantum mechanics, by setting up reciprocal relations between the statistical fluctuations affecting macroscopic quantities. Of course, the reciprocal limitation of the dynamical and thermodynamical modes of description has nothing to do with the quantum of action: the similarity of the two cases is on the logical level only; physically, they are completely independent of each other.

The fluctuations we are here considering are those which affect the macroscopic parameters defining the state of a system which is a small part of a larger, but finite, system; then, even intensive quantities, like the temperature, will exhibit fluctuations, for they are only exactly defined for infinite systems. It is easy to see that one must expect reciprocal relations between fluctuations of extensive and intensive quantities, that is, those quatities which (for a homogeneous system) are proportional to the number N of constituent elements of the system, and those which are independent of N. In fact, the fluctuation of an extensive quantity, the energy, for example, is proportional to $N^{1/2}$, while that of an intensive quantity, the temperature, say, is inversely proportional to $N^{1/2}$; the product of the two is therefore independent of N and cannot be made arbitrarily small by a change of the number of degrees of freedom of the system. A more detailed calculation shows that the products in question are of the form kC, where C is a quantity independent of the size, but depending on the physical properties of the system considered, and k is Boltzmann's universal constant. Thus, it is this constant which here plays a part analogous to that of Planck's constant in quantum theory: it is inversely proportional to Avogadro's number, which is thus seen to define a critical size of the system, for which the complementarity between dynamical and thermodynamical description becomes important.

This consideration is not as artificial as it may seem: it is easy to give examples of systems of this critical size, which can be considered from both the dynamical and the thermodynamical points of view. If we take as the systems under investigation droplets of emulsion distributed under gravity in a solution, we have a case in which the systems are large enough for us to see them and to follow their motion, which is a Brownian motion, and yet small enough for us to speak of their statistical distribution in the solution. As a representative thermodynamical quantity we may take the entropy S, which will fluctuate on account of this statistical distribution of the droplets. Now we need an intensive quantity characteristic of the individual motions of the droplets; for this let us choose the velocity v, which is subject to the well-known Brownian fluctuations due to the collisions with the molecules of the solution. A simple calculation gives in this case for the product of the two fluctuations:

$$\Delta S \cdot \Delta v = k v_0$$

where v_0 is a quantity of the order of the mean atomic velocities inside the droplet. We see that the pattern of complementarity applies to the present case in all details just as in quantum mechanics.

So far, therefore, we have found that the atomistic view of the world involves in its description two stages of complementarity. The first and most fundamental one occurs in the account of the properties of individual atomic systems, the second when we want to describe the behaviour of systems of large numbers of atoms. In the first case, a relation of complementarity exists between two sets of concepts which were used indiscriminately in classical theory: the space-time localization and the energy-momentum balance; it becomes significant when the mechanical action in the process considered is of the order of Planck's constant. At the second stage, we have complementarity between the dynamical and the thermodynamical behaviour of large systems, the critical size being given by Avogadro's number. In both cases, as we have seen, complementarity essentially implies a specific kind of statistical causality.

Let us now venture for a moment into the field of biology. We here meet with a very peculiar situation: what characterizes the biological behaviour, in contrast with the behaviour of inanimate matter, is the existence of a certain structural order, either definite spatial structures such as those of the protein molecules, or more formal, topological structures such as the interconnexions of nerve cells in the brain. In any event, we notice the appearance of some kind of order which has a remarkable stability, and which can be reproduced. Now, the great event of the past few years has been the discovery, at the very lowest stage at which the phenomena of life appear, of possible mechanisms by which the maintenance and perpetuation of such a type of order can be understood in simple terms of physics. There is certainly no reason to expect that, with the most elementary, also the most elaborate forms of biological behaviour will not eventually be reduced, in a certain sense, to the laws of physics and chemistry, that is, in the last instance, to quantum mechanics, since according to the famous saying of Dirac quantum mechanics explains a large part of physics and the whole of chemistry. Now we may perhaps add: the whole of biology.

Yet this is not the whole story. It is quite obvious that, when we are presented with such complicated structural problems, it is practically impossible, even with the most perfect knowledge of all the laws of physics and chemistry, to predict what kind of organism will result from a given structure of such complexity. We are here in the same situation as somebody who would know all about mechanics and is presented with a blueprint of some machine which he has never seen and which he does not know. It would be very hard for this person to guess the function of the machine, to understand why this particular combination of motions has been set up. We must realize that in order to get complete understanding of organisms we need the concept of function. We must somehow know beforehand what the function of a certain organ is in order to be able to understand its structure. We here meet with another type of causality, functional causality, which is obviously complementary to the ordinary causality of physics and chemistry, from which this concept of function is absent. Thus, we see that at the biological level we have another stage of complementarity, which again presents the same characteristics of mutual exclusiveness as the previous ones: on one hand, we have the complete description of organs in terms of physics and chemistry, on the other the functional mode of description, which will always be necessary for a complete account of such complicated arrangements of molecules as living beings. This complementarity is concerned with another aspect of atomic systems than just the size, although the size is rather critical, inasmuch as biological structures can only occur at a definite stage of molecular complexity, allowing on one hand for the necessary stability and on the other for the possibility of reproduction; but the main point here is the type of spatial ordering of the constituent atoms, which distinguishes biological systems from inanimate ones.

At this point I may perhaps be allowed to venture a little further still on the path of speculation, and raise the following question. The whole point of the logic of complementarity, in its application to the atomistic description of the world, is that the mutual limitations in the use of the classical concepts it implies, at each of the three stages we have recognized, are related in the last resort to the size of the sense organs of human beings. To be sure, there is no direct reference in physics or even in molecular biology to the human observer as such, but only to registering appliances which are entirely controllable physical systems. Still, the size of these appliances is dictated by the requirement that the complementary limitations be negligible in the account of their behaviour, and this effectively brings them to macroscopic size, that is, to a size adapted to our sense organs. There was a Greek philosopher, the sophist Protagoras, who used to say that "man is the measure of all things". There is a deeper truth in this than he could possibly fathom. This brings us to the question: Is the recourse to complementarity logic so to speak an accident, another of the many imperfections of human nature, and is it not conceivable that other rational beings, some kind of Maxwellian demons, could develop a different mode of description of Nature not affected by any complementary limitations?

The tentative answer I should like to submit is that there is nothing more accidental in the emergence of complementarity logic than in the emergence of man himself as a product of organic evolution. My point is that the concept of a thinking Maxwellian demon is self-contradictory. I suspect that the development of a computing and communication system like our brain demands about that complexity of organization which has been reached by our own species in the course of evolution. After all, it is an empirical fact that rational thinking has only arisen at the stage of mankind, presumably because it was only at this stage that a brain developed with a sufficient number of interconnexions between nerve cells to make possible this very peculiar behaviour which we call rational thinking. Hence the suggestion that no animal other than man or some being of equivalent complexity can possibly have any physics; and this degree of complexity, as we have seen, necessarily implies the introduction of complementarity in physical theory.

The view of physics, and in fact of science, which results from the development of quantum theory is very different from the scientific ideal of the nineteenth century; but it is by no means, or so it seems to me, inferior to it. The ideal for which such people as Huxley and Tyndall fought, the supremacy of human reason over all forms of irrationality, was based on the deterministic view of a Nature governed by inflexible laws, which the task of science was to discover and to formulate. These laws, by their very nature, would have absolute validity, and they would give a picture of the universe, in the literal sense of the world, a picture which could be contemplated, but from which the spectator, the human spectator, would be excluded, or at least reduced himself to an object of contemplation. The present conception, on the contrary, stresses the fact that at every stage the human observer intervenes in a very fundamental way by determining the very language in which the laws of Nature are formulated. The significance of the laws of Nature, from this point of view, is not that of drawing a picture of the world independent of who is looking at it, but rather of giving an analysis of the interaction between ourselves and the external world. At first sight it would seem that one thus introduces in science a subjective element. But the fear that the objectivity and rationality of the old view would be abandoned is entirely unfounded. True objectivity is simply the possibility of guaranteeing that the account of the phenomena will convey equivalent information to all observers, that it will consist of statements intelligible to all human beings. This objectivity is indeed ensured in quantum theory by the existence of transformations which allow you to pass at will from the point of view of one observer to that of another, who makes some other type of experiment. It is just the invariance of the laws of physics for such transformations that guarantees their objectivity, and nothing is changed in this respect by the advent of quantum theory.

Far from marking a retreat of human reason from the ambitious aim of the great pioneers of the preceding centuries, the new ideal of science appears as a tremendous advance towards this very aim. Compared with the old one, it is much broader of scope, much richer of content. It is more truly adapted to the actual situation that scientists have to face in their quest; the logic of complementarity enables us to cope with the dilemmas which this situation presents to us, and which were insoluble from the old point of view. I may therefore conclude on an optimistic note by assuring the younger generation that the instrument of rational analysis we are handing over to them contributes to a further increase of our power of understanding our position in Nature and of coping with the problems which it raises.

NOTE

* Substance of a lecture given at the University of Tokyo on October 6, 1960.

CHAPTER 10

THE EPISTEMOLOGICAL CONFLICT BETWEEN EINSTEIN AND BOHR

(Dedicated to Max Born on his 80th Birthday)

[1963c]

As our knowledge of atomic phenomena has grown and developed, we have been led to the conclusion that probability must play an essential part in our account of the unfamiliar behaviour of atoms. The classical framework of unrestricted determinism has proved too narrow and has given place to a more general relationship between the various aspects of phenomena. For this new relation Niels Bohr has coined the name "complementarity". But this momentous widening of our scientific and philosophical outlook was not achieved without a struggle. Classical causality had its staunch supporters, foremost among them Einstein. That Einstein, one of the most powerful minds that mankind ever produced, should thus refuse to accept the leading thought of the time is in itself an astonishing fact, which cannot fail to repay a closer analysis. But the interest of the case is further enhanced by certain features which at first sight look quite paradoxical.

It was Einstein who more than anyone else, certainly more than Max Planck, stressed the very aspects of quantum theory which ultimately led to Bohr's idea of complementarity. Einstein was the first to point out the dual nature of light, which manifests itself sometimes in the form of waves, sometimes in the form of particles. The point of Bohr's argument is just that these two aspects cannot be causally related in a determinate way, but must be regarded as complementary features of the phenomena. Moreover, the method used by Bohr in the analysis which led him to this conclusion was directly inspired by that followed by Einstein himself in the masterly analysis of the concept of simultaneity which forms the foundation of the theory of relativity. All the shrewd objections raised by Einstein were successfully countered by Bohr; the logical consistency of the complementarity idea and its adequacy in accounting for the empirical situation were demonstrated beyond any reasonable doubt; nevertheless, Einstein refused to the last to accept it as final, because, he said, it was repulsive to his innermost feelings.

The issue which was there at stake is in fact not a mere technicality of physics, but one of deep human interest. Einstein's opposition has not held up the development of atomic theory; it has rather acted as a stimulus to a fuller elucidation of Bohr's ideas about complementarity. But it does raise fundamental problems of the theory of knowledge. Why, we may ask, did Einstein lose faith in the method of thought by which he achieved his early triumphs? What is the nature of that stronger conviction which prompted him to appeal to feelings in a situation where, one would think, reason alone should decide? The answers to these questions affect our whole attitude to the function of science and the ultimate sources of human knowledge.

The theory of knowledge which guided both Einstein and Bohr in their work goes back to Mach. In the course of historical studies of past discoveries, Mach recognized and described the standard procedure by which, consciously or unconsciously, scientific concepts and laws are elaborated. It is, Mach points out, based on *Gedankenexperimenten*, that is to say on experiments not actually performed, but imagined in an idealized form suggested by some familiar experience. This means, essentially, that concepts are regarded as a kind of condensed code of action: they are defined by some prescribed sequence of manipulations of suitable apparatus, and are thereby referred, in the last instance, to definite relations between sense impressions of the observer. Thus Einstein, in order to analyse the concept of simultaneity, imagines two observers exchanging light signals; only by such a concrete action can simultaneity be ascertained and given a well-defined meaning. In the same way Bohr traces the lack of determinate causal connection in atomic phenomena to a property of the interaction between the atomic systems and the measuring instruments applied to them.

Mach's theory of knowledge, according to which all concepts are ultimately reduced to relations between sense impressions, undoubtedly contains an important element of truth. By laying emphasis on the interaction between the phenomena and the observer, it rightly stresses the fact that the function of science is not to give us a contemplative picture of the world, but to provide us with rules of action, which enable us to control the forces of nature. An obvious weakness of Mach's theory, however, is its oversimplification of the relation between concepts and sense impressions. This relation is very seldom an immediate one. Concepts are linked together by intricate nets of logical connections, which form such imposing edifices as, for instance, geometry. When we consider such logical constructions, an essential complication arises: it is not immediately clear whether sense impressions and the relations which we perceive between them determine uniquely the systems of concepts appropriate to their description. For example, one may ask the question whether the geometry of physical space is Euclidean or non-Euclidean.

Henri Poincaré, whose independent line of thought is strikingly akin to Mach's, had a sly way of escaping from this difficulty: he simply asserted that it does not matter. The choice of any particular system of concepts to account for some part of our experience is purely a matter of convention. It is dictated only by considerations of simplicity and convenience. Thus, says Poincaré, the postulates of Euclidean geometry are not the expression of experimental facts; we have just been led by certain physical observations to adopt them in preference to others. As compared with Mach's view, this opinion represents an essential alteration of the relation between concepts and sense impressions: the latter are deprived of the dynamical part assigned to them by Mach in the elaboration of the concepts; in fact, they are reduced to the passive role of helping to select a suitable system of concepts from some pre-existing assortment. Poincaré's conventionalism thus marks a fatal step into the blind alley of idealism.

It was not long before Einstein, in his great creative period, met with the same problem as Poincaré. Originally, he had followed the pattern described by Mach, and he was very appreciative of the pertinent and fruitful part of Mach's philosophy, from which he had received direct inspiration at decisive stages of his discoveries. The theory of relativity, however, grew naturally into one of those logical systems of concepts, condensed into mathematical symbols, which represent such a formidable concentration of mental power. Einstein's attention was gradually diverted from the concrete contents of the concepts he dealt with, and the further he proceeded from his starting point, the more be became impressed by the heuristic value of abstract mathematical thinking. It was then that, as I see it, his thought took a wrong turn.

Like Poincaré, he shifted the measure of values from the empirical and operational approach of Mach to a purely formal and idealistic view, dominated by such ill-defined criteria as simplicity and elegance. Certainly he could not be satisfied by Poincaré's conventionalism. His fine physical intuition could not fail to perceive that the harmony between the phenomena of the external world and our abstract conceptual constructions must have a deeper significance. He regards scientific concepts and theories as "free creations of the mind"; but he adds, taking here a firmer and truer view than Poincaré, that it must be possible in each case to find a definite formal theory which is better adapted to experience than any other and which is, in this sense, a true and unique representation of reality. Nevertheless, what he asserts is that the human mind is able, by its free activity, to build up a true picture of the world: and this, certainly, is just as much as Poincaré's more superficial view a somersault into idealism.

There is an apparent boldness in thus asserting the supremacy of the mind. Yet it is actually an attitude of despair. We shall not expect to learn from Einstein how this wonderful harmony between our mental constructions and the phenomena of the external world comes about; because this harmony, he declares, is a wonder "which we shall never be able to comprehend". Here, then, we have again the familiar picture: by making absolute the power of the mind, idealism is driven to the negation of this power and inevitably ends in pure mysticism. We now understand better the origin of Einstein's feelings about causality. Pygmalion-like, he is fascinated by the formal beauty of his own mental creation, which is undoubtedly the most complete embodiment of physical determinism. Not unnaturally, he exalts it to the rank of a universal pattern for the description of nature, and discounts as a temporary imperfection the failure of this pattern to apply to atomic phenomena. His faith in the deterministic mode of description loses the character of rational scientific belief; it becomes mystical.

How, then, are we to avoid this mystification? How does Bohr, for instance, face this problem? The most striking difference between the attitudes of Einstein and Bohr is that Einstein stresses the permanence, the finality of scientific thought, while Bohr insists on its unceasing development. This gives us the clue to the solution of the difficulty. It makes a tremendous difference whether one looks at a theory in its final state, or at the imperfect stages through which it passed in the course of its development. The harmony between form and content ultimately achieved may indeed appear wonderful; but it was much less so in the preliminary drafts. The mystical touch, the wonder, is an illusion resulting from a static approach; it vanishes as soon as the process of formation and growth is considered in its development. New ideas are, of course, creations of the mind; but they are by no means *free* creations. They only arise as a response to the always recurring necessity of adjusting our mental picture of the world to our changing empirical knowledge of it. In this perspective, the interaction between observer and phenomena appears as an inherent part of the process and thereby is also susceptible to rational description. We simply cannot afford to ignore the fact that science is a human activity tied up with the whole process of organic adaptation of man's mind to the external world. Or at least, those who choose to indulge in dreams of formal beauty must realize that they cut themselves off from the "solid ground of Nature".

CHAPTER 11

NIELS BOHR'S CONTRIBUTION TO EPISTEMOLOGY

[1963h]

When I was staying at Yukawa's Institute in Kyoto two years ago, I had occasion to discuss Bohr's ideas with the great Japanese physicist, whose conception of the meson with its complementary aspects of elementary particle and field of nuclear force is one of the most striking illustrations of the fruitfulness of the new way of looking at things that we owe to Niels Bohr. I asked Yukawa whether the Japanese physicists had experienced the same difficulty as their Western colleagues in assimilating the idea of complementarity and in adapting themselves to it. He answered, "No, Bohr's argumentation has always appeared quite evident to us;" and, as I expressed surprise, he added, with his aristocratic smile, "You see, we in Japan have not been corrupted by Aristotle."

If Yukawa had also mentioned Plato, his epigram would have given a complete characterization, which it would be difficult to make more pregnant, of the significance of Bohr's contribution to philosophical thought. Untrammelled by formal schooling, guided only by the sure intuition of the investigator of nature, Bohr rediscovered the dialectical process of cognition which had so long been obscured by the unilateral development of epistemology on the basis of Aristotelian logic and Platonic idealism. Indeed, when Newton formulated the new natural philosophy (as he called it) which was to give modern science its aim and its method, he was so scared at the prospect of the impious materialism to which exclusive reliance on human reason would lead that he injected into his philosophy a generous dose of mystical antidote of Platonic inspiration; and this uneasy alliance of rationalism and mysticism has since then been paralyzing scientific philosophy. Even those scientists of the nineteenth century who were most inclined to idealism were impervious (as the example of Oersted shows) to the Hegelian form of dialectics; as for the Marxist dialectics, which might have given the materialists among them a clue, it was effectively hidden from their view by an impenetrable social barrier.

When young Niels Bohr entered the scene, an antimaterialistic reaction was at its height. His father, an eminent physiologist, firmly rejected the superficial materialism of the Haeckel school; and this condemnation appeared the more significant as it came from a master in the quantitative investigation of the physical and chemical processes underlying the physiological functions. Just because of this first-hand experience, Bohr's father was in a good position to know how sterile materialistic speculations remained for the analysis of concrete cases, and to appreciate the indispensable part played in such analysis by specific biological concepts, such as that of function. Hence his insistence on the necessity of upholding, at least for methodical reasons, teleological viewpoints in biology.

This argumentation of his father, solidly founded on the conditions of actual scientific investigation, impressed Bohr much more than any teaching of traditional philosophy. The philosopher Høffding, a wise and broad-minded thinker, belonged to the circle of the father's nearest friends, and soon formed a trusting affection for the earnest and sensitive adolescent; but his influence on Bohr was more marked in the general attitude to human problems than in questions of epistemology. Although Høffding showed more appreciation than was common at the time of the role of science in the development of philosophical conceptions, he lacked the intimate knowledge of scientific thinking necessary for a detailed analysis of its philosophical implications. A little-known incident illustrates at the same time Høffding's shortcomings in the more technical aspects of philosophical analysis and the seriousness with which Bohr applied himself to their study: When he attended Høffding's course on formal logic, Bohr noticed some serious errors in the professor's exposition, which, with his wonted candor, he pointed out to him. Høffding took the student's censure quite philosophically, and he even submitted the proofs of the amended edition of his course to his young friend's approval. In later life, Bohr remembered the fact, but had forgotten what the errors in question actually were.

Høffding's lectures on the history of philosophy had, at any rate, the merit of not enforcing any system upon his audience. He was more concerned to present the problems than the solutions proposed by the various systems; for, he said, the systems come and go, but the problems remain. It was not, however, in Bohr's nature to be attracted by anything so vague as the classical problems of philosophy; his interest could only be awakened by some concrete, sharply outlined situation which would grip his imagination. He liked Spinoza's conception of the psychophysical parallelism, in which he later was tempted to see an endeavor to express the complementarity relations in the realm of psychological phenomena; eventually, however, when his conception of complementarity had reached full maturity, he abandoned this interpretation, and rejected the whole idea of a psychophysical parallelism in the spirit of Spinoza, as implying a kind of mutual determination and equivalence of the two parallel aspects which certainly does not belong to a true relation of complementarity. He read Kierkegaard, but admired the virtuosity of his style more than his highly strung meditations. What made a really deep and lasting impression on him, however, was the unpretentious 'Tale of a Danish Student', in which Poul Martin Møller has given such a delightfully humorous illustration of Hegelian dialectics.

Here at least, in the character of a soul-searching licentiate desperately struggling to keep his ideas together, he could see a vivid picture of the dramatic situation which might confront the unwary ponderer on the intricacies of human thinking:

> ... and then I come to think of my thinking about it; again I think that I think of my thinking about it, and divide myself into an infinitely retreating succession of egoes observing each other. I don't know which ego is the real one to stop at, for as soon as I stop at any one of them, it is another ego again that stops at it. My head gets all in a whirl with dizziness, as if I were peering down a bottomless chasm, and the end of my thinking is a horrible headache.

Habent sua fata libelli! Poul Møller had little imagined that his light-handed banter would one day start a train of thought leading to the elucidation of the most fundamental aspects of atomic theory and the renovation of the philosophy of science. Yet it is hardly an exaggeration to say that the perplexities of this licentiate, especially his struggle with his many egoes, were the only object lesson in dialectical thinking that Bohr ever received, and the only link between his highly original reflection and philosophical tradition.

Let us quote another characteristic scene, in which the licentiate tries to explain to his cousin Fritz why it takes him such a long time to write his dissertation:

> ... certainly I have seen before thoughts put on paper; but since I have come distinctly to perceive the contradiction implied in such an action, I feel completely incapable of forming

a single written sentence. And although experience has shown innumerable times that it can be done, I torture myself to solve the unaccountable puzzle, how one can think, talk, or write. You see, my friend, a movement presupposes a direction. The mind cannot proceed without moving along a certain line; but before following this line, it must already have thought it. Therefore one has already thought every thought before one thinks it. Thus every thought, which seems the work of a minute, presupposes an eternity. This could almost drive me to madness. How could then any thought arise, since it must have existed before it is produced? When you write a sentence, you must have it in your head before you write it; but before you have it in your head, you must have thought it, otherwise how could you know that a sentence can be produced? And before you think it, you must have had an idea of it, otherwise how could it have occurred to you to think it? And so it goes on to infinity, and this infinity is enclosed in an instant.

"Bless me," said Fritz with indifference, "while you are proving that thoughts cannot move, yours are proceeding briskly forth!"

"That is just the knot," replied the licentiate. "This increases the hopeless mix-up, which no mortal can ever sort out. The insight into the impossibility of thinking contains itself an impossibility, the recognition of which again implies an inexplicable contradiction."

Bohr must have recalled this passage as he manifested such amusement when Dirac, seeing him hesitate in the formulation of some remark, observed sententiously, "One should never start a sentence before knowing the end of it." Could it be that one reason why Bohr took such a fancy to the poor licentiate's embarrassment in writing a sentence or keeping the subject of his dissertation within bounds was that he saw in him a goodhumored caricature of himself? From the time when we were investigating together the complementarity relations of electromagnetic quantities, I remember that on one of those days when the solution of our problem seemed more elusive than ever, we took a bicycle ride in the country and came in sight of a beautiful mansion, called Naerumgaard, where Bohr's grandparents had lived. "It is here," said Bohr, "that I finished my first paper." (This was the determination of the surface tension of water that he carried out in his student years and that won him a gold medal of the Danish Academy of Sciences.) "I was working on it in Copenhagen, but the experiments had no end; I always noticed new details that I thought I had first to understand. At last my father sent me out here, away from the laboratory, and I had to write up the paper." Now it was just part of Bohr's greatness that he did finish his papers. He might pursue a problem for years on end, with dogged tenacity, looking at it from all angles, taking up the same point over and over again, but he knew when to stop. He vividly realized that our proud theories are but temporary resting places of the mind on the unending road to knowledge: Such resting places, however, there must be, where we may taste the joys of knowledge, feel that we have reached a certain harmony between our mental picture of the world and our experience of it.

It is remarkable that Bohr's first preoccupation with philosophical problems did not arise from his physical investigations, but from general epistemological considerations about the function of language as a means of communicating experience. The fictitious case of the licentiate trying to sort out his egoes was only an exaggeration of more common situations in which the same word is used, in different contexts, to denote aspects of human experience which are not only different, but even mutually exclusive; thus we currently use the same word when we speak of a state of our consciousness and of the concomitant behavior of our body. How to avoid ambiguity in such cases was the problem that worried Bohr. In his search for a solution, he was guided by a beautiful analogy with the concepts of multiform function and Riemann surface. The ambiguity of every word referring to our mental activity may be expressed by saying that it belongs to different 'planes of objectivity', just as the values of a multiform function are distributed on different Riemann planes. The use of words in everyday life must then be subject to the condition that they be kept within the same plane of objectivity; and as soon as we deal with words referring to our own thinking, we are exposed to the danger of gliding onto another plane. In mathematics, that highly sophisticated language, we are guarded from this danger by the essential rule never to refer to ourselves. But just as the gist of Riemann's conception lies in regarding all the branches of a multiform function as one single function, it is an essential feature of ordinary language that there is one word only for the different aspects of a given form of psychical activity. We cannot hope, therefore, to avoid such deep-rooted ambiguities by creating 'new concepts'. We must rather recognize the mutual relationships of the planes of objectivity as primitive, irreducible ones, and try to remain keenly aware of them.

As to the origin of this fundamental peculiarity of human language, it could be traced directly to man's position in the universe, which conferred on him the unique possibility of making himself the object of his own observation. Here again, we hear the licentiate describe the situation in his own way:

Thus on many occasions man divides himself into two persons, one of whom tries to fool the other, while a third one, who in fact is the same as the other two, is filled with wonder at this confusion. In short, thinking becomes dramatic and quietly acts the most complicated plots with itself and for itself; and the spectator again and again becomes actor.

Does not this last sentence call forth in our memories familiar echoes of one of Bohr's favorite reflections? Thus we find foreshadowed in these youthful meditations all the main themes that recur, steadily amplified, in his later epistemological work: the use of language for the objective communication of experience; the ensuing necessity of fixing the unambiguous meaning of words by reference to situations of common experience; and finally, the possible occurrence of a duality of aspects requiring special caution in the use of language to secure unambiguous communication.

Against this background it is easier to imagine Bohr's reaction to the dilemma about the dual nature of light, which had puzzled the physicists since Planck's discovery of the quantum of action had led Einstein to emphasize the corpuscular features of light in its interaction with matter, contrasting with the wave-like manner in which it spreads out in space. While the great masters were vainly trying to eliminate the contradiction in Aristotelian fashion by reducing one of the aspects to the other, Bohr realized the futility of such attempts; he knew that we had to live with this dilemma as with the others he had so deeply pondered over, and that the real problem was to refine the language of physics so as to provide room for the coexistence of the two conceptions of light, with suitable precautions in order to avoid any ambiguity in their application. Of course, this was just a program, and many years were to pass before it could be accomplished; but we see that the point of view of complementarity (to call it by its future name) was a constructive one, preventing effort from being wasted in sterile channels and exhibiting much more sharply than the current view the true significance of the quantum of action. There are still too many people who are prone to speak lightly of complementarity as 'just philosophy' because they fail to realize its importance as a guiding principle through the whole historical development of quantum theory.

In the decade following Planck's discovery, and especially after Rutherford's establishment of the existence of the atomic nucleus, it had become clear that the quantum of action intervened just as essentially in the constitution of atomic systems as in the nature of light; but there was here also a sharp contradiction between the mechanical properties of such systems and their quantal features. Just in the sharpness of this new contradiction, however, Bohr saw the hope of decisive progress; just because every other way out was barred, he boldly put forward his quantum postulates, feeling absolutely convinced that they were the expression of a fundamental truth. The quiet firmness of his attitude in the face of the general confusion then prevailing is a striking testimony to the strength he derived from his epistemological standpoint.

Most revealing of his dialectical turn of mind at this early stage is the concluding sentence of a lecture he gave in 1913 on his explanation of the Balmer formula; after stressing the contrast between the classical and quantal behavior of atomic systems, he ended with the remarkably prophetic statement: "Just by accentuating this contrast it will perhaps be possible in the course of time to bring about a certain consistency also in the new ideas."

When at last, with the advent of quantum mechanics, the long-sought logical foundation was established for a rigorous formulation of the complementary relationships of atomic theory, Bohr was not content to point out how this new development fulfilled his far-sighted expectations; he eagerly took up the task of refining the complementarity idea by a minute analysis of its multiple implications. In this work, he was to no small extent stimulated by Einstein's criticism; the account of this prolonged controversy, with its dramatic succession of shrewd objections and brilliant refutations, has been given by Bohr himself in a later paper which will remain a classic in the history of physics and in the history of philosophy.

The method to which the two giants had recourse in this memorable contest deserves special mention: in order to visualize more easily the meaning of the concepts at issue, they imagined more or less idealized experimental arrangements illustrating the way in which these concepts are actually applied. In devising such imaginary experiments, Einstein's inventiveness was supreme; but Bohr was unsurpassed for penetrating analysis of their implications. His greatest triumph was the successful refutation of the paradox which will remain part of the physical tradition under the name of the 'Einstein box'. Einstein had argued that it should be possible, in contradiction to the indeterminacy relation, to ascertain

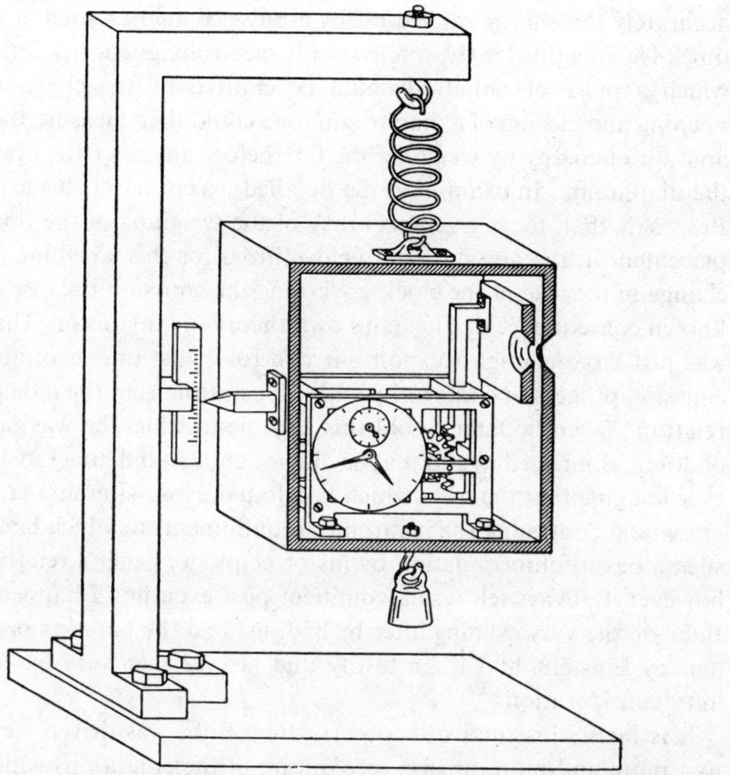

Bohr's 'realistic' representation of the 'Einstein box' shows a mock-realistic style of drawing intended to stress the classical nature of the measuring apparatus, as well as the necessity of carefully analyzing the measuring process and not being satisfied with too schematic arguments, which may be treacherous. The formulae

$$\Delta p \approx \frac{h}{\Delta x} < T \cdot g \cdot \Delta m$$

$$\frac{\Delta T}{T} = \frac{1}{c^2} g \Delta x$$

$$\Delta T > \frac{h}{c^2 \Delta m}$$

$$\Delta T \cdot \Delta E > h$$

summarize the analysis: in order to obtain an accuracy Δm in the determination of the mass, or $\Delta E = c^2 \Delta m$ in the determination of the energy, by weighing during a time T in a gravitational field of acceleration g, the accuracy of the determination of the momentum acquired accurately the energy carried off by a pulse of light emitted at a definite time. He imagined a box filled with electromagnetic radiation, from which a pulse of radiation could be emitted by the clock-controlled opening and closing of a diaphragm; one could then measure the emitted amount of energy by weighing the box before and after the operation of the diaphragm. In examining the detailed execution of this experiment, Bohr saw that, for a given accuracy of the weighing of the box, its displacement in the gravitational field utilized for this weighing implied a change in the rate of the clock governing the emission because of a well-known consequence of Einstein's own theory of gravitation. That change was just large enough to spoil our control of the time evolution of the emission process in conformity with expectation from the indeterminacy relation. When he later retold this argument, which he was never tired of doing, Bohr used to add that he did not understand how Einstein could ever imagine that the facts which had found expression in quantum theory would contradict the invariance requirements to which he had given such a beautiful formulation by his principles of general relativity. This, however, I always felt was a comment post eventum. I happened to see Bohr on the very evening after he had just had the paradox proposed to him by Einstein, and I can testify that he was then very far from such lofty considerations.

It is highly instructive to observe how Bohr was driven step by step to a more and more precise recognition of the conditions which had to be specified in order to achieve full harmony between the statements of the theory and the experiments devised to check them. The method reached its utmost refinement in the examination of the fundamental concepts of electromagnetic theory, in which I had the privilege of participating. In this case, one had to exploit to the last limit all possibilities of controlling the course of the measuring process before complete agreement could be reached with the theoretical predictions. Thus, suppose

---

by the box in this weighing must be Δp, implying an uncontrolled displacement Δx of the box. This in turn involves an error ΔT in the rate of the clock. Combining the first two relations, one obtains the third, or equivalently the fourth, which is the indeterminacy relation between time and energy. For full details, see N. Bohr, 'Discussion with Einstein on epistemological problems in atomic physics', in *Albert Einstein, Philosopher-Scientist* (Library of Living Philosophers, Evanston, Ill., 1949) p. 201 [reprinted in Niels Bohr, *Atomic Physics and Human Knowledge* (Wiley, New York, 1958)].

that the spacetime situation of two domains in which measurements of electromagnetic field quantities are performed is such that radiation from each of them is able to reach the other within the time taken by the measurements. Then the product of the measuring errors will not be reduced to the theoretical minimum unless we avail ourselves of the possibility of exchanging information between the two domains and establish between the two sets of test bodies an automatic signalling system enabling them to compensate their mutual perturbations in the most effective manner.

As the outcome of this patient and unrelenting effort of elucidation, there emerged an entirely novel view of the meaning and structure of physical description. Our experiments, to use Bohr's favorite phrase, are questions that we put to Nature; and in our theories we try to state what we have learned from her in a language ensuring unambiguous and objective communication. Hence the paramount importance of establishing a rigorous terminology, sufficiently general to make due allowance for the peculiar conditions under which we observe atomic systems. We let such systems interact with apparatus which is necessarily adapted to our means of immediate perception, and whose behavior is accordingly described in the language of classical physics. In the account of any atomic phenomenon, we must include a specification of the experimental arrangement fixing the conditions of observation, but a sharp distinction must be made between the part of the total system which forms the apparatus and the part which is the object of observations, and which may exhibit quantal properties. Moreover, the complete definition of the phenomenon must essentially contain the indication of some permanent mark left upon a recording device which is part of the apparatus; only by thus envisaging the phenomenon as a closed event, terminated by a permanent record, can we do justice to the typical wholeness of the quantal processes and steer clear of all the paradoxes to which a lax terminology may lead.

It is in this widened framework that the conception of complementarity in all its generality finds its natural place, as a logical relationship between two physical phenomena, both representing aspects of a physical system equally necessary for its complete description, but corresponding to mutually exclusive experimental conditions. The occurrence of such relationships in atomic physics has a momentous consequence for the causality relations between atomic phenomena: this causality can no longer be deterministic, as it is for large-scale events, but has to be fundamentally statistical. Moreover, the function of a physical theory appears itself in a new light: in view of the necessity of explicitly specifying, for each phenomenon, the conditions of its observation, the description of the phenomena is no longer, as in classical physics, a picture of events from which all reference to observation is eliminated; instead, it appears essentially as a rational and fully objective account of the interaction of the external world with human observers – a conception of science which obviously conforms much better to the part it actually plays in human society.

It is now thirty years since the clarification of the complementary features of quantum mechanics and quantum electrodynamics was essentially completed. On the face of it, physicists at large have reconciled themselves with this intrusion of dialectics into their traditional modes of thought. Who among us, however, could boast of having mastered all the intricacies of complementarity arguments to the point of being prepared for any situation? I would rather characterize our attitude in the words that I once heard from one of Bohr's most faithful and eminent disciples, Weisskopf, who, after having discovered, not without effort, the explanation of a tricky case of complementarity put to him by a doubting experimenter, exclaimed, "Bohr always wins!" Too few even now realize the earnestness of the epistemological issues with which Bohr had to contend single-handed; too few fathom the depth of the problems he had solved and appreciate the revolutionary significance of his contribution to a better understanding of the dialectic process and the development of a truly scientific philosophy.

Complementarity is no system, no doctrine with ready-made precepts. There is no via regia to it; no formal definition of it can even be found in Bohr's writings, and this worries many people. The French are shocked by this breach of the Cartesian rules; they blame Bohr for indulging in "clair-obscur" and shrouding himself in "les brumes du Nord". The Germans in their thoroughness have been at work distinguishing several forms of complementarity and studying, in hundreds of pages, their relations to Kant. Pragmatic Americans have dissected complementarity with the scalpel of symbolic logic and undertaken to define this gentle art of the correct use of words without using any words at all. Bohr was content to teach by example. He often evoked the thinkers of the past who had intuitively recognized dialectical aspects of existence and endeavored to give them poetical or philosophical expression; our only advantage over these great men, he would observe, is that in physics we have been presented with such a simple and clear case of complementarity that we are able to study it in detail and thus arrive at a precise formulation of a logical relationship of universal scope. The nature of this relation he regarded as sufficiently illustrated by his analyses of the limits of validity of classical physical concepts.

On the numerous occasions on which he addressed assemblies of specialists in the most varied disciplines, Bohr never failed to acquaint them with the idea of complementarity as it had taken shape in atomic physics, and to suggest possible applications in the realm of their own investigations. Psychical phenomena of course furnish striking examples of complementarity. In fact, in pointing to such examples, Bohr was returning to the very source of inspiration for his earliest dialectic meditations. He was always as eager to learn as to teach, however, and when much later, some time in the early thirties, his old friend the psychologist Rubin called his attention to William James' *Principles*, he joyfully recognized in this great book a general attitude akin to his own; he was particularly enthusiastic about James' brilliant description of the stream of consciousness. He also gave much thought to the complementary aspects of human relations. Above all he was concerned to point out that, in ethical and social questions the recognition of complementarity between seemingly irreconcilable points of view could help people to get rid of prejudices fostering intolerance; but the way in which hints of such prospects were generally received by church dignitaries and other pillars of society suggests that these problems are not yet ripe for an approach on so rational lines.

The complementarity relations in biology were constantly in Bohr's mind; it was a matter of particular satisfaction to him that he was able to take up his father's ideas and present them in the guise of a relation of complementarity between a purely physical and chemical account of biological phenomena and the use of the teleological concept of biological function equally indispensable for a complete description. Last summer, he had eagerly seized an opportunity to return to these considerations, in the light of the new discoveries in molecular biology. He

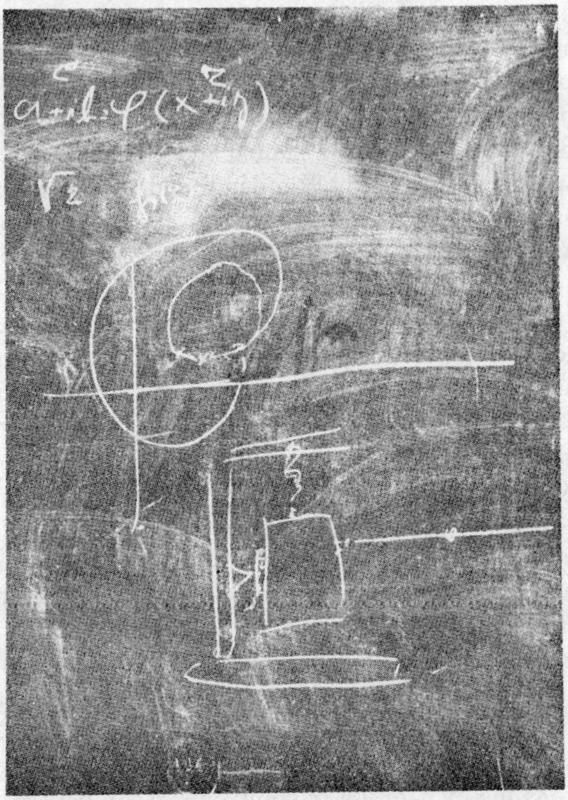

These rough drawings were made by Niels Bohr on the blackboard of his working room at Carlsberg in the course of conversation on the eve of his death. The upper one represents a contour in a complex plane, surrounding a singularity (at the origin) of the nonuniform function $\sqrt{Z}$. It illustrates the early considerations on the ambiguity of language mentioned in the text. The lower drawing is a sketch of the 'Einstein box', also described in the text. Fate has thus given us, by Bohr's own hand, a last symbolic record, as it were, of the beginnings and culmination of his thoughts on complementarity.

had accepted an invitation to deliver an address at the opening of a new institute of genetics in Cologne, and he worked very hard at the preparation of this lecture. He sought the expert help of his younger Copenhagen colleagues, who, coming for the first time in closer contact with the venerated master, were struck by the candor and humility with which he submitted to their criticism. This paper was the one he was most anxious to finish. We are only left, alas, with a preliminary version of it, which, imperfect as it is, embodies his last thoughts on these fundamental questions.

Bohr had great expectations about the future role of complementarity. He upheld them with unshakable optimism, never discouraged by the scant response he got from our unphilosophical age. On one of those unforgettable strolls during which Bohr would so candidly disclose his innermost thoughts, we came to consider that what many people nowadays sought in religion was a guidance and consolation that science could not offer. Thereupon Bohr declared, with intense animation, that he saw the day when complementarity would be taught in the schools and become part of general education; and better than any religion, he added, a sense of complementarity would afford people the guidance they needed. Utopian dreams? It is up to us to respond to the challenge and make of today's Utopia the reality of tomorrow. We owe Bohr a tremendous widening of our scientific and philosophical horizon; he has proposed to us a view of the world of greater wealth and deeper harmony, and a conception of the function of science which makes its scope more universal and its appeal more human. But his most precious gift to us who have known him is the shining example of a life so earnestly devoted to truth, so full of wisdom and humanity. His name is written in the annals of history near those of Newton and Einstein; his place in our hearts is among those that are dearest to us.

CHAPTER 12

THE MEASURING PROCESS IN QUANTUM MECHANICS

(On the 30th Anniversary of the Meson Theory by Dr. H. Yukawa, 1965)

[1965c]

> Erunt multi qui, postquam mea scripta legerint, non ad contemplandum utrum vera sint quae dixerim, mentem convertent, sed solum ad disquirendum quomodo, vel iure vel iniuria, rationes meas labefactare possent.
>
> G. GALILEI, *Opere* (Ed. Naz., vol. I, p. 412)
>
> There will be many who, when they will have read my paper, will apply their mind, not to examining whether what I have said is true, but only to seeking how, by hook or by crook, they could demolish my arguments.

The attempt to develop a consistent theory of atomic processes has disclosed reciprocal limitations of the applicability of classical concepts, the occurrence of which can only be understood by a deeper analysis of the definition of these concepts than was necessary in classical physics. This analysis has been carried out by Niels Bohr, to whom we owe the complete elucidation of the epistemological aspect of the theory. Bohr's approach is quite straightforward and does not appeal to any principle which is not either immediately obvious from everyday experience or well established by experiment; and his argumentation is conducted with full logical rigour. Nevertheless, it is understandable that in order to exhibit more directly the link between the physical concepts and their mathematical representation, a more formal rendering of Bohr's argument should have been attempted, as part of an abstract formulation of the whole theory by the method of axiomatization of modern logic. Such axiomatization of a physical theory would indeed be harmless if it were properly handled; but unfortunately this is still more difficult to achieve in physics than in mathematics, and in either domain, the attempt generally leads not to clarification, but to confusion.

The origin of this peculiar difficulty lies in the obvious fact that no formalization can be complete, but must leave undefined some 'primitive' concepts and take for granted without further analysis certain relations between these concepts, which are adopted as 'axioms': the concrete meaning of these primitive concepts and axioms can only be conveyed in a 'metalanguage' foreign to the formalism of the theory. However, just the proper choice of the primitive concepts and axioms and the proper use of the metalanguage demand a complete mastery of the actual physical situation which it is the purpose of the axiomatic analysis to clarify, and it is therefore a fatal illusion to hope that the really deep difficulties of this situation could be skipped or in any way alleviated by any process of axiomatization. It is for this very reason that Bohr never showed any interest in the endless discussions to which the axiomatization of quantum mechanics has given rise: in his view, formal logic could only help us to check whether we had not made any trivial mistake in our reasoning, and he rightly regarded his task as accomplished when he had exhibited in simple words the perfect adaptation of the principles of quantum mechanics to the empirical situation it purports to describe.

The following remarks are intended to dispell the impression, still prevailing in some quarters, that the translation of Bohr's argument into the formal language of the theory should present unrecognized difficulties or leave room for speculations deviating from the commonsense conclusions reached by Bohr. These misunderstandings, which go back to deficiencies in von Neumann's axiomatic treatment, have only recently been completely removed by the very thorough and elegant discussion of the measuring process in quantum mechanics carried out by Daneri, Loinger and Prosperi,[1] and all I shall have to do is to summarize and comment their work. Essentially, it consists in giving a detailed description by means of quantum mechanics of the entire course of a typical measuring process. This will not "prove" the impossibility of a completely deterministic description of the atomic phenomena, since this impossibility is embodied in the formalism of the theory, and any deduction of it from the axioms, such as that proposed by von Neumann, is therefore – as many sharp-sighted critics have observed – necessarily circular: but it will show us more clearly how the fundamental statistical predictions of the theory closely correspond to the possibilities of determination of the physical quantities compatible with the existence of the quantum of action.

It is necessary, in the first place, to insist on a point of method, disregard of which vitiated von Neumann's treatment of the measurement problem and has played havoc with many of the subsequent discussions.

When considering the epistemology of observation of physical phenomena, it may perhaps seem natural to include the observer into the field of investigation, and this was actually done by von Neumann: he followed up the indefinite chain of physiological and mental processes leading from the measuring apparatus through the sense organs and the brain of the observer to the obscure recesses of his 'consciousness'. The actual development of science, however, has consistently pointed to the advantage of the opposite procedure, namely the separate study of that part of our experience from which all features connoted with the word life are eliminated. This methodical elimination has not only been found possible, but even indispensable for the generality and simplicity of the description of nature, inasmuch as life belongs to rather special and complicated types of material structures. It has thus become a primary requirement of sound method in physics that all concepts should be defined by reference to purely physical processes, regarded as sufficiently familiar to be generally intelligible without further analysis. In particular, the measuring procedure by which the value of a physical quantity is ascertained has to be conceived in terms of some automatic recording device, subject only to the laws of physics and entirely analysable by means of them. The biological processes involved in human observation are perfectly irrelevant in this context; indeed, it is quite irrelevant whether or not any human observer is actually present, or even whether any reading is ever made of the recorded information. Only by upholding this 'detachment' of the observer is it possible to maintain the objective character of the physical description, i.e. its universal validity for all possible observers. In fact, as soon as the 'consciousness' of the observer is unduly mixed up in the argument, the impression is created that a 'subjective' element has been introduced in physics; but there is no objective truth in this, it is just a product of muddled thinking.

The muddle arises from not keeping the requirement of objectivity sufficiently distinct from another equally stringent one, namely that of definiteness: the information conveyed by the physical description must be formulated in such a way as to have, for each observer, a precise, unambiguous meaning; in the last analysis, it must refer to some well-defined disposition and manipulation of instruments by which the conditions leading to the appearance of the phenomenon are established. This requirement is so obvious that in classical physics it may simply be taken for granted, because the phenomena dealt with in this part of physics are of the same macroscopic character as the instruments used for their observation. In quantum physics, however, this is no longer so; a disparity arises between the atomic system we want to observe and the means of observation, which in order to fulfil their function are still necessarily bodies consisting of large numbers of atoms. This very circumstance, as Bohr showed, is of crucial importance for the clarification of the logical structure of quantum mechanics. Now, at this point the human observer, whom we have been at pains to keep out of the picture, seems irresistibly to intrude into it, since after all the macroscopic character of the measuring apparatus is imposed by the macroscopic structure of the sense organs and the brain. It thus looks as if the mode of description of quantum theory would indeed fall short of ideal perfection to the extent that it is cut to the measure of man. However, we are not facing here any deep philosophical issue, but the plain commonsense fact that it takes a complicated brain to do theoretical physics. Intelligent beings on other planets would not fare otherwise in this respect; they would have developed the same atomic physics as ours, and recognised the complementarity of position and momentum. They might have been spared axiomatics, however.

The part of the argument which has been a stumbling block for previous formal analyses of the measuring process is the design of a model of measuring apparatus sufficiently representative to illustrate the fundamental features and sufficiently simple to allow of a quantitative treatment. It is the merit of the Italian physicists that they have provided precisely such a model, in complete harmony with Bohr's ideas. We have to visualize the measurement as an interaction between the observed atomic system and a registering device, ultimately leading to the formation of some permanent record, uniquely related to a definite quantity characterizing the state of the atomic system. How does such a permanent mark come about? Since it is of macroscopic character, it cannot be a direct result of the initial atomic interaction; the latter is rather of the nature of a triggering effect. It starts off within the measuring apparatus a macroscopic reaction, terminating when a state of stable equilibrium is reached, in which the desired characteristic mark has appeared. The triggering process only lasts for a time interval of the order of magnitude typical of individual atomic reactions; the ensuing macroscopic process necessitates a "relaxation time" which, although it may be very short on the macroscopic time scale, will in any case involve a large number of atomic reactions. This is the decisive point: the formation of the permanent mark is a process of ergodic character, entailing the wiping out of all structural details of the initial state of the total system, known as the 'reduction' of this state.

The stationary states of a macroscopic body correspond to a discrete spectrum of energy values differing from each other by amounts extremely small on the macroscopic scale; they may be divided into 'shells,' each containing all the states within a given interval $(E, E+\Delta E)$ on a continuous energy scale, and any macroscopic change of the body will correspond to transitions between states of the same shell. The energy shells are then further subdivided into 'cells' with respect to conveniently chosen intervals of values of other quantities of interest. Each cell thus corresponds to a manifold of s_k orthogonal states Ω_{ki} $(i=1, 2, \ldots, s_k)$, which will not in general be eigenstates of the energy. There are a great many cells in each energy shell, nearly all of which contain numbers of states s_k very much smaller than the total number S of states of the shell; but some exceptional cells, corresponding to states of equilibrium of the body under specified external conditions, contain numbers of states of the same order of magnitude as S. Whatever the initial state of the body, its natural evolution will eventually carry it, after a time of the order of the relaxation time, into the cell corresponding to the equilibrium under the given conditions. It will then, at least within any reasonable time, remain in this cell, except for occasional excursions to smaller cells, corresponding to the usual fluctuations.

The mathematical description of this ergodic behaviour is straightforward. Macroscopically observable quantities are identified with the time averages of the corresponding atomic ones over intervals sufficiently long for the establishment of equilibrium: these are idealized as averages over an infinite time, more precisely by the operation

$$\mathfrak{M}\cdots \equiv \lim_{T\to\infty} \frac{1}{T}\int_0^T \mathrm{d}t\cdots.$$

Let us denote the Hamiltonian operator of the macroscopic body by $H=\sum_\rho E_\rho P_\rho$, where P_ρ is the projection operator of the eigenstate of energy E_ρ. This Hamiltonian defines a time evolution operator $\varepsilon(t) \equiv \exp[-iHt/\hbar]$, such that if the initial state of the body is Ω_{lj}, at time t it becomes $\varepsilon(t)\,\Omega_{lj}$. The probability that it will then be within the cell $\{\Omega_{ki}\}$ is

$$\sum_i |(\Omega_{ki}, \varepsilon(t)\,\Omega_{lj})|^2,$$

and the time average of this quantity is

$$\sum_\rho \sum_i |(\Omega_{ki}, P_\rho \Omega_{lj})|^2;$$

we expect this to be of the order of the statistical weight of the cell within the energy shell to which it belongs, i.e. s_k/S. Moreover, if we consider the quantity

$$\sum_i (\Omega_{ki}, \varepsilon(t)\,\Omega_{lj})\,(\Omega_{ki}, \varepsilon(t)\,\Omega_{l'j'})^*, \quad (l \neq l'),$$

which may be regarded as a measure of the correlation between probability amplitudes of different states, its time average, which is in absolute value

$$|\sum_\rho \sum_i (\Omega_{ki}, P_\rho \Omega_{lj})\,(\Omega_{ki}, P_\rho \Omega_{l'j'})^*|, \quad (l \neq l'),$$

will be expected to be much smaller than the statistical weight of the cell. The two 'ergodicity conditions'

$$(1) \quad \begin{aligned} &\mathfrak{M}\{\sum_i |\Omega_{ki}, \varepsilon(t)\,\Omega_{lj})|^2\} \approx s_k/S, \\ &|\mathfrak{M}\{\sum_i (\Omega_{ki}, \varepsilon(t)\,\Omega_{lj})\,(\Omega_{ki}, \varepsilon(t)\,\Omega_{l'j'})^*\}| \ll s_k/S, \quad (l \neq l') \end{aligned}$$

provide a sufficient basis for the whole argument. (They may even be replaced by weaker ones, but we shall not insist on this refinement.)

The next step consists in adding to this general description of a macroscopic system a specification characterizing it as a measuring apparatus. In order to fulfil this function, the system must be susceptible of a set of states Ψ_r, each of which is uniquely related, by the interaction between the body and some definite atomic system, to a definite state φ_r of the latter system. Let H_a denote the Hamiltonian of the atomic system, and let $\varepsilon_a(t) \equiv \exp[-iH_a t/\hbar]$; let us further assume that an interaction of energy V is established between the atomic system and the body during a time τ so short that the effect of the time evolution operators $\varepsilon_a(\tau)$, $\varepsilon(\tau)$

on the respective initial state vectors φ_r, Ψ_0 of the two systems is negligible. After the interaction has taken place, the initial state $\varphi_r \Psi_0$ has thus become, in sufficient approximation, $\exp[-iV\tau/\hbar]\ \varphi_r \Psi_0$, and the process is characterized as a measurement showing that the atomic system is in the state φ_r if the final state just obtained can be represented as $\varphi_r \Psi_r$. Then, indeed, if we start from any state

$$\psi_0 = \sum_r c_r \varphi_r$$

of the atomic system, the state of the total system after the interaction is terminated will be approximately expressed as

$$\sum_r c_r \varphi_r \Psi_r,$$

and at any later time t, as

$$\text{(2)} \qquad \sum_r c_r \varepsilon_a(t)\ \varphi_r \varepsilon(t)\ \Psi_r.$$

Thus, once the correspondence between the states φ_r and Ψ_r has been established during the short time τ by the interaction between the two systems, it subsists after they have again separated. We may now leave the atomic system to its fate, and all we have to do is to study the evolution of the measuring body and its very essential consequences for the nature of the information which the measuring process can give us.

This last stage, which may be described as the formation of a permanent record of the measurement, is represented in our model by the natural evolution of the macroscopic apparatus from the unstable state Ψ_r in which it has been brought by the interaction with the atomic system to the state of equilibrium which corresponds to the external conditions resulting from the measurement. In order to obtain a precise formulation of the situation, we must be a little more specific about the choice of the cell subdivision of the energy shell. Each state Ψ_r will in general be a superposition of the states Ω_{ki} defining the cells, but we may always choose the latter in such a way that a definite group of cells $\{\Omega_{ki}^{(r)}\}$ corresponds to each state Ψ_r, without overlap between these groups. Thus we may write, for every state Ψ_r,

$$\text{(3)} \qquad \Psi_r = \sum_k \sum_{i=1}^{s_k} (\Omega_{ki}^{(r)}, \Psi_r)\ \Omega_{ki}^{(r)}$$

with

(4) $$\sum_{ki} |(\Omega_{ki}^{(r)}, \Psi_r)|^2 = 1.$$

Then, we may characterize the state of equilibrium towards which the state Ψ_r will be tending (in the sense of ergodic theory) as the cell $\{\Omega_{e_r i}^{(r)}\}$ of maximum weight $s_{e_r} \approx S$.

The main quantity of interest is now the correlation coefficient

(5) $$C_{k,rr'} = \sum_i (\Omega_{ki}, \varepsilon(t)\, \Psi_r)\, (\Omega_{ki}, \varepsilon(t)\, \Psi_{r'})^*$$

between the probability amplitudes for any two states Ψ_r, $\Psi_{r'}$, and any cell $\{\Omega_{ki}\}$; in particular, $C_{k,rr}$ represents the probability of finding at time t the state evolving from Ψ_r in the cell $\{\Omega_{ki}\}$. Inserting in the expression (5) the representations of the type (3) for the states Ψ_r, $\Psi_{r'}$, we get

$$C_{k,rr'} = \sum_{lj} \sum_{l'j'} (\Omega_{lj}^{(r)}, \Psi_r)\, (\Omega_{l'j'}^{(r')}, \Psi_{r'})^* \times$$
$$\times \sum_i (\Omega_{ki}, \varepsilon(t)\, \Omega_{lj}^{(r)})\, (\Omega_{ki}, \varepsilon(t)\, \Omega_{l'j'}^{(r')})^*.$$

Now, applying the ergodicity conditions (1), and taking account of Eq. (4), we find for the time average

(6) $$\mathfrak{M} C_{k,rr'} \approx \delta_{rr'} \frac{s_k}{S}.$$

In this derivation, we have of course interpreted the second ergodicity condition as implying that the time averages of the terms with $r \neq r'$ could just be neglected. This leads us to the conclusion – indicated by the factor $\delta_{rr'}$ in Equation (6) – that the correlations between the different states Ψ_r, $\Psi_{r'}$ disappear when the state of equilibrium is reached; they are wiped out, so to speak, in the process of recording the result of the measurement. As to the average probability of the cell $\{\Omega_{ki}\}$ in the equilibrium state, it is

(7) $$\mathfrak{M} C_{k,rr} \approx s_k/S,$$

i.e. negligibly small except for the cell $\{\Omega_{e_r i}\}$, for which it practically amounts to certainty; in other words, we may write instead of Eq. (7)

(8) $$\mathfrak{M} C_{k,rr} \approx \delta_{ke_r}.$$

It is an easy matter, starting again from the ergodicity conditions (1), to estimate the average fluctuations of the values of $C_{k,rr'}$ around the average value (6); we shall not dwell on this point.

We are now in a position to discuss the fundamental prescription establishing the physical meaning of the formalism of quantum mechanics. Consider two quantities A, B pertaining to an atomic system, and let a_r, φ_r; b_s, χ_s denote their respective eigenvalues and eigenstates. Suppose that the initial state of the atomic system is $\psi_0 = \sum_r c_r \varphi_r$ and that a measurement of the quantity A is performed at time $t=0$. If a measurement of the quantity B is then performed at time t, the probability of finding the value b_s is given by

$$(9) \qquad \sum_r |c_r|^2 \, |(\chi_s, \varepsilon_a(t) \, \varphi_r)|^2 .$$

The fact that this expression only depends on the relative probabilities $|c_r|^2$ of the eigenstates φ_r, but not on the phases of the complex coefficients c_r, is referred to as the 'reduction' of the state ψ_0. In fact, if no measurement of A had been performed at $t=0$, the probability of finding the value b_s at time t would be given by

$$(10) \qquad |\sum_r c_r (\chi_s, \varepsilon_a(t) \, \varphi_r)|^2 ,$$

an expression differing from (9) by a correlation term

$$\sum_{r \neq r'} c_r c_{r'}^* (\chi_s, \varepsilon_a(t) \, \varphi_r) \, (\chi_s, \varepsilon_a(t) \, \varphi_{r'})^* .$$

It is the disappearance of this term as a consequence of the measurement of A that is formally denoted as the 'reduction' of the initial state, and the main purpose of the analysis of the measurement is to exhibit the physical process to which this formal 'reduction' corresponds.

To this end, let us consider the atomic system in the state ψ_0 and perform at time $t=0$ a measurement of the quantity A, in the way schematized in the previous discussion. At time t, the total system consisting of the atomic system and the apparatus used for the measurement of A will be in the state represented by the wavefunction (2). Let us now ask for the probability, if at that time t we should perform a measurement of the quantity B (by means of some other suitable apparatus), of finding the value b_s of this quantity, and of having the apparatus with which we had measured A in one of the states of the cell $\{\Omega_{ki}\}$. According to the law of quantum mechanics expressed by a formula of the type (10), this probability is

$$\sum_i |\sum_r c_r (\chi_s \Omega_{ki}, \varepsilon_a(t)\, \varphi_r \varepsilon(t)\, \Psi_r)|^2,$$

i.e. with the notation (5),

$$\sum_{r,r'} c_r c_{r'}^* (\chi_s, \varepsilon_a(t)\, \varphi_r)\, (\chi_s, \varepsilon_a(t)\, \varphi_{r'})^*\, C_{k,rr'}.$$

This is not yet, however, the answer to our problem, since we are not concerned with the state of the measuring apparatus on the atomic scale, but only with its macroscopic state after it has recorded the result of the measurement. In the last expression, we must therefore replace the coefficients $C_{k,rr'}$, pertaining to the measuring apparatus, by their ergodic averages $\mathfrak{M}C_{k,rr'}$ given by Equation (6). We see at once that this leads to the expression (9) prescribed by the laws of quantum mechanics, with the extra factor (cf. Equation (8)) δ_{ke_r}, which simply tells us that the measuring apparatus is in its state of equilibrium.

The advantage of the preceding analysis is to show quite clearly the origin and true character of the 'reduction' of the initial state of the atomic system, prescribed by the algorism of quantum mechanics. This prescription expresses in idealized fashion the effect of the recording of a piece of information about the atomic system upon the predictions we can make about the ulterior behaviour of the system on the basis of the information thus recorded. It entails an element of irreversibility in any measurement, which originates in the macroscopic character of the recording process, and results in the loss of all initial knowledge of correlations on the atomic scale. If at this stage any reader feels that such conclusions are so obvious that they could have been reached without operating the formidable machinery of quantum mechanics and ergodic theory, he is perfectly right. What else can be expected from an application of these theories to the present problem than a translation in mathematical language of a physical argument which can be more briefly stated in terms of simple concepts, referring directly or indirectly to commonplace experience? It is only fair to add that in performing this analysis so thoroughly, the Italian physicists have conclusively established the full consistency of the algorism, leaving no loophole for extravagant speculations.

There is perhaps one point here worthy of some comment. There should now remain no doubt that the reduction of the initial state of the atomic system has nothing to do with the interaction between this system and the measuring apparatus: in fact, it is related to a process taking place in the latter apparatus after all interaction with the atomic system has ceased. It would be easy (but pointless) to modify the discussion so as to give it the apparently paradoxical form to which the names of Einstein, Rosen and Podolsky are attached. Far from involving any real paradox, the kind of situation envisaged in the famous paper by these authors is, as we see, just the normal one; it arises (in principle) whenever we perform any measurement on an atomic system. The kind of measurement we decide to make irrevocably prescribes for all time the nature of the description we can give of the system. Since, however, we are not limited in the scope of the measurements we may choose to perform on many such atomic systems, we are able to explore all aspects of the behaviour of matter on the atomic scale.

NOTE

[1] Daneri, A., Loinger, A., and Prosperi, G. M., *Nucl. Phys.* **33** (1962), 297.

CHAPTER 13

STATISTICAL CAUSALITY IN ATOMIC THEORY (A GENERAL INTRODUCTION TO IRREVERSIBILITY*)

[1972d and 1974a]

Table of Contents:

1. Historical background of the concept of irreversibility

Although a naive feeling of a privileged direction of time has always been present in human thought (Fugit irreparabile tempus, Humpty Dumpty etc.), the scientific concept of irreversibility is relatively modern (19th century).

Modern physics can be said to have started with Stevin and Galileo in the 16th century, but at that time no attention was paid to irreversible phenomena because the primitive state of technology could permit only crude quantitative considerations. The main topic of scientific investigation was the analysis of planetary motions, which were characterized by periodicities exhibiting no irreversibility, and the definition and measurement of time also involved periodic motions, such as that of a pendulum.

Moreover, even though conservation laws were implicit in the laws of mechanics (Kepler's second law, for instance, expresses the conservation of angular momentum), they were not explicitly recognized by Newton and his immediate followers. It was only one century later that Laplace pointed out the relation between Kepler's second law and the conservation of angular momentum. The conservation law of mechanical energy was not formulated in full generality by the 18th-century physicists. The growing importance of the use of machines prompted Lazare Carnot to discuss the efficiency of such mechanical systems. His main conclusion was that maximum efficiency could be achieved by avoiding abrupt changes of velocity, so as to minimize the loss of 'living force'. This concept of 'living force' had been introduced by Leibniz as a measure of 'motion in opposition to Descartes' view, according to which this measure would be the absolute value of the momentum. The underlying idea was that all interactions between the particles of matter were contact interactions, i.e. transmission of motion by collisions, and that the total amount of 'motion' should be conserved, i.e. that the collisions were elastic. From this point of view, the 'living force', correctly defined by Leibniz as the product of the mass and the square of the velocity of the particles, is indeed conserved in all dynamical processes. However, it could be 'lost' in the cases considered by Carnot, involving production of heat.

Sadi Carnot, no doubt influenced by his father's work, tried to find the condition of maximum efficiency for the new source of power developed in the meantime, the steam engine. His masterly analysis showed that this condition was a reversible cycle of processes, consisting in an alternation of the only two reversible modes of transformation – isothermal and adiabatic. The Carnot principle of maximum efficiency could be applied quite generally to any transformation involving heat transfer. It made possible the analysis of experimental data concerning the thermal properties of gases and vapours and also made possible the definition of an absolute scale of temperature. Carnot's book was addressed to engineers and remained unnoticed by the physicists, who continued to regard friction and heat losses as a nuisance spoiling the regularity of the purely dynamical behaviour. It was only after W. Thomson had become aware of Carnot's work (through a paper by Clapeyron based on it) that the fruitfulness of Carnot's ideas was appreciated.

However, Thomson and Clausius were soon faced with an apparent contradiction between Carnot's principle and the discovery by Joule of the law of equivalence between heat and mechanical work. Indeed, Joule's principle implied the transformability of heat into motion, whereas Carnot's argument was based on the assumption that heat was a substance, the amount of which remained constant in all transformations. Clausius reconciled the two principles by regarding them as logically independent, and abandoning the idea of the substantiality of heat. Thomson came independently to the same conclusion about a year later.

Thus the two fundamental laws of thermodynamics were established in full generality. Clausius introduced the concept of entropy as a "measure of the transformability" (*Verwandlungsmass*) of a system: reversible transformations of an isolated system leave the entropy constant, irreversible transformations lead to an increase of its entropy.

The question then arose of deriving these principles of thermodynamics from the laws of mechanics. Indeed, since Descartes' time it had been quite generally accepted that material bodies were mechanical systems, consisting of elementary particles in motion. From this point of view, heat had to be regarded as a particular mode of motion of the particles and the thermal properties of bodies must appear as consequences of the purely dynamical behaviour of these particles. It was natural to assume that the motion of the particles was governed by the same laws as those that governed the bulk motions of macroscopic bodies. The first law of thermodynamics then resulted from an immediate application of the conservation law of mechanical energy. As to the second law, Clausius and Boltzmann, in an early attempt to derive it on a mechanical basis, both made the same mistake; they replaced at one stage of the argument the individual velocity of an atom by its average velocity in the steady state, thus unwittingly introducing a statistical element into the derivation. Maxwell was the first to realize the essentially statistical character of the second law; he was apologetic about the introduction of statistical considerations, which he excused by our ignorance of the exact initial conditions of the atomic motions. Boltzmann adopted the statistical method without at first realizing its full implications. Thus, his proof of the second law ('H-theorem') was based on the well-known '*Stosszahlansatz*', which is not valid when the system is far away from the steady state. He only reached a clear understanding of the situation when he had to face two serious objections. The first was put forward by Loschmidt. He argued that if a system is started in some difinite state, and if at some later time during its evolution the velocities of all particles were inverted, the system must retrace its steps back to the initial state. This implies that in this process the entropy should decrease, and therefore the H-theorem appears to be in contradiction with the laws of mechanics. Another paradox was pointed out by Zermelo, who referred to Poincaré's theorem stating that the motion of any isolated mechanical system of finite extension is quasi-periodic: such cyclic behaviour is at variance with the irreversibility expressed by the second law. The gist of these two objections was to emphasize the logical impossibility of deriving any form of irreversible behaviour from the strictly reversible laws of dynamics. In refuting the objections, Boltzmann did not deny this logical point, but argued that it was irrelevant, inasmuch as the derivation of the H-theorem, owing to its statistical character, was not strictly dynamical.

Thus the essential link between the occurrence of irreversibility and a statistical mode of description was clearly recognized. Boltzmann's views received further elaboration by Ehrenfest, who showed that the non-dynamical element was introduced by a passage from a strictly continuous to a 'coarse-grained' distribution of the states of the system, and by Gibbs, who compared the coarse-grained evolution of the system to a 'mixing' process.

In modern terms, we have here a typical relationship of complementarity between two modes of description of the phenomena, corresponding to conditions of observation which are mutually exclusive. Contradictions in the use of such complementary modes of description are avoided by specifying appropriate limitations of their validity. In the present case, these limitations refer essentially to the size of the system, more precisely to the number of degrees of freedom. The atomic mode of description requires an immense number of dynamical variables; the thermodynamic, macroscopic mode of description uses a very limited number of variables. The thermodynamic description applies to 'large' systems, i.e. to systems of spatial dimensions much larger than the atomic dimensions: such systems behave in general irreversibly, while small systems do not (as the case of Brownian motion shows). The critical size corresponds to typical dimensions of our sense organs, compared to those of single atoms. With respect to time, the evolution of the system is irreversible on a macroscopic time scale, long with respect to the atomic time scale defined by the average time of atomic interactions, and again corresponding to the relaxation times of the processes of reception of signals by our sense organs. Such reference to the process of observation does not imply (as has sometimes been stated, e.g. by G. N. Lewis, who called the entropy a 'subjective' concept) any departure from objectivity – if this word is used in its only sensible meaning as denoting the consensus of all observers placed in the same conditions.

The preceding considerations apply just as well to quantal systems, since the laws of quantum mechanics are also reversible. In this case, we meet a statistical element even at the level of individual atomic processes: this statistical element is logically independent of the one introduced in the description of large systems of atoms. Nevertheless, there is an indirect connection between the two owing to the fact that the variables describing the behaviour of single atoms are defined by measuring operations performed with the help of macroscopic apparatus.

2. The 'Arrow of Time' and the Irreversibility of Individual Processes

It was Boltzmann who first raised the possibility of associating the sense of the flow of time from past to future, the "arrow of time" (the expression is due to Eddington), with the second law, by defining the future as that direction of time in which the entropy of the universe increases. Although this procedure is logically acceptable if the direction of time is referred to a standard physical phenomenon (i.e. heat transfer), it is unsatisfactory because it does not express the real origin of the distinction between past and future. From a historical and a psychological point of view, human beings refer past and future to the evolution of their own life, and thus attach to the direction of time a much more profound significance than an increase of entropy of material systems can express. The direction of time is associated with the endeavour to achieve desirable aims, common to men and all animals, to orientate action towards such aims. Scientific observation, combined with rational thinking, is just a sophisticated example of this activity, inasmuch as knowledge is acquired in view of its use in orientating our behaviour. One ought therefore rather try to associate a direction of time with the process of observation. This can be done by realizing that an observation requires the reception of a signal, emitted from the outside world, and propagated with finite velocity. Hence, one can *define* a direction of time by stating that the instant of reception is *later* than the instant of emission. Having done this, the second law is given a well defined physical content. where entropy becomes in every case a measure of the transformability of the system. Moreover, this point of view makes it possible to avoid the paradoxes and confusion found in the literature. In fact, the invariance of the mechanical laws for time reversal is not contradictory with the distinction between past and future just introduced. When performing an observation the physicist normally prepares an experimental arrangement in order to observe what is going to happen. The consequences of such an observation are naturally expressed by solutions of the time-reversal invariant equations, but describe the evolution oriented towards the future (e.g. the retarded electromagnetic potentials).

However, the observer can also use the time-inverse solutions (e.g. advanced potentials) to infer by a 'retrodictive' process the events which may have led to the situation revealed by a present observation.

By establishing a direct connection between the sense of time and the process of observation, we not only satisfy the epistemological requirement of going back as far as possible to the immediate psychological roots of our scientific concepts, but we also emphasize the fact that irreversibility has a much wider scope than its occurrence in the second law of thermodynamics: an irreversible element may be attached to an individual atomic process, obeying a reversible law of evolution, by the conditions under which it is observed. As a typical example, we shall discuss the 'decaying states' of atomic systems, and more specifically those of atomic nuclei, which appear in the study of nuclear reactions. For simplicity, we shall consider reactions involving neutrons, in order to deal only with finite range interactions. This allows us to introduce the concept of *channel*, in the following way: outside the finite region of interaction, the fragments (projectile and target, or end products of the reaction) can be regarded as moving freely, and the asymptotic form of their wave-function can accordingly be factorized and written down explicitly – such an asymptotic wavefunction defines an entrance or exit channel. If we treat the process as stationary, i.e. occurring in both directions at a given energy out of the continuous energy spectrum, the channel wave-function plays the part of a boundary condition selecting a particular wave-function among the manifold of those belonging to the given energy.

Let us first consider the simple case of 'potential scattering' (i.e. scattering of the particle by a fixed potential of finite range): only one channel is possible, namely elastic scattering. We may decompose the wave function into 'partial waves', i.e. components corresponding to definite spherical harmonics with respect to the centre of the potential: the total cross-section will be the sum of the partial cross-sections. The partial wave components of the motion of the particle have the asymptotic form $I_c(k)\exp(-ikr)+O_c(k)\exp(ikr)$, with incoming and outgoing amplitudes $I_c(k)$, $O_c(k)$ depending on the wave-number k and a set of 'channel quantum numbers' c. In the absence of the potential, these amplitudes would be the same (if the phases are chosen appropriately); the effect of the scattering is then represented by the amplitude difference

$$1-S_{cc}(k)\equiv 1-\frac{O_c(k)}{I_c(k)},$$

and the cross-section is proportional to the absolute square of this quantity. The ratio $S_{cc}(k)$ expresses the asymptotic relation established by the potential between an incoming and an outgoing wave in the only available channel c.

In the more general case in which several channels c, c' are available we may still write channel wave-functions, of the same form as above, for each of them, with an extra factor, consisting of the wave-functions representing the internal states of the interacting fragments. The partial amplitudes will then be of the form

$$\delta_{cc'} - S_{cc'}(k), \quad S_{cc'}(k) = \frac{O_{c'}(k)}{I_c(k)},$$

where k is one of the wave-numbers, on which all the others depend by the condition of energy conservation $k_c^2 - \varepsilon_c = k_{c'}^2 - \varepsilon_{c'} = E$($\varepsilon_c$, $\varepsilon_{c'}$ being the internal energies in the various channels). The S-matrix thus introduced must satisfy two conditions: it must be unitary – this expresses the conservation of the total flux of particles, and it must be symmetrical, $S_{cc'}(k) = S_{c'c}(k)$, as a consequence of the invariance of the stationary processes occurring at a given energy for time-reversal: all reactions occur equally in both directions.

As a function of the wave-number k or the energy E, the cross-sections $\sigma_{cc'}(E)$ present experimentally for definite energies E_n a number of peaks which, when well separated, have definite 'widths' Γ_n: these are called 'resonances'. These can be interpreted physically in the following way: Let us introduce a fictitious infinite barrier around the interaction region. Within the barrier the energy spectrum is discrete, whereas outside it is continuous. Removing the barrier couples the two regions and allows the particle, which was in a perfectly sharp stationary state, to escape to the continuum. This process has associated with it a characteristic decay time: this is the physical origin of the 'resonance' phenomena. Therefore, an asymmetry in time is introduced even at this elementary level.

The decaying of the wave-function in time can be expressed by adding to the energy of a stationary state E_n, an imaginary term $-(i/2)\,\Gamma_n$; this gives a time factor

$$e^{-(i/h)(E \;\; -(i/2)\,\Gamma_n)t}$$

to the wave-function. How can one relate this concept of a 'decaying state' with the S-matrix? If we extend the definition of the functions $S_{cc'}(E)$ to the complex energy plane, a pole of $S_{cc'}(E)$ in the lower half plane will give rise to a term in its expansion of the form

$$\frac{R_n^{1/2}}{E-\left(E_n-\dfrac{i\Gamma_n}{2}\right)}$$

which gives a contribution to the cross-section of the form $R_n/[(E-E_n)^2-\frac{1}{2}\Gamma_n^2]$, correctly representing an isolated resonance peak. Of course, there must also exist poles in the upper half-plane, corresponding to unphysical situations which would be the time-inverses of the decaying states. We thus have no contradiction with reversibility. The observation process selects from all the poles those corresponding to the decaying states.

This interpretation is only valid if the S-matrix actually has the assumed analytic properties. The theory can be developed in full generality for the case of finite range potential scattering. There, one can prove that $S(E)$ is indeed meromorphic in the complex E plane. In the general case, no such proof can be given. In the first place, the wave-numbers are non-analytic functions of the selected independent variable k or E, and the S-matrix elements are therefore non-analytic functions with branch points on the real axis, corresponding to the 'opening' of the channels. These branch points necessitate the introduction of cuts in the complex E plane. However, it is then possible to extend the theory of elastic scattering by assuming a distribution of poles on the different Riemann sheets.

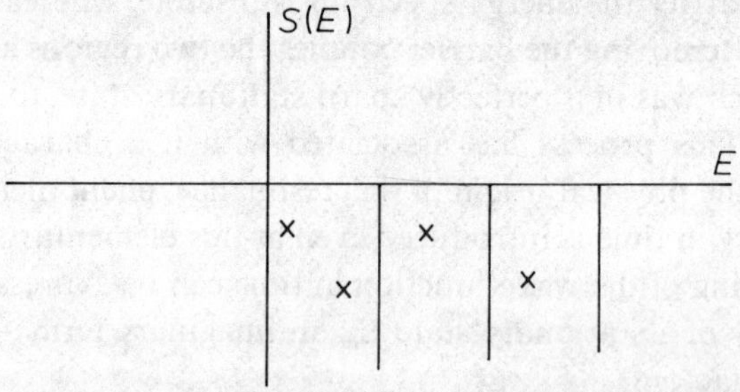

The idea of decaying states was first introduced by Gamow in the early days of quantum mechanics, to describe the spontaneous emission of α-particles from nuclei. It was received with much scepticism at the time, since the meaning of solutions of the Schrödinger equation with complex eigenvalues was unclear. Born elucidated their physical meaning by the method outlined above. Now, we realize that the complex poles are in fact eigenvalues of a non-Hermitian generator, which has been explicitly written down by Claude Bloch. Its non-Hermitian part, related to the irreversibility of the decaying state concept exhibits a direct dependence on the boundary conditions.

We meet here a situation for which the regular framework of quantum mechanics is obviously too narrow (as the early critics of Gamow's work correctly felt): this breakdown of the usual rules of quantum mechanics is shown by the fact that the wave-function corresponding to a decaying state is not normalizable. In fact, besides the time-factor expressing exponential decay in time, this function has a radial factor $\exp(ik_n r)$, with the complex wave-number $k_n = K_n - i\gamma_n (\gamma_n > 0)$, such that $k_n^2 = E_n - \frac{1}{2} i\Gamma_n$. This means that the amplitude of the wave-function increases exponentially with increasing distance from the interaction region. The physical meaning of this property was already explained by Gamow: the amplitude received at a given distance r of the interaction region was emitted at an anterior time $t-(r/v)$ (where v is the velocity of the outgoing motion), its variation with r must thus reflect the inverse of the variation with t. The non-renormalizability of the decay state wave-function is accordingly a direct consequence of the irreversibility in time of the process it describes. We expect, therefore, that the incorporation of irreversible processes in a quantal description will necessitate a widening of the formalism of quantum mechanics.

3. Aspects of the history of causality within atomic theory; ergodicity

The subject of my talk, or rather the choice of it, is also inspired by Sambursky, namely by a paper that he has written about the curious fact that in the ancient world there was no conception of statistics, of randomness – that was completely foreign, according to the evidence he has collected, to the mentality, to the outlook of the ancients. So it is a modern concept and even in modern science it has only appeared – the physicists, so to speak, dared to use it, with apology – in the 19th century. Now we have come to the point of being forced to realize that statistical causality is *the* form of causal link that dominates practically the whole of our description of the material world and very probably also of the organic world, and this turn of outlook has come about through the atomic theory – through the representation of the world as a huge system of atoms. If I may describe it like this – we have passed from the system of representation of deterministic trajectories of large bodies with which Newton was very much concerned, to the random walk of tiny atoms. The God of Newton held a firm hand over the planets – the God of the atomic physicist is intoxicated.

When one takes atoms seriously, one is necessarily led to a kind of consideration borrowed, as Maxwell put it, from the department of statistics, because it is quite essential – this is a point which is not always sufficiently emphasized – it is essential, for a rational atomic representation of the phenomena, to assume that the atoms are identical in structure, or at least that they consist of a few species of identical structures. Otherwise it would be quite impossible to make any coherent description of macroscopic bodies.

On the other hand they must be so small that they are inaccessible to direct observation. This creates a new epistemological situation, because we have somehow to describe the motions of the atoms and the forces with which they interact. Now, with the refinement of technology, as you know, we are now able to make observations of individual atomic processes, but of course any such observation always implies an amplification device, in order to bring the atomic phenomenon that we want to investigate to our level of observation, which is macroscopic. Therefore, what we actually observe are interactions between single atoms or atomic systems and macroscopic bodies or instruments. I prefer to speak in terms of fully automatic registration, so that we keep out the human observer and all the problems that are connected with that aspect of the problem. This is a quite novel situation in physics, and I need not remind you of the tremendous consequences for epistemology that have arisen out of it and which are expressed by the code word of 'complementarity', introduced by Niels Bohr to remind us of them.

This complementarity at the level of individual atomic processes introduces a dichotomy between the two fundamental aspects of 19th century physics, according to which we can describe phenomena: the one being the very old and primitive one of space-time localization (which by itself raises epistemologic problems – but let us accept it as a well-defined mode of description, always on the macroscopic scale), the other being the energetic aspect, which is a 19th century creation – the general conception that the various forms of energy are interchangeable and that there is a conservation of the energy (and, related with it, of the momentum) of any closed system.

Those two aspects are complementary, which means that when we concentrate on one aspect, we completely exclude the possibility of defining the other. Now this complementarity is of course not the result of an arbitrary choice on our part; it is based on a law of nature, namely, the existence of an element of discontinuity in the interactions between the atoms and the macroscopic instruments – the existence of the quantum of action, which limits the applicability of the features of continuity in the classical description: there can be no question of a trajectory, since a continuous variation of position would imply changes of action less than one quantum.

One very often hears the question: is this complementarity a property of the atoms themselves, or is it imposed by us on the atoms? This is a badly formulated question: the two partners – the atoms and ourselves – are inseparable. We are in the queer position of being ourselves macroscopic systems of atoms, and the only concepts with which we can communicate experience are concepts adapted to our scale, macroscopic concepts; we have also to describe, by means of these, experience derived from the interaction of those tiny atoms with our macroscopic receptors. To find fault with this situation is no help – we have to live with it. The concept of complementarity just reminds us of the necessity of always taking due account of the conditions in which complementary concepts may be used in the description of atomic phenomena. These conditions are embodied in a perfectly consistent mathematical theory, which allows us to cope with every possible type of phenomenon. Certainly, this formalism is adapted to observational situations 'only' – well, there are no other situations that we can speak about intelligibly.

Thus, quantum mechanics always refers to conditions of observation, which enter into the equations in two ways: first as what we call *external parameters* – the intensity of a magnetic field, for instance, which is generated by a big electro-magnet and measured in the usual, classical way; then as *boundary conditions*, by which we select that solution of the equations which is adapted to the experimental arrangement: this is always describable entirely in terms of classical concepts.

Here arises a problem of logical consistency. Because our instruments are themselves systems of atoms, their behaviour during the observation of an atomic process must also be describable by means of Schrödinger's equation, if we introduce in this equation a complete atomic description of the experimental arrangement and its interaction with the atom that we are observing. In principle we can find the solution of this problem, and it should lead to the same result as the procedure that I have just described, involving the summary description of the conditions of observation by external parameters and boundary conditions.

Clearly, this does not mean that we could dispense with the classical concepts, or 'reduce' them to quantal operators, since we already put them into the interpretation of the symbols with which we work in the Schrödinger equation: in order to interpret the solution for the whole system we have to use the same prescriptions that we use when we are interested in the behaviour of a single atom. All we can achieve is to demonstrate that the relation established in quantum mechanics between the formalism and the classical concepts describing macroscopic observation is *consistent*; this is a necessary ingredient in the formulation of this relation, since after all it is our duty to show that whatever results we express in classical terms are consistent with the formalism that we claim is capable of giving us those results.

Now, this brings us then to a definite problem in quantum mechanics, namely, that of the time evolution of systems with many degrees of freedom. This problem was first treated in the early days of quantum mechanics by von Neumann, who took as a model the similar treatment of classical systems which goes back to Maxwell and Boltzmann. But von Neumann's theory is not satisfactory; its failure is due to a deep-lying difficulty, which I must try to explain.

Let me first recall the argument of the classical ergodic theory. A state of the system is represented by a point in a multidimensional phase space, whose *coordinates* are the whole set of canonical coordinates and momenta of all the constituent elements, and whose *metric* is determined by the condition of being independent of the choice of canonical coordinates and momenta, and also invariant in time: Liouville's theorem shows that these requirements are satisfied by a simple Euclidian metric. A statistical distribution of states in a domain of phase space will be deformed and spread out in a complicated way in the course of time, but it will always occupy the same total volume of phase space: its evolution is comparable to an incompressible flow. As a result, if the evolution of the system is confined to a limited region of phase space, every bundle of trajectories starting from some domain within this region, however it may spread over it in the course of time, must eventually return to the initial domain, and not only once, but an infinity of times – this is the content of Poincaré's famous theorem about the cyclic character of the time evolution of a finite system; it strikingly expresses the reversibility of its dynamical behaviour.

How can we now pass from this dynamical behaviour at the atomic level to the level of macroscopic observation? Obviously, by an averaging process: we do not measure the value of a physical quantity for each single element of the system, but only its average value for the whole system at some instant of our macroscopic time, which means over a time interval very large on the atomic scale, involving many interactions between the constituent elements of the system.

In fact, we may just take the average over an infinite time – which is a convenient idealization from the mathematical point of view: for we know that on the macroscopic scale the system, after a certain relaxation time, short on that scale, reaches a state of equilibrium; and it does not matter if we extend the average over the infinity of Poincaré cycles during which (ideally, of course) the evolution of the system just repeats itself. At first sight, it would seem, however, that calculating such a time average is a hopeless undertaking, since it would require solving the equations of motion for the whole system; but we could by-pass this task if we only knew – or guessed – how often, on the average, the system occupies in the course of time each of its possible configurations: the time average would then be replaced by a statistical average over the part of phase space in which the system evolves.

Suppose we take a film of the successive configurations of our system: to obtain the time average of some physical quantity, we have to measure it on each picture and take the sum (divided by the number of pictures); obviously we get the same result if we cut out the single pictures and arrange them, regardless of the times at which they occur, in groups corresponding to the same value of the physical quantity in question: we have then to count how many pictures there are in each group, multiply these numbers by the corresponding values and sum the products (again divided by the total number of pictures).

Maxwell and Boltzmann guessed that a large system of interacting elements would pass equally often through each of its possible configurations: this is what Boltzmann called the 'ergodic hypothesis', and it leads, of course, to the simplest possible method for computing statistical averages. Mathematicians, however, frowned upon such a bold intuition, and at long last, after their tools were sufficiently sharpened, they managed to replace it by a forbiddingly rigorous theory, without, however, adding anything to its physical significance. Indeed, the rigorous definition of ergodicity does not provide us with any *practical* criterion helping us to decide whether a given system behaves ergodically or not.

Yet in one respect, ergodic theory did clarify our ideas. Gibbs had compared the endless winding of the trajectory in phase space to a mixing process – the stirring of milk in black coffee, leading to a mixture which has a *coarse* appearance of homogeneity: the mathematical analysis of this process, carried out by Hopf, revealed that it is logically independent of the ergodic character of the system – it only takes place if a further condition is satisfied. In fact, a mixing essentially depends, for its very definition, on the consideration of more than one trajectory – it turns out that it suffices to consider a pair of trajectories; this is conveniently done by representing a pair of phase space points, each belonging to one of the trajectories, by a single point in a 'superspace' with twice as many dimensions, in such a way that the pair of trajectories will appear as a single trajectory in superspace. Then, Hopf's mixing theorem simply states that the mixing process will occur if, and only if, the system is ergodic in superspace (as well as in the original phase space): this is a more restrictive ergodicity condition, essential for the 'normal' thermodynamic behaviour of the system. It is essentially a condition upon the *correlations* between pairs of configurations, and it thus emphasized the necessity of taking account of these pair correlations, in addition to the simple average distributions in phase space.

On this particular point, quantum theory has a great formal advantage over the classical description: it can operate, as von Neumann showed, with a density operator which comprehends *both* the average density distributions in the various states of the system (its diagonal elements in a matrix representation using a complete set of such states as a basis) *and* the correlation coefficients between pairs of states (its non-diagonal elements).

4. Irreversibility of Large Quantal Systems

If one attempts a general theory of large systems with the aim of analysing the origin of their irreversible behaviour, two different points of view offer themselves, both initiated by Boltzmann: the ergodic and the kinetic one. In the ergodic approach, one considers the evolution of finite systems, confined in the classical case in a finite manifold of their phase space. The cyclic character of the phase space trajectories makes it necessary to consider the whole motion in a global way, and the irreversibility appears in a 'coarse-grained' description, in which the ideally precise localization in phase-space is replaced by a statistical distribution over finite cells. Two conditions – 'metrical transitivity' and 'mixing' – must be satisfied in order to ensure an irreversible trend of the coarse-grained distribution towards an equilibrium distribution. However, these conditions have little practical value, since they are not directly related to the Hamiltonian and therefore do not provide us with any criterium to decide whether a given system satisfies them. Moreover, the quantal generalization of the coarse-graining concept gives rise to great formal complications.

The kinetic approach circumvents these difficulties if it is applied to an infinite system in the 'thermodynamic limit' (in which the extension and the number of degrees of freedom both become infinite in such a way that the density remains finite): for in this limit, the Poincaré cycle becomes of infinite duration, and the dynamical behaviour must therefore be expected to exhibit (in general) an irreversible trend in the asymptotic limit of sufficiently large, but finite, times. Thus, one obtains at the macroscopic level of description, a smoothed-out representation of the time evolution of the system, showing an approach to equilibrium on the macroscopic time scale: this is the content of Boltzmann's H-theorem. In the usual treatment, the definition of the asymptotic limit implies a departure from the strict dynamical description, taking the form of some statistical assumption – *Stosszahlansatz*, time averaging, etc. It is possible, however, as we shall see, to dispense with such explicit assumptions and obtain asymptotic solutions satisfying the dynamical equation of evolution exactly, but exhibiting an asymmetry in time which only results from the property of incoherent atomic correlations of limited range of having 'no memory', i.e. of dying out after times of atomic order of magnitude.

This method, due to the Brussels group led by I. Prigogine, has the further advantage of yielding a general criterium of 'dissipativity', directly expressed in terms of the Hamiltonian, and allowing one to decide whether a given system will show any irreversible approach to equilibrium on the macroscopic time-scale. The theory has recently been given by Prigogine, George and myself [1] a quite general form, based on the formalism of quantum mechanics, but representing a non-trivial extension of this formalism, enabling it to embody the treatment of irreversible processes. Let us now proceed to a brief account of this theory, valid for at any rate an extensive class of very large quantal systems considered in the thermodynamic limit.

In a superspace defined as the direct product of the Hilbert space and its dual, the density matrix appears as a supervector, varying in time according to Liouville's equation; the latter represents the rate of variation of the density supervector as resulting from a linear transformation of this supervector, effected by a linear superoperator L:

$$i\dot{\rho}(t)=L\rho(t).$$

The Liouville superoperator L may be expressed in terms of the Hamiltonian H of the system; to this end, let us introduce a convenient notation. Consider the special class of superoperators $O\equiv M\times N$ depending on a pair of supervectors M, N according to the definition $O\rho=M\rho N$ (we call them 'factorizable'); then we may write $L=H\times 1-1\times H$.

The Liouville equation describes the dynamical behaviour of the system on the atomic scale. In order to arrive at a description at the macroscopic level of observation, we ought – along the lines of the preceding classical argument – to introduce the analogue of the mixing process. In the classical case, this is simply done by a 'coarse graining' of the phase space, i.e. by averaging the distributions on the atomic scale over finite 'cells' of phase space (the shape of which is arbitrary). Here, then, is the stumbling block against which von Neumann's attempt foundered: a phase-space cell with a sharp boundary would contradict the indeterminacy relations, and the substitute von Neumann tried to formulate leads to hopeless complications. Later, it is true, a considerably better approach has been found: it consists in subdividing the Hilbert space into bundles of state vectors, which play the part of the cells; it is then necessary to perform an additional averaging over all such subdivisions, and the resulting formalism is still very cumbrous.

However, the introduction of coarse graining, which is the source of this difficulty, is solely a consequence of the cyclic character of the time evolution on the atomic scale, which forces us to extend averages over infinite time and represent the mixing effect of the correlations (which actually goes on all the time) by the schematic form of statistical average we call coarse graining. Now, the cyclic type of time evolution is a feature of finite systems, confined in a finite phase space extension if they are classical, or characterized by a discrete energy spectrum in the quantal case. Infinite systems, on the other hand (i.e. systems of infinite degree of freedom and infinite extension, but such that the number of elements within a given extension has a finite limit), show no cyclic return: the period of the cycle becomes infinite, their energy spectrum is essentially continuous; they may therefore be expected to exhibit directly, in the asymptotic limit of very large, but finite, macroscopic times, the approach to a state of equilibrium which corresponds to macroscopic observation.

This consideration invites us to look for the quantal form of Boltzmann's kinetic approach rather than to a generalization of ergodic theory: this is the line of investigation initiated by Prigogine and followed by him and his group in Brussels during the last decade. I came into it only recently, when Prigogine told me of the conclusion he had reached – a conclusion so striking that I did not quite believe it until I had repeated the calculation by a more direct method, which had been outlined, but not fully exploited, by the Brussels group.

In exploring virgin ground, one needs to test new ideas by applying them to concrete cases, and to seek guidance from such applications for further developments: the formalism useful to this end is not generally the one suited to give a systematic account of the main features of the theory. In the present case, it turned out that a different – though of course equivalent – mode of treatment led to a very compact, synthetic presentation of the line of argument and the essential results: it is this final form of the theory, developed in common by Prigogine, George and myself, that I am going to outline to you.

My late participation in the protracted investigations of the Brussels group allows me, perhaps, to take a more detached view of the subject, and I am not infringing the code of modesty about which scientists are sometimes so prejudiced, by telling you that the outcome of these investigations really represents a major advance in the strengthening of the conceptual frame of atomic theory: we have now a powerful mathematical tool to represent and analyse in a surprisingly transparent way the relationship of complementarity between the dynamical behaviour of large atomic systems and their macroscopic observation.

Our aim being to find out the long time effect of correlations, we must, to begin with, compare our system, defined by the Hamiltonian H, with a 'model' system H_0, from which the interaction energy V, responsible for the correlations, is removed, in such a way that $H=H_0+V$. The eigenstates of H_0 form a complete orthogonal basis of representation in Hilbert space, from which we construct a similar basis in superspace: the latter may be divided into two classes of supervectors, those built up of pairs of identical (or physically equivalent) eigenstates, and those built up of pairs of different eigenstates; they belong, respectively, to two orthogonal subspaces of superspace, characterized by projection superoperators P_0, P_c. Then, the projections $\rho_0=P_0\rho$ and $\rho_c=P_c\rho$ of the density supervector correspond, respectively, to the average distribution densities and the correlation amplitudes. Putting $L_{00}=P_0LP_0$, $L_{0c}=P_0LP_c$, etc. we obtain for ρ_0 and ρ_c the coupled Liouville equations

$$(1) \qquad i\dot{\rho}_0=L_{00}\rho_0+L_{0c}\rho_c, \quad i\dot{\rho}_c=L_{cc}\rho_c+L_{c0}\rho_0.$$

The next step is to extract from these equations the asymptotic forms $\tilde{\rho}_0(t)$, $\tilde{\rho}_c(t)$ of ρ_0 and ρ_c for large *positive* values of the time variable: these are expected to express our possibilities of prediction of the *future* evolution of the system on the macroscopic time scale.

We must here restrict the generality of the Liouville superoperator in order to characterize the class of systems which we expect to exhibit the 'normal' asymptotic behaviour, i.e. an approach to a state of equilibrium. To this end, we observe that the time evolution of the correlation density ρ_c is essentially governed by the superoperator $T_c=\exp(-iL_{cc}t)$, depending on the part of the Liouville superoperator which acts entirely in the correlation subspace. We assume accordingly that the asymptotic effect of this superoperator $T_c(t)$ upon any regular supervector which is not an invariant is to reduce this supervector to zero: $\lim_{t\to\infty} T_c(t)\,A=0$; we express by this assumption the fading of the system's 'memory' of its correlations.

This condition, which may also be expressed as an analyticity condition on the Laplace transform of $T_c(t)$, has first been verified in this form by a perturbation expansion, for infinite systems in whose description there enters a 'small' physical parameter such as the coupling constant or the density [3]. More recently, it has been shown that for soluble models, such as the Friedrichs model, the analyticity assumption is satisfied rigorously (i.e. independently of any perturbative approach) for a large class of interactions [4].

By means of this assumption, we readily derive from the second Liouville Equation (1) the following relation between the asymptotic densities:

$$\tilde{\rho}_c(t) = \int_0^\infty d\tau \, e^{-iL_{cc}\tau}(-iL_{c0})\, \tilde{\rho}_0(t-\tau); \tag{2}$$

it has the form of an integral equation, showing how the asymptotic correlations build up by sequences of processes starting from the average situations through which the system passes in the course of time.

Let us introduce at this stage an asymptotic time evolution operator by writing $\tilde{\rho}_0(t)$ in the form

$$\tilde{\rho}_0(t) = e^{-i\theta t}\tilde{\rho}_0(0); \tag{3}$$

the variable t in this formula represents the time measured on a macroscopic scale – to be *quite* rigorous, we ought to restrict the formula to some discrete sequence of values $t_1, t_2, \ldots, t_n$, which, however, could be chosen arbitrarily; let us forget about this restriction, which would complicate things without affording us more insight into the physical aspects. The advantage of the representation (3) is to reduce the integral Equation (2) to a simple linear relation between $\tilde{\rho}_0(t)$ and $\tilde{\rho}_c(t)$ taken at the same time:

$$\tilde{\rho}_c(t) = C\tilde{\rho}_0(t), \quad C = \int_0^\infty d\tau \, e^{-iL_{cc}\tau}(-iL_{c0})\, e^{i\theta\tau}. \tag{4}$$

The first Liouville Equation (1) then yields a functional equation for θ:

$$\theta = L_{00} + L_{0c}C, \tag{5}$$

which can be solved for θ by iteration.

The total asymptotic density $\tilde{\rho}=\tilde{\rho}_0+\tilde{\rho}_c$ thus obtained has the remarkable property of being an *exact* solution of the Liouville equation. According to Equation (4), it may be written in the form $\tilde{\rho}=P_a\tilde{\rho}$, with $P_a=P_0+C$. It is still more remarkable that this superoperator P_a has the characteristic properties of a projection operator: it defines a subspace in which the asymptotic density is confined. This subspace differs from the average subspace P_0 by the adjunction of a part of the correlation subspace P_c, namely that part which is specified by the superoperator C; the latter may be interpreted as representing the building up of correlations from *asymptotic* average situations (we call it the superoperator of correlation creation) – it sorts out those correlation processes which have a long time effect and accordingly manifest themselves at the macroscopic level.

Thus, the asymptotic density $\tilde{\rho}$ is not, as one might have expected, governed by a 'kinetic equation' different from the dynamical Liouville equation: it is an exact solution of the latter, and its asymptotic character is conferred upon it by its confinement to a subspace defined by the projector P_a. The expression for the superoperator of correlation creation C, which enters in the definition of P_a, is clearly unsymmetrical in time, and gives the projector P_a the expected bias towards a preferred direction of the time evolution. In fact, time inversion transforms P_a into a different projector $\bar{P}_a=P_0+D$, where the superoperator

$$D=\int_0^\infty \mathrm{d}\tau\, \mathrm{e}^{i\eta\tau}(-iL_{0c})\,\mathrm{e}^{-iL_{cc}\tau}$$

is the time-inverse of C; it contains the superoperator η which is the time-inverse of θ and obeys the equation $\eta=L_{00}+DL_{c0}$ derived from Equation (5) by time-inversion. In contrast to C, the superoperator D describes sequences of 'destructions' of correlations leading to asymptotic average situations. We thus see that in passing to the macroscopic level by means of the projector P_a, we have destroyed the time-reversal invariance of the dynamical behaviour on the atomic scale and imposed macroscopic time-directedness upon the asymptotic time-evolution.

An important element is still missing in the picture: we must establish a link between the asymptotic density supervector $\tilde{\rho}(t)$ and the arbitrarily chosen dynamical density supervector $\rho(0)$ from which the time-evolution is assumed to start. This is readily supplied, however, on the basis of a further remarkable property (easily derived) of the superoperator θ and its time-inverse η:

(6) $$P_a\theta = LP_a, \quad \eta\bar{P}_a = \bar{P}_a L.$$

With the notation $N_0 = 1 + DC$, it follows from Equations (6) and (3) that

$$N_0\tilde{\rho}_0(t) = N_0\,\mathrm{e}^{-i\theta t}\tilde{\rho}_0(0) = \mathrm{e}^{-i\eta t}N_0\tilde{\rho}_0(0)$$

and, on the other hand,

$$\bar{P}_a\rho(t) = \bar{P}_a\,\mathrm{e}^{-iLt}\rho(0) = \mathrm{e}^{-i\eta t}\bar{P}_a\rho(0).$$

This shows that $N_0\tilde{\rho}_0(t)$ and $\bar{P}_a\rho(t)$ have the same time-evolution, governed by the superoperator $\exp(-i\eta t)$: we may therefore equate them at any instant and in this way fix the correspondence between the dynamical and the asymptotic density. This gives the quite fundamental relation, valid at any time,

(7) $$\tilde{\rho}(t) = \tilde{\prod}\rho(t), \quad \text{with} \quad \tilde{\prod} = P_a N_0^{-1}\bar{P}_a,$$

from which follows

$$\tilde{\rho}(t) = \tilde{\sum}(t)\,\rho(0), \quad \text{with} \quad \tilde{\sum}(t) = \tilde{\prod}\,\mathrm{e}^{-iLt},$$

the answer to our last question, completing the theory.

The striking feature about the superoperator $\tilde{\prod}$ occurring in Equation (7) is that it is also a projector, and moreover one that is time-reversal invariant: we have now a time-symmetrical subspace of the superspace in which the asymptotic part of the time-evolution, starting from any given situation, remains confined, exhibiting the features observed at the macroscopic level; whereas the irregular fluctuations occurring on the atomic time scale are contained in the complementary subspace orthogonal to the asymptotic one. That such a clean separation between the two aspects of the atomic system could be effected was entirely unexpected: it could never have been found by a study of the evolution of the system in Hilbert space, for it can only be formulated in terms of the superspace formalism. The superoperator $\tilde{\prod}$ is not factorizable: one cannot ascribe any state vector to an asymptotic situation as we have defined it, but only a density supervector $\tilde{\rho}$.

In fact as appears from Equation (4), the correlation part $\tilde{\rho}_c$ of the density $\tilde{\rho}$ is directly derived from the part $\tilde{\rho}_0$ expressing the average probability distributions of the system; owing to this remarkable structure of the projector π, the evolution in $\tilde{\pi}$-space may be entirely described in terms of probabilities only. In particular, the 'reduction of the wave-packet' of an atomic system after its interaction with a measuring apparatus is a direct consequence of this property of the $\tilde{\pi}$-space description: the essential point being that the apparatus must necessarily belong to the macroscopic level of quantum mechanics, and its behaviour accordingly be described in terms of its $\tilde{\pi}$-space variables (loosely speaking, the behaviour of the apparatus has 'thermodynamical' character, inasmuch as variables pertaining to thermodynamic equilibrium or near-equilibrium states all belong to $\tilde{\pi}$-space). Any phase relations in the initial state of the atomic system are therefore wiped out (i.e. are rejected into the orthogonal subspace): this is the only meaning of the 'reduction' of the initial state of the atomic system resulting from the measurement. As to the human observer, his interaction with the apparatus is also entirely described in the $\tilde{\pi}$-subspace, and therefore without any influence whatsoever on whatever goes on in the orthogonal subspace.

One further point should be mentioned. The theory gives us a simple criterium to decide whether the system shows the normal macroscopic behaviour described by thermodynamics. The superoperator of asymptotic time-evolution θ is closely related to a 'collision superoperator' defined as the Laplace transform of the superoperator $L_{0c} T_c(\tau) L_{c0}$:

$$\psi(z) = \int_0^\infty \mathrm{d}\tau L_{0c} T_c(\tau) L_{c0} \, \mathrm{e}^{-z\tau};$$

indeed, Equation (5) may be written as a functional relation in terms of $\psi(z)$:

$$\text{(8)} \qquad i\theta = iL_{00} + \psi(-i\theta).$$

Now, a homogeneous system (i.e. a system for which $L_{00} = 0$) will obviously not exhibit any tendency towards equilibrium if the collision operator vanishes identically. Equation (8) then shows that it will exhibit an irreversible tendency towards equilibrium provided that θ itself does not vanish identically. This 'condition of dissipativity' is a practical one: it can be tested in concrete cases [4] by actual computation of θ.

I shall leave it at this summary sketch, except for one further point I wish to mention. Not all large systems show the normal macroscopic behaviour described by thermodynamics: how can we recognize those that do? Well, for the case of homogeneous systems at any rate, theory gives us a simple criterion: we have just to turn again to the superoperator of asymptotic time-evolution θ. It is obvious that any system for which θ happens to vanish identically will not show any irreversible tendency towards equilibrium; if the system is homogeneous, it can be seen that it will exhibit this tendency provided that θ does not vanish identically. This 'condition of dissipativity' is a practical one: it can be tested in concrete cases by actual computation of θ.

I need not insist on the epistemological significance of the neat, clear-cut representation we obtain for the complementarity between the two levels of description of atomic phenomena. In particular, the discussion of the measuring process in quantum mechanics can be reduced to a few words: the apparatus being a macroscopic system, the process is described in the asymptotic subspace, where any phase relations in the initial state of the total system are wiped out (i.e. they are rejected into the orthogonal subspace) – this is the simple meaning of the 'reduction' of the initial state of the atomic system resulting from the measurement. As to the human observer, his interaction with the apparatus is also entirely described in the $\tilde{\prod}$ subspace, and therefore without influence whatsoever on whatever goes on in the orthogonal subspace.

These brief indications may serve to illustrate how the disclosure of new formal relationships, provided that their adequacy to the physical situation is severely kept under control, can throw light on the epistemological aspects of atomic theory and bring out the deep harmony we have reached in our description of atomic phenomena.

NOTE

* [This paper is derived from two of Rosenfeld's papers which substantially borrowed from each other: 'General Introduction to Irreversibility' (1972d) and 'Statistical Causality in Atomic Theory' (1974a). The latter was contributed to a 1971 symposium in honor of S. Sambursky.]

This paper may serve also as Rosenfeld's introduction to the last exciting work of his life, the "major advance", as he saw it, which has been achieved by the Brussels group of Prigogine and others, with whom he was collaborating. The following paper expounds some of that achievement. – Ed.].

BIBLIOGRAPHY

[1] George, C., Prigogine, I., and Rosenfeld, L., *Det kgl. Danske Videnskabernes Selskab mat.-fys. Meddelelser* **38** (1972), No. 12, [partially reprinted in this volume, p. 571 – Ed.].

[2] Prigogine, I., George, C., and Henin, F., *Physica* **45** (1969), 418; see also Balescu, R. and Brenig, L., *Physica* **54** (1971), 504.

[3] Prigogine, I., *Non-Equilibrium Statistical Mechanics* (New York, Interscience Publ., 1962).

[4] Grecos, A. and Prigogine, I., *Physica* **59** (1972), 71; *Proc. Nat. Acad. Sci.* **69** (1972), 1629.

CHAPTER 14

THE MACROSCOPIC LEVEL OF QUANTUM MECHANICS*

[1972a and b]

(with C. George and I. Prigogine)

SYNOPSIS. A synthetic account is given of a general treatment of large quantal systems, allowing of a clear-cut characterization of the macroscopic level of description of such systems. The time-evolution of the density operator is given by a Liouville equation, which is written down in a superspace formed by the direct product of the Hilbert space with itself. It is shown how to construct a projector $\tilde{\Pi}$ in this superspace such that the subspace it defines contains the asymptotic time-evolution of the density operator for time intervals very large compared with those typical for atomic processes: this asymptotic subdynamics shows the characteristic features of macroscopic behaviour.

A quantitative criterion is formulated in superspace for the existence of a well-defined and unique macroscopic level in the sense just outlined of a separate subdynamics governing the asymptotic behaviour of the system. This 'condition of dissipativity' can be directly tested on the Hamiltonian of any given system.

In general, the subdynamics can only be formulated in superspace: it is not possible to return from the $\tilde{\Pi}$ subspace to a Hilbert space description of the system in terms of state-vectors. Thus, the scope of the latter description is clearly limited, and a precise formulation is obtained of the complementarity between the dynamical account of the system on the atomic scale and its description at the level of macroscopic observation.

The epistemological problems of quantum mechanics receive from the present point of view an especially transparent treatment. In particular, the consistency of the use of classical concepts for the account of quantal phenomena is obvious, since the macroscopic description operates directly with probabilities, all quantal interference effects being eliminated from the $\tilde{\Pi}$ subspace; thus, the rule of 'reduction' of the state-vector following a measurement performed upon an atomic system appears as an immediate consequence of the macroscopic character of the measuring process.

1. INTRODUCTION

In a detailed discussion of the epistemological problems of atomic physics, where explicit reference to the conditions of macroscopic observation is essential [1, 2], a decisive part is played by the analysis of the asymptotic approach to equilibrium of material bodies consisting of a very large number of interacting atomic constituents (and possibly also of a field of electromagnetic radiation). In the first place, such an analysis serves in its own right to establish the consistency of the atomistic description of macroscopic phenomena, by clarifying the relation of complementarity between the irreversible character of the macroscopic behaviour of the large system and the time-reversal invariance of its dynamical description at the atomic level [3]. In the second place, applied to the process of observation of an individual atomic phenomenon by means of a macroscopic apparatus interacting with the atomic object, it throws light on the conceptual foundations of quantum mechanics [4, 5].

Recent progress in the quantum theory of large systems allows us, as we intend to show in this paper, to treat these problems with more precision and more completely than was hitherto possible. The method we use was initiated by one of us [6] and developed in numerous publications by the Brussels group during the last decade. This method may be applied to arbitrary systems but becomes especially interesting and fruitful in the limit of large systems, whose energy spectrum is (at least in part) continuous. It is then possible to follow directly the time-evolution of observables depending on a finite number of degrees of freedom through the time evolution of the density operator, and in particular to study the asymptotic behaviour of the system for times of macroscopic order of magnitude. It turns out that this behaviour can be given a remarkably simple formal expression under very general assumptions about the correlations between the constitutive elements of the system: in physical terms these correlations should be of limited range and their effects should only persist for times of atomic order of magnitude. The asymptotic density operator will then exhibit the expected approach to equilibrium, provided that a further condition – playing here somewhat the same part as the condition of mixing in the theory of classical systems – is fulfilled by the interactions. In contrast with the ergodicity and mixing conditions for classical systems[1] this condition of 'dissipativity' can be explicitly tested for typical systems of actual physical interest. Of course, our method can also be applied to classical systems, for which it again leads, in the limit of large systems, to a condition of dissipativity.

This condition plays a fundamental role in our theory, inasmuch as its fulfilment guarantees the existence of a well defined macroscopic level of description of a large system, besides its dynamical description on the atomic level. This new mode of description contains the usual phenomenological account of the behaviour of the system in terms of thermodynamics, chemical kinetics and other macroscopic theories. Thus, the condition of dissipativity, when fulfilled, establishes the possibility of introducing in a well defined way two complementary levels of description of atomic systems.

To illustrate the significance of this remarkable result, let us consider the evolution of the temperature of a gas. We may proceed in two ways: either we use the statistical definition of temperature and solve the dynamical problem (which is possible, at least in principle, by means of an appropriate computer), or we use the Fourier equation of heat conduction. These two quite different procedures, so far as we know, give results in agreement with each other. This shows that the complete dynamical description contains elements which in fact are irrelevant for the evolution of such an observable as the temperature. Now, our method, in the form elaborated in recent papers of the Brussels group [9–18], allows us to define with precision the part of the dynamical description that is relevant and to discard in an unambiguous fashion the remainder.

As shown by Prigogine, George and Henin [14], the evolution of a mechanical system may be split into formally independent 'subdynamics', characterized by certain projection operators, as explained in Subsection 2.7 of [19].* One of these subdynamics, belonging to the projector $\tilde{\Pi}$ defined in that Subsection, contains all the information about equilibrium and linear transport properties, and the macroscopic level of description of a large quantal system is accordingly defined as that entirely expressed in terms of the variables corresponding to this $\tilde{\Pi}$-subdynamics. The consideration of the macroscopic level of a system thus entails an enormous reduction of its mode of description, since all the variables associated to the other subdynamics are excluded from it. In relation to the macroscopic description so defined, the condition of dissipativity plays a fundamental part. For non-dissipative systems, such as quantal systems with a discrete, non-degenerate energy spectrum, there is no dynamical evolution in the $\tilde{\Pi}$-space, and a macroscopic level of description cannot be defined. The consideration of large dissipative systems is therefore essential. There also exist large quantal systems, with continuous spectrum, which are not dissipative (an extreme example would be a superfluid system at zero temperature): for such systems there is no macroscopic description, in the above sense, of their dynamical evolution.

In the next section, we present a synthetic account of the formal framework of our theory, leading to the concept of subdynamics and the definition of the macroscopic level of quantum mechanics. Section 3 is devoted to a discussion of the physical content of the theory, with special emphasis on the epistemological problems of the atomic theory of material bodies. In section 4 the theory is applied to the analysis of the observation of an individual atomic process and its bearing on the epistemological aspects of quantum mechanics. This still leaves out of consideration a number of important problems upon which our approach throws new light, such as those related to transformation theory, the introduction of collective modes or quasiparticles, or the definition of unstable particles.[2] We hope nevertheless that the aspects discussed here are sufficient to show that the theory of large quantal systems is an essential part of quantum mechanics, both by its physical applications and its contributions to the questions of principle concerning the foundations of atomic theory. Indeed, it makes possible an incorporation in quantum theory of deep-lying, general properties of matter, which could not be achieved by means of the usual Hilbert space formalism of quantum mechanics.

2. The Problem of Relating the Quantum Mechanical Description of Atomic Processes to the Behaviour of Matter in Bulk is Resolved in a Conceptually Simple Treatment

[See Editorial Note, p. 596 – Ed.]

Atomic theory raises the epistemological problem of harmonizing the detailed dynamical description of atomic systems given by quantum mechanics and the description of individual atomic processes and of the behaviour of matter in bulk at the level of macroscopic observation. The logical side of this problem is completely elucidated by the recognition of relationships of complementarity between the two modes of description: on the one hand, there is the complementarity expressed by the indeterminacy relations, which governs the application of the macroscopic space-time localization and momentum-energy conservation to the individual atomic processes; on the other, there is the complementarity between the account of the behaviour of a large system of atomic constituents given by quantum mechanics (and electrodynamics) and the description of the same system as a material body in terms of the concepts of macroscopic mechanics, electromagnetism and thermodynamics.

The formal side of the problem, however, which consists in establishing the consistency of the rules connecting the formalism of quantum mechanics with the concepts used in the account of macroscopic observation, still leaves scope for a presentation more in accordance with the conceptual simplicity of the actual situation. A general treatment of the quantum theory of large atomic systems which satisfies this desideratum has been given in a recent article [19]. The purpose of this note is to outline the method and the principal results.

The work pursued during the past decade by the Brussels group has clearly shown that the method best adapted to the investigation of large atomic systems is not the ergodic but the kinetic approach of statistical mechanics, applied to the limiting case of infinite systems (that is, systems of infinite degree of freedom, but such that the number of elements within a given phase extension has a finite limit). Because such systems have essentially a continuous energy spectrum, the vexed question of 'coarse graining' can be ignored and the asymptotic limit of the density operator for times very large on the atomic scale, but finite, can be directly discussed. This discussion leads to the conclusion [14] that the time evolution of the system may be split rigorously into formally independent 'subdynamics', characterized, in a manner presently to be explained, by certain projection operators depending on the correlations between the constitutive elements of the system. One of these subdynamics belonging to the projector $\tilde{\Pi}$ given by equation (7) below contains all the information about equilibrium properties and linear transport properties. We may therefore define the macroscopic level of description of quantum mechanics as the reduced description in terms of the variables of the $\tilde{\Pi}$-subspace. That such a reduced description is at all possible results precisely from the fact that the $\tilde{\Pi}$-subdynamics is expressed by an independent equation of evolution. The application of this analysis to the case of a measuring apparatus, initially triggered off by an interaction of atomic duration with an atomic object, shows in a surprisingly simple way that its evolution in $\tilde{\Pi}$-space conforms to the "reduction" rule of quantum mechanics; this important conclusion follows directly from the mathematical structure of the $\tilde{\Pi}$-subspace of the apparatus, which entails the elimination of the initial phase relations between the components of the density operator of the atomic object. Our paper contains a simple and general derivation of these remarkable results, which shows more clearly how these methods not only lead to a deeper understanding of the epistemological problems of atomic theory, but even to a significant extension of the scope of quantum mechanics.

In a superspace defined as the direct product of the Hilbert space and its dual, the density operator $\rho(t)$ appears as a supervector, varying in time according to a Liouville equation of the form

$$i\dot{\rho}(t) = L\rho(t)$$

The Liouville superoperator L may be expressed in terms of the Hamiltonian H of the system. For this purpose, we may use a convenient notation for the special class of "factorizable" superoperators $O \equiv M \times N$ depending on a pair of supervectors M, N according to the definition $O\rho = M\rho N$; we may then write $L = H \times 1 - 1 \times H$. Our aim being to find the long time effect of correlations, we must, to begin with, compare our system, defined by the Hamiltonian H, with a 'model' system H_0, from which the interaction energy V, responsible for the correlations, is removed, in such a way that $H = H_0 + V$. The eigenstates of H_0 form a complete orthogonal basis of representation in Hilbert space, from which we construct a similar basis in superspace: the latter may be divided into two classes of supervectors, those built up of pairs of identical (or physically equivalent) eigenstates, and those built up of pairs of different eigenstates; they belong, respectively, to two orthogonal subspaces of superspace, characterized by projection superoperators P_0, P_c. Then the projections $\rho_0 = P_0\rho$ and $\rho_c = P_c\rho$ of the density supervector correspond, respectively, to the average distribution densities and the correlation amplitudes. Putting $L_{00} = P_0LP_0$, $L_{0c} = P_0LP_c$ and so on, we obtain for ρ_0 and ρ_c the coupled Liouville equations

$$(1) \qquad i\dot{\rho}_0 = L_{00}\rho_0 + L_{0c}\rho_c, \; i\dot{\rho}_c = L_{cc}\rho_c + L_{c0}\rho_0$$

The next step is to extract from these equations the asymptotic forms $\rho_0(t)$, $\rho_c(t)$ of ρ_0 and ρ_c for large positive values of the time variable: these are expected to express our possibilities of prediction of the future evolution of the system on the macroscopic time scale.

We must here restrict the generality of the Liouville superoperator in order to characterize the class of systems which we expect to exhibit the 'normal' asymptotic behaviour, that is an approach to a state of equilibrium. To this end, we observe that the time evolution of the correlation density ρ_c is essentially governed by the superoperator $T_c = \exp(-iL_{cc}t)$, depending on the part of the Liouville superoperator which acts entirely in the correlation subspace. We assume accordingly that the asymptotic effect of this superoperator $T_c(t)$ upon any regular supervector A which is not an invariant is to reduce this supervector to zero: $\lim_{t\to\infty} T_c(t)\, A = 0$; we express by this assumption the fading of the system's "memory" of its correlations. This condition, which may also be expressed as an analyticity condition on the Laplace transform of $T_c(t)$, has first been verified in this form by a perturbation expansion, for infinite systems in whose description there enters a "small" physical parameter such as the coupling constant or the density [20]. More recently, it has been shown that for soluble models, such as the Friedrichs model, the analyticity assumption is satisfied rigorously (that is, independently of any perturbative approach) for a large class of interactions [6]. By means of this assumption, we readily derive from the second Liouville equation (1) the following relation between the asymptotic densities:

$$\text{(2)} \qquad \tilde{\rho}_c(t) = \int_0^\infty d\tau e^{-iL_{cc}\tau} (-iL_{c0})\, \tilde{\rho}_0(t-\tau)$$

It has the form of an integral equation, showing how the asymptotic correlations build up by sequences of processes starting from the average situations through which the system passes in the course of time.

We now introduce an asymptotic time evolution operator by writing $\tilde{\rho}_0(t)$ in the form

$$\text{(3)} \qquad \tilde{\rho}_0(t) = e^{-i\theta t} \tilde{\rho}_0(0)$$

The advantage of the representation (3) is to reduce the integral equation (2) to a simple linear relation between $\tilde{\rho}_0(t)$ and $\rho_c(t)$ taken at the same time:

$$\text{(4)} \qquad \tilde{\rho}_c(t) = C\tilde{\rho}_0(t),\ C = C(\theta) = \int_0^\infty d\tau e^{-iL_{cc}\tau} (-iL_{c0})\, e^{i\theta\tau}$$

The first Liouville equation (1) then yields a functional equation for θ:

$$\text{(5)} \qquad \theta = L_{00} + L_{0c} C(\theta)$$

which can be solved by iteration.

The total asymptotic density $\rho = \rho_0 + \rho_c$ thus obtained has the remarkable property of being an exact solution of the Liouville equation. According to equation (4), it may be written in the form $\rho = P_a \rho$, with $P_a = P_0 + C$. It is again remarkable that this superoperator P_a has the characteristic properties of a projection operator in superspace, idempotency and 'adjoint symmetry' (that is, it is such that the projection $P_a A$ of a self-adjoint supervector A is self-adjoint). The projector P_a defines a subspace in which the asymptotic density is confined. This subspace differs from the average subspace P_0 by the adjunction of a part of the correlation subspace P_c, namely that part which is specified by the superoperator C; the latter may be interpreted as representing the building up of correlations from asymptotic average situations (we call it the superoperator of correlation creation) – it sorts out those correlation processes which have a long time effect and accordingly manifest themselves at the macroscopic level. Thus, the asymptotic density $\tilde{\rho}$ is not, as one might have expected, governed by a 'kinetic equation' different from the dynamical Liouville equation: it is an exact solution of the latter, and its asymptotic character is conferred upon it by its confinement to a subspace defined by the projector P_a.

The expression for the superoperator of correlation creation C, which enters in the definition of P_a, is clearly unsymmetrical in time, and gives the projector P_a the expected bias towards a preferred direction of the time evolution. In fact, time inversion transforms P_a into a different projector $\bar{P}_a = P_0 + D$, where the superoperator

$$D = \int_0^\infty d\tau e^{i\eta\tau} (-L_{0c})\, e^{-iL_{cc}\tau}$$

is the time-inverse of C; it contains the superoperator η which is the time-inverse of θ and obeys the equation $\eta = L_{00} + DL_{c0}$ derived from equation (5) by time-inversion. In contrast to C, the superoperator D describes sequences of 'destructions' of correlations leading to asymptotic average situations.

An important element is still missing in the picture: we must establish a link between the asymptotic density supervector $\tilde{\rho}(t)$ and the arbitrarily chosen dynamical density supervector $\rho(0)$ from which the time-evolution is assumed to start. This is readily supplied, however, on the basis of a further remarkable property (easily derived) of the superoperator θ and its time-inverse η:

(6) $$P_a\theta = LP_a,\ \eta\bar{P}_a = \bar{P}_a L$$

With the notation $N_0 = 1 + DC$, it follows from equations (6) and (3) that

$$N_0\tilde{\rho}_0(t) = N_0 e^{-i\theta t}\tilde{\rho}_0(0) = e^{-i\eta t}N_0\tilde{\rho}_0(0)$$

and, on the other hand,

$$\bar{P}_a\rho(t) = \bar{P}_a e^{-iLt}\rho(0) = e^{-i\eta t}\bar{P}_a\rho(0)$$

This shows that $N_0\tilde{\rho}_0(t)$ and $P_a\tilde{\rho}(t)$ have the same time-evolution, governed by the superoperator $\exp(-i\eta t)$: we may therefore equate them at any instant and in this way fix the correspondence between the dynamical and the asymptotic density. This gives the quite fundamental relation, valid at any time,

(7) $$\tilde{\rho}(t) = \tilde{\Pi}\rho(t) \text{ with } \tilde{\Pi} = P_a N_0^{-1}\bar{P}_a$$

from which follows

$$\tilde{\rho}(t) = \tilde{\Sigma}(t)\rho(0) \text{ with } \tilde{\Sigma}(t) = \tilde{\Pi}e^{-iLt}$$

the answer to our last question, completing the theory.

The striking feature about the superoperator $\tilde{\Pi}$ occurring in equation (7) is that it is also a projector in the extended sense defined above (which does not include the property of self-adjointness); moreover, it is time-reversal invariant: it defines a time-symmetrical subspace of the superspace in which the asymptotic part of the time-evolution, starting from any given situation, remains confined, exhibiting the features observed at the macroscopic level; whereas the irregular fluctuations occurring on the atomic time scale are contained in the complementary subspace orthogonal to the asymptotic one. That such a clean separation between the two aspects of the atomic system could be effected is an entirely unexpected property specific to the density supervector representation: it could never have been found by a study of the evolution of the system in Hilbert space, for it can only be formulated in terms of the superspace formalism.

The superoperator $\tilde{\Pi}$ is not factorizable: one cannot ascribe any state vector to an asymptotic situation as we have defined it, but only a density supervector $\tilde{\rho}$. In fact, as appears from equation (4), the correlation part $\tilde{\rho}_c$ of the density ρ is directly derived from the part $\tilde{\rho}_0$ expressing the average probability distributions of the system; owing to this remarkable structure of the projector $\tilde{\Pi}$, the evolution in $\tilde{\Pi}$-space may be entirely described in terms of probabilities only. In particular, the 'reduction of the wave-packet' of an atomic system after its interaction with a measuring apparatus is a direct consequence of this property of the $\tilde{\Pi}$-space description: the essential point being that the apparatus must necessarily belong to the macroscopic level of quantum mechanics, and that its behaviour must accordingly be described in terms of its $\tilde{\Pi}$-space variables (loosely speaking, the behaviour of the apparatus has 'thermodynamical' character, inasmuch as variables pertaining to thermodynamic equilibrium or near-equilibrium states all belong to $\tilde{\Pi}$-space). Any phase relations in the initial state of the atomic system are therefore wiped out (that is, they are rejected into the orthogonal subspace): this is the only meaning of the 'reduction' of the initial state of the atomic system resulting from the measurement. As to the human observer, his interaction with the apparatus is also entirely described in the $\tilde{\Pi}$-subspace, and therefore without any influence whatsoever on whatever goes on in the orthogonal subspace.

One further point should be mentioned. The theory gives us a simple criterion to decide whether the system shows the normal macroscopic behaviour described by thermodynamics. The superoperator of asymptotic time-evolution θ is closely related to a 'collision superoperator' defined as the Laplace transform of the superoperator $L_{0c}T_c(\tau)L_{c0}$:

$$\Psi(z)=\int_0^\infty d\tau L_{0c}T_c(\tau)L_{c0}e^{-z\tau}$$

Indeed, equation (5) may be written as a functional relation in terms of $\Psi(z)$:

(8) $$i\theta=iL_{00}+\Psi(-i\theta)$$

Now, an homogeneous system (a system for which $L_{00}=0$) will obviously not exhibit any tendency towards equilibrium if the collision operator vanishes identically. Equation (8) then shows that it will exhibit an irreversible tendency towards equilibrium provided that θ itself does not vanish identically. This 'condition of dissipativity' is a practical one: it can be tested in concrete cases [22] by actual computation of θ.

It is thus clear that the epistemological consistency problem raised at the beginning of this note is completely answered by the neat, clear-cut representation we obtain for the complementarity between the two levels of description of atomic phenomena. It need hardly be pointed out that the problem here discussed of the macroscopic level of quantum mechanics is just a simple illustration of a general method of representation in superspace, which actually amounts to an extension of the scope of quantum mechanics. Perhaps the most significant feature of the method is the introduction of generalized projectors, involving the replacement of self-adjointness by time-inversion invariance (which reduces to the former in the absence of dissipation, that is, for systems for which the collision operator vanishes) – a generalization which may be expected to find application to a large variety of problems.

3. Epistemological Problems of the Atomic Description of Macroscopic Phenomena

3.1. *The Historical Background*

The epistemological problems raised by the attempt to base the description of the directly observed phenomena on the atomic constitution of matter have forced themselves on the attention of physicists since the pioneering work of Maxwell and Boltzmann, but they could only be clearly formulated after the dynamical behaviour of the atomic constituents had found its definitive expression in quantum mechanics and the conceptual foundations of this theory had been elucidated. In the perspective of nineteenth-century physics it was natural enough to assume that the dynamics of the atoms was the same as that of large bodies, and accordingly to interpret the quantities characterizing the properties of these bodies as suitable averages over quantities pertaining to the constituent atoms. This averaging process seemed so far from being problematic that in the early papers of Clausius and Boltzmann it was not even mentioned explicitly.

However, the apparent contradiction between the irreversible evolution of macroscopic bodies and the time-reversal invariance of their description as atomic systems soon led Maxwell and his follower Boltzmann to their still fundamental analysis of the role of statistical causality in atomic physics. Of course, sharing the universal belief in determinism as the ultimate causal pattern of natural laws, they regarded any recourse to statistics as a *pis-aller*, but they correctly insisted on the fact that such a recourse was dictated by the very conditions of macroscopic observation. This point was variously elaborated both by Gibbs and by the Ehrenfests; the former forcibly pointed out that irreversibility at the macroscopic level results from the circumstances defining the corresponding mode of observation, while the latter introduced the notion of 'coarse-grained' distribution as a mathematical expression for a mode of observation not reaching down to the dynamical determination of the atomic system. On the other hand, it was clear to Maxwell and Boltzmann that the equilibrium distribution had to arise naturally from the dynamics, and they had the correct intuition (albeit incorrectly formulated) of ergodicity as the dynamical property primarily responsible for fixing the form of the asymptotic distribution. Gibbs wanted to express the same physical conception by his 'mixing' simile, which, however, turned out to be a stronger dynamical requirement than ergodicity.[3]

The development of quantum mechanics cannot affect the specific epistemological problems arising from the atomistic structure of matter, but it makes it easier to cope with them, by supplying us with adequate mathematical and logical tools for handling statistical averages of atomic quantities and probabilities of atomic processes. The statistical form of causality is now prevailing both in the account of individual atomic processes and of macroscopic phenomena; it is essential, however, to maintain a sharp distinction between these two forms of statistical causality, which are logically independent: the former has its origin in the existence of the quantum of action, the latter depends on the degree of freedom of the systems investigated, which may be characterized by some critical parameter, such as Avogadro's number, defining the order of magnitude we call 'macroscopic'. The formalism offers us uniform rules for dealing with both types of statistics and the relations of complementarity associated with each of them. In particular, as pointed out in the preceding section, the quantal density operator determines both the average distributions and the correlation coefficients of any system of interacting constituents, and accordingly yields a unified formal basis for the discussion of the two aspects of the asymptotic time-evolution – existence of the asymptotic distribution and 'mixing' effect of the correlations.

The ergodic approach leaves unanswered the question as to which physical characteristics of a dynamical system are decisive for its exhibiting a thermodynamic behaviour on the macroscopic time-scale; for the conditions of ergodicity and mixing are mere mathematical formulations of properties characteristic of such behaviour, but not explicitly related to the physical structure of the system. One lacks here, for distinguishing systems which allow of a thermodynamical description from those that do not, a criterion linking this property more directly to the Hamiltonian.

3.2. *Kinetic Approach vs. Ergodic Theory*

Even from the formal point of view, the extension of classical ergodic theory to quantal systems is far from straightforward: the stumbling block is the difficulty of finding a representation of the conditions of macroscopic observation of comparable simplicity to the classical concept of coarse-graining. In classical theory, it is permissible to replace a coarse-grained distribution, referred to some arbitrary subdivision of phase space into finite cells, by an idealized continuous distribution, and thus to make it independent of the mode of subdivision adopted. It is certainly possible, as the Italian school has shown [8], to introduce a 'cell' subdivision of the Hilbert space of state vectors and to arrive at certain quantal generalizations of the ergodic and mixing properties of classical theory, but both the enunciation of these results and the algorism leading to them – so long as one insists on mathematical rigour – are very cumbrous.

It must be observed, however, that the use of a coarse-grained cell subdivision is a consequence of the cyclic character of the time-evolution of the closed, finite systems dealt with in ergodic theory: in the classical case, this cyclic character is expressed by Poincaré's theorem; in the quantal case, the discreteness of the energy spectrum leads to an almost periodic time dependence of the density operator. As a result, the idealized definitions of macroscopic quantities can only use averages taken over an infinite time, and it is then the object of the theory to express these as statistical averages, necessarily coarse-grained in view of their physical meaning. The time-evolution of infinite systems, on the other hand, is, as we have pointed out in Subsection 2.2 [of the full paper, and sketched above – Ed.], radically different: the classical Poincaré cycle becomes of infinite duration, and the quantal energy spectrum, being continuous, allows of no almost-periodicity in time; but the correlations may lead to the occurrence of states idealized, as is usual in the theory of aperiodic processes, by (non-normalizable) state vectors with amplitudes decaying or building-up exponentially in time. This makes it possible, as we have seen, to study directly such one-sided effects of the correlations over long, but finite time intervals: the asymptotic form the density operator takes after a time of macroscopic order of magnitude can then be immediately interpreted as representing the conditions of macroscopic observation, without any need for the explicit consideration of coarse-graining. In this kinetic approach, the part of coarse-graining is played by the projection into the asymptotic subspace of our Hilbert superspace; besides its simplicity, this operation has the advantage of clearly exhibiting the fact that the macroscopic mode of description is uniquely fixed by the dynamics of the system.

3.3. *Main Features of the Macroscopic Aspect of the Kinetic Theory*

The form given to the kinetic theory in Section 2 [of the full paper, and sketched above – Ed.] has several noteworthy features. The most prominent is doubtless the clear-cut separation it effects, through the projection just mentioned, between the level of macroscopic observation, contained in the asymptotic subspace $\tilde{\Pi}$, and the fluctuating atomic behaviour of the system, which is fully accounted for by the density operator $\rho(t)$ describing the dynamical behaviour in the total superspace. If the system is initially in a pure state, its density operator at any time can be factorized as a dyadic product of state vectors: in the asymptotic subspace $\tilde{\Pi}$, however, no such factorization is possible, no analogue to a state vector can be defined. Indeed, the function of the projector $\tilde{\Pi}$ is to sort out the part of the density operator that subsists after time intervals long enough to 'wipe out' all initial phase relations between state vectors responsible for the fluctuations at the atomic level. In the asymptotic subspace, therefore, no 'interference' of probability amplitudes occurs, but the asymptotic density operator $\tilde{\rho}(t)$ embodies the description of any correlation effect observable at the macroscopic level.

Indeed, the formalism is not restricted to the study of the approach to thermodynamic equilibrium, but is applicable as well to systems presenting kinematic properties of macroscopic order of magnitude. It is important for the completeness of the theory that it should contain a formal criterium allowing us to recognize the type of asymptotic behaviour we may expect for a given system. This criterium is supplied by the condition of dissipativity, which gives a simple characterization of the class of systems exhibiting an irreversible approach to equilibrium. It is important to note that the dissipativity condition is amenable to explicit evaluation at least in simple cases, such as the Friedrichs model, and relates the dissipative character of the system directly to the structure of its energy spectrum.

For dissipative systems, a particularly clear picture is obtained of the origin of the irreversibility observed in their macroscopic behaviour. The projector $\tilde{\Pi}$, effecting the subdivision of the superspace into the subspaces which are the respective seats of the macroscopic and atomic evolution of the system, is invariant for time-reversal: this subdivision applies indeed to all systems, whether or not they are dissipative. The time evolution operator $\tilde{\Sigma}(t)$ in the asymptotic subspace is also time-reversal invariant; its form

(9) $$\tilde{\Sigma}(t)=e^{-iLt}\tilde{\Pi}=\tilde{\Pi}e^{-iLt}$$

if we use the spectral decomposition $\int \lambda \, dP(\lambda)$ of the superoperator L, just amounts to a Fourier integral representation $\int \exp(-i\lambda t) \, dP(\lambda) \, \tilde{\Pi}$ of the time-evolution. The expressions

(10) $$\tilde{\Sigma}(t)=P_a e^{-i\theta t} N_0{}^{-1} \bar{P}_a = P_a N_0{}^{-1} e^{-i\eta t} \bar{P}_a,$$

for $\tilde{\Sigma}(t)$ illustrate more explicitly how the asymptotic time-evolution starting from any given initial state proceeds by sequences of processes of 'creation' and 'destruction' of correlations between the elements of the system. The time direction in which one considers these processes is fixed by the choice of the conditions under which we want to observe the behaviour of the system: on the information supplied by observation at a single instant, we normally base expectation of future behaviour; but the symmetry in time of the two equivalent forms (10) of $\tilde{\Sigma}(t)$ shows what retrodiction we can also make, on the basis of the same information, about the evolution of the system, on the macroscopic time-scale, most likely to have led to its observed state: the sequences of creations and destructions of correlations appear inverted when, so to speak, we look back in time, and for dissipative systems we arrive at the well-known situation discussed by Ehrenfest in connection with the *H*-theorem.[4]

The collision superoperator $\Psi\mathscr{L}(+0)$ does not only govern the condition of dissipativity; it also clarifies, as we have seen, the role of the dynamical invariants. In particular, it allows us, by

$$(11) \quad [L_{00} - i\,\Psi\mathscr{L}(+0)]\Phi = 0$$

to characterize the regular invariants (which, at least in typical cases, coincide with the controllable ones). This solves one of the main riddles of the kinetic theory: it was indeed never clear why the 'collision operators' occurring in the usual formulations of the theory, such as the Boltzmann operator, were only in simple relation with the controllable invariants, independently of the possible existence of other invariants. At the same time, we see how decisive in this respect is the distinction between regular and singular invariants, – a distinction which can only be made for infinite systems.

Finally, it must be stressed that the macroscopic description we arrive at by projection into the $\tilde{\Pi}$ subspace is not the 'classical' one: it still contains Planck's constant wherever quantal effects occur on a macroscopic scale, as in the spectral distribution of electromagnetic radiation in thermal equilibrium, or in superconductivity and superfluidity. The laws of classical physics only appear as a limiting case of this macroscopic description in which all quantal effects are neglected, i.e. formally where $\hbar$ is treated as infinitesimal.

Nevertheless, it would be a serious logical error to imagine that the procedure we have been following – starting from the quantal description of the system, obtaining by the projection $\tilde{\Pi}$ its macroscopic description, and letting in the latter $\hbar$ tend to zero – would be a deduction of classical physics from quantum mechanics. For we must not forget that the quantal description from which we start has to be made in terms of the very concepts of classical physics. All we have done, therefore, is to prove the logical consistency of the rules by which a connection is established between the mathematical formalism and the classical description, upon which our account of macroscopic observation must ultimately rest. Obviously, there can be no question of any formal axiomatization of such a scheme: it must be based upon some set of classical concepts (e.g. space-time localization and momentum-energy) which remain unanalyzed and are treated as 'primitive'. These concepts (and these only) must be explained by processes of measurement, i.e. prescribed manipulations of specially designed apparatus, entirely describable in classical terms. Typically quantal concepts, on the other hand, like Planck's constant or the electron spin, cannot be related directly to such purely classical measuring apparatus, – but the occurrence of quantal phenomena at the macroscopic level allows us to establish indirect connections, involving more than one classical measurement, between quantal parameters and classical quantities, and thereby to determine with arbitrary accuracy the numerical values of the quantal parameters.[5]

4. Observation of Individual Atomic Processes

4.1. *The Consistency Problem of Quantum Mechanics*

Quantum Mechanics presents essentially two epistemological problems. The one, just recalled, concerns the consistency of the rules of interpretation by which the formalism is brought into relation with macroscopic observation, the other the relations of complementarity between different conditions of observation. There is nothing to add to Bohr's analysis of these relations, and we shall therefore confine ourselves to a comment, from the point of view developed in this paper, on the consistency problem. Briefly restated, this problem arises from the fact that the basic concepts used in the formulation of the laws governing individual atomic processes necessarily belong to the classical modes of description of direct macroscopic observation, which it is the aim of atomic theory to relate to the very laws of atomic behaviour. This means, as mentioned above, that these basic concepts must be regarded, from a strictly logical point of view, as 'primitive'. However, there is nothing to prevent us from describing on the atomic scale the process of measurement by which a value of the physical quantity denoted by any of these concepts is ascribed to an atomic system, as a dynamical process involving interactions between the atomic system observed and all the atomic constituents of the measuring apparatus, and obeying the laws of quantum mechanics. Such an analysis will only lead to the same conclusion as that resulting from the direct application of the rules of interpretation of the formalism if these rules are consistent. A test of this consistency is especially desirable with respect to the rule usually designated as the 'reduction' of the wave function representing the initial state of the observed atomic system, i.e. its replacement after the measurement by another wave function expressing the information obtained by this observation. Although such a rule is obviously consonant with the statistical form of causality inherent in quantum mechanics, its relation with the dynamical process of measurement can only be elucidated by a careful consideration of the specific function of the measuring apparatus and its implications for the course of the measuring process.

The decisive step was made by N. Bohr [1], who pointed out that any measurement is essentially the codified registration of some characteristic signal arising from the interaction between the observed atomic system and a suitable recording device of macroscopic dimensions. Indeed, he emphasized that the very definition of a phenomenon must contain a specification of the experimental conditions under which it is observed, including the apparatus recording some kind of permanent mark allowing us to identify the process observed. Until the process has been terminated by the registration of its permanent record, we have no basis for the use of the classical concepts corresponding to such record, and accordingly no possibility of giving any well-defined account of the process. In other words, it is precisely the recording by the apparatus which establishes the necessary link between the behaviour of an atomic system and its description in terms of concepts referring to our possibilities of observing it. Now, it is clear that the formation of a permanent mark on a recording device is an irreversible macroscopic process, retaining no other trace of the original state of the atomic system than the specific feature corresponding to the construction of the apparatus; this information modifies the conditions of observation upon which statistical predictions about the behaviour of the atomic system must henceforth be based, and it is precisely this modification which is expressed in the formal language of the theory by the assignment to the atomic system, according to the rule of 'reduction', of a new wave-function, appropriate to the new conditions of observation.

Once it is realized that the wave-function of the atomic system after the measurement is a component factor in an expression representing the asymptotic state of the whole system including the measuring apparatus, it can hardly be doubted that it will have the form prescribed by the reduction rule, since the latter precisely corresponds to this asymptotic situation. As already mentioned, we are only concerned with the consistency of the use of the small set of classical concepts we have called 'primitive'. In quantum mechanics, these are the complementary sets defining space-time localization and momentum-energy exchange; the completion of the above argument in this case only requires, as Bohr showed, the consideration of simple dispositions of fixed or moving diaphragms. It is sufficient to discuss these, as he did, by the methods of classical optics, since the latter use an idealized representation of diaphragms which directly accounts for the irreversible modification of a wave pattern arising from its interaction with such material bodies.[6] In this connection, it must be remembered that, although the apparatus fixing the conditions of observation, like any other macroscopic bodies, may in principle be described as large quantal systems, they must be kept outside such a description in order to fulfil their special function (otherwise, they would become themselves objects of observation); it is for this deep-lying reason that their intervention is represented in the formalism of quantum mechanics by parameters denoting 'external' forces and prescriptions for the selection of appropriate solutions of the fundamental equations– among them the reduction rule. Disregard of this circumstance is the most frequent source of misunderstanding concerning the foundations of quantum mechanics.

The general theory of large quantal systems obviously offers another possibility of dealing with the consistency problem: one has only to apply this theory to the system formed by the atomic system under observation and the measuring apparatus described as a large assembly of atoms. This was done by the Italian physicists [4] on the basis of the form of quantal ergodic theory they had previously developed: [8][7] their result is, of course, in full harmony with Bohr's argument [5]. The kinetic approach outlined in the preceding sections allows us to discuss the issue in a more direct and simpler manner and to throw further light on the role of the measuring process in the epistemological analysis of physical theory.

4.2. *Discussion of the Measuring Process*

In order to analyze the course of a measuring process, we may use the schematization discussed in detail in ref. [5]. The main point is to arrive at a sufficiently simple and general formulation of the restrictions that have to be imposed upon the structure and dynamical behaviour of a macroscopic body in order that it may fulfil its function of measuring a specific property of a given atomic system. Its mode of interaction with the atomic system must be such as to bring it into a state observable at the macroscopic level and uniquely related (at least approximately) to the specific atomic state of interest. Without loss of generality, we avoid unessential complications by assuming that the interaction between the atomic system and the measuring apparatus only lasts for a time short enough to allow us to disregard any dynamical change of the atomic system as a result of this interaction. The apparatus, on the other hand, reaches, at the termination of the interaction, a dynamical state which must be such as to fix the ultimate outcome of the measurement. For from the moment the interaction ceases, the two parts of the total system evolve independently, and we expect the apparatus to settle down, at the macroscopic level of observation, into a situation providing a 'permanent' record[8] of its specific interaction with the atomic system. This requirement imposes essential restrictions upon the constitution of the measuring apparatus and upon the interactions of its constituent elements with the atomic system under investigation. If the aim of the measurement is to ascertain whether the atomic system is in an eigenstate $|\varphi_s\rangle$ of the operator representing some physical quantity attached to the system, there must exist a complete orthogonal basis $|sm\rangle$ of dynamical states of the measuring apparatus such that (i) only a definite set $\{|sm\rangle\}$ of such states (denoted by a common index s) interacts with the atomic system when the latter is in a definite state $|\varphi_s\rangle$; (ii) the correlations between any two states $|sm\rangle$, $|s'm'\rangle$ are much weaker when they belong to different such sets ($s \neq s'$) than when they belong to the same set ($s = s'$). (As an example, we may think of the formation of a particle 'track' in a bubble chamber or a photographic emulsion.[9])

In order to show how these conditions ensure the desired functioning of the measuring device, let us examine the asymptotic time-evolution of its density supervector in its own superspace. We define the subspace P_o by the projector

$$P_o = \sum_s P_o^{(s)}, \quad P_o^{(s)} = \sum_m P_{sm} \times P_{sm},$$

and, accordingly, the correlation subspace by the projector

$$P_c = \sum_{ss'} P_c^{(ss')}, \quad P_c^{(ss')} = \sum_{mm'} P_{sm} \times P_{s'm'},$$

the summation Σ' extending over all values of m and m' if $s \neq s'$ and all different values of m and m' if $s = s'$. The superoperator $\Psi(t)$ given by

(12) $$\Psi(t) = L_{oc} T_c(t) L_{co}.$$

and the asymptotic superoperators θ, η derived from it by

(13) $$\theta = L_{oo} - i \int_0^\infty d\tau \Psi(\tau) e^{i\theta\tau}, \quad \eta = L_{oo} - i \int_0^\infty d\tau e^{i\eta\tau} \Psi(\tau),$$

will essentially effect transitions from any subspace $P_o^{(s)}$ to the same subspace: for other transitions, on account of the condition (ii) above, will be much less probable. Moreover, if we consider the mutually orthogonal subspaces defined by the projectors $P^{(ss')} = P_o^{(s)} \delta_{ss'} + P_c^{(ss')}$, the leading terms of the superoperators of destruction and creation of correlations will be those linking any subspace $P^{(ss)}$ with itself (i.e. the subspace $P_o^{(s)}$ with the correlation subspace $P_c^{(ss)}$); next in importance will be the links between subspaces $P^{(ss')}$ and $P^{(ss'')}$ for values of s' and s'' in a small interval around s. Therefore, if we start from any initial density supervector $\rho^{(M)}$ of the measuring apparatus, and only retain the leading terms, the asymptotic time-evolution of its projection $P^{(ss')} \rho^{(M)}$ will take the form

(14) $$\tilde{\Sigma}(t) P^{(ss')} \rho^{(M)} = \delta_{ss'} P^{(ss)} \tilde{\rho}^{(M)}(t) \equiv \delta_{ss} \tilde{\rho}_s^{(M)}(t):$$

this formula expresses a clear-cut asymptotic trend of the measuring system towards definite states $\tilde{\rho}_s^{(M)}(t)$ uniquely associated with the eigenstates $|\varphi_s\rangle$ which are the objects of the measurement.[10]

After this preparation, the discussion of the measuring process is readily performed. Let the initial state of the atomic system be represented by a superposition $\Sigma_s c_s |\varphi_s\rangle$ of the eigenstates $|\varphi_s\rangle$; the resulting density supervector

$$\rho^{(S)} = \sum_{ss} \rho_{ss'}^{(S)}, \quad \rho_{ss'}^{(S)} = c_s c_{s'}^* |\varphi_s\rangle\langle\varphi_{s'}|$$

exhibits the correlation between the states $|\varphi_s\rangle$ arising from definite phase relations between the coefficients $c_s, c_{s'}^*$. The initial form of the density supervector of the total system formed by the atomic system and the apparatus (i.e., let us recall it, the form this supervector takes immediately after the two constituents of the total system have interacted) is easily set up in conformity with our condition (i): to each component $\rho_{ss}^{(S)}$ there corresponds a component $P^{(ss')}\rho^{(M)}$ of the apparatus density supervector. The total density supervector is thus initially

$$\sum_{ss'} \rho_{ss'}^{(S)} \cdot P^{(ss')} \rho^{(M)},$$

and each factor evolves independently in its own superspace: the time-evolution of the atomic system is governed by its Liouville superoperator $T^{(S)}(t) = \exp\{-iL^{(S)}t\}$; as to the apparatus, we are interested in its evolution at the macroscopic level, described in its asymptotic subspace by the superoperator $\tilde{\Sigma}(t)$. Adopting the idealized situation expressed by equation (14), we therefore obtain at time t the total density supervector

$$\sum_s T^{(S)}(t)\, \rho_{ss}^{(S)} \cdot \tilde{\rho}_s{}^{(M)}(t).$$

From this expression a density supervector making explicit reference only to the atomic system may be derived by averaging it over the apparatus superspace, i.e. by taking its trace with respect to the apparatus basis; we may simply ascribe the same limiting value to all the tr $\tilde{\rho}_s{}^{(M)}(t)$ and normalize it to unity:

$$\lim_{t\to\infty} \operatorname{tr} \tilde{\rho}_s{}^{(M)}(t) = 1.$$

This gives us for the density supervector of the atomic system, after a time sufficient for the recording of a permanent mark on the apparatus, the expression

$$(15) \qquad \sum_s |c_s(t)|^2 |\varphi_s\rangle\langle\varphi_s|,$$

from which every explicit reference to the apparatus has disappeared, and whose form agrees with the prescription of the reduction rule.

The way in which this result has been derived makes its meaning quite clear. In the first place, it is essentially an asymptotic result, and there is no question of its contradicting any consequence of the time-dependent Schrödinger equation. Secondly, the only interactions involved in the whole process are the specific interaction of short duration between the atomic system and the apparatus and the long sequence of interactions it triggers off between the constituent atoms of the apparatus; there is no question whatsoever of any non-physical intervention of any kind upon the atomic system: the whole measuring process is a purely automatic registering operation, obeying only the laws of quantum mechanics, and its outcome is accordingly in full harmony with these laws. Let us here emphasize again that the treatment of the measuring apparatus as a quantal system we have performed has no other purpose than exhibiting the consistency of the conceptual framework of quantum mechanics, in which formalism and rules of interpretation in terms of classical concepts form an inseparable whole. Thirdly, the difference between the 'reduced' density (15) and that of the system before the measurement simply reflects the change in our information about the system brought about by this measurement; the occurrence of such a difference is no peculiarity of quantum mechanics (though the particular form it takes is of course a specifically quantal one) – it is common to all statistical situations and in fact an essential component of the concept of probability.

4.3. *The Role of the Observer*

Statistical causality has often been misrepresented as implying the intrusion of a 'subjective' element into the description of the phenomena; this epistemological error (which was not made by the founders of the theory of probabilities) arises from an insufficient analysis of the conditions under which knowledge of the phenomena is obtained. The inclusion of a specification of the conditions of observation into the account of the phenomena is no arbitrary decision, but, as we have insistently stressed, a necessity imposed by the very laws governing the course of these phenomena and the mechanism of their observation, and thereby an indispensable part of their objective description, since it ensures that such a description will be common to all observers placed in the specified conditions. It is in order to make this objectivity quite apparent that we have analyzed, in the preceding subsection, the observation process as a purely physical one, limited to the automatic registration of a record, which need not even be read.

Obviously, such an analysis touches only one side of the process of acquisition of objective knowledge. This process is not complete until a reading of the record has actually occurred, i.e. until the information it contains has been stocked into the brain of some observer. This 'psychological' side of the cognitive process is obviously just as objective as the other, since it is the same for any conceivable observer, but the question arises whether it can legitimately be separated as sharply as we have done from the purely physical registration process. It would seem that our understanding of the fundamental biological processes has now reached a stage allowing us to give a definite answer to this question.

We must above all realize that there is a large class of physical and chemical dissipative processes, occurring far from the thermodynamic equilibrium conditions, which present structural inhomogeneities quite foreign to the familiar physical phenomena observed at or near equilibrium, but analogous to typical features of biological systems [25, 26]. It has been shown, for example, that chemical reactions essential for the metabolism of living systems, as well as for the regulation of genetic and evolutionary processes, can be treated in detail in terms of usual chemical kinetics, and involve branches of the solution of the non-linear kinetic equations which do not belong to the classical thermodynamic description of near-equilibrium situations, but only appear under conditions far from equilibrium [27, 28]. Now, thermodynamics, both in its classical and its more general form just mentioned, is a mode of description adapted to the behaviour of macroscopic systems in the asymptotic $\tilde{\Pi}$ subspace; indeed, it has been explicitly established [29] that the thermodynamic description can be derived from the kinetic equation associated with the time-evolution operator $\tilde{\Sigma}(t)$. It is natural to conclude from these considerations that the distinction between 'living' and 'non-living' systems appears to be part of their description in $\tilde{\Pi}$-space, and accordingly cannot be based on any quantal description on the atomic scale. An analogy which may perhaps clarify the significance of this remark is the distinction between laminar and turbulent motion, which correspond to two types of solution of the Navier-Stokes equation: as this equation belongs to the macroscopic level – it is a consequence of the kinetic equation – we cannot characterize the difference between these two types of flow by means of the formalism of quantum mechanics. Likewise, it would seem that the characteristics of 'life' cannot be formulated at the atomic level but are essentially macroscopic. Incidentally, this circumstance is sufficient to dismiss all 'paradoxical' situations (such as the famous example ascribed to Schrödinger) which one would allegedly encounter when attempting to treat living systems by the methods of quantum mechanics. In such cases, just as in the general problem of observation, a correct epistemological analysis can only be developed at the macroscopic level of description.

If we accept this general inference from our present knowledge of biology, we do not see any special role to be attributed to a 'living' observer in the discussion of the consistency problem of quantum mechanics: the only condition the intervention of such an observer has to fulfil is to be amenable to physical description at the macroscopic level. Now, we may confidently assume that the sense organs of an animal register signals from the physical environment in essentially the same way as material apparatus, and that the storage of these signals and their incorporation into sensory-motor schemes is also the result of physical and chemical processes of the type commonly observed in biology.[11] Since all the organs involved are of macroscopic order of magnitude, their activity can in principle be entirely described in the asymptotic subspace $\tilde{\Pi}$ of the appropriate superspace, like the functioning of any physical measuring apparatus.

5. General conclusion

Perhaps the most significant general result of our approach is the introduction of generalized projection superoperators, made possible through the consideration of superspace. This generalization involves the concept of time-reversal invariance which replaces the usual one of self-adjointness, to which it reduces in the absence of dissipation. These new projectors allow us to give a precise characterization of the macroscopic level of description of general quantal systems, and to formulate quantitatively the conditions under which a given system will exhibit the properties belonging to this level: a formulation obviously susceptible to experimental test (for example, we may verify whether a system is in thermodynamical equilibrium). These questions actually amount to an extension of the scope of quantum mechanics, which has only been outlined in this paper and deserves a more detailed treatment. In a sense, the question of the definition of a macroscopic level of quantum mechanics may be considered as a simple illustration of this general method, which applies as well to a large class of other problems.

We are grateful to our colleagues A. Jauch, M. Kac, L. Lanz, L. A. Lugiato and G. Ramella for epistolary and verbal discussions. We also wish to thank all the members of the groups of statistical mechanics in Brussels and Austin for suggestions and discussions, especially R. Balescu, R. W. Gibberd, A. Grecos, F. Henin and G. Nicolis. Finally we want to acknowledge the support of the Fonds de la Recherche Collective (Belgium) and of the Welch Foundation (Houston, U.S.A.).

NOTES

* [Editorial Note: we print here as one text two papers which had the same title: *The Macroscopic Level of Quantum Mechanics*. They were the work of Rosenfeld in collaboration with I. Prigogine and C. George of the Université Libre in Brussels and the Center for Statistical Mechanics and Thermodynamics, University of Texas, Austin. We are grateful to Professor Prigogine for kindly approving of this reprint, and of the arrangement. The papers were published in *Nature* **240** (1972), 25–27, and *Det kgl. Danske Videnskabernes Selskab. mat.-fys. Meddelelser* **38** no. 12 (1972), 1–44. Here Part 1 and Parts 3–4–5 come from the publication of the Royal Danish Academy, while our Part 2 is the complete text of the *Nature* article which contains a summary of the technical Part II of the Danish paper.

[1] For a discussion of these conditions and their extension to finite quantal systems, see in particular refs. [7, 8].

[2] A monograph by I. Prigogine, C. George and F. Henin, dealing with these problems, is in preparation.

[3] See, e.g. ref. [7].

[4] It must be observed, in this connection, that while both $\tilde{\rho}(t)$ and $\rho(t)$ appear as functions of the same time variable t, and in fact as solutions of the same Liouville equation, their time variations may be extremely different: $\rho(t)$ exhibits variations at the rate of atomic processes, whereas the variations of $\tilde{\rho}(t)$ may look quite smooth in comparison. This circumstance may be made quite explicit, as shown by L. Lanz, L. A. Lugiato and G. Ramella [22], at the unavoidable cost of formal complication: one must then renounce using the asymptotic time-displacement superoperator θ acting at any time, but introduces a similar superoperator, whose action is only defined for an arbitrary finite sequence of separate instants.

[5] Thus, Millikan's determination of Planck's constant from the study of the ejection of electrons from metallic surfaces by light of various frequencies involves essentially two measurements: that of the light frequency and that of the corresponding kinetic energy of the ejected electrons.

[6] Somewhat less elementary is the case of quantum electrodynamics, owing to the necessity of taking account of all retardation effects in the measurement of a field component; in this case also, the consistency of the formalism could be established by a detailed analysis [2], in which advantage was taken of the smallness of the coupling between electromagnetic fields and distributions of electric charge and current.

[7] More recently, they have given some consideration to the problem from the point of view of their own kinetic theory, but made no attempt at a detailed examination of the issues involved [23].

[8] By "permanent" we do not imply that the state in question is one of actual thermodynamic equilibrium, but only that it lasts long enough for macroscopic observation.

[9] The interaction of a particle with the medium gives rise at a certain point (corresponding to a state $|\varphi_s>$ in which the particle has a definite "position") to a local fluctuation (represented by the set $\{|sm>\}$) around which a bubble or a spot develops (leading to the asymptotic density $\tilde{\rho}_s^{(M)}(t)$ defined on p. 591. The condition (ii) expresses the requirement that the

successive bubbles or spots formed by the passage of the particle through the medium be sufficiently distinct from each other for a recording of the corresponding positions of the particle.

[10] It should be stressed that the approximation (14) is sufficient for the analysis of the most general measuring process of physical interest. The case envisaged by Wigner's theorem [24], in which the system formed by the atomic object and the measuring apparatus has an additive invariant, only occurs if the quantity to be measured is precisely such an invariant, or commutes with it, and an idealization of the type (14) is then in accordance with the theorem. If, however, the quantity to be measured does not commute with the additive invariants, Wigner's theorem has no relevance, for it is an obvious physical requirement that any such invariance should be destroyed by coupling the apparatus to a suitable external system of infinite extension. For instance, while a momentum measurement of infinite accuracy may be performed by means of an elastic collision of the atomic object with a freely moving test-body, the spatial localization of the object requires the test-body to be rigidly attached to some body of infinite mass serving as a spatial system of reference. It is worth pointing out how well adapted the $\tilde{\Pi}$-space representation is to a concise and general account of the process of measurement.

[11] Of great interest in this respect is the experimental evidence showing that the memory of a specific sensory-motor scheme can be stored in the form of a coded macromolecule.

BIBLIOGRAPHY

[1] N. Bohr, *Dialectica* **2** (1948), 312.

N. Bohr, 'Discussion with Einstein on Epistemological Problems in Atomic Physics' in *Albert Einstein: Philosopher-Scientist* (Library of Living Philosophers, Evanston, Ill., 1949).

N. Bohr, 'Quantum Physics and Philosophy: Causality and Complementarity' in *Philosophy in the Mid-Century, a Survey*, ed. R. Klibansky (Firenze 1958).

[2] N. Bohr and L. Rosenfeld, *Dan. Vid. Selsk. mat.-fys. Medd.* **12**, no. 8 (1933) [See this volume, p. 357 – Ed.]; *Phys. Rev.* **78** (1950), 794; [See this volume, p. 401 – Ed.].

L. Rosenfeld, 'On Quantum Electrodynamics' in *Niels Bohr and the Development of Physics* (Pergamon Press, London, 1955). [See this volume, p. 413 – Ed.]

[3] N. Bohr, Faraday Lecture in *J. Chem. Soc.* (1932), 349.

L. Rosenfeld, 'Questions of Irreversibility and Ergodicity' in *Rendiconti Scuola Int. Fisica*, Corso XIV, Varenna 1960 (Zanichelli, Bologna, 1962). [See this volume, p. 808 – Ed.]

[4] A. Daneri, A. Loinger, and G. M. Prosperi, *Nucl. Phys.* **33** (1962), 297; *Nuovo Cimento* **44B** (1966), 119.

A. Loinger, *Nucl. Phys.* **A108** (1968), 245.

[5] L. Rosenfeld, 'The Measuring Process in Quantum Mechanics' in *Suppl. Progress Theor. Physics*, Commemoration issue, 1965, p. 222. [See this volume, p. 536 – Ed.]

[6] I. Prigogine, *Non-Equilibrium Statistical Mechanics* (Interscience Publ., New York, 1962).

[7] L. Rosenfeld, *Acta Physica Polonica* **14** (1955), 3. [See this volume, p. 762 – Ed.]

[8] A. Loinger, 'A Study of the Quantum Ergodic Problem' in *Rendiconti Scuola Int. Fisica*, Corso XIV, Varenna 1960 (Zanichelli, Bologna, 1962).

J. E. Farquhar, *Ergodic Theory in Statistical Mechanics* (Interscience Publ., New York, 1964), chap. VIII.

[9] C. George, *Physica* **30** (1964), 1513; G. Severne, *Physica* **31** (1965), 877; also ref. [10].
[10] M. Baus, *Bull. Cl. Sc., Acad. Roy. Belg.* **53** (1967), 1291, 1332, 1352.
[11] J. Rae, *Bull. Cl. Sc., Acad. Roy. Belg.* **55** (1969), 980, 1040.
[12] R. Balescu, P. Clavin, P. Mandel, and J. W. Turner, *Bull. Cl. Sc., Acad. Roy. Belg.* **55** (1969), 1055.
[13] I. Prigogine, 'Dynamic Foundations of Thermodynamics and Statistical Mechanics' in *A Critical Review of Thermodynamics*, ed. B. Stuart, B. Gal-Or, and A. J. Brainard (Mono Book Corp., 1970).
[14] I. Prigogine, C. George, and F. Henin, *Physica* **45** (1969), 418.
[15] I. Prigogine, C. George, and F. Henin, *Proc. Nat. Acad. U.S.A.* **65** (1970), 789; the same with P. Mandel and J. W. Turner, *ibid.*, **66** (1970), 709.
[16] A. Grecos, *Physica* **51** (1970), 50.
[17] A. Grecos and I. Prigogine, *Physica* **59** (1972), 77.
[18] R. Balescu and J. Wallenborn, *Physica* **54** (1971), 477.
[19] C. George, I. Prigogine, and L. Rosenfeld, *Det Kgl. Danske Videnskabernes Selska, mat.-fys. Meddelelser* **38** (1972), 12. [See Editorial Note in this volume on p. 596 – Ed.]
[20] R. Balescu and L. Brenig, *Physica* **54** (1971), 504.
[21] A. Grews and I. Prigogine, *Proc. US Natl. Acad. Sci.* **69** (1972), 1629.
[22] L. Lanz, L. A. Lugiato, and G. Ramella, *Physica* **54** (1971), 94; see also L. Lanz and G. Ramella, *Physica* **44** (1969), 499; L. Lanz and L. A. Lugiato, *Physica* **44** (1969), 532 and **47** (1970), 345.
[23] L. Lanz, G. M. Prosperi, and A. Sabbadini, *Nuovo Cimento* **2B** (1971), 184.
[24] E. Wigner, *Z. Physik* **133** (1952), 101.
[25] I. Prigogine, in *Theoretical Physics and Biology* (1967), ed. M. Marois (North Holland Publ. Co., Amsterdam, 1969).
[26] P. Glansdorff and I. Prigogine, *Stability, Structure and Fluctuations* (Wiley Interscience, New York, 1971; French edition, Masson, 1971).
[27] I. Prigogine, R. Lefever, A. Goldbeter, and M. Herschkowitz-Kaufman, *Nature* **223** (1969), 913.
[28] M. Eigen, 'Self-Organization of Matter and the Evolution of Biological Macro-molecules', *Naturwissenschaften* **58** (1971), 465.
[29] P. Résibois, *J. Stat. Phys.* **2** (1970), 21.

CHAPTER 15

QUANTUM THEORY AND GRAVITATION*

[1966a]

The incorporation of gravitation into a general quantum theory of fields is an open problem, because the necessary empirical clues for deciding the question of the quantization of the gravitational field are missing. It is not so much a matter here of the mathematical problem of how one should develop a quantum formalism for gravitation, but rather of the purely empirical question, whether the gravitational field – and thus also the metric – evidences quantum-like features. In the absence of relevant observations, we can only attempt to shed light on such a question from the epistemological side; and we cannot hope thereby to reach any sort of final conclusions, since epistemological considerations can indeed help in investigating the logical structure of a given theory, but not its conformity to phenomena.

1. Quantum mechanics and classical gravitational theory

First of all we wish to recall the famous discussion between Einstein and Bohr concerning the uncertainty relation between energy and time. Bohr illustrated the meaning of this uncertainty relation through the following example: The emission of a pulse of light during a certain time interval, the endpoints of which are precisely known, can be effected with the help of a shutter arrangement controlled by a clockwork; this arrangement consists of two parallel screens, one of which is set in motion by the clockwork and thus can control the opening, during the prescribed time interval, of a hole that has been bored into the other. In this way a section is cut out of a light ray falling on the first screen, which can afterwards be absorbed by means of a suitable contrivance so that its total energy can be determined with arbitrary exactness. However, to infer from this the energy emitted by the light source during the given time interval, one must also take into account as accurately as possible the exchange of energy between the light ray and the shutter arrangement; and it is precisely at this point that the quantum-like character of this interaction makes itself felt. The energy change of the shutter could simply be deduced from the change in momentum of the moveable screen, since the velocity v of the latter is precisely known. An error Δp in determining the momentum results then in an error $\Delta E = v\Delta p$ in determining the energy exchange. Now, since the location at which the interaction between the radiation and the shutter takes place is precisely determined only up to the diameter d of the hole in the fixed screen, the error in the measurement of the momentum exchange cannot be smaller than $\hbar/d$ and consequently

(1) $\Delta E > \hbar v/d.$

Now one might think of reducing this inaccuracy by choosing the ration v/d to be very small. But then one impairs the knowledge of the instant at which the opening of the hole begins, for this instant is obviously determined by the arrangement only up to an inaccuracy of order of magnitude

(2) $\Delta t \approx d/v.$

As one sees, the requirements placed on the arrangement in order to facilitate a precise determination of the energy on the one hand, and on the other a precise determination of the time, are incompatible with one another; and the inaccuracies (1) and (2) stand in the reciprocal relation

(3) $\Delta E \Delta t \gtrsim \hbar.$

The epistemological significance of this analysis of Bohr's lies in the fact that on one hand it depends directly on the function of the clock as an apparatus which links the instant at which some physical event occurs with an invariable time scale fixed by the clockwork; while on the other hand it shows with full clarity how this function of the clock as chronometer involves an unavoidable exchange of energy with the physical system in question, which can be taken into account only subject to a limitation imposed precisely by this function. Nevertheless, Einstein thought it possible to escape the force of the argument by means of an artifice. Let us, he suggested, enclose the light source and the whole shutter arrangement in a box; by weighing the box twice – once before the beginning and once after the end of the emission process triggered by the clock – we can quite simply determine with arbitrary exactness the energy radiated by the light source without affecting the determination of the time of the process. In this way he thought he could avoid the necessity of taking into account the exchange of energy between the radiation and the clock and thus also the uncertainty relation (3) following from it.

To clarify this apparent paradox, we must first of all investigate the suitability of the proposed arrangement as a weighing device. Suppose the weighing takes place, by suspending the box on a spring and reading its vertical displacement q along a fixed scale (Figure 1). If this reading is

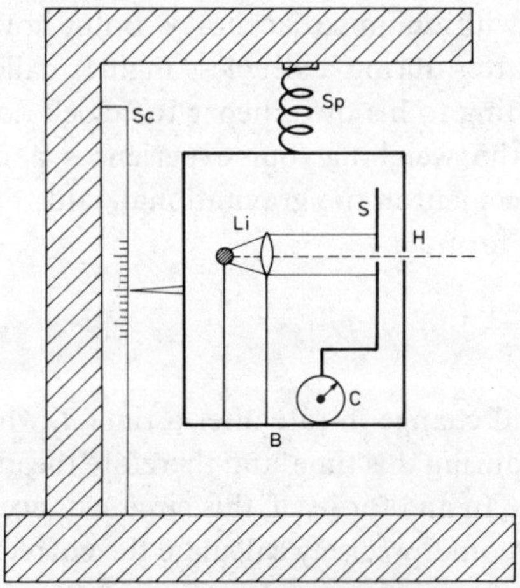

Fig. 1. The 'Einstein Box'.
Sp Spring, B Box, H Hole, Li Light source,
S Shutter, Sc Scale, C Clockwork.

taken with an inaccuracy of $\lesssim \Delta q$, then the determination of the conjugate momentum component p of the box will be subject to an inaccuracy $\Delta p \gtrsim \hbar/\Delta q$, and consequently the determination of mass with an inaccuracy

$$\Delta m \gtrsim \frac{\Delta p}{gT} \gtrsim \frac{\hbar}{gT\Delta q}, \tag{4}$$

where g is the acceleration due to the gravitational field at point q and T is the duration of the weighing. Formula (4) shows that with an arbitrarily small inaccuracy Δq in the reading it is possible to arbitrarily reduce the inaccuracy Δm by making the duration of the weighing sufficiently long. The apparatus is thus, in fact, suited to an arbitrarily accurate determination of the energy mc^2 emitted during the given time interval; and indeed, if the second weighing at the time t of the closing of the hole begins and lasts a time T, the error in the energy measurement ΔE will be connected with the error in the reading Δq by the formula

$$(5) \qquad \Delta E \gtrsim \frac{\hbar c^2}{gT\Delta q}.$$

But now how about the determination of the time fixed by the end of the emission process? Einstein thought that this time could be simply fixed by the clock with unlimited accuracy. Bohr, however (after having reflected on the matter during a sleepless night!), called his attention to the fact that according to his own theory the clock does not remain undisturbed during the weighing, but experiences a change in its rate through the displacement in the gravitational field; this change is given by the well-known formula

$$\frac{\delta T}{T} = \frac{g\delta q}{c^2}.$$

Here δT is the total change in rate after a time T, where the box is displaced a height δq during this time, and therefore the gravitational potential changes by $g\delta q$. In and for itself, this circumstance would be without significance, if one could precisely calculate the correction δT: this, however, would require precise knowledge of the shift δq. An error Δq in the reading of q results, on the contrary, in an unavoidable error Δt in knowledge of the rate change δT and thus in knowledge of the time t of the conclusion of the emission process, and this error

$$(6) \qquad \Delta t \approx \frac{gT\Delta q}{c^2}$$

proves to be greater, the greater the duration T of the weighing is. Here the expected complementarity between the possibilities of determining the conjugate quantities E and t now shows itself; indeed one again finds by comparison of formulas (5) and (6) that it is precisely the product $\Delta E \Delta t$ that is independent of the duration T of the weighing and must necessarily be $\gtrsim \hbar$.

What can one now conclude from the analysis of this *Gedankenexperiment* concerning the relationship between quantum theory and gravitational theory? To answer this question one must, above all, note that gravitational theory was only applied to a description of the behavior of the box and the clockwork, but not to the system being investigated, the radiation field. What Bohr showed is only that the apparent contradiction with quantum theory was actually based on an inconsistent application of classical gravitational theory to a macroscopic system, and disappears upon consistent consideration of all the circumstances. Determining the energy of a radiation field by weighing depends, that is to say, not only on the equality of energy and mass, but also on the equivalence principle; it is thus logically necessary to introduce all the consequences of this principle in analyzing the measurement process, and thus also the influence of the gravitational field on the rate of clocks. Proof of the uncertainty relation $\Delta E \Delta t \gtrsim \hbar$ results by reducing this relation to the corresponding relation $\Delta p \Delta q \gtrsim \hbar$ which refers to the determination of the exchange of momentum between the radiation field being studied and the spring on which the box in the gravitational field hangs. But, the gravitational field plays, so to speak, only an intermediary role in this process, and its effect is completely described by the classical theory.

2. Measurability of the Quantized Gravitational Field

To progress further we must now make some sort of assumption about the supposed quantum properties of the gravitational field. It is certainly simplest to pursue the analogy with the electromagnetic radiation field as far as possible; this amounts actually to considering a solution of the linear approximation to the gravitational equations, which may represent gravitational radiation of weak intensity propagating in empty space. It will be instructive to establish where such an analogy breaks down. Let us consider then a component $\mathfrak{F}$ of the gravitational radiation field, which we define as the force exerted in a given direction on a test body of mass unity. As a consequence of this definition, the corresponding energy density is expressed by $\mathfrak{F}^2/G$, where G designates the usual gravitational constant. As the critical field magnitude we can, as in electrodynamics, select the mean field strength $\mathfrak{F}$ of a gravitational quantum of wavelength L, which we can estimate from the condition that its energy $\hbar c/L$ is contained within a volume of linear dimensions L; this yields

(7) $$\mathfrak{F} = \sqrt{G\hbar c}/L^2.$$

Just as in electrodynamics, we can also determine this mean field strength temporally averaged over a period L/c, by means of its ponder-motive effect on a test body with total mass M that uniformly fills the volume in question. With an indeterminacy Δx in the displacement of the test body in the measurement process, there results for the measurement of the field an uncertainty of order of magnitude

(8) $$\Delta\mathfrak{F} \approx \hbar c / ML\Delta x.$$

For the measurement to yield a meaningful result, it should in any case satisfy the modest requirements

$$\Delta\mathfrak{F} < \mathfrak{F}, \Delta x < L;$$

through substitution into equation (8) in light of equation (7), there follows from this for the mass of the test body the condition

(9) $$M > M_0,$$

where

(10) $$M_0 \equiv \sqrt{\hbar c/G} \approx 10^{-5} \text{ g}$$

represents a critical mass for quantized gravitational theory.

The analogous condition in electrodynamics, e.g. that which occurs in the measurement of an electrical field strength, implies only that the total charge of the test body must consist of a great number (at least $\sqrt{(\hbar c/e^2)}$) of elementary charges: in principle such a condition can always be fulfilled, because pure electromagnetic radiation theory (which contains only the two universal constants $\hbar$ and c) fixes no space-time dimensions. But the situation is different in quantized gravitational theory, which in addition depends on the gravitational constant and in which, consequently a definite scale for physical quantities can be defined: Besides the mass M_0, the critical length $L_0 = \hbar/M_0 c = \sqrt{G\hbar c}/c^2 \approx 10^{-33}$ cm and the critical time interval L_0/c appear. To have before us an analogous situation from quantum electrodynamics, we must proceed to the next approximation and introduce quantization of the current and charge distributions that interact with the radiation field. It appears, then, that the occurrence of a definite scale (say given by the proton mass m, the corresponding Compton wave length $\lambda_p = \hbar/mc$ and the time λ_p/c) for the measurability of the total currents and charges contained in a space-time region, indeed, does not entail a logical restriction, though it does certainly entail a fundamental practical restriction. Such a restriction comes about in that the measurement results, according to quantum theory, are subject to statistical fluctuations and consequently are practically useful only in so far as these fluctuations remain sufficiently small on the average in comparison to the mean values. Now this practical requirement cannot be fulfilled for arbitrarily small linear dimensions of the space-time regions in which the currents and charges to be measured are contained, because the corresponding mean fluctuations grow beyond all limits when the limitation of the space-time region becomes infinitely sharp. Consequently, in order that the fluctuations not become dominant, these regions can only be defined with a finite uncertainty whose linear dimensions cannot be smaller than, say, the Compton wave-length λ_p. With this, naturally, a limit is also given for the linear dimensions of the space-time region itself.

It is certainly questionable to what extent such an analogy can be used for guidance in the case of gravitational theory. As long as an actually completed quantum theory of the gravitational field does not exist, the physical meaning of the critical quantities M_0, L_0, and thus also of conditions (9) and (10) will remain uncertain. Yet one can point out that a test body of mass M_0, should it be constructed of baryons, would assume a volume whose linear dimensions would be of an order of magnitude not less than 10^6 times the range of the so-called strong interactions, i.e. approximately 10^{-7} cm. This circumstance would allow the conjecture that quantization of the gravitational field could have no testable consequence in regions of atomic dimensions; on the other hand, since such a quantization can obviously play no important role in the macroscopic or even cosmological domain, one would then be inclined to conclude that the gravitational field possesses no quantum-like properties whatever and is subject to purely classical laws. As already emphasized, such an inference is, however, in no way imperative; the above considerations can only serve to indicate circumstances that in principle may complicate the interpretation of a quantized gravitational theory.

As concerns the further question of how a classical gravitational theory might then be united with a quantum theory of the remaining fields which interact with the gravitational field, electrodynamics again provides an example of such a union, namely the limiting case of classical electromagnetic fields interacting with atomic systems. The operators of current and charge density naturally do not occur as sources of the potentials determining such fields; but rather the expectation values of these operators in the state characterizing the source distribution. Further, in order not to come into contradiction with the consequences of the non-commutativity of the variables which describe the sources, one must also agree that only fields considerably stronger than the critical field strengths, given by the mean fields of light quanta, exert an observable effect on currents and charges. In electrodynamics, to be sure, the latter convention appears only as the neglect of real quantum effects of the radiation field. It is conceivable, however, that a similar restriction of the field effects on masses in gravitational theory could have a fundamental character. The critical field strengths in question could be, say, those given by the above equation (7) which, according to equation (10), can also be expressed as GM_0/L^2; however, we lack any clue for indulging in speculations about the possible form of such a restriction on the field concept. The above considerations merely had the purpose of showing that no logical compulsion exists for quantizing the gravitational field; rather the question of whether and how such a quantization is to be carried out can only be decided empirically.

Appendix

In the discussion, Professor Hönl produced an instructive example that points to an apparent contradiction between the predictions of classical gravitational theory and those of quantum theory. Let us consider the passage of a particle through a lattice distribution of particles, which act on the falling particle through their gravitational field. According to quantum theory one obtains, for the deflection of the particle, a probability prediction, represented by the diffraction pattern of the de Broglie wave, which depends on the mass of the particle. According to classical gravitational theory, however, the deflection of the particle is independent of its mass. In order to accentuate the apparent paradox more sharply, we wish to modify the problem somewhat. We can simply replace the grating by a screen with two holes: then the well-known diffraction pattern will occur on a second screen, consisting of bands at a distance $l\lambda/d$ from each other; where l denotes the distance between the screens, d the distance between the holes and λ the de Broglie wave-length of the particle. Now one could place a mass close to one of the holes which would give to the particle, if it passes through this hole, an additional deflection that would indeed noticeably shift the point where the particle strikes the intercepting screen; however, by less than the distance between the bands. One could thus think he had at hand a means of determining which of the two holes the particle had passed through, in direct contradiction to the fact that the existence of bands of possible places where the particles strike the intercepting screen is essentially dependent upon the impossibility of such a determination, since, the diffraction pattern depends on the presence and mutual position of both holes.

Closer consideration shows that the stratagem mentioned cannot have the intended result. The displacement δH on the intercepting screen caused by the additional deflection may be easily estimated. Let p be the momentum of the particle perpendicular to the first screen, δp the change in momentum parallel to this screen which the particle undergoes due to the gravitational effect of the mass introduced. One then obviously has

$$\delta H = l\delta p/p = l\lambda\delta p/h$$

if one also uses the relation $p = h/\lambda$ between the momentum and the de Broglie wave-length of the particle. The requirement that δH should be smaller than the band interval $l\lambda/d$ thus means that the change in momentum of the deflecting mass, which except for sign also equals δp, should certainly be smaller than h/d. But this is impossible to attain if, simultaneously, it must be determined with certainty that the deflecting mass is situated near the first and not by the second hole. This last requirement means, namely, that the location of the mass in the direction parallel to the screen must be known with at most an uncertainty $\Delta x < d$. But then its momentum in this direction can only be determined with an uncertainty $\Delta p > h/\Delta x > h/d$. Placing the mass next to a hole will thus necessarily result in a complete blurring of the diffraction pattern, and a contradiction can never arise between the consistent application of quantum theory on one hand and of classical gravitational theory on the other.

NOTE

* A talk at the Einstein Symposium of November 2–5, 1965, in Berlin.

CHAPTER 16

QUESTIONS OF METHOD IN THE CONSISTENCY PROBLEM OF QUANTUM MECHANICS

[1968f]

ABSTRACT. Attempts to set up on an axiomatic basis an abstract 'theory of measurement' in quantum mechanics lead to imaginary difficulties and suggest speculations about non-physical effects which have no objective justification.

The immediate occasion for the following remarks is the publication of two papers (refs. [1, 2]) in which the well-understood analysis of the foundations of quantum mechanics[1] is again questioned, and speculations put forward, to which the conclusions of this analysis would give no support. Such doubts are seen to have their origin in a radical difference of conception (going back to von Neumann) concerning the method of approach to epistemological problems, and can therefore only be dispelled by a critical re-examination of these problems from the methodical point of view; more specifically, we shall be concerned with the problems of measurement connected with the observation of atomic phenomena.

The analysis of measurements performed upon atomic systems is not carried out for its own sake. Its only purpose is to illustrate the consistency of the description of atomic phenomena given by quantum mechanics. Since the quantities associated with the physical concepts used in this description are determined by measurements, the obvious way to understand the physical meaning of the formal relations between the operators representing such quantities is to translate them in terms of concrete physical operations. There are two ways in which this straightforward argument can be spoiled: (i) by attempting to "formalize" it [1], (ii) by introducing into the argument the "consciousness of the observer" [2].

It is clear that one cannot formalize the crucial point of the consistency argument, namely the part of it in which the account of the phenomenon under discussion is ultimately referred to those classical concepts that are not further analysed, but taken as expressions of immediate experience to form the basis of our means of communication of unambiguous information. In particular, any 'axiomatization' of the theory can at best help to avoid trivial contradictions or redundancies in its formal apparatus, but is incapable of throwing any light on the adequacy of the theory as a mode of description of experience. This last problem belongs to what the scholastics pointedly called 'real logic', i.e. logic of things, in contradistinction to 'formal logic'.

On the other hand, it is perfectly legitimate, in the analysis of some measuring process, aiming at a consistency test of the theory, to make use of the formalism of the theory and to apply it both to the atomic system and to the bodies constituting the experimental apparatus; the latter are also composed of atoms and interact with the atomic system under investigation by the usual physical forces according to the laws of quantum mechanics. This was done, for instance, in the discussion of the measurability of electromagnetic fields [5], and more recently, in an examination of the statistical aspects of the interaction between the atomic system and the macroscopic measuring apparatus [6]. All that such detailed studies intend to accomplish is to bring out more clearly the essential features of the complicated physical processes on which the consistency argument is based, but they do not pretend to be a substitute for the argument itself, and do not impair its logical simplicity. It is not superfluous to insist on this point, since failure to realize it has induced people to evoke the picturesque vision of an infinite sequence of 'observers', each of whom observes what the preceding one is doing; a prospect that puts one in the uncomfortable state of mind of the licentiate in the 'Tale of a Danish student' [7].

Kept within proper bounds, axiomatization of the formalism would not make any difference to this situation. When, however, exaggerated claims are made about its powers, disastrous results follow. The abstract concepts with which our theories operate are idealizations of actual physical situations, and their application is accordingly subject to limitations, which cannot be ignored with impunity. Thus, it is evident that any definition of a measuring instrument, constituted by an assembly of many atoms, must allow for the possible occurrence of those improbable atomic configurations which do not conform to the ideal behaviour characterized by the laws of thermodynamics. If a set of axioms purporting to give a formal characterization of measuring instruments turns out to make such exceptional behaviour absolutely impossible, the simple-minded conclusion would seem to be that this particular axiomatization does not adequately fulfil its purpose. Its promoters [1], however, draw a quite different conclusion: they maintain that the exceptional behaviour must be eliminated, even though they are only able to suppress it (in the Carrollian sense of the word) by a principle which is 'outside and contrary to the present doctrine of quantum mechanics'; they thus get involved in a tangle which is entirely of their own making. Indeed, by taking the improbability of the exceptional situations properly into account with the help of the known methods of statistical physics, it is readily shown [6] that the measuring process, far from presenting any difficulty, can be completely described within the framework of quantum mechanics.

The authors of the pilloried paper [1] do not feel the weight of this last point because they believe that a measurement does not necessarily involve the interaction of the atomic object with a macroscopic body. Here they are again misled by disregard of the requirements of real logic. They want to build up an abstract 'theory of measurement', which would presumably encompass such non-classical quantities as spin (which they take as an example in their argumentation). Such a theory, if it existed, would be of no use for the real epistemological problems of quantum mechanics. These problems are only concerned (let us repeat it) with the application of certain sets of classical concepts to the description of atomic systems, and these classical concepts are *defined*, in their quantitative aspect, by macroscopic apparatus interacting with the atomic systems and recording the result of the interaction. As to non-classical quantities (like the quantum of action, the spin, the direction of an angular momentum, etc.), it would be inconsistent to require them to be measurable; in fact, it can be checked that they are not measurable, but only indirectly related to measurable quantities.

Now, as a concrete example of 'measurement' in which the atomic system is allegedly 'left undisturbed' by the apparatus, the authors quote a so-called 'negative result measurement' discussed by Renninger [7]. This example is particularly unfortunate, since the fallacy of Renninger's considerations is immediately obvious; in fact, it is a relapse, with a slight variation in form, into the Einstein-Rosen-Podolsky fallacy. Like the latter, it is a consequence of not devoting sufficient care to the examination of the actual physical conditions required to make the proposed procedure workable. Let us send photons onto a screen pierced with two holes, and absorb them in another screen placed behind the first. The distribution of the spots they leave on this second screen is radically changed if we somehow ascertain that the photons have passed through one of the two holes. A familiar way to achieve this change is to stop the other hole. A more gentle method, according to Renninger, would be to work out at what time after its emission by the source a photon would reach one of the holes; if at that time a counter placed at this hole does *not* register the passage of the photon, we may conclude that it has passed through the other hole, and our expectation of the further fate of this photon is thus changed by an observation which (so it is asserted) has not affected its state. The argument, however, should not be allowed to end at this verbal stage; for as soon as one asks how an apparatus accomplishing such a feat would look, one realises that it would have to include a clockwork timing the instant of emission of the photon and correlating it with the instant of observation, with the result that an interaction of the usual type necessarily takes place between the photon and this clockwork.

The preceding indications, however fragmentary, should suffice to show that the difficulties under which the adepts of von Neumann's method appear to labour are only consequences of this particular method, and do not occur if the problem is properly treated. Above all, there is no substance in the doubt they cast upon the well-established fact that the epistemological analysis of the foundations of quantum mechanics can be carried out on a purely physical basis, without any appeal to non-physical interventions of any kind. Measuring apparatus, in particular, are completely automatic registering devices, so contrived as to provide the kind of information needed for the objective description of the phenomena. Of course, the conditions of objective description themselves are determined by the activity of human 'observers'; but the essential point is that for the purpose of the analysis in question it is sufficient to express these conditions in physical terms, namely as the requirement to base the description on classical concepts, quantitatively defined by macroscopic measuring apparatus. The physiological and mental processes leading to the formation of these concepts constitute an independent object of study, equally susceptible to the rigorous method of scientific investigation, resting on experiment and rational analysis. It is conceivable (at least for people inclined to mysticism) that the epistemological problem of quantum mechanics could not have been solved without recourse to an influence of the consciousness of the observer on the physical phenomena; just as it was conceivable for Kepler that the positions of the planets in the heavens could influence human affairs. But nowadays there is as little justification for the former speculation as for the latter.

NOTE

[1] See in particular two recent review essays by R. Peierls [3], in which the true situation in quantum mechanics is admirably summarized, and which also contain pertinent remarks about general aspects of scientific method. In this last respect, the terse and lucid essays of M. Born [4] give inspiring examples of the sound attitude of the physicist towards philosophical issues.

BIBLIOGRAPHY

[1] Jauch, J. M., Wigner, E. P., and Yanase, M. M., *Nuovo Cim.* **48B** (1967), 144.
[2] Wigner, E. P., 'Remarks on the Mind-Body Question', in *The Scientist Speculates*, I. J. Good, ed. (Heinemann, London and Basic Books, New York, 1961) p. 284.
[3] Peierls, R., *History of Science* **4** (1965), 124; *Scientific American* **216** (1967), 137.
[4] Born, M., *Physik in Wandel meiner Zeit*, Vieweg, Braunschweig, 1957; 'Symbol and Reality', in *Natural Philosophy of Cause and Chance* (Oxford University Press, Oxford, 1949; Dover reprint, New York, 1964).
[5] Bohr, N. and Rosenfeld, L., *Mat. fys. Medd. Dan. Vid. Selsk.* **12** (1933), No. 8. [English translation: this volume, p. 357].
[6] Daneri, A., Loinger, A., and Prosperi, G. M., *Nuclear Physics* **33** (1962), 297; *Nuovo Cim.* **44B** (1966), 119; Rosenfeld, L., *Suppl. Progr. Theor. Phys.*, extra number (1965), 222. [See p. 536 of this volume].
[7] Cf. Rosenfeld, L., *Physics Today* **16** (1963), 47 (English translation of relevant passage p. 48). [See p. 522 of this volume].
[8] Renninger, M., *Z. Phys.* **158** (1960), 417.

THE METHOD OF PHYSICS

[1968g]

NOTICE
(with apology to Mark Twain)

Persons attempting to find materialism in this report will be prosecuted; persons attempting to find idealism in it will be banished; persons attempting to find positivism in it will be shot.

BY ORDER OF THE AUTHOR

(Cf. the preamble to *The Adventures of Huckleberry Finn*)

1. Introduction

Scientific activity is distinguished from those other creative efforts we call art by its tremendous efficiency: it aims at controlling the forces of nature and it achieves this control with steadily increasing assurance both in width and in depth. The feeling of certainty we associate with scientific knowledge does not arise only from its successful application to the satisfaction of practical needs, but finds a more direct justification in its internal coherence and power of prediction; ultimately, of course, these two sources of scientific certainty boil down to the unsophisticated statement: "we believe it is true because it works". In any case, by providing us with such deep insight into the workings of nature, science has asserted itself as a fundamental – if not the most fundamental-element of human culture. In its very essence universal, the method of rational analysis of the natural phenomena which characterizes the scientific approach must affect all forms of culture developed by human communities.

What has just been said of science as a whole applies of course to physics, that part of science which excludes from consideration, for purely practical reasons, the manifestations of organic life. This restriction of its scope considerably simplifies the task of physics, and has made possible a corresponding refinement in the elaboration of its methodical tools. Thus, in the course of history, physics has always served as a model in this respect to the other sciences, and there is at present no evidence that the extension of the method of physics to other domains would meet any essential limitation. The odds are, therefore, that a discussion of the method of physical investigation may exhibit at least the basic features of the universal method of science. However this may be, the study of the physical phenomena on the atomic scale, by confronting the physicists with new conditions of observation and interpretation, has necessitated a deepgoing revision of the underlying methodological and epistemological problems, which alone would make a re-examination of these problems of general interest – not least to the physicists themselves.

How acute this need is for the younger generation of physicists was recently called to my attention by a talented research student, who solicited an interview in order to discuss some difficulties he had concerning the epistemological aspects of atomic physics: "I am spending most of my time", he explained, "in the computer room carrying out complicated calculations of some detail of structure of a small family of atomic nuclei. I am doing this in order to obtain results whose publication may secure me a position (excuse me to be so frank), but I also wish to devote some little time to thinking about physics". This reminded me of an anecdote told by the renowned accompanist Gerald Moore [1]: it is about a double-bass player who on his day off decided to hear the performance of the opera in which he was daily playing his part in the orchestra; what was his surprise to discover that while he was just producing a monotonous sequence of two or three notes "marvellous tunes were going on on the stage"! Physicists are indeed threatened by a similar disharmony between the everyday drudgery imposed by the increasing complexity of analytic investigation and the occasional enjoyment of critical and synthetic reflexions. All is well with the bright youngsters who are aware of the danger and take steps to overcome it; but how many risk remaining double-bass players all their life?

However serious their consequences for the form and organization of physical investigation, the increased complication and specialization of experimental and theoretical techniques are not the most significant methodical changes brought about by the modern development of atomic physics (we thus denote not only the study of atoms and molecules, but also that of atomic nuclei and of the still deeper structural properties of matter commonly called "elementary particles"). It is by the novel epistemological problems it has presented us that this development has profoundly affected our whole view of the meaning of physical science. The novelty arises from the disparity in scale between the atomic processes we want to study singly and our means of observation, which are necessarily systems of material bodies composed of large numbers of atoms. This means that fluctuations in the course of the interaction between the observed system and the apparatus, which would be negligible if the object of observation were itself a large system, may essentially affect the conclusions we can draw from the observation of a single atomic system. The full implications of this situation cannot be properly appreciated without first discussing the physics of large systems, since its knowledge is required for the analysis of the process of observation itself. Indeed, the whole body of concepts and relations by which we describe the behaviour of matter in bulk and electromagnetic fields of usual intensities, and which we denote as 'classical' or 'macroscopic' physics, must be established in its own right, without any appeal to atomistic features; for, according to what has just been said, it must also form the basis (with appropriate modifications) of our account of atomic phenomena.

2. Method of classical physics

The traditional characterization of physics as an inductive science – indeed, as the inductive science *par excellence* – remains as adequate as ever so far as it goes, but it does not go very far. No one will quarrel with the statement that observation of the phenomena, through the channel of sensory perception, is the primary source of our knowledge, and that we derive from these data of observation, by the logical process of induction, those regular features of the phenomena we call the "laws" of nature; but the crux of the matter is the detailed understanding of the relation between these two activities, for neither of them is normally exercised without a substantial intervention of the other, so that the sharp distinction made between them in the traditional view is at least very artificial. Clarity in this respect can only be expected from a more searching analysis, going back to the physiological basis of perception and thinking. Locke's pioneering work and its continuation by Hume and Condillac were steps in the right direction, and played a constructive part – especially the last – in the first elaboration of the method of physics and chemistry in the XVIIIth century; but these philosophers were hampered by the rudimentary state of physiology at that time, and their speculations, for all their sagacity, could not be free from arbitrariness. In fact, the problems of epistemology are themselves of scientific, not philosophic, character and can only be solved by the very method their solution purports to elucidate. Thus, the decisive progress made in the XIXth century was exclusively due to scientists, familiar with the actual conditions of experimental research and theoretical thinking both in physics and in physiology.

The chief merit of Helmholtz was to clear the ground of two erroneous beliefs, fatal to the philosophical systems in which they are entertained: the first is that our sensations would somehow be 'faithful copies' of the things perceived, the other that our mind would be endowed with '*a priori* synthetic ideas'. He pointed out that our sensations are by no means directly related to the physical phenomena that normally produce them, but are mere symbols of qualities – colours, sounds, etc. – associated with these phenomena. He further demonstrated that it is possible to construct our most fundamental abstract concepts – space and time, number and magnitude – entirely on the basis of definite correlations of sensations, without recourse to any *a priori* element. By thus establishing the fundamental identity of sensations and ideas, he achieved a considerable unification of our view of the mental processes underlying the method of investigation of the physical phenomena. To the scope of this method Fechner, whose great work seldom receives the full appreciation it deserves, added the whole domain of the phenomena of perception themselves, by showing that sensations have – within limits – the character of magnitudes and that their relations to the physical stimuli are therefore susceptible to quantitative representation.

On the broad basis of the analysis of sensations thus raised to the scientific status, Mach carried out a penetrating study of the processes of theoretical reasoning leading to the formulation of physical laws; in particular, he critically examined the true nature of those general 'principles' sometimes set up as 'self-evident axioms', and clearly brought out their dependence on unanalysed preconceptions originating from experience, and accordingly in the last instance from sensations. Mental processes unsupported by sensory perception were thus banished from the whole realm of rational thinking, and the validity of the method of physics in all its aspects ultimately referred to sensory experience. The general conclusions we have here recalled were either drawn from detailed experimental results or arrived at by careful logical argumentation, and they have proved to be of lasting value. All that subsequent investigation has led to is to outline more sharply the limitations of their validity, without impairing their fundamental significance.

Yet the whole approach of Helmholtz and Mach still suffers from an inherent weakness, in as much as they only dealt with the ultimate stage of rational thinking reached by modern science; even in his historical studies, Mach did not go further back than the XVIth century (or more accurately than Archimedes, who was taken as a model by the pioneers of modern science). They were accordingly confronted with all the intricacy of the fully developed connections between sensations and ideas that constitute our science, and in trying to reconstruct the processes by which these connections might have been gradually established, they could only present a plausible picture, which would suffice logically to prove their case, but could not claim to correspond to the actual development of the human mind. It is to this last problem that a fundamental contribution has been made in the present century by Piaget and his school. From extensive experimental studies on children of various ages, Piaget derived a complete account of the successive elaboration of our mental representation of the world, as well as of the concomitant evolution of affective life and the formation of personality. The 'genetic epistemology' resulting from Piaget's work not only confirms the main points of Helmholtz and Mach, but forms a very necessary complement to their work: one may say that it clinches their argument by removing from it any remaining arbitrariness.

The genetic approach is indeed the most appropriate to exhibit the essential features of the method of physics, as they step by step emerge from the natural mental development of the individual members of human communities. It may seem far-fetched to start from the new-born infant's first reactions; but it is the simplest way to demonstrate the complete continuity of mental evolution from the earliest reflexes to the last refinements of scientific thinking. Continuity, of course, does not mean smooth progress; in fact, Piaget has been led to distinguish in this evolution a number of successive stages, each of which represents a new effort of the subject to adapt himself to a wider and wider environment: each time, the subject passes from an attitude of extreme egocentrism to the recognition of new aspects of the external world, until he reaches a temporary equilibrium between the two opposite tendencies. Also the fully-fledged physicist has to go through such a dialectic process if he wants to arrive at a balanced judgment of his activity.

Before the acquisition of language, during the two first years of its life, the child accomplishes, entirely on the sensory-motoric level, a tremendous feat of adaptation, at the end of which he has acquired the mode of coordination of his sensations and motions that is going to form the framework for the whole representation of the world he will successively construct: he has consolidated sensory-motoric schemes – much in the way imagined by Helmholtz – corresponding to the basic concepts expressing relations of space and time, and he has realized the limits of his own body and the necessity of distinguishing sensations associated to it from those related to a world external to it. Initially, the child's endowment consists of a few built-in reflex arcs, which he learns by trial and error to combine into more and more complicated schemes; in these schemes he incorporates those sensations which he experiences as significant, as being signs useful for the orientation of his behaviour.

This symbolic character of sensations is abundantly evidenced by the mechanism of conditioned reflexes and the typical modes of animal behaviour analysed by Lorenz, Tinbergen and others. A striking example is the utilization by some species of insects of the polarization of the light from the sky as a substitute for the image of the sun for purposes of spatial orientation; precisely the same functional equivalence of two widely heterogeneous physical phenomena has been discovered by the Norse navigators in the Viking period – with the difference, typical for the human species, that the determination of the polarization was made with the help of a tool, the 'sun stone', and not by a special built-in organ. Without recurring to such complex forms of behaviour, however, one may invoke, as Helmholtz did, Johannes Müller's law of the 'specific energy' of the sensory nerves, according to which the specific excitation transmitted to the brain by a sensory nerve is independent of the physical nature of the stimulus. It is clear that this property rules out any close correspondence between the physical phenomena and the sensations to which they give rise; there is of course some correspondence between the phenomena and the sense organs at the physical level, but it is usually an indirect one: thus the sensations of colour are not directly associated with the frequencies of the light waves, but depend on these frequencies only indirectly, through the absorption properties of the pigments of the retina. We must therefore regard the sensory schemes as a kind of code system, whose function is to signal, and thereby to signify, the physical phenomena.

Language in its primitive form is another code system, transposing, so to speak, the individual sensory-motoric schemes onto the social plane: primarily it serves as a vehicle for communication of experience from individual to individual. This function raises the problem of objectivity: sensations perceived by different subjects cannot be compared in isolation, but they can be brought together in certain associations that are recognized by all subjects; these are the sensory-motoric schemes that can be communicated and thus ascribed objective significance – there is no other meaningful way of defining the objectivity of our description of experience. In the words which correspond to the sensory-motoric schemes no distinction is made between the varying degrees of abstraction of these schemes, i.e. the varying amounts of sensations of different qualities entering into them: this is to be expected in view of the fundamental uniformity of formation of the schemes in question, whether they refer to what we call "concrete objects" or "abstract ideas" expressing, for instance, relations of spatial orientation or temporal succession.

This form of language is learnt by the child in the second stage of his development, between the ages of two and seven. It is at this stage that the links between the words and the sensory-motoric schemes are consolidated: the words are incorporated into these schemes as additional signs belonging to them. In his thinking, which is extremely active, the child does not dissociate the words from the schemes, but reasons directly in terms of the schemes themselves, by extending their scope or combining them in imagination. This type of reasoning, which Piaget calls intuitive, is dominated by finalistic motivations and proceeds by analogies. When the aim is to devise an actual operation, the success or failure of the latter provides some control of such analogies; but this is not the case when the analogy is intended to account for some natural phenomenon: it then mostly takes an animistic form. This is a pre-logical stage of human thinking, traces of which are noticeable in the cosmological systems of the earliest pre-socratic philosophers: to become logical, it lacks an essential element, the inversion of operations.

This is the main progress realized by the child at the beginning of the next stage of his growth, at the age of seven or eight. Since every sensory-motoric scheme represents an operation leading irreversibly from an initial to a definite final state, the two operations we call inverse of each other appear at first as radically different schemes, and it takes the child a considerable effort of abstraction to realize that there is between two such operations a relation of quite general character, independent of the concrete form of these operations. This momentous step enables the child to compare objects or situations of the most various kinds with the help of operations connecting them in both directions. The composition, or successive performance, of operations of the same kind and their inversion lead him to a 'grouping' of these operations which, while not quite attaining the logical simplicity of the mathematical concept of group, is a sort of prefiguration of it.

By using such groupings of operations he rapidly masters the problems of ordering objects with respect to different qualities, and of collecting them into classes, and in particular he arrives at the notion of conservation of definite properties of bodies – substance, weight, volume, in this order of difficulty – when these bodies are transformed by certain operations. At the end of this period, at the age of twelve, he has acquired familiarity – albeit in incomplete and unsystematic fashion – not only with the main features of the logic of relations and classes, but also with the basic concepts of arithmetics, geometry and kinematics. It is of paramount importance to realize that all these 'cognitive schemes' have been built up at the intuitive level, by actual operations governed by sensory-motoric schemes and with little help from the use of language. We are here witnessing the formation of those conceptual frames that bear the whole edifice of physics, and we see that they have their root in sensation and action.

There is a far cry, however, from the twelve-year old's representation of the world to that of modern science; it rather resembles the Aristotelian outlook. The missing element – which was being discovered by the geometers in Aristotle's time – is the accuracy of analysis and expression obtained by rational thinking. This latest phase in man's mental development appears, like the preceding ones, as the natural unfolding of sensory-motoric activity; this activity is now concentrated on exploiting to the utmost the resources of language, now recognized as a symbolic representation of the sensory-motoric schemes and of the newly discovered relations between them, as a code system of incomparably greater flexibility and capacity than that constituted by the sensations themselves. At the age of twelve, the child is becoming aware of the advantage of dissociating the words from the sensory-motoric schemes to which they belong and using them as representatives for these schemes; the emergence of cognitive schemes, requiring new words for their expression, accelerates the process of transition from intuitive to verbal thinking.

With this new technique, the child is now able to grasp and express logical relations in their full generality and rigour, and within a few years the adolescent has enriched his vocabulary with a large quantity of words representing concepts called "abstract" in the sense that they are formally defined in terms of other concepts and thus only indirectly referred to sensory-motoric schemes. He has learned from the adults how to handle this marvellous instrument which can give his thought such sharp precision and such security of deduction; all too often, he has also learned from them how to misuse it in empty speculations. The right use of abstract rational thinking demands indeed an awareness of its pitfalls and limitations, which only painstaking analysis can reveal. We may here take our guidance from Mach's discussion of the formation of the concepts and theories of physics, which are the most refined products of this process of abstraction and logical construction.

The economy of thought, in which Mach rightly sees the *raison d'être* of such a process, can only be achieved by a voluntary schematization of the description, retaining from the wealth of sensory experience only those few features which are deemed essential for unique specification. The selection of such elements is, like all activities of living beings at any level of organization, a matter of trial and error. It is guided, of course, by the requirement of objectivity, which we found operating already at the intuitive level of thinking, where it gives prominence to coordinations of sensations persisting under changing conditions of observation. The process of induction and the conception of natural law can be traced back to this intuitive level; in particular, the notion of causality arises spontaneously from the occurrence of invariable sequences of sensations under similar circumstances. At the abstract level, this selection of objective features of the observed phenomena becomes explicit; the conditions of observation are symbolized by groups of operations, and objective significance is ascribed to the 'invariants' of such groups; these invariants may represent permanent properties of physical systems in stable states or permanent relationships between such properties in systems undergoing change. Induction appears, from this point of view, as a search for invariants of a given group of operations, and the formulation of a law of nature as the description of such a group of operations and its invariants: the principle of special relativity, for example, enunciates the invariance of all phenomena with respect to the group of uniform rectilinear motions of the observers.

An extreme degree of abstraction is reached with the concept of magnitude, which allows us to attach a number to a physical property and opens the way to the most precise mode of description we have achieved, the language of mathematics. The character of magnitude can be ascribed to sensory schemes when it is possible to establish between them, without changing their quality, relations of equivalence and additivity (the latter amounting to possibility of increase and decrease). The mathematical formulation of physical relations expressing what is called the "quantitative" aspect of the phenomena, ensures the highest degree of unambiguity of information, but it only achieves this at the cost of neglecting qualitative aspects which are just as important for a complete account of experience. By placing excessive emphasis (as sometimes happens) on mathematical theory, one runs the risk of losing contact with the sensory-motoric schemes whose objective description is the ultimate purpose of the theory. It may be helpful to visualize the situation by recalling the two code systems with which our representation of the world operates, one of them formed by the sensations and their coordination into sensory-motoric schemes, the other by the elements of language – the words of everyday language as well as the sophisticated terminology of mathematics and physics. The latter code system is entirely conventional and only acquires meaning by being referred to the former. No physical description is therefore meaningful if it does not contain the necessary linkage to sensory experience.

The ease with which abstract concepts can be handled – and which gives abstract theories such power as instruments of analysis and prediction of physical phenomena – encourages the illusion that sheer logical necessity could impose a definite form upon natural laws, that the whole content of a physical theory could be extracted by way of purely logical consequence from some general principle. Mach's sharpsighted examination of many principles of this kind proposed in the course of history has exposed the fallacy of this tenacious belief; he has shown how in all such cases unanalysed sensory experience has unwittingly guided abstract thinking in the formulation of the principle. He has especially stressed the role of *Gedankenexperimente*, 'imagined experiments', in this respect: the principle in question may be suggested by the consideration of fictitious processes, the outcome of which has to be imagined; but in order to achieve the desired result, imagination must appeal – consciously or unconsciously – to actual sensory experience. There is in fact no essential difference between imagined and actual experiments in the part they play in establishing the indispensable link between abstract theory and sensory experience. Rightly understood, imagined experiments – i.e. appeal to previous experience stored at the intuitive level – actually afford a most effective guidance in the edification of physical theories.

The assignment of the character of magnitude to a physical property, in particular, is by no means a matter of arbitrary definition: the fulfilment of the required conditions by the property in question must be ascertained by observation. Establishing the correspondence between a physical magnitude and its abstract representation by a number, its 'measurement', is again a well-defined procedure, the modality of which is determined by some natural law fulfilled by the property considered: either we establish a direct comparison with a given standard of reference, or we bring the magnitude envisaged in relation with some other magnitude which we already know how to measure. For instance, the measurement of length depends on the possibility of constructing solid rods whose state is sufficiently invariable, or sources emitting light of sufficiently constant wavelength; the measurement of an electric current by a galvanometer depends on Ampère's law of electrodynamics. Sensations of a given kind satisfy to some extent the criteria of magnitude; as Fechner showed, their indirect measurement can be based on Weber's law, which states that when a sensation is excited by a stimulus of given intensity, the change in this intensity required to produce the least perceptible change in the sensation is proportional to the total intensity of the applied stimulus; accordingly – within the limits of validity of this law – sensations can be graduated by the fractional variations of intensity of the corresponding stimuli.

These examples clearly illustrate a quite general feature of the quantitative determinations of magnitude, namely that they necessarily involve an idealization of the measuring procedure: in order to give these determinations the sharpness they must have to be amenable to mathematical treatment, we must neglect, within assigned limits, the unavoidable fluctuations in the structure of the measuring apparatus and the performance of the measuring operations. The estimate of the accuracy of a definite measuring process is usually quite a complicated problem, necessitating a precise knowledge of the physical laws on which this process depends. Consequently, no definition of a physical magnitude can be self-contained in the sense that it could be immediately referred to the sensory scheme it represents: it always implies the indirect intervention of a considerable body of theory about entirely different phenomena. The same can be said, *a fortiori*, of any experiment purporting to establish a quantitative relation between physical magnitudes: to believe that the law expressing such a relation is founded only on direct sensory experience, is just as illusory as to believe that physical principles could be founded on purely logical necessity. Empirical positivism and logical formalism are the Scylla and Charybdis between which we have to steer a cautious course.

If we keep in mind the symbolic character of both sensory schemes and abstract concepts which constitute our two correlated code systems, we should have no difficulty in understanding the actual situation that confronts us. Sensory-motoric schemes and concepts both appear as idealizations; to say that the former are more 'concrete' than the latter, in the sense that they are more directly related to the phenomena, is not relevant for our purpose: the important point is that both kinds of idealizations are indispensable because they mutually support each other. There is no circularity in this mutual relationship, because neither sensations nor concepts are formed at once in a single process of perception or abstraction, but on the contrary are the outcome of a multiplicity of such processes, leading by successive approximation to sensory-motoric schemes of increasing delicacy and to more and more refined determinations of the concepts. At each stage of approximation, the mutual relationship between sensory experience and theoretical thinking is perfectly well-defined on the basis of the approximation already reached.

A direct consequence of this state of affairs is that all our idealizations can only have limited validity. On the one hand, they must be sharply defined in order to ensure that our 'coding' of the phenomena should convey unambiguous information; but, as we have seen, this sharpness can only be obtained by ignoring minor variations in the phenomena, which *eo ipso* are excluded from the scope of application of the idealization in question. A physical theory is therefore nor merely a set of concepts and of laws expressing definite relations between these concepts; it is not complete without a specification of its domain of validity, which (strictly speaking) is only derived with precision from the next stage of approximation. As soon, however, as a clear criterion of applicability is formulated, a theory that has stood the test of experience acquires, within the domain thus circumscribed, perennial validity. Thus, Newtonian mechanics remains for all time the adequate mode of description of the slow motions of material bodies; the criterion of slowness is supplied by the more accurate theory of relativity, the scope of which includes motions of any velocity.

Once one has understood the origin of the inherent limitations of all abstract concepts, one realizes that these limitations do not imply any arbitrariness in their definition, but are dictated by the practical use to be made of them. The suggestion that the choice of the basic concepts of a physical theory would be a mere matter of convention disregards the decisive motivation of this choice: a theory is only deemed satisfactory when the concepts it uses are fitted as closely as possible to the aspects of sensory experience which the theory describes. This, as we have seen, is normally achieved by a process of gradual adjustment leaving no room for arbitrary steps.

Yet, the formal constructions of logical thinking, especially in its mathematical form, have such a power of proliferation that among the innumerable products of this autonomous activity, it has repeatedly happened that the physicist found ready-made those providing him with the adequate mode of expression for quantitative laws he had discovered by induction from experiment. Could such coincidences now indicate that our physical conceptions would be, as Einstein in his later years expressed it, "free creations of the mind" that in some "miraculous" way would reflect the true laws of nature? There can be no greater delusion than such a belief. In the rare cases presenting this startling harmony between abstract mathematics and natural law, one is dealing with some theory of very general scope – like Einstein's own theory of gravitation – and consequently with the invariants of some group presenting a high degree of abstraction: it is anything but a miracle that such groups should find application to more than one aspect of the phenomena, and in the first place to the geometrical and kinematical aspects studied by the mathematicians; thus, the group of the theory of gravitation is the extension to four dimensions of that governing the geometry of surfaces. The true content of Einstein's theory, however, is not the requirement of general invariance with respect to space-time description, but the 'equivalence principle' leading to the identification of metric and gravitation; and this is a genuine physical statement, based directly on the experimental fact that all bodies fall with the same acceleration.

The study of the genetic development of our present method of investigation of the physical phenomena should guard us from the temptation to ascribe to rational thinking a singular position: we have seen it emerging as a natural outcome of the unfolding of potentialities entirely contained in our sensory-motoric organization. It arises – both in the mental development of the individual and in the cultural evolution of human communities-primarily in response to social pressure; on the one hand, the need to make more efficient use of the forces of nature, on the other – of more immediate interest for our discussion – the need of more accurate description and communication of experience. The determining influence of social needs on this process is strikingly illustrated by the fact that no human community had proceeded much beyond the intuitive level of thinking until the rise of industrial civilization provided the incentive and the means for the rapid progress of the rational outlook characteristic of modern science.

In every respect, the inseparable union of accurate experimentation and logical analysis which constitutes the method of physics is "un outil de merveilleux service", all the more wonderful when one compares it to its origin; but wonder is indeed the usual impression we get when we reflect upon any biological process. Most remarkable is perhaps the fact that this method has enabled us to build up a representation of the universe free from the limitations imposed upon our sensory experience by the structure of our sense organs; to give only one example, while our sensory perception of light is limited to about one octave of frequencies, we have at our disposal a whole arsenal of apparatus allowing us to make use of the complete spectrum of electromagnetic radiations for the exploration of phenomena ranging from the atomic to the cosmological scale. This confers upon scientific knowledge a degree of objectivity transcending human communication. Recently Prof. Yang raised the question whether inhabitants of some more favoured planet could not have achieved an understanding of the laws of nature so much deeper than ours that it would be as inaccessible to us as our science is to small children. On the strength of the preceding considerations, I am confident that we would be able to converse with those intelligent beings on equal terms, and to learn from them whatever answers to our unsolved problems their higher organization might have enabled them to find before us.

3. Method of Atomic Physics

The conception of the atomic constitution of matter is much older than the actual identification of individual atomic processes. It is suggested by mere qualitative observation of numerous phenomena evidencing the divisibility of matter and the transmutations it undergoes in chemical reactions, although such indications are by no means sufficient for a rigorous process of induction. The strongest motivation, however, for the formulation of the atomic hypothesis was the endeavour to reduce as much as possible the multiplicity of qualities offered by the phenomena: the programme of atomic theory was to explain all qualitative appearances in terms of properties of the individual atoms, which one could hope to limit to a few characteristic magnitudes. This is a very sophisticated idea, which is not found, in the child's mental development, at the intuitive level of thinking. At that level, observation of a solution of sugar in water, for example, may well suggest to the child the divisibility of the matter of sugar into tiny, invisible bits, but he does not dissociate from these bits the qualities of sugar in bulk: the conception he elaborates may be compared to Anaxagoras' *homoioumeria* rather than to atomism[2].

Indeed, an atomic theory at the scientific level implies much more than just the possibility of 'explaining' the manifestations of matter in bulk as the result of atomic motions: such an explanation can only have scientific value if it is quantitative and sufficiently free from arbitrariness. In view of the lack of direct guidance from experience, the edification of an atomic theory must proceed by trial and error; one assumes some specific 'atomic model' of material bodies, the specification including the laws assumed to govern the behaviour of the atoms, and one tries to compare the properties of the model, following from these assumptions, with various observed properties of the material bodies; if all goes well, one can derive from such comparisons quantitative determinations of the atomic magnitudes.

Every atomic model of the constitution of matter raises general problems of great epistemological importance. In the first place, it is obviously necessary to assume that even the smallest grains of matter in which we can recognize the qualities of matter in bulk contain extremely large numbers of atoms. This immediately implies the necessity of applying to the analysis of atomic systems representing macroscopic bodies the method of statistics, based on the calculus of probabilities. This is not merely a practical, but a logical necessity: the only sharply defined quantities resulting from the immensely complicated motions of the atoms are statistical averages over atomic magnitudes, and only such averages can be compared with macroscopic magnitudes; at the same time, the statistical fluctuations around these averages provide us with a precise estimate of the approximation involved (as explained above) in the very definition of the macroscopic magnitudes. Atomic theory, therefore, throws new light on the nature of the idealization of the concepts with which macroscopic physics operate; in particular, from the atomistic point of view, the deterministic form of the macroscopic laws no longer appears as a primary feature, but only as a consequence of the statistical 'law of large numbers'.

To justify the application of statistical methods large numbers are not sufficient; the elements subjected to statistics must also present a certain uniformity of structure. Thus, it is a necessary condition for the atoms of matter that they should only exist in a limited number of species, and that all atoms of a given species should have identical structures, and consequently be indistinguishable and interchangeable. Obvious as it is, this point is worth emphasizing, for it is quite essential for a proper assessment of the meaning we can attach to statements concerning single atoms: these statements clearly do not pertain to the fate of a specified individual atom, but to the behaviour of any atom of the given species, picked at random. In consequence, the experiments from which laws determining such behaviour of single atoms can be derived are of a distinctly statistical type, called "sampling": one picks out, under specified conditions, a sufficient sample out of a much larger collection of atoms subjected to the same conditions, and one examines the distribution of its properties. The experiment may perhaps destroy the sample, or make it unsuited for other experiments: this matters little, since other samples may be extracted from the large collection without sensibly affecting its state.

The statistical method of investigating assemblies of atoms in order to ascertain the basic laws of their behaviour does not in any way prejudge the form of causality that these laws will exhibit. Statistics may be applied just as well to collections of elements themselves governed by statistical laws (as we now know is actually the case for the atoms) as to systems of deterministic behaviour. Before one had any technical means of attaining the atomic laws experimentally, it was natural to assume that they had the same deterministic form as the corresponding laws of macroscopic systems; but there is no logical compulsion to do so: there is no need of a 'deterministic substratum' to justify the use of statistics. In fact, from the epistemological standpoint we have by now reached the hypothesis of deterministic causality on the atomic scale looks very artificial, nay inconsistent. For the laws of atomic behaviour must necessarily be formulated in terms of classical, macroscopic concepts, and therefore partake, at least to that extent, of the statistical elements entering, as we have just seen, into the definition of these concepts. Accordingly, we must conclude that even independently of the assumptions specifying the atomic model, the fundamental type of physical causality is the statistical one.

Atomic theory thus offers us a description of nature dominated at all levels by randomness: "Democrito, che il mondo a caso pone", according to Dante's lapidary formula. This description extends even without any noticeable discontinuity to the biological phenomena and includes in particular the processes of formation of sensory-motoric schemes and abstract concepts in the brain. Everywhere we meet the same remarkable aspect of the dialectic of nature: determination and order emerging from randomness, achieved by trial and error. In contrast to a misconception popular in the XIXth century and still lingering in our own days, randomness does not lead to chaos: both the deterministic and the finalistic types of causality derive from statistical causality. In a sense one may say that the former types are 'reduced' to the latter, but one must then be careful not to confuse this reduction with an elimination of the concepts thus 'reduced'. It is clear that the statistical foundation of classical physics does not eliminate deterministic causality from it, just as little as it eliminates the qualities of the classical phenomena by reducing them to atomic motions; likewise, the finalistic concept of function will always remain indispensable in biological thinking, where it helps to orient towards fruitful aims the investigation of the complicated molecular processes underlying the biological phenomena. Only, by tracing their origin to random processes on the microscopic scale, atomic theory divests these logical 'categories' of all mystical colouring.

The most conspicuous manifestation of the dialectical relationship between the macroscopic phenomena and the atomic processes is the irreversible character of the former, expressed by the second law of thermodynamics, and contrasting with the reversibility of the laws of time evolution on the microscopic scale. The synthesis of the two points of view is brought about by statistical causality: macroscopic irreversibility indicates a statistical trend in the course of the phenomena; deviations in the direction opposite to the normal one cannot strictly be excluded, but are extremely improbable as soon as the system is sufficiently large. Irreversibility thus has its root in the conditions of macroscopic observation, in which the immense number of modes of variation of the system on the atomic scale escapes the observer's control. There are, however, bodies large enough to be accessible to direct observation and yet small enough to be sensibly affected by interactions with single atoms: such bodies, emulsion droplets in solution, for example, show hardly any irreversible tendency in the coordination of their motions, but unceasing fluctuations giving rise to the phenomenon of 'Brownian motion', which reflects to some extent the reversibility of the underlying atomic processes. Classical irreversibility and Brownian motion appear, therefore, as two extreme idealizations, referring to mutually exclusive conditions of observation. The former entails complete neglect of the statistical fluctuations, the latter is entirely dominated by these fluctuations.

We are here confronted with a logical relationship between two modes of description of physical systems, the clear recognition of which is undoubtedly the most important advance in scientific epistemology since the days of Newton. Although historically it was ushered in by the problem of irreversibility in the way we have just outlined, it was only imperfectly understood even by the greatest pioneers of atomic theory, Clausius, Maxwell, Boltzmann, Gibbs, who were unable to find its proper formulation. For this one had to wait until the discovery of the quantum of action revealed at a much deeper level a similar logical situation which was eventually elucidated by Bohr's introduction of the general conception of 'complementarity'. Relations of complementarity arise whenever a given body of experience can be approached from two standpoints which turn out to offer mutually exclusive, though equally significant, accounts of experience when the concepts with which they operate are applied in their strict, idealized form. By appropriately relaxing the sharpness of their definition, it is possible to use complementary concepts concurrently without incurring contradiction; one then expects some reciprocal limitation of their respective domains of validity.

In this sense, the macroscopic and atomistic modes of description are complementary to each other; the statistical fluctuations affecting the definition of the macroscopic properties under specified conditions assign a limit to their applicability in these conditions, and there are indeed reciprocal relations between such fluctuations which are independent of the size of the system and therefore quite generally fix the extent to which the two complementary descriptions may give a synthetic account of the phenomena. The quantitative expression for these reciprocal limitations contains a scale parameter connecting atomic and macroscopic dimensions, and indicating the size of physical system for which the complementary relation is of practical significance; one may take for this purpose Avogadro's number, which is the number of molecules of any species forming the macroscopic unit amount, the 'mole', in terms of which the amounts of the components entering into chemical reactions are expressed by simple integers.

As an instrument of dialectical logic, complementarity has many possibilities of application to all branches of science; in biology, for instance, fruitless controversies are eliminated by recognizing that the concept of function, which resorts to finalistic causality, can be upheld as complementary to a purely physical mode of description of the biological phenomena. The most fundamental and thoroughly explored case of complementarity, however, is that arising as a consequence of the existence of the quantum of action; for it cuts right through the whole of physics – classical as well as atomic physics. The reason why it was discovered so late is that the quantization of action has easily observable effects only on the atomic scale.

The concept of action has the peculiarity of bringing together, in its definition, quantities representing the two points of view under which we essentially envisage the physical processes: on the one hand their localization in space and time, and on the other the exchanges of momentum and energy they involve; the ideal space-time localization is symbolized by the concept of 'particle', the exchange of momentum and energy by the concept of 'field' or agent of transmission of the interaction between particles. The classical description of the motion of particles along definite trajectories and of the wave-like propagation of fields of force between them is governed by a general principle in which the two aspects are united in the very concept of action: this principle picks out the actual trajectory and the actual mode of propagation in a given space-time region as those involving the least quantity of action. It is clear, therefore, that the existence of a minimum quantity of action, expressed by Planck's constant, must set a limit to such a classical description in terms of trajectories and waves: to assign this limit is precisely the role of Heisenberg's indeterminacy relations. We learn from these relations that the limitation in question is a reciprocal one: there is no absolute limit to the possibility of either localizing a process or ascertaining its momentum-energy balance, but these two aspects turn out to be mutually exclusive, and can only be considered synthetically with the minimum reciprocal limitation imposed by the quantum of action.

The physical origin of the quantal complementarity can be traced in detail by an analysis of the process of measurement by which we are able to attach to any atomic object those classical characteristics-localization in space and time, or the complementary determination of momentum and energy – which establish the necessary connection between the processes on the atomic scale and our sensory experience. By reason of the quantization of action, atomic processes cannot be subdivided in the same fashion as the macroscopic phenomena, but must be treated as a whole; they are not well defined until they are terminated by leaving some permanent record in a registering device (a spot on a photographic plate, the definite position of a pointer, the discharge of a counter, etc.). The definition of an atomic phenomenon must accordingly include the detailed specification of the experimental arrangement fixing the conditions of its production and of the measuring apparatus recording its outcome. It is this permanent recording that represents the characteristic feature of measurements performed on atomic objects, and that gives rise to their complementary limitations. The measuring device must amplify the atomic process, transfer it so to speak from the atomic to the macroscopic scale: the interaction of the atomic object with the macroscopic apparatus is too small to produce immediately detectable effects, but by disturbing the state of equilibrium of the registering apparatus, it triggers off within this apparatus a chain of atomic processes tending to restore the perturbed equilibrium, and this change in the state of the registering apparatus is a macroscopic effect. It is a typical irreversible phenomenon on the macroscopic scale, which necessarily involves, owing to its statistical character, a loss of control over the underlying atomic processes. With regard to the quantity which is being measured, this is just the usual indetermination affecting all classical measurements; it is of purely practical character and can be reduced at will by refining the construction of the apparatus. The quantization of action, however, imposes a limitation of principle upon the determination of the aspect of the process complementary to that represented by the quantity measured, and this loss of control of the complementary aspect is unavoidable: under the best experimental conditions, it cannot be reduced below an amount reciprocal to the instrumental indetermination of the measured quantity, in conformity with the indeterminacy relations.

The statistical form of the causal relations between atomic processes is an immediate consequence of the fundamental complementarity embodied in the quantization of action. It affects not only the type of theoretical prediction we can make about these processes, but also the type of experimental technique suited to their investigation. Especially in the study of processes of extremely rare occurrence, we must accumulate very large samples of individual events in order to extract from their statistical analysis the fundamental atomic laws of which they are a manifestation. The handling of such large amounts of statistical data requires a more and more extensive automatization of the means of observation and of the mathematical elaboration of the results of measurements: the function of these automatic devices is not only the usual one of eliminating the direct intervention of the sensory and mental activity of individual observers, and thereby ensuring the best conditions of objective analysis, but above all that of overcoming the slowness of the physiological mechanisms by substituting for them the use of atomic processes of tremendously greater speed. In the last resort, however, all this technological display – however indispensable for further progress in domains so far exceeding the powers of our sensory and mental organization – is strictly dependent on this very organization; in spite of the disparity of the underlying physical processes, the workings of our most ingenious auxiliary devices are entirely similar to those of our sense organs and our brain: their amplifying the range of our perceptions does not modify the qualitative distinctions between them which have their origin in direct sensory experience, their recording function is strikingly akin to our memory, and even the mathematical feats of electronic computers are produced by the very methods of trial and error which, as we learn from genetic epistemology, have given birth to our mathematical concepts.

The outlook of modern physics seems to be radically different from the aims professed by the XIXth century physicists, whose genius opened the way to the modern developments; but this apparent difference is due more to their insufficient realization of the epistemological background than to any real difference in the tasks that confronted them. They laboured under the illusion that the description of the world they were drawing was like a picture that a human spectator could contemplate without being involved in its making, but in reality they were actively engaged, just as much as we are, in building up the material and mental tools that would ensure us a steadily firmer control of the forces of nature, and they knew – at least the most thoughtful among them – that this task required an attack on two fronts, leading to a deeper insight into the operation of man's sensory and mental apparatus as well as into the laws of physical processes. By taking clear consciousness of man's ambiguous position as spectator and actor, and placing the emphasis on the interaction between the human observer and his environment as the true object of its quest, modern physics brings us a decisive step forward in our striving after a comprehensive and harmonious view of the universe.

NOTES

[1] Gerald Moore, *The Accompanist Speaks* (lecture-recital), Angel Records, New York, No. 35262.

[2] I differ on this point from Piaget's own interpretation.

CHAPTER 18

SOME REFLECTIONS ON KNOWLEDGE

[1971e]

To Louis D'Or, in memory of our peripatetic Zurich conversations of forty years ago.

In setting up as a principle the apparently banal observation that "man is the measure of all things", Protagoras freed philosophy from the impasse into which Parmenides had led it by opposing the eternal immutability of a rational truth of divine origin to the "fragile opinions of mortals." Moreover, his religious scepticism suited the moderation of his theory of knowledge: as to the gods, he said further, we do not really know whether they exist or not, because we lack the data to decide the question. It would be tempting (if such a comparison were not far too artificial) to compare the epistemological discussions of physics in our own century to this ancient debate. It is not by chance that the dominant ideas of contemporary scientific thought are those proposed and defended by Niels Bohr and Max Born, two physicists whose attitudes are distinguished as much by their balanced approach to questions of epistemology as by the humanism inspiring their views on the social relations of science. They conceived of the rational analysis of phenomena in a forthright way, free of any dogmatism, and their very mastery of formal methods clearly showed them their limits. As for the younger generation, although for the last few years it has appeared to be in the throes of a new Parmenidean delirium, it too will eventually find its Protagoras who will break the spell of fruitless formalisms and lead it back to a human measure of things.

Like all philosophical maxims, that of Protagoras greatly needs to have its import made precise. One is tempted at first, and not mistakenly, to dispute its application to knowledge of the physical world (that is, a world from which the manifestations of life are excluded for methodological reasons): the processes of this physical world, it is said, unfold according to their own laws, independently of the purely accidental existence of beings who are curious to know these laws. Mixing up this so-called subjective preoccupation with a description of physical phenomena could only adulterate its purity. It must be noted that this objection is modern; it did not exist for the ancients and even less for the men of the middle ages who believed the world had been created for them. The conception of an autonomous universe, separated from humanity, is a product of the Copernican revolution. By one of those paradoxes in which the dialectic of thought abounds, this enlargement of the cosmic scale, while leading Newton to the development of a method of analyzing physical phenomena which is still ours, suggested to him an ideal view of the objectivity of this method which concealed the position of the human observer in the picture which this method permitted him to sketch. Newton made no mistake about the need to specify the point of view to which the picture was related: but he never doubted that this point of view was that of an eternal and omnipresent God–the same God, moreover, who for good or ill, regulated human affairs. In addition, Newton thereby created for himself the illusory problem of identifying some intermediary medium, some ether, which would allow God to watch over his material creation, and to act on it from time to time, as needed, to correct its faults.[1] Let us not laugh at this oddity: Newton, relating the course of things directly to his divine reference frame, was more consistent than certain modern cosmologists who would like to convince us that the sense of time is fixed by the expansion of the universe, and that if the latter changed into a contraction we would live in a world in reverse[2]: unrepentant Platonists, they do not admit that the fixing of the sense of time is an integral part of the observation of phenomena. Man is assuredly very insignificant on the cosmic scale, but one cannot let him tend toward zero with impunity.

All the pioneers of modern science, Newtonians and Leibnizians alike, agreed in attributing a determinist character to the causality of physical phenomena. This is not, as is sometimes asserted, because this mode of causality was suggested to them by the laws of motion of macroscopic bodies, particularly the stars. For they did not abstain from speculations on the atomic structure of these bodies, which implied the idea of disorganized motions or arrangements of the atoms; they were in possession of the theory of probability and the basis of its practical application, the law of large numbers: and they had before their eyes the success of its application in calculating life insurance based on mortality tables. If they nevertheless did not introduce the statistical method in their reflections on atoms, it is because they were inhibited by their transcendental conception of the objectivity of physical laws: in a few phrases of extraordinary clarity, Bernoulli[3] stresses the subjective character of estimates based on the law of large numbers and contrasts it with the objectivity of rational truths which constitute part of the plan of Creation. That such a plan leaves nothing to chance seems so manifestly obvious that even Laplace or, closer to us, Einstein still accepted it without question. However, without the idea of a personal God who is its author, this cosmic plan loses all substance. Far from saving it, Laplace's expedient of replacing the demiurge with an 'intelligence', which he envisioned as finite though 'vast', only added an incongruity: for to fulfil the task which Laplace assigned it, this intelligence would have to make abstraction of itself, keep itself somehow 'outside' of the cosmos that it claims to embrace in its totality. Laplace would have done better to call openly on the goddess invoked by Parmenides in his poem; however Laplace was not a poet, but rather a strict atheist. Thus, in the absence of an act of faith, which would clash with modern science because it introduces an arbitrary element, which eludes all control, into the description of phenomena, we are obliged to 'redescend' to the human scale. But then we are obliged to recognize, as Bernoulli had very accurately perceived, that the mode of causality adapted to the human condition is not determinism, but rather statistical causality.

Of course, this conclusion does not oblige us at all to adopt Bernoulli's opinion on the subjective character of probability judgments. On the contrary, this opinion was justified only by the hypothetical existence – which we must set aside – of a norm of certitude transcending human experience and to which our judgments had to be compared. If we decide to remain within the framework of human observations, the only possible criterion of the objectivity of a proposition is the agreement of all the observers with regard to it; if they cannot call upon any superhuman arbiter, the observers have no other resource than to agree among themselves. Here we see the fundamental link between the problem of objectivity and that of communication. For a proposition to be objective, it is essential that it relates to an experience which is communicable to any observer in terms which enable the latter to repeat the experience and to verify the proposition in question. Our experience of the external world meets this requirement, as does a good portion of our mental activity, which allows us to establish that knowledge achieved by introspection is common to all men. From this point on, the term subjective applies only to certain uncommunicable experiences of the emotional life which, by this very fact, are not the object of science. Undeniably, the conception of scientific knowledge which has just been sketched differs radically from the one bequeathed to us by the pioneers and which was still current in the nineteenth century, when the great theories which constitute what we call classical physics were being developed; but it is important to stress that these theories, though evolved in another spirit, are perfectly compatible with modern ideas on the meaning of their objectivity; in fact, this change of perspective affects them in no way, except by ridding them of the illusion of a transcendental source of objectivity. But this illusion is tenacious and those who still cling to it try, by subtle casuistry, to distinguish it from the new conception by inventing for the latter the name 'intersubjectivity': a purely verbal distinction, echoing useless regrets.

However rational it may appear to us now, after long use, the concept of statistical causality, objective in the new sense of the word, but more fundamentally irreducible to an underlying determinism, could not have been elaborated except by an intellectual effort which only intellects remarkably free of prejudice were capable of making. The effort may be discerned in the hesitations of the creators of quantum mechanics who took some time to adapt to the new perspectives. At the Solvay Congress in 1927, Dirac said that 'nature makes a choice' among the various possible behaviors of an atomic system. Without lending himself to such flights of language, Bohr himself spoke at this time of a 'renunciation of the exigencies of causality', thus confusing causality and determinism; and in presenting the recourse to statistical concepts as a 'renunciation', he involuntarily suggested that it was simply an attitude of retreat and that it was giving us only an imperfect view of the phenomena. These early *faux-pas* today occasion nothing more than the prattling of philosophers: the physicists have long since forgotten them. Indeed, Bohr had long recognized the necessity of rigorously determining the terminology of atomic physics so as to render it compatible with the limitations imposed on the classical concepts by the existence of the quantum of action. The indeterminacy relations provided him with the theoretical schema required for completing this program. Within the framework of complementarity, statistical causality takes its natural place in a dialectical logic: any relationship of complementarity in effect implies a statistical element, since a deterministic causality would leave no room for the option between mutually exclusive situations which is just what this complementarity envisions.

Conversely, it may be seen that on the two levels where statistical causality is introduced into atomic physics, it is in relation to situations of complementarity. The typical situation is that of the description of an individual atomic phenomenon where we find the well known complementarity of the two modes of macroscopic observation which we possess: spatio-temporal localization and energy-momentum balance. The other case we encounter is that of describing macroscopic systems composed of material atoms or quanta of radiation: Here the alternatives are a dynamical description of these systems or their thermodynamic description, modes of description demanding mutually exclusive observation conditions. In these two cases we can see very well that if the choice of conditions of observation is left to our decision, the form itself of these conditions is independent of our will and fixed objectively by our position with regard to the physical universe: on one hand, the obligation to work with macroscopic instruments is simply a requirement of our own constitution which forces us to relate atomic dimensions to the scale of our normal sensory-motor activity: on the other, the complementarity of the aspects of the phenomena which we can grasp with the aid of these instruments is the consequence of a natural law – the existence of the quantum of action.

Hence, it was inevitable that the exploration of the atomic domain led physicists to effect a self-examination, a reflection on their own activity – always a disagreeable procedure because it is difficult and disruptive of routines – but whose recompense is the attainment of a more profound harmony between nature and its mental representation, of greater clarity concerning the significance of this representation. Laplace's cold contemplative 'intelligence' is succeeded by an idealization on the human scale, that of the concerted action of men upon nature. From this point of view, science has the function of analyzing our interaction with the world of which we are part and codifying the results of this analysis so as to render them communicable: thus it appears as an extension of our sensory-motor activity in the direction of an abstraction which prodigiously increases its efficacy. To the purely perceptive codification of our experience is added a codification of another order of powerfulness – that of abstract language, the instrument of rational thought.

Far from being an innate 'faculty' of the human species, rational thought developed in the course of history only in those societies whose industry could not remain satisfied with more primitive mental representations, based principally on the perceptive code. Similarly, the mental development of the individual in such a society attains rational thought only after having first passed through a stage which is dominated by the concrete representations of sensory-motor schemes. Now, the conception of chance, as shown by the history of scientific thought as well as by the study of individual psychogenesis, is a very late acquisition – so strong is the tendency to attribute an apparent absence of determinacy to ignorance of some regularity governed by a higher will or which a sufficiently vast intelligence could comprehend. Before a roll of the dice, the reactions of children differ from those of Bernoulli or Laplace only in form: the adult who questions them might 'know the trick'. To reach the understanding that there is no trick to know, it was necessary for us to take account more correctly of the conditions and of the very nature of knowledge.

NOTES

[1] Cf. Rosenfeld, L., 'Newton and the Law of Gravitation', *Archive for History of Exact Sciences* **2** (1965), 365. [See this volume, p. 58.]

[2] See T. Gold, ed., *The Nature of Time* (Cornell University Press, Ithaca, 1967), p. 145.

[3] Bernoulli, Jac., *Ars conjectandi* (Thurnisi, Basileae, 1713), p. 210.

CHAPTER 19

EPISTEMOLOGY ON A SCIENTIFIC BASIS

[1971a]

To My Friend Harald Wergeland

For a physicist to venture onto the treacherous ice of epistemology requires some justification. So I shall begin by relating why I have come to concern myself with these questions, because this will form the most natural introduction to what follows.

The development of atomic physics in this century has led to a drastic change in the conceptual structures that physicists use to describe experience. To achieve a logically consistent and complete description of the properties of atoms, physicists were forced to impose upon the two modes of description that together exhaust experimental possibilities – space-time description and energy-impulse conservation – a reciprocal limitation according to definite rules inbuilt into quantum mechanics itself. This curious situation created an epistemological problem of a similar depth to that previously encountered in the theory of relativity, and once again it could be solved in the same way, viz. by widening the logical framework for the use of concepts: in the same way as Einstein in that case had generalized space-time co-ordination, so now Niels Bohr introduced the logical concept of complementarity in order to make room for a non-contradictory use of conceptions that are subject to quantum-theory limitations. In both these cases solutions were achieved within physics itself without appealing to other than physical laws (the invariability of the velocity of light and the existence of the quantum effect) and to quite general (although slightly idealized) physical experience. One would imagine that as soon as such new conceptual frameworks, firmly based on assured experience, were introduced, physicists would agree to accept them and enter into them, but that is not what is occurring. The idea of complementarity, in particular, still meets with strong opposition not only, as is to be expected, from older scientists who find it difficult to abandon their habitual lines of thinking, but also from some of the most capable of younger physicists. What can be the cause of this resistance? It cannot be brushed aside lightly; investigation must be made to find out where the difficulty lies in approaching this new situation.

First of all, one discovers that the objections do not come from physics; the physical situation is, of course, quite plain: quantum mechanics offers a finely honed and reliable method capable of accounting for all experience within its clearly defined area of competence; no one disputes this. Whence then the opposition? It comes from another side of the scientist's view of his theories, precisely that which concerns epistemology. Any physicist, any scientist indeed (I am naturally building on my experience as a physicist, but biologists are in the same situation), conceives the theories he is using and constantly trying to refine from a dual viewpoint. First of all he views them as technical aids that permit him to describe and analyze experience according to very precise rules: optimally, using mathematics, which is the most precise language we have. But he also believes that the exacting labour entailed in the investigation of natural phenomena ought to be seen in a broader human perspective, that it leads to cognition. This does, of course, lead on to possible practical applications to social tasks, but in any case one important motivation for scientific work is that it contributes to knowledge of nature and of our position as men in nature – over against nature, if one will. The value of a theory, not to say its dignity, will therefore be highly dependent on whether it is thought to satisfy this striving towards a deeper knowledge of the regularities of nature.

To reach clarity on this aspect of theories ought to be the aim of epistemology, but here lies the difficulty, because many physicists (and in this respect biologists are often even more amateurish) are not aware that their attitude to epistemological questions is far from unprejudiced, since it all too often rests upon arbitrary assumptions in which they fairly haphazardly and unquestioningly place their trust. Whereas in the discussion of experimental results and theoretical calculations they manifest the greatest meticulousness, they plunge into epistemological observations with incredible lightheartedness. When it is a matter of evaluating such quite new features of the description of nature as complementarity, they attack the matter from obsolete viewpoints whose inadequacy is apparent from the development of physics itself. Even so penetrating a thinker as Einstein objected to the epistemological consequences of quantum mechanics because (quite correctly) he found them incompatible with a certain postulate about 'elements of physical reality' that he himself had propounded, and which he claimed to be universally valid without being able to give reasons for this 'hunch'.

However, Einstein's criticism was at least constructive precisely in pointing to a necessary modification of the idea of 'physical reality' in its application to atomary objects, but what is to be said of the many amateurish attempts to maintain the deterministic form of causality in atomic theory! In all such attempts one is obliged to differentiate between the empirical world in which, under given experimental conditions, we observe the behaviour of varied atomic processes distributed according to statistical laws, and an ideal world (or perhaps several such) in which the individual processes occur in a determined way, which, however, escapes our observation. How then can one imagine that there could be any connexion between these different worlds? In no other way – as Bell recently demonstrated by an ingenious analysis of these curious notions – than by assuming something reminiscent of Leibniz's "prästabilierte Harmonie". All that the innovators of quantum mechanics have to offer us, then, is a return to the theological world-view of the baroque age.

For Newton and Leibniz, of course, determinism in natural laws was an almost obvious assumption: the world was God's creation and the laws according to which it was to develop had been predetermined by Him without leaving anything to chance (although Newton considered that God might well have overlooked some consequences of these laws!). From this point of view one understands too what Newton meant by the introduction of absolute space and absolute time: by the word 'absolute' he was referring to the framework established by God Himself – he even added the provision that the centre of gravity of the solar system was a fixed centre at rest to which all motions could be referred. Laplace retained the Newtonian conception completely, but as he was not so pious, he spoke of a "being", a finite being it is true, but whose intelligence is so comprehensive that it can survey the state of motion and the interactions of all the particles in the world at a particular time and from them calculate the state of the world at any time. For him too, then, the idea of a determinism of the laws of nature was inseparably linked to the idea of a world description from a point of view that was, if not divine, then at least superhuman. But what we have learnt from the theory of relativity and even more from the quantum theory is precisely that we are obliged to abandon the existence of any absolute system of reference; there is no higher being to refer to as the guarantor of the consistency of our conceptual formations – we can only rely on ourselves. Newton had set himself the life's task of deciphering the purposes of God; Pascal showed a deeper insight into our position in the world when he said that God "conceals himself" from us. A concern with 'concealed' variables cannot offer any evasion of an inevitable consequence of scientific development.

This example – and several others I could have cited – show not only into what byways one can wander when one does not take sufficient account of the foundations of epistemology, but also that one cannot fathom these foundations if one limits oneself, like Einstein and Bohr, to an analysis of the description of physical phenomena. Our task here is concerned with the other aspect of elaborating such a description: the answering of the justified questions about the cognitive content and consistency of natural science demands a discussion of the way in which our concepts are formed and used – of the origin and nature of the whole of rational thinking. Such considerations have been the starting-point for my preoccupation with these problems. The first question one asks is: what method is one to use to avoid pitfalls like those I have just tried to illustrate? The answer is straightforward: since the time of Newton science, and especially physics, has developed a methodology that has been thoroughly tested and has proved so fruitful that, purely pragmatically, it must be seen as the most reliable we have. According to the classical scheme propounded by Newton, one proceeds from experience: from observations and experiments one learns by what laws processes occur; then one establishes the concepts and logical relationships that provide a rational account of the regularities by which we are able to predict the phenomena that correspond to given conditions of observation. Now, the processes that lead to the formation of ideas and concepts and the setting up of conceptual structures (i.e. the processes that are the bases of our cognition) are, of course, themselves natural phenomena; even though they occur in our interior world, they are just as susceptible of analysis by the scientific method as our external phenomena. We can observe the behaviour of both animals and men; by appropriate experiments we can even extend our insight into the development of thought-processes and, from these accumulated experiences, hope to form well-founded notions of the laws of cognition.

This appears to be a large programme, but it is, of course, not the first time it has been put forward. In the history of epistemology one can, in fact, trace a plain line marking a clear development of epistemology in the scientific direction I have just outlined. As is to be expected, this line parallels the development of science itself, a little delayed in time, for new insights concerning cognition usually follow upon scientific advances. In this way the danger, so prominent in our time, arises that epistemology may completely lose contact with science. Yet it seems to me that we have now reached a point at which, with vigorous scientific work from both physics and epistemology, we could achieve a synthetic treatment of the physical and psychological aspects that would form the basis of a more harmonious world view.

It will be not without profit to survey briefly the successive stages in the historical development to which I have just referred. The very first beginning comes from Newton himself, who more systematically than Galileo and with greater practical experience and deeper insight than Descartes and Pascal tried to formulate the main rules for the physical induction method and, in addition, to work out new mathematical tools appropriate to a rational, quantitative account of phenomena. He also formed some ideas about epistemological problems proper, without developing them further. However, his friend John Locke, who understood well how to analyze and reproduce the thinking that had developed with the advent of science, completed this task. Locke's systematic presentation of the epistemology that the new inductive method led to was rightly regarded as quite revolutionary and immediately provoked indignation among philosophers like Berkeley and Leibniz, who could see plainly that it could not but undermine the ideological foundation on which the established social system rested. Just as progressive spirits of the day regarded science as a tool for creating an industrial economy, so they found in Locke's demystifying of man's mental life a powerful weapon in their battle for a reform of the social order. Locke's main thesis that all our concepts and the very laws of our thinking are derived from experience overturned the traditional notion, according to which abstract concepts are imprinted upon our minds right from birth. The mind of the baby was thought to be fully equipped with the faculty of logical thought, and the role of experience was merely to assist this innate ability to full development. But Locke maintained that all logical conceptual structures had to be formed from sense perceptions and had no other origin. Locke's successors, Hume and Condillac, carried this thinking to its final conclusion. Hume pointed out the empirical character of the principle of causality itself, and, in his refutation of Berkeley's idealism, Condillac demonstrated that the movements of the body constituted an essential element in our knowledge of an external world independent of our thinking.

And yet the opposition that these bold theories provoked shows that they were not yet well enough grounded to hold their own in the way that the newly discovered physical laws had done. Unlike these latter, the epistemological assertions were not built on any exact empirical knowledge but only on broad, insufficiently analyzed experiences. The first step in the direction we now regard as the correct one was as yet so hesitant that it is understandable that even some of the greatest thinkers of the eighteenth century did not find it adequate. It is instructive to note the tenor of their opposition. The most consistently thought out was Leibniz's conception, already mentioned, of "prästabilierte Harmonie" – the soul (or mind) has no contact at all with the body, which behaves like a material system according to its own laws and the soul pursues its own existence independently of it – but God Himself provides that at every moment the movements of the soul are completely harmonized with those of the body. This was the singular theory to which a strict logician like Leibniz was actually driven when he did not want to recognize experience as the ultimate basis for the formation of our ideas as well as the origin of our thinking.

But Leibniz lacked the subtle intuition for the phenomena of the material world that Euler possessed to the highest degree. Euler saw the unreasonableness of Leibniz's extreme standpoint and adopted a midway position, which however d'Alembert thought he ought to be ashamed of. Like Leibniz, Euler did indeed declare that the connexion between mind and body is a miracle that God has brought about and which surpasses human understanding; thus far d'Alembert's reproaches were no doubt justified, in any case in the sense that as a scientist he ought not to have abandoned in advance any possibility of explaining even the most recondite matters. But apart from this, when we turn to the details of how Euler describes (without any attempt at an explanation, which he rejects) psychological phenomena – how our thinking occurs, what influences the mind receives from its contact with the external world through the sense organs, then we must admit that his purely descriptive psychology is completely on a par with what we know today. Here we have a typical example of the scientist's strict method of procedure and of its superiority to the still groping method of Locke or Hume. As a well-trained scientist Euler understands how problems should be formulated, even though he considers that such problems are not susceptible of rational solution.

Just as Euler could not approve of Leibniz's denial of any connexion between soul and body, so he disliked the attempts of the French materialists to reduce the soul to matter, to refer psychological phenomena to material properties or movements that had to remain hypothetical. He realized that if one wished to achieve a rational description both of the material world and of the world of the soul (or mind), one had to begin by defining the main concepts as sharply as possible. His psychology therefore distinguishes itself by its clearcut formulations. The dividing line between mind and matter is established by quite precise definitions. The characteristic properties of matter are, of course, in accordance with the physics of the time – impenetrability, inertia, localization in time and space. More original in his concept of the spiritual aspect: he introduces individual spirits – 'esprits' – to which he proceeds to assign very definite qualities selected in harmony with his description of mental phenomena. In particular, he stresses that space-time localization is not applicable to spirits: the connexion between a particular spirit and a particular human or animal body occurs exclusively by the will of God. However, he develops a detailed description of the way in which this connexion functions: in the brain the spirit is receptive to the sense impressions that are registered there; under the influence of these, the spirit forms a conceptual notion of the state of the outside world, upon which it reacts again upon the body by sending through the nerves signals to muscles and causing movements – as we see, a logically quite coherent description of what we now call sensory-motor schemes. With the means available at that time there was no possibility of approaching nearer to the intermediate link 'the spirit': Euler's only, but serious, weakness was that he capitulated to the temporary difficulty of this problem.

Not until the nineteenth century did experimental sense physiology (in which the physicist Fechner made a great contribution by introducing quantitative measurements) make a serious beginning. That this firmer basis gave rise to a fresh advance in epistemology was due to the intervention of the most outstanding scientist of the period, Helmholtz, who was both a physiologist and a physicist. His analysis of the bases in sense physiology for the concept of space is a classic example of the penetrating combination of analysis and synthesis that he applied to these problems. He started out by using pure logic to unravel the characteristic properties of the concept of space; he placed the main stress on the transformational properties of space, which are closely linked to the movements of bodies. Then he attempted to show how these properties could be deduced from various sense impressions, particularly sight and touch experience, either separately or in combination. He thought that he had thereby finally refuted Kant's idealistic conception of space as an *a priori* given form of our perception. Like Locke, he saw in scientific method the way to the liberation of thought from metaphysical prejudices. Yet, we must again question whether this ingenious reconstruction of the formation of the space concept really corresponds to the mental processes that lie behind it. Helmholtz was not able to answer this question, for he still lacked sufficient insight into the functioning of abstract thinking. For we know that all that logic can give us is a network of coherent statements: it cannot unambiguously determine by what paths our thought will in fact proceed through this network from its starting-point in our experience to the final abstract conclusions.

The same criticism can be made of Ernst Mach's psycho-physical analyses and attempts to root the ultimate origin of physical concepts in sense impressions. In comparison with Helmholtz, one might even say that Mach undervalued the share that purely abstract theoretical thinking has in the formation of scientific concepts. On one extremely important point, however, Mach's epistemological investigations were trailblazing: from an analysis of the information that the history of science provides about how the great physicists set about making their discoveries, he was able to demonstrate certain common features in this creative activity; in particular, he found that the discoverer often builds upon general experience, which he then applies to "Gedankenexperimente", i.e. imagined, idealized experiments suitable for illustrating the supposed new law.

What can now be said about the present situation? Some physicists still live in the Kantian thought-world, known as critical idealism notwithstanding that it is the least critical attitude imaginable. Even the depredations of Platonism are still apparent among mathematicians and often affect our young theorists. But even though most people will agree that physical concepts have their origin in sense perceptions, they cannot free themselves from the assumption or the belief that some mental faculty peculiar to the brain and having its own laws must be involved: they are thus adopting the same position as Euler. Our survey of the historical development shows us clearly that this confusion may be due to the lack of a theory in the spirit of Locke or Mach, free from any suspicion of arbitrariness. The way forward is therefore quite plain: we must look around for areas of experience that may be able to shed light on the actual process of concept-building – not in Helmholtz's imagination but in living people.

What experience have we available? Above all, when we are thinking of the development of scientific concepts, we have the whole history of science: this is the material that lies nearest to hand and which Mach employed to great profit. The historical documents (and we might wish that these were more abundant) can inform us about the motivation and the thought that actually led to a fruitful formation of scientific concepts: in this, historical analysis has a similar significance to the observation of phenomena as against their analysis in experiments under controlled conditions. If such experimental information is also available, then historical information provides valuable material for comparison, but it cannot in itself give us any exhaustive answer to such complicated questions. For we must realize that the scientists whose behaviour we are observing are adults who from their childhood have been passing through a mental development, about which we are, in most cases, insufficiently informed; they possess experiences that in many and various ways have exerted influence on their attitudes to physical problems, in brief, they are not purely experimental objects.

If we wish to find pure minds, we shall have to go to children. Here we are in the fortunate position of having available the results of the investigations made by the school of Piaget. Piaget, who began his career as a biologist, early on moved from an interest in shellfish to an interest in children (via what intermediate stages I know not) and has since devoted his whole life to investigating the mental development of children by purely scientific empirical methods. For this purpose he had first of all to develop a reliable technique yielding reliable information about the reactions of children when faced with certain problems and when asked what they think in such situations. The soundness of the method and the objectivity of the conclusions is strengthened by the fact that children exposed to the same test react in definite ways corresponding to a few regular stages of development, which are found in the most varied tests. In this way a generally valid picture is achieved of the way in which the thinking faculty of the child arises and develops, and the conclusions that can be drawn acquire, for the first time in the history of epistemology, a truly scientific character. Piaget's genetic epistemology can thus with justice be designated the beginning of the scientific age in the study of human cognition.

It is not, of course, necessary to delay investigating children's behaviour until the age at which they can speak and understand questions; from the very earliest stages, from birth to an age of three or four, we are limited to the methods employed in animal psychology: a study which above all shows clearly that there is complete continuity between animals and man and finally disposes of the myth of man's privileged mind equipped with the innate faculty of rational thought. The baby enters the world with a few built-in reflexes without which it simply could not survive, but apart from these its brain is completely empty, like a computer without a programme. The programme is filled in, though, at a fantastic rate: some sensory-motor schemes are developed as early as the first hours after birth, and altogether the first eighteen months of human life are the period in which, with a good conscience, we can claim to have done some work. The essential feature of this building up of nerve connexions in the brain – a feature that Piaget strongly stresses – is that the various nerve schemes are both sensory and motor: the sense perceptions received are always preserved in connexion with signals to motor reactions; right from the beginning the set-up of the brain aims at activity and not at looking on passively.

This is confirmed by the child's behaviour in the next stage of development: the child begins spontaneously to combine sensory-motor schemes in sequences of operations by which to reach the more remote goals that an extended field of experience places before his activity. From the actions governed by the separate sensory-motor schemes the child forms concrete conceptions, and it is this ability to combine concrete conceptions with a definite goal that is the first step to logical thought – a logic that is still inhibited by the finite range of ideas and is marked by a preponderance of action and finality.

Whether rational thought will proceed to grow into the fully developed form we know, is no longer the decision of the individual child: it depends completely on the influence the child receives from society. In what are known as primitive societies, the thinking of the adults has largely halted at the stage of concrete conceptions, for they have no need of involved abstractions. Even in our educated society, the use of sophisticated thought-forms of science are an exception: Michotte's experiments on adult subjects' spontaneous experience of simulated interactions between moved and stationary objects reveal an unexpected survival of pre-Galilean mechanics in their conceptual world. Through language children are given access to fully developed rational thought, and this they learn from adults. Language is a code system, a system of acoustic or visual signals linked to the sensory-motor schemes and representing these just as the concrete conceptions do – in fact, there is only a difference of degree between words, which we regard as abstract signals, and conceptions, which we call concrete because they are built upon direct sense perceptions: the latter are of course also signals we receive from the world around us. In language, however, abstraction is driven to the limit: whether language signs are heard or seen, or felt by the blind, they are expressions of the same linguistic structure. Not until a late stage of development, around ten or eleven, do children discover the superiority of language as a tool of thought: at that age they realize that the thought operations they used to carry out by means of concrete ideas can be performed with words, and that thinking in words literally opens up boundless prospects for them, for language offers the possibility of comprising in a finite sentence the idea of the infinite continuation of a series of operations.

Having reached this stage in understanding the origin of rational thinking in the brain of the individual child, we also understand better how, under varying social orders, science has taken the development that history shows. I shall here merely cite a single example. When we read the fragments we have from the pre-Socratic philosophers (without being influenced by the widespread but exceedingly ill-founded belief in the amazing superiority of the Greeks), we cannot avoid the impression that they are using child language, that their cosmological speculations are on the same level as the explanations of natural phenomena that children weave with bold imagination from their limited world of ideas. I would assert that we have here not merely an analogy, but an actual identity of intellectual development: in the simple social conditions of the past, people could make do with concrete ideas and correspondingly primitive thought-forms. When we examine the historical documents from this point of view, we can, in fact, observe when they became aware of the necessity for abstract thinking: Aristotle stands, we might say, between two worlds; his powerful thought still moves partially at the concrete stage, but he has already begun to analyze explicitly the abstract form of thinking – for one of his greatest achievements is that he established the laws of logic.

Thus, in the development of society we can recognize the same stages that are found in the mental development of the individual child; by combining these two areas of experience we may therefore hope to achieve an adequate account of the origin, continued growth, and extension of human cognition, an account that would have the same degree of certainty as that obtained in science. On such an empirical basis one would be able to demystify epistemology, sweep away all these arbitrary assumptions, these philosophical prejudices inherited from dead systems, and really work with the same assured methods that have brought science to its unique position. By this path we can expect to achieve complete harmony in a world-view that would embrace both natural science, which describes our experience of the external world, and epistemology, which analyzes our experience of our inner world of thought. If the same strict methods are applied to the two areas of experience, there should be no reason for any conflict between our ideas about these two aspects of human existence.

Translated by David Stoner

CHAPTER 20

CONDILLAC'S INFLUENCE ON FRENCH SCIENTIFIC THOUGHT

[1972f]

There is a widespread tendency among historians of science to look down upon the eighteenth century as a period of stagnation in the scientific development, contrasting with the triumphant achievements of the great creative period which preceded it. Newton's towering personality is indeed such as to overshadow investigators of nature who would otherwise have ranked as the foremost of their time and on a par with their great followers of the nineteenth century and of our own time. The main achievement of Newton, if we look at it in a wider historical perspective than could be done by the popularizers of the time, is not what impressed them so much – his formulation of the law of universal gravitation, his experimental discoveries in optics, his vindication of a theological conception of the laws of nature. Above all, he initiated a new approach to nature, a new method of investigation of natural phenomena. This inductive method, starting from the direct observation of the phenomena by painstaking, accurate experiments, and leading to a rational account of them, has since then demonstrated its unfailing efficiency and power. It is the only method in human endeavour that leads to an unending accumulation of ever improved, ever expanding knowledge, which never suffers any set-back. This is not to say, of course, that it is the only way in which we establish relationships of lasting value with nature, but it is unique in this cumulative character, and the resulting certainty, of its acquisitions. For Newton, the source of such certainty was not in doubt: brought up in the Puritan ideology, with its deeply religious mode of expression, he naturally regarded the concept of a personal creator of the world as inherent in his natural philosophy, and saw in the divine origin of human reason the ultimate justification of the uncompromising rationalism that characterizes his whole approach.

From this point of view, one may say that the time of Newton, and above all Newton's own contribution, marks the birth of modern science, of science as we understand it, as the rational description of our experience of nature. No doubt Newton had a precursor in Galilei, who had a clear vision of the meaning of experiment as the source of our knowledge of natural phenomena and of the power of human reason to discover the laws of these phenomena; but Galilei was still groping for the right way of experimenting, his mathematics had not the needed sophistication, and on both counts Newton was truly the innovator. As to Descartes' influence on Newton, deep as it was, it was more in the nature of an inspiration and a challenge than of a direct guidance. If any philosopher may be called a rationalist, it is certainly Descartes, but he lacked balance in his judgment: he tended to give rational thinking too large a part in the elaboration of a coherent view of the world, and to neglect the role of the information derived from observation. As a result, his theoretical constructions about natural phenomena soon proved complete failures; the real point at issue, however, between Newton and the Cartesians was not, despite appearances, the contrast between their conceptions of the laws of nature, but above all the question of the method leading to the discovery and formulation of these laws.

With regard to the fundamental physical principle underlying the description of the world, there was no opposition between Newton and the followers of Descartes: all accepted the view that the material universe was a huge mechanism, a system of various kinds of particles moving in various ways and interacting only by direct contact. This mechanistic conception remained in fact unchallenged throughout the eighteenth and the nineteenth century, until it finally foundered against the problem of the nature of light, which was already puzzling Newton, About the necessarily hypothetical modalities of the mechanical world picture there was much arbitrariness, and consequently much quarrelling: Newton dodged such fruitless speculations by his insistence on proceeding inductively to the stage which observation would lead to, and provisionally renouncing complete mechanical explanation rather than indulging in unverifiable hypotheses. It is on this methodical issue that Newton took his stand against the Cartesian advocacy of a purely rational construction of the mechanism underlying the phenomena. The paramount importance of this issue was keenly felt by Newton's contemporaries and those that upheld the Newtonian tradition in the eighteenth century. The latter were no mere epigones: their boundless admiration for Newton was perfectly genuine; it was prompted by a clear recognition of the superiority of his scientific method.

As to Newton's opponents – those I just described as Cartesians in the wide sense of the word – I must especially mention Leibnitz, whose mathematical conceptions – quite independently of the philosophy underlying them – proved even more successful than Newton's. The notorious clash between the two giants about the discovery of the calculus should not be judged on its superficial course; if this conflict arose and took such proportions, it must have been more deeply motivated than by a mere quarrel of priority. Indeed, we again see that it did not really concern the facts of the case: the contestants and their supporters knew perfectly well that the two methods, the Newtonian method of fluxions and the Leibnitzian differential calculus, were essentially the same – otherwise, there would have been no question of possible plagiary. The real cause of the conflict is not difficult to establish, not by conjecture, but on the basis of actual documents, some of them only recently published: it was the opposition of two powerful temperaments, both of them imbued by the belief in rational thinking as the instrument of scientific inquiry, but disagreeing about the way in which this instrument should be handled in order to achieve a quantitative description of the phenomena. Newton was of the intuitive type – his whole conception of fluxion was derived from a visual picture of the motion of a particle on its trajectory – whereas Leibnitz was of the abstract logical type, laying the emphasis on the practical advantage of a neat formal representation of the various aspects of the phenomena by carefully chosen symbols subject to precise mathematical and logical rules. The next generation, having assimilated these apparently conflicting approaches, was quite happy to use both of them and nobody would have dreamt of arguing which of the two was more fruitful or more powerful. But for the great pioneers, the issue was one aspect of the deeplying questions of method concerning which they held irreconcilably opposite views.

Thus was the scene set, at the beginning of the second quarter of the eighteenth century, for an exploration of nature stimulated by the impulse that Newton and Leibnitz had given to the development of adequate methods of investigation. The history of the following period may conveniently be divided into two stages. The second quarter of the century was a time of consolidation, so to speak, during which the natural philosophers learned to handle the new mathematical tools and refined them, while the experimental method opened up the new domains of electricity and 'pneumatic' chemistry (the discovery of various kinds of gases, paving the way to a quantitative study of chemical reactions). Then, in the second half of the century, we see a new generation at work, fully trained in the use of the new methods, both as regards their technical aspects and their philosophical implications. There was no lack among them of powerful and original thinkers: the greatest perhaps was Euler, who came from the Leibnitzian school, but carried further and systematized Newton's ideas; of comparable eminence were the French 'geometers' who professed to be Newton's disciples – Clairaut and d'Alembert, followed by the latter's brilliant pupils Lagrange and Laplace – and the French chemists under Lavoisier's leadership. To get the measure of their achievements, one has only to compare Newton's *Principia* with the *Mécanique céleste* of Laplace, or the chemical ideas of Stahl with those of Lavoisier: the progress accomplished in those fifty or sixty years is quite comparable in importance and scope with the progress we have witnessed in the first half of our own century.

Now, after this outline of the general background, let us come to the special subject I have chosen to illustrate in more detail: the epistemological aspects whose importance for the development of eighteenth century science I have tried to emphasize. Against this background, Condillac is indeed a natural choice, for he is quite representative of the evolution of ideas about scientific method during the two stages I have distinguished – the period of consolidation and the period of expansion and progress. Condillac is not a scientist in the modern sense of the term, he did not make any contribution to the scientific activity he was witnessing; but he did observe this activity with deep understanding and managed in a masterly way to analyse and formulate the general methodical principles which guided his philosophical friends in their scientific endeavours.

The conceptions of Newton entered France rather late, because they had to overcome the formidale obstacle of the Cartesian tradition, whose influence was still so strong by the middle of the century that Voltaire himself, that knight errant of progressive ideas, had to champion the Newtonian philosophy in full armour. He was of course not alone in this fight, not indeed the most influential. His role as a clever propagandist should not be underestimated, but cannot be compared with Clairaut's much weightier intervention: not only had Clairaut a hand in the French translation of the *Principia*, but he was the first to make a significant contribution to a problem discussed in the *Principia* – the problem of the shape of the earth – which went further than Newton's own treatment.

The Cartesians had one supporter of tremendous vitality, Fontenelle, who, as late as 1752, published a theory of the solar system based on an improved version of the Cartesian 'tourbillons'. Much more important, however, was the influence of Malebranche – I am not concerned here with Malebranche's theology, but with his scientific views, which are often underrated: he saw clearly the insufficiency of Descartes' physics and did his best to improve it, without abandoning its fundamental principles. Now, the Oratorian order, to which Malebranche belonged, introduced this better brand of Cartesianism into their schools, in which more place was given to the teaching of science than in most Jesuit colleges: this circumstance accounts for the prolonged hold of Cartesianism on French thought. Indeed, even such a staunch Newtonian as d'Alembert betrays in his philosophical attitude a greater influence of his Oratorian schooling than he would have liked to acknowledge: in the preface to his *Traité de Dynamique*, for instance, one perceives a distinct undercurrent of Cartesian rationalism.

The interest of the French 'philosophers' for Newton was not purely scientific; of course, they were able to appreciate his scientific achievements, but (as clearly appears from Voltaire's *Lettres philosophiques*) they above all regarded his natural philosophy as a part of a wider intellectual and political movement they had good reasons to contrast with the oppressive regime they were subjected to; hence their keenness in promoting Newtonian studies and Newton's natural philosophy. In the turbulent company of the 'philosophes', Condillac, modest and retiring, stands out as a sharpsighted and acute thinker, grasping the essential points and expounding them with unequalled lucidity. His role in the philosophical debate was to discuss the epistemological issues raised by the Newtonian view of science. In England, this role had been fulfilled by Locke, whose *Treatise on Human Understanding* offered a systematic exposition of a theory of knowledge in harmony with Newton's conceptions. Condillac's first contribution, the *Exposition des origines des connaissances humaines*, had no other pretension than to introduce Locke to the French public – though, as we shall soon see, it achieved much more. It was soon supplemented by a *Traité des systèmes*, which exposed the weaknesses of the metaphysical attitude common to Descartes and Leibnitz and contrasted them with the firm guidance afforded by Newton's inductive method.

These two slender teatises, written with sharp precision and luminous simplicity, won him immediate recognition as the spokesman of the 'philosophes' in matters of epistemology and psychology; but these qualities did not by far exhaust their significance. Locke had traced the origin of our knowledge of the laws of nature to sensations and abstract ideas, but he had left undecided the question of the nature of the latter; Condillac took this last step and showed that our abstract ideas could themselves be referred to sensations. This was a bold step indeed, which carried epistemology from what I called the stage of consolidation to that of innovation, – implying as it did a definitive break with the idea of the divine origin of rational thinking, a possibility that Locke had still left open. Yet, Condillac's thesis did not at first appeal to the materialists among the 'philosophes': Diderot pointed out the danger implicit in it of bringing support to Berkeley's idealistic opposition to Newton. If, he argued, our ideas come from our sensations, they are somehow elaborated in our brain and in this process become separated from the sensations and belong to our internal world; hence our rational thinking is a purely internal activity, without direct contact with the external world, as Berkeley asserted. It was this paradoxical conclusion that prompted Diderot to denounce idealism as the scandal of philosophy, because it was so difficult to prove its absurdity.

Well, Condillac took up the challenge, and his answer to Diderot's query was the *Traité des sensations*, his masterpiece and a landmark in the history of epistemology. The argumentation he developed in this book is extremely remarkable, both by its logical subtlety and by the deep insight it reveals in an essential feature of the perceptive process. In a boldly dialectical way, he started by carrying Berkeley's thesis to the extreme, by means of his famous artifice – we would now say "Gedankenexperiment" – of the statue endowed with sensibility: beginning with the sensation of smell, and proceeding successively to the other senses, he showed that the statue could very well develop a kind of rational thinking, without being aware of any relation of this thinking to an external world. There is one exception, however: the sense of touch – we would speak of haptic sensations. The preceding argument holds only so long as the statue does not move, remains passive: this is its essential difference from a living being. As soon as we add the possibility of touching objects, which implies an active exploration of the environment, the spell of solipsism is broken: the thinking subject realizes that he is acting upon an external world. This tremendous intellectual achievement is all the more impressive when it is brought into relation with the modern development of genetic psychology: one of the fundamental features brought out by Piaget's studies of the mental development of children is precisely the inseparable union of the passive reception of a sensation and of the motor reaction it leads to, the building up of a sensory-motor scheme, which is the true unit with which our thinking, both concrete and abstract, operates.

The next work of Condillac was a direct continuation of the trend of argument by which he had established that there is nothing in our rational thinking that does not come from a purely material interaction with the external world. Buffon at that time had developed speculations about the evolution of the animal world, which seemed to suggest a separation of principle between animals and man, the animals being deprived of the power of building up abstract ideas. In his *Traité des animaux*, Condillac examines this question, and concludes against Buffon that there is no recognizable discontinuity between the behaviour of higher animals and that of man: it is in animals that the beginnings of consciousness and of our mental properties can be perceived. This was a contribution of great importance to the development of the materialistic view of the world, which Diderot and his friends were attempting, but from which Condillac himself kept aloof. In fact, Diderot's materialism remained rather ambiguous, since he was forced, as a result of Condillac's analysis, to extend the fundamental property of 'sensibility' to all the constitutive elements of his material world.

In order to appreciate the scientific value of Condillac's psychology from the point of view of his own time, it is interesting to compare it with the views of one of the most powerful contemporary thinkers, Euler. As a mathematician and a theoretical physicist (as we would now call him), he dominates the eighteenth century in the same way as Newton dominated the second half of the seventeenth: the mathematical tools and the concepts of classical mechanics found in our textbooks and taught to students from generation to generation have the form that was given them by Euler. It is, therefore, important to know what opinions a man like Euler would have about the origin of the ideas he was so instrumental in shaping. We are fortunate in having from him a popular exposition of his view of the world, which he wrote in the form of letters, addressed to a German princess whose preceptor he was. Now, the psychology developed in these letters is very interesting, and certainly does not deserve the contempt which d'Alembert expresses about it. In fact, Euler goes even further than Condillac in trying to put psychology on a biological basis, by discussing the physiological processes underlying our thinking. Of course, such a discussion must remain very rudimenditary and speculative, but at any rate he puts the question of the relation between our ideas and our nervous system. He fully accepts the thesis that our ideas have their origin in sensations which are located somewhere in the brain; he mentions the *corpus callosum* as the likely seat of sensations, the organ where they are collected and from which signals for motor reactions go out – thus a physiology perhaps a bit primitive, but essentially correct (we might rather speak of the cerebellum, but this does not change the general picture). When he comes to the concept of idea, however, he is very embarrassed, because as a mathematician, he feels the need for a sharp definition; and since he has no basis for identifying any part of the nervous system with what we call idea, he introduces the abstract concept of "esprit" as an immaterial, individual element, whose role is to elaborate ideas and combine them according to the rules of thought. With mathematical rigour, he specified the properties which differentiate these 'esprits' from matter – foremost among them the impossibility of localizing them in space and time.

The dark point in this system was of course the problem of how the connection between the 'esprit' and the sensations comes about. Euler plainly declares that the establishing of such a connection is a 'miracle': this was the statement that aroused d'Alembert's wrath. What it amounts to, however, is just a confession of ignorance; unable to imagine any material way in which sensations could be related to ideas, Euler introduces the 'esprit' as a conceptual (not to say fictitious) representative of this necessary, but unknown link. What has Condillac to say about this problem, which he had also to face (even though he does not explicitly inquire about the processes going on in the brain)? He seems to dismiss it by the remark that talking about the 'mind' is just confused thinking. This means, actually, that he felt as keenly as Euler the need for some clear conception of the relationship between sensations and ideas, and like the latter was at a loss what picture of the process to imagine. Unlike Euler, he was reluctant to operate with such elusive concepts as 'esprit' or 'âme', which do not provide any picture at all. This difference of attitude illustrates once more the fundamental opposition between the Cartesian procedure which Euler followed when he introduced the purely ideal, but sharply defined, concept of 'esprit', and the Newtonian method which Condillac favoured. At any rate, the comparison shows that the French philosopher was not unequal to the leading scientist of the time.

In the last quarter of the century, the intellectual movement inspired by the 'philosophes' was on the wane, but a brilliant constellation of young scientists had taken over the tradition they had initiated and was carrying on the exploration of nature with undiminished vigour. The impact of Condillac's theory of knowledge upon this new generation was profound, although it is not easy to document – they had little occasion to quote him (with one exception soon to be mentioned); but his influence is noticeable in all their manner of thinking. In the field of psychology, in which he was a pioneer, Condillac had direct followers in the 'idéologues' of the period of the Directoire and the Empire, but there was then hardly anything to add to his work – in fact, it is only in our own time that his views can be fruitfully taken up again in the light of the increased knowledge of psychical phenomena derived from experimental investigation. Incomparably more significant, however, was the influence of Condillac's didactic writings – the course of instruction he wrote up for his pupil the duke of Parma. These volumes, in which he develops his epistemological considerations with the same wonderful simplicity, straightforwardness and clarity as in his original treatises, were the source from which the French scientists, all through the years of the Revolution, the Empire and the Restauration, received guidance and inspiration.

The most explicit evidence we have of an immediate application of Condillac's ideas to a scientific problem concerns Lavoisier's elaboration of a rational chemical nomenclature. This was a major advance in the development of chemistry – much more important than the analogy with Linné's system of classification of the three 'reigns' of nature might suggest: for Linné's nomenclature was purely conventional, while Lavoisier's embodied his fundamental ideas – the precise definition of a chemical element and the theory of the acidifying principle, – which mark the beginning of modern chemistry. It was quite natural for Lavoisier, faced with a chaotic mass of arbitrary denominations, to feel the need for a more orderly designation of the substances whose composition he was able to determine. Nevertheless, the emphasis he laid on the significance of this task and the methodical way in which he accomplished it are directly inspired by Condillac's teaching. Both in the preface to the *Méthode de Nomenclature chimique* and in the introduction to the *Traité de chimie*, Lavoisier invokes the theory of the function of language presented in Condillac's *Logique* for a deeper justification of the characteristic features of the nomenclature he proposes. In the framework of Condillac's psychology, language occupies a prominent position. Emanating somehow from the sensations, the ideas and the relations between ideas have in turn to be represented by material signals: these constitute the language, which therefore provides us not only with a means of communication of our perceptions, but with an exquisitely flexible tool for the logical combination of ideas. Hence, it is a true analytical method, by means of which we proceed from the known to the unknown, just as mathematicians do. A language, describing a certain order of phenomena can even attain, ideally, the same precision as mathematics: it then becomes what we denote as the 'science' of these phenomena – 'a science is a well-contrived language'. Mathematics itself is a language – "la langue des calculs", as Condillac calls it. When one realizes that his conception of language, with all its implications, was vividly present to Lavoisier's mind, one understands the full import of his concern with nomenclature and his insistence on the active part it could play in the investigation of chemical processes.

There is in Lavoisier's considerations on the role of language a remark of striking modernity: the scientist, he says, is like a child. The child has also to learn the language, and how to use it. In this learning process, he makes mistakes, which he corrects by experience – let the scientist take a lesson from the child, and do the same. We know more, thanks to Piaget's studies, about the acquisition of language by the child, than Condillac and Lavoisier could imagine, and this more precise knowledge only goes to support Condillac's conception of the nature and function of language, and to strengthen the comparison of the scientist with the child. The learning of the language proceeds in two steps: in the first stage, the child has to learn the correspondence between the words and the sensory-motor schemes they symbolize; but the great discovery the child makes, in a second stage occurring at the average age of eleven, is that the word-symbols can be detached, so to speak, from the concrete representations of the sensory-motor schemes and used by themselves in an infinity of logical combinations. This abstract mode of thinking pays for its tremendous increase in speed and scope with the danger of losing in the process the necessary contact with the sensory-motor experience – a danger to which children and scientists are indeed equally exposed. Lavoisier has a lapidary formula to summarize the profound insight into our thinking which he owed to Condillac and had made his own: "faits, idées, mots sont comme trois empreintes d'un même cachet". Facts, ideas, words – three imprints of the same seal: is this not the wisest legacy of the enlightenment?

CHAPTER 21

UNPHILOSOPHICAL CONSIDERATIONS ON CAUSALITY IN PHYSICS

[1971b]

1. Causality in classical physics

Historically, the elaboration of the scientific concept of causality as a fundamental element of rational thinking is a by-product of the development of the formulation of the laws of motion of material bodies and their successful application to the dynamics of the solar system. Especially this last circumstance strongly influenced early epistemological reflection in encouraging radical idealizations, such as the law of inertia and the resulting notion of force as producing acceleration and thereby uniquely determining the motion. Thus, the concept of physical causality became endowed right from the start with the connotations of necessity and determinism. The belief in the adequacy of this deterministic causality was strengthened by the initial success of the attempt to reduce all physical phenomena to mechanical processes; such a view of the physical world was not necessarily atomistic, but it accepted the basic idea that all forces were essentially contact interactions between elements of matter. It is true that the force of gravitation, and later those of electricity and magnetism, were treated as long range interactions. This, however, was regarded as a phenomenological description, which ought to be reduced to some local form of interaction between the elements of ordinary matter and those of some subtle medium through which such forces could be propagated even in the absence of gross matter. Although this radically mechanistic view had to be abandoned, and the transmission of electromagnetic and gravitational interactions ascribed to autonomous agents, the basic causal structure of physical theory in its final classical form was fully upheld in this theory: all interactions of the material atoms and atomic constituents with the fields of electromagnetic and gravitational force are strictly local and lead to causal relations of a deterministic character.

Let us at once emphasize how remote this abstract form of physical causality is from the immediate perception of causal relations manifested by adult subjects when they are confronted, as in the classical experiments of Michotte and his school [1, 2], with moving visual patterns simulating the collision of solid bodies. Not only is a causal relationship perceived even when the collision represented is dynamically impossible, but this relation is felt to arise before the contact and subsist after it within finite intervals of separation of the colliding bodies – their 'action radii' – depending on their velocities. When the body to which momentum has been transferred has moved away from the other farther than its action radius, its motion is perceived as autonomous and no longer related to the collision (unless the velocity it acquires is considerably larger than the initial velocity of the other body). Indeed, the descriptions of the process perceived, given spontaneously in common language by the subjects, are strikingly reminiscent of an ancient and medieval conception of the causality of motion, of which we find the earliest historical evidence in the work of the Byzantine scholar, Johannes Philoponos [3]. Just as the latter argued that the motion of a stone or an arrow is due to an incorporeal 'kinetic power' (later called 'impetus') communicated to the mobile by the hand or the bow, the subject speaks of 'motion' as of something with which a moving body is endowed and which may pass by contact onto another body. The finite action radius of perceptive causality corresponds to the view expressed by Philoponos that the supply of kinetic power given to the mobile in the initial contact with the mover is finite and can only maintain the motion over a finite distance.

A further feature common to perceptive causality of the transmission of motion and to the medieval conception is the complete absence of any notion of inertia: the transmission of motion is conceived as a one-sided process, in which one body acts upon another to set it in motion, or modify its motion; the latter body thereby remaining entirely passive and without influence on the process. This last peculiarity of the type of perceptive causality exhibited by adults is the more remarkable as it is in regression from the causal anticipations and explanations of mechanical processes to which children are led in the course of their spontaneous mental development. In fact, the investigation of the formation and elaboration of schemes of causal relations by children, currently pursued by Piaget and his school,[1] shows that children do take account of the fact that a body exerts more or less "resistance" to being set in motion – an attitude quite in harmony with the general predominance, in the child's thinking, of experience derived from his activity over purely sensory elements. A study by R. Droz,[2] in particular, reveals that this notion of resistance of a body to being moved, which may be regarded as a prefiguration of the Newtonian concept of inertia, is even incorporated by some children in their spontaneous description of the perception of Michotte visual patterns simulating transmission of motion. It would seem, however, that this link between the perception of causal relations and an improved dynamical representation of the processes does not persist, and that the adult reverts to a more primitive form of perceptive causality, dominated by the sensory pattern.

An essential feature of this pattern, emphasized by Michotte, is its global character; as soon as a break occurs in the continuity of the process represented, any perception of causal relation disappears. This global character of perceptive causality is the opposite of the analytical mode of description of classical mechanics, in which the motion is indefinitely subdivided into elements characterized by instantaneous positions and momenta of the moving bodies. Here the relation of causality expresses the existence of a necessary and unique dynamical link between neighboring elements of the motion. Necessity and uniqueness are foreign to perceptive causality: a slight variation of the circumstances suffices to wipe out the perception of a causal link in the global pattern observed. There is no question, therefore, of representing (as some physicists uncritically do) the abstract form of deterministic causality of classical physics as corresponding to some deep-lying requirement of our mental setup. Certainly, the concepts of classical physics may be said in a loose way to originate from 'common experience'; at any rate, the systematic experimental studies of child behavior initiated by the Piaget school have already traced some of them, in a rudimentary form, to their origin in the sensorimotor schemes built up by the child in the course of his mental development; but their final elaboration demands the highest power of abstraction of which human minds are capable. They represent idealizations exquisitely adapted to a symbolic, codified description of the domain of experience ultimately accessible to us through sensory perception; but such a refined code system is not the one we need for the guidance of our day-to-day behavior in the physical world: for this purpose, unsophisticated sensorimotor schemes still suffice in our scientific age, as they sufficed for the millennia of human history preceding the emergence of modern science.

In a sense, Lagrange was right to describe 'analytical mechanics' as a branch of mathematics; but he made a fatal error of judgment – shrewdly pointed out by his contemporary Poinsot [4] – in ascribing to mathematical reasoning the power of establishing the necessary truth of physical laws of absolute generality. He underestimated the danger inherent in the idealizations that make the mathematical formulation of such laws possible and give them such sharp precision and supreme elegance: this result is only achieved by fixing the scope of the idealizations according to mathematical criteria, at the risk of extending it beyond the limit at which these idealizations cease to give an adequate representation of the physical phenomena they purport to describe. Increased experimental knowledge of the phenomena may thus force us to assign limits to the validity of a physical concept; such a step is not only a progress in scientific theory at the practical level, but also a major epistemological advance, since no physical concept is sufficiently defined without the knowledge of its domain of validity. A case in point is the law of inertia itself, whose simple Newtonian enunciation, far from being universal, is now recognized as only the asymptotic form of a more general one, which is part of a theory taking into account physical conditions for spatiotemporal localization, ignored in Newton's mechanics. From this point of view, the deterministic type of causality of classical physics also appears as an idealization rooted – not (as we have seen) in any psychical experience, still less in any innate property of the mind (that will-o'-the-wisp of philosophers) – but in highly sophisticated experience of the physical phenomena. Accordingly, the determination of its domain of validity is not a matter of arbitrary decision, but entirely one of inference from experiment.

2. Conservation laws

So far, we have discussed the causal relations of classical physics as they appear when one looks at the evolution of the phenomena in space and time. There is, however, another aspect of causality which is not concerned with spatiotemporal localization of a physical process, but with the balance of energy and momentum exchanged between different parts of the system in the course of the process. In its full generality, including the phenomena of heat and electromagnetic radiation as well as purely mechanical processes, this accountant's view of causality is surprisingly modern. During the middle ages and the Renaissance, the dream of a *perpetuum mobile* was entertained even by the most skillful and experienced craftsmen – like that thirteenth-century church builder, Villard de Honnecourt, who jotted down the sketch of such a contraption in his notebook [5]. When it was banished from the realm of mechanics in the eighteenth century, the secret hope of discovering some inexhaustible source of power persisted in less well-known domains of physics. Thus Joule himself spent much effort in constructing electromagnets because he originally expected that, with increasing intensity of the exciting current, the power he could gain from the magnet would increase more rapidly than that needed by the battery to produce the current [6].

Even in the domain of purely mechanical phenomena, the long and confused controversy during the eighteenth century about the conservation of 'motion' in collision processes had not thrown sufficient light on the role of the forces operating the transmission of motion between interacting bodies, and the existence of a common measure for force and motion, regulating the transmutation from the one into the other, remained quite unfamiliar down to the middle of the nineteenth century. Accordingly, the idea of an equivalence between heat and mechanical work was also foreign to the then prevailing mechanistic conception of the physical phenomena, even for those physicists who (like Laplace) were perfectly aware of the possible identification of heat with molecular motion. When the young engineer Colding submitted to Oersted his thoughts about the production of heat by friction, which he wanted to test by large-scale experiments, Oersted's reaction betrays a confusion characteristic of this general attitude: there can be no relation between the force applied and the heat produced, he objects, since, assuming heat to be due to some kind of motion, the total momentum of this motion (which he takes as its measure) is always zero; no applied force can be 'lost' – this was d'Alembert's conception – except by being compensated by the reaction of some constraint. (Nevertheless, let it be said to his and his colleagues' credit, they did give Colding all the support and help he needed to perform and publish his experiments!) [7].

This unpreparedness of the mechanistic school for the aspect of the physical processes expressed by conservation laws is very instructive: it convincingly shows that this aspect was quite a novel one, of which the most experienced masters of rational mechanics were not conscious. Of course, once attention was focused onto these conservation laws, it was easy enough to exhibit them as logical consequences of the laws of motion, and to define the concepts of kinetic and potential energy in order to describe the transmutation of energy continually going on between work and motion in any dynamical process; the fact remains, however, that the existence of this transmutation had to be pointed out to the chevroned physicists by young outsiders whose vision was not obscured by too much knowledge. From the epistemological point of view, this means, obviously, that in spite of the trivial (and, as we shall soon see, misleading) circumstance that they are formally interrelated, the two aspects of the causality of physical processes arising from the consideration of their evolution in space and time, on the one hand, and of the exchanges of momentum and energy taking place in them, on the other, are actually independent of each other and correspond to two different ways of looking at the phenomena, both equally necessary for obtaining a complete picture of them.

3. Complementarity in Quantum Physics

Within the scope of classical physics, the logical relation between the two aspects of causality we have discussed does not seem to present any problem; and indeed, as soon as mechanistic physicists of the nineteenth century realized that the equivalence of work and heat was a simple consequence of the interpretation of heat as the kinetic energy of molecular motion, they were confident of having 'reduced' the laws of heat phenomena to the Newtonian laws of motion. The extension of these laws to the atomic domain, where their applicability could not be directly tested, was not regarded as problematic, since the unlimited validity of the idealized concepts of mechanics was taken for granted. Mach's opposition to atomism, although (like all long-range predictions of prophets) it appears misguided to our increased wisdom, was actually based on a pointed questioning of this very extrapolation of our usual space-time description to atomic systems [8, 9]. We are not justified, he argued, in attributing to these hypothetical constituents of matter, which we neither can see nor touch, a location or a displacement in space, since these notions only acquire a meaning by being referred to sensations of sight and touch, in which they originate. We should therefore not expect the parameters needed to specify the state of an atomic system to be the same as those describing a system of ordinary material bodies. The causality governing atomic systems need not be of the spatiotemporal type; it could have the more general form (of which the conservation law of energy is an example) of a set of relations between parameters of a different kind. This argument suffers from Mach's general tendency to envisage the relationship between physical concepts and sense impressions as more direct than it actually is. Anyhow, it was too inconclusive to make an impression on the advocates of a representation of the phenomena that had proved so successful. They went on in their own naïve way, taking their description of the molecular motions as an account of the spectacle that would offer itself to a human observer who, like Alice, would have eaten a bit of the right-hand side of the mushroom.

The causality problem took an unexpected turn, however, after the discovery of the quantum of action and the subsequent development of a theory of atomic constitution and atomic processes incorporating in a rational way the content of the quantum postulates and the correspondence principle. This theory operates with the same physical idealizations – space-time localization and momentum-energy exchange – as classical physics, for the simple reason that the conditions under which individual atomic processes are observed are essentially the same as those of ordinary observation. All the information we are able to obtain about atomic processes has ultimately to be referred to the indications recorded by experimental devices whose functioning is entirely described in classical terms. Now, the new feature introduced by the existence of the quantum of action is the occurrence of a mutual limitation in the use of the two modes of description just mentioned, which in classical physics, when quantal effects are ignored, become logically compatible without any restriction. It is important to realize that the limitation in question is not due to any imperfection of the classical idealizations: within its scope (excluding the domain of phenomena in which the structural properties of nucleons and mesons come into play), quantum theory is just as exquisitely adapted to the account of the experiments as classical theories are within their own scope. There is no limit to the separate application of either space-time localization or momentum-energy conservation to an atomic system, and there is no question of eliminating either of them, since both correspond to significant and distinct aspects of the system.

However, the conditions of observation allowing us to localize the system and those allowing us to set up the momentum-energy balance of the processes taking place in it are no longer compatible – they are mutually exclusive.

We are here confronted with a peculiar epistemological situation, which is not of our own choice, but is imposed upon us as a straightforward consequence of the existence of the quantum of action. In order to remind us of its implications, the concept of 'complementarity' has been introduced: it denotes a relation between two physical phenomena, both exhibiting aspects of a given system indispensable for a complete account of its behavior, but whose conditions of observation are mutually exclusive. It is necessary, in view of the occurrence of this new type of relation, to include in the definition of a 'phenomenon' the complete specification of the experimental arrangement by means of which it is observed; an essential item of such an arrangement is a device in which the characteristics of the process are registered in some codified form, thus ensuring the complete objectivity of its description. It cannot be too strongly emphasized that the concept of complementarity (like all our concepts) is just an element of a convenient code system, serving to warn us of the caution to be exercised in the use of classical physical concepts in order to avoid logical contradictions. Far from implying any deviation from ordinary logic, it helps us to keep our terminology in harmony with the strict logical deductions of the mathematical formalism. If one chooses to speak of 'contradiction' in connection with the mutual exclusion of complementary phenomena, then one can only mean a 'dialectical' contradiction; in fact, one may regard the occurrence of complementarity relations in quantum theory as a clear and precise example of what is implied by the general conception of 'dialectical process.' The quantum theory, which comprehends the complementary aspects in a logically consistent whole, represents the synthesis resolving the dialectical opposition between these aspects.

The two complementary modes of description of quantum phenomena correspond in classical physics, as we saw, to two independent types of causality, both deterministic. In order to achieve the synthesis of these complementary aspects, quantum theory abandons the excessive idealization of determinism and replaces it by the more comprehensive and flexible type of causal relations provided by the mathematical formalism of probability theory. From the logical point of view, this step is in harmony with the dialectical character of the theory; the logical formulation of a dialectical contradiction, according to Apostel's analysis [10], essentially involves an element of probability. From the physical point of view, the use of a statistical mode of description is immediately suggested by our most common experience in the investigation of atomic processes. When we have fixed all the controllable circumstances of an experiment, we usually observe, not one definite type of process, but a great diversity of phenomena; under such conditions, we can expect to find regularities only in the statistical distribution of the various properties observed. Thus, it was evident since the earliest formulation of the quantum postulates that radiative transitions between stationary states of atomic systems could only be characterized by the rate of probability for their occurrence, since there is nothing in their definition that could determine the time at which such a transition would take place.[3] Indeed, the concepts of stationary state and monochromatic radiation are extreme idealizations excluding every time determination, and the situation referred to in the quantum postulates is strictly complementary to the time evolution of radiation processes.

4. Structural identity of atomic systems

The statistical method necessarily excludes from consideration individual differences between the objects with which it deals; it is concerned with the properties of species, not of individuals. It is this feature that most obviously distinguishes causal relations of the statistical type from classical causality. That a statistical method should be appropriate to the analysis of the properties of matter in bulk in terms of atomic processes, is merely a consequence of the fact that the conditions of observation of these properties exclude all possibility of determining the underlying dynamic behavior of the atoms. We have here two modes of description that are complementary, not on account of quantum limitations (which play no part in this argument), but because of the definitions of the quantities characterizing the macroscopic states of a physical system. In addition the formulation of the laws governing such states involves the neglect of all effects of the deviations of individual atoms from their average behavior under the given external conditions. Accordingly, the atomistic description of the phenomena (as was clearly recognized by Maxwell and Boltzmann) has to assign all the constituent atomic systems of a given species identical structures and to treat them as indistinguishable and interchangeable [11, 12].

In this idealization of identical atoms there is in the first instance an element of arbitrariness, inherent in the statistical method: How must we define the atomic or molecular species, whose members shall have identical structures? Obviously, this definition will depend on the experimental possibilities for establishing such specific distinctions. The history of chemical 'philosophy' since Lavoisier displays a succession of advances in structural representations of molecules, each reminding us of the relativity of the concept of molecular species and of the pitfalls into which many a skillful chemist stumbled, under the illusion that he was sticking to 'facts' and shunning 'hypotheses.' Thus, Stas believed that his accurate determinations of atomic weights disproved the hypothesis of a common building unit for the atoms of the chemical elements. Kekulé then showed a better grasp of the problem when he objected that the results of Stas were equally compatible with the hypothesis of a statistical distribution of the masses of the individual atoms of the same element. He added, furthermore, that it would then be conceivable that reactions could be found leading to a segregation of the heavier and lighter atoms – a pure speculation, he was anxious to stress, but it was a speculation based on sound logic [13].

Atomic theory, however, has now reached the stage in which the conditions of observation of individual atomic processes have put an end to speculation and removed all arbitrariness. We may now regard as firmly established the identity of the constituents of atomic systems, electrons and nucleons, and the identity of structure of the atomic systems themselves, in all processes involving momentum and energy exchanges small enough not to affect the stability of the constituent particles. Indeed, in conjunction with the quantum laws to which atomic particles and systems are subjected, the identity of these elements acquires a fundamental significance, by leading to definite limitations of the possible symmetry properties which an ensemble of such interchangeable elements exhibits in its various quantal states. The specification of these symmetry properties is the key to the interpretation of types of interaction and correlation entirely outside the reach of classical ideas, and playing an essential role in the economy of nature: on these interactions and correlations depend, among others, the most important molecular bonds, the properties of the metallic state, as well as the striking phenomena of superconductivity and superfluidity appearing at low temperatures. The foundation of this vast body of knowledge would be wiped away if some 'hidden parameter,' embodying any further specification of atomic states than that given by quantum theory, crept out into the open.

The statistical causality of quantum theory refers to a deeper level of analysis, logically independent of the statistics of matter in bulk we have just been considering, and it affects the classical notion of a particle in a much more radical way. In the space-time representation of the motion of a particle, there is no place any longer for the classical picture of a continuous temporal succession of positions of the particle along a spatial trajectory; the link between successive observations of the spatial position of the particle is purely statistical. This not only eliminates from the quantal description of an atomic process any possibility of individualizing the particles taking part in it, but even of ascribing to these particles the permanence they would still retain in a classical description. Indeed, the process is completely characterized by indicating the numbers of particles of each species initially present in the various states they can occupy (which may, for instance, be spatially localized) and the changes these numbers undergo in the process. In the domain of atomic physics, these changes normally conserve the total number of particles of each species, but this is a much weaker statement than is implied by the classical notion of permanence of the particles. There is no controllable way in which the changes in the occupation of the states could be ascribed to transitions of definite particles from one state to another. A noncommittal way of accounting for the immediate data of observation would be to say that particles 'appear' in certain states and 'disappear' from others; one prefers to speak of 'creations' and 'annihilations' of particles in the states in question. The connotation of activity contained in this terminology has the merit of recalling that the observed appearances and disappearances are due to actual dynamical interactions. However much this quantal outlook departs from the classical picture of the evolution of systems of material bodies in space and time, it fulfills in the domain of atomic physics exactly the same function as the latter in its own domain: it is the only mode of objective conceptual representation of the phenomena that is accurately adapted to the experimental conditions.

The impossibility of attaching to the constituents of the atomic world the attribute of permanence inherent in the concept of material body used in classical physics will appear in its proper light when we realize how extreme an idealization this abstract notion of permanence involves. Here again, it will be instructive to turn to the psychology of perception; one aspect only of the very complex problem of perceptive permanence[4] need detain us. The typical situation exhibiting the perception of permanence of motion is the 'screen effect': if the subject is shown a red square, along a side of which a white narrow band is made to appear and expand gradually in the direction perpendicular to this side, what the subject perceives is a solid white band emerging from behind the red square that hid it from view. A corresponding response is obtained with the inverse pattern, in which the white band initially present alongside the red square is made to contract and vanish: the subject perceives the band gliding solidly behind the screen. Now, one may combine the two patterns along two opposite sides of the red square, in such a way as to simulate the entry of the white band on one side and its exit on the other; one then obtains a remarkable 'tunnel effect.' If the motions of entry and exit are the same, and the time interval between them has the right order of magnitude, the subject perceives the motion of the band as permanently continued inside the tunnel; if the entry-exit time interval increases slightly, the impression of permanent motion persists – only, there is a 'hitch' somewhere inside the tunnel, at a place which, curiously enough, the subject indicates with precision. However, this perception of the permanence of the motion disappears rather sharply when the entry-exit time interval reaches a critical value (depending on the other circumstances of the experiment). The subject then finds that the motion of entry 'dies out' and that a new motion starts somewhere near the exit side, with an 'empty space' in between. Graphical representations of the perceived motions, drawn spontaneously by some subjects, show in the former case a continuous trajectory, in the latter two branches of trajectory with an intermediate gap. Continuity of the motion, therefore – as a direct or indirect[5] component of the sensory pattern – is an essential condition for ascribing permanence to it in ordinary perception. If this element of continuity is not present under the conditions of observation, the unsophisticated observer reacts in much the same way as the atomic physicist.

5. By way of conclusion

The assessment of the epistemological novelties of quantum theory is often obscured by the belief that the causal structure of classical physics is such a compelling exigency of our *Anschauung* that it is the mold in which every physical theory should necessarily be cast. The experimental study of perception clearly shows that this belief is unfounded: the needs of ordinary perception of causality are satisfied by a mental representation of the motion of bodies much more primitive than the abstract conceptual framework of classical mechanics, which in fact has to respond to the exacting requirements of a rather singular mode of interaction of man with his environment. Nevertheless, the two responses have a common origin, and it is important to understand how such a sharp differentiation between them could arise, lest one draw unwarranted conclusions from its presence.

The decisive point is that mental development does not proceed along one uninterrupted line, but by cycles or stages, each of which leads to temporary harmony between the child's limited environment and the system of mental operations enabling him to cope with it. The ultimate stage is that ending with the acquisition of the 'formal operations' which remain the tools of adult logical thinking. The attainment of this stage is prepared by the preceding one, the stage of 'concrete operations,' based on the direct use of the sensorimotor schemes of perception.

The characteristic difference between the two stages lies in the function of language. At the stage of concrete operations, language is simply a means of communication of sensorimotor experience: words are incorporated into the significant combinations of sensorimotor schemes, and constitute a symbolic representation of the latter. Since the sensorimotor schemes are themselves symbols of the aspects of experience retained as significant, one may say without excessive schematization that two correlated sets of symbols, or code systems, are used for the registration and communication of experience [14, 15]. At the formal stage, however, the verbal code system pursues an autonomous development by purely abstract derivations of new concepts without immediate correspondence in the sensorimotor field. It is this formal refinement that makes scientific thinking possible, while it is obviously of little use in daily perceptive experience: hence, in particular, the discrepancy we noted between perceptive and physical causality. This detachment of the formal code system from the concrete one does not proceed without raising its problems, above all the ever recurring one of the adequacy of conceptual constructions whose link with sensorimotor experience is only an indirect one. The only way to clarify such questions is to go back to those basic idealizations which remain firmly anchored in perception, and to examine to what extent they can be consistently related to the other formal elements of the theory. In this respect, classical and quantal theories are in the same situation and the arguments by which their respective domains of validity have been bounded follow exactly the same pattern. Only, the way from idealization to sensory experience is a bit more roundabout for an atom than for a planet.

NOTES

[1] These problems were recently discussed at the thirteenth symposium on genetic epistemology held in Geneva in June 1968. I take this opportunity of expressing my gratitude to Professor Piaget for enabling me to attend this symposium and report on the epistemological situation in modern physics. The present paper is a version of this report, revised in the light of the new information discussed at the symposium.

[2] This study was presented at the thirteenth symposium on genetic epistemology mentioned in the preceding note.

[3] This was repeatedly mentioned by Niels Bohr in conversation.

[4] A thorough discussion of the whole problem, including the recent investigations of the Michotte school, will be found in the works by A. Michotte and his collaborators, cited above.

[5] This refers to the distinction proposed by Michotte and Burke between the 'modal' and 'amodal' data of sensation: the former are part of the immediate sensory content, the latter are only derived from it, as for instance the invisible part of the motion in the screen and tunnel effects. Piaget's genetic investigations indicate that such amodal elements may arise from the coordination of sensorimotor schemes in the child's early mental development: the screen effect is manifest in the behavior of ten-month-old infants.

BIBLIOGRAPHY

[1] Michotte, A., *La perception de la causalité*, Publications Universitaires de Louvain, 1954.

[2] Michotte, A. *et al.*, *Causalité, permanence et réalité phénoménales*, Publications Universitaires de Louvain, 1962.

[3] Dijksterhuis, E. J., *Val en Worp*, Noordhoff, 1924, pp. 36–40.

[4] Poinsot, L., 'Théorie générale de l'équilibre et du mouvement des systèmes', *Journal de l'Ecole Polytechnique* (1805); reprinted in L. Poinsot, *Eléments de statique*, 9e édition, Bachelier, 1848.

[5] Album de Villard de Honnecourt (Bibliothèque Nationale ms. fr. 19093, f°5r°);

reproduced in P. Mérimée, *Etudes sur les arts du moyen-âge*, édition Flammarion, 1967, p. 237. See further L. Whyte, *Medieval Technology and Social Change*, Oxford University Press, 1962, pp. 129 ff. and related notes.

[6] Joule, J. P., *Scientific Papers*, Taylor and Francis, London, 1884, vol. 1, p. 14 (from a letter published in 1839).

[7] Marstrand, V., *Ingeniøren og Fysikeren Ludvig August Colding*, Danmarks Naturvidenskabelige Samfund, 1929, pp. 23–24.

[8] Mach, E., *Die Geschichte und die Wurzel des Satzes von der Erhaltung der Arbeit*, Calve, 1872. [English translation by P. E. B. Jourdain, *History and Root of the Principle of the Conservation of Energy* (Open Court, Chicago and Kegan Paul, Trench, Trübner & Co., London, 1911) – Ed.].

[9] Mach, E., *Principien der Wärmelehre*, J. A. Barth, 1896. [English translation in process: *Vienna Circle Collection* (Dordrecht-Holland and Boston-U.S.A.) – Ed.].

[10] Apostel, L., 'Logique et dialectique', in *Logique et connaissance scientifique* (ed. by J. Piaget), Encyclopédie de la Pléiade, vol. 22, Gallimard, Paris, 1967.

[11] Rosenfeld, L., 'On the Foundations of Statistical Thermodynamics', *Acta Phys. Polon.* **14** (1955), 3. [This volume, p. 762].

[12] Rosenfeld, L., 'Questions of Irreversibility and Ergodicity', *Rend. Scuola Intern. Fis. "Enrico Fermi"* **14** (1962), 1. [This volume, p. 808].

[13] Gillis, J., 'Auguste Kekulé et son oeuvre réalisée à Gand de 1858 à 1867', *Acad. Roy. Belg., Classe Sci., Mém.*: Collection in 8° **37** (1966), fasc. 1.

[14] Rosenfeld, L., *The Method of Physics*, UNESCO report, 1968. [This volume, p. 614].

[15] Pavlov, I. P., 'An Attempt to Understand the Symptoms of Hysteria Physiologically' (1932), in *Conditioned Reflexes and Psychiatry*, International Publishers, New York, 1941, pp. 113–114.

CHAPTER 22

IRREVERSIBILITY – A LAY SERMON

(On the Occasion of Professor K. Bleuler's Sixtieth Birthday, as a Token of Esteem and Friendship)

[1977b]

I am not sure whether a lay sermon demands a text, but if so, I would choose the report (*Acta* **17**, 16–34) of Paul's preaching to the Athenians. Curious of novelties as they were, they quietly let him talk – until he came to mention the resurrection of the dead. Then they laughed at him and went away, saying: "We'll listen to this another time". Religious people will not appreciate the Athenians' humour; Marxists, for instance, will severely tax them with positivism. Yet, did they not just apply the supreme scientific criterium of truth, or plausibility, namely common sense? as physicists do, or ought to do. Now, the physicists' reaction to the flow of time is not as simple-minded as the Athenians': while they acknowledge the irreversibility of all processes they directly observe, they do assume strict reversibility in 'another world' – the world of atoms, of which until recently they had no better knowledge than the apostle could have of the paradise or the other place. This ambiguous attitude has a long history, on which it will not be irrelevant to look back.

The time concept, as a means of ordering the succession of events in the world, including our own part in them (to which we attach such disproportionate importance), is by no means a primitive one: the Eskimo, for instance, have no root for it in their elaborate language; they do, however, form a compound word, *piv-fik*, from the root *piv* denoting 'an event' and the location suffix *fik*, by which they aptly express the idea of a general frame in which 'things happen'. There is no use for greater precision in a society leading a monotonous existence in an almost unchanging environment.

A real need for a more accurate control of time arises in agricultural communities, and we have very early evidence of such peoples taking advantage of celestial bodies' cyclic motions as natural clocks. In the Mesopotamian creation myth 'Enuma elish', the first act of the creator god, after separating earth from heaven, is to make a calendar by assigning the celestial bodies their regular routes; one of them, Nibir (Jupiter), is even appointed as a kind of 'watchdog', seeing to it that the others do not wander from their courses – a detail revealing the preoccupation of ensuring the conservation of this celestial calendar. The reality to which this myth gives the consecration of a divine origin is the actual construction of a more or less serviceable method of prediction of the return of important seasonal phases, based on long sequences of observations, carefully recorded and combined – by simple linear interpolation – in an endeavour to bring out periodicities. Concurrently with this practical goal, the more precarious hope was entertained of reading in the stars the fate of men – the emphasis in this elusive guess being laid on unique features irrevocably leading to an irreversible evolution.

Three-sided relief. Cast with proposed restoration. 5th century B.C. Courtesy, Museum of Fine Arts, Boston, H. L. Pierce Fund.

Thus, right from the start, those observers of the heavens were unwittingly confronted with the relation of complementarity we now recognize between the irreversible flow of time, on the one hand, and, on the other hand, the measure of time, which is based on periodic, i.e. essentially timeless, processes, governed by conservation laws; the most primitive form of such laws being the reproduction of a definite pattern of celestial phenomena – some conjunction of planets, or the position of some celestial body with respect to a constellation, or similar recognizable configurations. We know that a complementarity situation cannot be represented in a single picture; yet Greek imagination managed to do it: on the famous throne in the Boston museum there is an illustration of the myth of Adonis which embodies the two aspects of time in all their ambiguity. Between the seated deities of life and death, facing each other, stands a winged figure holding a lever (which was originally protruding and is now broken off); at each end of this lever hangs a weight on which a human body is depicted: on the side of Persephone it is collapsing in the pangs of agony, on the side of Aphrodite it is surging to life with uprised arms. Thus, at each swing of the lever, a new being is lifted from the deep to the smile of the life-giving goddess, while an old one sinks down to the realm of the mourning queen of Hades. But who is the winged god that wields the lever and thus rules the succession of life and death? A pessimist will recognise him as death-bringing Thanatos; an optimist will say that it is the love-god Eros: was it the artist's intention to leave us with this doubt?

The transition from the mythical expression of cosmic problems to their modern scientific formulation took place in Western Europe under a strong incentive from the expansion of the overseas trade, with the attending problems of long-range navigation. The problem of the determination of the longitude in the open sea put exacting demands on timekeeping methods of high accuracy. Its solution was sought concurrently in two directions: the traditional one of using some celestial clock, and the novel one of building a material clock, essentially a pendulum, sufficiently well regulated and yet sturdy enough to resist the shaking of a stormy sea. This was the ultimate source of tremendous progress both in astronomy and mechanics, culminating in Huygens' supremely elegant (though technically too delicate) solution of the pendulum problem, and in Newton's all-embracing cosmology. However, the very emphasis on the measurement of time gave this development a bias of far-reaching consequence towards the reversible types of motion: the emerging world-picture became one of a huge mechanism, the elementary pieces of which, the atoms, were assumed to be governed by strictly reversible laws of motion. Curiously enough, it took a long time – and some acrimonious controversy – before the conservation laws inherent in such a system were recognized; but its determinism was never questioned. It is true that Newton did not believe the pattern of the solar system could be stable: it would have to be restored from time to time by a direct intervention of the Creator; but he took it for granted that it was the Creator's intention to maintain this pattern eternally repeating itself, as a good-time-keeper ought to do.

Yet there were other circumstances to remind people of the frailty and decay of human fortune. Overseas trade was profitable, but hazardous, and the merchant-adventurers felt the urge to protect themselves against the risks they incurred, and also to make more quantitative estimates of their own resources and of those of their competitors. The response to these preoccupations gave birth to a new science, which we now call statistics, operating with a new concept, probability – a quantification of the vague notion of chance. 'Statistic' is one of those pedantic words the Germans are so fond of: it dates from the beginning of the XIXth century and is supposed to denote the study of 'states' i.e. of the political entities so called. Sir William Petty, when he made the first attempt to compare the resources of France and England 'by number weight and measure', had given his book the title 'Political arithmetic', a more homely expression of the same idea. As to the 'doctrine of chances', it was fully worked out in Holland and in England, with the aim of computing fair insurance premiums, as well as annuities based on life-expectations derived from carefully compiled tables of mortality. The crowning piece to the edifice was Bernoulli's formulation of the law of large numbers, which establishes the concept of probability on an empirical basis – by deriving its value in each specific case from the observation of sufficiently long sequences of events – and thus immensely widens its scope.

The discovery that even casual events could be reduced to mathematical laws of their own would seem at first sight a challenge to the determinism of the mechanistic world-view; but the prevailing theological approach to the theory of knowledge could all too easily fit them into a comprehensive system. As Bernoulli explained, occurrences can either be necessary or 'contingent': the former result from deterministic laws ordained by God, the latter depend on man's free will and thus apparently escape the bond of necessity; however, the gift of free will to mankind is also part of God's creation and its exercise cannot therefore, disturb the harmony of the world. Such endeavours at reconciling 'contingency' with necessity by appealing to an 'act of God' (this is the phrase still used by insurance people to designate imprevisible events) entailed a subordination of the former type of causality to the latter, which effectively concealed the true relationship between them, which we now know to be no other than the complementarity between timeless conservation laws and the irreversible time evolution of the phenomena. This was a high price to pay for the sake of a formal consistency which, after all, was very fragile, depending as it did on the assumed solicitude of a personal God. When Laplace tried to formulate the deterministic ideal of causality without this assumption, he realized the need to substitute for it, as an absolute frame of reference for his world-picture, the fiction of an 'intelligent being' capable of encompassing its totality at some instant, from which both its future and its past course would be uniquely determined; in his intention, this 'being' (however much his powers exceeded human capacities) would just be a representative of ordinary mankind, divested of any divine attributes, and himself a part of this universe which would have no secret for him. This, however, is plainly impossible: he would be doomed to keep outside it, as a mere shadowy spectator, not unlike the mythical character of the 'Enuma elish' who bears the 'tablet of destinies', on which are inscribed the fates of all men.

CHAPTER 23

BERKELEY *REDIVIVUS*

(Review of W. Heisenberg's *Natural Law and the Structure of Matter* [Rebel, London, 1970])

[1970c]

In 1919, during the street fighting in Munich, while the din of shooting filled the air, a schoolboy called Werner Heisenberg, installed on the roof of the seminar building, was reading the *Timaios* – a near enough approximation to the prophet receiving the law on the mountain amid thunder and lightning. In 1964, with a prodigious career of creative work behind him, Werner Heisenberg chose to ascend another height, this time the very Hill of Pnyx, to proclaim from this hallowed ground the final triumph of Plato. This sermon has been preached before, by a famous bishop of Cloyne, among others, who thought he had routed atheism, just as Heisenberg now fancies he has liquidated its modern avatar, dialectical materialism. Berkeley's judgment was dimmed by his new theory of vision: Heisenberg is led astray by his new theory of matter.

Persiflage won't do, however, with a man of Heisenberg's stature. Every one of his pronouncements deserves close scrutiny – though not every one will bear it. In trying to expose the weak point of his plea for Platonism, I am reminded of Diderot's dictum about idealism being the scandal of philosophy, not so much because it is absurd as because it is hard to refute by rational argument. How could one find fault with Heisenberg's brilliant sketch of the course of Greek philosophy, or his penetrating characterization of the essential difference between it and modern science? Yet, on closer examination, the whole picture, dazzling as it is, shows itself utterly unreal, just because it is drawn from a radically idealistic point of view, with complete disregard of the material circumstances which alone gave the events and people substance and life. From the dramatic struggle of mankind for the control of nature he makes a shadow-play of idle philosophizing – thus demonstrating, much in spite of himself, the not very novel truth that disembodied ideas are mere *flatus vocis*, signs of things, not essentially different from material sensations.

When Heisenberg comes to modern physics, it is again hard to contradict him when he reminds us of the limitations of ordinary language for the description of atomic phenomena, or when he stresses, by contrast, the universality of an abstract mathematical description. Why, however, does he stop there? Why doesn't he adduce at this point the all-important fact that our mathematical concepts are also human inventions and accordingly bear the mark of their material origin? Why does he pretend to ignore the fact that the most abstract mathematical ideas only gain meaning from reference to sensory-motoric schemes? There is nothing, alas, in the Platonic trend he perceives in atomic physics but the pertinacious illusion of wonderful harmony in the mathematical structure of the laws of nature: a harmony that is very real, but appears less wonderful when it is realized that it only pertains to a deliberately schematized description of the phenomena. "The wonder is no wonder," said the old Stevin, who was an engineer.

I am sorry to conclude that there was really very little urgency to climb the Pnyx only to conjure up such weary ghosts. To this shadowy company, by the way, belongs as well the bogy of dialectical materialism that Heisenberg seems to have invented for the pleasure of slaying it. The conflict between materialism and idealism, under various forms, has played its historical part. If Heisenberg had been content to state that it has no longer any part to play in modern science, there would be no quarrel between us. Modern science operates with modern concepts – the most powerful of which, complementarity, is the direct outcome of the greatest discovery we owe to Heisenberg's genius.

CHAPTER 24

THE WAVE-PARTICLE DILEMMA

[1973b]

The dilemma between wave and particle aspects of the various constituents of matter and radiation arose after Planck's discovery of the quantum of action, and Einstein's quite independent proposal of a sort of revival of the corpuscular theory of light, again including the concept of the quantum of action in the form of energy or light quanta.

A famous confrontation of the two points of view took place in Salzburg in 1909 between their respective protagonists. The attitude of the two great masters, Planck and Einstein, is very characteristic of two tendencies that were then competing, and equally uncomfortably, in facing the quite unexpected dilemma that the theory of radiation presented. Planck took the conservative side. He took it openly by saying that, "in introducing novelties in physics one must proceed in as conservative a manner as possible." That was his outspoken principle: one must be very reluctant to take any new step which is not logically connected with the preceding one. In other words, Planck was willing and daring enough to overcome a prejudice, to use Dirac's pointed way of expressing the progress of physics, but he was uncomfortable about it.

Planck felt uncomfortable about abandoning the classical theory of radiation. And the discomfort, or hesitation, that he felt was an inhibiting feature in the progress of his own thinking. However, he had good reasons to feel uncomfortable and he expressed those reasons at Salzburg in very strong terms. He pointed out that the classical theory of radiation based on the field conception of Maxwell, or rather of Hertz to be more accurate, was supported by a great mass of experimental evidence about interference and diffraction phenomena, phenomena which were based on the superposition principle and which, therefore, could not, except by very forced assumptions, be incorporated in any corpuscular description of the propagation of light.

Einstein, on the contrary, represented the radical attitude. In fact, when he introduced the concept of the light quantum, he probably was not quite aware of the connection between his conception and Planck's introduction of the fundamental constant that bears his name. He thought that certain phenomena, like the photoelectric effect and the photochemical reactions, indicated that in the interaction between matter and radiation, the exchange of energy (and also, he added later, momentum) took place according to the laws of collisions of particles, so that radiation in such interactions acted as if it were composed of photons. However, when you treat those photons as independent particles you do not get Planck's radiation law; you get Wien's law, which is only a limiting case valid for high frequencies or for high values of the ratio $h\nu/kT$.

Very soon, however, Einstein realized that by abandoning the idea that photons composing a flow of radiation were independent, he could then recover Planck's law. His argument was essentially based on the consideration of fluctuations in the radiation field. He found the surprising result that the fluctuation in the radiation field, which is governed by Planck's law, was the sum of the fluctuations that would arise from a classical radiation field and those that would arise from an assembly of independent photons. As the outcome of this work, one had the very uncomfortable situation that the two aspects with which one had tried to describe radiation phenomena were now concurring in producing the result that was known to be the right one from Planck's analysis.

In Salzburg Einstein tried to defend nevertheless the radical assumption of the corpuscular structure of radiation by suggesting, although it could not be more than a qualitative suggestion, that the photons, or the light quanta as they were called then, were some kind of singularity, of concentration of energy and momentum inside a radiation field. The radiation field would so to speak guide the photons in such a way as to produce also the interference and diffraction phenomena that Planck wanted, with good reasons, to see reproduced by the radiation theory. Of course, none of these two opposing attitudes was satisfactory. Planck was perhaps too conservative and Einstein too radical.

Now one may ask at this stage what Bohr thought of the whole matter. Bohr was then still a student, but he was an uncommonly thoughtful student, and he was very eagerly following all those discussions. In this particular confrontation he took sides with Planck, but not entirely. He did not reject Einstein's conception of the light quantum, because he saw that this conception had its justification simply from the fact that it was a useful one, that it could form the basis of a method of analysis of the interaction between radiation and matter, which led to consistent results in a very simple way that the classical theory of radiation, of course, could not achieve. That was especially evident in the photoelectric effect. But why did he then not treat the two aspects of light on the same footing? Why did he give a sort of preference to Planck's view of the electromagnetic field as being in some sense more fundamental than the photon concept?

This is a point of view that Bohr never abandoned. The reason reveals the first germ in his mind of an attitude which was in the sequel extremely fruitful, and which he then called the correspondence argument. To put it in somewhat anachronistic terms, as we see it after the event, he attached reality, i.e. reality as it was defined by him, to those aspects that could be directly observed in certain limiting circumstances, by direct macroscopic observation. And, of course, in the case of radiation it is clear that direct observation in the limiting case of small values of $h\nu/kT$ gives the usual classical wave description of Hertz and Maxwell. As to the photon or the light quantum concept, introduced by Einstein, Bohr regarded it as a useful but an *auxiliary* concept, one which he later called symbolical, meaning thereby that it was not an aspect of the radiation phenomena which could be directly observed as such. It could only be observed through the bias of an application and verification of the conservation laws of energy and of momentum. Bohr always made this distinction between the two aspects of radiation.

By mentioning the word 'correspondence' in this context, at this early stage, I am of course being a bit unfaithful to Bohr's later idea of correspondence which was much deeper and also much bolder. In fact, the point that for low values of $h\nu/kT$ one has to expect the classical theory of radiation to be valid had already been made by Lord Rayleigh in 1900, when he pointed out that for low frequencies the application of the equipartition theorem gave the law that bears his name. Rayleigh also suggested that it could not be valid generally as it led to a divergence, but should be valid at least for the modes of low frequency.

When I mentioned to Bohr that this was perhaps the first germ of the correspondence argument, he protested. He said emphatically that "it is not the correspondence argument. The requirement that the quantum theory should go over to the classical description for low modes of frequency, is not at all a principle. It is an obvious requirement for the theory." Nevertheless that was the first beginning of it.

Bohr also realized at that time the importance of the quantum of action as a stabilizing element for the Rutherford atom. As soon as he became acquainted with the Rutherford model of the atom, he realized that this model had a number of consequences of the greatest importance. There was no doubt that this model had a serious grain of truth in it, since after all it was introduced by Rutherford to explain results of the experiments on the scattering of alpha particles, which would not be understandable in any other way. He pointed out already at that early stage that this model would, for the first time, allow a clear distinction between an atom and a molecule, a molecule being characterized by the fact that it has a structure with more than one nucleus and the appropriate number of electrons. He also made at that time a large amount of calculations on the stability of certain ring structures of electrons which would bind the nuclei of a diatomic or polyatomic molecule together.

Bohr also realized that radioactive phenomena, on the other hand, would be nuclear phenomena. It was for the first time that such a clear distinction was made, and this led him to formulate what were later known as the displacement laws of radioactivity, which on the view of the Rutherford model, were rather visible consequences. When he presented those considerations to Rutherford, the latter gave him a severe lesson by condemning the impatience of youth in publishing too quickly speculations which could only be justified by some more experimental material than was then available. Later on, when this material accumulated within the next few months, the displacement laws were recognized by Fajans and Soddy. It is rather curious to note that Fajans, when he published those displacement laws, asserted that they were an argument *against* Rutherford's model. Well, this is an aside.

The main thing was that Bohr was convinced right from the start that the stability of Rutherford's model was, as he expressed himself, of 'non-mechanical' origin. Rutherford's atom was essentially unstable, and it could only be stabilized by the introduction of an element which was not part of classical physics, and it was obvious that it should be the quantum of action. It was known already, not only by Bohr but also by other people, that the orders of magnitude were just about right. Nicholson, especially, had made some very fancy speculations about the radiation of atoms being governed by Planck's quantum hypothesis.

I shall pass over rather rapidly on Bohr's formulation of the quantum postulates only to point out that the formulation of those postulates represented an extremely daring step and a typical example of overcoming prejudice. And the prejudice that was overcome in those postulates was the idea that the frequency of radiation emitted by a vibrating system of charges should be the same as the frequency of vibration of the charges, which is of course not the case according to the idea of transition between stationary states.

Einstein's reaction was very positive to the news that experiments had confirmed Bohr's ascription of the Pickering and Fowler series to helium. Hevesy wrote in a letter to Bohr that when he (Hevesy) told Einstein that it was established with certainty that the Pickering-Fowler spectrum belongs to helium, Einstein remarked: "Then the frequency of the light does not depend at all on the frequency of the electron. And this is an *enormous achievement*. The theory of Bohr must then be right." In a letter to Rutherford, Hevesy wrote: "When I told him of the Fowler Spectrum the big eyes of Einstein looked still bigger and he told me 'Then it is one of the greatest discoveries.'"

In the derivation of Balmer's formula from the postulates which was Bohr's triumph at that time, he again made use of this argument which was only one part, and not the main part, of the correspondence argument. He argued, namely, that if you go to the orbits containing many quanta, the very excited orbits, then the difference of energy between neighbouring orbits becomes smaller and smaller, and the frequency of the transition between neighbouring orbits in that region approaches the mechanical frequency, so that you will recover the classical picture in the limit of high quantum numbers. That is, the phenomena involving many quanta are therefore not distinguishable from the classical description.

But again, this was not yet the correspondence argument. This was only a part of it which Bohr had described at that time. He had not yet used the word correspondence. He used the word analogy. He said that it was a beautiful analogy with classical electrodynamics. He was extremely concerned about deciding by very careful argumentation as to which parts of the classical theory he could retain and which parts he had to abandon. Of course, he had to abandon the classical mechanism of emission and absorption of radiation. These were described as transition processes, which were quite outside the scope of classical description. But the stationary states themselves, i.e. the calculation or the estimation of the energy for stationary states, he argued, were still amenable to classical theory, involving only the Coulomb field, the statical part of the electromagnetic field.

Bohr tried very hard to find ways of determining the energies of stationary states for given systems. His idea, which by the way was also developed independently at the same time by Ehrenfest under the name of adiabatic invariant theory, was simply to start from the only mechanical system for which the quantization was well-known and verified, i.e. the harmonic oscillator. Then he argued that if you transform the system, if you change the forces acting in the system in a very slow way, then no quantum transition was possible, and the system would gradually evolve into some other system. Whatever invariants belonged to the first system would keep the same value for the ultimate system. He called this the principle of mechanical transformability, by which he then tried to determine the stationary states of given systems of electrons and nuclei.

The success, and even the possibility, of formulating the postulates and of defining the idea of stationary state as he did, was only due to the weakness of the radiation coupling. If it had not been for that weakness there would not have been sharply defined, or reasonably sharply defined, energies of the excited states. They would have widths and that would have disturbed the picture. Bohr was, however, very conscious about it and he used repeatedly the weakness of the fine structure constant in order to make a sharp separation, in an ideal way of course, between those parts of the theory that were governed by the Coulomb field, by the static force only, and which he thought at that time could still be managed by classical physics, while the other parts which represented the pure radiation field were governed by the postulates and by transition processes which escape any possibility of classical analysis.

Even at that early stage the fundamental ideas of quantum mechanics were already contained in this picture, though they were not clearly expressed in the first publications; especially the fact that the radiation field is entirely outside any classical picture and must be treated by postulates, by the concept of transition processes, which are not analyzable in classical terms. Any attempt to analyze the mechanism of emission or absorption of radiation would involve a continuous succession of processes, each of which would involve an exchange, or modification of action less than the quantum, and therefore were quite excluded by this model.

However, Bohr thought at that time that processes that depended only on static forces, were still amenable to classical analysis.

The next stage in the development of the theory was first of all accelerated in a very essential way by Sommerfeld's intervention, Sommerfeld's extension of Bohr's quantum conditions, derived from the principle of mechanical transformability or the use of adiabatic invariants, to more than one degree of freedom. Bohr had used until 1916, until Sommerfeld's work, the idea of electrons distributed in rings. When I asked him why he did that, why he used a rather improbable picture like that, he said, "Well, that was what was used then, that was what we were used to." It had come, of course, from Thomson's attempts to describe an atomic model with stable rings of electrons inside a distribution of positive charge.

Sommerfeld's paper, in which he introduced the quantization in three dimensions, made a deep impression on Bohr. Fortunately he had then the help of Kramers to refresh his mathematics and mechanics. Both of them undertook a deeper analysis of atomic structures based on the Hamiltonian scheme of equations and on the theory of multiply periodic systems. It is very important to remember that the whole framework of quantum mechanics has its origin in the consideration of multiply periodic systems, and this was quite essential.

The problem was then to analyze the spectroscopic data that had accumulated in the mean time. The spectroscopists were stimulated to obtain a deeper understanding of the regularities of the spectra that they had been so accurately measuring. But they did not quite know what to do with it. In the exciting environment of the German physics of that time, this was done in a very systematic way at different places, especially in Tübingen. The problem arose of accounting, not only for the series of stationary states called the spectroscopic terms that could be disentangled from the measurements of the frequencies of the spectra, but also for the intensities of the spectral lines, which seemed to be quite capriciously varying from one transition to the other.

It was here that Bohr developed the correspondence argument in an essential way. He said that the correspondence with classical physics, with the classical radiation theory, is valid not only in the classical limit, in the limit of many quanta, but is also approximately valid in a semi-qualitative way even in the domain, far away from the classical limit, in which the radiation processes are governed by the transition postulate.

In the classical theory of multiply periodic systems, the mechanical frequencies of the electrons are described by definite modes of frequencies and their harmonics, and then rules were established by trial and error, by just fumbling between those classical harmonics and the transitions between pairs of stationary states. This was not a simple correspondence, since it was a problem of associating a single number in the classical description, the number of the harmonic, with two numbers, or two sets of quantum numbers corresponding to the two states between which the transition was proceeding. A number of people concentrated their efforts on that problem, and it is amazing to think of the tremendous amount of work that was put into that in a few years' time. Reasonably clear understanding of optical spectra was obtained on the basis of the correspondence argument. I say, the "*correspondence argument.*" That is the phrase that Bohr liked to use rather than the 'correspondence principle', because he was a bit afraid of giving the name of principle to a rule the foundation of which was so shaky.

In the course of this work Bohr and Kramers were led more and more to the conviction that the classical picture, the classical model of atoms as systems of electrons revolving in orbits around the nucleus, orbits which were not simple Keplerian ellipses then, but which were turning around the nucleus in a complicated way, was breaking down in a very serious way. As we know now, after the event, the missing element was the spin, of course, that showed up by the necessity of introducing half integral quantum numbers, whereas there was no justification for such a half integral quantization. The feeling arose that those models were after all very unreliable, and that the only element on which finally the analysis of spectra was based was the system of empirical rules that were developed by various people for calculating the intensities of the various transitions in terms of the quantum numbers involved.

There was a curious divergence at that stage between Sommerfeld and Bohr, Sommerfeld keeping his faith in the literal validity of the models at a time when Bohr had already abandoned this idea. Bohr was always amused to recall that when he had sent reprints of his first papers of 1913 to Sommerfeld, he received in reply from Sommerfeld a very flattering comment on the content of the papers with the remark that "I doubt very much in the reality of models."

Now, at that time, it was in 1924, an incident occurred which induced Bohr and Kramers to crystallize their opinion. It was the arrival in Copenhagen of a young American, Slater, who had ideas of his own about the possible reconciliation of the two aspects of the radiation field. He had the idea that the classical electromagnetic wave would be a sort of guiding wave for the photons, and he realized that the connection between the emitting or absorbing atoms and the classical radiation field could not be upheld in the classical way. He had therefore conceived the idea that one had to replace the actual atoms by a system of oscillators, which could play an equivalent role in the mechanism of emission and absorption to the real atoms. Bohr and Kramers took up that idea, but only partially. It was quite in keeping with what I said before about Bohr's attitude towards the photon concept, that he rejected from Slater's proposal the photon altogether. He kept only the idea of the virtual oscillators creating a virtual radiation field, containing potentially all the possible frequencies that could be exchanged between the oscillators representing the atoms. Slater was very dissatisfied with that turn of events. He felt that what he considered as his main idea had been just left aside, and that of his own conception there remained only the grin of the Cheshire cat. And yet, he did nothing else than to let Kramers write up the paper and put his name together with those of Bohr and Kramers underneath. This paper of Bohr, Kramers and Slater represents the precursor stage to what emerged finally as the conceptual framework of quantum mechanics.

In the first place, it was a completely statistical description. The function of the radiation field was now reduced to transmitting interaction by the mechanism of absorption and emission in this particular model. But it was, as the name virtual field indicates, not a real field in the sense of being directly observed. It was a virtual field, a field containing in itself all the elements, all the frequencies, that could be absorbed or emitted by the various atoms. And the atoms themselves were also reduced to virtual systems of oscillators. Thus it was a rather abstract theory, eliminating all that remained of the classical description, even eliminating, and that was the culminating point, the validity of the conservation laws of energy and momentum. Since there was no element that could decide, after an atom, or a virtual oscillator, had emitted its corresponding radiation, when this radiation could be absorbed again, the conservation of energy and momentum could only be upheld in a statistical sense on the average.

Now this was a challenge in the first place to the experimentalists. Bothe and Geiger, within a very short time, and Compton and Simon in another way, tested this very revolutionary consequence of the statement of the famous paper of Bohr, Kramers and Slater, and found that it was not true in the case of the Compton effect. In the Compton effect they could show that there was a coincidence between the absorption and the emission of radiation and the recoil of the electron.

One would normally expect that Bohr and Kramers, and Slater also, would have abandoned those ideas in view of these experimental results, but Bohr's attitude was quite different. It did not influence him in the least. Not that he despised experimental results, far from it. But he said that those results only showed that we have not yet got to the depth, to the crux, of the matter, and that the difficulty, the dilemma of the two aspects of light, lies still deeper than we thought; and there was no cheap way of avoiding the dilemma, not even by renouncing a deterministic relation between the acts of absorption and emission. That did not mean that the fundamental conception of this representation of the phenomena by means of virtual oscillators and virtual radiation between them should not be a way, although very imperfect, and it ought to be pursued further. You may call it the resilience of Bohr, which was a very typical characteristic of his optimism in all matters.

At that time, even before the arrival of Slater in Copenhagen, Kramers had already started work on the theory of dispersion. Bohr's postulates coped with emission and absorption, and there remained the phenomena of dispersion, which were much harder nuts to crack from the point of view of the correspondence argument. Kramers made the first step in that direction, and then he and Bohr had the good fortune of Heisenberg's coming to their rescue. Heisenberg not only completed Kramers' work on the dispersion theory, but he was led from that to the crucial discovery of the actual mathematical formalism that was the correct expression for all those concepts that were only vaguely expressed in the paper of Bohr, Kramers and Slater. Heisenberg has himself testified that this paper made a great impression on him, and was very helpful in leading him in the right direction.

It is a bit embarrassing for me to talk about Heisenberg in his presence at this symposium, but I do that on good authority I think, i.e. on Heisenberg's own testimony in his writings. Heisenberg developed a matrix theory, operating only with observable quantities, the latter being the transition probabilities and the energies of the stationary states, and he firmly maintained that his theory was complete. It was true at least for multiply periodic systems in the framework of quantum mechanics. And then, another great pioneer who unfortunately is no longer among us, Pauli, a close friend of Heisenberg, was a bit more critical. He said, "Well, after all the moon also revolves along a Keplerian orbit, and we have no difficulty in localizing the moon and defining its trajectory. So there is an element in the description which is still missing after all, in spite of the formal completion of the matrix scheme." Who knows, perhaps Pauli and Heisenberg by their own exertions would have completed the scheme, but they were helped anyway, unwittingly, by Schrödinger's publication of wave mechanics.

Schrödinger's wave mechanics was a direct onslaught on quantum mechanics, on quantum theory altogether. He wanted to get rid of Bohr's postulates. He wanted to reinstall the classical description of radiation. He showed how, by means of the wave functions that he had determined as eigenfunctions of the eigenvalues of a mechanical system, he could construct charge and current densities, which were almost periodic functions with just the periodicities that would produce in a classical way the correct frequencies observed in the spectra. But Heisenberg immediately pointed out that that was an illusion, because any classical mechanism of emission and absorption of radiation of whatever frequencies would never lead to Planck's distribution law, but to the classical distribution.

As a result Heisenberg, at that time, was very sharply sensitized against Schrödinger's theory. Pauli, on the other hand, recognized that although Schrödinger's proposal was untenable, nevertheless Schrödinger's wave conception and the conception of a wave packet could just help to provide the missing element of a description, comprising all the elements of observation – not only those which had been contemplated by Heisenberg originally, but all the other elements of a spatial localization of charge and matter as well as a treatment of aperiodic phenomena, such as scattering processes. That programme was very quickly developed by Born, and the crowning piece to the edifice was provided, at about the same time, by Jordan and Dirac. That was the transformation theory, showing that you can pass by a canonical transformation from any mode of description, based on the choice of variables, to any other.

The very rapid sketch which I have given of the development shows that the impression that the pioneers of quantum mechanics, especially Heisenberg, found a formalism and then wondered about how to interpret it, is completely wrong. They were guided all the time by physical ideas, and at every step they were very clear about the physical implications of the symbols, operators, etc., that they were introducing. That was not the problem. The problem that arose very rapidly, and it was recognized by Heisenberg, was that this comprehensive description provided by the more flexible formalism arising from the fusion of the matrix idea, of the operator idea of Dirac, with the wave function, wave vector, state vector, or ket vector as Dirac called it, that this whole formalism implied limitations in the application of the classical concepts.

There was never any doubt in the minds of the people working on the elaboration of the theory that the symbols that they used had to be associated with classical concepts. That was quite clear at that time; although it does not seem to be so clear nowadays, I do not know why. But at that time nobody had any qualms about that. However, what appeared immediately, or very soon, was that this application of classical concepts was subject to limitations. And the real problem that Heisenberg had to face was to account for these limitations, to understand them. At that stage Bohr intervened again actively. Bohr was always, of course, in the background. Although he did not take part in the elaboration of the new theory directly, at this critical stage concerning the understanding of the limitations in the application of the classical concepts, Bohr intervened again very forcefully in collaboration with Heisenberg, and this intervention presented a very interesting and instructive feature.

Heisenberg argued that since his scheme was complete, it had to contain also the answer to the question on the origin of the limitations. Heisenberg tells us in his wonderful book *Der Teil und das Ganze* [*Physics and Beyond*], that when he was pondering about that problem he was reminded of a remark that Einstein had made to him when he presented to Einstein his ideas about using only observable quantities. Einstein said: "What quantities are observable should not be our choice, but should be given, should be indicated to us by the theory." That was a very wise remark which Einstein ought perhaps to have remembered later on. Be that as it may, it was extremely fruitful in this particular case and it led Heisenberg to the discovery of the uncertainty relations.

The question then was: What to do with those uncertainty relations? What was their significance? That was a matter for discussion, and Bohr fumbled very much for the right way of formulating the meaning of the uncertainty relations. And then, Bohr finally saw in these relations, in this peculiar situation, the solution of the dilemma that had hovered over the whole theory all the time. The dilemma of light, with which I began, had been doubled by the similar dilemma of matter, arising from Schrödinger's introduction of matter waves. Here again we have the two opposing aspects of particles and waves, applicable to matter as well as to radiation. Bohr reflected that the uncertainty relations were a consequence of the commutation rules, the only element in the theory which was not classical, in which Planck's constant was introduced with the coefficient *i*, the imaginary unit. Bohr was always fond of pointing out this feature which excluded any immediate classical interpretation of these relations. This situation was regarded by Bohr as the feature of the theory that opened up new prospects and eliminated all logical difficulties.

He did that by introducing the notion of complementarity as a logical relation of which we had so far no example in physics. This was a way of accommodating concepts, or phenomena as he called them later, which were mutually exclusive, but nevertheless were part of a comprehensive description of the phenomena. It is curious to point out in this connection that the situation between the complementary aspects of light, the wave aspect which is in direct correspondence with the classical description, and the concept of photon which is symbolical, which is only an expression for the exchange of energy and momentum between matter and radiation, is not paralleled, as de Broglie thought, by the two wave and particle aspects of matter, but it is just the other way around. For matter, the aspect which is in correspondence with classical observation is the particle aspect, of course; whereas the wave aspect is a symbolical one, and is the basis for the statistical description of the processes that can occur, and which Bohr called *individual* processes, meaning thereby that they are not divisible, that they are not analyzable in the classical way. Complementarity necessarily involves statistics. This is obvious, because if you want to accommodate two concepts which are mutually exclusive, if you want to use them together, you can only do that by renouncing the precise determination to a certain extent, so that you have to introduce a statistical element.

This was not a surprise for any of the Copenhagen group, because statistical causality was already contained in the paper by Bohr, Kramers and Slater, and was accepted as a natural form of causal relations in atomic physics, that was quite well adapted to the conditions of observation. There was no mystery about it. Here I may respectfully disagree with the statement made yesterday by Dirac, and I do not think that there is any reason to be uncomfortable about the use of statistical causality in physics. As a matter of fact, I do not believe that causality poses any problem at all in physics at any stage, for the simple reason that causality is not a matter of our choice. Causality is inherent in the form of the description that we develop, and we develop this manner of description in reference to the conditions of observation. If the conditions of observation are such that they allow for several possibilities, well, statistics is the indicated method. It is a method created for just such purposes. In fact, the first to introduce statistics in physics – I disregard of course Maxwell and statistical mechanics because that is another thing – but in fundamental atomic processes, were Rutherford and Soddy when they formulated the law of radioactive transformations. Now anyone who has seen Rutherford remembers a force of nature, and not a very deep, subtle or intricate thinker. As a matter of fact, he expressed his own character so gently in a famous lecture in which he said, "I like simple pictures because I am a simple person myself." So, if Rutherford introduced a statistical law for radioactive transformations, he did not have any deep epistemological qualms about it. He simply used statistics in the same way as an insurance agent uses statistics to calculate premiums. And that is the right way to do it; statistics is just made for that kind of problem. Moreover, it is not strictly true to say that quantum mechanics has eliminated determinism, because after all the laws of conservation of energy and momentum are just as deterministic as ever, and they have universal application.

Well, I must leave it at that. I hope to have described to you a very exciting period, one of the most exciting periods in the history of physics, which exhibits a remarkable continuity in the logical development of a trend of ideas. There has been very little of external influences disturbing the course of events. The only external influence was the advent of Schrödinger's wave mechanics, and that was a very helpful one, although Schrödinger himself was rather sorry for the use, rather the misuse, as he thought, that was made of his wave ideas.

DISCUSSION

F. Bopp: I have a question concerning the action integral. It seems to have been a curious development. The action integral appears in Bohr's famous theory of 1913. I came across this integral in Sommerfeld's famous lecture in 1911 at Karlsruhe. Do you know the point to which the ideas were directed in the action integral up to 1910? There is the wave quantization of energy, and suddenly the action integral comes up. There is the possibility that perhaps Sommerfeld invented it and Bohr knew of this development. At the Solvay conference in 1911, when Lorentz put his question concerning adiabatic invariance, Einstein answered it. Was this the origin of this transformation from quantum energies to the connection of the action integral with Planck's constant?

L. Rosenfeld: I think that these two concepts, the action integral and the adiabatic invariance, were a bit earlier. The action integral was introduced for the first time by Planck in an effort to generalize the quantum condition to many degrees of freedom. He then concentrated on the statistical aspect of the quantum of action and tried to establish a connection with the cells in phase space. That was, I think, about 1910.

H. Rechenberg: That was already in 1906.

L. Rosenfeld: Already in 1906? Did he really do that in 1906 for *n* degrees of freedom? I don't think so. For one, yes, but then he pursued this line and it was by 1910 that he tried to introduce it in a general way.* Both Sommerfeld and Bohr knew Planck's work and were interested in it. The adiabatic invariants were introduced by Ehrenfest in connection with the statistical theory of radiation. It was a very brilliant piece of work, though it is not such a well-known paper unfortunately. Ehrenfest tried to see where the discontinuity introduced by Planck's constant occurred, and how necessary it was. There were anticipations of other things in that paper, in particular the notion of adiabatic invariance. When Ehrenfest learned about Bohr's papers he was well prepared to think of this idea as a natural one. He knew that the phase integral in particular was an adiabatic invariant for periodic systems and so it was natural for him to try to use it to determine the stationary states of other systems.

F. Bopp: The most quoted paper of Ehrenfest was later than this.

L. Rosenfeld: Yes. Later on, Burgers, a pupil of Ehrenfest's had a thesis devoted to the theory of adiabatic invariance.

F. Hund: I believe that the introduction of the phase integral came about more gradually. It is difficult to say who at first had noticed its consequences. Planck, in his lectures of the winter term in 1905, which were essentially the same as his book of 1906 [*Theory of Heat Radiation*], had a footnote – he didn't emphasize it – that in a harmonic oscillator, the stationary state or the energy used, is expressed by the integral; but *only* for the harmonic oscillator. At the Solvay conference in 1911 it was somewhat more complicated. I think it was Nernst who asked the question as to how to quantize the rotator. Lorentz gave an answer, which was not quite correct, and Hasenöhrl then came up and pointed out the difference between harmonic oscillator and rotator. In a harmonic oscillator, equal steps of energy are also equal steps of the phase integral, but in the rotator this is not the case. Then he said that one has the choice of either taking equal steps of the energy or taking equal steps of the phase integral, and he [Hasenöhrl] thought that "the second choice would be in the sense of Planck." I think this was the first clear conception of a generalization, but some pages later he confused it totally for reasons that are not so important. Three weeks later, at the *Naturforscherversammlung* at Karlsruhe, Hasenöhrl used that integral as the essence of a quantum theory, but he gave an unhappy example afterwards. Again, at the Solvay conference in 1911, and in 1912, Planck took the phase integral as the central consideration for the harmonic oscillator, but only for the harmonic oscillator. In April 1913, at the so-called *Kinetischen Gas Kongress* at Göttingen [organized by David Hilbert], Debye gave the action integral as the real generalization of Planck's ideas without citing Hasenöhrl. He didn't know of him, I think. And, in the same year [1913], I think in May, Ehrenfest gave the right quantization of the rotator, but he touched upon the adiabatic principle in a few words only. So it was in 1913 that Debye and Ehrenfest have first understood that this was the sound foundation. It was, as I said, a somewhat gradual and complicated development. [*Spontaneous and sustained applause from the audience.*]

L. Rosenfeld: Thank you very much. May I just say one word more about Ehrenfest. Bohr criticized Ehrenfest's application of the adiabatic principle in the case of the rotator, because if you try to go from an oscillation to a rotation, then you get a singularity when you pass to the rotation.

F. Hund: Ehrenfest tried another approach which was connected with that discontinuity, but it was afterwards.

B. L. van der Waerden: Rosenfeld quoted what Einstein said to Heisenberg about what is observable, but he quoted only the second half of what Einstein said. But in his book [*Physics and Beyond*], Heisenberg also told the beginning of the story, approximately as follows: Einstein expressed his doubts about his opinion that a theory should contain only statements about facts that are observable in principle. Heisenberg was astonished and said: "But you yourself propagated this principle." To which Einstein replied, "Yes, maybe I have made use of this kind of philosophy, but still it is nonsense (Quatsch). *The theory alone decides what is observable and what not*".

* *Author's Note added in proof*

In my reply to Bopp's question I gave (from memory) the correct date 1910 (anterior, of course, to the dates of publication) for Planck's non-trivial conception of the statistical meaning of the quantum of action related to the phase integral, and for Ehrenfest's first application of the adiabatic invariance of the phase integral for periodic systems (his paper was published in *Ann. Physik* **36** (1911), 91). For more details, see L. Rosenfeld, *Max-Planck Festschrift* (Berlin 1959), p. 203 and *Osiris* **2** (1936), 149. [This volume, pp. 235 and 193].

For the subject of the lecture, see also L. Rosenfeld, *Archive for Hist. of Exact Sc.* **7** (1971), 69. [This volume, p. 266].

CHAPTER 25

A VOYAGE TO LAPLACIA

[1955d]

> Spirit of Nature! all-sufficing Power,
> Necessity! thou mother of the world!
>
> *Queen Mab*, c.VI.

How did we ever, of all places, stumble upon that one? What unfathomable part in the predetermined concatenation of events which animate the scene of the universe did this trip fulfil? What may its necessary link have been with the tumultuous collisions of galaxies in the distant realms of space or the discreet embraces of earthworms in the cold recesses of the soil? When faced with such anxious queries, the tragic poet's familiar lines will come naturally to the reader's mind:

> Διὸς ἵμερος οὐκ εὐθήρατος ἐτύχθη
> παντᾷ τοι φλεγέθει
> κἀν σκότῳ μελαίνᾳ ξὺν τύχᾳ
> μερόπεσσι λαοῖς.
>
> (Aeschylos, *Suppliant Maidens*, 87–90)

"Nature's striving is not easy to trace: everywhere it flares, but in the gloom of events obscure to mortal men."

On the face of it, things looked simple enough: never trust appearances, however. It was in the autumn of 1931: as he was passing through Belgium, Bohr suggested to me that we might spend a few days in the countryside to go on with our work. Only some weeks before we had enjoyed such a spell of rustic seclusion in an isolated villa on the shore of the lake of Como. Nothing favours inspiration like the necessity of crossing a lake to post a letter; once we were caught in a storm and nearly drowned in the process. We found it equally stimulating to have to cope with angrily expostulating peasants, when their cattle had been disturbed by the landlady's unmanageable dog, which we all too hopefully took with us in our strolls. No wonder that we should have longed to repeat the experience of such a quiet retreat at the first occasion.

It occurred to me that Spa would be the ideal spot for our studious purposes. The bustle of the tourist season would be over and we would have the delightful woods all to ourselves. I ought perhaps to point out that this town does not owe its name to the fact that it is a watering place: the rigid arrow of causality points in the reverse direction. Spa is, or was (in its heyday in the 18th century), the paragon of watering places. However much it may have decayed from this position, it still does boast of an easily accessible post-office and of pastures well-fenced against depredations by unruly dogs. However, these are, after all, but minor failures in point of romantic wilderness. Bohr welcomed my proposal most warmly, and off we went to Spa.

Portents of forthcoming events are constantly hovering around us, but we heed them not. (I am *not* alluding here to preprints.) As the train slowed down along the platform in Spa station, I casually (or so it seems) glanced at the station clock as the big hand, with a silent jerk, was moving to the next minute: the time it showed exactly matched the predetermined figure which I had just looked up in the time-table! The singularity of such a coincidence did not then impress me in the least: I only recalled it long afterwards, when pondering upon the past. Thus Jupiter blinds the unfortunate on whom his doom has fallen.

We had our first inkling of something queer about the place on arriving at the hotel. Bohr had decided this time to shun the surprises of romantic inns and to pin his faith to the palatial comfort of vast establishments: with what success we shall presently see. Our arrival at this late time of the year startled the population of errand boys and porters idling about in the deserted hall. They eyed us with tip-assessing curiosity and formed a silent background to the ensuing scene, like the chorus of a Greek tragedy while the heroes are discussing their problems in alternate rhythmical lines. The protagonist soon made his appearance; he was presumably some sort of manager. We explained that besides bedrooms we wished to dispose over some drawing room in which we could do our writing. The reply was perhaps not very rhythmical, but it was lapidary enough: writing-rooms were not available in this hotel. On our mild expression of surprise, the man relented somewhat and grudgingly suggested a suite of two bedrooms with a 'salon' between them. The climax came when we enquired about the price of this princely accommodation. He mustered us with a stern look and, after quoting a figure, added in the same breath: "*payable in advance.*" Speechless with amazement I looked at the chorus: their stony faces expressed, if anything, complete detachment. Bohr wanted me to intimate to the man that he must be labouring under some misunderstanding; but, curt as I tried to be, I could not hope to beat him in that line, and he departed with flying standards, abandoning us to the unfeeling chorus.

The 'salon' turned out to be a late 19th century impression of what an 18th century boudoir might have looked like. This was not a particularly cheerful sight, and we both felt a strong desire to take the air. It was a gloomy pair that left the hotel under the receptionist's strangely inquisitive gaze. The scene outside was not encouraging either. The large square in front of us was the exclusive abode of a medley of dogs, looking fully as forlorn as their kindred of neighbouring Aix-la-Chapelle:

Zu Aachen langweilen sich auf der Strass'
Die Hunde, sie flehn untertänig:
"Gib uns einen Fusstritt, o Fremdling, das wird
Vielleicht uns zerstreuen ein wenig.

(Heine, *Deutschland*, Kap. III.)*

Eventually, as we proceeded, we found something to divert *us* from our depressed feelings. As we came in sight of the Casino, we noticed a human crowd, and we hastened to join it and see what was going on. When we tried to enter the building, however, we were obsequiously requested to produce our member's tickets, and readily informed, on our puzzled enquiry, that nobody was admitted to the 'salon des jeux' without such a ticket. This disclosure was a flash of light in the darkness of the night, illuminating the hidden threads out of which Necessity had spun the web of our recent fate. The tourists gone, the gamblers came: that was obviously the ebb and flow regulating the life of this town. From this simple principle the hotel manager's attitude immediately derived a perfectly clear meaning.

Elated by this discovery, we entered a near-by 'patisserie' to have some coffee. There we had occasion to make another observation, which essentially deepened our understanding of the situation. Presently a man came in and sat down at a table near ours. He took out of his pocket a sheet of paper covered with long columns of figures. Feverishly he scanned these figures and started performing calculations which he scribbled all over the paper. The interpretation of this unusual behaviour was not in doubt: we were beholding, for the first time in our lives, a *real* gambler, that is, a gambler with a 'system'.

I hope, gentle reader, that thou dost not share the common fallacy of associating gamblers with hazard. The real gambler scorns the idea that roulettes should revolve at the whim of some blindfolded damsel delicately poised upon a winged wheel. His nights are not abandoned to slumbering trust in chance, but consumed in ardent search of law. Dame Fortune's bland bosom is not his pillow; the god he adores is that capacious Laplacian mind before whose eye past and future are spread out with the same shadowless clarity. His own mind being only a feeble adumbration of such a perfect being, he will owe that the 'system' he has managed to work out may not be infallible: but of the existence of an ideal system he is unrockably convinced. The real gambler is the pure, unspoiled type of the determinist.

This, then, was the root of the matter: we had unwittingly landed in the midst of Laplacia; and we had now to make the best of it. The first decision we took was to flee from that inhospitable host who at the mere sight of us had taken such a poor view of our system. On leaving, the next morning, I muttered some vague apology about unforeseen circumstances forcing us to change our plans. I shall never forget the knowing look with which the man received this intelligence. A true Laplacian does not know of unforeseen circumstances; and our precipitate departure fitted only too well into the deterministic interpretation of our behaviour which he had been elaborating for himself.

We found a more congenial dwelling-place at 'Annette et Lubin', an inn of the romantic variety, as the rococo flavour of the name sufficiently indicated. Not that we were able to dispel the deep distrust with which the landlord greeted guests to whose apparition he felt unable to assign a ground satisfying his sense of causality. When you come to think of it, this suspicious state of mind is a natural consequence of a Laplacian philosophy of life. If you go seriously about it, you must be asking yourself all the time: "Didn't I overlook some causal connection?"

Our landlord was remarkably thorough in this respect. When we mentioned writing in the drawing-room, he became very grave, and said: "Yes, but not with ink." The point was that a former guest had once spilled some ink on a Persian carpet. We could not help feeling that the assumption of a non-Newtonian attraction of Persian carpets for ink, which seemed to underlie his argument, showed excessive disregard of the great truth that we are not only spectators, but actors in the drama of existence. By the time, however, we had already become sufficiently adapted to the Laplacian environment to receive the intimation of this taboo with perfect composure.

Yet, in some subtle and insidious way, the spirit of Laplacia was nagging at our very souls. I had a sudden revelation of this on a gloomy evening ending a gloomy day, during which we had struggled in vain with our problems of measurability of quantized fields. As we were dejectedly musing by the fire, at length Bohr broke the silence and muttered wearily: "We must look the situation in the face ... Could it be ... could it be that this quantization is all nonsense?" To be sure, this was only a passing mood; there was no trace of it the next day. Still, it was a warning not to be ignored. The decision to leave the Laplacians to their fate did not require long deliberation. And as the train gathered momentum along the platform, while the clock was marking the prescribed time of departure, we waved a hearty farewell to Laplacia – *for ever*!

NOTE

* Milling around on Aachen's streets
The dogs were humbly imploring
"Oh stranger, give us a little kick!
Life has become so boring."
[tr. Aaron Kramer in *The Poetry and Prose of Heinrich Heine*, ed. Frederic Ewen, p. 185 (Citadel Press, N.Y., 1948).]

PART III

THEORETICAL PHYSICS

CHAPTER 1

ON THE ENERGY-MOMENTUM TENSOR

[1940b]

The work [1] of Lorentz, Hilbert, De Donder, F. Klein and Weyl has established the close relation which connects the energy-momentum tensor of an arbitrary system of physical entities, such as material particles or electromagnetic fields, to the gravitational field. In principle, this relation leads automatically to a well-determined, symmetric form of this tensor, as soon as the Lagrangian of the system under consideration has been expressed in a way which is invariant under arbitrary transformations of the space-time coordinates. But it may appear at first sight that this general procedure for deriving the energy-momentum tensor will run into practical difficulties if it is applied, since it seems to require in each case special considerations and calculations, relating to the gravitational potentials, i.e. on variables not directly related to the problem, and whose influence in general is negligible empirically. It is for this reason that procedures which are less direct, but which can be applied immediately to a Lagrangian invariant only under the Lorentz group, have often been preferred. On the other hand, these procedures also require special considerations in each case, especially to ensure that a symmetric form of the tensor is found [2], so that no practical advantage compensates for the unsatisfactory neglect of the profound relation which exists between the energy-momentum tensor and the general invariance of the Lagrangian.

It appears, however, that the problem has a solution which does not suffer from the difficulties recalled above. Indeed, it can be seen that precisely by using as fully as possible the invariance properties of the Lagrangian, one can obtain in a very simple and direct way an expression for the energy-momentum tensor, the 'special' form of which, that is the form invariant only under the Lorentz group, can be immediately specified in each particular case as soon as the 'special' form of the corresponding Lagrangian is known. The object of the present work is to deduce this expression, under sufficiently general conditions to include in particular the cases which have occurred in physics up to now.[1]

1. Definition of the Energy-Momentum Tensor

We shall begin by recalling how the energy-momentum tensor is introduced in the derivation of the gravitational field equations from the Lagrangian. The gravitational field can be specified in two equivalent ways, which must both be considered: In the first, the field variables are the components of the metric tensor g_{lm}; in the second [3], a 'tetrad', or set of four orthogonal vectors $h_j^{(\mu)}$ ($\mu = 1, 2, 3, 4$), is introduced at each point, and the metric tensor is defined, with their help, by[2]

$$(1)\quad \begin{cases} g_{lm} = g_{(\mu\nu)}\, h_l^{(\mu)}\, h_m^{(\nu)} \\ \text{where} \quad g_{(\mu\nu)} = \begin{cases} 0 \text{ for } \mu \neq \nu \\ 1 \text{ for } \mu = \nu = 1, 2, 3 \\ -1 \text{ for } \mu = \nu = 4. \end{cases} \end{cases}$$

Whenever it is unnecessary to specify which of these descriptions is being used, we will designate the gravitational variables by $Q\gamma$, where γ is an arbitrarily fixed indexing number. Since, in the present context, these variables play only an auxiliary role, we will not need to envisage their quantization, and in all that follows, they will be treated as simple parameters ('c – numbers').

The Lagrangian of a system composed of the gravitational field and of some physical agencies capable of producing this field is the sum of a term referring to the gravitational field alone and of another term depending on the physical agencies and their interaction with the gravitational field; these two terms are the integrals over the space-time region occupied by the system of, respectively, scalar densities $\mathscr{G}$ and $\mathscr{L}$ which depend on the variables Q_γ and their derivatives. Representing by $Q_{\bar{\gamma}}$ an arbitrarily chosen linear combination of the Q_γ's, homogeneous of first degree, we obtain the gravitational field equations by requiring that the Lagrangian of the total system be an extremum for arbitrary variations of the $Q_{\bar{\gamma}}$'s vanishing at the limits of integration. With an integration by parts, this condition:

$$\delta \int (\mathscr{G} + \mathscr{L})\, \mathrm{d}w = 0, \quad (\mathrm{d}w = \mathrm{d}x^1 \mathrm{d}x^2 \mathrm{d}x^3 \mathrm{d}x^4),$$

takes the form

$$\int \frac{\delta(\mathscr{G} + \mathscr{L})}{\delta Q_{\bar{\gamma}}} \delta Q_{\bar{\gamma}}\, \mathrm{d}w = 0,$$

where the coefficient of $\delta Q_{\bar{\gamma}}$ defines the 'variational derivative' [4] with respect to $Q_{\bar{\gamma}}$ of the density function $\mathscr{G}+\mathscr{L}$; the field equations then follow

$$(2) \qquad \frac{\delta \mathscr{G}}{\delta Q_{\bar{\gamma}}} = -\frac{\delta \mathscr{L}}{\delta Q_{\bar{\gamma}}}.$$

In this form, the first member contains only the gravitational variables, while the second, which represents just the role of the physical agencies as sources of the gravitational field, determines the 'tensor density of energy-momentum' of the system under consideration.

If the variables specifying the nature of the system do not include spinor magnitudes, such as the wave functions of elementary particles of spin $\frac{1}{2}$, we can adopt the usual description of the gravitational field with the help of the g_{lm}'s; in the opposite case, the function $\mathscr{L}$ depends on the tetrads $h_j^{(u)}$ in a way not rationally reducible to the g_{lm}'s, and it is advantageous to choose these tetrads to represent the field. In the first case, equations (2) directly yield the energy-momentum tensor density in symmetric form

$$(3) \qquad \mathscr{T}^{lm} \equiv \mathscr{T}^{ml} \equiv -\left(\frac{\delta \mathscr{L}}{\delta g_{lm}} + \frac{\delta \mathscr{L}}{\delta g_{ml}}\right).$$

To obtain the generalization of this definition for the case where $\mathscr{L}$ depends essentially on the $h_j^{(\mu)}$'s, it suffices to observe that for any function (such as $\mathscr{G}$) which depends on the $h_j^{(\mu)}$'s only through the g_{lm}'s, we have [4]

$$(4) \qquad \frac{\delta \mathscr{G}}{\delta h_j^{(\mu)}} = \frac{\delta \mathscr{G}}{\delta g_{lm}} \frac{\partial g_{lm}}{\partial h_j^{(\mu)}};$$

for which, since by (1)

$$(5) \qquad \frac{\partial g_{lm}}{\partial h_j^{(\mu)}} = g_{(\mu\nu)}\left\{\delta_l^j\, h_m^{(\nu)} + \delta_m^j\, h_l^{(\nu)}\right\},$$

we obtain

$$(6) \qquad \frac{\delta \mathscr{G}}{\delta h_j^{(\mu)}} h_i^{(\mu)} = \left(\frac{\delta \mathscr{G}}{\delta g_{jm}} + \frac{\delta \mathscr{G}}{\delta g_{mj}}\right) g_{mi}.$$

Thus it can be seen that the energy-momentum density is given in this case by

$$(7) \qquad \mathscr{T}_{.i}^{j} = -\frac{\delta \mathscr{L}}{\delta h_j^{(\mu)}} h_i^{(\mu)},$$

an expression which, with the help of (6), reduces to (3) if $\mathscr{L}$ contains the $h_j^{(\mu)}$'s only through the g_{lm}'s. The symmetry condition on the indices is no longer identically satisfied by expression (7); we shall come back to this point later.

2. THE FUNDAMENTAL IDENTITIES

Our goal is now to show how the direct calculation of the variational derivatives $\delta\mathscr{L}/\delta Q_{\bar{\gamma}}$ occurring in formulas (3) or (7) can be avoided by using the fundamental identities which result from the invariance of the Langrangian

$$\int \mathscr{L}\,\mathrm{d}w$$

under arbitrary coordinate transformations. Though these matters are well known, we shall quickly review the derivation of these identities, while stating precisely the conditions under which we will operate, and using a notation which emphasizes certain auxiliary quantities which we shall need later on.

The variables specifying the physical agencies under consideration are the components of certain tensors $(Q^{(x)})$, $x=1, 2, \ldots$; to fix our ideas, we shall use only covariant components, and these will be designated collectively by Q_α, where the index α represents, besides the indexing number x, a set of values of the indices $i_1, i_2, \ldots\, i_n$ of the tensor $Q^{(x)}_{i_1, i_2, \ldots, i_n}$. We may adjoin to this system of variables Q_α the gravitational variables Q_γ, which either make up a tensor $(Q^{(0)}) \equiv (g)$, or a system of four tensors $(Q^{(0,\,\mu)}) \equiv (h^{(\mu)})$; we shall designate the total system of variables Q_α and Q_γ by the global notation Q_ω: Any expression containing a summation with respect to the index ω (that is in which the index occurs twice) will then be decomposable into a sum of two analogous expressions containing, respectively, a summation on α and a summation on γ.

If we now consider an arbitrary, infinitesimal coordinate transformation defined by

$$\delta x^i \equiv \overline{x^i} - x^i = \xi^i(x), \tag{8}$$

the ('substantial') variation of the variable Q_ω will be given, by virtue of the tensor character of the variables, by an expression of the form

(9) $$\delta Q_\omega \equiv \bar{Q}_\omega(\bar{x}) - Q_\omega(x) = c^j_{\omega,i}\frac{\partial \xi^i}{\partial x^j};$$

the auxiliary quantities $c^j_{\omega,i}$ are the components of a system of tensors related respectively to each type of variable $(Q^{(x)})$ by the following definition:

(10) $$c^{(x)j}_{i_1,i_2\ldots i_n,i} = -\sum_{p=i}^{n}\delta^j_{i_p}\,Q^{(x)}_{i_1\ldots i_{p-1}\,i\,i_{p+1}\ldots i_n}.$$

For instance, for the gravitational variables we have:

(i) If the tetrad variables $h^{(\mu)}_l$ are used

$$c^{(o,\mu)j}_{l,i} = -\delta^j_l\,h^{(\mu)}_i,$$

(ii) If the metric tensor g_{lm} is used,

$$c^{(o)j}_{lm,i} = -(\delta^j_l\,g_{mi} + \delta^j_m\,g_{li}).$$

By comparison with formulas (3) and (7), it can be seen that in any case the energy-momentum tensor density can be written, with the help of this notation, as

(11) $$\mathcal{T}^{\,j}_{\cdot i} = \frac{\delta\mathcal{L}}{\delta Q_\gamma}c^j_{\gamma,i},$$

and it is precisely this expression which will appear in the fundamental identities. We will again use the 'local' variations

(12) $$\delta^* Q_\omega \equiv \bar{Q}_\omega(x) - Q_\omega(x) = \delta Q_\omega - \frac{\partial Q_\omega}{\partial x^j}\xi^j,$$

for which we obviously have

(13) $$\delta^*\frac{\partial Q_\omega}{\partial x^j} = \frac{\partial}{\partial x^i}\delta^* Q_\omega.$$

The invariance condition for the Lagrangian under the transformation (8), which will be assumed satisfied for an arbitrary region of integration, can be written as

(14) $$\delta\int \mathcal{L}\,\mathrm{d}w = \int \delta^*\mathcal{L}\,\mathrm{d}w + \int \frac{\partial}{\partial x^j}(\mathcal{L}\xi^j)\,\mathrm{d}w = 0.$$

We shall now assume that the function $\mathscr{L}$ depends only on the variables Q_ω and on their first derivatives

$$Q_{\omega|j} \equiv \frac{\partial Q_\omega}{\partial x^j};$$

as the derivatives of the gravitational variables occur only through the covariant derivatives of the other variables Q_α, it would in fact be sufficient to formulate this hypothesis for the Q_α alone. If we ignore the quantization of these latter variables as well as that of the Q_γ's, we will thus have

$$(15)\qquad \delta^*\mathscr{L} = \frac{\partial \mathscr{L}}{\partial Q_\omega}\,\delta^* Q_\omega + \frac{\partial \mathscr{L}}{\partial Q_{\omega|j}}\,\delta^* Q_{\omega|j};$$

It will be seen that even for all the quantized fields of Q_α which have been introduced in physics, the Lagrangian depends on the Q_α's and $Q_{\alpha|j}$'s in a sufficiently simple manner that equation (15) can still be preserved – on condition that the latter is 'symmetrized', that is, that the second term is interpreted as the half-sum of the quantity written, and its hermitian conjugate.[3] Thus, it is the restrictive condition (15) which determines the range of validity of the calculations below.

We can first, using (13), write (15) in the form

$$(16)\qquad \delta^*\mathscr{L} = \frac{\delta \mathscr{L}}{\delta Q_\omega}\,\delta^* Q_\omega + \frac{\partial}{\partial x^j}\left(\frac{\partial \mathscr{L}}{\partial Q_{\omega|j}}\,\delta^* Q_\omega\right),$$

where the variational derivatives reduce under our assumptions to

$$(17)\qquad \frac{\delta \mathscr{L}}{\delta Q_\omega} = \frac{\partial \mathscr{L}}{\partial Q_\omega} - \frac{\partial}{\partial x^j}\left(\frac{\partial \mathscr{L}}{\partial Q_{\omega|j}}\right),$$

Introducing (16) into (14), and using (12) and (9), we obtain the invariance condition

$$(18)\qquad \begin{aligned} -\int &\left\{\frac{\partial}{\partial x^j}\left(\frac{\delta \mathscr{L}}{\delta Q_\omega}\, c^j_{\omega,i}\right) + \frac{\delta \mathscr{L}}{\delta Q_\omega}\, Q_{\omega|i}\right\} \xi^i \,\mathrm{d}w \\ +\int &\frac{\partial}{\partial x^j}\left\{\left(\mathscr{L}\delta^j_i - \frac{\partial \mathscr{L}}{\partial Q_{\omega|j}}\, Q_{\omega|i} + \frac{\delta \mathscr{L}}{\delta Q_\omega}\, c^j_{\omega,i}\right)\xi^i \right. \\ &\left. + \frac{\partial \mathscr{L}}{\partial Q_{\omega|j}}\, c^k_{\omega,i}\, \frac{\partial \xi^i}{\partial x^k}\right\} \mathrm{d}w = 0. \end{aligned}$$

Considering a transformation ξ^i which vanishes at the limits of integration, we first obtain the identical vanishing of the coefficient of ξ^i in the first integral of (18)

$$\frac{\partial}{\partial x^j}\left(\frac{\delta \mathscr{L}}{\delta Q_\omega} c^j_{\omega,i}\right)+\frac{\delta \mathscr{L}}{\delta Q_\omega} Q_{\omega|i}\equiv 0, \tag{19}$$

then the vanishing of the second integral for an arbitrary choice of the region of integration and of the transformation ξ^i; from this we deduce, since the coefficients of the ξ^i's and their first and second derivatives vanish separately, two new sets of independent identities:

$$\mathscr{L}\delta^j_i-\frac{\partial \mathscr{L}}{\partial Q_{\omega|j}} Q_{\omega|i}+\frac{\delta \mathscr{L}}{\delta Q_\omega} c^j_{\omega,i}+\frac{\partial \mathscr{R}_i{}^{kj}}{\partial x^k}\equiv 0 \tag{20}$$

and

$$\mathscr{R}^{jk}_i+\mathscr{R}^{kj}_i\equiv 0, \tag{21}$$

where we have put

$$\mathscr{R}^{jk}_i=\frac{\partial \mathscr{L}}{\partial Q_{\omega|j}} c^k_{\omega,i}. \tag{22}$$

By virtue of the way in which they were derived, the fundamental identities (19), (20), (21) are covariant. It will be advantageous, for the clarity of the following calculations, to make this covariance formally manifest by introducing covariant derivatives instead of the partial derivatives $Q_{\omega|j}$.

Before this, however, for purposes of orientation, let us consider the special relativistic point of view. Taking identity (20) and formula (11) and putting $Q_\gamma|_j=0$, it can then be seen that the identity reduces the calculation of $\mathscr{T}^j{}_{.i}$ to operations solely on the non-gravitational variables Q_α, as long as the tensor density $\mathscr{R}^{jk}_i$ is known. On the other hand, expression (22) for this tensor density shows that it contains gravitational variables explicitly only through the derivatives $\partial\mathscr{L}/\partial Q_{\gamma|j}$, which reduce at once to the $\partial\mathscr{L}/\partial Q_{\alpha|j}$, since the $Q_{\gamma|j}$'s occur only through the covariant derivatives of the Q_α's. The whole question thus reduces to eliminating the $\partial\mathscr{L}/\partial Q_{\gamma|j}$'s from $\mathscr{R}^{jk}_i$, and we shall see that this elimination can be very easily accomplished. Since consideration of the spinor variables requires some supplementary developments, entirely parallel to the preceding ones, on the invariance of the Lagrangian under orthogonal transformations of the tetrads, we shall, for the sake of greater clarity, first treat the case separately in which only properly tensor variables are involved.

3. Tensor variables

We shall now distinguish among the variables Q_α between those which are purely tensorial, to be collectively denoted by Q_τ, and the spinor variables, to be called Q_σ; we shall provisionally assume that all variables of the system under consideration are of the first type. For the latter, an expression for the covariant derivative $Q_{\tau\|l}$ in our notation is easily obtained by noting that a parallel displacement at a point (x) is represented by formulas like (8), (9), (12) if we impose on the ξ^i's at that point the condition

$$\frac{\partial \xi^i}{\partial x^j}+\Gamma^i_{jl}\xi^l=0,$$

where the Γ^i_{jl} are the components of the affine connection [5]; the covariant derivative $Q_{\tau\|l}$ is then defined as the coefficient of ξ^l in the corresponding expression for $-\delta^* Q_\tau$, that is:

$$(23)\qquad Q_{\tau\|l}=Q_{\tau|l}+c^j_{\tau,i}\Gamma^i_{jl}.$$

If the gravitational field is described by the metric tensor, we will have on the other hand, with an analogous notation,

$$(24)\qquad Q_{\gamma\|l}=0.$$

Finally, the tensor densities which occur through their partial derivatives in the identities (19) and (20) have, respectively, the divergences:

$$(25)\qquad \left(\frac{\delta\mathscr{L}}{\delta Q_\omega}c^j_{\omega,i}\right)_{\|j}=\frac{\partial}{\partial x^j}\left(\frac{\delta\mathscr{L}}{\delta Q_\omega}c^j_{\omega,i}\right)-\frac{\delta\mathscr{L}}{\delta Q_\omega}c^j_{\omega,k}\Gamma^k_{jl}$$

and

$$(26)\qquad \mathscr{R}^{kj}_{i\|k}=\frac{\partial\mathscr{R}^{kj}_i}{\partial x^k}-\mathscr{R}^{kj}_l\Gamma^l_{ki};$$

in (26), we have used the antisymmetry of $\mathscr{R}^{kj}_i$ with respect to its superscripts, as expressed by the identity (21).

Using formulas (23), (25), (26) and (22), we can write identities (20) and (19) in covariant form:

$$(27)\qquad \frac{\delta \mathscr{L}}{\delta Q_\omega} c^j_{\omega,i} \equiv \frac{\partial \mathscr{L}}{\partial Q_{\omega|j}} Q_{\omega\|i} - \mathscr{L}\delta^j_i - \mathscr{R}^{kj}_{i\|k},$$

$$(28)\qquad \left(\frac{\delta \mathscr{L}}{\delta Q_\omega} c^j_{\omega,i}\right)_{\|j} + \frac{\delta \mathscr{L}}{\delta Q_\omega} Q_{\omega\|i} \equiv 0.$$

With (11) and (24), formula (27) gives the desired expression for the energy-momentum tensor density:

$$(29)\qquad \boxed{\mathscr{T}^{j}_{.i} \equiv \frac{\partial \mathscr{L}}{\partial Q_{\tau\|j}} Q_{\tau\|i} - \mathscr{L}\delta^j_i - \mathscr{R}^{kj}_{i\|k} - \frac{\delta \mathscr{L}}{\delta Q_\tau} c^j_{\tau,i}.}$$

To exhibit more clearly the covariant form of this expression, we have written

$$\frac{\partial \mathscr{L}}{\partial Q_{\tau\|j}} \text{ instead of } \frac{\partial \mathscr{L}}{\partial Q_{\tau|j}}$$

As for equation (28), it acquires a simple and well-known meaning, if we take into account the field equations which must be satisfied by the variables Q_τ. To write these equations, we shall form, with the help of Q_τ, a set of independent variables $Q_{\bar{\tau}}$, and, to avoid any complications due to the quantization of the variables, we shall assume that the $Q_{\bar{\tau}}$ are linear and homogeneous functions of the Q_τ. The field equations are then

$$(30)\qquad \frac{\delta \mathscr{L}}{\delta Q_{\bar{\tau}}} = 0.$$

Furthermore, it can be seen that

$$(31)\qquad \frac{\delta \mathscr{L}}{\delta Q_\tau} c^j_{\tau,i} = 0$$

and

$$(32)\qquad \frac{\delta \mathscr{L}}{\delta Q_\tau} Q_{\tau\|i} = 0,$$

as follows from the equations (30). The identity (28) thus signifies, through (11) and (24), that

$$(33) \qquad \boxed{\mathscr{T}^{j}_{.i\|j}=0}$$

by virtue of the field equations: this is the general relativistic expression for the conservation of energy-momentum.

It should be noted that equation (31) can be used to simplify definition (29) of the energy-momentum density without affecting the conservation theorem (33): indeed, we can put, instead of (29)

$$(34) \qquad \mathfrak{T}^{i}_{.j}=\frac{\partial \mathscr{L}}{\partial Q_{\tau\|j}}\,Q_{\tau\|i}-\mathscr{L}\,\delta^{j}_{i}-\mathscr{R}^{kj}_{i\|k}.$$

However, the symmetry condition (3) is no longer necessarily identically satisfied by $\mathfrak{T}^{ij}$, but only as a consequence of the field equations. By putting

$$(35) \qquad d^{ij}_{\omega}=\frac{1}{2}\left(c^{i}_{\omega,k}\,g^{kj}-c^{j}_{\omega,k}\,g^{ki}\right)$$

in general we can write, instead of (3), by comparing (29) and (34)

$$(36) \qquad \mathfrak{T}^{ij}-\mathfrak{T}^{ji}=2\,\frac{\delta\mathscr{L}}{\delta Q_{\tau}}\,d^{ij}_{\tau}.$$

The antisymmetric tensor d^{ij}_{ω}, defined by (35) and which will soon play an essential role, has a very simple meaning in the case of *special* relativity, where we assign the constant values $\mathring{g}_{lm}$ to the variables g_{lm}. Indeed, it can be seen from (8) and (9) that an infinitesimal Lorentz transformation corresponds to linear and homogeneous ξ^{i}:

$$(37) \qquad \mathring{\xi}^{i}=\varepsilon^{i}_{.k}\,x^{k}.$$

with the orthogonality condition

$$(38) \qquad \varepsilon_{ik}+\varepsilon_{ki}=0,\ \text{setting}\ \varepsilon_{ik}=\mathring{g}_{il}\,\varepsilon^{l}_{.k}.$$

Thus, for such a transformation, we have precisely

$$(39) \qquad \mathring{d}Q_{\omega}=\delta^{ij}_{\omega}\,\varepsilon_{ji}.$$

Finally, we must now express the $\mathscr{R}_i^{kj}$ occurring in (29) or (34), and defined by (22), as functions of only the $\partial\mathscr{L}/\partial Q_{\tau\|j}$. The elimination of the $\partial\mathscr{L}/\partial Q_{\gamma|j}$ is immediately achieved if we note that for the components Q_γ of the metric tensor we have, from (35),

$$(40) \qquad d_\gamma^{ij}=0.$$

Thus it follows that the quantities

$$(41) \qquad \mathscr{D}^{k;\,ij} \equiv \frac{1}{2}\left(\mathscr{R}_l^{ki}\, g^{lj} - \mathscr{R}_l^{kj}\, g^{li}\right) = \frac{\partial\mathscr{L}}{\partial Q_{\tau\|k}}\, d_\tau^{ij}$$

do not depend anymore on operations explicitly involving the gravitational variables. On the other hand, by virtue of their antisymmetry property (21), the $\mathscr{R}_l^{jk}$ can be expressed in terms of $\mathscr{D}^{k;\,ij}$ only, in the form:

$$(42) \qquad \mathscr{R}_l^{jk}\, g^{li} = \mathscr{D}^{j;ki} - \mathscr{D}^{i;jk} + \mathscr{D}^{k;ij}.$$

It can be seen that the explicit form of the energy-momentum density given by (29) or by (34), (41), (42) was obtained by using general consequences of the invariance of the Lagrangian function as expressed by the fundamental identities (20), (21) while the third identity (19) led to the conservative theorem (33). The extension of the preceding considerations to the case where the Lagrangian contains spinor variables will similarly require the use of identities devolving from the invariance of the Lagrangian under orthogonal transformations of the tetrads. This will be shown in the following paragraph.

4. Spinor variables

We now assume that the variables of the system under consideration are either tensors and are denoted by Q_τ, or are spinors, written Q_σ, and we take as the gravitational variables Q_γ the tetrads $h_i^{(\mu)}$. An infinitesimal orthogonal transformation of this tetrad is represented by

$$(43) \qquad \delta h_i^{(\mu)} = d_{i(\lambda\nu)}^{(\mu)}\,\varepsilon^{(\lambda\nu)}, \qquad (\varepsilon^{(\lambda\nu)} + \varepsilon^{(\nu\lambda)} = 0)$$

where the coefficients $d_{i(\lambda\nu)}^{(\mu)}$ are simply related, through the equations

$$(44) \qquad d_{i(\lambda\nu)}^{(\mu)} = g_{(\lambda\rho)}\, g_{(\nu\sigma)}\, h_j^{(\rho)}\, h_k^{(\sigma)}\, d_i^{(\mu)jk}$$

to the $d_i^{(\mu)jk}$ defined by the general formula (35); explicitly, we have

$$(45)\quad \begin{cases} d_i^{(\mu)jk} = \tfrac{1}{2}(\delta_i^k\, h^{(\mu)j} - \delta_i^j\, h^{(\mu)k}), \\ d_{i(\lambda\nu)}^{(\mu)} = \tfrac{1}{2}(\delta_\lambda^\mu\, g_{(\nu\sigma)}\, h_i^{(\sigma)} - \delta_\nu^\mu\, g_{(\lambda\sigma)}\, h_i^{(\sigma)}). \end{cases}$$

Under such a transformation, the spinor variables undergo variations of the form

$$(46)\quad \begin{cases} \delta Q_\sigma = d_{\sigma(\lambda\nu)}\, \varepsilon^{(\lambda\nu)}, \\ \text{with } d_{\sigma(\lambda\nu)} = -d_{\sigma(\nu\lambda)}. \end{cases}$$

If, in particular, Q_σ represents the set Ψ of two-spinors of rank corresponding to an elementary particle of spin $\frac{1}{2}$, we have [3]

$$(47)\qquad d_{(\sigma)\lambda\nu} = -\tfrac{1}{8}(\gamma_{(\lambda)}\,\gamma_{(\nu)} - \gamma_{(\nu)}\,\gamma_{(\lambda)})\, Q_\sigma,$$

where the matrices $\gamma_{(\lambda)}$ satisfy the relations

$$(48)\qquad \gamma_{(\lambda)}\,\gamma_{(\nu)} + \gamma_{(\nu)}\,\gamma_{(\lambda)} = -2g_{(\lambda\nu)};$$

for the associated variable [6] $Q_{\sigma'} \equiv \Phi = \Psi^\dagger \beta$ ($\Psi^\dagger$ being the Hermitian conjugate of Ψ, and β^{-1} a matrix transforming the $\gamma_{(\nu)}$ into their Hermitian conjugates), we have as a consequence

$$(49)\qquad d_{\sigma'(\lambda\nu)} = \tfrac{1}{8}Q_{\sigma'}(\gamma_{(\lambda)}\,\gamma_{(\nu)} - \gamma_{(\nu)}\,\gamma_{(\lambda)}).$$

The invariance of the Lagrangian under transformations (43), (46) is expressed by the condition

$$\int \left\{ \frac{\delta\mathcal{L}}{\delta h_i^{(\mu)}}\, \delta h_i^{(\mu)} + \frac{\delta\mathcal{L}}{\delta Q_\sigma}\, \delta Q_\sigma \right\} \mathrm{d}w + \int \frac{\partial}{\partial x^j} \left\{ \frac{\partial\mathcal{L}}{\partial h_{i|j}^{(\mu)}}\, \delta h_i^{(\mu)} + \frac{\partial\mathcal{L}}{\partial Q_{\sigma|j}}\, \delta Q_\sigma \right\} \mathrm{d}w = 0$$

and yields two sets of independent identities:

$$(50)\quad \begin{cases} \dfrac{\delta\mathcal{L}}{\delta h_i^{(\mu)}}\, d_{i(\lambda\nu)}^{(\mu)} + \dfrac{\delta\mathcal{L}}{\delta Q_\sigma}\, d_{\sigma(\lambda\nu)} \equiv 0 \\[2ex] \dfrac{\partial\mathcal{L}}{\partial h_{i|j}^{(\mu)}}\, d_{i(\lambda\nu)}^{(\mu)} + \dfrac{\partial\mathcal{L}}{\partial Q_{\sigma|j}}\, d_{\sigma(\lambda\nu)} \equiv 0. \end{cases}$$

From (44), we have

$$h^{(\lambda)j}\, h^{(\nu)k}\, d_{i(\lambda\nu)}^{(\mu)} = d_i^{(\mu)jk};$$

if we write similarly (except for a difference in sign the meaning of which will be apparent later on)

$$h^{(\lambda)j}\, h^{(\nu)k}\, d_{\sigma(\lambda\nu)} = -d_{\sigma}^{(h)jk} \tag{51}$$

[The index (h) being added to avoid any confusion with the quantity d_{σ}^{jk} defined by (35)], we can put identities (50) in the equivalent form

$$\begin{cases} \dfrac{\delta \mathscr{L}}{\delta h_{i}^{(\mu)}}\, d_{i}^{(\mu)jk} - \dfrac{\delta \mathscr{L}}{\delta Q_{\sigma}}\, d_{\sigma}^{(h)jk} \equiv 0 \\[2ex] \dfrac{\partial \mathscr{L}}{\partial h_{i\|j}^{(\mu)}}\, d_{i}^{(\mu)lk} - \dfrac{\partial \mathscr{L}}{\partial Q_{\sigma|j}}\, d_{\sigma}^{(h)lk} \equiv 0. \end{cases} \tag{52}$$

With equation (45) and the definition (7) of $\mathscr{T}^{j}_{\cdot i}$, the first identity (52) gives

$$\mathscr{T}^{jk} - \mathscr{T}^{kj} \equiv 2\, \frac{\delta \mathscr{L}}{\delta Q_{\sigma}}\, d_{\sigma}^{(h)jk}. \tag{53}$$

If it is assumed again that the independent spinor variables $Q_{\bar{\sigma}}$ are linear and homogeneous functions of the Q_{σ}, it can be seen that we can write

$$\frac{\delta \mathscr{L}}{\delta Q_{\sigma}}\, d_{\sigma}^{(h)jk} = 0 \tag{54}$$

by virtue of the field equations

$$\frac{\delta \mathscr{L}}{\delta Q_{\bar{\sigma}}} = 0. \tag{55}$$

Formula (53) thus shows [3] that the tensorial density of energy-momentum $\mathscr{T}^{jk}$ is still symmetric with respect to the indices j, k, if not identically at least as a consequence of the field equations (55).

Going back to the identities (19) and (20), we can first of all write them in covariant form [analogous to (27), (28)]

$$\begin{cases} \dfrac{\delta \mathscr{L}}{\delta Q_{\omega}}\, c^{j}_{\omega,i} \equiv \dfrac{\partial \mathscr{L}}{\partial Q_{\omega|j}}\, Q_{\omega;i} - \mathscr{L}\,\delta^{j}_{i} - \mathscr{R}^{kj}_{i\|k} \\[2ex] \left(\dfrac{\delta \mathscr{L}}{\delta Q_{\omega}}\, c^{j}_{\omega,i}\right)_{\|j} + \dfrac{\delta \mathscr{L}}{\delta Q_{\omega}}\, Q_{\omega;i} \equiv 0, \end{cases} \tag{56}$$

where we have put, for all variables,

$$(57)\qquad Q_{\omega;\,l} = Q_{\omega|l} + c^{j}_{\omega,\,i}\, \Gamma^{i}_{ji}$$

and also taken into account (25), (26). For purely tensor variables, $Q_{\tau;\,l}$ is, of course, just the covariant derivative $Q_{\tau\|l}$ defined by (23). But for the spinor variables, the covariant derivative $Q_{\sigma\|l}$ differs from $Q_{\sigma;\,l}$ by a term [3] which occurs because a parallel displacement of the tetrad is equivalent to an orthogonal transformation which affects the Q_σ. If the parallel displacement is defined by a vector ξ^j (constant at the point under consideration), this equivalence can be expressed, from (43), in the form

$$\xi^{(\lambda\nu)}\, d^{(\mu)}_{i(\lambda\nu)} = -h^{(\mu)}_{i;\,j}\, \xi^j;$$

From this we obtain, using (45), for the parameters $\xi^{(\lambda\nu)}$ of the equivalent orthogonal transformations:

$$(58)\qquad \begin{cases} \xi^{(\lambda\nu)} = \Gamma^{(\lambda\nu)}_{j}\, \xi^j \\ \text{with}\quad \Gamma^{(\lambda\nu)}_{j} = -\tfrac{1}{2}\{h^{(\lambda)}_{i;\,j}\, h^{(\nu)i} - h^{(\nu)}_{i;\,j}\, h^{(\lambda)i}\}; \end{cases}$$

it follows that for the covariance derivative of Q_σ:

$$(59)\qquad Q_{\sigma\|l} = Q_{\sigma;\,l} + d_{\sigma(\lambda\nu)}\, \Gamma^{(\lambda\nu)}_{l}.$$

With the above, the first identity (50) yields

$$-\frac{\delta\mathscr{L}}{\delta h^{(\mu)}_{l}}\, h^{(\mu)}_{l;\,i} + \frac{\delta\mathscr{L}}{\delta Q_\sigma}\, d_{\sigma(\lambda\nu)}\, \Gamma^{(\lambda\nu)}_{i} \equiv 0,$$

whence

$$\frac{\delta\mathscr{L}}{\delta h^{(\mu)}_{l}}\, h^{(\mu)}_{l;\,i} + \frac{\delta\mathscr{L}}{\delta Q_\sigma}\, Q_{\sigma;\,i} \equiv \frac{\delta\mathscr{L}}{\delta Q_\sigma}\, Q_{\sigma\|i};$$

similarly, from the second identity (50)

$$\frac{\partial\mathscr{L}}{\partial h^{(\mu)}_{l\|j}}\, h^{(\mu)}_{l;\,i} + \frac{\partial\mathscr{L}}{\partial Q_{\sigma|j}}\, Q_{\sigma;\,i} \equiv \frac{\partial\mathscr{L}}{\partial Q_{\sigma\|j}}\, Q_{\sigma\|i}.$$

Carrying these results into (56), we obtain

$$(60)\qquad \begin{cases} \dfrac{\delta\mathscr{L}}{\delta Q_\omega}\, c^{j}_{\omega,\,i} \equiv \dfrac{\partial\mathscr{L}}{\partial Q_{\alpha\|j}}\, Q_{\alpha\|i} - \mathscr{L}\,\delta^{j}_{i} - \mathscr{R}^{kj}_{i\|k} \\ \left(\dfrac{\delta\mathscr{L}}{\delta Q_\omega}\, c^{j}_{\omega,\,i}\right)_{\|j} + \dfrac{\delta\mathscr{L}}{\delta Q_\alpha}\, Q_{\alpha\|i} \equiv 0, \end{cases}$$

where the symbol Q_α includes all tensor variables Q_τ and spinor variables Q_σ. The first formula (60) can be written by (11),

$$(61) \qquad \mathcal{T}^j_{.i} \equiv \frac{\partial \mathcal{L}}{\partial Q_{\alpha\|j}} Q_{\alpha\|i} - \mathcal{L}\delta^j_i - \mathcal{R}^{kj}_{i\|k} - \frac{\delta \mathcal{L}}{\delta Q_\alpha} c^j_{\alpha,\, i},$$

giving a direct generalization of (29).

By taking into account especially (54) and (59), we find that the field equations (30) and (55) imply

$$(62) \qquad \begin{cases} \dfrac{\delta \mathcal{L}}{\delta Q_\alpha} c^j_{\alpha,\, i} = 0 \\[2ex] \dfrac{\delta \mathcal{L}}{\delta Q_\alpha} Q_{\alpha\|i} = 0, \end{cases}$$

from which the conservation theorem (33) follows as before. With (62), we can simplify even further the definition (61) by taking for the energy-momentum tensor density

$$(63) \qquad \mathcal{T}^j_{.i} = \frac{\partial \mathcal{L}}{\partial Q_{\alpha\|j}} Q_{\alpha\|i} - \mathcal{L}\delta^j_i - \mathcal{R}^{kj}_{i\|k}.$$

By virtue of (35) and of (53), we then have

$$(64) \qquad \mathcal{T}^{ji} - \mathcal{T}^{ij} = 2 \frac{\delta \mathcal{L}}{\delta Q_\alpha} s^{ji}_\alpha$$

if we put

$$(65) \qquad \begin{cases} s^{ji}_\tau = d^{ji}_\tau, \\ s^{ji}_\sigma = d^{ji}_\sigma + d^{(h)ji}_\sigma; \end{cases}$$

by (62) and (54), relation (64) shows that the symmetry of the tensor density $\mathcal{T}^{ji}$ is a consequence of the field equations.

As for the tensor density $\mathcal{R}^{kj}_i$, it is still calculated by (42); but we now have, instead of (41),

$$\mathcal{D}^{k;\, ij} = \frac{\partial \mathcal{L}}{\partial Q_{\omega|k}} d^{ij}_\omega,$$

or again, from the second identity (52), and with the notation (65),

$$(66) \qquad \mathcal{D}^{k;\, ij} = \frac{\partial \mathcal{L}}{\partial Q_{\alpha\|k}} s^{ij}_\alpha,$$

a formula in which the gravitational variables do not occur explicitly anymore.

In the context of *special relativity*, it can immediately be seen that, in the case where there are spinor variables, the s_α^{ij} are the natural generalization of the $\mathring{d}_\alpha^{ij}$ appearing in formula (39). In particular, we can take

$$\mathring{h}_i^{(\mu)}=\delta_i^\mu,$$

in which case, the orthogonal transformation (43) of the tetrad is identical to the space-time Lorentz transformation defined by (37). With equations (39), (46), and (51), it can be seen that the s_α^{ij} given by (65) reduce in this case to the coefficients of the variations

$$\text{(67)} \qquad \mathring{\delta} Q_\alpha=\mathring{s}_\alpha^{ij}\,\varepsilon_{ji}$$

corresponding (because of the choice of covariant tensor components for the variables Q_α) to the Lorentz transformation *contragradient* to the transformation (37) of the space-time variables. [It is this relation of contragradience which is the cause of the minus sign introduced in formula (51).]

5. Summary and Applications

Formulas (63), (42) and (66) express the general result which we have derived for a system described by tensor and spinor variables Q_α and having a Lagrangian which – except for some supplementary restrictions required by the quantization of the variables – is assumed to contain only the Q_α and their (covariant) first derivatives $Q_{\alpha\|j}$. Putting together these formulas, we can thus see that the energy-momentum tensor density of such a system can be calculated from the Lagrangian density $\mathscr{L}$ with the help of formula

$$\text{(I)} \qquad \mathscr{T}^j_{.i}=\frac{\partial\mathscr{L}}{\partial Q_{\alpha\|j}}\,Q_{\alpha\|i}-\mathscr{L}\,\delta_i^j-\mathscr{R}^{kj}_{i\|k},$$

by using

$$\left\{\begin{aligned} &\text{(II)} && \mathscr{R}_l^{kj}\,g^{li}=\mathscr{D}^{j;\,ik}-\mathscr{D}^{i;\,kj}+\mathscr{D}^{k;\,ji}\\ &\text{(III)} && \mathscr{D}^{k;\,ij}\equiv\frac{1}{2}\left(\mathscr{R}_l^{ki}\,g^{lj}-\mathscr{R}_l^{kj}\,g^{li}\right)=\frac{\partial\mathscr{L}}{\partial Q_{\alpha\|k}}\,s_\alpha^{ij},\end{aligned}\right.$$

where the s_α^{ij} are some very simple linear combinations of the Q_α. The components $\mathcal{T}^{ji}$ are symmetric with respect to their indices by virtue of the field equations for the Q_α, and these same equations also imply the conservation theorem

$$\text{(IV)} \qquad \mathcal{T}^{j}_{.i\|j}=0.$$

It can be seen that, with the exception of the term $-\mathcal{L}\delta^j_i$, the expression for $\mathcal{T}^j_i$ depends solely on the derivatives of the Lagrangian with respect to the derivatives of the variables Q_α.

As formulas (I)–(III) do not explicitly involve the way in which the Lagrangian depends on the gravitational variables, they are directly suited to the usual case where it is desired to calculate the expression for the energy-momentum tensor, within the bounds of *special* relativity, from the 'special' form (that is invariant only under the Lorentz group) of the Lagrangian.

In this case, the covariant derivatives reduce to the corresponding partial derivatives, and the quantities s_α^{ij} acquire a particularly simple meaning: They are the coefficient which determine the variations

$$\text{(V}_1\text{)} \qquad \mathring{\delta} Q_\alpha=\mathring{s}_\alpha^{ij}\,\varepsilon_{ji}$$

of the variables Q_α under an infinitesimal Lorentz transformation

$$\text{(V}_2\text{)} \qquad \begin{cases} \mathring{\delta} x^i=\varepsilon^i_{.k}\,x^k \\ (\varepsilon^i_{.k}=\mathring{g}^{il}\,\varepsilon_{lk}\,;\ \mathring{g}^{il}=0\ (i\neq l),\ 1\ (i=l=1,\,2,\,3),\ -1\ (i=l=4)) \\ \varepsilon_{il}+\varepsilon_{li}=0. \end{cases}$$

The divergence of the term $\mathcal{R}^{kj}_{i\|k}$ occurring in the tensor density $\mathcal{T}^j_{.i}$ can be written in general, because of the antisymmetry of $\mathcal{R}^{kj}_i$ with respect to the indices k, j.

$$\text{(68)} \qquad (\mathcal{R}^{kj}_{i\|k})_{\|j}=\tfrac{1}{2}\mathcal{R}^{kj}_h\,\mathsf{R}^{\;h}_{ijk},$$

$\mathsf{R}^{\;h}_{ijk}$ being the Riemann-Christoffel tensor. In the case of *special* relativity, this expression vanishes (as can indeed be seen directly, since it reduces to the second partial derivatives of antisymmetric components). In this case, therefore, the divergence $\mathcal{R}^{kj}_{i\|k}$ could be omitted from the definition (I) of $\mathcal{T}^j_{.i}$ without affecting the validity of the conservation theorem (IV); inversely, a specialized procedure for looking for the special-relativistic expression for $\mathcal{T}^j_{.i}$, starting simply from the condition that the expression sought must satisfy equation (IV) in virtue of the field equations, could not lead unambiguously to the adjunction of this term $\mathscr{R}^{kj}_{i\|k}$. Nevertheless, if this term is omitted, the remaining expression for $\mathscr{T}^{ji}$ no longer generally has the essential property of being symmetric with respect to the indices j, i (by virtue of the field equations).

Still remaining within the bounds of *special* relativity, it can be seen that the term $\mathscr{R}^{kj}_{i\|k}$ does not affect the definition of the total energy-momentum vector of the system:

$$(69) \qquad G^i = -\int \mathscr{T}^{4i}\, dv, \ (dv = dx^1\, dx^2\, dx^3)$$

but, by contrast, it appears in an essential way in the expression for the total angular momentum:

$$(70) \qquad M^{ij} = -\int (x^i \mathscr{T}^{4j} - x^j \mathscr{T}^{4i})\, dv. (i, j = 1, 2, 3)$$

If, in effect, one puts

$$(71) \qquad \mathscr{P}^\alpha \equiv \frac{\partial \mathscr{L}}{\partial Q_{\alpha|4}},$$

the definition (69) (where the $-$ sign has been chosen to conform to the usual conventions) yields[4] for the total momentum

$$(72) \qquad \mathbf{G} = -\int \mathscr{P}^\alpha \operatorname{grad} Q_\alpha\, dv$$

and for the total energy

$$(73) \qquad G_4 = \int \left\{ \mathscr{P}^\alpha Q_{\alpha|4} - \mathscr{L} \right\} dv.$$

But, because of the presence of the divergence term $\mathscr{R}^{kj}_{i\|j}$ in the form (I) of the energy-momentum tensor, definition (70) very naturally leads to a distinction between 'orbital angular momentum'

$$(74) \qquad \mathbf{M}_o = -\int \mathscr{P}^\alpha (\mathbf{x} \wedge \operatorname{grad})\, Q_\alpha\, dv,$$

which corresponds directly to the momentum density actually occurring in expression (72), and 'intrinsic angular momentum' of 'spin'

$$(75) \qquad M_s^{ij} = \iint \left\{ x^i \frac{\partial(\mathscr{R}_l^{k4} \mathring{g}^{lj})}{\partial x^k} - x^j \frac{\partial(\mathscr{R}_l^{k4} \mathring{g}^{li})}{\partial x^k} \right\} dv,$$

which is due precisely to the divergence term. Of course, this separation of the total angular momentum into two terms:

$$(76) \qquad M^{ij} = M_o^{ij} + M_s^{ij},$$

only has direct physical sense for physical agencies endowed with inertia, to which a reference system can be attached that is at rest with respect to them. By means of integration by parts, the intrinsic angular momentum (75) can be put in more simple form:

$$M_s^{ij} = \int \left\{ \mathscr{R}_l^{j4} \mathring{g}^{li} - \mathscr{R}_l^{i4} \mathring{g}^{lj} \right\} dv,$$

or again, according to (III),

$$(77) \qquad M_s^{ij} = 2 \int \mathscr{P}^{\alpha} s_{\alpha}^{ij} \, dv.$$

Discussion of the passage to the Hamiltonian form and of the quantization of the variables, by means of (71), (73), is beyond the scope of the present work. In this respect, we shall content ourselves with some additional remarks, limiting ourselves to the 'regular' case where formulas (71), for the independent variables $Q_{\bar{\alpha}}$ do not give rise to any identities between these variables and their canonically conjugate momenta $\mathscr{P}^{\bar{\alpha}}$. We note that if the independent variables $Q_{\bar{\alpha}}$ are linear and homogeneous combinations of the Q_{α}, the summations over all the components Q_{α}, appearing in the formulas of this paragraph can be replaced by the corresponding summations over the independent variables $Q_{\bar{\alpha}}$. If, then, in the 'regular' case which we are considering, appropriate commutation relations are defined between the $Q_{\bar{\alpha}}$ and the $\mathscr{P}^{\bar{\alpha}}$, the well-known connection between the total momentum operator and the translation group is immediately recognized in formula (72), and in formula (76) [with (74) and (77)], it is the analogous connection between the total angular momentum and the rotation group which is found. Moreover, it can be easily verified that the components of orbital angular momentum on the one side, and the components of spin on the other, are related among themselves by the commutation relations characteristic of the rotation group, while each component of orbital angular momentum commutes with each component of spin. These properties complete the justification for the distinction between orbital angular momentum and spin for physical agencies of finite mass.

6. Examples

In conclusion we shall consider, as examples, the case of elementary particles with spin $\frac{1}{2}$, and then the case of spin 1 fields such as vector meson fields or the electromagnetic field.

A. *Spin* $\frac{1}{2}$

Using the notations introduced above for formulas (47), (48), (49) we can take

$$\text{(78)} \qquad \mathscr{L} = -\operatorname{Re}\left\{\frac{\hbar c}{i}\Phi\gamma^j\Psi_{\|j} + \Phi\mu\Psi\right\},$$

as the Lagrangian density, where $\gamma^j = h^{(\nu)j}\gamma_{(\nu)}$ and Re designates the real part of the expression following it. The operator μ is related in the usual way to the mass of the particles under consideration. To be consistent with the usual representation, we shall choose $\beta = \gamma^4$, and introduce the current matrices

$$\text{(79)} \qquad \alpha^j = \gamma^4\gamma^j$$

and the spin matrices

$$\text{(80)} \qquad \sigma^j = i\gamma^k\gamma^l;$$

(j, k, l, being permuted cyclically); with this choice of β, the $\gamma^j (j=1, 2, 3)$ are anti-Hermitian and the matrices β, α^j, σ^j and μ Hermitian.

The momenta conjugate to Ψ and $\Phi = \Psi^\dagger\beta$ are respectively

$$\text{(81)} \qquad \mathscr{P}_\Psi = -\frac{1}{2}\frac{\hbar c}{i}\Phi\beta, \quad \mathscr{P}_\Phi = \frac{1}{2}\frac{\hbar c}{i}\beta\Psi.$$

Therefore the total momentum is:

$$G_i = -\int \left\{ \mathscr{P}_\Psi \, \Psi_{\|i} + \Phi_{\|i} \, \mathscr{P}_\Phi \right\} dv$$

(82)
$$= \mathrm{Re} \frac{\hbar c}{i} \int \Phi \beta \, \Psi_{\|i} \, dv$$

$$= \int \Psi^\dagger \, p_i \, \Psi \, dv,$$

introducing the operator

(83)
$$\boldsymbol{p} = \frac{\hbar c}{i} \, \mathrm{grad}.$$

Similarly, the total energy becomes

(84)
$$G_4 = \int \left\{ \mathscr{P}_\Psi \Psi_{\|4} + \Phi_{\|4} \mathscr{P}_\Phi - \mathscr{L} \right\} dv$$

$$= \int \Psi^\dagger \left\{ \sum_{j=1}^{3} \alpha^j \, p_j + \beta \mu \right\} \Psi \, dv.$$

The angular momentum is composed of the orbital angular momentum

(85)
$$\mathbf{M}_0 = \int \Psi^\dagger (\boldsymbol{x} \wedge \boldsymbol{p}) \, \Psi \, dv$$

and of the spin

$$M_s^{ij} = 2 \int \left\{ \mathscr{P}_\Psi \, s_\Psi^{ij} + s_\Phi^{ij} \, \mathscr{P}_\Phi \right\} dv;$$

From (47), (48), (49), (51) we have

$$2 s_\Psi^{ij} = \frac{1}{2} \gamma^i \gamma^j \Psi,$$

$$2 s_\Phi^{ij} = \frac{1}{2} \Phi \, \gamma^i \gamma^j,$$

hence

$$M_s^{ij} = \frac{1}{2} \mathrm{Re} \int \mathscr{P}_\Psi \, \gamma^i \gamma^j \, \Psi \, dv,$$

or finally, with the notation (80)

$$\mathbf{M}_s = \frac{\hbar c}{2}\int \Psi^\dagger\, \boldsymbol{\sigma}\, \Psi\, \mathrm{d}v. \tag{86}$$

B. *Spin 1*

We shall use as variables the covariant components Q_i of the four-vector potential – i.e. the components of the vector potential **U** and the negative of the scalar potential, $-V$. We shall introduce the field tensor

$$F_{ik} = \frac{\partial Q_k}{\partial x^i} - \frac{\partial Q_i}{\partial x^k} \tag{87}$$

representing the two spatial vectors **H** and **E**, defined respectively by

$$\begin{cases} H_i = F_{kl}\,; \text{ (with cyclic permutations of } i, k, l,) \\ E_i = F_{i4},\ i = 1, 2, 3. \end{cases} \tag{88}$$

We shall take for the Lagrangian density

$$\mathscr{L} = -\frac{1}{4} F^{ik} F_{ik} - \frac{1}{2}\kappa^2 Q^i Q_i$$

$$= -\frac{1}{2}(\mathbf{H}^2 - \mathbf{E}^2) - \frac{1}{2}\kappa^2(\mathbf{U}^2 - V^2), \tag{89}$$

where κ is, as is well known, a constant related to the mass of the particles associated with the field under consideration.

The momenta conjugate to the components of **U** are, from (89) the corresponding components of $-\mathbf{E}$ but the momentum conjugate to V is identically zero. Hence it happens that the field equation expressing the vanishing of the variational derivative $\delta\mathscr{L}/\delta V$ reduces to a form containing no time derivative:

$$\operatorname{div} \mathbf{E} + \kappa^2 V = 0. \tag{90}$$

The total energy can easily be expressed as

$$G_4 = \frac{1}{2}\int \left\{ \mathbf{H}^2 + \mathbf{E}^2) + \kappa^2(\mathbf{U}^2 + V^2) \right\} \mathrm{d}v$$

$$- \int V(\operatorname{div} \mathbf{E} + \kappa^2 V)\, \mathrm{d}v; \tag{91}$$

As is well known, in the quantization of the variables (**U**, $-$**E**), can be chosen as canonical variables, but two essentially different cases occur depending on whether κ is equal to zero or not: In the latter case, it is necessary to consider equation (90) as a definition for the operator V, and the last integral in (91) is identically zero; in the other case (that of electrodynamics), equation (90), which reduces to div $\mathbf{E}=0$, is a 'constraint', by virtue of which the last term in (91) vanishes.

As for the total momentum and the orbital and intrinsic angular momenta, they can be written in the following form independent of κ:

$$\mathbf{G}=\sum_{i=1}^{3}\int E^i \operatorname{grad} U_i \, dv, \tag{92}$$

$$\mathbf{M}_0=\sum_{i=1}^{3}\int E^i(\boldsymbol{x}\wedge \operatorname{grad})\, U_i \, dv, \tag{93}$$

$$\mathbf{M}_s=\int \mathbf{E}\wedge\mathbf{U}\, dv. \tag{94}$$

Instead of (92) it is possible to write

$$\mathbf{G}=\int \mathbf{E}\wedge \operatorname{rot}\mathbf{U}\, dv+\int (\mathbf{E}\operatorname{grad})\,\mathbf{U}\, dv$$

$$=\int \mathbf{E}\wedge\mathbf{H}\, dv-\int \mathbf{U}\operatorname{div}\mathbf{E}\, dv, \tag{95}$$

which for a pure electromagnetic field reduces to the usual expression for the Poynting vector, using the constraint div $\mathbf{E}=0$. In the same manner, starting from (93) and (94), the corresponding expression for the total angular momentum is found

$$\mathbf{M}=\int \boldsymbol{x}\wedge(\mathbf{E}\wedge\mathbf{H})\, dv-\int (\boldsymbol{x}\wedge\mathbf{U})\operatorname{div}\mathbf{E}\, dv. \tag{96}$$

However, to establish this result in the case of the electromagnetic field, it is necessary to go back to the initial form (75) for the term $\mathbf{M}_s$. In this case it can be seen that expression (96) does indeed possess the property of gauge invariance[5] in virtue of the condition div $\mathbf{E}=0$, while the distinction between orbital angular momentum and spin no longer has an unambiguous significance.

The present work was done (in April 1939) at the Institute for Advanced Study in Princeton, during a research trip to the United States made possible by a grant from the Belgian American Educational Foundation. [– L. R.]

NOTES

[1] When the present work (the publication of which has been subjected to various delays) was completely finished, essentially identical results to those of Section 5 below were published by F. Belinfante (*Physica* **6** (1939), 887). The latter has kindly informed me, after becoming aware of my manuscript, that he has himself very recently rederived his results by a method very analogous to the one discussed in this work, but which nevertheless contains a few interesting differences. Given the didactic form of my exposition, its present publication without any modifications may still be useful, as it presents an overall view from the standpoint of method of the problem, the solution to which Belinfante first published.

[2] We adopt the usual summation convention where repetition of an index in a term signifies summation over that index.

[3] This 'symmetrization' convention applies to all following formulas. It should be noted that it does not apply to the gravitational variables.

[4] We put $x^4 = ct$. The expression for **G** and **M** represent the corresponding physical quantities multiplied by c.

[5] If we are considering an electromagnetic field interacting with a system of charges with density ρ, the constraint becomes div $\mathbf{E} - \rho = 0$. The last terms on the right-hand sides of (95) and (96) then combine with additive terms corresponding respectively to the momentum and to the angular momentum of the system of charges, in such a way that the final expressions do possess gauge invariance.

BIBLIOGRAPHY

[1] Lorentz, H. A., *Amsterdam Akad. Verslag* **23** (1915), 1073; **24** (1916), 1389, 1759; **25** (1916), 468, 1380.
Hilbert, D., *Gött. Nachr.* (1915), 395.
De Donder, Th., *Amsterdam Akad. Verslag* **25** (1916), 153; **26** (1917), 101; **27** (1918), 432; *Archives du Musée Teyler* **3** (1917); *Bull. Acad. Roy. Belg.* **5** (1919), 201, 317; **7** (1922), 371, 420; **8** (1923), 129; **10** (1924), 188; *La Gravifique einsteinienne* (1921) (*Compléments*, 1922); *C.R. Acad. Sci. Paris* **174** (1922), 1288; **176** (1923), 1700; **177** (1923), 106, 254.
Klein, F., *Gött. Nachr.* (1917), 469; (1918), 235.
Weyl, H., *Ann. d. Phys.* **54** (1917), 117; *Raum, Zeit, Materie* (1918; 4th edition 1921, §28). [See English translation (1922) by H. L. Brose, *Space–Time–Matter*, reprinted Dover, New York, 1950, with a new preface by the author – Ed.]

[2] Cf. for example W. Pauli, *Handb. d. Phys.* **24**/1 (1933), 235.

[3] Weyl, H., *Zs. f. Phys.* **56** (1929), 330.
Fock, V., *Zs. f. Phys.* **57** (1929), 261.

[4] Cf. De Donder, Th., *Théorie invariantive du Calcul des variations* (new edition, Gauthier-Villars, Paris, 1935). See in particular pp. 8 and 84.
[5] Cf. for example Weyl, H., *Raum, Zeit, Materie*, *op. cit.*, §14.
[6] Pauli, W., *Ann. Inst. Poincaré* **6** (1936), 109.

CHAPTER 2

ON THE DEFINITION OF SPIN FOR A RADIATION FIELD

[1942e]

Using the general expression (1) for the energy-momentum tensor of any field, it is possible to define the spin or intrinsic angular momentum of the field, contained in a given region. But it is clear that this definition is ambiguous due to the fact that a certain integral over the surface of the domain under consideration can be retained or omitted at will in this definition; and, for a radiation field, this integral might very well be different from zero. Let us define a radiation field to be a field the amplitude of which at a distance R from a fixed point does not fall off more rapidly than R^{-1}. The purpose of this note is simply to show that in this case the flux of spin carried by the radiation field – which is in fact the quantity having a direct physical significance – is not affected by the ambiguity in question. For a better understanding, we will summarize, as briefly as possible, the entire problem.

I

It is well known[1] that the energy-momentum tensor for any system, the field equations of which are derived from a variational principle, can be written (in special relativity) as follows:

$$(1)\qquad \mathcal{T}^{ik}=\bar{\mathcal{T}}^{ik}+\frac{\partial}{\partial x^j}\mathcal{R}^{ij,k},\ (i, j, k=1, 2, 3, 4),$$

where the tensor $\mathcal{R}^{ij,k}$ is antisymmetric with respect to the indices i, j. Because of this property, it can be seen at once that the last term does not play any role in the expression of the energy-momentum conservation law, which can be expressed as well by

$$(2)\qquad \frac{\partial \mathcal{T}^{ik}}{\partial x^i}=0 \text{ as by } \frac{\partial \bar{\mathcal{T}}^{ik}}{\partial x^i}=0.$$

In particular, the total momentum of the system can be defined by the integral, over the volume V occupied by the system,

$$(3) \qquad P^i = -\int \mathcal{T}^{4i}\, dv \; (i=1, 2, 3)$$

as well as by

$$(4) \qquad \bar{P}^i = -\int \bar{\mathcal{T}}^{4i}\, dv,$$

as is usually done. The difference

$$(5) \qquad p^i = P^i - \bar{P}^i = -\int \frac{\partial \mathcal{R}^{4j,\, i}}{\partial x^j}\, dv$$

converts to a flux integral

$$(6) \qquad p^i = -\int n_j\, \mathcal{R}^{4j,\, i}\, dS$$

over the surface S containing the volume V. In this formula the n_j ($j=1, 2, 3$) are the components of the external normal to the surface element dS.

II

But the presence of the divergence term in expression (1) has the effect of making the tenor for $\mathcal{T}^{ik}$ symmetric (in contrast to $\bar{\mathcal{T}}^{ik}$) with respect to indices i, k (either identically or because of the field equations), and this symmetry property, as is well known, is essential to guarantee the conservation of the total angular momentum of the system. Setting (for $i, k=1, 2, 3$)

$$(7) \qquad \mathcal{M}^{j,\, ik} = x^k \mathcal{T}^{ji} - x^i \mathcal{T}^{jk},$$

one can deduce from (2) and from the symmetry property of $\mathcal{T}^{ji}$ that:

$$(8) \qquad \frac{\partial \mathcal{M}^{j,\, ik}}{\partial x^j} = 0,$$

an equation that can be interpreted as the expression for the conservation of total angular momentum for the system:

$$(9) \qquad M^{ik} = \int \mathscr{M}^{4,\, ik}\, \mathrm{d}v$$

Indeed, equation (8) means that the time derivative of M^{ik} will reduce to the flux of angular momentum

$$-\int \sum_{j=1}^{3} n_j \mathscr{M}^{j,\, ik}\, \mathrm{d}S$$

crossing the surface S per unit time. (The velocity of light is taken as unity.)

III

From (1), the angular momentum (9) will separate into two terms:

$$(10) \qquad M_0^{ik} = \int (x^k \mathscr{T}^{4i} - x^i \mathscr{T}^{4k})\, \mathrm{d}v$$

and

$$(11) \qquad M_s^{ik} = \int \left(x^k \frac{\partial \mathscr{R}^{4j,\, i}}{\partial x^j} - x^i \frac{\partial \mathscr{R}^{4j,\, k}}{\partial x^j} \right) \mathrm{d}v\,;$$

For a system of non-zero mass, these two terms correspond respectively to the orbital angular momentum and to the intrinsic angular momentum or spin of the system,[2] using the necessary and sufficient condition

$$(12) \qquad p_i = 0,$$

at least in the limiting case where the surface S goes to infinity; for according to (5), the expression (11) is then independent of the choice of the origin with respect to which the angular momentum has been defined. It can be seen that this condition is just the one which indicates that the total momentum of the system reduces to the usual form (4).

IV

Defining

$$(13) \qquad \mathscr{S}^{j,\, ik} = \mathscr{R}^{ji,\, k} - \mathscr{R}^{jk,\, i},$$

the expression (11) for the spin can be written

$$M_s^{ik} = \bar{M}_s^{ik} + m_s^{ik} \tag{14}$$

with

$$\bar{M}_s^{ik} = \int \mathscr{S}^{4,\, ik}\, \mathrm{d}v \tag{15}$$

and

$$\begin{aligned} m_s^{ik} &= \int \frac{\partial}{\partial x^j} (x^k \mathscr{R}^{4j,\, i} - x^i \mathscr{R}^{4j,\, k})\, \mathrm{d}v \\ &= \int n_j (x^k \mathscr{R}^{4j,\, i} - x^i \mathscr{R}^{4j,\, k})\, \mathrm{d}S. \end{aligned} \tag{16}$$

Even if condition (12) is satisfied, the surface integral m_s^{ik} might very well be different from zero. An example of this peculiarity is given, as J. Serpe[3] has indicated, by the (neutral) meson radiation field.

V

Nevertheless, it can easily be seen that it is always possible to take $\bar{M}_s^{ik}$ as a definition for the spin instead of M_s^{ik}. Indeed, setting

$$\mathscr{M}_0^{j,\, ik} = x^k \bar{\mathscr{T}}^{ji} - x^i \bar{\mathscr{T}}^{jk}, \tag{17}$$

in a fashion analogous to (7),

$$\frac{\partial \mathscr{M}_0^{j,\, ik}}{\partial x^j} = -\frac{\partial}{\partial x^j} (\mathscr{R}^{kj,\, i} - \mathscr{R}^{ij,\, k}) = -\frac{\partial \mathscr{S}^{j,\, ik}}{\partial x^j}, \tag{18}$$

is obtained from (8) and from the antisymmetry of $\mathscr{R}^{ij,\, k}$, and by using (13). From (18) and using (10) and (15), it follows that the time derivative of $M_o^{ik} + \bar{M}_s^{ik}$ does indeed reduce to a flux integral.[4]

VI

In the face of this ambiguity in the definition of spin, it is important to observe first that it does not occur in the usual case for which the field decreases sufficiently rapidly at large distances from the sources; and for which consequently the surface integral m_s^{ik} vanishes. Then, formula (15) gives an unambiguous definition of the total spin of the system. On the other hand, in the cases where the ambiguity can occur, corresponding to the radiation at large long distances from the sources, the total spin contained inside a given (spherical) wave surface is not particularly interesting. What matters then is the flux of the spin carried through this surface per unit time, or even rather the time average of that quantity. If (15) is taken as the definition of the spin, this flux, is equal by (18) to

$$\phi_s^{ik} = -\int \sum_{j=1}^{3} n_j \mathscr{S}^{j,\,ik}\, \mathrm{d}S, \tag{19}$$

while, starting from definition (11) it would obviously be

$$\phi_s^{ik} - \frac{\partial m_s^{ik}}{\partial x^4}.$$

Now the two expressions coincide when time averaged, with the necessary and sufficient condition that

$$\overline{\frac{\partial m_s^{ik}}{\partial x^4}} = 0, \tag{20}$$

the sign $\overline{}$ meaning time average.

VII

Now, for a radiating system, the field amplitudes and consequently the expression m_s^{ik} are almost periodic functions of time (in the sense of H. Bohr), and it is easy to see that for such a function the condition (20) is indeed satisfied. It can in fact be deduced immediately from the fact[5] that any almost periodic function is bounded:[6] Since from

$$|f(t)| \leqslant C \text{ for } |t| < \infty,$$

it follows that

$$\left|\frac{1}{T}\int_0^T f'(t)\,\mathrm{d}t\right| = \left|\frac{f(T) - f(0)}{T}\right| \leqslant \frac{2C}{T},$$

and hence

$$\lim_{T\to\infty} \frac{1}{T} \int_0^T f'(t)\, dt = 0.$$

Thus it can be seen in a very general way that the ambiguity in the definition of spin, due to the existence of the quantity $m_s^{ik} \neq 0$, has no influence on the average flux of spin radiated by the system.

In the particular case of the meson field discussed by Serpe,[3] the quantity m_s^{ik} is independent of time; hence, the condition (20) is satisfied automatically and the flux of the spin is accordingly given by (19) in any case. Serpe shows in addition that this flux through a wave surface during a determined time interval is equal to the total spin contained in the volume enclosed between the initial and final positions of the wave surface, this spin being calculated precisely using the definition (15), i.e. leaving m_s^{ik} out of account.

VIII

On the other hand, the above considerations do not apply to the electromagnetic field. For that field, indeed, the distinction between angular momentum and spin is meaningless, since those terms are not separately gauge invariant. For the same reason, it is necessary to keep the surface integral m_s^{ik} in the expression for angular momentum. It is clear, however,[7] that choosing – as usual – the vector potential in such a way that it varies as R^{-1} with the distance R from the source of radiation, condition (20) is fulfilled, and the calculation of the flux of total angular momentum can formally still be done without considering the term m_s^{ik}.

NOTES

[1] F. J. Belinfante, *Physica* **7** (1940), 449; L. Rosenfeld, *Mem. Acad. Roy. Belg.* (*Cl. Sciences*) **18,** fasc. 6 (1940). Cf. §5, formula (I) [this volume, p. 711]. This present note supplements these papers.

[2] It can be verified (cf. L. Rosenfeld, *loc. cit.* (1), pp. 24–25) that these two terms have the non-commutative properties required for this interpretation.

[3] J. Serpe, *Physica* **8** (1941), 748.

[4] The non-commutative properties of $\bar{M}_s^{ik}$ and M_s^{ik} are obviously identical.

[5] H. Bohr, *Fastperiodische Funktionen*, §46, theorem I. [See English translation, *Almost Periodic Functions* (Chelsea Publ., New York, 1947) – Ed.]

[6] I owe this remark to M. D. J. Bouman.

[7] J. Humblet, work in preparation. [We have not traced this work.–Ed.]

ON THE BEHAVIOR OF A CANONICAL ENSEMBLE DURING AN ADIABATIC TRANSFORMATION

[1942f]

It is well known that the concepts developed by Gibbs in his famous treatise on statistical mechanics [1] were the object of a penetrating critique by P. and T. Ehrenfest [2]. Concerning the general way in which a non-stationary distribution tends to a distribution in statistical equilibrium, recent progress in ergodic theory [3] now permits us to present Gibbs' ideas in a form which is safe from these criticisms. Nevertheless, in discussions of the effect of quasistatic transformations on an equilibrium distribution there is a point that, to my knowledge, remains obscure. While the treatment of isothermal transformations obviously does not present any difficulties, the same cannot be said of adiabatic transformations. If, starting with a canonically distributed ensemble, we suddenly vary the external parameters by an infinitesimally small amount, after which we keep the parameters at their new values, the systems of the ensemble remaining thermally isolated, we would expect – in order to be able to establish in complete generality the statistical analogue of the concept of entropy – that the new equilibrium distribution, to which the ensemble we are considering 'tends' (in the sense of ergodic theory), would again be a canonical distribution corresponding to the new values of the external parameters and to a suitably modified value of the modulus. P. and T. Ehrenfest correctly remark [4] that this property, implicitly assumed by Gibbs when he discusses thermodynamic analogies in chapter XIV of his book, is neither demonstrated nor even stated in the preceding chapters. Lorentz [5], in the very clear exposition he gave of Gibbs' theory, slides too quickly over this question. He confines himself to briefly alluding to the equivalence of the microcanonical and the canonical distribution for systems having a large number of degrees of freedom. It is evident that in this case considerations of this kind are insufficient. In fact, it is of importance to know if, on thermodynamic grounds, the distribution finally achieved would or would not be changed if the systems of the ensemble under consideration were put in contact with a thermostat at the temperature that should be attributed to the system after the adiabatic transformations: for such a change, which would occur if the final state reached were not canonical, would imply an increase in entropy during this contact, in contradiction with thermodynamics.

Of course, the problem we are discussing has a well-determined solution, which should emerge automatically from the analysis of the process by Gibbs' methods. This is what we want to show briefly here. We first consider a canonical distribution with canonical modulus θ_0. We will treat only one external parameter, whose initial value is a_0. The density in phase space will therefore be initially given by $e^{H(p,q,\theta_0,a_0)}$ with

$$(1) \qquad H(p, q; \theta_0, a_0) = \frac{F(\theta_0, a_0) - E(p, q; a_0)}{\theta_0},$$

F representing the free energy, E the energy of a system of the ensemble at a point (p, q) in phase space.[1] If the external parameter undergoes a sudden variation da and subsequently remains at the value $a = a_0 + \mathrm{d}a$, the previous distribution will not be immediately modified, but the 'surfaces' of constant energy in phase space will be deformed, so that the distribution will no longer be uniform in the shells included between two infinitesimally close surfaces $E(a)$, $E(a) + \mathrm{d}E$. Since the systems of the ensemble stay energetically isolated, this distribution, by virtue of the ergodic theorem, will evolve separately in each of these shells towards a uniform distribution; the density of the latter is none other than the microcanonical mean for the shell in question of the initial distribution. But

$$e^{H(p,q,\theta_0,a_0)} = e^{H(p,q;\theta_0,a)} \left\{ 1 - \frac{\mathrm{d}a}{\theta_0} \left(\frac{\partial F}{\partial a} - \frac{\partial E}{\partial a} \right) \right\}$$

$$= e^{H(p,q;\theta_0,a)} \left\{ 1 - \frac{\mathrm{d}a}{\theta_0} (A - \bar{A}) \right\},$$

where $A \equiv -\partial E/\partial a$ designates the generalized force relative to the parameter a, and $\bar{A}$ its mean value in the initial canonical distribution.[2] Noting that $e^{H(p,q;\theta_0,a)}$ and $\bar{A}$ are constant in the shell $E(a)$, $E(a) + \mathrm{d}E$, we now obtain for the final distribution a density

$$(2) \qquad e^{H(p,q;\theta_0,a)}\left\{1-\frac{da}{\theta_0}(\bar{A}\rceil_{E(a)}-\bar{A})\right\},$$

the symbol $\bar{A}\rceil_{E(a)}$ designating the microcanonical mean of A for the energy $E(a)$. The point now is to compare this distribution with a canonical distribution $e^{H(p,q;\theta,a)}$ with a suitably chosen modulus $\theta=\theta_0+d\theta$.

If we limit ourselves to the case of systems possessing a large number of degrees of freedom (the only one which permits a comparison with thermodynamics), it will be sufficient to study the distribution (2) in the neighborhood of the mean energy $\bar{E}$ (the bar always designates the canonical mean). We can then write, keeping the value of the parameter a fixed

$$\bar{A}\rceil_E=\bar{A}\rceil_{\bar{E}}+\frac{\partial\bar{A}\rceil_E}{\partial\bar{E}}(E-\bar{E})+\frac{1}{2}\frac{\partial^2\bar{A}\rceil_{\bar{E}}}{\partial\bar{E}^2}(E-\bar{E})^2$$

and

$$\bar{A}=\bar{A}\rceil_{\bar{E}}+\frac{1}{2}\frac{\partial^2\bar{A}\rceil_{\bar{E}}}{\partial\bar{E}^2}\overline{(E-\bar{E})^2},$$

whence we get for distribution (2)

$$e^{H(p,q.\theta_0,a)}\left\{1-\frac{da}{\theta_0}\frac{\partial\bar{A}\rceil_{\bar{E}}}{\partial\bar{E}}(E-\bar{E})-\frac{1}{2}\frac{da}{\theta_0}\frac{\partial^2\bar{A}\rceil_{\bar{E}}}{\partial\bar{E}^2}\left[(E-\bar{E})^2-\overline{(E-\bar{E})^2}\right]\right\}.$$

Let us now define $d\theta$ by

$$(3) \qquad \frac{d\theta}{\theta_0}=-da\frac{\partial\bar{A}\rceil_{\bar{E}}}{\partial\bar{E}};$$

we can then replace

$$e^{H(p,q;\theta_0,a)}\left\{1+\frac{d\theta}{\theta_0^2}(E-\bar{E})\right\}$$

by

$$e^{H(p,q;\theta,a)} \text{ (with } \theta=\theta_0+d\theta),$$

since

$$\frac{\partial}{\partial\theta_0}\left(\frac{F}{\theta_0}\right) = -\frac{\bar{E}}{\theta_0^2}.$$

The distribution (2) finally reached after the infinitesimal adiabatic transformation considered thus takes the form

$$(4) \qquad e^{H(p,\, q;\, \theta,\, a)}\left\{1 - \frac{1}{2}\frac{\mathrm{d}a}{\theta_0}\frac{\partial^2 \bar{A}|_{\bar{E}}}{\partial \bar{E}^2}\,[(E-\bar{E})^2 - \overline{(E-\bar{E})^2}]\right\};$$

It therefore differs by the last term from the canonical distribution with modulus θ, relative to the final value a for the variable parameter.

Nevertheless, it is easy to see that this factor plays only a negligible role for systems with a large number of degrees of freedom.[3] The correction to the canonical mean of any quantity G is indeed (equating θ to θ_0, which is permissible here)

$$-\frac{1}{2}\frac{\mathrm{d}a}{\theta}\frac{\partial^2 \bar{A}|_{\bar{E}}}{\partial \bar{E}^2}\,\overline{(G-\bar{G})(E-\bar{E})^2},$$

or else, by a general formula of Gibbs,[4]

$$-\frac{1}{2}\frac{\mathrm{d}a}{\theta}\frac{\partial^2 \bar{A}|_{\bar{E}}}{\partial \bar{E}^2}\,\theta^2\frac{\partial}{\partial\theta}\left(\theta^2\frac{\partial \bar{G}}{\partial\theta}\right);$$

one can see that, as the number of degrees of freedom increases, this quantity tends to zero even faster than do the mean energy fluctuations (and barring exceptional circumstances, the mean fluctuations of the other quantities) in the canonical ensemble. The same can also be said about the correction made by the factor under discussion to the mean fluctuation of the ensemble quantities. Thus, we arrive at the conclusion that the final distribution indeed reduces to a canonical one.

Let us return to formula (3) which gives the variation of the modulus. We can replace $\partial\bar{A}|_{\bar{E}}/\partial\bar{E}$ in it by $\partial\bar{A}/\partial\bar{E}$, or equivalently $(\partial\bar{A}/\partial\theta)/(\partial\bar{E}/\partial\theta)$. It then takes the well-known form of the thermodynamic equation for adiabatic transformations, derived from the constancy of the entropy.

NOTES

[1] The letters p, q represent the coordinates $p_1, p_2 \ldots p_f; q_1, q_2 \ldots q_f$ of phase space.

[2] If we limit ourselves to first order terms in da, it is permissible to replace a_0 by a in the functions H and A appearing in the coefficient of da.

[3] In the case of a perfect gas, or a pure radiation field, or of another system for which $\partial^2 \bar{A}|_E / \partial E^2 = 0$, one would have to consider a factor depending on higher order terms in $E - \bar{E}$ which would be negligible *a fortiori*.

[4] This is formula (257) (chapter VII, p. 87). It is enough to write it for $h=1$ and $h=2$ and to suitably combine the formulas obtained. The quantity G is here supposed to be independent of the modulus.

BIBLIOGRAPHY

[1] J. W. Gibbs, *Elementary Principles in Statistical Mechanics* (Yale University Press, 1902).

[2] P. and T. Ehrenfest, 'Begriffliche Grundlagen der statistischen Auffassung in der Mechanik', *Enzykl. d. math. Wissensch.* **IV** 2ii (1911), 32 [English translation: *The Conceptual Foundations of the Statistical Approach in Mechanics*, transl. by M. J. Moravcsik (Cornell University Press, Ithaca, N.Y., 1959)].

[3] See e.g. E. Hopf, 'Ergodentheorie', *Ergebn. d. math. Wiss.* **V** (1937), 2.

[4] P. and T. Ehrenfest, *loc. cit.*, §23b.

[5] H. A. Lorentz, 'Über den zweiten Hauptsatz der Thermodynamik', *Abhandlungen über theoretische Physik* **I** (1907), 292, §81.

CHAPTER 4

ON THE ISOLATED AND ADIABATIC SUSCEPTIBILITIES

[1961c]

ABSTRACT. A general discussion of the relation between the thermodynamical and statistical treatments of adiabatic transformations is carried out both on classical and quantal lines. In classical theory, the equivalence of the two treatments, and consequently the equality of the isolated and adiabatic susceptibilities, is established for very large systems, the deviations from exact equality, for any quantity, being of the order of fluctuations. In quantum theory, the similar proof of asymptotic equivalence requires an additional assumption of continuity. Conditions for exact equivalence are also discussed.

1. INTRODUCTION

In 1951, L. J. F. Broer [1] called attention to a paradoxical result met with in the theoretical discussion of the magnetization of certain crystalline powders by an oscillating magnetic field. Under appropriate experimental conditions, it seemed reasonable to regard the process as adiabatic, and accordingly to expect for the magnetic susceptibility the value derived from simple thermodynamical considerations. On the other hand, a direct computation of the susceptibility, based on the study of the way in which the statistical distribution of the elementary magnetic moments of the system was modified by a small variation of the external magnetic field, seemed to lead to a different result.

For a classical system, the occurrence of such a discrepancy between a statistical analysis of the process and the consequences of usual thermodynamics seemed definitely excluded; for a quantal system, however, the situation was not so clear, since the considerations by which the connexion between statistics and thermodynamics can be established in classical theory [2] could not (as we shall see) be extended to the quantal case without further assumption. Still, it was difficult to see how Broer's use of perturbation theory in his direct estimate of the susceptibility could be justified, and it could therefore be doubted whether the discrepancy he had found was genuine.

In the months following the publication of Broer's paper, I exchanged some correspondence about it with Dr. Broer himself, Dr. L. J. Smits and

Dr. D. Polder. While this discussion led to the clarification of some misleading statements in the paper (which for certain reasons was rather hurriedly written) it remained inconclusive regarding the main point at issue, because none of us was apparently disposed to attempt the more accurate direct evaluation of the susceptibility, which we recognised was needed to settle the question. Now, precisely such an evaluation has recently been carried out by W. J. Caspers [3, 4], whose thorough discussion of the problem may be said to elucidate the situation in all essential aspects.

The way is thus clear for a synthetic presentation of the whole problem from a more general point of view, especially emphasizing its relation to the foundation of the statistical thermodynamics of quantal systems. A renewed discussion on these lines is perhaps not superfluous, inasmuch as Broer's treatment, which Caspers takes as a starting point without comment, is in need of improvement [1].

2. Adiabatic transformation of a classical mechanical system

It will be convenient to start with an analysis of the problem on classical lines, since in this case the situation is quite transparent and leads to clear-cut conclusions of a perfectly general character. The case of quantal systems will then be taken up and its special features illustrated by comparison with the classical theory.

The first step in the argument consists in establishing a connection between the thermodynamics of adiabatic transformations and their statistical treatment. Although this question is obviously an essential and by no means trivial part of any logical exposition of statistical thermodynamics, it is surprisingly neglected in most textbooks; I shall therefore repeat here, with some improvements, the discussion presented in my previous note on the subject [2].

Let us consider a mechanical system, whose dynamical behaviour is described, in terms of canonical coordinates (p, q), by a Hamiltonian $H(p, q; a)$, depending also on some external parameter a (which may be, e.g., an external magnetic field). A quasi-static adiabatic transformation of this system may be idealized as a succession of very small instantaneous variations δa of the parameter a, separated by extremely long intervals during which the system is kept isolated and allowed each time to settle down to a stationary state under the new external conditions.

It is clearly sufficient to consider a single such variation of the parameter from a value a to $a+\delta a$. In the initial state, the system is assumed to have a definite temperature, which means, statistically, that it may be represented by a canonical distribution of given modulus, the inverse of which we denote by β; the density of this distribution has the form

$$(1) \qquad n(p, q; a) = \exp \eta(p, q; a, \beta)$$

with a probability exponent η given by

$$(2) \qquad \eta(p, q; a, \beta) = \beta[F(a, \beta) - H(p, q; a)]$$

in terms of the Hamiltonian $H(p, q; a)$ and the inverse modulus β; the normalization function $F(a, \beta)$ represents the free energy of the system. We must now study how this distribution will be modified after the change $a \to a+\delta a$ has taken place and the contact of the system with the thermostat of modulus β^{-1} has been severed, and see how far it can be said to be equivalent with a new canonical distribution of inverse modulus $\beta+\delta\beta$, the variation $\delta\beta$ being related to the initial values a, β and the variation δa in the way expected from thermodynamics.

The instantaneous change from a to $a+\delta a$ will not at first affect the distribution of the system in phase space. In the usual case, in which the parameter a only enters into the potential energy, it is in fact clear that the system will at first remain in the same phase (p, q). If, however – as in the case of a magnetic field, – the parameter a occurs in the expression of the momenta in terms of the velocities:

$$(3) \qquad p_k = \mu_k \dot{q}_k + \mathscr{A}_k(q; a)$$

(where the coefficients μ_k are constant, whereas the parameter-dependent terms $\mathscr{A}_k$ are functions of the coordinates), the phase (p, q) will be displaced by an amount

$$(4) \qquad \delta_0 q_k = 0, \quad \delta_0 p_k = \frac{\partial \mathscr{A}_k}{\partial a}\, \delta a,$$

implying no change in the velocities, and consequently in the energy: in other words, the initial phase will move to another phase of the initial energy surface, in which the distribution density (which is in any case uniform in each energy shell) is the same. We may therefore quite generally[2] conclude that, immediately after its isolation under the new external conditions fixed by the value $a+\delta a$ of the parameter, the statistical distribution of the system will just be the initial one of density $n(p, q; a)$.

This distribution, however, will no longer correspond to statistical equilibrium under the new conditions and if the ergodicity conditions are satisfied will therefore evolve, during the period of isolation of the system, towards a new equilibrium distribution, which will be ergodic on each of the energy surfaces

$$(5) \qquad H(p, q; a+\delta a) = E(a+\delta a)$$

corresponding to the new value of the external parameter.

If we denote in general by $\langle G(p, q; a)\rangle_{E(a)}$ the ergodic average of the phase function $G(p, q; a)$ over the energy surface $H(p, q; a) = E(a)$, the density of the new equilibrium distribution may therefore be expressed as

$$(6) \qquad n_{\mathrm{ad}}(p, q; a+\delta a) = \langle \eta(p, q; a)\rangle_{E(a+\delta a)}:$$

this is the formula underlying the whole theory of adiabatic processes.

3. Adiabatic and Canonical Distributions

According to our assumption (1), expressing that we start from a canonical distribution, the adiabatic distribution (6) takes the form

$$(7) \qquad n_{\mathrm{ad}}(p, q; a+\delta a) = \langle e^{\eta(p, q; a, \beta)}\rangle_{E(a+\delta a)}.$$

We now want to compare this with a canonical distribution of modified modules

$$(8) \qquad e^{\eta(p, q; a+\delta a, \beta+\delta\beta)}.$$

The necessary and sufficient condition for the identity of the two distributions (7) and (8), which may be expressed as

$$\langle e^{\eta(p, q; a-\delta a, \beta-\delta\beta)}\rangle_{E(a)} = e^{\eta(p, q; a, \beta)},$$

can simply be written as

$$(9) \qquad \left\langle \frac{\mathrm{d}\eta}{\mathrm{d}a} \right\rangle_{E(a)} = 0 \quad \text{for all} \quad E(a);$$

in this formula, the inverse modulus β is regarded as a function of the parameter a.

On account of the explicit form (2) of $\eta(p, q; a, \beta)$ the condition (9) can be expanded as

$$\frac{\mathrm{d}\beta}{\mathrm{d}a}\left[\frac{\partial(\beta F)}{\partial\beta}-H\right]+\beta\left[\frac{\partial F}{\partial a}-\left\langle\frac{\partial H}{\partial a}\right\rangle_E\right]=0. \tag{10}$$

Let us denote by

$$A(p, q; a)\equiv-\frac{\partial H}{\partial a} \tag{11}$$

the phase function associated with the external parameter a in the expression $-A\delta a$ for the change in the energy of the system on varying a by δa; and let us use the notation

$$\bar{G}\equiv\int G(p, q; a)\, e^{\eta(p, q; a, \beta)}\,\mathrm{d}\omega \tag{12}$$

for the canonical average of the phase function $G(p, q; a)$. In virtue of the well-known relations

$$\frac{\partial(\beta F)}{\partial\beta}=\bar{E}, \quad \frac{\partial F}{\partial a}=-\bar{A}, \tag{13}$$

our condition (10) becomes

$$\frac{1}{\beta}\frac{\mathrm{d}\beta}{\mathrm{d}a}=\frac{\bar{A}-\langle A\rangle_E}{\bar{E}-E} \quad \text{for all} \quad E. \tag{14}$$

In other words, the difference $\bar{A}-\langle A\rangle_E$ should be strictly proportional to the difference $\bar{E}-E$ for all values of the energy. Then, the dependence of the modulus on the parameter a is determined by equating $\mathrm{d}\log\beta/\mathrm{d}a$ with the coefficient of proportionality.

In my previous discussion of the problem [2], I did not formulate this condition explicitly, because it seemed too restrictive to serve as a basis for a general foundation of the thermodynamics of adiabatic processes. Some time after the publication of Broer's paper, the precise analogue of the condition (14) for the case of quantal systems was pointed out by M. Klein [5]. We shall return to it later; for the moment, remaining within the limits of classical theory, we shall pursue the discussion on the assumption that the condition in question is not generally fulfilled: we must then examine whether the thermodynamical relations can nevertheless be established with sufficient accuracy under some weaker condition.

4. Adiabatic and Canonical Averages

Let us form the 'adiabatic average' of the arbitrary phase function $G(p, q; a)$, i.e. the average calculated by means of the distribution density n_{ad} given by Equation (7):

$$\langle G(p, q; a+\delta a)\rangle_{ad} = \int G(p, q; a+\delta a)\, \langle e^{\eta(p, q; a, \beta)}\rangle_{E(a+\delta a)}\, d\omega = \int e^{\eta(p, q; a, \beta)}\, \langle G(p, q; a+\delta a)\rangle_{E(a+\delta a)}\, d\omega,$$

i.e. for a general value a of the external parameter,

$$(15) \qquad \langle G(p, q; a)\rangle_{ad} = \int e^{\eta(p, q; a-\delta a; \beta-\delta\beta)}\, \langle G(p, q; a)\rangle_{E(a)}\, d\omega.$$

Expanding the exponential factor as before yields the desired comparison of this average with the canonical average (12) of variable inverse modulus $\beta(a)$ in the form

$$(16) \qquad \langle G(p, q; a)\rangle_{ad} - \bar{G} = -\delta a \int e^{\eta(p, q; a, \beta)}\, \langle G(p, q; a)\rangle_{E(a)} \frac{d\eta}{da}\, d\omega,$$

where $d\eta/da$ has the value given by the left-hand side of eq. (10), i.e.

$$(17) \qquad \frac{d\eta}{da} = (\bar{E} - E)\frac{d\beta}{da} - (\bar{A} - \langle A\rangle_E)\, \beta.$$

In order to express in a concise and physically significant way the difference $\langle G\rangle_{ad} - \bar{G}$ given by Equation (16), we introduce the general definition of the *correlation* between the phase functions $G(p, q)$, $K(p, q)$ in a canonical ensemble:

$$(18) \qquad \Gamma(GK) \equiv \overline{(G-\bar{G})\,(K-\bar{K})} = \overline{GK} - \bar{G}\bar{K} = \overline{G(K-\bar{K})}.$$

With this notation, Equation (16) becomes, when the value (17) of $\mathrm{d}\eta/\mathrm{d}a$ is inserted,

$$(19)\qquad \langle G\rangle_{\mathrm{ad}}-\bar{G}=-\beta\delta a\left[\Gamma(\langle G\rangle_E\langle A\rangle_E)-\frac{1}{\beta}\frac{\mathrm{d}\beta}{\mathrm{d}a}\Gamma(GH)\right].$$

Now[3], one way to choose the value of $\mathrm{d}\beta/\beta\,\mathrm{d}a$ as a function of a, and thereby the function $\beta(a)$ characteristic of adiabatic transformations, short of the complete equivalence of the adiabatic and canonical distribution densities, is to require the equality of the adiabatic and canonical averages of some quantity G of thermodynamical significance. It is natural to select for this purpose the entropy, i.e. to impose the condition

$$(20)\qquad \langle\eta\rangle_{\mathrm{ad}}=\bar{\eta}$$

upon the exponent of probability $\eta(p, q; a)$ given by Equation (2).

From the form of the right-hand side of eq. (19) it is immediately clear that this amounts to imposing the same condition upon the energy:

$$(21)\qquad \langle E\rangle_{\mathrm{ad}}=\bar{E}.$$

This gives

$$(22)\qquad \frac{1}{\beta}\frac{\mathrm{d}\beta}{\mathrm{d}a}=\frac{\Gamma(AH)}{\Gamma(H^2)},$$

and this condition (22) ensures the equivalence of the adiabatic and canonical averages of the entropy and the energy of the system. The other quantities necessary for a complete thermodynamical description are those of the type $A=-\partial H/\partial a$ (which enter into the fundamental formula for the differential of the thermodynamic potential, in our case the free energy): for these quantities we shall not expect exact equivalence between the two averages; there will subsist a difference

$$(23)\qquad \langle A\rangle_{\mathrm{ad}}-\bar{A}=-\beta\delta a\left\{\Gamma(\langle A\rangle_{E^2})-\frac{[\Gamma(AH)]^2}{\Gamma(H^2)}\right\}.$$

It is easy to evaluate this difference, or more generally the difference

$$(24)\qquad \langle G\rangle_{\mathrm{ad}}-\bar{G}=-\beta\delta a\left\{\Gamma(\langle G\rangle_E\langle A\rangle_E)-\frac{\Gamma(GH)\,\Gamma(AH)}{\Gamma(H^2)}\right\},$$

derived by combining Equations (19) and (22). To this end, all we need are the readily established formulae

$$(25)\qquad \Gamma(GH)=-\frac{\partial\bar{G}}{\partial\beta},\ \Gamma(GH^2)=\frac{\partial^2\bar{G}}{\partial\beta^2},$$

$$\Gamma(GH^3)=-\frac{\partial^3\bar{G}}{\partial\beta^3}+3\frac{\partial\bar{H}}{\partial\beta}\frac{\partial\bar{G}}{\partial\beta},\cdots$$

Remembering that $\beta^{-1}=kT$ (where T denotes the absolute temperature and k Boltzmann's constant), we see that $\Gamma(GH)$, besides a factor of macroscopic order of magnitude, contains a factor k, which is of the order of the molecular dimensions. Likewise, $\Gamma(GH^2)$ and $\Gamma(GH^3)$ are 'of order k^2'; it is readily verified that the further correlations $\Gamma(GH^n)$ for $n>3$ are of higher order in k than k^2.
Expanding $\langle A\rangle_E$ in powers of $e\equiv E-\bar{E}$, we may write

$$(26)\qquad \langle A\rangle_E=A^{(0)}+A^{(1)}e+\tfrac{1}{2}A^{(2)}e^2+\tfrac{1}{6}A^{(3)}e^3+\cdots$$

with the notation

$$(27)\qquad A^{(n)}\equiv\frac{\partial^n\langle A\rangle\bar{E}}{\partial\bar{E}^n}.$$

We then get, up to the order k^2,

$$\Gamma(\langle G\rangle_E\langle A\rangle_E)=\Gamma(G\langle A\rangle_E)=A^{(1)}\Gamma(GH)+\tfrac{1}{2}A^{(2)}\Gamma(GH^2)+ +\tfrac{1}{6}A^{(3)}\Gamma(GH^3)+\cdots,$$

$$\Gamma(HA)=\Gamma(H\langle A\rangle_E)=A^{(1)}\Gamma(H^2)+\tfrac{1}{2}A^{(2)}\Gamma(H^3)+ +\tfrac{1}{6}A^{(3)}\Gamma(H^4)+\cdots,$$

and therefore

$$\Gamma(\langle G\rangle_E\langle A\rangle_E)-\frac{\Gamma(GH)}{\Gamma(H^2)}\Gamma(AH)= =\tfrac{1}{2}A^{(2)}\left[\Gamma(GH^2)-\frac{\Gamma(GH)}{\Gamma(H^2)}\Gamma(H^3)\right]+ +\tfrac{1}{6}A^{(3)}\left[\Gamma(GH^3)-\frac{\Gamma(GH)}{\Gamma(H^2)}\Gamma(H^4)\right]+\cdots:$$

the terms of order k have cancelled each other; moreover, in the coefficient of $A^{(3)}$, a similar cancellation of the terms of order k^2 occurs, according to (25), so that this coefficient is actually of order k^3. The coefficient of $\frac{1}{2}A^{(2)}$ may be written

$$\frac{\partial^2\bar{G}}{\partial\beta^2}-\frac{\partial\bar{G}/\partial\beta}{\partial\bar{E}/\partial\beta}\frac{\partial^2\bar{E}}{\partial\beta^2}=\frac{\partial\bar{E}}{\partial\beta}\frac{\partial}{\partial\beta}\left[\frac{\partial\bar{G}/\partial\beta}{\partial\bar{E}/\partial\beta}\right]=$$

$$=\frac{\partial\bar{E}}{\partial\beta}\frac{\partial}{\partial\beta}\frac{\partial\bar{G}}{\partial\bar{E}}=\left(\frac{\partial\bar{E}}{\partial\beta}\right)^2\frac{\partial^2\bar{G}}{\partial\bar{E}^2};$$

as to $A^{(2)}=\partial^2\langle A\rangle_{\bar{E}}/\partial\bar{E}^2$, it may be replaced by $\partial^2\bar{A}/\partial\bar{E}^2$, the difference between these two quantities being obviously at least of order k. We thus finally obtain for the difference (26) the expression

$$\text{(28)}\qquad \langle G\rangle_{\text{ad}}-\bar{G}=-\tfrac{1}{2}\beta\delta a\frac{\partial^2\bar{A}}{\partial\bar{E}^2}\left(\frac{\partial\bar{E}}{\partial\beta}\right)^2\frac{\partial^2\bar{G}}{\partial\bar{E}^2},$$

which is of order k.

In particular, the difference (23) between the adiabatic and canonical averages of the fundamental quantities A will be of order k, and may accordingly be regarded as vanishingly small for macroscopic systems, for which alone thermodynamic concepts are meaningful. Returning to the relation (22), we can write it, on account of the first eq. (25), in the form

$$\text{(29)}\qquad \frac{1}{\beta}\frac{\mathrm{d}\beta}{\mathrm{d}a}=\frac{\partial\bar{A}}{\partial\bar{E}},$$

which is readily seen to be identical with a well-known thermodynamical relation characteristic of quasi-static adiabatic processes. This completes the proof that (to the approximation indicated) an adiabatic transformation leads from a canonical distribution to another one, of suitably varying modulus.

5. Adiabatic and isolated susceptibilities

The equivalence of the adiabatic and isolated susceptibilities, for classical systems, may be proved quite generally in a few lines; it is in fact an immediate consequence of the relation (28). In general, the susceptibility of a material system for a given kind of phenomenon may be defined as the rate of change of some extensive quantity characterizing the phenomenon with respect to some intensive parameter belonging to the class of phenomena in question. The extensive quantity $G(p, q; a)$ and the intensive parameter a might be associated in the usual way (i.e. G might be the quantity A associated, by Equation (11), with the parameter a), but for the following argument we may just as well treat G as an arbitrary function of a.

The *isolated susceptibility* will be defined by the actual rate of change of the quantity G in the isolated system, i.e. by

$$\chi_{is}=\frac{1}{\delta a}[\langle G(a+\delta a)\rangle_{ad}-\overline{G(p, q; a)}], \tag{30}$$

since the initial value of the macroscopic quantity corresponding to G is the canonical average $\bar{G}$, whereas the final one is the average $\langle G(a+\delta a)\rangle_{ad}$ given by the preceding statistical derivation. On the other hand, the *adiabatic susceptibility* will be expressed by the rate of change of the latter average

$$\chi_{ad}=\frac{1}{\delta a}[\langle G(a+\delta a)\rangle_{ad}-\langle G(a)\rangle_{ad}], \tag{31}$$

since it is the difference occurring in Equation (31) which directly and strictly represents the change of the corresponding macroscopic quantity in an adiabatic transformation, as it is considered in thermodynamics.

The difference of the two susceptibilities thus defined is, according to Equation (28),

$$\chi_{ad}-\chi_{is}=-\frac{1}{\delta a}[\langle G\rangle_{ad}-\bar{G}]=\tfrac{1}{2}\beta\frac{\partial^2\bar{A}}{\partial\bar{E}^2}\left(\frac{\partial\bar{E}}{\partial\beta}\right)^2\frac{\partial^2\bar{G}}{\partial\bar{E}^2}, \tag{32}$$

and therefore negligible for sufficiently large systems.

6. Extension to Quantal Systems

We must now turn our attention to quantal systems and examine to what extent the preceding considerations remain valid when the appropriate changes required by quantum theory are introduced. In the first place, our considerations can only be significant for ergodic systems, i.e. systems satisfying the conditions[4] which ensure the existence of time averages of the physical quantities independent of the initial state in which the system is observed, and guarantee that in the course of time the isolated system will reach a state of equilibrium represented by a statistical distribution with equal weights over all the eigenstates of the total energy within a macroscopically infinitesimal energy interval, or 'energy shell'. The first condition requires that the (discrete) energy levels of the system are non-degenerate; the second that they are so densely distributed that each 'shell' can be subdivided into a large number of 'cells', or groups of levels, each containing itself many levels. The fulfilment of these conditions restricts the argument to systems of a very large number of degrees of freedom, such as those representing macroscopic bodies, and in which the individual elements interact in such a way as to lift all possible degeneracies in the total energy [5].

The canonical distribution is defined by a density operator which has again the form (1), (2), the Hamiltonian $H(a)$ being now treated as an operator. Denoting by $E_i(a)$, $\varphi_i(a)$ the eigenvalues and eigenfunctions of this operator, we may write the canonical density operator in the form

$$e^{\eta(a)}=\sum_i e^{\eta_i(a)}D_{\varphi_i(a)},\quad \eta_i(a)\equiv\beta[F(a,\beta)-E_i(a)],\tag{33}$$

where $D_{\varphi_i(a)}$ represents the projection operator of the (non-degenerate) eigenstate $\varphi_i(a)$. Since the instantaneous change $a\to a+\delta a$ does not at first affect the eigenfunctions, the operator (33) will still represent the distribution density immediately after the system has been isolated and left to follow its time evolution determined by the Hamiltonian $H(a+\delta a)$. In the energy shell formed by the S states $(\varphi_{j_1}, \varphi_{j_2}, \ldots, \varphi_{js})$, the new equilibrium distribution ultimately reached by the system, provided that it satisfies the ergodicity conditions recalled above, will then be defined by the density operator

$$\frac{1}{S}\sum_{\nu=1}^{S}\operatorname{tr}[e^{\eta(a)}D_{\varphi_{j_\nu}(a+\delta a)}]\,D_{\varphi_{j_\nu}(a+\delta a)},$$

or, changing $a+\delta a$ into a,

$$\frac{1}{S}\sum_{\nu=1}^{S}\operatorname{tr}[e^{\eta(a-\delta a)}D_{\varphi_{j_\nu}(a)}]\,D_{\varphi_{j_\nu}(a)}.$$

The density operator of the new distribution is accordingly

(34) $$D_{ad}=\sum_j \operatorname{tr}[e^{\eta(a-\delta a)}D_{\varphi_j(a)}]\, D_{\varphi_j(a)},$$

the summation extending over *all* eigenstates $\varphi_j(a)$.

This may be written, assuming the inverse modulus β, in the probability exponent $\eta(a-\delta a)$, to be a function of the parameter a,

$$D_{ad}=\sum_j e^{\eta_j(a)}[1-\operatorname{tr}(\delta\eta_j(a)\, D_{\varphi_j(a)})]\, D_{\varphi_j(a)},$$

or, introducing the notation

(35) $$\langle G(a)\rangle_{E_j(a)}=\operatorname{tr}(G(a)\, D_{\varphi_j(a)})$$

and dropping the argument (a),

(36) $$D_{ad}=e^{\eta}[1-\sum_j \langle\delta\eta\rangle_{E_j} D_{\varphi_j}].$$

Equation (36) shows that the 'adiabatic' distribution of density D_{ad} reduces exactly to a canonical one, of density $\exp\eta[a, \beta(a)]$ with variable modulus, if and only if

(37) $$\left\langle\frac{d\eta}{da}\right\rangle_{E_j}=0 \quad \text{for all} \quad E_j.$$

The quantity $\langle G\rangle_{E_j}$ defined by equation (35), and representing the expectation value of the operator G in the eigenstate of energy E_j, appears here as the quantal analogue of the ergodic average $\langle G\rangle_E$, so that the condition (37) is the analogue of the classical formula (9).

The calculation transforming Equation (9) into the Klein conditions (14) can be repeated, starting from Equation (37). The canonical averages are now defined as

(38) $$\bar{G}=\operatorname{tr}(e^{\eta}G)$$

and the Klein conditions take the form

(39) $$\frac{1}{\beta}\frac{d\beta}{da}=\frac{\bar{A}-\langle A\rangle_{E_j}}{\bar{E}-E_j} \quad \text{for all} \quad E_j;$$

the quantity A is again defined, as an operator, by Equation (11). More generally, we may repeat the calculation leading to the formula (19) for the difference between the 'adiabatic' average

$$\langle G\rangle_{\rm ad}=\operatorname{tr}(D_{\rm ad}G)$$

and in the canonical average. In this formula, the correlation functions Γ are still defined by Equation (18); in making use of this notation, however, it is necessary to introduce operators to represent the ergodic averages:

$$\langle G\rangle_E\equiv\sum_j \langle G\rangle_{E_j} D_{\varphi j}. \tag{40}$$

We may further proceed with the argument leading to Equation (22) for the adiabatic variation of temperature as a function of the external parameter, and to Equation (24) for the remaining deviation of the adiabatic averages from the canonical ones.

The evaluation of this deviation, however, can no longer be carried out in full generality as in the case of classical systems. It is true that we may allow ourselves to replace the discrete energy spectrum by a continuum, since the assumption of an extremely large number of energy levels in any macroscopically infinitesimal energy shell is part of the ergodicity conditions for the large systems we are considering; we may thus replace Equation (40) by

$$\langle G\rangle_E=\int \langle G\rangle_\varepsilon \,\mathrm{d}D(\varepsilon), \tag{41}$$

introducing a continuous distribution $\mathrm{d}D(\varepsilon)$ (in the sense of von Neumann) of the energy levels and a function $\langle G\rangle_\varepsilon$ of the energy representing the ergodic average[6] of G in the state of energy ε. But the possibility of expanding the function $\langle A\rangle_\varepsilon$, in analogy to Equation (26), as a power series in $e=\varepsilon-\bar{E}$ requires an additional assumption, independent of the ergodicity conditions, and concerning more specifically the way in which the Hamiltonian depends on the parameter a; quite precisely, we must assume[7] that the ergodic average $\langle A\rangle_\varepsilon$ is a continuous function of ε, differentiable a sufficient number of times for $\varepsilon=\bar{E}$. Granting this, we may just take over the classical argument and reach the same final conclusions[8]. It may further be noticed that the Klein conditions then also reduce to the classical form.

7. Application to Systems of Magnetic Ions

A system of magnetic ions (such as a paramagnetic crystal powder) in an external magnetic field modulated with a period intermediate between the spin-spin and spin-lattice relaxation times offers an instructive example of a behaviour approximating that of an ideally isolated system, which may be regarded as ergodic. In a suitable range of temperatures, it is possible to reduce the consideration of the energy spectrum to a finite number of the lowest levels, and the formulae for the isolated and adiabatic susceptibilities then take the simple forms on which Caspers' work [3, 4] is based. However, it turns out, as shown by Caspers [4], that in these conditions, and if the additional assumption mentioned above about the behaviour of

$$\langle A \rangle_{E_i} \equiv -\partial E_i / \partial a$$

as a function of the energy for very large systems is satisfied, the asymptotic dependence of the energy eigenvalues E_i on a is such as to satisfy the Klein conditions. While this result disposes of Broer's paradox, it also exhibits the rather special character of this class of systems. The assumption about $\langle A \rangle_{\varepsilon}$, which is sufficient to ensure the consistency of the statistical interpretation of thermodynamics, appears to be much weaker than Klein's condition, but it would be of interest to investigate more closely how restrictive either of these assumptions actually is.

This paper was written during a stay at the School of Physics of Sydney University. I wish to thank Prof. H. Messel and his colleagues for a most cordial hospitality.

NOTES

[1] After this paper was completed, I saw a recent article by Yamamoto [9], whose considerations, however, appear rather superficial.

[2] In the note mentioned above [2], the case in which the relation between velocity and momentum involves the parameter a was not explicitly treated. However, the simple argument of the text shows that in order to include this case, there is no need to resort to the indirect reasoning proposed by Broer [1].

[3] The following treatment, which is more elegant than the argument developed in my previous paper [2], was suggested to me by that which Prof. C. Møller uses in his lectures, and which he kindly communicated to me in 1942.

[4] For a discussion of these conditions, see, e.g., refs. [6, 7].

[5] In discussing the application of the theory to systems of magnetic ions, Caspers [4] rightly insists on these restrictions.

[6] In this sense, the ergodic average $\langle G \rangle_e$ appears, as Broer [1] points out, as an average over the ergodic averages $\langle G \rangle_{E_j}$ in the actual states within the distribution interval

d$D(\varepsilon)$. It is clear from the preceding analysis, however, that the derivation of the fundamental formula (36) does not require any such averaging; the attempt at a similar analysis in Broer's thesis [8] suffers from a misunderstanding of this point.

[7] The necessity to introduce such an assumption was pointed out in ref. [6] (footnote 13, p. 32), and later, independently, by Caspers [4].

[8] In the section of Broer's paper [1] dealing with quantal systems, the expression for the isolated susceptibility (formula preceding his Equation (17)) is incorrectly written (although he uses a correct expression for it in the subsequent application to paramagnetic bodies). This slip unfortunately vitiates the important formula (17), which ought to be the same as (11), and utterly confuses the issue. The desirability of dispelling this confusion may justify my insistence on points of detail in the above exposition.

BIBLIOGRAPHY

[1] Broer, L. J. F., *Physica* **17** (1951), 531.
[2] Rosenfeld, L., *Proc. Ned. Akad. Wetensch.* **45** (1942), 970. [This volume, p. 742].
[3] Caspers, W. J., *Physica* **25** (1959), 43
[4] Caspers, W. J., *Physica* **25** (1959), 645.
[5] Klein, M., *Phys. Rev.* **86** (1952), 807.
[6] Rosenfeld, L., *Acta Phys. Polonica* **14** (1955), 3. [This volume, p. 762].
[7] Rosenfeld, L., *Scuola Intern. Varenna*, 1960. [1960a].
[8] Broer, L. J. F., 'Beschouwingen en metingen over paramagnetische relaxatie', Thesis, Amsterdam 1945.
[9] Yamamoto, T., *Phys. Rev.* **119** (1960), 701.

CHAPTER 5

ON THE FOUNDATIONS OF STATISTICAL THERMODYNAMICS*

[1955a]

To Professor Rubinowicz as a token of high esteem and respectful friendship.

ABSTRACT. The aim of this essay is to survey the logical structure of statistical thermodynamics and to realize how faithfully this theory reflects the peculiar dialectical character of the physical situation. The author tries to outline very briefly the foundations of the atomistic interpretation of the laws of thermodynamics, laying the emphasis on the physical meaning of the formal arguments. The points that he wants to make are no doubt familiar to those physicists who have cared to look into the matter; but there is still widespread confusion concerning some of them, which may afford some justification for taking up the question from a synthetic point of view. In this connexion, it is not at all superfluous to discuss the classical approach as well as the quantal one: the true significance of the latter can only be perceived by contrasting it with the former.

The atomistic interpretation of thermodynamics, on a statistical basis, was elaborated in friendly emulation by two of the deepest thinkers of the XIXth century, Maxwell and Boltzmann. After Maxwell's untimely death, Boltzmann had to carry on single handed a gallant fight against the rising tide of idealistic and positivistic criticism, which for a time swept even the solid bastions of physics. This reaction was the – hardly conscious – reflexion in the minds of physicists of the turmoil agitating the bourgeoisie during the period of transition to the imperialist phase of capitalistic society. In the field of physics, the effect of such a critical mood was, on the whole, far from detrimental: even an idealistic approach is not barren when it is controlled by close contact with concrete reality. In fact, the work of the critical school was a necessary phase of the dialectical movement which liberated the materialist conception of the physical world from its narrow mechanistic frame and carried it to the wider viewpoint of quantum theory.

Boltzmann himself remained to the last on the ground of mechanistic materialism. He did not live to witness the unfolding of the new synthesis and enjoy the final triumph of his views. But in his controversy against the school of 'energetics' and the other opponents of atomism he was right in every point. With sure intuition he emphasized those fundamental aspects of the atomic theory, which were bound to survive even the most drastic changes in our conception of the structure of matter. Admittedly, a robust faith was needed to uphold atomistic views in front of the accumulating evidence which pointed to a severe limitation of the validity of the equipartition theorem. Boltzmann felt, however, that while the breakdown of equipartition undoubtedly demonstrated our ignorance of deep-lying features of atomic constitution, there simply could not be anything wrong in the general physical ideas underlying the statistical treatment of atomic motions and interactions. The fruitfulness of these ideas, their power of coordinating qualitatively different phenomena, was for him a warrant of their conformity with essential elements of reality.

To appreciate the superiority of Boltzmann's philosophical standpoint, it is instructive to contrast it with the timid and ambiguous attitude of Gibbs. While Boltzmann remained undaunted, Gibbs retreated before the difficulties. He endeavoured to make use only of those parts of the formal structure of statistical mechanics which were susceptible to a fully rigorous treatment with the methods then available. This means that he shunned the whole problem of the ergodic hypothesis, in spite of its fundamental importance, and that with regard to the physical interpretation he took refuge to the crudest Platonistic conception of mechanical 'analogies' of the thermodynamical laws. If one takes this view, the real physical basis of the theory, and especially the essential character of uniqueness of the atomistic interpretation, is utterly lost. In fact, the paradoxical outcome of such an idealistic treatment is an atomic theory of heat from which the concept of atom is wellnigh banished. The exquisite formal elegance of Gibbs' treatise should not blind us to the fundamental inadequacy of the underlying philosophy.

We now know that Gibbs' mathematical scruples were beside the point. Lebesgue's theory of the measure of sets and extension of the concept of integral have supplied mathematical tools perfectly adapted to the needs of statistical thermodynamics. With their help it has become possible to give a precise formulation of the conditions under which the ergodic theorems are valid; and these mathematical conditions correspond to an easily visualizable physical situation. The failure of the classical treatment is therefore not due to any formal inconsistency, but to a real departure of the atomic processes from the behaviour of classical mechanical systems.

The extension of the methods of statistical thermodynamics to the radiation field by Rayleigh and Planck made it conspicuous that the difference between atomic and classical laws is a radical and deep-lying one. For a radiation field, the atomistic elements whose statistical behaviour is analyzed are the proper electromagnetic oscillations of which it is composed. The breakdown of equipartition in such a simple and transparent situation brought about the crisis out of which the new synthetic conceptions of quantum theory eventually emerged. Quantum theory has removed all the uncertainties and contradictions which had hampered the completion of statistical thermodynamics on a classical basis. Moreover, by forcing upon our consciousness the dialectical character of physical laws, expressed by the idea of complementarity, it has enabled us to get a clearer view, or at least a clearer formulation, of the complementary relationship underlying the ergodic theorems.

We are thus now in a position to survey completely the logical structure of statistical thermodynamics and to realize how faithfully this theory reflects the peculiar dialectical character of the physical situation. To bring this out is the aim of this essay, in which I shall try to outline very briefly the foundations of the atomistic interpretation of the laws of thermodynamics, laying the emphasis on the physical meaning of the formal arguments. The points that I want to make are no doubt familiar to those physicists who have cared to look into the matter; but there is still widespread confusion concerning some of them, which may afford some justification for taking up the question from a synthetic point of view. In this connexion, it is not at all superfluous to discuss the classical approach as well as the quantal one: the true significance of the latter can only be perceived by contrasting it with the former.

1. The basic assumptions of atomic theory

The most immediate assumption underlying any atomistic interpretation of phenomena of matter in bulk is that these phenomena are the outcome of processes involving immense numbers of atoms. This is of course the basis for the representation of macroscopic physical quantities by statistical averages taken over corresponding quantities pertaining to the atomic system, the fluctuations of the latter quantities becoming negligible as the number of degrees of freedom of the sytem increases. Above all, it is a necessary requirement for understanding the emergence of macroscopic irreversibility from the reversible phenomena on the atomic scale; for the statistical irreversibility expressed by the ergodic theorems has an essentially asymptotic character. Even this elementary point is obscured by the formalistic approach of Gibbs, who lays a misplaced emphasis on the derivation of exact formulae valid for any number of degrees of freedom. This pointless elegance irritated Boltzmann: 'I can understand statistics applied to a gas, he remarked [1], but I see no point in applying it to a sewing machine."

No less essential than the large number of the atoms entering into the constitution of ordinary bodies is the identity of their structure. The importance of the uniformity of atomic structure for a coherent account of large scale regularities was properly emphasized by Maxwell; he compared molecules with "manufactured articles" [2]. Incidentally, he carried the metaphor so far as to assert that the existence of these molecules, all 'manufactured' on the same pattern, pointed to the existence of a Manufacturer: – a delightful example of the indirect, but very substantial impact on physical thought of the new social outlook dominated by the development of mass production methods; natural philosophy no longer culminates into the apotheosis of the craftsman – potter or clockmaker – but in that of the selfsatisfied factory owner.

The full incorporation of the structural identity of the atomistic elements (particles or radiation waves) into the description of systems of such elements has only been achieved within the framework of quantum theory, where it is embodied in the requirement that any state vector pertaining to a system of identical elements be either symmetrical or antisymmetrical in the sets of variables defining the elements. This condition, by which the degeneracy arising from the various permutations of the elements is removed from the start has also a direct physical meaning: it expresses a fundamental property of the elements, corresponding to their classification into 'bosons' and 'fermions', which has important repercussions upon their thermodynamical behaviour.

In the classical treatment, no such new physical principle is recognized, of course; but from the logical point of view the situation is in no way different: here also one has very essentially to take account of the identity of the elements in a manner appropriate to the formalism. I shall go into this point in some detail, since it is still currently misrepresented, in spite of Boltzmann's crystal-clear analysis. We have simply to consider that a physical state of a system of identical elements does not uniquely correspond to an ordered sequence of the sets of variables defining the states of the elements, but rather to the whole set of such sequences taken in any order. A physical quantity, expressed in terms of the element variables, is accordingly invariant with respect to the permutations of these variables; in other words, it is a function, not of any particular sequence of variables, but actually of the set of all the sequences. In Boltzmann's (1877) terminology, an ordered sequence of element variables, i.e. a point in the phase space of the total system (more precisely, a small neighbourhood around such a point) is a 'complexion'; for the set of complexions arising from each other by permutations of the element variables, and thus representing the same physical state of the system, Boltzmann himself uses the word 'Zustandsverteilung', Ehrenfest (1911, p. 37) the more picturesque 'Stern', which may perhaps best be rendered as 'constellation'.

Now, for the computation of statistical averages of physical quantities, it is obviously convenient to extend the required integrations over the whole phase space of the system, i.e. over all complexions, whereas the physical average should be taken over the constellations. To reduce the integral over phase space to a physical average, we must therefore divide it by the number of permutations of the elements. Thus, the mathematical representation of the system in phase space, which operates with complexions, is not immediately adapted to the physical situation, and the adjustment needed to restore the connexion with physically meaningful quantities is directly derived from the identity of the elements.

Confusion on this point may easily arise from a rash assimilation of the element variables entering into the complexions to the elements themselves. It is clear, however, that the concept of complexion is a purely mathematical device, and that individuality of the element variables, far from corresponding to a similar property of the elements, is a spurious formal feature which has to be removed. The abstract character of the complexions is aptly expressed by Boltzmann's (1877, p. 171 and p. 200 in Wiss. Abh.) graphic account of the statistical calculations: he works with "labels" (Zettel) on which sets of atomic variables are inscribed, and imagines these labels put into an urn from which they can be drawn at random to form complexions. He never indulges in the misleading – and perfectly superfluous – fiction of attaching the labels to the atoms, which would be tantamount to regarding the atoms themselves as distinguishable from each other. On the contrary, the whole trend of Boltzmann's argument can only be understood on the assumption that the atoms are essentially undistinguishable.

Gibbs' (1902, Chapt. XV) discussion of 'systems composed of molecules' in the last chapter of his treatise is in essential agreement with Boltzmann's, but as usual he rather spoils the tale by undue reticence. His distinction between 'specific' and 'generic' phases is identical with Boltzmann's distinction between complexions and constellations, and the relation of the concept of generic phase to the assumption of the identity of the molecules is clearly indicated (Gibbs 1902, p. 188). Unfortunately, there is no hint of the fundamental significance of this assumption: the very fact that it is only mentioned, almost casually, at the end of the book shows how reluctant Gibbs was to commit himself to an atomistic philosophy.

Before leaving the question of the structural identity of the atomic elements, it will not be superfluous to dispel a possible misunderstanding. In classical theory, the individuality of the elements is actually maintained, in spite of their structural identity, during the course of their interactions: in fact, the trajectory in phase space, which represents the temporal evolution of the system, establishes a definite connection between the successive complexions it takes up. Such a connection is, of course, entirely lost in quantum theory, in which the concept of trajectory is replaced by a mode of description of the interaction processes most consistently upholding the impossibility of individualizing the elements. This radical difference between the classical and quantal treatment with regard to the permanence of the individuality of the elements in no way conflicts with their essential similarity in respect of the statitical behaviour of these elements. The temporal aspect of the phenomena is in fact quite irrelevant for the statistical properties of the state of equilibrium; the whole purpose of the ergodic theorems is precisely to replace averages taken over the temporal succession of complexions by purely statistical averages defined by a stationary distribution in phase space. In Boltzmann's discussion of the H-theorem, in particular, the statistical aspect is introduced by the "Stosszahlansatz", which does not express the true variation in time of the distribution, but only the asymptotic average of this variation. The averaging process, of course, removes any trace of temporal succession and, accordingly, of individuality of the elements.

2. The central problem of statistical thermodynamics

The identification of macroscopic quantities with averages over their atomistic counterparts involves only a trivial use of statistical conceptions. But in the analysis of heat phenomena the most fundamental features of statistical theory are brought into play and the statistical aspect appears as a deep-lying element of the whole atomistic conception of matter. Heat is, in fact, the macroscopic manifestation of a *disordered* form of motion or propagation of the elements constituting the material bodies and radiation fields; by which is meant that the parameters describing this motion or propagation escape the control of the macroscopic observer, whose knowledge of the state of the system is restricted to a few mechanical and thermal characteristics. The statistical approach to thermodynamics is thus immediately related to the fact that the very definition of the macroscopic states of the system entails a limitation of our control of the microscopic behaviour of its elementary constituents. This is indeed the dialectical contradiction which arises in the development of a consistent atomistic picture, and which resolves itself into the synthesis of statistical thermodynamics.

The central problem is the statistical interpretation of the second law of thermodynamics, in its twofold aspect, concerning the behaviour of a system under controlled (quasistatic) and spontaneous transformations, respectively. In the first place, we have the statement of the existence of a quantity called *entropy* which belongs to each state of thermal equilibrium of the system, and whose variation when passing to some other equilibrium state in a controlled way is a well-defined function of the mechanical and thermal parameters characterizing the transformation. The statistical meaning of entropy is well illustrated by such simple processes as the mixing of two gases. For the increase of entropy in this case, when considered from the atomistic point of view, can only be a function of the numbers of atoms involved in the mixing process: in fact, it is simply related to the greater probability of the disordered state represented by the mixture, as compared with that of the separated gases. Quite generally, entropy appears as a statistical measure of the degree of 'disorder' of the microscopic state of the system. Alternatively, to use a fashionable terminology, we may say that it is a measure of our 'information' pertaining to the microscopic state; but we must not forget (a point which is not always properly appreciated in the so-called theory of information) that we are here referring to a very special type of information, viz. that which is obtained or lost by physical operations involving controlled changes of mechanical and thermal parameters, and that the entropy is not merely a mathematical concept, but itself a physical quantity, directly related to the quantity of heat exchanged between the system and other bodies during the process.

The other part of the second law, according to which the entropy of a thermally isolated system increases when the system undergoes a spontaneous transformation from one state of equilibrium to another, expresses the essentially irreversible character of physical processes on the macroscopic scale. From the atomistic point of view, this means that the thermal evolution of an isolated system, when its elements are left to follow the course determined by their own laws, tends statistically towards states of greater and greater probability. The derivation of this fundamental property is the object of the ergodic theorems, which form the backbone of statistical thermodynamics. Asymptotic irreversibility is indeed a very general feature of statistical 'chains' of events: it is exhibited by results of lotteries just as well as by assemblies of atoms. It is especially remarkable in the latter case, however, in view of the reversibility of the elementary laws of motion: the conciliation (in a statistical sense) of the dialectical opposites of macroscopic irreversibility and microscopic reversibility is the essential achievement of the statistical interpretation of the second law, strikingly illustrating the synthetic character of this interpretation.

From the preceding outline it is quite clear that the introduction of the quantal description of the elementary constituents as a basic assumption instead of the classical picture does not make the least difference to the fundamental structure of statistical thermodynamics; for the quantal laws, just as the classical ones, are reversible with respect to time, and the problem of establishing the macroscopic irreversibility by taking account of the statistical element involved in the concept of macroscopic observation remains unchanged and is again solved by ergodic theorems. The issue has been obscured by the fact that quantum theory itself, in contrast to classical theory, introduces a statistical element at the microscopic level; and it has sometimes been confusedly argued that it is this elementary quantal statistics which provides the basis of macroscopic irreversibility. In reality, we have here two completely distinct statistical features, which are not only logically independent of each other, but also without physical influence upon each other. The question whether the elementary law of change is deterministic (as in classical physics) or statistical (as in quantum theory) is entirely irrelevant for the validity of the ergodic theorems. (This purely mathematical property of ergodic chains shows, incidentally, how groundless is the opinion that the application of probability considerations to physical phenomena should have to be 'justified' by the existence of some 'deterministic substratum'.

The two quite independent statistical features which pertain to the behaviour of the quantal elements on the one hand, and to the definition of the macroscopic states of a system of such elements on the other, both arise from a synthetic relationship of complementarity between mutually exclusive aspects of the phenomena. The complementarity between the macroscopic and microscopic modes of description of an atomistic system presents indeed (cf. Bohr, 1932) a striking formal analogy with the familiar pattern of reciprocal limitation in the determination of quantal variables: a complete knowledge of the microscopic state of the elements precludes the use of such macroscopic concepts as temperature and entropy; conversely, the macroscopic description implies a widegoing renunciation of the knowledge of the microscopic state. But of course, this complementarity must be carefully distinguished from the entirely different dialectical situation of quantum theory.

While it does not alter the main logical issue, quantum theory has, of course, an important physical significance for the establishment of statistical thermodynamics. In the first place, the physical consequences of quantal statistics, which contain those of the classical theory as limiting cases, are free from the inconsistencies of the latter. The equipartition of energy, which was the source of these inconsistencies, is reduced to a law of asymptotic validity; moreover, as already noted, the identity of the elements is more correctly taken into account, so as to cope with even such extreme cases of non-classical behaviour as that of the pure radiation field or the metallic state of matter. Above all, quantum theory makes possible, in a much more compendious way than classical theory, a unified treatment of both matter and radiation (and in fact of any physical system whatsoever). This is, of course, a consequence of the elimination of the dualism between matter and force and its replacement by the conception of the complementary manifestations of particle and field for every constituent of the physical world.

A logical building up of statistical thermodynamics starts with the consideration of isolated systems and the establishment of the ergodic theorems which define the corresponding equilibrium distribution. The systems of actual physical interest are usually not isolated, but in interaction with their surroundings. However, the derivation of the equilibrium distribution under such conditions can be reduced to the ergodic case by conceiving the system under investigation as embedded in a much larger system, which may be treated as isolated to any desired approximation (just by making it large enough). The second step in the development of the statistical theory is therefore the study of the distribution of a small part of an isolated system. The final step consists in linking up the equilibrium distribution with thermodynamical quantities; this is achieved by setting up, in each case, a function of the statistical distribution which has all the properties of the corresponding characteristic function in thermodynamics. For the discussion of the ergodic theorems it is necessary to give separate treatments for the classical and quantum theory, because the formulation has to be adapted to the radically different conceptions of the atomistic states of the system in the two theories. Once this is done, and a formal correspondence between classical and quantal expressions for statistical operators is established, the remaining steps of the argument can be presented in a unified manner.

3. Classical statistics of isolated systems

Ergodic theorems can be enunciated for mechanical systems of the most general description, and it is advisable to begin by discussing them with this high degree of generality and abstraction, in order to bring out the part played by the special characteristics of atomistic systems in the adaptation of the mathematical formalism to physical requirements. But before coming to the exposition of this formalism, let us try to gain an orientation into the problem by approaching it from a semi-empirical point of view.

Let us consider the temporal evolution of an isolated system with a very large number of degrees of freedom, from a certain instant at which all connexions with the surroundings have been severed. We know from general experience that after a period of adjustment, or 'relaxation', it will tend to a stationary state in which it will persist indefinitely. From the atomistic point of view, the value of any macroscopic quantity pertaining to the system will be interpreted as the average of a definite microscopic quantity over the interval of time necessary for the measurement. Such a time interval may be extremely short in the macroscopic scale and yet cover the passage of the system through a large number of successive microscopic configurations. In the present case, however, the time average will be the same over any interval, except perhaps during the period of relaxation; the macroscopic value pertaining to the stationary state of the system may thus be defined, without appreciable error, by an average extending over an infinite time. The introduction of this idealized conception of the time average is the essential step which opens the way to the application of statistical methods. For such an average, extending over all the microscopic configurations actually taken by the system, is obviously independent of the order in which they have been successively taken; it is therefore equivalent to a definite *statistical* average, i.e. an average defined by a distribution law indicating the relative frequencies of occurrence of the various configurations of the system in the course of its evolution. Such a statistical distribution law, from which the time has disappeared, completely characterizes the stationary state of the isolated system, and the fundamental problem of the statistical thermodynamics of isolated systems is the derivation of this distribution law.

The above argument, however, is incomplete in as much as it takes for granted the empirically observed irreversible tendency of the isolated system to a stationary state. This behaviour, as already pointed out, is by no means obvious in view of the reversibility of the elementary law governing the change of configuration of the system with time. Our task is therefore not only to find the conditions under which the time average of any quantity exists independently of the initial state of the system, but also to define the precise meaning of the macroscopic irreversibility, i.e. to determine in what sense we may say that any initial distribution of states tends towards the statistical distribution expressing the time average. This is the object of the ergodic theorems.

Using the convenient geometrical representation of the microscopical configurations of the system by points of a symbolic 'phase space', we may visualize its temporal evolution from some initial phase as a trajectory in this space. The macroscopic time evolution, however, is not truthfully represented by any such single trajectory, nor even by any bundle of trajectories whose initial phases have a known distribution: for the density of this distribution, by Liouville's theorem, will remain the same as we follow it up along the trajectories, and will therefore not exhibit any tendency towards any limiting distribution. Of course, this is just the irreversibility paradox in a new guise: we cannot hope to conjure up any irreversible trend from purely mechanical considerations.

The way out has been described with great clarity by P. and T. Ehrenfest (1911, pp. 60–62) in their masterly critical survey of the problem: we must distinguish between a "fine" distribution of phases, which is defined on the microscopic scale and is transported without alteration along the trajectories, and a 'coarse' distribution, corresponding to a subdivision of phase space into finite cells (which may be of arbitrary size and shape). The physical meaning of such a distinction is clear: it expresses the fact that the macroscopic determination of the state of the system is represented, not by a single phase, but by an unknown distribution of phases, which we may in the most favourable case imagine reduced to a finite cell surrounding a given phase. The coarse distribution is thus the mathematical element by which the standpoint of the macroscopic observer is introduced into the analysis; and it is just this coarse distribution which reflects the macroscopic appearance of irreversibility. In fact, on this 'coarse' scale, any initial distribution spreads out gradually through the whole region of phase traversed by the trajectories; in the same way as a little milk poured into a cup of coffee is diluted by stirring until a homogeneous beverage is obtained: yet all the time, the space occupied by the milk, although deformed into thinner and thinner streamers, keeps the same volume. This simile so aptly illustrates the 'mixing process' in phase space which explains macroscopic irreversibility that even Gibbs (1902, p. 145) did not forsake it in the obscure pages he grudgingly devoted to the problem in his book.

We are now in a position to turn to the mathematical aspect of the theory, and to examine how it fits the physical situation. The mathematical relationships are best illuminated from the point of view of group theory[3]. We conceive the time evolution as an automorphism of phase space, the infinitesimal transformation of which is expressed by the Hamiltonian equations of motion. Trajectories are thus regarded as the sets of all successive transforms P_t, by this group, of some initial phase P_{t_0}, and it is the invariance of such sets with respect to the group that governs the global properties of the trajectories which are of interest for statistical considerations. Above all, it is essential that it should be possible to define a measure in phase space, invariant for the 'Hamiltonian group' (as we shall call the automorphism in question). The existence of such an invariant measure is guaranteed by Liouville's theorem, and its simplest expression is obtained by using canonical coordinates p_i, q_i and a Euclidean metric in phase space. The volume element $d\omega = \Pi_i dp_i dq_i$ has indeed the required invariance, and it is easy to derive from it an invariant element of measure $d\sigma$ on the 'energy surface' which contains all the trajectories of an isolated system of given energy.

No other property of the motion than the measure invariance expressed by Liouville's theorem is needed for the proof of *Birkhoff's fundamental ergodic theorem*[3a], which applies to any conservative mechanical system, provided only that all its trajectories are contained within a finite domain of phase space, – an assumption obviously fulfilled by all systems of physical interest. Birkhoff's theorem asserts the existence of the time average

$$(1) \qquad F(P_{t_0}) = \lim_{T \to \infty} \frac{1}{T} \int_{t_0}^{t_0+T} f(P_t)\, dt$$

of any phase function $f(P)$ for *almost all* trajectories, i.e. almost all initial phases P_{t_0}. The value $F(P_{t_0})$ of the average is, of course, independent of the initial instant t_0, but does in general depend on the trajectory (represented by one of its phases, P_{t_0}). It is easy, however, to indicate a general condition necessary and sufficient to ensure the constancy of the time average almost everywhere on the energy surface: it is the *metrical indecomposability* of this surface, i.e. the impossibility of decomposing it into two invariant sets (with respect to the Hamiltonian group) both of positive measure. Moreover, if this condition is satisfied, the constant value F of the time average can readily be found by considering the integral of the time average over all initial phases, i.e. over the whole energy surface. For such an integration, taken in the sense of Lebesgue, the occurrence of exceptional trajectories, for which formula (1) breaks down, is immaterial, since they form a set of measure zero. This is just an example of how exquisitely the theory of measure meets the requirements of statistics. The existence of exceptional trajectories must of course remain a stumbling block for any attempt at a rigorous treatment of the ergodic problem unless full use is made of the resources of Lebesgue's algorism. The resulting expression for F is

$$F = \frac{1}{\Sigma} \int_S f(P)\, \mathrm{d}\sigma, \tag{2}$$

where $\Sigma = \int_S \mathrm{d}\sigma$ is the invariant measure of the energy surface.

Although the argument leading to formula (2) is based on the consideration of a definite trajectory, it holds just as well if we start from any coarse distribution of initial phases. The quantity F given by formula (2) thus represents the statistical average equivalent to the time average, i.e. the macroscopic quantity corresponding to the phase function $f(P)$ in the state of equilibrium. The equilibrium distribution, called 'ergodic' (or in Gibbs' terminology 'microcanonical'), may simply be described as uniform over the energy surface with respect to the invariant measure on this surface. We may still be somewhat more realistic, and introduce also some uncertainty in the determination of the energy of the system, which amounts to considering an 'energy shell' of finite thickness instead of an energy surface. We must then assume metrical indecomposability for almost all the energy surfaces contained in the shell, and replace (2) by a formula of the same type,

$$F = \frac{1}{\bar{\Sigma}} \int_{\bar{S}} f(P)\, \mathrm{d}\omega, \tag{3}$$

in which the integration now extends over the energy shell $\bar{S}$ whose volume is denoted by $\bar{\Sigma}$; in other words, we get a coarse ergodic distribution uniform over the energy shell.

So far Birkhoff's theorem. It establishes the existence – and the form – of the statistical equilibrium distribution, but it sheds no light on the 'mixing process' by which it is reached asymptotically. The condition of metrical indecomposability of the energy surface tells us that the bundle of trajectories issuing from any initial cell will eventually 'fill up' the whole energy surface or shell, but it is not sufficient to ensure that the initial coarse distribution will itself in the course of time fill the surface or shell uniformly. A rather trivial example, which, however, sufficiently illustrates the issue, is that of a rigid rotation around a fixed axis: any distribution of phases will in this case be bodily displaced within the energy shell with the velocity of the rotation. Such a behaviour is by no means of mere academic interest. To understand its physical significance, it will be necessary to look a little more closely at our mixing process in phase space. Like every mechanical process, it is not, of course, truly irreversible; on the contrary, it exhibits the peculiar quasi-periodicity, arising from its invariance for time reversal, which is expressed by Poincaré's theorem. In the long run, any deviation from uniformity in the distribution, however large, is bound to recur again and again; the average periods of recurrence, however, increase so rapidly with the number of degrees of freedom of the system as to exclude the possibility of ever observing the recurrence of large fluctuations on the macroscopic scale. Now, this is just the point with respect to which our rigid rotator is anomalous. For this system, fluctuations of arbitrary amplitude around the ergodic average of any phase function occur with the frequency of the motion itself; the ergodic average will then not have the clear-cut physical interpretation we wish to ascribe to it.

The mathematical analysis of the mixing process has been carried out by E. Hopf[4], and one of his results, which admirably answers the physical requirements, fully deserves the dignity of *second ergodic theorem.* Roughly speaking, it supplements Birkhoff's theorem by giving the additional condition under which fluctuations average out in the course of time, thus guaranteeing the adequacy of the ergodic average as representative of macroscopic quantities. We are here concerned, of course, not with the instantaneous values $f(P_t)$ of phase functions, but with the 'coarse estimates' of such functions, obtained by smoothing them out over some arbitrary phase distribution, e.g. a finite cell around some phase, specified by a density function $D(P)$. At time t, the coarse estimate of $f(P)$ will be given by the integral $\int f(P)\, D(P_t)\, d\omega$, where the transform P_t of P has to be expressed as a function of P and t. The mixing process may then be described by saying that while the estimate just defined may in the course of time repeatedly deviate by any amount from its ergodic average F, the time average of these fluctuations (taken absolutely, e.g. quadratically) will tend to zero as it is extended over longer and longer times:

$$(4) \qquad \lim_{T\to\infty} \frac{1}{T} \int_0^T \left[\int f(P)\, D(P_t)\, \mathrm{d}\omega - F \right]^2 \mathrm{d}t = 0.$$

Hopf's ergodic theorem states the necessary and sufficient condition for the validity of formula (4). In order to formulate this condition, let us consider the phase space representing pairs of phases (P, P') and the automorphism defined by the direct product of the Hamiltonian group transformations applied to P and P', respectively. A trajectory in this product space will thus represent a pair of trajectories of the original system, contained in the two energy surfaces defined by the phases P, P'. By taking all possible pairs of such energy surfaces, within the energy shell defined by the distribution $D(P)$, we define an energy shell in the product space which contains all the trajectories in this space corresponding to the product distribution $D(P)\, D(P')$. Now, it is a simple matter to transform the time average occurring in formula (4) into a time average in the product space, which reduces the proof of (4) to an application of Birkhoff's theorem to the whole energy shell in the product space. The condition for the validity of (4) is therefore the metrical indecomposability, with respect to the product Hamiltonian group, of almost all energy surfaces of the shell in the product space. Thus, while the existence of the ergodic average depends on a property of single trajectories, the mixing process involves a further condition applying to pairs of trajectories. The meaning of this somewhat abstract condition may be elucidated by our example of the rotating motion around an axis. If the angular velocity is not a constant, but a definite function $\omega(J)$ of the angular momentum, the mixing process will in general occur, the initial distribution becoming more and more distorted and diluted; it is only required to exclude any rational proportionality between the angular velocities (except possibly for a set of pairs $\omega(J)$: $\omega(J')$ of measure zero), and this is precisely equivalent to Hopf's condition.

The last, but not the least interesting, question to examine is that of the fulfilment of the condition of metrical indecomposability for actual physical systems. This condition obviously plays the part of the original 'ergodic hypothesis' of Maxwell and Boltzmann, the inadequacy of which had been proved, as soon as the methods of the theory of sets became available, by Rosenthal (1913) and Plancherel (1913). The great advantage of the exact formulation of this hypothesis is to make it amenable to more detailed analysis. Such an analysis is also very much needed, since, as we shall see, the notion of metrical indecomposability requires an important modification to adapt it to the cases of physical interest [5].

Let us first mention a rather trivial correction. Strictly speaking, the energy surface can never be metrically indecomposable, since (as is easily seen) any time-independent integral of the motion, distinct from the energy integral, can be used to effect a decomposition of the surface into two invariant sets of positive measure, in which the integral in question is, respectively, larger and not larger than some suitably chosen value. Thus, the manifold containing a given trajectory must be defined by assigning a definite value, not only to the energy integral, but also to those other time-independent integrals (such as momentum and angular momentum) which can be fixed by external conditions and are thus, in principle, under the control of the macroscopic observer. However, for most practical purposes, it is permissible to assume that those other 'controllable' integrals have a constant value over the greatest part of the energy surface (e.g. that the system is connected with some auxiliary body so heavy that it remains practically at rest), so that the identification of the reduced manifold with the entire energy surface only involves a negligible error.

In general only a few of the distinct time-independent integrals, which together determine the shape of the trajectory, are 'controllable'; this is intimately connected with the circumstance that the same physical state may correspond to a whole set of phases, and that phase functions representing physical quantities must accordingly be uniform, in the sense that they must have the same value for all equivalent phases: only uniform integrals can be regarded as controllable. We have already discussed the case of systems of identical elements, with the distinction of specific and generic phases it entails; another example is afforded by angular coordinates, which are only determined physically *modulo* 2π, so that only phase functions which are periodic in these coordinates can have a physical meaning. The system formed by two uncoupled rigid rotators with different fixed axes illustrates in a simple way the occurrence of both uniform and non-uniform integrals.

Now, a non-uniform integral will also give rise to a decomposition of the energy surface (or of the reduced manifold defined by the uniform integrals) into invariant parts of positive measure. But if this decomposition is such that no two phases of any part are physically equivalent, it will not impair the validity of the ergodic theorems provided that their application be restricted to uniform phase functions. In the example of the two rotators, we clearly get a complete picture of the physical situation by restricting the range of the angular variables to the interval $(0,2\,\pi)$ and imagining the successive segments of the trajectory referred, by appropriate translation, to the part of the energy surface corresponding to these intervals. In order, therefore, to cope with the very general occurrence of equivalent phases, we must substantially modify the condition of metrical indecomposability and the scope of the ergodic theorems. We shall call a decomposition of the energy surface into two invariant parts of positive measure *essential* when all physically equivalent phases belong to the same part. The surface will then be metrically indecomposable *in the physical sense* when it does not allow of any essential decomposition. It is this property which guarantees the applicability of the ergodic theorems to uniform phase functions.

Considering the most important case of atomistic systems composed of identical elements, the modification just indicated of the ergodic theory is exactly what is needed to establish the relation between mathematical expressions involving complexions and their interpretation in terms of constellations. How essential this reduction is for the consistency of the whole atomistic picture will appear, in particular, if we ask for the definition of the relaxation period necessary for the macroscopic equilibrium state to establish itself. To visualize this process, the concept of constellation is clearly indispensable: for otherwise we should be led to the absurd conclusion that it would require several of the recurrence periods defined by Poincaré's theorem. We must, however, imagine the trajectory issuing from the domain of phase space which represents the initial constellation and traversing successively the domains of the most varied constellations, which form a chaotic entanglement; the constellations corresponding to the equilibrium state, occupy such an overwhelming part of the energy shell that, however improbable the initial constellation, the trajectory is bound to enter very soon some "prong" of an equilibrium constellation; in fact, if the number of elements and the energy are taken to correspond to usual physical conditions, this will happen in a time of the order of magnitude of the observed relaxation periods.

4. Quantal Statistics of Isolated Systems

We shall now review the ergodic problem on the basis of quantum theory and try to bring out the complete parallelism with the classical case. From the formal point of view, the correspondence between the classical and quantal concepts and laws which come into play is exhibited by the following comparison:

A classical state of the system is represented by a phase P. A distribution over phases is defined by a density function $D(P)$, and the average of any phase function $f(P)$ over this distribution by the integral

$$\int f(P)\,D(P)\,d.$$

The time evolution of the state P is represented by a trajectory defined by the Hamiltonian group transformation

$$P \to P_t.$$

Any distribution density $D(P)$ obeys the differential equation, expressing Liouville's theorem,

$$\frac{\partial D}{\partial t} + (H, D) = 0,$$

where (H, D) denotes the Poisson bracket.

A quantal state of the system is represented by a state-vector ψ. An arbitrary statistical distribution is defined by a density operator[6] D, and the expectation value of any operator f for this distribution by the trace

$$\operatorname{tr}(fD).$$

The time evolution of the state ψ is represented by a 'rotation' of the state-vector defined by the transformation

$$\psi \to \psi_t = e^{-(i/\hbar)Ht}\psi, \tag{5}$$

where H is the Hamiltonian operator.

The density operator D obeys the differential equation, equivalent with Schrödinger's equation,

$$\frac{\partial D}{\partial t} + \frac{i}{\hbar}[H, D] = 0, \tag{6}$$

where $[H, D]$ denotes the commutator.

For the discussion of the conditions of validity of the ergodic theorems, it will be necessary to assume (just as in the classical case) that the system considered is enclosed within a finite volume; as a consequence of this hardly restrictive assumption, the energy spectrum E_j is discrete and defines a complete orthogonal system of eigenvectors φ_j with associated density operators D_{φ_j}. The analogue of Birkhoff's theorem is here an immediate consequence of the fundamental Equation (6). According to this equation, the time dependence of any density operator D, expressed by

$$\begin{aligned} D(t) &= e^{-(i/\hbar)Ht} D(0)\, e^{(i/\hbar)Ht} \\ &= \sum_{j,k} e^{-(i/\hbar)(E_j - E_k)t} D_{\varphi_j} D(0)\, D_{\varphi_k}, \end{aligned}$$

is almost periodic (in the sense of H. Bohr), and the expectation value $\mathrm{tr}[D(t)\, f]$ of any operator f has accordingly a time average. In general the time average will depend on the detailed structure of the initial distribution $D(0)$; but if (and only if) the energy spectrum is non-degenerate, the time average $\langle D \rangle$ of the distribution takes the simple form

$$\begin{aligned} (7) \qquad \langle D \rangle &= \sum_j D_{\varphi_j} D(0)\, D_{\varphi_j} \\ &= \sum_j \mathrm{tr}\,[D_{\varphi_j} D(0)]\, D_{\varphi_j}\,. \end{aligned}$$

In this formula, the initial distribution occurs only through the coefficients $\mathrm{tr}\,[D_{\varphi_j} D(0)]$, which represent the statistical weights with which the various energy states enter into the composition of the initial distribution. Thus, the operator $\langle D \rangle$ defines a statistical distribution formed by a 'mixture' (in v. Neumann's terminology) of energy states with statistical weights which can be chosen arbitrarily: this is indeed the closest possible analogue to the ergodic distribution over an energy shell, and the condition of *non-degeneracy of the energy* is the pendant of the metrical indecomposability of the energy surface.

The element of uniformity introduced by the time average is well brought out by considering the special case of the distribution $D\psi$ corresponding to some statevector ψ, which we can expand in the form $\psi = \sum_j c_j \varphi_j$. While such a distribution expresses in its most elementary form (as a 'pure case' according to v. Neumann's nomenclature) the statistical feature introduced in quantum theory on the microscopic scale, its time average has a more macroscopic character, inasmuch as it only depends on the probabilities $\pi_j \equiv |c_j|^2$ and no longer on the complex amplitudes $c_j = \pi_j^{1/2} \exp(i\alpha_j)$ themselves. In fact, the whole time dependence of the distribution $D\psi$ is contained in the arguments α_j, and the time average is identical with an average over these arguments, assuming that they are distributed at random. It is this uniform distribution of the arguments α_j which corresponds to the uniform phase distribution in the classical energy shell.

The discussion of the 'mixing process' leading to the ergodic distribution $\langle D \rangle$ has been carried out by v. Neumann (1929). The first problem is how to construct the analogue of a 'coarse' subdivision of phase space. To this end, we may try to replace non-commuting canonical variables p, q by a pair of commutable operators P, Q which approximate them with reciprocal latitudes not exceeding considerably the limit indicated by the indeterminacy relation; it can be shown that this is actually possible. We obtain in this way a set of commuting variables, by means of which we redefine all operators of physical interest, including the Hamiltonian: the new operators may be regarded as 'coarse estimates' of the actual ones. Some suitably selected set of "coarse" operators can now be used to fix a complete orthogonal system of state-vectors ω_k, such that to each distinct set of eigenvalues of our operators there corresponds a finite subset of state-vectors, defining a 'cell' characterized by a density operator $d = s^{-1} \sum D_{w_k}$, where the sum extends over the s state-vectors of the cell. To a definite eigenvalue of the 'coarse' Hamiltonian there will correspond a 'shell' with the density operator $D_E = S^{-1} \sum D_{\varphi_j}$, comprising a subset of S eigenvectors φ_j. Since the φ_j and ω_k are linear combinations of each other, we may also write $D_E = S^{-1} \sum D_{\omega_k}$ and thus consider the shell as an assembly of a certain number N of cells:

$$D_E = S^{-1} \sum_{\nu=1}^{N} s_\nu \, d_\nu .$$

Any initial distribution D within the energy shell has the time average D_E, provided that (as we shall henceforth assume) the energy is nondegenerate.

We may now formulate *von Neumann's theorem*, which indicates under which conditions the time average of the quadratic fluctuation

$$\lim_{T \to \infty} \frac{1}{T} \int_0^T \{\operatorname{tr}[D(t)\, F] - \operatorname{tr}(D_E F)\}^2 \, dt, \tag{8}$$

for any 'coarse' operator F, can be said to vanish. There is, as in the classical theory, an additional necessary and sufficient condition on the Hamiltonian, viz. that there be no 'energy resonances', i.e. no two energy differences $E_j - E_k$ equal to each other; the parallelism with Hopf's theorem need not be stressed. The vanishing of the time average (8), however, can only be established in the sense of probability; i.e. one shows that the probability for it *not* to vanish can be made arbitrarily small by suitably fixing the coarse subdivision of the energy shell.

A brief outline of the proof of von Neumann's theorem will make this last point clearer. In the first place, an upper bound for the expression (8) can be derived by a straightforward calculation, in which use is made of the condition of no resonances to eliminate the structure of the initial distribution D; one gets

$$tr(D_E F^2)\cdot\left\{\sum_j \pi_j M_j + \sum_{j\neq k} \pi_j \pi_\kappa M_{jk}\right\},$$

with

$$M_j = \sum_{\nu=1}^{N} \frac{S_\nu}{S}\,[S\, tr(D_{\varphi_j} d_\nu) - 1]^2,\quad M_{jk} = \sum_{\nu=1}^{N} s_\nu S\, \mathrm{tr}(D_{\varphi_j} d_\nu D_{\varphi_k} d_\nu).$$

In this expression, the operator F only enters through the ergodic average of its square, and the initial distribution D through the probabilities π_j of the energy states. The coefficients M_j, M_{jk}, on the other hand, depend on the choice of the cell subdivision symbolized by the d_ν. All we can do is to study the distribution of the values of the M_j, M_{jk} as the set of eigenvectors ω_k is rotated with respect to the φ_j, and to estimate the probability that the upper bound of these coefficients be larger than some assigned value ξ. This is a delicate and tedious investigation [7], which finally yields for the probability just mentioned an expression of the form

$$\text{const.}\exp\{-b\sqrt{\xi S/N} + 2\log S\},$$

where b is a number of the order of unity; this estimate is valid for $\xi > \{2N/(S-2)\}$. To make this probability vanishingly small, we have only to take

$$S/N \gg (2\log S)^2,$$

i.e. to ensure that each cell of our coarse subdivision contains a large number of eigenstates of the Hamiltonian: a quite reasonable requirement.

Summing up the results reached so far, we conclude that in spite of the deeplying disparities in the fundamental atomistic conceptions, classical and quantum theory lead to essentially the same description of the basic statistical features which account for the macroscopic behaviour of isolated systems. The ergodic theorems are hardly of any direct practical use, but, as we shall proceed to show, they provide the natural starting point for the derivation of statistical laws adapted to the treatment of all the thermodynamical processes.

5. Statistics of non-isolated systems

The interactions of a physical system composed of a large number of elementary constituents with its surroundings are of two kinds: there are dynamical interactions, involving exchange of energy, momentum and angular momentum; there is also a flow of the atomic constituents themselves into and out of the system, with an accompanying exchange of mass and perhaps electric charge or other substantial properties. Strictly speaking, the two types of exchange cannot be separated; in particular, according to quantum theory, no physical system in interaction with others can be regarded as having a fixed number of elementary constituents, since, with an unlimited supply of energy from the surroundings, we cannot exclude the occurrence of creation and annihilation processes. However, it is of course a quite useful approximation in many cases to neglect such processes, and it is convenient to distinguish *closed* and *open* systems according as the number of elements is fixed or variable, and to treat separately these two classes of non-isolated systems. Moreover, it will be sufficient to restrict the analysis of dynamical interactions to the exchange of energy, which is the only one of interest for the theory of heat, and which is anyhow typical of the kind of argument to be applied to other dynamical processes if necessary.

It is in fact on the process of energy exchange that the concept of *temperature* is based. If we want to ascribe any system a definite temperature, we must bring it into thermal contact with a 'thermostat', i.e. allow it to exchange energy with a system whose heat capacity is so large that it maintains its temperature constant to any desired approximation. The concepts of energy and temperature thus refer to mutually exclusive situations, which afford an especially clear example of the general complementary relationship between the dynamical and thermodynamical aspects of the phenomena. The synthetic character of the atomistic conception reveals itself strikingly in the existence of a quantitative relation between the temperature and the *average* energy of the system: we here see again how the introduction of a statistical element provides the necessary link between the complementary aspects.

The statistical distribution characterizing a closed system of given temperature reflects this dialectical situation by an essential difference from the ergodic distribution pertaining to systems of given energy: it is not limited to an energy shell, but extends over the whole domain of possible states of the system and, in particular, assigns definite statistical weights to the different energy shells. To derive this law, the system we must consider is formed by the association of the body which we describe as a closed system [8] and the thermostat with which it exchanges energy. The latter must be supposed to be so large that the total system may always be regarded as isolated, and accordingly subjected to an ergodic distribution law. Our aim must be, then, to separate from this distribution of the total system that of the body by eliminating the variables which describe the state of the thermostat, so that in the final result, the influence of the thermostat is only represented by one parameter, related to the temperature. This can be achieved by suitably idealizing the conditions of the problem.

Although it is quite essential for the whole argument that there should be an interaction between the body and the thermostat, it is immaterial how large this interaction is. It is therefore permissible to assume it to be infinitesimal in comparison with the energies of the body and the thermostat considered separately. This allows us to treat the total system formally as a mere juxtaposition of body and thermostat, the total Hamiltonian being just the sum of the Hamiltonians of the two components. We may go a step further, and avail ourselves of the arbitrariness in the specification of the thermostat to introduce more symmetry into the question: we assume that the thermostat itself is a juxtaposition of a very large number of components (each of them of quite arbitrary structure), which are only bound together by negligibly small interactions; the body under investigation then appears as just one of the many units composing the total system. Our problem is thus reduced to a form of great simplicity and generality: if an isolated system is made up of a very large number of components with negligible interactions, to find the statistical distribution of any one of the components. The formulation and solution of this fundamental problem is one of Boltzmann's (1871) great achievements; he had especially in view the derivation of the distribution law for an ideal gas of molecules of complex structure[9], but he was well aware of the universal scope of the argument.

It is instructive to go a little more closely into the technical details of the analysis. Let us start with the system composed of the body and the thermostat. A state of this system will be represented

in classical theory by a phase

$$P_s \equiv (P, P_\theta)$$

of the 'product space' obtained from the phases P, P_θ of body and thermostat.

in quantum theory by a state vector

$$\psi_s = \psi \times \psi_\theta$$

which is the direct product of two state-vectors ψ, ψ_θ pertaining to body and thermostat, respectively.

The Hamiltonians H, H_θ and

$$H_s = H + H_\theta$$

define in the respective phase spaces sets of energy surfaces $S(E)$, $S_\theta(E)$ and $S_s(E)$ with elements of invariant measure

$$\mathrm{d}\sigma(E),\ \mathrm{d}\sigma_\theta(E),\ \mathrm{d}\sigma_s(E).$$

It is easy to see that the element of measure $\mathrm{d}\sigma_s(E)$ which serves to compute the ergodic average on the surface E of the total system can be expressed as

$$\mathrm{d}\sigma_s(E) = \int \mathrm{d}\varepsilon \mathrm{d}\sigma(\varepsilon)\, \mathrm{d}\sigma_\theta(E-\varepsilon);$$

this formula means that one must

define sets of eigenstates φ, φ_θ, φ_s with corresponding 'decompositions of unity'[10]

$$\mathrm{d}U_s(E),\ \mathrm{d}U(E),\ \mathrm{d}U_\theta(E).$$

It is easy to see that the operator $\mathrm{d}U_s(E)$ which represents the density operator of the ergodic distribution in the energy shell $(E, E+\mathrm{d}E)$ can be expressed as

$$\mathrm{d}U_s(E) = \int \mathrm{d}U(\varepsilon) \times \mathrm{d}U_\theta(E-\varepsilon),$$

the integral being interpreted in the

associate in all possible manners pairs of energy surfaces in the body and thermostat spaces in such a way that their energies have the sum E, integrate first over each pair and then over all values of the energy ε characterizing each pair.

sense of Stieltjes, as explained by v. Neumann, and extended over the whole range of values of the energy ε.

In this way, any 'coarseness' in the distribution of the total system is entirely attributed to the thermostat, while the body has a 'fine' distribution in energy.

The ergodic average F of any quantity pertaining to the body, and accordingly represented by

a function $f(P)$ in the body space, | an operator f acting upon the ψ's,

may thus be written in the form

$$F = \int d\varepsilon F(\varepsilon)\, u(\varepsilon) \frac{u_\theta(E-\varepsilon)}{u_s(E)}; \tag{9}$$

in this formula, $F(\varepsilon)$ represents the ergodic average for the value ε of the energy of the body, a quantity given by

$$F(\varepsilon) \sum (\varepsilon) = \int_{S(\varepsilon)} f(P)\, d\sigma \qquad \Bigg| \qquad F(\varepsilon)\, \mathrm{tr}[dU(\varepsilon)] = \mathrm{tr}[f\, dU(\varepsilon)]$$

and completely independent of the thermostat; the factors $u(E)$, $u_\theta(E)$, $u_s(E)$ are defined by expressions of the type

$$u(E) \equiv \sum(E) \qquad \Bigg| \qquad u(E)\, dE \equiv \mathrm{tr}[dU(E)].$$

One has

$$u_s(E) = \int d\varepsilon\, u(\varepsilon)\, u_\theta(E-\varepsilon), \tag{10}$$

and it is only the ratio $u_\theta(E-\varepsilon)/u_s(E)$ which represents the influence of the thermostat on the statistical weight of the energy value ε.

Boltzmann's original argument proceeds from this point to a direct estimate of the weight factor $u_\theta(E)$ for a very large system of weakly interacting components. If the number of components is N, the expression $u_\theta(E)$ depends on the energy by the factor $E^{3/2N}$. Writing $E=\frac{3}{2}N\theta$, we therefore get, asymptotically,

$$u_\theta(E-\varepsilon)\sim E^{3/2N}\left(1-\frac{\varepsilon}{\frac{3}{2}N\theta}\right)^{3/2N}\sim E^{3/2N}\exp(-\varepsilon/\theta).$$

Since further, by (10),

$$u_s(E)\sim E^{3/2N}Z, \quad \text{with} \quad Z=\int \mathrm{d}\varepsilon\; u(\varepsilon)\exp(-\varepsilon/\theta),$$

we finally get the familiar result

$$\frac{u_\theta(E-\varepsilon)}{u_s(E)}=Z^{-1}\exp(-\varepsilon/\theta).$$

This part of the argument, however, has recently been refined in a most interesting way by Khinchin (1949, Chapt. IV and V), whose beautiful little treatise presents a penetrating analysis of the mathematical structure of the classical theory. Besides being more rigorous than Boltzmann's, and thereby providing the basis for a direct treatment of fluctuation problems, Khinchin's derivation is remarkable because it illuminates the intimate connexion between Boltzmann's distribution law and the very foundation of the calculus of probabilities.

Khinchin's argument is based on the remark that if we consider the total system of body and thermostat as composed of $N+1$ non-interacting systems, we have between the corresponding quantities $u_i(E)$ ($i=0, 1, 2, \ldots, N$) and that $u_s(E)$, belonging to the total system, a relation

$$(11)\qquad u_s(E)=\int \mathrm{d}\varepsilon_1 u_1(\varepsilon_1)\int \mathrm{d}\varepsilon_2 u_2(\varepsilon_2)\int\cdots\int \mathrm{d}\varepsilon_N \mu_0\left(E-\sum_{i=1}^{N}\varepsilon_i\right),$$

precisely of the same form as the 'law of composition' expressing the probability density $u_s(E)$ of the sum $E=\sum_{i=0}^{N}\varepsilon_i$ of $(N+1)$ stochastic variables with probability densities $u_i(\varepsilon_i)$. The u's cannot be immediately identified with such probability densities because they cannot be normalized, but this is readily remedied by multiplying them with exponential factors of the form $\exp(-\beta\varepsilon_i)$, where β is an arbitrary positive parameter; i.e. the quantities

$$\pi_i(\varepsilon_i)=\{Z_i(\beta)\}^{-1}u_i(\varepsilon_i)\exp(-\beta\varepsilon_i), \tag{12}$$

with

$Z_i(\beta)=\int \mathrm{d}\varepsilon \sum_i(\varepsilon)\exp(-\beta\varepsilon)$ $= \int \mathrm{d}\omega \exp(-\beta H_i)$,	$Z_i(\beta)=\int \mathrm{tr}[\mathrm{d}U_i(\varepsilon)]\exp(-\beta\varepsilon)$ $=\mathrm{tr}[\exp(-\beta H_i)]$,

are properly normalized probability densities satisfying a composition law of the type (11). Now, to these densities, formally 'associated' to our systems by the relations (12), we may apply the central theorem of probability theory, which gives an asymptotic form for large N of the probability density $\pi_s(E)$, valid for arbitrary densities $\pi_i(\varepsilon_i)$. Using this form both for the density $\bar{\omega}_s(E)$ associated to the total system and that, $\pi_\theta(E-\varepsilon)$, associated to the thermostat, and returning to the u's by means of (12), we get (with neglect of unimportant fluctuation factors)

$$\frac{u_\theta(E-\varepsilon)}{u_s(E)}\approx\{Z(\beta)\}^{-1}\exp(-\beta\varepsilon), \tag{14}$$

with the normalization factor $Z(\beta)$ defined by (13) in terms, of course, of the Hamiltonian H of the body. All that remains to be done to establish (14) as (in Gibbs' terminology) the 'canonical' distribution law is to relate the formally introduced parameter β to the temperature; but we need not dwell upon this point.

Let us now turn to the open systems. This is the new field opened up by Gibbs, and in which his powerful methods reign supreme. Not only did he develop singlehanded the whole thermodynamical treatment of these systems, dominated by the concept of *chemical potential*, but he also gave the key to their statistical analysis by his definition of 'grands ensembles' obeying a distribution law with respect to the numbers of constituent elements, built up on the same pattern as the canonical distribution in energy. Faithful to his Platonistic attitude, however, he did not attempt to link up the new conceptions directly with the physical situation they were intended to describe, but relied for their justification upon the wide range of their successful application. Yet such a direct link is by no means difficult to establish. From the thermodynamical point of view, Brønsted (1933) showed by a simple and elegant argument, involving only the usual manipulation of idealized pistons and semipermeable membranes, that the concentrations of the various constituents of a chemical system can be changed in a quasi-static way without modifying either the entropy or the other thermodynamical variables; thus justifying Gibbs' procedure of treating these concentrations as independent variables in his definition of the chemical potentials. As regards the statistical theory of the 'grand ensembles', it is clear that it should find its foundation in a relationship similar to the body-thermostat interaction, but involving in this case an exchange of matter between the body under consideration and some very large reservoirs of the various species of elementary constituents whose numbers are allowed to vary. The way in which such an argument can be develop has been sketched by Tolman (1940); who, however, confined himself to classical theory and, moreover, left out any discussion of the points raised by the identity of the elementary constituents. It will therefore not be superfluous to re-examine the problem.

To avoid unessential complications, we shall just consider a homogeneous body exchanging elementary constituents (which we shall call 'molecules') with a large reservoir of the same kind. The total system of body and reservoir will be assumed to contain a fixed number of molecules, and we shall moreover take it to be in thermal equilibrium at a given temperature. As to the mechanism by which molecules are exchanged between the body and the reservoir, we may idealize it arbitrarily without altering the ultimate equilibrium distribution. The macroscopic state of the reservoir will depend on the mechanisms of molecule exchange, but this is immaterial for the argument; it is only necessary to realize that, for a given exchange mechanism, the state of the reservoir is fixed by that of the body. We shall imagine that molecules are only allowed to pass so slowly between body and reservoir that there is no appreciable interaction energy between these two components of the system, and that the variation of the number of molecules in the body appears as a succession of independent chance events. Since the body is very small compared with the reservoir, the *a priori* probability of a given partition $N+N'$ of the total number N_s of molecules between the body and the reservoir is accordingly expressed by a Poisson distribution law

(15) $$p(N)=\frac{\bar{N}^N e^{-\bar{N}}}{N!}$$

around some average value $\bar{N}$, which is part of the external conditions. The use of the Poisson distribution in this case is entirely similar to the application of the central theorem to the partition of energy between body and thermostat.

Now, for any given partition, characterized by the number N of molecules in the body, the macroscopic value of an atomistic quantity pertaining to the body is (owing to the additivity of the Hamiltonians of body and reservoir) the usual canonical average F_N, which of course depends on the number N of molecules. For the 'grand ensemble' formed by all possible partitions, however, we must take a weighted average of the F_N's taking account of the presence of the reservoir; our task is to reduce the weight factors to a simple form in which the reservoir is represented by only one parameter, related to the chemical potential. To explain how this programme is carried out, we must specify a little more closely the formal treatment of atomistic quantities relating to systems with a variable number of molecules; such quantities are of the general type

(16) $$f(P_N)=\sum_{i=1}^{N} f^{(1)}(P_i)+\sum_{i=1}^{N}\sum_{j=1}^{N} f^{(2)}(P_i, P_j)+\cdots$$

where, in classical theory, the phase P_N is formed of the individual phases $P_i (i=1, 2, \ldots N)$ of the molecules.

where, in quantum theory, the 'configuration space' P_N is formed of sets of variables $P_i (i=1, 2, \ldots N)$ defining the individual states of the molecules.

If $H(P_N)$ denotes the body Hamiltonian, the canonical average F_N may be written

$$F_N=\{Z(N)\}^{-1}\int \mathrm{d}\omega_N f(P_N)\times \times e^{-\beta H(P_N)},$$

with

$$Z(N)=\int \mathrm{d}\omega_N\, e^{-\beta H(P_N)}.$$

$$F_N=\{Z(N)\}^{-1}\,\mathrm{tr}_N(fe^{-\beta H}),$$

with

$$Z(N)=\mathrm{tr}_N(e^{-\beta H}),$$

The integrations are taken over *specific* phases; but the passage to generic phases, effected by a division of each integral by $N!$, does not affect the average F_N itself.

the trace being taken with respect to a subset of symmetrized or antisymmetrized eigenvectors corresponding to the given number N of molecules.

When we take the canonical average for the total system, the reservoir contributes a weight factor $Z_r(N_s - N)$, defined in the same way as $Z(N)$ in terms of the reservoir variables, and referring to the number $N_s - N$ of molecules in the reservoir. The statistical average over the 'grand ensemble' thus takes the form

$$(17) \qquad F = \sum_N w(N)\, F_N Z(N) \frac{Z_r(N_s - N)}{Z_s(N_s)},$$

with

$$(18) \qquad Z_s(N_s) = \sum_N w(N)\, Z(N)\, Z_r(N_s - N);$$

the structure of these formulae is fully analogous to that of the formulae (9) and (10) referring to the energy partition over the body-thermostat system.

The *a priori* weight factor $w(N)$ has, in the classical case, just the Poisson expression $p(N)$ given by (16).

In the quantal formula, we may take $w(N) = 1$, since the *a priori* weighting is included in the restriction to symmetrized or antisymmetrized eigenvectors in the computation of the traces.

The elimination of the reservoir variables rests upon the remark that the quantity $\log Z_r(\bar{N}')$, taken for the equilibrium value $\bar{N}'$ of the molecule number in the reservoir, is directly related to an *extensive* thermodynamical function and is therefore (in our simple case of an homogeneous system) proportional to $\bar{N}'$. For sufficiently small deviations from the equilibrium state, it will therefore be permissible to write

$$(19) \qquad \log \frac{Z_r(N')}{Z_r(\bar{N}')} = (N' - \bar{N}')\, z = (\bar{N} - N)\, z,$$

where the reservoir function z can be ultimately expressed in terms of the thermodynamical variables, including $\bar{N}$, which specify the macroscopic state of the body; large fluctuations of N, on the other hand, are so seldom that there is no appreciable error in applying formula (19) to the whole range of values of N. Introducing a parameter ζ by the definition

$$\beta\zeta = -z + \log \bar{N}, \qquad \Bigg| \qquad \beta\zeta = -z,$$

and disregarding factors independent of N, we may therefore re-write the formulae (17), (18) in the form given by Gibbs:

$$(20) \quad F = \frac{\sum_N F_N \frac{Z(N)}{N!} e^{\beta\zeta N}}{\sum_N \frac{Z(N)}{N!} e^{\beta\zeta N}}; \qquad \Bigg| \qquad F = \frac{\sum_N \mathrm{tr}_N(fe^{-\beta H})\, e^{\beta\zeta N}}{\sum_N \mathrm{tr}_N(e^{-\beta H})\, e^{\beta\zeta N}};$$

the parameter ζ can be identified with the chemical potential.

In the classical expression (20), the explicit occurrence of the factor $\{N!\}^{-1}$, which has the effect of reducing the weight factor $Z(N)$ to generic phases, appears here as a natural consequence of the Poisson distribution of the partitions of the molecules. It once more illustrates the essential part played by the consideration of generic phases, which accounts for the identity of the elementary constituents, for the interpretation of the formalism of classical statistics. In fact, the factor in question is quite indispensable to secure the fulfilment of the product relation $W(N_1+N_2) = W(N_1)\, W(N_2)$, which must hold between the weight factors corresponding to a homogeneous system of N_1+N_2 molecules and its subdivision into two pieces containing N_1 and N_2 molecules, respectively.

In quantum theory, the constellations are introduced from the start by the process of symmetrization or antisymmetrization of the state-vectors, and there is no need to make any other explicit reference in the formalism to the distinction between complexions and constellations. In particular, it is possible to treat number variables themselves as quantal operators by expressing them in terms of suitable annihilation and creation operators, and to base on such operators a mode of representation of all physical quantities pertaining to systems of identical elements, in which this identity finds an immediate expression.

By considering the variable N in the quantal formula (20) as the eigenvalue of an operator (denoted by the same letter), we readily obtain the more compact expression

$$(21) \qquad F = \frac{\mathrm{tr}\,(fe^{-\beta H + \beta\zeta N})}{\mathrm{tr}\,(e^{-\beta H + \beta\zeta N})},$$

where the trace is now extended to the full set of eigenvectors covering the whole ensemble of configuration spaces for all possible eigenvalues of N. The immense advantage of this formula over its classical counterpart lies, however, in its invariance with respect to the mode of representation of the operators, also in their dependence on the number variables. Thus, if we use the representation just alluded to in terms of annihilation and creation operators, we arrive at a very simple treatment of open systems, especially of those with negligible interactions between the molecules: indeed, the distribution laws of ideal gases of bosons or fermions are then immediate consequences of the formula (21).

6. Statistics and thermodynamics

The knowledge of the equilibrium distribution law suffices, in principle, to derive all the macroscopic properties of the system in terms of the macroscopic variables which fix the external conditions. In thermodynamics, the properties of the system are similarly concentrated in a single *characteristic function* of the independent variables chosen to define the state of the system. Thus, for isolated systems, for which the energy (or enthalpy) is an independent variable, the characteristic function is the entropy; for non-isolated systems of given temperature, it is the free energy, which occurs in several variants according to the choice of the nonthermal independent variables. The establishment of the link between statistics and thermodynamics therefore reduces to finding in each case a statistical interpretation of the characteristic function. It is well-known that the solution of this problem is not unique, but that for large systems the ambiguity does not matter: this situation has been elucidated in a masterly fashion by Lorentz (1916), who called it the 'insensibilité' of the entropy definition.

The proof that a given expression built up by means of the statistical distribution law may be identified with a characteristic function consists of two parts, corresponding to the behaviour of the latter function under quasi-static and irreversible transformations, respectively. A controlled change by $\mathrm{d}a_i$ of the independent variables a_i results in a well-defined change $\mathrm{d}G = \sum A_i \,\mathrm{d}a_i$ of the characteristic function G. If, on the other hand, a spontaneous transformation brings the system from one state of equilibrium to another, one can only establish an inequality $G(a') - G(a) < o$ (say) between the initial and final values of the characteristic function; such inequalities can all be reduced to the corresponding entropy increase of the isolated system including the system investigated and its surroundings. It is well-known from Sadi Carnot's pioneering analysis (Carnot (1824); p. 23, sqq.) that any quasi-static transformation can be decomposed into a succession of infinitesimal steps, each of which is either an isothermal or an adiabatic transformation. The latter is idealized as a sudden infinitesimal change of some non-thermal variable, after the system has been severed from the surroundings, followed by an infinitely long period during which the system is allowed to reach the state of equilibrium corresponding to the new value of the variable in question. Conversely, all irreversible transformations can be reduced to two types: either an adiabatic change in which the periods of recovery following each infinitesimal step are of finite duration, or a process of exchange of heat between bodies of different temperatures.

The discussion of quasi-static transformations is not so straightforward as it looks. Thus, if we are dealing with isolated systems, we have no means of defining the temperature except in a purely formal way, as the inverse of the partial derivative $\partial S/\partial E$ of the very entropy function S for which we want to establish the expression in statistical terms: the suitability of such an expression must therefore be tested by the verification of equations of the form

$$-A_i = \frac{\partial S/\partial a_i}{\partial S/\partial E}$$

for the non-thermal variables a_i; the A_i being directly defined as the ergodic averages of the partial derivatives $\partial H/\partial a_i$ of the Hamiltonian with respect to these variables. It is often overlooked, however, that this verification is by itself not yet sufficient for a logically complete proof: it is further required to show that the *form* of the statistical distribution law is not modified by quasi-static transformations of the external conditions.

Let us take up again the two points just made for the more interesting case of non-isolated systems of given temperature. For simplicity, we shall only consider a closed system; we have then to test the identification

$$(22) \qquad F = -\theta \log Z$$

of the free energy F with an expression depending on the weight factor Z of the canonical distribution.[11] The verification is immediate for isothermal transformations, for which the variation dF directly expresses the work done on the system by the external forces. Moreover, the formula (22) readily yields for the macroscopic energy E, which is the canonical average of the Hamiltonian H, the expression

$$(23) \qquad E = \frac{\partial(\beta F)}{\partial \beta}$$

known as Helmholtz' formula; we use the notation $\beta \equiv \theta^{-1}$. For the variation $\mathrm{d}F \equiv -S\mathrm{d}\theta$ corresponding to a quasi-static change of temperature $\mathrm{d}\theta$, we therefore get

$$(24) \qquad -S = \beta(F - E),$$

which *formally* coincides with the thermodynamical relation between entropy, free energy and energy. From our present point of view, this relation must logically be regarded as a statistical *definition* of the entropy, in harmony with the thermodynamical formalism. However, this argument tacitly assumes that in discussing a quasistatic change of temperature we may treat the equilibrium distribution as a canonical one with variable temperature. Since the transformation in question, when reduced to the two basic types, involves an adiabatic step, our statistical interpretation of F and S will not be completely justified until the *permanence* of the canonical distribution under quasi-static adiabatic transformations is established.

To this end,[12] let us start from a state of the system characterized by a temperature θ_0 and the value a_0 of some non-thermal variable. With the help of the Gibbsian notion of 'probability exponent,'

$$(25) \qquad \eta \equiv \beta(F - H),$$

we can represent the normalized distribution density by $\exp \eta$. Let us imagine that at a certain instant the contact with the thermostat is interrupted and the value of the non-thermal variable suddenly changed by an infinitesimal amount δa from a_0 to $a = a_0 + \delta a$. This operation will not at first modify the microscopic state of the system, and the ensuing evolution will be that of the isolated system under the external conditions defined by the new parameter value. The density $D(a)$ of the equilibrium distribution eventually reached is accordingly given by the ergodic average of the initial one,

$$D(a)\,\mathrm{d}\omega = \langle e^{\eta(\beta_0, a_0)} \rangle_{\varepsilon(a)} \mathrm{d}\sigma_{\varepsilon(a)} \mathrm{d}\varepsilon\,;$$

$$D(a) = \sum_j \langle e^{\eta(\beta_0, a_0)} \rangle_{E_j(a)} D_{\varphi_j(a)}\,;$$

here $\langle f \rangle_{\varepsilon(a)}$ denotes the ergodic average of f,

$$\langle f \rangle_{\varepsilon(a)} = \frac{1}{\sum(a)} \int\limits_{S(a)} f(P)\,\mathrm{d}\sigma,$$

taken over the *new* energy surface $\varepsilon(a)$ containing the initial phase.

$$\langle f \rangle_{\varepsilon(a)} = \mathrm{st}\,[f D_{\varphi_j(a)}], \quad \text{for } \varepsilon(a) \equiv E_j(a),$$

taken with respect to the *new* Hamiltonian $H(a) = \sum_j E_j(a)\, D_{\varphi_j(a)}$.

We may first compare this expression for $D(a)$ with the density of the canonical distribution of temperature θ_0 for the new Hamiltonian $H(a)$. Representing by $\delta\eta$ the corresponding variation of η, we readily get

$$D(a) = e^{\eta(\beta_0, a)}\,[1 - \langle \delta\eta \rangle_{\varepsilon(a)}].$$

$$D(a) = e^{\eta(\beta_0, a)}\,[1 - \sum_j \times \langle \delta\eta \rangle_{E_j(a)} D_{\varphi_j(a)}].$$

For the ergodic average of $\delta\eta$ we may write (dropping now the argument a)

$$\langle \delta\eta \rangle_\varepsilon = \beta_0 (A - A_\varepsilon)\, \delta a,$$

where A and A_ε denote the canonical and ergodic averages of $\partial H/\partial a$, respectively. Now, if the system is so large that quadratic effects of energy fluctuations can be neglected,[13] we may identify the canonical average A with the ergodic average A_E taken for the average value E of the energy, and replace the difference $A_E - A_\varepsilon$ by the first term of its expansion in powers of E − ε. We thus see that

$$\langle \delta\eta \rangle_\varepsilon \approx (E-\varepsilon)\, \beta_0 \frac{\partial A_E}{\partial E}\, \delta a$$

is nearly proportional to the energy difference $E-\varepsilon$. It can therefore be compensated by a suitable alteration $\delta\beta$ of the inverse temperature to a value $\beta=\beta_0+\delta\beta$; for, by Helmholtz' formula (23), there corresponds to such a change of temperature an average variation

$$\langle \delta'\eta \rangle_\varepsilon = (E-\varepsilon)\, \delta\beta$$

likewise proportional to $E-\varepsilon$. In other words, we may describe the final distribution $D(a)$, for the new external conditions fixed by the parameter a, as being again a canonical distribution, with a temperature slightly different from the initial one. The condition required for the compensation connects the change of inverse temperature $\delta\beta$ with the change δa of the parameter a_0 by the relation

$$\delta\beta + \beta \frac{\partial A_E}{\partial E}\, \delta a = 0,$$

which indeed coincides with a thermodynamical equation characteristic for quasistatic adiabatic transformations.

Turning to the consideration of irreversible transformations, I may be very brief, since this aspect of the theory has been thoroughly laborated by Gibbs (1902, Chap. XI), and the transposition of the argument to quantum theory effected by Klein (1931). For systems of given temperature, the statistical expression (24) for the entropy can be described, according to (25), as minus the average probability exponent of the canonical distribution. If the external conditions are altered, the distribution may take an arbitrary form, but the notion of average probability exponent can still be applied to it. If the normalized distribution density is represented by exp η, its average probability exponent is defined, in classical theory, by $\int \eta e^\eta \, d\omega$ in quantum theory, by $\mathrm{tr}(\eta e^\eta)$. The whole discussion of irreversible transformations from the statistical point of view can be based on some simple inequalities which these expressions satisfy on account of the concavity of the exponential curve.

The case of adiabatic transformations is covered by the following lemma: "The average probability exponent of a distribution with uniform density in each energy shell is not larger than that of any other distribution with the same total probability for each shell." The limiting case of equality of the initial and final entropies is obtained with quasi-static transformations, since, as we have just seen, in each infinitesimal step δa of such a transformation the first order variation δS vanishes. In any other case, the successive steps will produce irregular variations of the average probability exponent until the establishment, with the last step, of an equilibrium distribution of uniform density in each energy shell; the lemma then tells us that the corresponding entropy is larger than that of the initial state.

To cope with transformations involving heat exchange between different parts of the system, we have a second lemma: "If a system is divided into two parts, the sum of the average probability exponents of these parts is not larger than the average probability exponent of the total system. The equality holds if, and only if, the two parts are simply juxtaposed without interaction." The two preceding lemmas are logically sufficient to exhaust all forms of irreversible thermodynamical transformations. If, however, one wants a more detailed analysis of the role of the temperature of the bodies involved in exchange of heat, a third lemma proves useful: "If θ is a positive parameter, the sum of the average energy and θ times the average probability exponent is a minimum for the canonical distribution of temperature θ (for which the quantity in question represents the free energy)." By combining the second and third lemmas one readily derives the fundamental theorems of Carnot and Clausius about cycles of transformations in which a body exchanges heat with various reservoirs.

I shall not conclude this survey of the logical structure of statistical thermodynamics without some discussion of the novel type of irreversible process with which we are confronted in quantum theory, viz. the process of *measurement*. Strictly speaking, the consideration of this kind of irreversibility falls outside the scope of thermodynamics, since it is connected with (ideal) processes affecting single atomistic elements. Still, it is instructive to compare the two situations, which, while referring to entirely different problems of observation, exhibit a striking similarity in their statistical aspect. The feature of irreversibility involved in the measuring process is intimately related to the peculiar 'wholeness' of quantal phenomena, which is their most fundamental character; indeed, a quantal phenomenon is only well-defined when it is brought to completion by some measuring intervention of the macroscopic observer. Just as in thermodynamics, though of course at a deeper level of observation, such an intervention introduces an irreversible element into the observer's experience.

In a measurement performed on a quantal system, the interaction between the system and the macroscopic bodies serving as measuring instruments partly escapes the control of the experimenter; this has consequences similar to those of the averaging process implied in the definition of macroscopic quantities. Even if the system was originally in a well-defined state, its state after the interaction with the measuring apparatus will be a statistical mixture of all the eigenstates of the quantity measured: the density operator D_φ, which contains, so to speak, all the statistical potentialities inherent in the 'pure' quantal state ψ, is transformed into an operator of the general form

$$D = \sum_i \operatorname{tr}(D_\varphi D_{\varphi i})\, D_{\varphi i}, \tag{26}$$

in which one of these potentialities is actually expressed. One may look at this situation from the point of view of the theory of information: the measurement entails the loss of the information about the previous state of the system; to regain an equivalent amount of information, one has to perform a 'reading' of the measuring instrument, i.e. sort out the mixture described by the distribution (26), and this costs a certain minimum amount of physical entropy.

It is interesting to note that Gibbs' conception of the average probability exponent is adequate also in this case to give the measure (apart from the sign) of the entropy associated in the way just indicated with the process of the measurement. In fact, v. Neumann (1927, p. 273; see also 1932, p. 191 sqq.) has shown that this entropy is connected with the distribution density (26) by the relation

$$S = -\operatorname{tr}(D \log D). \tag{27}$$

According to this formula, all pure states have the entropy zero, while for a mixture of pure states like (26), with 'concentrations' $\pi_i \equiv \operatorname{tr}(D_\varphi D_{\varphi i})$, one gets the familiar expression for the entropy of mixing, $-\sum_i \pi_i \log \pi_i$. The justification of the definition (27) requires in the first place the proof that it is possible to transform any pure state into any other by quasistatic adiabatic operations involving suitable measurements, but no sorting out; a sequence of such operations yielding the desired result has been most ingeniously contrived by v. Neumann. Once this statistical equivalence of all pure states is established, the case of mixtures is easily understood. Let us start from an ensemble of N systems all in the state ψ, and subdivide it into subsets containing $N\pi_1$, $N\pi_2 \ldots$ systems, respectively: we can then transform the systems of the first subset into the state φ_1, those of the second into the state φ_2, etc. without any change of entropy: all that remains to be done is then to mix up all these subsets, an operation whose minimum cost in entropy is just given by the usual expression quoted above.

This argument clearly shows that the relation (27) exhibits more than a purely formal analogy between the statistical features of two different orders of phenomena: it expresses the necessary connexion always to be expected between the abstract appraisal of the information obtainable about such phenomena and the thermodynamical properties of the macroscopic physical processes by which the information may in principle be obtained.

7. Dialectical aspects of statistical thermodynamics

Looking back on the preceding analysis of the atomistic theory of heat, I should like, in conclusion, to stress once more the epistemological principles which find such a striking illustration in the development of this theory. Above all, we realize that only the patterns of dialectical logic are flexible enough to formulate adequately the peculiar relationship between the atomistic picture and the macroscopic laws it purports to explain. Any attempt to follow rigidly a line of deduction exclusively based on atomistic conceptions, without paying due regard to the limitations imposed upon such deductions by the requirements of the macroscopic mode of description, inevitably leads to the impasse characterized by the famous paradoxes of Loschmidt and Zermelo [14]: the behaviour of atomistic systems does not exhibit any irreversibility. The second law of thermodynamics therefore appears, if one attributes it absolute validity, as the very negation of atomic theory. Such a dilemma cannot be resolved by any superficial patching up of the antithetic conceptions, but only by the introduction of a radically new point of view, from which these conceptions can be indissolubly integrated into a true synthesis. The unfolding of this dialectical process can be clearly discerned in the historical development of the subject. It was actually Clausius himself, the discoverer of the second law, who gave the first impulse to a renewed analysis on atomistic lines of the thermal properties of matter; and this re-assertion of atomic theory found its completion in the emergence of the statistical viewpoint, which confers on it its new synthetic character. The protagonists of this dialectical drama were blissfully ignorant of the arcana of Hegelian logic, but they keenly realized the novelty of the momentous step they had taken: "The modern atomists", wrote Maxwell (1873, p. 373) "have adopted a method which is, I believe, new in the department of mathematical physics, though it has long been in use in the section of Statistics."

Yet they were not aware of any break in the development of the theory of heat: for the basic idea which inspired the pioneers of thermodynamics (with the solitary exception of Robert Mayer) had always been the atomistic conception of heat. They thus experienced the dialectical transformation of thermodynamics into a statistical theory as a perfectly natural evolution within the general framework of atomism. Only at a late stage of the discussion were the laws of thermodynamics played up against atomic theory, when an attempt was made to revive a pluralistic conception of continuous matter, reminiscent of Anaxagoras' "homoiomeria," and involving, like the latter, an idealistic view of the 'cosmos'. Mach (1871), in particular, the most consistent advocate of such a conception, developed on the model of the logical structure of macroscopic thermodynamics a formal idea of causality, according to which the causal connexions of physical phenomena would be exhaustively described by functional relationships between macroscopic quantities; he opposed this type of direct causal relations outside of space and time to the determinism of classical mechanics, which was then thought to apply also to atomic phenomena. Statistical thermodynamics, however, introduces a new form of causality, which transcends the two others and eliminates their apparent opposition. It is remarkable that in quantum theory statistical causality again appears in a similar synthetic role, as part of a mode of description embodying the complementary aspects of space-time coordination and energy-momentum conservation.

I have already pointed out that in statistical thermodynamics just as in quantum theory, the notion of complementarity adequately expresses the logical relationship between the mutually exclusive modes of description which are synthesized in the theory. The existence of a relation of this kind did not escape the sagacity of Maxwell (1879), when he remarked, for instance, that "as soon as we begin to see the molecules distinctly, heat becomes resolved into motion." From this point of view, phenomena such as the Brownian movement appear in an interesting light. They represent intermediate situations, in which the two complementary modes of description can be applied with reciprocal latitudes. Thus, if we observe an emulsion of very fine droplets through the microscope, we may partially follow the motion of an individual droplet, as well as count the average numbers of droplets in thin layers at various depths: the individual motion will deviate by irregular fluctuations from the course determined by gravity and viscosity, and its character will change from complete randomness to complete dynamical regularity as the size of the droplet increases; the variation of droplet density with depth, on the other hand, which for ultramicroscopic droplets is accurately characterized by a canonical distribution of definite temperature, will deviate more and more from this simple form with increasing droplet size. Examples like this illustrate the essential point, repeatedly emphasized in the preceding discussion, that the mutual limitation of the complementary modes of description is a direct consequence of the difference of scale between the physical systems investigated and the tools of human observation.

In this respect, the particularly clear and cogent situation met with in quantum theory and so acutely elucidated by Niels Bohr has been of paramount importance in helping us to realize the insufficiency of current ideas about the relations between phenomena and their observation. For the idealist (I am only speaking of the most moderate variants of this attitude), scientific concepts are 'free creations of the mind' projected onto the physical world; for the mechanistic materialist, they are the 'reflexions' in the mind of external events. Common to both conceptions is the assumption that the interaction between the material world and the mind does not appreciably disturb the course of their separate existences. We have learned, however, that the peculiar 'wholeness' of atomic processes, expressed by the law of the existence of a quantum of action, makes such a separation of atomic systems from the conditions of their observation impossible: a quantal phenomenon can only be defined in terms of these conditions. The occurrence of complementary quantal phenomena, as well as the statistical character of the predictions which can be made about results of observations on quantal systems, have their origin in this fundamental fact. We are therefore now in a much more unprejudiced position to re-examine the problem of observation in statistical thermodynamics.

The contrast between the macroscopic mode of description of the thermal properties of matter and the dynamical analysis of the underlying atomic motions found a picturesque, but not altogether harmless, expression in Maxwell's (1871, Chap. XXII *in fin.*) fiction of 'demons' with exquisitely refined senses, enabling them to perceive individual atoms and act upon them. Like the all-embracing 'intelligence' in which Laplace had symbolized the deterministic ideal of mechanistic materialism, Maxwell (1873, p. 374) credited his demons with a more perfect kind of knowledge than that attainable by human observation: "when we pass from the contemplation of our experiments to that of the molecules themselves, we leave the world of chance and change, and enter a region where everything is certain and immutable." The statistical approach was represented as a *pis-aller*, and the macroscopic laws deduced by such means could not 'pretend to absolute precision.' Once it is recognized, however, that statistical relationships can logically subsist on their own right and provide, without any 'deterministic substratum,' a consistent and exhaustive account of a field of experience, there is no longer any point in asserting with Maxwell that while such relations may be true 'for all practical purposes,' they lack the finality attributed to 'the laws of abstract dynamics'; or in imagining like the less subtleminded epigones, that atomic theory 'reduces' – if not in practice, yet 'in principle' – the account of all experience to its atomistic aspect. Indeed we are here dealing with a typical situation in which any such distinction between 'practice' and 'principle' utterly vanishes: for the decisive role of the conditions of observation in determining the mode of description of experience is just as much a matter of 'principle' as of 'practice.'

This point was already appreciated very clearly by Clausius (1879, pp. 315–316), whose robust realism rebelled against the metaphysical misuse of Maxwell's demons: he acutely pointed out that the statistical description of the kind of molecular motion which we call heat is a matter of *definition*, directly arising from the actual impossibility of discerning individual atomic processes with the usual means of observation; the intervention of demons destroys the very conditions which give molecular motion the peculiar character to which the concept of heat refers. Only the undisputed acceptance in his time of the mechanistic dogma of determinism prevented him from fathoming the full epistemological significance of the statistical theory of heat, as we are able to assess it, now that we have come to realize the essential *autonomy* of statistical causality.

As rightly stressed by Clausius, the explicit consideration of the human observer in framing the concepts and methods of thermodynamics is a logical requirement for a proper definition of these concepts. Such an abstract outlook, however, may easily mislead people muddled by some idealistic philosophy to the belief that a subjective element is thereby introduced into the theory. Thus, no less an authority than G. N. Lewis (1930) could make the extraordinary statement that entropy "corresponds merely to a loss of information with regard to a state of the system and is thus a purely subjective concept." Clausius had a healthier view of the matter: he perfectly understood that, quite on the contrary, we are here dealing with an essential condition of *objective* description, imposed upon us by the laws of nature which govern the process of observation itself; entropy, he tersely remarked, "is not concerned with what heat can do with the help of demons, but with what it can do by itself." It is indeed remarkable that the development of the atomistic conception of the physical world has so insistently, in the theory of heat and again in quantum theory, reminded the physicists that science is not a detached contemplation of Nature, but is itself a natural process in which man is actively engaged. Our conception of the structure of scientific theories is thus more concretely brought into harmony with the social function of science. The founders of thermodynamics had no doubt that the task of science is – as Marx then expressed it with youthful enthusiasm – "to change the world." Deeper insight into the meaning of scientific knowledge, gained from their achievement, gives a new impetus to the idea of human progress which inspired them in their endeavour.

NOTES

[1] This utterance was reported in conversation by Ehrenfest. It certainly has the characteristic Boltzmannian touch.

[2] Maxwell, 1873 [*Sci. Papers*, Vol. II, p. 376]. See also Maxwell 1875 [*Sci. Papers*, Vol. II, p. 483] and 1871, Chap. XXII *in fin.* Maxwell attributes the comparison to Sir John Herschell.

[3] This was emphasized by B. O. Koopman (1931).

[3a] G. D. Birkhoff (1931). The first impulse to the modern ergodic theory was given by J. v. Neumann; see the historical survey by B. O. Koopman and G. D. Birkhoff (1932).

[4] Hopf (1932). See also Hopf (1934), especially p. 92, and Hopf (1937), p. 37.

[5] For all the following discussion see Khinchin (1949), Chap. III.

[6] This is v. Neumann's projection operator U, normalized to unit trace, i.e. $D=[\mathrm{tr}U]^{-1}U$. Cf. v. Neumann (1932), p. 41.

[7] This part of v. Neumann's analysis has been simplified and improved by Pauli and Fierz (1937).

[8] It may be a material body or a radiation field within a finite enclosure.

[9] In the case, the components are identified with the individual molecules, while the gas itself plays the part of the thermostat. It would seem at first sight that we here run into the inconsistency of assigning to the gas both a definite temperature and a definite energy; but the latter must, strictly speaking, be thought of as infinite.

[10] See v. Neumann (1932), p. 59 sqq. In our case of a discrete energy spectrum, we have (*ibid.* p. 65) $\mathrm{d}U(E)=\sum D_{\varphi_i}$ the sum extending over all the eigenstates in the interval $(E, E+\mathrm{d}E)$.

[11] Strictly speaking, in order to make the argument of the logarithm dimensionless, we ought, in classical theory, to divide all integrals over phase space, including the function Z, by some arbitrary 'unit' volume ω, for which we may take the volume of a cell of the coarse distribution. In quantum theory, the function Z, defined as a trace, is dimensionless if the state vectors used to compute it are normalized in the usual way; instead of the arbitrary cell volume ω, there appears a factor h^f, where f is the number of degrees of freedom of the system.

[12] The following proof was given for the classical case by L. Rosenfeld (1942); it is here carried out for the quantal case also. The discussion of the latter case in J. Broer's thesis (1945), is not quite satisfactory.

[13] In the quantal case, we further assume that the energy spectrum is so narrowly spaced that we may treat the ergodic average A_ε as a continuous function of the energy ε, as in the classical case.

[14] See P. and T. Ehrenfest (1902), pp. 22–23.

BIBLIOGRAPHY

Birkhoff, G. D., *Proc. Nat. Acad. Sci. U.S.A.* **17** (1931), 650, 656.

Bohr, N., 'Faraday Lecture', *J. Chem. Soc.* (1932), 376.

Boltzmann, L., *Wien. Ber.* **63** (1871), 679 [*Wiss. Abh.*, vol. 1, p. 284]; *Wien. Ber.* **76** (1877), 373 [*Wiss. Abh.*, vol. 2, p. 169].

Broer, J., *Beschouwingen en metingen over paramagnetische relaxatie*. Thesis, Amsterdam 1945.

Brønsted, J., *K. Danske Vidensk. Selsk., mat-fys. Medd.* **12** (1933), Nr. 6.

Carnot, S., *Réflexions sur la puissance motrice du feu*, 1824.

Clausius, R., *Die mechanische Wärmetheorie*, 2, Aufl. 2. Bd. (1879).

Ehrenfest, P. and T., *Enz. Math. Wiss.*, IV–32, (1911). [English translation by M. J. Moravcsik: *The Conceptual Foundations of the Statistical Approach in Mechanics* (Cornell University Press, Ithaca, N.Y., 1959) – Ed.].

Gibbs, J. W., *Elementary Principles in Statistical Mechanics*, Yale University Press, New Haven, 1902.

Hopf, E., *Proc. Nat. Acad. Sci. U.S.A.* **18** (1932), 204; *J. Math. Phys.* **13** (1934), 51; *Ergodentheorie*, 1937.

Khinchin, A. I., *Mathematical Foundations of Statistical Mechanics*, Dover, New York, 1949.

Klein, O., *Z. Phys.* **72** (1931), 767.

Koopman, B. O., *Proc. Nat. Acad. Sci. U.S.A.* **17** (1931), 315.

Koopman, B. O. and Birkhoff, G. D., *Proc. Nat. Acad. Sci. U.S.A.* **18** (1932), 279.

Lewis, G. N., *Science* **71** (1930), 569.

Lorentz, H. A., *Les théories statistiques en thermodynamique*, Teubner, Leipzig-Berlin, 1916.

Mach, E., *Die Geschichte und die Wurzel des Satzes von der Erhaltung der Arbeit*, 1871; 2. Aufl., 1909. [English translation by P. E. B. Jourdain: *History and Root of the Principle of the Conservation of Energy* (Open Court, Chicago, and Kegan Paul, Trench, Trübner & Co., London, 1911) – Ed.].

Maxwell, J. C., *Theory of Heat*, 1871; 'Molecules', *Nature* **8** (1873) [*Sci. Papers*, vol. II, p. 361]; Atom, *Encyclopaedia Britannica*, 1875 [*Sci. Papers*, vol. II, p. 445]; *Cambr. Phil. Trans.* **12** (1879), 547 [*Sci. Papers*, vol. II, p. 727].

v. Neumann, J., *Göttinger Nachr.* (1927), 273; *Z. Phys.* **57** (1929), 30; *Mathematische Grundlagen der Quantenmechanik*, 1932. [English translation by R. T. Beyer: *Mathematical Foundations of Quantum Mechanics* (Princeton University Press, Princeton, 1955) – Ed.].

Pauli, W. and Fierz, M., *Z. Phys.* **106** (1937), 572.

Plancherel, M., *Ann. Phys.* [*Leipzig*] **42** (1913), 1061.

Rosenfeld, L., *Proc. K. Ned. Akad. Wetensch.* **45** (1942), 970. [This volume, p. 742].

Rosenthal, A., *Ann. Phys.* [*Leipzig*] **42** (1913), 796.

Tolman, R., *Phys. Rev.* **57** (1940), 1160.

CHAPTER 6

QUESTIONS OF IRREVERSIBILITY AND ERGODICITY

[1962b]

1. Introduction

The aim of the following considerations is to point to some problems, partly of epistemological, partly of more technical character, raised by the concept of irreversibility and its position in the framework of statistical thermodynamics. Opinions concerning such questions are divided; but the origin of the confusion still prevailing in this domain must be sought more in the insufficient care taken by writers on the subject to analyse the epistemological basis of their point of view than in some inherent obscurity of the situation. I shall therefore try in the following to develop one consistent line of argument, which has the advantage of making use of well-established epistemological conceptions, namely those which have proved decisive in the elucidation of the foundations of quantum theory. I hope in this way to provide at least a convenient starting point for further discussion.

I shall first point out that irreversibility is introduced in the description of physical phenomena at a very early stage, and is not directly connected with the degree of complexity of the physical systems under consideration. Then, the well-known dilemma of statistical thermodynamics with respect to irreversibility on the macroscopic scale will be discussed from the point of view of complementarity. Finally, the connexion between macroscopic irreversibility and ergodic properties of mechanical systems will be briefly reviewed.

2. The 'arrow of time'.

It is clear that since the structure of the 'elementary' laws of physics (and this includes the laws of quantum theory as well as those of classical mechanics, electromagnetism and gravitation) is essentially reversible, i.e. invariant for a reversal of the sense of flow of time, any irreversible behaviour of a physical system must have its origin in some circumstance preventing a complete analysis of this behaviour in terms of such laws.

Within the framework of classical physics it was natural to seek the source of irreversibility in the statistical element introduced in the description of macroscopic systems by the complexity of their atomistic structure. The prevailing view was that there is in principle no limit to a strictly dterministic account of the phenomena, whose evolution in time is necessarily reversible; any observed irreversibility can therefore only be 'apparent', in the sense that it refers only to the most probable variation of the average distribution density of the atomic constituents of the system under consideration; the underlying reversibility on the atomic scale is then revealed by the character of the statistical fluctuations around the average variation. Characteristic of this conception is the endeavour to secure objectivity by eliminating all reference to the processes by which we may gain access to the states of the system entering the description of its observed behaviour.

In accordance with this 'mechanistic' philosophy, Boltzmann [1] suggested that the sense of the 'arrow of time' is imposed upon us by the direction of increasing entropy on a cosmical scale in the part of the universe which is accessible to our observation. He imagined that large parts of the universe may have been brought by statistical fluctuation into states far removed from thermal equilibrium: to an observer living in any one of these regions there will be a trend towards equilibrium which will define the sense of time. The lower limit for the time-scale of such an evolution and the size of the region taking part in it give an idea of the magnitude of the kind of fluctuation one has to assume. Moreover, one must realize that if at any time a large fluctuation is present, it is overwhelmingly probable that it has arisen from a still larger one. Even without leaving the ground of XIX-th century mechanicism, therefore, the conjecture of a thermodynamical origin of irreversibility in nature leads to cosmological hypotheses which, to say the least, are not very plausible.

The development of quantum theory, however, has taught us that the whole epistemology of mechanicism is untenable and has to be replaced by another conception, embodying a deeper-going analysis of the relation between the description of the phenomena and their observation. Although the role of the process of observation for the very definition of the concepts in terms of which the physical laws are formulated has been emphasized in the discussion of typical quantum effects, it is clearly of universal significance and just as imperative, from the epistemological point of view, in classical physics, even though in this domain its quantitative implications may be negligible. Now, it turns out that this analysis of the observation process also throws light on the question of 'time's arrow'.

Every observation of a physical system involves an interaction between this system and the apparatus which fixes the experimental conditions and thereby defines the phenomenon which is observed, or the physical quantity which is measured in the process. The observation or measuring process is closed by the registration of some permanent mark left by the system on the appropriate part of the apparatus; this may be a spot on a photographic plate, a stylus tracing or some similar record. Owing to the finite velocity of propagation of all interaction, it is always possible, in principle, to establish a definite order in the time-sequence of the physical process under investigation and its recording by the apparatus; and this offers a natural basis for the definition of the 'earlier-later' relationship. In plain words, if we look at an object, we say – by definition, or convention – that a photon which strikes our retina has been emitted *earlier* by the object. Thus, the sense of time is referred to the propagation of 'signals' from the object of observation to the detectors by which we observe it [1].

The preceding analysis stresses the role played by a property of the registering device which may be called 'memory': the production of a permanent mark (i.e. a mark whose duration is at least of the order of physiological relaxation times). This memory has no immediate relation (except for the lower time limit just indicated) with that of a human observer, although the ultimate prototype of the observation process is of course the direct sense perception and its recording in the brain. This physiological aspect of the matter, however, is by no means involved in our argument, which stops at the purely automatic physical recording of the phenomenon and thereby makes it clear that our fixation of the direction of time fulfils every reasonable requirement of objectivity. One must take account, in this connexion, of the alteration of our conception of objectivity which results from the recognition of the necessity of including a complete specification of the conditions of observation in the definition of the phenomena.

We further realize that the existence of a definite sense of time, or time-directedness, is a universal character of all physical processes, and in particular independent of the nature of the system considered, which may just as well be a single atomic object as a macroscopic body. There is always, however, an element of macroscopic complexity in this time-directedness, inasmuch as the registering device, in order to provide unique determinations of the quantities under consideration, must be describable in classical terms. To that extent, the irreversibility implied in time-directedness is again of statistical character; it shares this character with all concepts directly referring to our immediate experience and used without further analysis in the description of the phenomena; but this does not mean that one could contemplate any return to the thermodynamical arguments outlined above. Indeed, if one would try to analyse the thermodynamical behaviour of the registering device in its interaction with the observed system, in the hope of thereby obtaining further insight into the irreversibility of the recording process, this would amount to treating the registering device as a physical system under observation, whereby it would no longer fulfil its intended function of providing a *unique* specification of the phenomenon.

In fact, no amount of shifting the distinction between system observed and measuring apparatus can ever elude the necessity of maintaining this fundamental distinction, by which alone a connexion is established between the system and the concepts by which it is described. The concept of time-directedness appears therefore as a primary one, in the sense that it is immediately related to the process of observation, at a deeper epistemological level than the type of statistical irreversibility occurring in thermodynamics. The latter can be analysed in its own right (with the help of the H-theorem and the consideration of the fluctuations, as outlined above), but it must be realized that in this analysis the sense of time is already determined independently of any statistical consideration. The statement that the entropy of an isolated system in the final state of any process is larger than in the initial state has accordingly in *every* case a well-defined physical content, and can never be regarded as an indirect definition of the sequence 'initial-final'. Such a conclusion is in harmony with the fact that any irreversible change in the state of an isolated system must be initiated by an external intervention: the removal (without expenditure of work) of some constraint, which can only lead to a state which is less determined and therefore has a larger entropy. Gibbs had no doubt this situation in mind when he wrote [5]: "... while the distinction of prior and subsequent events may be immaterial with respect to mathematical fictions, it is quite otherwise with respect to the events of the real world." In fact, he refers the time-directedness in the 'real world' to an essential difference in our estimation of probabilities of subsequent and anterior events, which he describes in the at first sight sybilline sentence: "while the probabilities of subsequent events may often be determined from the probabilities of prior events, it is rarely the case that probabilities of prior events can be determined from those of subsequent events, for we are rarely justified in excluding the consideration of the antecedent probability of the prior events."

The fundamental significance of time-directedness, in the sense defined above, is illustrated by the fact, pointed out by Kronig [6] many years ago, that it is equivalent to the existence of dispersion relations for the Fourier amplitudes of the field whose propagation is considered: a function satisfying dispersion relations

$$f(\omega)=\frac{1}{\pi i}\,P\int_{-\infty}^{\infty}\frac{f(\omega')\,\mathrm{d}\omega'}{\omega'-\omega}$$

is in fact the Fourier transform of a function vanishing for negative values of its argument, and which therefore (if its argument is the time variable) may represent the propagation of a signal. In the recent development of the theory of dispersion relations, such functions are usually described as obeying a 'causality condition'; this terminology is unfortunate since causal relations are time-symmetrical; what is actually meant is that the functions in question express the time-directedness of the field propagation [2].

3. Complementarity of Dynamical and Statistical Descriptions

The analysis of macroscopic irreversibility on the basis of the H-theorem has resolved the apparent contradiction between this type of behaviour and the reversibility of the underlying dynamical laws of time evolution by introducing a statistical element into the passage from the atomistic to the macroscopic picture. In this way, the two modes of description – dynamical and statistical – are incorporated into a synthetic account as two complementary aspects of the behaviour of large assemblies of atoms[3]. This situation is even an especially simple and instructive example of the peculiar kind of logical relationship to which the name of complementarity may be applied.

Above all, it is characterized by the mutual exclusiveness of the two descriptions: conditions allowing of a complete mechanical (and electrodynamical) description of a system exclude the possibility of applying to the system any of the typical thermodynamical concepts; and conversely, the definition of the latter requires conditions of observation under which the mechanical parameters essentially escape our control. Thus, in order to assign to the system a definite temperature, it is necessary to allow it to exchange energy with a thermostat, and it is thereby impossible to assign to its energy any definite value; conversely, in order to keep its energy at a constant value, one must isolate the system, and the concept of temperature cannot be applied to it.

This complementarity is often overlooked owing to the fact that for very large systems it is permissible to disregard the fluctuations of the macroscopic quantities around their averages, and, for practical purposes, to assimilate a canonical ensemble of given modulus to the microcanonical ensemble corresponding to the average energy. Such a procedure may be compared to the disregard of the quantal limitations imposed by the uncertainty relations when one is dealing with phenomena involving many quanta of action. A closer examination of the statistical fluctuations, however, reveals reciprocal relationships between them, closely analogous to the uncertainty relations of quantum theory.

Let us consider a small part (but still of macroscopic dimensions) of a very large system, and study the deviation of this part-system from equilibrium, characterized by changes of the thermal variables, entropy S and temperature T, and of a typical pair of other macroscopic quantities, consisting of an extensive variable a and an intensive variable A, specified by their contribution $-A\,\mathrm{d}a$ to the change of the total energy. The (root mean square) fluctuation of all thermodynamical quantities pertaining to the part-system, in such random deviations from equilibrium, may be calculated by a method due to Landau and Lifschitz [10]. These authors show that the fluctuations of the pairs of variables (T, a) and (S, A) are statistically independent, and that (k denoting Boltzmann's constant)

$$(1) \qquad \begin{cases} \langle \Delta T^2 \rangle = kT^2 C_a^{-1}, & \langle \Delta a^2 \rangle = kT D_T; \\ \langle \Delta S^2 \rangle = k C_A, & \langle \Delta A^2 \rangle = kT D_S^{-1}. \end{cases}$$

In these formulae, C_a, C_A represent the heat capacities at constant a or A, and D_T, D_S the absolute values of the isothermal and adiabatic rates of change of a on variation of A:

$$(2) \qquad D_T = -\left(\frac{\partial a}{\partial A}\right)_T, \quad D_S = -\left(\frac{\partial a}{\partial A}\right)_S;$$

all these factors C and D are extensive, i.e. (for a homogeneous body) proportional to the average number N of atoms (or other elementary constituents) of the part-system.

From Equations (1) it is therefore apparent that the fluctuations of extensive variables are proportional to $N^{1/2}$, whereas those of intensive variables are proportional to the inverse of this quantity. In particular, if we assume for simplicity that the volume of the part-system is invariable, the square of the energy fluctuation will be

$$(3) \qquad \langle \Delta E^2 \rangle = kT^2 C_a,$$

and it is clear that we cannot simultaneously reduce the fluctuations of both temperature and energy of the part-system by increasing its size. In fact the product of the two fluctuations is independent of this size:

$$(4) \qquad \langle \Delta E^2 \rangle^{1/2} \langle \Delta T^2 \rangle^{1/2} = kT^2.$$

For a given temperature (defined by the very large total system), the order of magnitude of this product is governed by the universal constant k, or the inverse of Avogadro's number; this is the atomistic parameter which plays a part similar to the quantum of action in fixing the quantitative importance of reciprocal limitations in the use of different modes of description.

In the present case it is of course the size, more precisely the number of degrees of freedom, of the system which is decisive for the applicability of either dynamical or thermodynamical concepts to the analysis of its behaviour; in particular, the Landau-Lifschitz formulae (1) clearly show that no definite temperature or pressure can be assigned to a system of only a few degrees of freedom. There is no point in saying that the statistical definition of these concepts is still formally valid for such systems: for the smallness of the fluctuations is an essential condition for the practical usefulness of statistically defined quantities[4]. Here, as always, we find that the character of the physical description is directly imposed by the conditions of observation: the statistical character of the thermodynamical description arises from the large number of degrees of freedom of macroscopic systems, and its meaningfulness is accordingly limited to such systems; the preceding discussion of fluctuations has no other aim than to illustrate this situation by more quantitative considerations.

It may be instructive to pursue the matter somewhat further, and consider one of those intermediate cases in which one is dealing with a system, like an emulsion droplet, large enough to be accessible to direct individual observation, and yet small enough to exhibit large fluctuations. Such an object may then be considered from either of the complementary points of view, and the two corresponding modes of description may be applied to it with reciprocal limitations: we may on the one hand ascribe to this body an individual motion, and on the other hand treat it as a member of an assembly characterized by a statistical distribution and the resulting thermodynamical quantities. However, the motion will not have a purely dynamical aspect: it will be a 'Brownian' motion; neither will the statistical distribution be so well defined as the ideal ones which correspond to systems with very large numbers of constituents. To the translational motion of the barycentre of the droplet there correspond an intensive variable, the velocity v, and an extensive one, the momentum p; they are of course connected by the relation $p = Mv$ if M denotes the mass of the droplet, and their contribution to the energy differential is $v\,\mathrm{d}p$. Application of Equations (1) to this case gives, in particular,

$$(5) \qquad \langle \Delta S^2 \rangle^{1/2} \langle \Delta v^2 \rangle^{1/2} = k v_0 ,$$

where

$$(6) \qquad v_0 \equiv \sqrt{\frac{CT}{M}}$$

is of the order of the mean atomic velocities inside the droplet. The formula (5) expresses the reciprocal limitation (independent of the size of the droplet) of the statistical and dynamical descriptions of its behaviour.

In spite of the widegoing parallelism between the complementary situations in statistical thermodynamics and quantum theory, people are often reluctant to admit that the two cases are equally fundamental: while in quantum theory one is clearly confronted with the inescapable necessity of using two complementary sets of concepts in order to achieve an exhaustive account of experience, it would seem that in thermodynamics the dynamical description, applied to the systems of atoms constituting the bodies under investigation, is sufficient in all cases to account for the observed phenomena: the proper thermodynamical quantities would merely appear as convenient auxiliaries, derived by statistical procedures from the dynamical quantities; no doubt, their use might be subject to restrictions, but these would not have the same radical character as the quantal limitations expressed by the uncertainty relations.

The flaw in such an argument is that it does not sufficiently take into consideration the peculiar epistemological problems arising from the fact that all our knowledge about the atomic constitution of matter and the behaviour of atomic systems is derived from macroscopic observations and can only be formulated in terms of concepts referring to such observations. The kind of conceptual difficulty arising from this fact is illustrated by the necessity, when discussing the dynamical description at the atomic level, of appealing to fictitious observers, who are called 'demons' precisely because they are required to perform feats incompatible with the limitations of human observation. The concepts of thermodynamics have been constructed, historically, to cope with a definite situation characterized by a certain type of macroscopic observations; as Clausius said, they are "not concerned with what heat can do with the help of demons, but with what it can do by itself". The dynamical mode of description applies to phenomena defined by different conditions of macroscopic observation which are in a relation of mutual exclusion, or complementarity, to those of thermodynamics. This is a perfectly objective relationship, reflecting different modes of behaviour, under different external conditions, of systems composed of a large number of atomic constituents; there can therefore be no question of eliminating it. The logical role of the atomistic picture underlying statistical thermodynamics is rather, by providing a common foundation for the two complementary aspects, to allow a more precise determination of the conditions under which they appear.

It is instructive to reflect that we are facing an entirely similar epistemological situation in biology [12]: here there is a relationship of complementarity between the mode of description of molecular biology, in which the processes of life are accounted for in terms of ordinary physical and chemical concepts and laws, and that of macroscopic physiology, based on the concepts of biological function and organization. Here again, we find the two complementary aspects distinguished by different types of causality: in molecular biology we have just the statistical causality characteristic of quantum theory, whereas concepts like function and organization imply a consideration of final causes – not, of course, with any suggestion of an underlying finality arising from some principle fundamentally different from those governing inanimate matter, but simply in a formal methodological sense. Just as the atomistic conception is the common basis of the two complementary pictures of thermal phenomena, the structural properties of macromolecules, entirely accounted for by the known laws of quantum chemistry, are a sufficient basis for the understanding of all biological processes (at any rate, there is in our present experience no indication of any possible limitation in this respect). This, however, does not mean that the traditional functional approach to these processes would thereby become superfluous; not primarily because the excessive complexity of most living organisms prevents a detailed analysis of their behaviour at the macromolecular level from being actually performed, but rather on purely logical grounds: it is quite obvious that without the formal concept of some organization adapted to certain functions, or even several such concepts, corresponding to the successive stages of increasing complexity of biological evolution [13], it is impossible to bring order into the wealth of structures and behaviours which is the most striking character of life. By emphasizing the complementarity of the two contrasting points of view one preserves the possibility of doing equal justice, in a rational way, on the one hand to the unity of molecular structure and behaviour underlying all biological processes, and on the other to the qualitative diversity they exhibit at the same time.

We have thus surveyed three great subdivisions of a general account of natural phenomena, corresponding to increasing complexity or specialization of structure: quantum theory, which deals with the simplest atomic systems, statistical thermodynamics, whose main concern is the atomic constitution of matter in bulk, and biology, which describes the properties of peculiar macromolecular structures and the resulting macroscopic aggregations, which present characters of organization of increasing refinement. At each stage, we meet with a typical complementarity between mutually exclusive modes of description, the origin of which is ultimately the same: the necessity of applying essentially macroscopic concepts to the account of experience concerning processes at the atomic level. This necessity is probably very deeply rooted in the evolution of living organisms, for it is difficult to believe that the communication system which we call thinking could develop in an organ of less complexity than the human brain.

4. Irreversibility and Ergodicity

The physical basis of the atomistic interpretation of macroscopic irreversibility within the framework of classical mechanics was laid in all essentials by Boltzmann's conception of the ergodic character of the trajectories in phase space representing the time evolution of a large mechanical system. This idea was most instructively visualized by Gibbs' famous simile of the mixture of two liquids of different colours, which at the same time – as Ehrenfest especially emphasized – illustrated the fact that all statements of a statistical nature (and thereby exhibiting irreversible features) refer to distributions defined by a *coarse-grained* density in phase space. A rigorous mathematical formulation of these conceptions had to wait for the elaboration of appropriate methods, but was eventually achieved on the basis of the concept of metric transitivity.

The extension of the argument to quantum theory raises a delicate problem: whereas in classical theory the coarse-grainedness of the subdivision of phase space symbolizing the macroscopic specification of the state of the system does not in any way affect the functional representation of the physical quantities, this is no longer so in quantum theory, owing to the uncertainty relations between canonically conjugate mechanical variables. A general method for representing coarse-grained distributions of quantal systems, and for defining the corresponding 'coarse' operators, was proposed by J. von Neumann [14]. With its help, he was able to give a quantal treatment of the mixing process presenting a wide-going analogy to the classical discussion (cf. on this point ref. [9]). However, it proved impossible to eliminate completely from the result all reference to the structure of the coarse distribution: the asymptotic statistical distribution could only be established in the sense of probability, and the proof accordingly involved, in its last stage, an averaging over all possible coarse distributions.

A few years ago, Fierz [15] raised doubts about the adequacy of this averaging process, and his paper stimulated others to a critical re-examination of the problem. It soon became clear that if the averaging was performed from the start (and not as the last stage of the argument, as in von Neumann's proof), every reference to the structure of the system, such as entailed by ergodicity conditions [16], and even to its time evolution [17], was completely swept away from the resulting statistical properties; the latter, which we may call 'pseudo-ergodic', were therefore too general to be of any physical significance [5]. It must be emphasized, however, that neither Fierz nor the other critics have brought forward any direct objection to von Neumann's original approach, which in my opinion still stands as an acceptable foundation for the statistics of quantal systems.

Nevertheless, the critical investigations just mentioned, in which an Italian group of physicists had taken a prominent part, had the happy consequence of prompting these physicists to explore new possibilities, which led them to interesting results. They [20, 21] in fact discovered a class of statistical inequalities very similar in form to the pseudo-ergodic ones derived by an all-too-sweeping application of von Neumann's average, but quite different in mathematical character as well as physical meaning. In this approach, no average is taken over the coarse distribution; instead, one considers statistical averages over the initial-state vectors of the system, specified by the introduction of a definite measure in Hilbert space; and the derivation of the inequalities in question does require that the time evolution of the system be governed by a unitary operator. These inequalities reduce asymptotically to the desired statement of the mixing process when the number of degrees of freedom of the system is sufficiently large. Clearly, no ergodicity conditions are implied in such considerations since the possibility of exceptional behaviour is allowed for by the mathematical formulation; nevertheless, it is easily shown that the approach to the canonical distribution for any system coupled to a thermostat is ensured as a consequence of this coupling. It is therefore apparent that we have here an alternative treatment of the fundamental problem of statistical thermodynamics which is equally satisfying from the mathematical and the physical point of view. It is further instructive to note, as has been pointed out by the Italian physicists [22], that this treatment can be closely paralleled in classical theory, on the basis of a remark originally made by Khinchin. There are thus at the moment two rival modes of analysis of the statistical foundations of thermodynamics; we shall now take up in turn the main points of the general argument and examine in what light they appear from these two points of view.

4.1. *The General Argument*

It would hardly be necessary to recall the well-known chain of reasoning which leads to the definition of the concepts of temperature and associated canonical distribution (cf. *e.g.* ref. [9]), if it were not for the fact that ever recurring criticisms levelled against it on insufficient grounds invite some brief comment. It will be convenient to distinguish three steps in the argument:

(1) Birkhoff's theorem for isolated systems (and its quantal analogue);

(2) the introduction of a coarse-grained density distribution and the 'mixing process' leading to the asymptotic equilibrium distribution;

(3) the derivation of the canonical distribution for a system weakly coupled to a very large 'thermostat'.

In the third step, the total system formed by the given system and the thermostat is treated as isolated, and it then appears that the establishment of the canonical distribution for the given system is a direct consequence of the mere existence of a coupling, however weak, between this system and the thermostat; the coarse-grainedness of the (microcanonical) distribution of the total system has no influence on the result since it concerns only the thermostat variables. These simple features of an analysis going back to Boltzmann suffice to show how unfounded are the doubts recently expressed by Blatt [11] about the adequacy of the idealizations of isolated system and coarse-grained distribution.

The notion of isolated system is an abstraction of exactly the same type as any other physical concept, and its use in a given situation is merely a matter of practical convenience. In a mode of exposition adapted as closely as possible to the physical situation, it is the canonical distribution, and not the microcanonical one, which forms the natural basis for the statistical interpretation of thermodynamics. The necessity of a lack of isolation for bringing about thermal equilibrium, on which both Blatt

[11] and Fierz [15] insist, is indeed directly illustrated by the role of the coupling to the thermostat in the establishment of the canonical distribution; but it can in no way be regarded as a valid objection to the consideration of isolated systems as an auxiliary step in the mathematical derivation of results with direct physical application.

Likewise, the idea of coarse-grained density is a useful mathematical device to describe in a simple and efficient way how incomplete determination of the variables defining the state of the system affects statements which can be made about this state. Since statements of this type about the 'mixing process' which leads to the microcanonical distribution do not require any other presupposition than the law of time evolution of an isolated system, there ought not to be any qualm about their objective character; confusion in this respect can always be traced to some loose use of language or unwarranted intrusion of what Blatt [11] calls 'philosophical' standpoints.

Obviously – as Blatt [11] has rightly emphasized by pointing to the beautiful case of the 'spin echo' experiment – situations may arise in which a more or less 'fine-grained' behaviour of the system is observed. As pointed out in Section 3, such cases are complementary to the typical thermodynamical behaviour and call for an appropriate dynamical treatment. It is not 'philosophical' to conclude from their occurrence to the inadequacy of the introduction of the coarse-grained density to deal with thermodynamical problems.

The arbitrariness in the definition of the coarse-grainedness is only reflected in the expression for the entropy, and it is well known that, for large systems, the thermodynamical behaviour of this quantity is insensitive to its precise statistical definition. For the discussion of the approach to the state of thermal equilibrium and of the fluctuations around this state, along the lines of the H-theorem or by the equivalent consideration of the mixing process, it is clearly necessary, as emphasized by Lorentz [23] in a criticism of Gibbs' entropy definition, to reduce the coarseness of the distribution sufficiently to be able to individualize states of the system deviating in some particular way from the equilibrium configurations. It is therefore difficult to understand Fierz' [15] contention that von Neumann's entropy definition is based on too detailed a specification of the configurations of the system. Again, it should be recognized that the choice of the appropriate degree of coarseness is indicated by the circumstances of the problem under investigation, and is not restricted by any *a priori* considerations.

4.2. *The Ergodic Theorem*

The role of Birkhoff's theorem for classical mechanical systems is to prove the *existence* of a time average of every uniform phase function for almost all initial phases, and to formulate the condition – metric transitivity of the energy surface[6] – under which this time average is independent of the initial phase. For quantal systems, the analogous existence theorem is quite trivial; in any state characterized by the projection operator $D(t)$, the expectation value $\mathrm{Tr}\, D(t)\, f$ of the operator f has a time average because it is an almost periodic function of time, and the condition for this time average to be independent of the initial phases of the energy eigenfunctions in the expansion of the wave function of the system with respect to these eigenfunctions is obviously the non-degeneracy of the energy.

The 'ergodicity' condition of metric transitivity of the energy surface or non-degeneracy of the energy is a very natural one, and its suitability from the point of view of the physical interpretation does not raise any particular query. It is true that we do not know any general criterion by which we could recognize from inspection of the Hamiltonian whether the ergodicity condition is satisfied; but when we are dealing with simple models (not necessarily with few degrees of freedom: linear chains of coupled mass-points, e.g.), the ergodicity condition does prove modestly useful. In any case it is not this condition which has directly been called in question in the recent discussions, but rather the similar one underlying the analysis of the mixing process.

In view of this last problem it is important to realize that we can dispense entirely with the ergodicity condition, when discussing very large systems, by taking advantage of the relatively rare occurrence of appreciable deviations from the equilibrium values of the physical quantities pertaining to such systems. As especially stressed by Khinchin (ref. [24], Section 13), this means in the classical case that even if there is a certain integral of the motion taking different values in different regions of positive measure, one of these regions will have an overwhelmingly larger measure than the others; accordingly, the value of the integral in the latter region will not differ appreciably from the ergodic average taken over the whole energy surface. While the fulfilment of the ergodicity condition guarantees that the statistical weights of the exceptional values, in such an average, are strictly zero, the point is that even if it is not fulfilled, these weights are relatively so small that the practical result is the same.

It is possible to systematize this point of view by a quite general argument, developed by Bocchieri and Loinger [20] and their collaborators [21, 22]. In its classical aspect [22] it may be presented as follows: if the ergodicity condition is satisfied, the time integral $M_t f(P_t)$ of the phase function $f(P_t)$ is independent of the initial phase P, and one immediately obtains its value by taking its ergodic average with respect to the initial phase and permuting the order of the averages: owing to the time invariance of the ergodic measure $d\sigma_P$ on the energy surface, this gives (with $\Sigma = \int d\sigma_P$)

$$(7) \qquad M_t f(P_t) = \frac{1}{\Sigma} \int f(P)\, d\sigma_P,$$

i.e. the equality of time average and ergodic average. If we do not assume the ergodicity of the system, however, we ought to consider as the macroscopic physical quantity corresponding to the function $f(P)$ an integral

$$(8) \qquad F_D \equiv \int F(P)\, dD(P)$$

of the time-average $F(P) \equiv M_t f(P_t)$ over a certain distribution $dD(P)$ (in the sense of Stieltjes) which comprehends only the non-exceptional initial phases: but as the distribution of these phases is not known, the best we can do is to take all possible distributions and average in some definite way over them; for a large system, the dominating contribution to such an average will anyhow come from the distribution over the non-exceptional phases, and the 'distribution average' $\mathscr{D}F_D$ will accordingly represent the corresponding physical quantity with a negligible error. Now, if the averaging process $\mathscr{D}$ over distributions is defined in such a way as to give equal probability to all densities satisfying the normalization condition $\int dD(P) = 1$, it is readily found that $\mathscr{D}dD = d\sigma_P/\Sigma$. This, by Liouville's theorem, again leads for $\mathscr{D}F_D$ to the ergodic average:

$$(9) \qquad \mathscr{D}F_D = \frac{1}{\Sigma} \int f(P)\, d\sigma_P.$$

Passing to the quantal case, in which we deal with time averages of the form $F_D \equiv M_t \operatorname{Tr} D(t) f$, all we have to do is to perform an additional averaging over all possible projection operators $D(0)$; now, this is a process formally (but by no means physically) identical with the von Neumann averaging over coarse distributions, and the techniques elaborated by von Neumann himself [14], Pauli and Fierz [14] and the other physicists who made a critical study of his procedure [16, 17] may be taken over and put to better purpose. For the present problem, denoting this average over the projection operators again by $\mathscr{D}$, we find at once

$$\mathscr{D} F_D = \frac{1}{S} \sum_i \operatorname{Tr} D_i f, \tag{10}$$

where D_i is the projection operator of an eigenstate of the energy, belonging to the eigenvalue E_i, and the summation extends over the S energy eigenstates constituting the energy shell; Equation (10) holds irrespectively of any degeneracy of the energy operator: if the eigenvalue E_i is degenerate, the summation extends over a set of independent eigenstates, each of which gives its contribution $\operatorname{Tr} D_i f$, and the number S is the total number of such independent eigenstates in the energy shell. The right-hand side of Equation (10) is of course closely analogous to the ergodic average occurring in Equation (9).

The preceding considerations may be looked upon as presenting two different mathematical idealizations of the physical concept of macroscopic observation. In the 'ergodic' approach, this concept is represented by the time average, and the ergodicity condition guarantees that the result is independent of the initial phase. The alternative view, renouncing ergodicity conditions, eliminates the dependence on the initial phase by modifying the operation by which the macroscopic quantities are derived from the corresponding phase functions: one has to perform not only a time average, but also a distribution average. In neither case, however, is the physical interpretation just discussed sufficiently justified by a mere existence proof: for such a proof entails no indication of the relative magnitude of the possible errors involved by the fluctuations of the system around the equilibrium configurations in the course of its time evolution. The discussion of these fluctuations, leading to an estimate of such errors, is precisely the object of the second step in the general argument, the analysis of the 'mixing process'.

4.3. *The Mixing Process*

A classical system, even though it satisfied the ergodicity condition on each energy surface, might nevertheless possess some intrinsic periodicity, independent of its energy; in such a case, any initial coarse distribution in an energy shell would reproduce itself after each period, and there would accordingly be no 'mixing', i.e. no tendency towards the establishment of a uniform coarse distribution in the shell. To exclude this type of exceptional system, a further condition is necessary: it is provided by the study of the quadratic fluctuation

$$(11) \qquad M_t\left[\int f(P)\,\mathrm{d}D(P_t) - \frac{1}{\Sigma}\int f(P)\,\mathrm{d}\sigma_P\right]^2,$$

and embodied in Hopf's theorem, according to which the expression (11) vanishes provided that in the phase space corresponding to pairs of phases $P \times P'$ of the system considered the energy surfaces are metrically indecomposable. The quantal analogue of Hopf's theorem has been proved, at least in the sense of probability, by von Neumann [14]; for quantal systems, Hopf's ergodicity condition takes the form of a requirement that there be no 'resonance', i.e. no equalities of energy *differences*, in the set of energy eigenvalues of the system.

Whereas Hopf's theorem is a direct consequence of Birkhoff's theorem and has the same degree of generality (i.e. holds for almost all composite phases), its quantal analogue can only be established in the sense of probability with respect to the coarse distributions. It is conceivable that a more restrictive definition of the coarse distribution could lead to a sharpening of the enunciation of the theorem; attempts in this direction [26] have not so far given any clear answer to the question. However this may be, I should like to stress once more that in its present form von Neumann's theorem does provide a sufficient basis for the derivation of the canonical distribution of systems coupled to a thermostat, which is the decisive result to be achieved.

Let us now turn to the method of distribution averages. As pointed out above, it is essential for the justification of the physical interpretation of the compound averages $\mathscr{D}F_D$ that the fluctuations of the quantities $\int f(P)\,\mathrm{d}D(P_t)$ or $\mathrm{Tr}\,D(t)\,f$ around these averages should only very seldom exceed the order of magnitude of the average itself. Theorems establishing the fulfilment of this requirement have been proved by Bocchieri and Loinger and their collaborators in both the classical and the quantal case. The classical result [22]

$$(12)\qquad \mathscr{D} M_t\left[\int f(P)\,\mathrm{d}D(P_t)-\frac{1}{\Sigma}\int f(P)\,\mathrm{d}\sigma_P\right]^2=0$$

holds for systems of any number of degrees of freedom; the quantal theorem, on the other hand, holds only asymptotically for very large systems, in the following sense [20]. The coarse distribution is characterized by a division of the energy shell into N cells ($\nu=1, 2, \ldots, N$) each defined by a set of s_ν states with projection operators $D_{\omega\nu j}$; the probability of finding the system in the cell ν at the time t is

$$(13)\qquad u_\nu \equiv \sum_{j=1}^{s_\nu} \operatorname{Tr} D(t)\, D_{\omega\nu j},$$

and it suffices[7] to consider its relative quadratic fluctuation

$$(14)\qquad \Delta(t)_\nu \equiv \frac{(u_\nu-(s_\nu/S))^2}{(s_\nu/S)}.$$

For the distribution average of this fluctuation one has the inequality

$$(15)\qquad \mathscr{D}\Delta_\nu(t)<\frac{1}{s_\nu},$$

from which follows

$$(16)\qquad \mathscr{D} M_t \sum_\nu \Delta_\nu(t)<\sum_{\nu=1}^{N}\frac{1}{s_\nu}.$$

For the relative fluctuation of the time average $M_t u_\nu(t)$,

$$(17)\qquad \bar{\Delta}_\nu \equiv \frac{(M_j u_\nu-(s_\nu/S))^2}{(s_\nu/S)^2},$$

similar results are at once derived by application of Schwarz' inequality:

$$(18)\qquad \mathscr{D}\bar{\Delta}_\nu<\frac{1}{s_\nu}$$

and

$$(19) \qquad \mathscr{D} \sum_{\nu} \bar{\Delta}_{\nu} < \sum_{\nu=1}^{N} \frac{1}{s_{\nu}}.$$

The right-hand side of the inequalities (16) and (19) is extremely small if only the s_{ν} may be taken much larger than N. We may then conclude that for almost all initial states of the system, the probability $u_{\nu}(t)$ of finding the system in any given cell is very seldom different from s_{ν}/S, and that its time average $M_t u_{\nu}(t)$ does not differ appreciably from this quantity.

This last statement is all that is needed for the establishment of thermodynamics; but, as shown by Prospèri and Scotti [21], it is possible to analyse the mixing process further by dispensing with the averages and estimating instead the probability of any of the fluctuations $\Delta_{\nu}(t)$ exceeding a given value, or of the time during which this occurs being larger than an assigned value: it is found that these probabilities are negligibly small provided that the smallest s_{ν} is very much larger than $\log S$. This kind of analysis, which could be carried out in the classical case as well, represents an extension to systems of the most general character of the considerations commonly applied to systems of weakly interacting molecules in order to elucidate the meaning of the H-theorem.

When one tries to compare the two methods outlined in the foregoing, one gets the impression that there is little to choose between them. One may incline to think that the restrictive ergodicity conditions have the advantage of excluding from consideration exceptional systems specified by structural properties, such as absence of coupling between different degrees of freedom (black-body radiation) or periodicities (rotators or uncoupled oscillators): such cases are admittedly 'academic', but it is nevertheless instructive, from the epistemological point of view, to have a precise characterization of those structural features which counteract the tendency to thermodynamic irreversibility. In this connexion it must be observed that the significance of the ergodicity conditions could probably[8] also be demonstrated by the consideration of distribution averages.

NOTES

[1] This argument has been developed by Niels Bohr many years ago. I had recently occasion to present it in a course of lectures [2] and in a discussion [3]. See also a previous article by v. Weizsäcker [4], whose philosophical comments, however, especially regarding

the alleged subjectivity of quantum theory, are quite at variance with Bohr's own views.

[2] Cf. on this point ref. [2].

[3] This was first pointed out by Niels Bohr in his Faraday lecture [7] in 1932. See also refs. [8] and [9].

[4] In a recent paper by J. Blatt [11], this point is not sufficiently taken into account.

[5] This critical phase, as well as the kinetic aspects not touched upon here, are surveyed in ref. [18]. For the subsequent developments and further references to the whole problem, see ref. [19].

[6] To avoid unessential complications we assume that the energy integral is the only uniform one; otherwise the energy surface must be replaced by the manifold defined by the values of all the uniform integrals. We recall that a uniform phase function is one which takes the same value for all phases corresponding to the same physical state (e.g. in a molecular system, phases differing only by permutations of the molecules). Metrical transitivity (or, in Khinchin's terminology: indecomposability) is understood in the physical sense, i.e. as the impossibility of a decomposition of the energy surface into two parts of positive measure such that all physically equivalent phases are in the same part. See Khinchin's text-book [24] or ref. [9].

It must not be forgotten that an essential condition for the validity of the whole argument is, in the classical case, the finite extension of the energy surface (a condition paralleled in the quantal case by the discreteness of the energy spectrum). In a recent paper, P. Résibois and I. Prigogine [25] have shown explicitly how the abandonment of this condition may lead to a breakdown of the ergodic character of the system. In this connexion they raise the question whether the usual conception of the approach to equilibrium, embodied in the ergodic theorems, is sufficiently realistic, inasmuch as it implies the intervention of some physical processes at the boundary of the system which, for large systems, might require too much time. This circumstance would not seem, however, to invalidate the physical picture, since the time necessary for reaching equilibrium does not depend on the absolute size of the system, but only on its density.

[7] A typical 'coarse' operator is of the form

$$f = \sum_{\nu=1}^{N} f_\nu \sum_{j=1}^{s_\nu} D_{\omega_{\nu j}},$$

and its quadratic fluctuation

$$\left[\operatorname{Tr} D(t) f - \frac{1}{S} \sum_i \operatorname{Tr} D_i f\right]^2 = \left[\sum_\nu f_\nu \left(u_\nu - \frac{s_\nu}{S}\right)\right]^2$$

is never larger, by Schwarz' inequality, than

$$\frac{1}{S} \sum_i (\operatorname{Tr} D_i f^2) \sum_\nu \Delta_\nu(t).$$

[8] Thus, as Drs. Bocchieri and Loinger have kindly pointed out to me, the non-degeneracy of the Hamiltonian appears to be connected with an extremum property of the fluctuations Δ_ν of the average cell-probabilities $M_t u_\nu(t)$ with respect to variations of the coarse division into cells. More precisely, if $\mathscr{M}$ denotes the von Neumann average over coarse distributions, the quantity $\mathscr{M}\mathscr{D} \sum_\nu \Delta_\nu$ will take its minimum value if the Hamiltonian is non-degenerate; and this value will then, by the usual estimates, be negligibly small. This kind of argument is quite suggestive; but it would seem very difficult to make it rigorous.

BIBLIOGRAPHY

[1] Boltzmann, L., *Nature* **51** (1895), 413; *Wied. Ann.* **60** (1897), 392; *Math. Ann.* **50**

(1898), 325 [reprinted in *Gesammelte Abhand.*, vol. 3, nos. 112, 120, 128]; *Vorlesungen über Gastheorie*, Teil 2, Leipzig 1898, § 90. Cfr. also A. Grünbaum: *American Scientist* **43** (1955), 550.

[2] Rosenfeld, L.: *Cours de l'Ecole d'Eté des Houches*, 1957.

[3] *La structure et l'évolution de l'univers, XIe Conseil de physique Solvay*, Bruxelles 1958, pp. 81–95.

[4] von Weizsäcker, C. F., *Ann. d. Phys.* **36** (1939), 275.

[5] Gibbs, J. W., *Elementary Principles in Statistical Mechanics*, New Haven 1902, p. 150.

[6] Kronig, R., *Journ. Opt. Soc. Amer.* **12** (1926), 547.

[7] Bohr, N., *Journ. Chem. Soc.*, (1932), 376.

[8] Rosenfeld, L., *Norske Vidensk. Selsk. Forhandl.* **31** (1958), Nos. 9, 10. [This volume, p. 484].

[9] Rosenfeld, L., *Acta Phys. Pol.* **14** (1955), 3. [This volume, p. 762].

[10] Landau, L. D. and Lifschitz, E. M., *Statistical Physics*, 1st ed., Oxford 1938, § 40, p. 114; 2nd ed., New York 1958, § 111, p. 350.

[11] Blatt, J., *Progr. Theor. Phys.* **22** (1959), 745.

[12] Bohr, N., in *Symposium on Models in Biology*, Bristol 1959. [In this connection, see Bohr's essay 'Light and Life Revisited' (1962) in *Essays 1958-1962 on Atomic Physics and Human Knowledge* (Vintage Books, New York, 1963) – Ed.].

[13] Rosenfeld, L., *Anais da Acad. Brasiliera de Ciências* **26** (1954), 47.

[14] von Neumann, J., *Zeits. Phys.* **57** (1929), 30; Pauli, W. and Fierz, M., *Zeits. Phys.* **106** (1937), 572.

[15] Fierz, M., *Helv. Phys. Acta* **28** (1955), 705.

[16] Farquhar, I. E. and Landsberg, P. T., *Proc. Roy. Soc., A* **239** (1957), 134.

[17] Bocchieri, P. and Loinger, A., *Phys. Rev.* **111** (1958), 668.

[18] Caldirola, P. and Loinger, A., in *Max Planck-Festschrift 1958*, B. Kockel, W. Macke and A. Papapetrou, eds. (Berlin, Deutscher Verlag der Wissenschaften, 1959), p. 225.

[19] Caldirola, P., *Nuovo Cimento* **14** (1959), 260.

[20] Bocchieri, P. and Loinger, A., *Phys. Rev.* **114** (1959), 948.

[21] Prosperi, G. M. and Scotti, A., *Nuovo Cimento* **13** (1959), 1007.

[22] Albertoni, S., Bocchieri, P. and Loinger, A., *Journ. Math. Physics* **1** (1960), 244.

[23] Lorentz, H. A., *Abhandlungen über theoretische Physik*, Leipzig 1906, No. 11, § 83.

[24] Khinchin, A. I., *Mathematical Foundations of Statistical Mechanics*, Dover, New York, 1949.

[25] Résibois, P. and Prigogine, I., *Bull. Acad. Roy. Belg.* (5) **46** (1960–62), 53–60.

[26] Ludwig, G., *Zeits. Phys.* **150** (1958), 346; **152** (1958), 98.

CHAPTER 7

PART I

DYNAMICAL THEORY OF NUCLEAR RESONANCES

In Memory of Racah

[1968b]

1. Introduction

The physical concept of resonance takes its name from the theory of sound, but it has found a wide range of application in all aperiodic quantal processes at the atomic, nuclear and subnuclear levels. The elucidation of the dymamical features responsible for an effect common to such different types of interaction is a problem that would have appealed to Racah's turn of mind, and it would no doubt have afforded him particular satisfaction to see that methods developed in atomic and molecular physics by one of his nearest collaborators, U. Fano, proved extremely well suited to the analysis of the corresponding nuclear phenomena. It is in this last domain that the mechanism of resonance may be most completely studied. Optical resonances are not particularly informative because of the smallness of the interaction of atomic systems with the radiation field. In collisions between atoms and electrons or in chemical reactions, in which the interactions have no such clear-cut characteristics as the short range of the nuclear forces, resonance effects only become prominent at high energy resolution, or in rather special circumstances, like predissociation; this may explain why the need for a precise definition of the resonant states of atomic and molecular systems was not acutely felt. As to the processes involving so-called elementary particles, on the other hand, where such resonant states play a dominant role, our present inability to treat the relevant interactions hampers their detailed analysis.

A full clarification of the nature and mechanism of nuclear resonances requires the solution of two distinct problems, which may be conveniently called the structural and the dynamical problem. The former consists in establishing the typical resonance structure of the energy dependence of the elements of the *S*-matrix as a consequence of a precise definition of the resonant states: it can be shown that an adequate definition for this purpose is obtained by associating the resonant states with certain complex poles of the S-matrix regarded as a function of the complex energy variable in a specified domain of analyticity. The task of the dynamical problem then is the derivation of the resonance poles of the S-matrix from the Hamiltonian which describes the compound system of interacting nucleons; it is this determination that affords an insight into the mechanism leading to the occurrence of resonant states in such a system. In principle, the dynamical problem, for a given Hamiltonian, is a straightforward one in quantum mechanics; it is in fact a simple generalization of the usual eigenvalue problems to a more extended type of boundary conditions; but it is only with the utmost schematization of the actual interactions by fixed potentials (allowing for no other processes than elastic and inelastic scattering) that such a direct approach can be successful. For a general analysis recourse to a more elaborate 'model' of the nuclear system is unavoidable; this model is also characterized by some schematization of the Hamiltonian, but this is ultimately corrected by the explicit consideration of the 'residual' interactions. The dynamical problem then naturally divides itself into two successive steps: the first is the derivation of an expression for the elements of the S-matrix in terms of the energy, the second is the determination of the resonance poles of these functions in the complex energy plane – a problem which reduces to the solution of a transcendental equation in the energy.

In order to clarify the logical situation, it should be realized that the first step alone does not involve any concept of resonance at all: it leads to a representation– complete in principle – of the energy dependence of the reaction cross-sections, which exhibits the various maxima or other features associated with resonance effects, but it says nothing about the precise relation of such features to resonant states of the system. The answer to this last question can of course first be given when the resonant states of the system have been given a precise characterization, which is the object of the structural problem. Some theories of nuclear reactions which have received wide currency are in fact limited to what we have called the first step of the dynamical problem and are therefore (apart from other defects) essentially incomplete with respect to the analysis of resonance effects; it is true that users of these theories indulge in loose talk about 'resonances' and list under the name 'resonance parameters' certain numbers to which no precise physical meaning can be attached, but such practices are no substitute for logical thinking. In this connexion it must be stressed that resonant states are conceived to be intrinsic properties of the nuclear systems, whose definition must be independent of any arbitrariness in the description of the latter, such as the arbitrary separation of the Hamiltonian into a model Hamiltonian and a residual interaction operator. This is achieved in the structural analysis by characterizing these states with the help of a mathematical idealization; thus, it is ensured that the final step of the dynamical problem, for a given Hamiltonian, will always yield the same values for the resonance parameters, irrespectively of the choice of the model Hamiltonian in the first step of the dynamical analysis. Of course, the values obtained for the resonance parameters will depend on the specification of the total Hamiltonian, involving a phenomenological description of the nuclear interactions which has no secure theoretical foundation and is by no means unique; however, since all other sources of arbitrariness have been eliminated, the parameters defining resonant states obtained by the rational procedure just outlined, provide – like those defining bound states – unambiguous information about the properties of the nuclear forces schematically represented by the phenomenological interaction operators; this cannot be said of the unreliable 'resonance parameters' still in current use.

The present article, in which I shall describe the main features of the dynamical analysis of resonance effects in nuclear reactions, is a sequel to the outline I have recently given of the structural point of view [1]. I shall endeavour to make the exposition self-contained, but must refer the reader to the previous paper for a fuller explanation of the structural aspects. I adopt in all essentials, with some changes which – I hope – are improvements, the method developed by Weidenmüller and his collaborators [2] in Heidelberg, which is directly inspired by Fano's work on cognate atomic problems, alluded to above, and the principle of which was in fact set out by Dirac in one of his early papers. I should like to point out, however, that an equally powerful method, based on a variational principle, is being applied to atomic and molecular problems by Herzenberg and his group [3] in Manchester. It is noteworthy that in spite of all specific differences between atomic and nuclear systems, the treatment is qualitatively the same for both, owing to the analogous roles played by the field of the atomic nuclei in the former case and the model potential in the latter.

2. MODEL HAMILTONIAN AND RESIDUAL INTERACTIONS

Whatever method is used, the basis of the argument is the separation of the total Hamiltonian into a 'model' part and a 'residual interaction': the model is supposed to describe some average behaviour of the system, while the residual interaction is responsible for fluctuations around this average, as well as for aperiodic processes such as the 'reactions' in which the total system of nucleons passes from one 'channel' configuration (in which it is separated into definite fragments in specified relative motion) to another such configuration. The choice of the model is not crucial; every model is in the last resort derived from the shell-model, based on the conception of individual nucleon states in some potential representing the average interaction of any nucleon with all the others (including, when necessary, the Coulomb interaction). The shell-model configurations characterized by the occupation of given individual states by the nucleons of the system, in accordance with the exclusion principle, form an adequate starting-point for a more accurate description of the compound system of nucleons, taking account of the residual interactions. The effect of the latter will be to couple together the shell-model stationary states. The complete spectrum of these states consists of a discrete and a continuous part, forming two complementary subspaces characterized by projection operators B and C, respectively, with $B+C=1$, $BC=CB=0$; the residual interaction operator V is accordingly decomposed into the terms $V_B \equiv BVB$, $V_C \equiv CVC$ and $V_{BC} \equiv BVC+CVB$, which bring about couplings within the discrete subspace, within the continuum, and between the discrete states and the continuum, respectively.

The first coupling does not change the discrete or continuous character of the two subspaces; all it does is to modify the orthonormal basis of the discrete subspace, as well as the shape of the discrete energy spectrum. In particular, it includes the 'pairing' interaction between correlated nucleons, which may have considerable effect on the properties of the bound states of the system. Moreover, it is the handling of the coupling V_B that gives scope for the introduction of various nuclear models; since my aim in this article is only to present the main line of argument, I shall not insist on this point, although it is of great practical interest, but simply take as orthonormal basis of the discrete subspace the set of discrete eigenfunctions ψ_i of the operator $\hat{H} \equiv \hat{H}^{(0)} + V_B$ (where $\hat{H}^{(0)}$ is the shell-model

Hamiltonian); the corresponding eigenvalues of the energy will be denoted by ${}^0E_i$.

Besides this discrete spectrum, the Hamiltonian $\hat{H}$ has a continuous spectrum, unaffected by the residual interaction V_B; an orthonormal basis is conveniently built up of states which asymptotically represent the various channel configurations; each basis function $\phi_c(E)$ will thus be an eigenfunction of the Hamiltonian $\hat{H}^{(0)}$, belonging to the eigenvalue E and further specified (we shall presently explain more fully how) by its asymptotic behaviour, defined in terms of a set of channel quantum numbers c. The effect of the residual coupling V_C upon the continuous spectrum – like that of the coupling V_B on the discrete spectrum – is to reshuffle the basis states $\phi_c(E)$ coupling together the different channels. One may then construct a new orthonormal basis out of eigenfunctions $\psi^{(c)}(E)$ of the Hamiltonian ${}^0H \equiv \hat{H} + V_C$, each of which is again associated with a definite channel c by the asymptotic condition of having an incoming component in channel c only. The bases of discrete eigenfunctions ψ_i and of continuum eigenfunctions $\psi^{(c)}(E)$ of the Hamiltonian ${}^0H$ form together a complete orthonormal set of functions by means of which we can represent all states of the compound system, and analyse the effect of the remaining residual interaction V_{BC}, coupling the bound states with the continuum.

This last coupling will give rise to resonance phenomena whenever the discrete spectrum of bound states of the model system described by the Hamiltonian ${}^0H$ overlaps with the continuous spectrum. This situation, not unfamiliar in atomic and molecular physics, is of general occurrence for nuclear systems. Indeed there will always be shell-model configurations in which one or more nucleons from the inner shells are excited to individual states unoccupied in the ground-state configuration, in such a way that the excitation energy, if it were concentrated on a single nucleon of the uppermost shell, would raise this nucleon to an unbound individual state, thus producing a shell-model state of the continuum. It is the coupling of such a bound state 'embedded in a continuum' to the neighbouring continuum states that establishes a 'resonance' between the former and the latter. As a result, transitions between the continuum states will be enhanced, producing the typical peaks in the cross-sections. The bound state itself – if we imagine for a moment that the residual coupling has been suddenly switched on and look at the time evolution of the process – loses its stationary character and decays essentially according to an exponential law. It thus gives rise to a 'broad' state of the nuclear system, with a 'width' inversely proportional to its life-time; this resonant state formally appears as a complex pole of the S-matrix. In the nuclear case, processes of the type just described are designated as 'compound resonances', because they involve the intermediary of metastable configurations of the whole compound system.

Resonance effects, however, are even exhibited by the shell-model described by the Hamiltonian $\hat{H}^{(0)}$, ignoring all residual couplings. These processes are usually called 'single-particle resonances', since they concern behaviour of a single nucleon in a fixed potential field. The elastic scattering amplitude in any channel c is expressed as $\hat{S}_{cc}^{(0)}-1$ in terms of the element $\hat{S}_{cc}^{(0)}$ of the S-matrix associated with the Hamiltonian $\hat{H}^{(0)}$; regarded as a complex function of the energy, it has in general complex poles corresponding to a typical resonance behaviour of the cross-section. These resonances can in fact be ascribed – albeit somewhat artificially – to a mechanism not essentially different from that outlined above. Imagine all the nucleons of the compound system confined within a finite region of space by an infinitely high potential barrier. The stationary states of this fictitious system have a purely discrete energy spectrum, whereas in the region exterior to the barrier, a nucleon has a continuous spectrum of stationary states at its disposal. Let us now reduce the fictitious potential barrier to its actual shape, given by the shell-model Hamiltonian; this effectively amounts to introducing a coupling between the discrete bound states of the fictitious system and the adjacent continuum states in the exterior region. Thereby the stationary character of the bound states is destroyed and they are transformed into resonant states of the shell-model Hamiltonian. The effect of adding to the shell-model Hamiltonian the residual interaction between the continuum states will be – as we shall see in detail in a later section – merely to displace the single-particle resonance poles on the complex energy surface, without changing their character.

We thus expect that both the bound states and the single-particle resonant states of the shell-model Hamiltonian will give rise to resonant states of the nuclear system, the former leading to compound resonances, the latter to single-particle resonances. From the structural point of view, there is no difference between these two types of resonance, and we have just seen that they have ultimately the same dynamical origin. The distinction between them refers only to the character of the shell-model states from which they originate, and accordingly depends very much on the choice of the parameters defining the model. Thus, a slight change in the strength of the shell-model potential may suffice to change a bound state into a single-particle resonance or vice-versa. Still, once the underlying model Hamiltonian is fixed, the distinction is a clear-cut one, and necessitates separate formal treatments of the two types of resonance.

In the following sections, we shall accordingly proceed by successive steps. In the first place, it will be necessary to set up explicitly the two bases which, as mentioned above, may be applied to the description of the continuum, and to establish the relation between them, which is the starting point for the discussion of the single-particle resonances. It will be more convenient, however, to postpone this last discussion, and to go over first to the analysis of the compound resonance mechanism, in which the essential features appear more clearly and completely. The treatment of the single-particle resonances and of the related non-resonant processes of direct interaction will then round-off this general survey of the dynamical theory of nuclear reactions.

3. Stationary States in the Continuum and Residual Interactions Between Them

In deciding how to define channel configurations, in which the compound system is divided into two or more fragments, we are faced at the outset with a formal difficulty connected with the application of the exclusion principle. From the physical point of view, the most natural procedure would be to include in the description of the individuals states of the fragments as much as possible of the effect of the residual interactions; however, this leads to complications in the antisymmetrization of the total wave-function. These complications are avoided if in the specification of the channel configuration one leaves all the nucleons in their individual shell-model states, and takes account of their binding explicitly in the course of the calculation of the elements of the S-matrix; but this procedure raises other practical problems, inherent in any treatment of systems of more than two particles in relative motion, with regard to the most judicious choice of coordinates. This is a technical point of serious concern, since it is after all the labour of the computations that will set a limit to the usefulness of the one or other approach. The matter being as yet undecided, we shall here dodge the issue by confining the discussion to the well-studied case of channels composed of a target or residual nucleus and a single nucleon in relative motion to it. This will cover both elastic and inelastic scattering of nucleons, as well as (p, n) and (n, p) reactions, – an extensive and varied class of reactions.

For a channel of this restricted type, it is convenient, in constructing the eigenfunction of the residual nucleus, to take into account the effect of the residual interaction between its constituent nucleons. The difficulties we want to avoid would again arise if this eigenfunction would include shell-model configurations containing nucleons in unbound individual states; we explicitly exclude such states of the residual nucleus from consideration. Then, the residual interactions operating in the residual nucleus are included in the Hamiltonian $\hat{H}$, and all the matrix elements of the remaining residual interaction between states of the continuous spectrum will be finite. The asymptotic form of the basis wave-function $\phi_c(E)$ may then be written as the antisymmetrized product of two factors $R_c(r_c, k_c)$ and $\varphi_c(\Omega_c)$. The first factor $R_c(r_c, k_c)$ represents a stationary spherical wave, depending on the radial variable r_c – the distance of the unbound nucleon from the barycentre of the residual nucleus – and the wave number k_c, related to the kinetic energy E_c of the relative motion of the fragments by the expression $E_c=\hbar^2k_c^2/2M_c$ (where M_c denotes the reduced mass of the fragments). The second factor $\varphi_c(\Omega_c)$ is a function of all the other variables (symbolized as Ω_c) which describe the internal configuration of the residual nucleus and the state of the unbound nucleon; it includes the eigenfunction of the residual nucleus, corresponding to an energy ε_c, and the wave-function of the angular variables of the relative motion and the spin and isospin variables of the nucleon.

The total energy E of the compound system determines the kinetic energy of the relative motion by the equation

$$(1) \qquad E_c=E-\varepsilon_c,$$

and it will often be convenient to use E_c instead of E as the energy variable in channel c; the relation (1) will then always be understood to hold. The energy ε_c represents the threshold at which the channel opens, and the continuous spectrum of real energy values extends from ε_c upwards. We fix the zero of the energy E at the threshold of the channel of least internal energy; all other threshold energies ε_c are accordingly positive. The continuous spectrum of the compound system is thus conceived as a superposition of a number of continua, corresponding to the various channels, and starting from the respective thresholds ε_c.

Returning to the radial factor $R_c(r_c, k_c)$, we write it more explicitly (omitting an inessential numerical factor) as

$$(2) \qquad R_c(r_c, k_c) \propto \frac{1}{v_c^{1/2} r_c} \left[e^{-i\delta_c(k_c)} I_c(r_c, k_c) - e^{i\delta_c(k_c)} O_c(r_c, k_c) \right];$$

the functions I_c, O_c are those solutions of the free particle radial equation (including if necessary the Coulomb potential) that behave asymptotically (for real values of k_c) as an incoming and an outgoing wave, respectively. The phase-shift $\delta_c(k_c)$ is due to the distortion of the waves by the shell-model Hamiltonian. As to the factor $v_c^{-1/2}$, where v_c is the velocity of the unbound nucleon, it is required in order to ensure the normalization

$$(3) \qquad \langle \phi_{c'}(E'_{c'}) \mid \phi_c(E_c) \rangle = \delta_{c'c} \delta(E'_{c'} - E_c).$$

We may observe that the form (2) of the radial factor (involving only one wave-number k_c) and the orthonormality relation (3) would cease to hold rigorously if we admitted in the residual nucleus eigenfunction any admixture of unbound shell-model configurations.

Let us now introduce the residual interaction V_C and the corresponding orthonormal basis $\psi^{(c)}(E_c)$ of eigenfunctions of the Hamiltonian ${}^0H$. These functions may be expressed in terms of the basis $\phi_c(E_c)$, to which they are related by a unitary transformation of the form

$$(4) \qquad \psi^{(c)}(E_c) = \sum_{c'} \int dE'_{c'} \phi_{c'}(E'_{c'}) \langle E'_{c'} | \, {}^0a | E_c \rangle .$$

The expansion coefficients $\langle E'_{c'} | \, {}^0a | E_c \rangle$ are solutions of the equations

$$(5) \qquad (E' - E) \langle E'_{c'} | \, {}^0a | E_c \rangle + \\ + \sum_{c''} \int dE''_{c''} \langle \phi_{c'}(E'_{c'}) | \, V | \phi_{c''}(E''_{c''}) \rangle \langle E''_{c''} | \, {}^0a | E_c \rangle = 0.$$

As one would expect, these equations do not determine the coefficients completely, but leave open the possibility of adding to any solution terms proportional to $\delta(E'-E)$; this is just the freedom needed to satisfy the asymptotic boundary conditions imposed upon the functions $\psi^{(c)}(E_c)$. We want the asymptotic radial factor

$$(6) \qquad \sum_{c'} \int \mathrm{d}E'_{c'} R_{c'}(r_{c'}, k_{c'}) \langle E'_{c'} | \, {}^0a | E_c \rangle$$

of $\psi^{(c)}(E_c)$ to contain only outgoing waves, except in channel c. This is achieved by taking $\langle E'_{c'} | \, {}^0a | E_c \rangle$ in the form

$$(7) \qquad \langle E'_{c'} | \, {}^0a | E_c \rangle = \frac{\langle E'_{c'} | \, {}^0T | E_c \rangle}{E' - E^+} + \delta_{c'c} \delta(E'-E),$$

where the notation E^+ indicates that the pole $E'=E$ on the real axis of the E'-plane is displaced by an infinitesimal amount into the upper half of the plane; such a displacement, as is well-known, has the effect of eliminating the incoming wave-components from all the terms in the expression (6) containing the factor $(E'-E^+)^{-1}$. The asymptotic form of the wave-function $\psi^{(c)}(E)$ thus becomes (if $\mathscr{A}$ denotes the antisymmetrization operator, and the summation extends over the open channels c')

$$(8) \qquad \psi^{(c)}(E) \sim \mathscr{A} \Bigg\{ R_c(r_c, k_c)\, \varphi_c(\Omega_c) - \\ - \sum_{c'} \frac{2\pi \mathrm{i}}{v_{c'}^{1/2} r_{c'}} \mathrm{e}^{\mathrm{i}\delta_{c'}(k_{c'})} O_{c'}(r_{c'}, k_{c'})\, \varphi_{c'}(\Omega_{c'}) \langle E_{c'} | \, {}^0T | E_c \rangle \Bigg\},$$

in conformity with our requirement.

This asymptotic expression exhibits a very simple relation between the elements of the matrix $\langle E_{c'} | \, {}^0T | E_c \rangle$ corresponding to continuum states "in the energy shell" (i.e. with $E'=E$) and the elements ${}^0S_{c'c}(E)$ of the S-matrix for the reaction leading from channel c to channel c' at the energy E when the interaction is described by the Hamiltonian ${}^0H$. Indeed, the latter matrix element appears in the asymptotic wave-function when it is written as

$$(9) \qquad \psi^{(c)}(E_c) \sim \mathscr{A} \Bigg\{ \frac{1}{\sqrt{v_c}} \mathrm{e}^{-\mathrm{i}\delta_c(k_c)} \Bigg[\frac{1}{r_c} I_c(r_c, k_c)\, \varphi_c(\Omega_c) - \\ - \sum_{c'} \sqrt{\frac{v_c}{v_{c'}}} \, {}^0S_{c'c} \frac{1}{r_{c'}} O_{c'}(r_{c'}, k_{c'})\, \varphi_{c'}(\Omega_{c'}) \Bigg] \Bigg\},$$

and the comparison with the expression (8), together with Equation (2), shows that

$$(10) \qquad {}^0S_{c'c} = \mathrm{e}^{\mathrm{i}[\delta_{c'}(k_{c'})+\delta_c(k_c)]}\{\delta_{c'c}+2\pi\mathrm{i}\langle E_{c'}|\,{}^0T|E_c\rangle\}.$$

The matrix $\langle E'_{c'}|\,{}^0T|E_c\rangle$, according to its definition (7), is determined as the solution of the inhomogeneous integral equation, replacing Equation (5),

$$(11) \qquad \langle E'_{c'}|\,{}^0T|E_c\rangle+\sum_{c''}\int \mathrm{d}E''_{c''}\langle\phi_{c'}(E'_{c'})|\,V|\phi_{c''}(E''_{c''})\rangle\frac{1}{E''-E^+}\times$$
$$\times\langle E''_{c''}|\,{}^0T|E_c\rangle+\langle\phi_{c'}(E'_{c'})|\,V|\phi_c(E_c)\rangle=0.$$

We see that – just as we can obtain the bound shell-model states ψ_i by taking the residual interaction V_B separately into account – we can treat separately the effect of the residual interaction V_C within the continuum by solving the integral Equation (11). We shall assume for the time being that this last operation is accomplished, and pass directly to the discussion of the residual coupling V_{BC} between the discrete and the continuous spectrum, respectively defined by the orthonormal bases ψ_i and $\psi^{(c)}(E_c)$.

4. Compound Resonances

We are now interested in an eigenfunction $\psi(E)$ of the total Hamiltonian

$$H={}^0H+BVC+CVB$$

belonging to an energy E in the continuum of the compound system, and specified by given amplitudes $y_c(E)$ of the incoming waves in the various channels open at the energy E. We expand this eigenfunction in terms of the complete orthonormal set ψ_i, $\psi^{(c)}(E_c)$:

$$(12) \qquad \psi(E)=\sum_i\psi_i a_i(E)+\sum_{c'}\int \mathrm{d}E'_{c'}\psi^{(c')}(E'_{c'})\,\langle E'_{c'}|\,a|E\rangle,$$

and obtain for the expansion coefficients the following set of coupled equations:

$$(13) \qquad ({}^0E_i-E)\,a_i(E)+\sum_{c'}\int \mathrm{d}E'_{c'}\langle i|\,V|E'_{c'}\rangle\,\langle E'_{c'}|\,a|E\rangle=0,$$

$$(14) \qquad (E'-E)\langle E'_{c'}|\, a|E\rangle + \sum_k \langle E'_{c'}|\, V|k\rangle\, a_k(E) = 0;$$

the matrix elements of the residual interaction are expressed in the representation defined by the bases ψ_i, $\psi^{(c)}(E_c)$.

By the same argument as in the preceding section, we replace $\langle E'_{c'}|\, a|E\rangle$ by

$$(15) \qquad \langle E'_{c'}|\, a|E\rangle = \frac{\langle E'_{c'}|\, \alpha|E\rangle}{E'-E^+} + y_{c'}(E)\,\delta(E'-E)$$

in order to satisfy the asymptotic conditions characterizing the eigenfunction $\psi(E)$. The Ansatz (15) transforms Equations (13) into a set of inhomogeneous linear equations, the terms not containing the unknown quantities being $\sum_{c'} y_{c'}(E)\langle i|\, V|E_{c'}\rangle$. This means that the unknown quantities themselves are linear combinations of the arbitrarily given amplitudes $y_c(E)$, of the form

$$(16) \qquad a_i(E) = \sum_c y_c(E)\, b_i(E, c),$$

$$\langle E'_{c'}|\, \alpha|E\rangle = \sum_c y_c(E)\langle E'_{c'}|\, T|E_c\rangle.$$

The new coefficients are now determined by the equations

$$(17) \qquad ({}^0E_i - E)\, b_i(E, c) + \sum_{c'}\int \mathrm{d}E'_{c'}\langle i|\, V|E'_{c'}\rangle \frac{1}{E'-E^+}\times$$
$$\times \langle E'_{c'}|\, T|E_c\rangle + \langle i|\, V|E_c\rangle = 0,$$

$$(18) \qquad \langle E'_{c'}|\, T|E_c\rangle + \sum_k \langle E'_{c'}|\, V|k\rangle\, b_k(E, c) = 0.$$

Moreover, the asymptotic form of the eigenfunction becomes, on account of Equations (12), (15), (16) and (9),

$$\psi(E) \sim \mathscr{A}\sum_c y_c(E)\left\{\psi^{(c)}(E_c) + \sum_{c'}\int \mathrm{d}E'_{c'}\psi^{(c')}(E'_{c'})\frac{\langle E'_{c'}|\, T|E_c\rangle}{E'-E^+}\right\}$$

$$\sim \mathscr{A}\sum_c y_c(E)\left\{\psi^{(c)}(E_c) - \sum_{c'} 2\pi \mathrm{i}\langle E_{c'}|\, T|E_c\rangle\, \mathrm{e}^{-\mathrm{i}\delta_{c'}(k_{c'})}\right.$$

$$\left.\times \sum_{c''}\frac{1}{v_{c''}^{1/2} r_{c''}}\, {}^0S_{c''c'} O_{c''}(r_{c''}, k_{c''})\, \varphi_{c''}(\Omega_{c''})\right\};$$

by comparing this form with its expression in terms of the S-matrix belonging to the total Hamiltonian H,

$$\psi(E) \sim \mathscr{A} \sum_c y_c(E) \frac{1}{\sqrt{v_c}} e^{-i\delta_c(k_c)} \left[I_c(r_c, k_c)\, \varphi_c(\Omega_c) - \right.$$

$$\left. - \sum_{c'} \sqrt{\frac{v_c}{v_{c'}}}\, S_{c'c} \frac{1}{r_{c'}} O_{c'}(r_{c'}, k_{c'})\, \varphi_{c'}(\Omega_{c'}) \right],$$

we find for the matrix element $S_{c'c}$ the expression

$$S_{c'c} = {}^0S_{c'c} + 2\pi i\, e^{i[\delta_{c'}(k_{c'}) + \delta_c(k_c)]} \times \tag{19}$$
$$\times \sum_{c''} e^{-i[\delta_{c'}(k_{c'}) + \delta_{c''}(k_{c''})]}\, {}^0S_{c'c''} \langle E_{c''}|\, T\, |E_c\rangle .$$

This formula shows that, besides ${}^0S_{c'c}$, we have to determine the elements in the energy shell of the matrix $\langle E'_{c'}|\, T\, |E_c\rangle$. The latter are immediately given by Equation (18) in terms of the $b_k(E, c)$, and we may therefore write for $S_{c'c}$, instead of Equation (19),

$$S_{c'c} = {}^0S_{c'c} - 2\pi i\, e^{i[\delta_{c'}(k_{c'}) + \delta_c(k_c)]} \sum_k \langle \tilde{E}_{c'}|\, V\, |k\rangle\, b_k(E, c), \tag{20}$$

where we have introduced the dual of the state vector

$$|\tilde{E}_c\rangle = \sum_{c'} |E_{c'}\rangle\, {}^0S^*_{c'c}\, e^{i[\delta_{c'}(k_{c'}) + \delta_c(k_c)]}. \tag{21}$$

The physical meaning of this new state vector is readily derived from its asymptotic form. According to Equation (9), and in virtue of the unitarity of the matrix ${}^0S$ (in the subspace of open channels, to which the summations are confined), we have

$$|\tilde{E}_c\rangle \sim -\mathscr{A} \left\{ \frac{1}{\sqrt{v_c}} e^{i\delta_c(k_c)} \left[\frac{1}{r_c} O_c(r_c, k_c)\, \varphi_c(\Omega_c) - \right. \right.$$

$$\left. \left. - \sum_{c'} \sqrt{\frac{v_c}{v_{c'}}}\, {}^0S^*_{c'c} \frac{1}{r_{c'}} I_{c'}(r_{c'}, k_{c'})\, \varphi_{c'}(\Omega_{c'}) \right] \right\},$$

i.e. $|\tilde{E}_c\rangle$ is essentially the transform of $|E_c\rangle$ by time-reversal; it may differ from the time-reverse by its sign, according to the choice of the phases of the factors $\varphi_c(\Omega_c)$, but this sign is in any case the same for all channels which are coupled with each other, and is therefore without interest.

There only remains to determine the coefficients $b_k(E, c)$ from the set of linear equations

$$(22) \qquad \sum_k d_{ik}(E)\, b_k(E, c) + \langle i|\, V\,|E_c\rangle = 0,$$

$$(23) \qquad d_{ik}(E) \equiv ({}^0E_i - E)\,\delta_{ik} - \sum_{c'} \int \mathrm{d}E'_{c'}\, \frac{\langle i|\, V\,|E'_{c'}\rangle\,\langle E'_{c'}|\, V\,|k\rangle}{E' - E^+},$$

derived from Equations (17), (18) by elimination of the $\langle E'_{c'}|\, T\,|E_c\rangle$. The algebraic solution of these equations may be expressed as

$$(24) \qquad b_k = -\frac{1}{D(E)} \sum_i D_{ki}(E)\,\langle i|\, V|E_c\rangle,$$

where $D(E)$ denotes the determinant of the matrix $\{d_{ik}(E)\}$ and $D_{ki}(E)$ the minor of the element $d_{ik}(E)$. We may now re-write Equation (20), combined with Equation (10), in the form

$$(25) \qquad S_{c'c} = \mathrm{e}^{\mathrm{i}\,[\delta_{c'}(k_{c'}) + \delta_c(k_c)]}\,\{\delta_{c'c} + 2\pi\mathrm{i}\,\langle E_{c'}|\, {}^0T|E_c\rangle + \\ + 2\pi\mathrm{i}\,\langle \tilde{E}_{c'}|\, {}^1T|E_c\rangle\},$$

where, according to Equation (23), we have

$$(26) \qquad \langle \tilde{E}_{c'}|\, {}^1T|E_c\rangle = \sum_{i,k} \langle \tilde{E}_{c'}|\, V|k\rangle \frac{D_{ki}(E)}{D(E)} \langle i|\, V|E_c\rangle.$$

Equations (25), (26) solve the first part of the dynamical problem, the determination of the S-matrix as a function of the energy. They will form our starting point for the second part of the problem, the elucidation of the relationship between the singularities of the S-matrix which govern the resonance phenomena and the residual couplings we are considering. In the previous paper [1] dealing with the analytic structure of the S-matrix, it is shown that resonances are associated with a certain family of complex poles $\mathscr{E}_j = E_j - \frac{1}{2}\mathrm{i}\Gamma_j$ of the S-matrix, situated below the real E-axis. In view of the double determination of all channel wave-numbers k_c for each value of the energy, one must consider the distribution of the poles on the whole Riemann surface of the energy and specify by appropriate cuts, starting from the branch-points of energies $E = \mathscr{E}_c$ on the real axis, those portions of the various sheets of this surface that contain the poles of physical interest. We need not repeat here the details of this discussion, but may summarize its conclusion in a sufficiently definite way by saying that the resonance poles are those most easily accessible from the real axis on a 'physical' energy-plane in which cuts are drawn from the branch-points parallel to the imaginary axis.

If we examine from this point of view the analytic properties of the matrix element (25), we may in the first place leave out of consideration the factor $\exp[\mathrm{i}\delta_{c'}(k_{c'})+\mathrm{i}\delta_c(k_c)]$, which represents the effect of the 'potential scattering' by the average field of the shell-model Hamiltonian. It is true that the phase-shift $\delta_c(k_c)$, related to the matrix-element $\hat{S}^{(0)}_{cc}$ of the shell-model S-matrix by the equation $\hat{S}^{(0)}_{cc}=\exp[2\mathrm{i}\delta_c(k_c)]$, has complex poles in the k_c-plane corresponding to the shell-model single-particle resonances; but in the expression (25), it only occurs for a real value of its argument, and has therefore no influence on the integrated cross-sections. The threshold values $k_c=0$, $k_{c'}=0$, which are also critical points, can be approached as closely as we wish, owing to the continuity of the functions $\delta_c(k_c)$. The resonance contributions are contained in the factor between brackets of the expression (25), and the two matrix-elements of ${}^0T$ and ${}^1T$, according to their origin, contain the respective contributions of the single-particle resonances and the compound resonances. We cannot yet discuss the former, but we may analyse the latter with the help of the explicit form (26) of the matrix-element $\langle \tilde{E}_{c'}|\,{}^1T|E_c\rangle$. Indeed, Equation (26) immediately shows that the poles of the S-matrix corresponding to compound resonances are found among the roots of the equation

$$(27)\qquad D(E)=0;$$

it is not superfluous to stress that this is not an algebraic equation of the 'secular' type, but a transcendental equation in the energy, since this variable also appears in each element $d_{ik}(E)$ in the indirect manner exhibit by Equation (23).

Let us examine more closely the form of this equation. The complex integrals occurring in the expression (23) for the matrix-element $d_{ik}(E)$ may be evaluated by replacing the pole $E'=E$ on the real E'-axis and deforming the path of integration in its neighbourhood into an infinitesimal half-circle in the lower half-plane; each integral is then expressed as the sum of its principal part and the contribution from the pole, which is $\mathrm{i}\pi G^{(c')}_{ik}(E)$, with

$$(28)\qquad G^{(c')}_{ik}(E)\equiv\langle i|\,V\,|E_{c'}\rangle\,\langle E_{c'}|\,V\,|k\rangle.$$

Since the matrix $\{G_{ik}^{(c')}\}$ is Hermitean, we thus obtain a separation of the Hermitean and the anti-Hermitean part of $\{d_{ik}\}$:

$$(29) \qquad d_{ik} = ({}^0E_i - E)\,\delta_{ik} - R_{ik}(E) - \mathrm{i}\pi G_{ik}(E),$$

where

$$(30) \qquad R_{ik}(E) = P\int \mathrm{d}E' \sum_{c'} \frac{G_{ik}^{(c')}(E')}{E' + \varepsilon_{c'} - E}, \quad G_{ik}(E) = \sum_{c'} G_{ik}^{(c')}(E);$$

the symbol P indicates that the principal part of the following integral has to be taken. It may further be observed that the $G_{ik}(E)$ are real quantities, and accordingly symmetrical with respect to the indices i, k; this follows from the invariance of the operator V and the state-vectors $|k\rangle$ for time-reversal, which allows us to write, e.g.

$$\langle E_{c'}|\, V\,|k\rangle^* = \langle \tilde{E}_{c'}|\, V\,|k\rangle,$$

and from the relation

$$\sum_{c'} |E_{c'}\rangle\,\langle E_{c'}| = \sum_{c'} |\tilde{E}_{c'}\rangle\,\langle \tilde{E}_{c'}|,$$

which results from Equation (21) and the unitarity of the matrix ${}^0S$.

A considerable simplification occurs if the energy dependence of the $G_{ik}^{(c')}$ is weak enough to be neglected, for in this case the R_{ik} vanish, and Equation (27) takes the form of a secular equation

$$\det\{({}^0E_i - E)\,\delta_{ik} - \mathrm{i}\pi G_{ik}\} = 0.$$

There is then a one-to-one correspondence between the complex roots $\mathscr{E}_j$ and the bound-state energies ${}^0E_i$, and by imagining the residual coupling, absent at first, to increase gradually to its full strength, one can visualize the formation of a resonant state from a shell-model bound state as a displacement on the energy-plane of the point of the real axis corresponding to the latter state towards a final position in the lower half of the physical plane, representing the resonant state. In the general case, a correspondence can still be established between the shell-model states and the complex roots of Equation (27) by following the movement of the latter on the energy surface as the strength of the residual coupling varies. There may then correspond more than one root $\mathscr{E}_j$ to each bound-state energy ${}^0E_i$; but starting from a given resonance pole situated in the physical region of the cut energy-plane and letting the residual coupling decrease, we can uniquely assign this resonant state to a definite bound state of the chosen shell-model.

One may approach the question from a somewhat different angle. The matrix $\{d_{ik}(E)\}$ is a representation, in the basis ψ_i of the discrete subspace, of the operator ${}^1H(E)-E$ acting in this subspace, where ${}^1H(E)$ is given by

$$(31) \qquad {}^1H(E)={}^0H-VC\frac{1}{{}^0H-E^+}CV.$$

The solution of the set of Equations (22) therefore amounts to the inversion of this non-Hermitean operator. This is performed algebraically in the way expressed by Equation (24); alternatively, one may proceed by diagonalizing the operator ${}^1H(E)$ and its adjoint ${}^1H(E)^\dagger$, which is obtained from the expression (31) by replacing E^+ by E^-, i.e. by displacing the pole E into the lower half of the energy plane. The two sets of eigenvectors $|n\rangle$, $|\tilde{n}\rangle$ of ${}^1H$ and ${}^1H^\dagger$ are transforms of each other by time reversal, and form together a biorthogonal set of eigenfunctions to the complex eigenvalues ${}^1\mathscr{E}_n(E)=\varepsilon_n(E)-\frac{1}{2}\mathrm{i}\gamma_n(E)$ and ${}^1\mathscr{E}_n(E)^*$ of the two operators; this means that

$$\langle\tilde{n}\mid m\rangle=\delta_{nm}\langle\tilde{n}\mid n\rangle.$$

The determing factor $\langle\tilde{E}_{c'}|\,{}^1T|E_c\rangle$ in the term of the S-matrix we are discussing becomes, according to Equation (26) and the meaning just explained to the operator ${}^1H(E)$,

$$(32) \qquad \langle\tilde{E}_{c'}|\,{}^1T|E_c\rangle=\langle\tilde{E}_{c'}|\,VB\frac{1}{{}^1H(E)-E}BV|E_c\rangle.$$

If we go over to the basis $|n\rangle$, $\langle\tilde{n}|$ in the discrete subspace, it takes the form

$$(33) \qquad \langle\tilde{E}_{c'}|\,{}^1T|E_c\rangle= =\sum_n\langle\tilde{E}_{c'}|\,V|n\rangle\frac{1}{[\varepsilon_n(E)-E-\frac{1}{2}\mathrm{i}\gamma_n(E)]\,\langle\tilde{n}\mid n\rangle}\langle\tilde{n}|\,V|E_c\rangle.$$

Such a form may give the misleading impression that resonances could be accounted for by means of the complex eigenvalues ${}^1\mathscr{E}_n(E)$ of the operator ${}^1H(E)$, without further consideration of the analytical behaviour of the S-matrix. If we start from the representation $\{d_{ik}(E)\}$ of the operator ${}^1H(E)-E$, the eigenvalues ${}^1\mathscr{E}_n(E)$ are obtained from the equation

$D(E)=0$, provided that one solves it as if it were a secular equation, keeping E fixed in the matrix elements of ${}^1H(E)$; it thus appears, by the same argument as above, that the eigenvalues ${}^1\mathscr{E}_n(E)$ quite generally stand in a one-to-one correspondence to the shell-model bound states ${}^0E_i$. Since the matrix elements $\langle i|\, V|E_c\rangle$ of the residual interaction are the same as those of the total Hamiltonian H, the operator ${}^1H(E)$, restricted to the discrete subspace, may be written in the form $H-HG(E)\,H$, with

$$G(E)=\int \mathrm{d}E' \sum_{c'} \frac{|\psi^{(c')}(E')\rangle\,\langle\psi^{(c')}(E')|}{E'+\varepsilon_{c'}-E^+},$$

which shows that it only depends on the choice of the model Hamiltonian ${}^0H$ through the specification of the continuum basis $|\psi^{(c)}(E)\rangle$. This implies, however, that the eigenvalues ${}^1\mathscr{E}_n(E)$ may depend to that extent on the choice of the model; at any rate, their dependence on the energy deprives them of fundamental significance. The complex poles $\mathscr{E}_j$ of the S-matrix, on the other hand, do not, of course, exhibit any model or energy dependence; they are related to the eigenvalues ${}^1\mathscr{E}_n(E)$ by being roots of the equations ${}^1\mathscr{E}_n(E)-E=0$, equivalent to Equation (27); even if these equations look different for different choices of the shell-model Hamiltonian, their roots $\mathscr{E}_j$ will necessarily be the same. There will be cases when some eigenvalue ${}^1\mathscr{E}_n$ will turn out to be practically energy-independent, and thus accidentally coincide with some pole of the S-matrix, but one has obviously to use the latter for an unambiguous and universal characterization of resonant states as intrinsic properties of the compound system.

This is not to say that the states represented by the eigenfunctions and eigenvalues of the operator ${}^1H(E)$ are entirely deprived of physical meaning. Within the framework of the adopted model description of the various configurations of the compound system, these states represent the outcome of the residual coupling of such configurations to the continuum, – a coupling which changes configurations that would be bound without it into metastable states decaying into the various open channels of the continuum. From this point of view, the 'width function' $\gamma_n(E)/\hbar$ represents the total probability per unit time of all these decay processes, whereas the difference $\varepsilon_n(E)-{}^0E_i$ (for corresponding states) can be interpreted as a mean 'shift' in energy of the shell-model state ${}^0E_i$ owing to the residual coupling.

Our analysis, however, does not stop at this pattern of metastable states, changing with the energy of the compound system, but shows that its formation is governed by an invariable set of resonant states $\mathscr{E}_j$, to which they are related by the identities ${}^1\mathscr{E}_n(\mathscr{E}_j)=\mathscr{E}_j$. We rather regard the shell-model states "embedded in the continuum" as the limiting forms taken by the intrinsic resonant states when the residual coupling is fictitiously made to vanish; as the coupling gradually decreases, each complex pole $\mathscr{E}_j$ in the physical energy plane moves along a definite path towards some point of the real axis representing one of the shell-model states ${}^0E_j$. This gives a clear picture of the connection between the resonant states and the shell-model states, without any intervention of the metastable states. Nor is the determination of these states helpful as an intermediate step for the computation of the resonant state parameters $\mathscr{E}_j$: for this purpose, a direct solution of Equation (27) is more expeditious.

A somewhat more detailed insight into the mechanism of the resonance phenomena can be obtained if one makes more definite assumptions about the form of the nuclear interactions. In particular, it is interesting to pursue the consequences of the commonly adopted assumption that the couplings between pairs of nucleons play the dominant part, whereas direct interactions involving more than two nucleons are negligible. Let us then go back to the eigenstates of the shell-model Hamiltonian $\hat{H}^{(0)}$, i.e. the bound configurations of nucleons in uncoupled individual shell-model states. We can make a distinction among these states according as they are or are not directly coupled by pair interactions to the channel states of the continuum. Only those of the first class are immediately involved in the resonance mechanism; the others take part in the resonance indirectly, by being coupled to the states of the first class through the residual interactions of the type V_B; hence the picturesque name of 'doorway states' given to the states of the first class by Feshbach [4], who (albeit without regard to the intrinsic resonant states) was the first to call attention to their special role in the production of resonance phenomena. The coupling of the states of the second class to the doorway states may itself proceed by a chain of any number of pair interactions, and we thus arrive at a description of the resonance phenomena involving a whole network of coupled shell-model configurations which all concur more or less indirectly to the process through the intermediary of some doorway state directly coupled to the continuum. This is the counterpart in quantal terms of Bohr's classical analogy picturing the formation of the compound nucleus by multiple collisions among the nucleons. It is for this reason that the resonant states so far discussed, which result from the residual coupling between bound shell-model states and the continuum, are called 'compound resonances'.

For a formal treatment of the doorway state mechanism, all we need is a slight modification of the preceding formalism. We now understand our basis in the discrete subspace to consist of the eigenstates of the Hamiltonian $\hat{H}^{(0)}$; we distinguish the doorway states $|\alpha\rangle$, $|\beta\rangle$,... from those of the second class denoted by $|a\rangle$, $|b\rangle$,... and call the respective eigenvalues $\hat{E}^{(0)}_\alpha$,...; $\hat{E}^{(0)}_a$,.... We must then introduce in the definition (23) of the coefficient $d_{ik}(E)$ the additional term $\langle i|\,V\,|k\rangle$, and the equation giving the resonant states becomes, instead of Equation (27),

$$\left|\begin{array}{c|c} \delta_{\alpha\beta}(\hat{E}^{(0)}_\alpha - E) + \langle\alpha|\,V\,|\beta\rangle - \left\langle \alpha\right|\, VC\dfrac{1}{\hat{H}^{(0)} - E^+}CV\,\left|\beta\right\rangle & \langle\alpha|\,V\,|b\rangle \\ \hline \langle a|\,V\,|\beta\rangle & \delta_{ab}(\hat{E}^{(0)}_a - E) + \langle a|\,V\,|b\rangle \end{array}\right| = 0.$$

In this form, the equation clearly exhibits the special role of the doorway states and the indirect coupling of the other states to the continuum by the intermediary of the doorway states.

5. Single-particle resonances and direct interaction processes

We may now turn to the integral Equation (11) for the matrix $\langle E_{c'}'|\,{}^0T\,|E_c\rangle$, which we have still to solve. This will give us the contribution of the continuum couplings to the S-matrix element given by Equation (25). Regarded as a mathematical problem, the solution of this equation is not straightforward, since the usual iteration procedure will not converge for every real value of E. This difficulty is not a deep one; it can be circumvented in various ways. However, our problem is not just the calculation of the S-matrix, but the analysis of the way in which resonances occur; to this particular end, it is not indifferent how one treats the coupling between states of the continuum. Thus, the method adopted by the French group headed by Bloch and Gillet [5] cannot be commended from this point of view: they just replace the continuum by an arbitrary discrete sequence of states, and thus reduce the problem to purely algebraic operations, which are abandoned to a computer. Such a treatment 'à la hussarde' (an approach which may relieve a pressing need, but does not afford deeper satisfaction) leaves in the dark interesting aspects which a more systematic investigation brings out. We shall therefore follow a slower course. Our first task will be to determine the domain of convergence, on the energy surface, of the iteration series for the inverse of the operator $1+V_C(\hat{H}-E^+)^{-1}$ in the continuum subspace; it will turn out that this discussion will give us the means of carrying out the inversion in the exceptional regions as well.

Let us first consider the operator $V_C(\hat{H}-E^+)^{-1}$ in the upper-half of the physical energy plane, i.e. in the region characterized by positive imaginary parts of all channel wave-numbers k_c. In this region, we can define the eigenfunctions of the operator, which behave asymptotically in all channels as outgoing waves only; and regarding them, as well as the associated complex eigenvalues, as functions of the energy, it will prove possible to continue their definition to the lower half of the physical plane (including the real axis), which is the region of interest to us. It will be convenient to start with the 'left' eigenfunctions $\langle\tilde{\psi}_n|$; the tilde serves to indicate that the dual state vectors $|\tilde{\psi}_n\rangle$ are the transforms by time-reversal of a set of state vectors $|\psi_n\rangle$ which likewise asymptotically consist of outgoing waves. Calling the corresponding eigenvalues $-\lambda_n$, we have, for each value of the energy in the region indicated,

$$(34) \qquad \langle\tilde{\psi}_n(E)|\left\{\lambda_n(E)+V_C\frac{1}{\hat{H}-E^+}\right\}=0.$$

Writing this equation in the form

$$(35) \qquad \langle\tilde{\psi}_n(E)|\,\{\hat{H}-E+\eta_n(E)\,V_C\}=0, \quad \eta_n(E)\equiv\frac{1}{\lambda_n(E)},$$

we see that we can interpret $\langle\tilde{\psi}_n(E)|$ as an eigenfunction, belonging to the eigenvalue E, of the Hamiltonian corresponding to the non-Hermitean interaction operator $\eta_n(E)\,V_C$, and further specified by the asymptotic condition of comprising only outgoing waves. Taking the adjoint and the time-reverse of Equation (35), on the assumption that $\hat{H}$ and V_C are self-adjoint and invariant for time-reversal, we get

$$\{\hat{H}-E+\eta_n(E)\,V_C\}\,|\psi_n(E)\rangle=0, \tag{36}$$

which may also be written

$$\left\{\lambda_n(E)+\frac{1}{\hat{H}-E^+}\,V_C\right\}|\psi_n(E)\rangle=0. \tag{37}$$

Acting from the left with the operator V_C, and putting

$$|\phi_n(E)\rangle=V_C|\psi_n(E)\rangle, \tag{38}$$

we derive from Equation (37)

$$\left\{\lambda_n(E)+V_C\frac{1}{\hat{H}-E^+}\right\}|\phi_n(E)\rangle=0. \tag{39}$$

On the assumption that the operator V_C is such that no state vector $|\phi_n(E)\rangle$, defined by Equation (38), identically vanishes, we conclude from Equation (39) that these vectors $|\phi_n(E)\rangle$ are the set of 'right' eigenfunctions of our operator $V_C(\hat{H}-E^+)^{-1}$, belonging to the same eigenvalues $-\lambda_n(E)$. Assuming further that these eigenstates present no degeneracy, it is easy to prove the biorthogonality of the sets $|\phi_n\rangle$, $\langle\tilde{\psi}_n|$:

$$\langle\tilde{\psi}_m(E)|\,\phi_n(E)\rangle=\delta_{mn}V_n(E), \tag{40}$$

with the normalization factor

$$V_n(E)=\langle\tilde{\psi}_n(E)|\,V_C|\psi_n(E)\rangle. \tag{41}$$

Now, since in our case the operator V_C is of limited range, and may even be taken to vanish outside a finite region of space, its matrix elements $\langle\tilde{\psi}_m(E)|\,V\,|\psi_n(E)\rangle$ are finite for any value of the complex energy variable, and the biorthogonality relations (40), (41), whose validity was originally restricted to the upper half of the physical energy plane, may therefore be continued to the whole physical plane. The same continuation may be achieved for the eigenvalues $\lambda_n(E)$; indeed, by multiplication from the left with $\langle\tilde{\psi}_m(E)|$, we derive from Equation (39) the relation

$$\lambda_n(E)\,V_n(E)\,\delta_{nm}+\left\langle\tilde{\psi}_m(E)|\,V_C\frac{1}{\hat{H}-E^+}\,V_C|\psi_n(E)\right\rangle=0, \tag{32}$$

in which the matrix element of the operator $V_C(\hat{H}-E^+)^{-1}V_C$ is finite on the whole physical plane, excepting isolated points which are poles of the functions $\lambda_n(E)$.

We are now prepared to discuss the integral Equation (11), which we write in operator form, in the continuum subspace,

$$(43)\qquad \left(1+V_C\frac{1}{\hat{H}-E^+}\right){}^0T+V_C=0.$$

Let us single out some group of eigenstates $|\phi_n(E)\rangle$, $\langle\tilde{\psi}_n(E)|$ by means of its projection operator

$$(44)\qquad R(E)=\sum_n R_n(E),\qquad R_n(E)=\frac{1}{V_n(E)}|\phi_n(E)\rangle\,\langle\tilde{\psi}_n(E)|,$$

and decompose the operators ${}^0T$ and V_C into two parts:

$$(45)\qquad {}^0T={}^0T_R+{}^0T_D,\qquad V_C=V_R+V_D,$$

with

$$(46)\qquad {}^0T_R=R\,{}^0T,\qquad V_R=RV.$$

Equation (43) yields separate equations for each component of ${}^0T_R$ and for ${}^0T_D$:

$$(47)\qquad (1-\lambda_n(E))\,R_n(E)\,{}^0T+R_n(E)\,V=0,$$

$$(48)\qquad \left(1+V_D\frac{1}{\hat{H}-E^+}\right){}^0T_D+V_D=0.$$

Whatever the value of $\lambda_n(E)$, Equation (47) immediately solves the inversion problem for ${}^0T_R$. At the same time, it shows that the iteration procedure (which here amounts to expressing $(1-\lambda_n)^{-1}$ as a power expansion) will converge provided $|\lambda_n|<1$, i.e. $|\eta_n(E)|>1$. In other words, the exceptional regions on the energy plane are those mapping the interior of the circles $|\eta_n|=1$ in the η_n-planes; and among such regions, only those which include some interval of the real axis require attention. In any case, the exceptional regions will correspond to a finite number of eigenstates $|\phi_n\rangle$, $\langle\tilde{\psi}_n|$; it will suffice to project these out according to the above pattern and obtain their contributions to ${}^0T$ by means of equations of the form (47). The Equation (48) for the remaining part ${}^0T_D$ will then be soluble by iteration. It may be advisable to project out more terms than strictly necessary, in order to improve the convergence of the iteration by reducing the remaining coupling V_D.

The distribution of the exceptional regions on the physical energy plane is easily discussed. The centres of the circles $|\eta_n|=1$ correspond to roots of the equations $\eta_n(E)=0$ situated on this part of the energy surface. As shown by Equation (36), these are the single-particle resonant states $\hat{\mathscr{E}}_i$ of the model system of nucleons described by the Hamiltonian $\hat{H}$, and appearing as poles of the scattering amplitudes $\hat{S}_{cc}-1$, in the manner explained in Section 3. Together with the bound states ${}^0E_i$ they form the set of poles of physical interest of the S-matrix associated with the Hamiltonian $\hat{H}$, from which the resonant states of the Hamiltonian ${}^0H$ originate. Indeed, as the residual coupling V_C is introduced and increases to its full strength, the pole $\hat{\mathscr{E}}_i$ moves, according to Equation (36), towards a new position ${}^0\mathscr{E}_l$, which is a root of the equation $\eta_n(E)=1$ and a resonant state of the Hamiltonian ${}^0H$. The boundary of the exceptional region around $\hat{\mathscr{E}}_i$ passes through the point ${}^0\mathscr{E}_l$, and although no general statement can be made about the range of variation of the distances $|{}^0\mathscr{E}_l-\hat{\mathscr{E}}_i|$ or the shape of the boundaries, examples indicate that no considerable fluctuations in the size of the exceptional regions or irregularities in their shape should be expected. Accordingly, only those exceptional regions around poles near the real axis – which are those giving rise to marked resonance effects – are likely to include segments of the real axis. We thus see that the term ${}^0T_R$ separated from the operator ${}^0T$ by the procedure described above corresponds precisely to the most important resonant states of the single-particle type.

We are now in a position to write down the solution of the integral Equation (11) in a form susceptible to direct physical interpretation. We select a set of eigenstates $|\phi_n(E)\rangle$, $\langle\tilde{\psi}_n(E)|$ embodying (in the sense just explained) the important single-particle resonances, and leaving over in the residual interaction V_C a remainder V_D small enough to ensure a rapid convergence of the iteration series solving Equation (48). According to Equation (47) – which has to be written in the representation of basis $\phi_c(\mathrm{E}_c)$ in order to conform with Equation (11) – and Equations (44), (45), we get immediately

$$(49)\qquad \langle E_{c'}'|\,{}^0T_R\,|E_c\rangle = -\sum_n \langle\phi_{c'}(E_{c'}')|\,V\,|\psi_n(E)\rangle \times \\ \times\frac{1}{[1-\lambda_n(E)]\,V_n(E)}\langle\tilde{\psi}_n(E)|\,V\,|\phi_c(E_c)\rangle.$$

As to ${}^0T_D$, we shall here just retain the lowest approximation

$$\text{(50)} \qquad \langle E'_{c'}|\, {}^0T_D \,|E_c\rangle \approx -\langle \phi_{c'}(E'_{c'})|\, V_D \,|\phi_c(E_c)\rangle .$$

A slight transformation of Equation (49) is desirable in order to isolate the contributions from single-particle resonances from a non-resonant background on which they are superposed. In fact, writing

$$\frac{1}{1-\lambda_n} = \frac{\lambda_n}{1-\lambda_n} - 1 = \frac{1}{1-\eta_n} - 1,$$

we get

$$\text{(51)} \qquad \langle E'_{c'}|\, {}^0T_R \,|E_c\rangle = -\langle \phi_{c'}(E'_{c'})|\, V_R \,|\phi_c(E_c)\rangle +$$

$$+ \sum_n \langle \phi_{c'}(E'_{c'})|\, V \,|\psi_n(E)\rangle \frac{1}{[1-\eta_n(E)]\, V_n(E)} \langle \tilde{\psi}_n(E)|\, V \,|\phi_c(E_c)\rangle .$$

Now, writing Equation (36) in the form

$$\text{(52)} \qquad ({}^0H - E)\,|\psi_n\rangle = (1-\eta_n(E))\, V_C \,|\psi_n\rangle ,$$

and using Equations (38) and (44), we derive from it

$$\text{(53)} \qquad |\psi_n\rangle \frac{1}{(1-\eta_n)\, V_n} \langle \tilde{\psi}_n| = \frac{1}{{}^0H - E^+} R_n = \tilde{R}_n^{\dagger} \frac{1}{{}^0H - E^+} R_n .$$

This allows us to express the last term in Equation (51) as the matrix element of an operator:

$$\text{(54)} \qquad \langle E'_{c'}|\, {}^0T_{R_1} \,|E_c\rangle = \langle \phi_{c'}(E'_{c'})|\, V\tilde{R}^{\dagger} \frac{1}{{}^0H - E^+} RV \,|\phi_c(E_c)\rangle .$$

In the operator form (54), the matrix element in the energy shell $\langle E_{c'}|\, {}^0T_{R_1} \,|E_c\rangle$, which enters into the expression (25) for the S-matrix of the total Hamiltonian, is entirely parallel to the compound resonance term (32). In its more explicit form given by Equation (51), it can be compared to the corresponding expression (33). The complex poles ${}^0\mathscr{E}_l$ defining the single-particle resonances are roots of the equations $\eta_n(\mathrm{E}) = 1$, corresponding to the equations ${}^1\mathscr{E}_n(E) - E = 0$ which determine the compound-resonance poles. The background term $-\langle \phi_{c'}(E_{c'})|\, V_R \,|\phi_c(E_c)\rangle$ can be combined with the remaining term $\langle E_{c'}|\, {}^0T_D \,|E_c\rangle$, which has the same form, and arises both from non-resonant processes and from broad resonances corresponding to poles far away from the real axis, not included in the subspace R of the selected states. To the approximation (50), the general background effects are thus given by the simple expression

$$(55) \qquad \langle E_{c'}| \, {}^0T_{D_1} \, |E_c\rangle = -\langle \phi_{c'}(E_{c'})| \, V \, |\phi_c(E_c)\rangle.$$

Thus, our analysis, which in this respect resembles an argument first put forward by Mac Donald, clarifies the relationship between resonance and direct interaction processes and indicates how the treatment of the latter, if necessary, can be carried out to higher approximations by means of the further terms of the iteration series solving Equation (48) for ${}^0T_D$. The symmetry of the S-matrix with respect to the entrance and exit channels is readily verified by transforming the expressions (32), (54) and (55) for its different terms according to the scheme

$$\langle \tilde{c}'| \, O \, |c\rangle = \langle c'| \, O^\dagger \, |\tilde{c}\rangle^* = \langle \tilde{c}| \, O \, |c'\rangle.$$

6. Applications

Actual calculations according to the method here outlined proceed in three stages: (1) the choice of the shell-model and the construction of the basis ψ_i, $\phi_c(E)$; (2) the construction of the basis $\psi^{(c)}(E)$; (3) the calculation of the S-matrix and the determination of its resonance poles. The first stage needs no comment. The second stage requires the calculation of the whole ${}^0T$ matrix (and not only of its elements in the energy shell); this cannot be carried out directly by following the ideal pattern of the preceding section, since finding the eigenstates $|\phi_n\rangle$, $\langle\tilde{\psi}_n|$ would be tantamount to solving an eigenvalue problem of the type (36), involving an interaction operator V_C of complicated structure, which is precisely what we want to avoid. Accordingly, one will replace in practice V_C by some simple potential $V_C^{(0)}$, for which the eigenstates $|\phi_n^{(0)}\rangle$, $\langle\tilde{\psi}_n^{(0)}|$ and their eigenvalues $\eta_n^{(0)}$ can be evaluated in terms of the total energy E. This gives an approximate determination of the single-particle resonances of interest by appropriate roots of the equations $\eta_n^{(0)}(E)=1$.

We still decompose ${}^0T$ and V_C into two parts:

$$(56) \qquad {}^0T = {}^0T^{(0)} + {}^0T^{(1)}, \quad V_C = V_C^{(0)} + V_C^{(1)},$$

and we directly identify $V_C^{(0)}$ with $R^{(0)}V_C^{(0)}$, where

$$(57) \qquad R^{(0)} = \sum_n R_n^{(0)}, \quad R_n^{(0)} = \frac{1}{V_n^{(0)}} |\phi_n^{(0)}\rangle \langle \tilde{\psi}_n^{(0)}|,$$

$$(58) \qquad |\phi_n^{(0)}\rangle = V_C^{(0)} |\psi_n^{(0)}\rangle, \quad V_n^{(0)} = \langle \tilde{\psi}_n^{(0)}| V_C^{(0)} |\psi_n^{(0)}\rangle.$$

Now, ${}^0T^{(0)}$ is defined as the solution of the equation

$$(59) \qquad \left(1 + V_C^{(0)} \frac{1}{\hat{H} - E^+}\right) {}^0T^{(0)} + V_C^{(0)} = 0;$$

it is of the form $R^{(0)}\,{}^0T^{(0)}$ (but this is no longer $R^{(0)}\,{}^0T$), and the components $R_n^{(0)}\,{}^0T^{(0)}$ are given by equations similar to Equation (17),

$$(60) \qquad (1 - \lambda_n^{(0)}) R_n^{(0)}\,{}^0T^{(0)} + R_n^{(0)} V_C^{(0)} = 0.$$

For the part ${}^0T^{(1)}$ we now obtain by some simple operator algebra, using Equations (43), (45) and (59),

$$(61) \qquad \left(1 + V_C^{(0)} \frac{1}{\hat{H} - E^+}\right) {}^0T^{(1)} + V_C^{(1)} \left(\frac{1}{\hat{H} - E^+}\,{}^0T + 1\right) = 0;$$

moreover, by taking the adjoint and the time-reversed of Equation (59) we derive an equation for the transformed matrix ${}^0T^{(0)T}$:

$${}^0T^{(0)T} \left(1 + \frac{1}{\hat{H} - E^+} V_C^{(0)}\right) + V_C^{(0)} = 0,$$

from which it follows that

$$\left(1 + {}^0T^{(0)T} \frac{1}{\hat{H} - E^+}\right)\left(1 + V_C^{(0)} \frac{1}{\hat{H} - E^+}\right) = 1,$$

and that therefore Equation (61) yields

$$(62) \qquad {}^0T^{(1)} = -\left(1 + {}^0T^{(0)T} \frac{1}{\hat{H} - E^+}\right) V_C^{(1)} \left(\frac{1}{\hat{H} - E^+}\,{}^0T + 1\right).$$

The matrix-element $\langle E'_{c'}|\,{}^0T^{(1)}\,|E_c\rangle$ has to be calculated from Equation (62) in the representation of basis $\phi_c(E_c)$. According to Equations (4) and (7), one has

$$(63) \qquad \left(\frac{1}{\hat{H} - E^+}\,{}^0T + 1\right) |\phi_c(E_c)\rangle = |\psi^{(c)}(E_c)\rangle;$$

similarly, we define a new basis $\chi^{(c)}(E_c)$ in the continuum subspace by

$$(64) \qquad \left(\frac{1}{\hat{H}-E^{+}}\,{}^{0}T^{(0)}+1\right)|\phi_c(E_c)\rangle=|\chi^{(c)}(E_c)\rangle;$$

this gives

$$(65) \qquad \langle\tilde{\chi}^{(c)}(E_c)|=\langle\phi_c(E_c)|\left(1+{}^{0}T^{(0)T}\frac{1}{\hat{H}-E^{+}}\right),$$

and accordingly, from Equation (62),

$$(66) \qquad \langle E'_{c'}|\,{}^{0}T^{(1)}\,|E_c\rangle=\langle\tilde{\chi}^{(c')}(E'_{c'})|\,V_C^{(1)}\,|\psi^{(c)}(E_c)\rangle.$$

This is an exact expression, generalizing Equation (50).

Equation (66) cannot be used directly, since it involves the unknown basis $\psi^{(c)}(E_c)$; however, it may be made the framework of an iteration process leading to an approximate determination of increasing accuracy of both ${}^{0}T^{(1)}$ and the basis $\psi^{(c)}(E_c)$. With the help of the basis $\chi^{(c)}(E_c)$ calculated by Equation (64), we can indeed start from the approximate value

$$\langle E'_{c'}|\,{}^{0}T^{(1)}\,|E_c\rangle\approx\langle\tilde{\chi}^{(c')}(E'_{c'})|\,V_C^{(1)}\,|\chi^{(c)}(E_c)\rangle.$$

The corresponding approximation of the total matrix ${}^{0}T$, inserted in Equation (60), yields a first approximation to the required basis $\psi^{(c)}(E_c)$. If necessary, this first result may be used to obtain from Equation (66) an improved value of ${}^{0}T^{(1)}$, which again will lead, through Equation (63), to a more accurate determination of the basis $\psi^{(c)}(E_c)$.

I shall leave it at these general indications; only the study of the papers describing the application of the method to concrete problems can enable the reader to appreciate its power. A particularly impressive case is that of elastic and inelastic neutron scattering on ^{15}N, treated by the Heidelberg group [6]. An analysis of elastic neutron scattering on ^{12}C by Lovas [7], carried out by a somewhat different method, but essentially on the same principles, affords an interesting example of the comparison between metastable, energy-dependent states and corresponding poles of the S-matrix; in Lovas' case, one of the metastable states results from a compound mechanism, the other is of the single-particle type. The most extensive application of the method at Heidelberg so far is the very thorough investigation [8] of the analogous bound and resonant states in the spectra of isobaric nuclei, differing by substitution of neutrons for protons and vice-versa. Mention should also be made, of course, of the results obtained by the French group [5], as well as of independent investigations along essentially the same lines carried out by Mac Donald, Balashov and many other physicists [9]; a discussion of this work will be found in papers of the Heidelberg group [2, 8, 10]. The theory has recently received two important extensions, which, when fully explored, will give the account of nuclear reaction processes a much-needed completion. The first aspect calling for renewed examination in the light of the analysis here outlined is the statistical distribution in energy of the large number of resonances revealed by high-resolution techniques; the search for the adequate formulation of the statistical properties underlying the observed features, extensively pursued from various points of view by Moldauer [11], has been taken up again by Mahaux *et al.* [12], and some interesting conclusions have already been reached. A related problem is then the formulation of a simplified theory of the processes observed at moderate resolution, when the effects of many resonances are superposed to give a smoothed out variation of the cross-sections with energy; considered in their time evolution, such processes are those involving direct interaction of short duration between projectile and target. An extreme model representation of these direct interaction processes is suggested by the optical analogy of the complex refractive index, which includes the absorption by the scatterer: it is found that the addition of a small imaginary part to the shell-model potential is sufficient to reproduce the general features of the direct reactions. For a more accurate description, however, the optical model needs to be supplemented by contributions from compound nucleus formation, with the correct phase relationships between them. This can only be achieved by a theory starting from first principles, and in which a rational definition of the optical model potential is built in. The first to give a clear formulation and an elegant discussion of the problem was Mac Donald [9], who, however, did not pursue the matter in detail. A renewed analysis by the Heidelberg group [12] on the basis of the techniques here described shows that they lead to a complete solution, allowing of practical computations, and giving the precise justification and limitation of the *ad hoc* procedures widely used in the interpretation of experimental data, but hitherto lacking a firm foundation. The elucidation of the still outstanding questions concerning the extension of the method to many-fragment channels will reveal its full scope and possible limitations.

BIBLIOGRAPHY

[1] Rosenfeld, L., in *Proc. Intern. Conference on Elementary Particles, Kyoto, 1965*, p. 182.

[2] Weidenmüller, H. A., *Nucl. Phys.* **75** (1966), 189;
Weidenmüller, H. A. and Dietrich, K., *Nucl. Phys.* **83** (1966), 332;
Glöckle, W., Hüfner, J., and Weidenmüller, H. A., *Nucl. Phys.* **A 90** (1967), 481.

[3] Herzenberg, A. and Mandl, F., *Proc. Roy. Soc. (London)* **A 270** (1962), 48; **A 274** (1963), 253;
Herzenberg, A., Kwok, K. L., and Mandl, F., in *Proc. Third Intern. Conference on the Physics of Electronic and Atomic Collisions, London, 1963*, North-Holland Publ. Co., Amsterdam, 1964, p. 98;
Bardsley, J. N., Herzenberg, A., and Mandl, F., in *Proc. Third Intern. Conference on the Physics of Electronic and Atomic Collisions, London, 1963*, North-Holland Publ. Co., Amsterdam, 1964, p. 415;
Herzenberg, A., Kwok, K. L., and Mandl, F., *Proc. Phys. Soc. (London)* **84** (1964), 345, 477;
Herzenberg, A., Sherrington, D., and Süveges, M., *Proc. Phys. Soc. (London)* **84** (1964), 465;
Kwok, K. L. and Mandl, F., *Proc. Phys. Soc. (London)* **86** (1965), 501;
Mandl, F., *Proc. Phys. Soc. (London)* **87** (1966), 871;
Phythian, R., *Proc. Phys. Soc. (London)* **87** (1966), 171;
Bardsley, J. N., Herzenberg, A., and Mandl, F., *Proc. Phys. Soc. (London)* **89** (1966), 305, 321;
Mandl, F., *Proc. Phys. Soc. (London)* **90** (1967), 913;
Bardsley, J. N., *Proc. Phys. Soc. (London)* **91** (1967), 300.

[4] Feshbach, H., in *Comptes-rendus du Congrès intern. de physique nucléaire, Paris, 1964*, Vol. 1, pp. 227–251, especially p. 248.

[5] Bloch, C. and Gillet, V., *Phys. Letters* **16** (1965), 62; **18** (1965), 58;
Gillet, V., Melkanoff, M. A., and Raynal, J., *Nucl. Phys.* **A 97** (1967), 631;
Raynal, J., Melkanoff, M. A., and Sawada, T., *Nucl. Phys.* **A 101** (1967), 369.

[6] Ebenhöh, W., Glöckle, W., Hüfner, J., and Weidenmüller, H. A., *Z. Physik* **202** (1967), 302.

[7] Lovas, I., *Nucl. Phys.* **81** (1966), 353.

[8] Weidenmüller, H. A., *Nucl. Phys.* **85** (1966), 241;
Mahaux, C. and Weidenmüller, H. A., *Nucl. Phys.* **89** (1966), 33;
Mahaux, C. and Wiedner, C. A., *Nucl. Phys.* **A 93** (1967), 327;
Mahaux, C. and Weidenmüller, H. A., *Nucl. Phys.* **A 94** (1967), 1;
Weidenmüller, H. A., *Nucl. Phys.* **A 99** (1967), 269, 289;
see also Lipperheide, R., *Nucl. Phys.* **A 105** (1967), 545.

[9] Mac Donald, W. M., *Nucl. Phys.* **54** (1964), 393; **56** (1964), 636;
Balashove, W. W., *et al.*, *J. Nucl. Phys. (USSR)* **2** (1965), 643;
Bauer, M. and Prats, F., *Nucl. Phys.* **89** (1966), 230;
other references cited in refs. [2], [8], [10] and [13].

[10] Mahaux, C. and Weidenmüller, H. A., *Nucl. Phys.* **A 97** (1967), 378.

[11] Moldauer, P. A., *Phys. Rev.* **135** (1964), B 642; *Rev. Mod. Phys.* **36** (1964), 1079; *Phys. Rev.* **136** (1964), B 947; *Phys. Letters* **8** (1964), 70; in *Proc. Intern. Conference Nucl. Physics, Gatlinburg, 1966*; in *Annual Meeting American Physical Society, New York, Jan. 1967*; *Phys. Rev. Letters* **18** (1967), 249; *Phys. Rev.* **157** (1967), 907.

[12] Mahaux, C. and Weidenmüller, H. A., *Nucl. Phys.* **A 91** (1967), 241;
Lejeune, A. and Mahaux, C., *Z. Physik* **207** (1967), 35.
[13] Hüfner, J., Mahaux, C., and Weidenmüller, H. A., *Nucl. Phys.* **A 105** (1967), 489.

NOTE ADDED IN PROOF

The reader's attention is called to a forthcoming paper by K. Dietrich and K. Hara, in which the 'random phase approximation' method is extended so as to allow the inclusion in the model Hamiltonian of the ground state correlations and the collective features of the excited configurations.

PART II

COUPLING BETWEEN COMPOUND AND SINGLE-PARTICLE RESONANCES

Addendum to 'Dynamical Theory of Nuclear Resonances'

[1968c]

Although it proved possible to develop the calculation of the general reaction amplitude (25) in such a way that the contributions from compound and single-particle resonances can be treated separately, the equivalent expression (19) for this amplitude clearly shows that the two kinds of resonance are actually coupled with each other, and it is interesting to exhibit this coupling more explicitly.

To this end, we shall in the first place ascertain that the single-particle resonances $\hat{\mathscr{E}}_i$ of the shell-model Hamiltonian are completely eliminated from the final expression (19) for $S_{c'c}$ as well as from the expression (10) for the amplitude ${}^0S_{c'c}$ on which $S_{c'c}$ depends. To begin with, it is convenient to absorb the phase factors $e^{i\delta_c}$, $e^{-i\delta_c}$ into the respective wave-functions $|\phi_c\rangle$, $\langle\phi_c|$; $|\psi^{(c)}\rangle$, $\langle\psi^{(c)}|$ by writing

$$|\bar{\phi}_c\rangle = e^{i\delta_c}\,|\phi_c\rangle, \quad \langle\bar{\phi}_c| = e^{-i\delta_c}\langle\phi_c|,$$

$$\text{(1a)} \qquad |\bar{E}_c\rangle \equiv |\bar{\psi}^{(c)}\rangle = e^{i\delta_c}\,|\psi^{(c)}\rangle, \quad \langle\bar{E}_c| \equiv \langle\bar{\psi}^{(c)}| = e^{-i\delta_c}\,\langle\psi^{(c)}|.$$

The formulae for $S_{c'c}$ and ${}^0S_{c'c}$ may then be rewritten as

$$\text{(2a)} \qquad S_{c'c} = \sum_{c''} {}^0S_{c'c''}\,\{\delta_{c''c} + 2\pi i\,\langle\bar{E}_{c''}|\,T\,|\bar{E}_c\rangle\}$$

and

$$\text{(3a)} \qquad {}^0S_{c'c} = e^{2i\delta_{c'}}\,\{\delta_{c'c} + 2\pi i\,\langle\bar{E}_{c'}|\,{}^0T\,|\bar{E}_c\rangle\}.$$

In these formulae, we have, on account of Equations (18) and (24),

$$\text{(4a)} \qquad \langle\bar{E}_{c''}|\,T\,|\bar{E}_c\rangle = \sum_{i,k} \langle\bar{E}_{c''}|\,V\,|k\rangle\,\frac{D_{ki}(E)}{D(E)}\,\langle i|\,V\,|\bar{E}_c\rangle,$$

whereas we only need to consider the part ${}^0T_R$ of ${}^0T$, which contains the important single-particle resonance contributions and whose matrix-elements are given by Equation (49):

$$\text{(5a)} \qquad \langle \bar{E}_{c'}| \, {}^0T_R | \, \bar{E}_c \rangle = -\sum_n \frac{\langle \bar{\phi}_{c'} | \, V \, | \psi_n \rangle \langle \tilde{\psi}_n | \, V \, | \bar{\phi}_c \rangle}{(1-\lambda_n) \, V_n} = \\ = \sum_n \frac{\langle \bar{\phi}_{c'} | \, V \, | \psi_n \rangle \langle \tilde{\psi}_n | \, V \, | \bar{\phi}_c \rangle \, \eta_n}{(1-\eta_n) \, V_n},$$

the last form resulting from the replacement of λ_n by $1/\eta_n$.

Taking first a non-diagonal amplitude ${}^0S_{c'c}$ (with $c' \neq c$), we see at once from the last Equation (5a) that its residue at any pole $\hat{\mathscr{E}}_i$ of $e^{2i\delta_{c'}}$ will vanish on account of the factors $\eta_n(\hat{\mathscr{E}}_i)=0$ occurring in the different terms of the function $\langle \bar{E}_{c'}| \, {}^0T_R \, |\bar{E}_c\rangle$ taken for $E=\hat{\mathscr{E}}_i$. As to the amplitude ${}^0S_{cc}$, it may be written in the form

$${}^0S_{cc} = \frac{e^{2i\delta_c}}{\prod_n (1-\lambda_n) \, V_n} \begin{vmatrix} (1-\lambda_n) \, V_n \, \delta_{nm} & \langle \tilde{\psi}_n | \, V \, | \bar{\phi}_c \rangle \\ 2\pi i \langle \bar{\phi}_c | \, V \, | \psi_m \rangle & 1 \end{vmatrix},$$

where, according to Equation (42), the underdeterminant $\det\{(1-\lambda_n) \times \times V_n \delta_{nm}\}$ may be replaced by $\det\{V_n \delta_{nm} + \sum_{c'} W_{nm}^{(c'+)}\}$, with

$$W_{nm}^{(c'\pm)} \equiv \int \mathrm{d}E' \, \frac{\langle \tilde{\psi}_n(E') | \, V \, | \bar{\phi}_{c'}(E') \rangle \langle \bar{\phi}_{c'}(E') | \, V \, | \psi_m(E') \rangle}{E' + \varepsilon_{c'} - E^{\pm}} = \\ = P \int \mathrm{d}E' \, \frac{\langle \tilde{\psi}_n | \, V \, | \bar{\phi}_{c'} \rangle_{E'} \langle \bar{\phi}_{c'} | \, V \, | \psi_m \rangle_{E'}}{E' + \varepsilon_{c'} - E} \pm \\ \pm i\pi \langle \tilde{\psi}_n | \, V \, | \bar{\phi}_{c'} \rangle_E \langle \bar{\phi}_{c'} | \, V \, | \psi_m \rangle_E .$$

By adding to the mth column ($m=1, 2, \ldots$) the terms of the last column multiplied with $-2\pi i \langle \bar{\phi}_c | \, V \, | \psi_m \rangle$, we obtain

$${}^0S_{cc} = \frac{e^{2i\delta_c}}{\prod_n (1-\lambda_n) \, V_n} \det \left\{ V_n \delta_{nm} + \sum_{c' \neq c} W_{nm}^{(c'+)} + W_{nm}^{(c-)} \right\}.$$

Now, $W_{nm}^{(c-)}$ – in contrast to $W_{nm}^{(c+)}$ – is regular at the poles of $e^{2i\delta_c}$, and the same is then obviously true of ${}^0S_{cc}$. Finally, since according to Equation (9) the phases are completely eliminated from the asymptotic forms of $|\bar{E}_c\rangle$ and $\langle \bar{E}_c|$, the amplitudes $S_{c'c}$, as shown by Equation (2a), are also regular at the poles of $e^{2i\delta_c}$.

On the other hand, the amplitudes $S_{c'c}$ do depend on the single-particle poles ${}^0\mathscr{E}_n$ of ${}^0S_{c'c''}$, which also affect the matrix elements $\langle \bar{E}_{c''} | \, T \, | \bar{E}_c \rangle$, and we must now investigate more closely the behaviour of the function

$S_{c'c}(E)$ in the neighbourhood of one of these poles ${}^0\mathscr{E}_n$; indeed, only poles near the real axis, corresponding to narrow single-particle resonances, will be expected to have a marked effect on the reaction amplitudes, and if there is more than one such pole, the argument presently to be developed for a single one can be successively applied to each of them. We write the amplitude $S_{c'c}$ in the form

(6a) $$S_{c'c}(E)=N(E)/D(E),$$

where $D(E)=\det\{d_{ik}(E)\}$ and the numerator $N(E)$, according to Equation (4a), is readily put into the form

(7a) $$N(E)=\sum_{c''} {}^0S_{c'c''}(E)\begin{vmatrix} d_{ik}(E) & \langle i|\,V\,|\bar{E}_c\rangle \\ -2\pi i\langle \bar{E}_{c''}|\,V\,|k\rangle & \delta_{c''c}\end{vmatrix} = \begin{vmatrix} d_{ik}(E) & \langle i|\,V\,|\bar{E}_c\rangle \\ -2\pi i\sum_{c''} {}^0S_{c'c''}\langle \bar{E}_{c''}|\,V\,|k\rangle & {}^0S_{c'c}\end{vmatrix}.$$

Let us now determine the leading terms of the elements of the determinant (7a) in the neighbourhood of the pole ${}^0\mathscr{E}_n$ of ${}^0S_{c'c}$. According to the structure theory of the S-matrix, the residue of the function ${}^0S_{c'c}$ is of the form $(v_c v_{c'})^{1/2} g_{cn} g_{c'n}$ (where the modulus of the complex quantity g_{cn} is essentially the square root of the partial width for channel c). In the neighbourhood of the pole ${}^0\mathscr{E}_n$, the asymptotic form (9) of $|\bar{\psi}^{(c)}\rangle$ is dominated by its outgoing components, which occur in the combination

(8a) $$-\mathscr{A}\sum_{c'} g_{c'n} O_{c'}({}^0\mathscr{E}_n)\,\varphi_{c'}/r_{c'},$$

multiplied by the coefficient $v_c^{1/2} g_{cn}(E-{}^0\mathscr{E}_n)^{-1}$; the asymptotic condition (8a) characterizes the eigenfunction $|\Psi_n({}^0\mathscr{E}_n)\rangle$ of the resonant state ${}^0\mathscr{E}_n$ corresponding to the Hamiltonian ${}^0H$, and we have therefore

(10) $$|\bar{E}_c\rangle \approx \frac{v_c^{1/2} g_{cn}}{E-{}^0\mathscr{E}_n}\,|\Psi_n\rangle.$$

Likewise, if we choose the sign in the definition of the time-reversal in such a way as to ascribe to the dual function $\langle\tilde{\Psi}_n|$ the asymptotic form (8a) with the opposite sign, we find (cf. Equation (21) and the following one)

(11a) $$\langle \bar{\bar{E}}_{c'}| = \sum_{c'} {}^0S_{c'c''} \langle \bar{E}_{c''}| \approx \frac{v_{c'}^{1/2} g_{c'n}}{E - {}^0\mathscr{E}_n} \langle \tilde{\Psi}_n|.$$

On the other hand, since the density operator $|\bar{E}_{c'}\rangle \langle \bar{E}_{c'}|$ has also a pole at $E = {}^0\mathscr{E}_n$, the corresponding integral occurring in the expression (23) for $d_{ik}(E)$ takes the approximate form of a constant $C_{ik}^{(c')}$ times the integral

$$\int \frac{\mathrm{d}E'_{c'}}{[E'_{c'} - ({}^0\mathscr{E}_n - \varepsilon_{c'})]\,(E'_{c'} + \varepsilon_{c'} - E^+)} = \int \frac{\mathrm{d}E'}{(E' - {}^0\mathscr{E}_n)\,(E' - E^+)} =$$
$$= \frac{1}{E - {}^0\mathscr{E}_n} \int \mathrm{d}E' \left\{ \frac{1}{E' - E^+} - \frac{1}{E' - {}^0\mathscr{E}_n} \right\} = \frac{2\pi i}{E - {}^0\mathscr{E}_n},$$

since the path of integration passes below the pole E and above the pole ${}^0\mathscr{E}_n$ of the respective integrands; the function $d_{ik}(E)$ has therefore a pole in $E = {}^0\mathscr{E}_n$, whose residue $-2\pi i \sum_{c'} C_{ik}^{(c')}$ is readily found to be

$$-2\pi i \langle i| V |\Psi_n\rangle \langle \tilde{\Psi}_n| V |k\rangle.$$

Thus,

(12a) $$d_{ik}(E) \approx ({}^0E_i - E)\,\delta_{ik} - \frac{2\pi i \langle i| V |\Psi_n\rangle \langle \tilde{\Psi}_n| V |k\rangle}{E - {}^0\mathscr{E}_n}.$$

On account of Equations (10a), (11a), (12a), the leading term of the determinant (7a) becomes

$$N(E) \approx \left| \begin{array}{c:c} ({}^0E_i - E)\,\delta_{ik} - \dfrac{2\pi i \langle i| V |\Psi_n\rangle \langle \Psi_n| V |k\rangle}{E - {}^0\mathscr{E}_n} & \dfrac{v_c^{1/2} g_{cn} \langle i| V |\Psi_n\rangle}{E - {}^0\mathscr{E}_n} \\ \hdashline -\dfrac{2\pi i v_{c'}^{1/2} g_{c'n} \langle \Psi_n| V |k\rangle}{E - {}^0\mathscr{E}_n} & \dfrac{v_c^{1/2} v_{c'}^{1/2} g_{cn} g_{c'n}}{E - {}^0\mathscr{E}_n} \end{array} \right|$$

$$\approx \frac{v_c^{1/2} v_{c'}^{1/2} g_{cn} g_{c'n}}{E - {}^0\mathscr{E}_n} \left| \begin{array}{c:c} ({}^0E_i - E)\,\delta_{ik} - \dfrac{2\pi i \langle i| V |\Psi_n\rangle \langle \Psi_n| V |k\rangle}{E - {}^0\mathscr{E}_n} & \dfrac{\langle i| V |\Psi_n\rangle}{E - {}^0\mathscr{E}_n} \\ \hdashline -2\pi i \langle \Psi_n| V |k\rangle & 1 \end{array} \right|$$

which by an obvious transformation becomes

(13a) $$N(E) \approx \frac{(v_c v_{c'})^{1/2} g_{cn} g_{c'n}}{E - {}^0\mathscr{E}_n} \det\{({}^0E_i - E)\,\delta_{ik}\}.$$

The denominator $D(E) = \det\{d_{ik}(E)\}$ becomes, with the approximation (12a) for the $d_{ik}(E)$,

$$\text{(14a)}\qquad D(E)\approx\left|\begin{array}{c:c}({}^0E_i-E)\,\delta_{ik} & \dfrac{\langle i|\,V\,|\Psi_n\rangle}{E-{}^0\mathscr{E}_n}\\ \hdashline 2\pi i\langle\Psi_n|\,V\,|k\rangle & 1\end{array}\right|\approx-\frac{1}{E-{}^0\mathscr{E}_n}\,\Delta(E)$$

with

$$\text{(15a)}\qquad \Delta(E)=\left|\begin{array}{c:c}({}^0E_i-E)\,\delta_{ik} & \langle i|\,V\,|\Psi_n\rangle\\ \hdashline -2\pi i\langle\Psi_n|\,V\,|k\rangle & {}^0\mathscr{E}_n-E\end{array}\right|.$$

The Equations (13a), (14a), (15a) show that the pole ${}^0\mathscr{E}_n$ is eliminated from the function $S_{c'c}(E)=N(E)/D(E)$ but replaced by another pole, given in our approximation by the root of the equation $\Delta(E)=0$ nearest to ${}^0\mathscr{E}_n$. In other words, the single-particle resonant state ${}^0\mathscr{E}_n$ of the Hamiltonian ${}^0H$ is coupled to the bound states ${}^0E_i$ by the residual interaction V_{BC} in much the same way as the bound states between themselves. The single-particle resonant states of small width we have been considering can only occur near the thresholds, and a small change in the choice of the Hamiltonian ${}^0H$ would suffice to change them into bound states: it was therefore to be expected that there would be no sharp distinction their behaviour and that of bound states, nor consequently between the single-particle and compound resonances of the total Hamiltonian to which they respectively give rise.

CHAPTER 8

THE STRUCTURE OF QUANTUM THEORY

[1968d]

1. Non-relativistic Quantum Mechanics

1.1. *Wholeness of Quantal Processes*

The object of non-relativistic mechanics is the description of the behaviour of systems of particles under the influence of interactions specified by potential functions of the coordinates.

In classical mechanics, the motion of the system can be considered from a local or a global point of view, and the two descriptions are equivalent. The instantaneous state of the system is represented by a point in a Euclidean phase-space (whose coordinates are a set of canonical coordinates and conjugate momenta). The local point of view leads to a unique temporal sequence of phases, forming a unique trajectory in phase-space, completely determined by any one of its constituent phases. The global point of view fixes uniquely the segment of trajectory joining any two phases by a variational principle, the principle of least action. This principle expresses the fact that the actual trajectory is singled out from the manifold of curves joining the two phases on the same energy surface by the property of corresponding to the least total quantity of mechanical action.

The existence of a minimum quantum of action radically changes all this mode of description. The evolution of the system cannot form a continuous sequence of phases, since the difference of action between any two successive states must at least amount to a quantum: hence the representation of states by phases and of changes of state by a continuous trajectory in phase-space are no longer meaningful. The concept of transition between states (of the same total energy) subsists with a global meaning which is now quite fundamental: such a transition process, in which action is transferred by quanta from one part of the system to another, cannot be analysed or subdivided. Moreover, since the choice between the possible transition processes from any given state of the system is no longer decided in a unique way by a local law of change, the occurrence of such transitions is governed by statistical causality, i.e. by laws governing the probabilities of their occurrence under given external conditions.

1.2. *Physical Magnitudes as Operators*

Physical magnitudes have a double function. In stationary states of the system, they have constant values which serve to characterize the state. Changes of the state of the system are specified by the particular way in which the physical magnitudes characterizing it change their values. In classical physics, the latter changes are consequences of the local laws of motion: the values of the physical magnitudes appear as continuous functions of time. In quantum physics, they must be connected with the transition processes themselves and thus depend on both the initial and the final state of any such process: they are therefore adequately represented by operators in a space whose elements are representative of the states of the system.

This representation links, more directly than the classical one, the physical magnitudes to their origin in measurements performed on the physical systems: such measurements are indeed operations aiming at the determination of the parameters which enter the quantitative description of the states of the system, and thereby establishing the connexion between this abstract description and our sensory perception of the systems.

1.3. *Superposition Principle and State-Vectors*

In the classical mode of description the phases contain all relevant information about the states of the system: the values of the physical magnitudes in a given state can be calculated in terms of the phase, as well as their variation when the state changes. Likewise, the quantal elements representing the states must allow us to compute the corresponding statistical information, viz. the probability of presence of the system in a given state and the probabilities of transition to other states. The adequate mathematical structure of the quantal state space is fixed by the physical principle of superposition, according to which the probability of presence of the system in any one of a set of states is not in general the sum of the probabilities of presence in each of them, but contains additional correlations between different states, of the same type as optical interference. This leads to the well-known Hilbert space structure for the state-vectors and the corresponding dyadics as probability operators; the average value of any operator is then given by the trace of the product of the operator and the probability operator of the state under consideration.

The structure of the Hilbert space of state-vectors is invariant for the group of unitary transformations, which physically correspond to the choice of different modes of definition of the states of the system, resulting from different standpoints of observation. All physically meaningful relations must be invariant with respect to the unitary group; the trace of an operator is such an invariant property.

1.4. *Non-Commutative Operator Algebra and Complementarity*

The algebra of linear operators in Hilbert space is non-cummutative. In particular, the coordinates and their conjugate momenta do not commute, but satisfy canonical commutation relations, which are invariant for unitary transformations, and correspond to the classical Poisson-brackets between these quantities. It is through the commutators between coordinates and momenta that the quantum of action, expressed by Planck's constant, is introduced in the formalism. Physically, the non-commutation of two operators indicates that the magnitudes they represent stand in a relationship of complementarity: their determinations in any state of the system are subject to reciprocal latitudes, the product of which exceeds the average value of the commutator in the state in question.

The origin of these reciprocal latitudes can be traced to the quantal limitations of the measuring operations involved in the determinations of either of the complementary magnitudes. Every measuring process consists in a quantal interaction between the atomic system and macroscopic apparatus, and partakes of the character of wholeness of all quantal processes: it is only well-defined when the apparatus has reached a state of equilibrium, in which the outcome of the measurement is permanently recorded. This last stage of the measuring process is essentially irreversible, and the loss of information about the original state of the atomic system it entails reduces, to the extent stated above, the possibility of determining magnitudes not commuting with the measured one.

Since the observation of quantal process always amounts to some appropriate measurement, the concept of process or phenomenon in quantum theory requires, according to the preceding argument, the precise specification of all macroscopic apparatus serving to fix the conditions of its production and registration. Phenomena whose characterization involves the measurement of non-commuting magnitudes will accordingly exhibit relationships of complementarity.

1.5. *Dynamical Behaviour and Temporal Evolution*

Transition processes between states of given energy are 'timeless', in the sense that a measurement of the time of occurrence of a process, i.e. the establishing of a connexion between this process and a standard clock, is complementary to the determination of its energy. It is possible, however, to describe the time evolution of an atomic system, i.e. the way in which the interactions between its constituent particles bring about a variation of the state-vector in the course of time. The transition from a given state-vector to the state-vector at some other time is a unitary transformation belonging to a continuous group, whose infinitesimal operator is essentially the Hamiltonian. Instead of treating the state-vectors as variable in time, one may equivalently regard the state of the system as fixed by the invariable external conditions and apply the unitary time transformation to the operators representing the coordinates and momenta (as well as other magnitudes depending on these canonical variables). One then obtains for these operators equations of motion which have the same form as the classical ones, provided that the classical Poisson brackets are replaced by the corresponding commutators. However, this local form of quantal dynamics – in contrast to the classical one – is not equivalent, but complementary to the global mode of description based on the consideration of transition processes at given energy.

The typical course of an atomic process starts from a configuration of the system in which it is separated in various parts whose mutual interaction is negligible; likewise, the final stage of the process is again a configuration of a similar type, consisting of different non-interacting parts, or the same parts in different states. Such asymptotic configurations are called 'channels', and the process may be described either as an evolution in time leading from an initial to a final channel, or, in a complementary fashion, as a transition at given energy between the states of the system defined by the condition of taking the asymptotic form of the former and latter channel. Either mode of description leads to the construction of a collision operator transforming channels into each other and determining the probabilities of all transitions between them. Whereas this collision operator contains less information than the Hamiltonian about the local dynamics of the system, it can completely replace the Hamiltonian in the global account of the transition processes.

1.6. *Spin and Exclusion Principle*

The states of the constituent particles of atoms and nuclei, the electrons and the nucleons, present a duplicity indicative of some entirely non-classical structure, called 'spin'. This is represented by additional dynamical variables having only two distinct values. Such dichotomic operators can be expressed in terms of three standard spin variables, which form a two-dimensional representation of the group of spatial rotations, and accordingly represent the components of a spatial vector. The spin vector has to be added to the orbital angular momentum vector in order to fulfil the angular momentum conservation law. Thanks to the isomorphism of the two-dimensional unitary group and the group of spatial rotations, the spin duplicity of states is thus formally associated with an intrinsic angular momentum of the particles; the spin of charged particles further defines an intrinsic magnetic moment of the particle. However this is not a physical correspondence: the spin of a free particle cannot be measured as an angular momentum or a magnetic moment.

Particles with spin satisfy the exclusion principle, stating that in an assembly of such particles, no two particles can occupy the same individual state. This is one of the most fundamental structural principles of nature. Not only does it form the basis for the "shell structure" of systems of particles subject to this exclusion, but it leads to the occurrence, in these systems, of peculiar types of non-classical interactions between pairs of particles, which are essential for the explanation of a large number of the most conspicuous properties of matter. In its most immediate application, it accounts for the sequence of atomic structures reflected in the periodic table of the elements, for the similar type of regularity of nuclear shell structures, as well as for the behaviour of the electron assemblies responsible for the properties of metals. The "exchange" interactions between pairs of electrons, occurring in addition to the classical ones as a consequence of the exclusion principle, are in particular responsible for the homopolar molecular bonds as well as for the spin correlations of ferromagnetic metals. Another type of correlation occurs between electrons excited to unoccupied individual states and the holes they leave within the set of occupied states: especially strong interactions occur between particle-hole pairs of equal and opposite momenta and spins, and the resulting long range correlations give rise to the striking phenomena of the superconductive state. Similar pairing interactions between nucleons and holes in the nuclear shells have a conspicuous influence on the properties of nuclei with even numbers of protons and neutrons.

2. Relativistic Quantum Theory

2.1. *The Electromagnetic Field*

In order to take account of the finite velocity of propagation of all interactions, one has to introduce fields as agents of transmission of these interactions between the particles. Leaving out of consideration the gravitation field, classical physics offers only the example of the electromagnetic field, transmitting interaction between charged particles. Thanks to the correspondence with classical electromagnetic theory, it is possible to quantize the electromagnetic field by a straightforward extension of the methods of non-relativistic quantum mechanics, at any rate to the approximation in which the sources of the field are treated classically. The only new structural element appearing at this stage is the introduction of field operators which are continuous functions of (non-quantal) space-time coordinates.

Since the coupling between the field and the charged particles is linear in the components of the potential four-vector operator, the effect of applying the latter to a state-vector will be to lead to a new state in which the number of radiation quanta of a given mode has been increased or decreased by one unit. The field operators may therefore be regarded as linear combinations of creation and annihilation operators for radiation quanta. Such creation and annihilation operators may be referred to points of space-time or to any other set of parameters specifying the state of the single radiation quanta; they belong to a non-commutative algebra with simple rules of non-commutation. With their help, one can define number operators, giving the numbers of radiation quanta in the various states, and one can construct any state-vector of the radiation field from the state-vector for the vacuum by successive application of the appropriate creation operators.

2.2. *Relativistic Particles*

The relativistic extension of the quantal description of free electrons or nucleons leads to a remarkable physical consequence: besides the spin duplicity, the requirement of relativistic invariance introduces another one, the particle-antiparticle conjugation, exhibited by the occurrence of processes in which pairs consisting of a particle and an antiparticle are created or annihilated. The technique of creation and annihilation operators is of course very convenient to describe such processes – the antiparticle being formally represented by a hole in a fictitious distribution of particles in the states of negative energy, to which the vacuum state-vector is associated. The rules of non-commutation of these pair creation and annihilation operators are different from those for the radiation operators, since they embody the exclusion principle, but they have also a very simple form. If the operators are referred to points of space-time, they are the same functions of space-time coordinates as the state-vectors of a single particle: they can thus be formally compared with the quantized operators of the radiation field; conversely, the radiation quanta can formally be compared with the particles and denoted as the radiation particles, or photons, associated with the radiation field. However, both the field aspect of the electrons and nucleons and the particle aspect of the radiation field have no direct physical meaning, since they both vanish in the classical limit.

With respect to the group of spatial rotations, the field operators of electrons and nucleons transform according to the simplest two-valued representation of this group, whereas those of the radiation field transform according to the one-valued representation of the group allowing of a maximum intrinsic angular momentum of one quantum. Formally, they thus appear as the prototypes of two distinct families of fields, called fermions and bosons, respectively, which are associated to the successive two-valued and one-valued irreducible representations of the rotation group, and are accordingly characterized by all successive values of the intrinsic spin – integral numbers of quanta (starting from zero) for the boson family, half-integral numbers of quanta for the fermion family. A number of representatives of the two families are actually observed in their aspect of free particles; all bosons other than photons, which are called mesons, have finite masses and accordingly a normal kinematical behaviour. All fermions obey the same rules of non-commutation which entail an exclusion principle; likewise, all bosons obey the same canonical commutation rules as the radiation field, which imply that they enter symmetrically in all systems in which they occur.

2.3. *Types of Interaction*

All particles of non-vanishing mass except the electrons and the protons are unstable; they decay into various associations of other particles by many different channels. They can also appear in various associations as a result of mutual collisions at high energies. These processes reveal the existence of three types of coupling between the fields associated with the particles. The best known is the electromagnetic interaction, which is transmitted by the electromagnetic field between all other fields; this coupling is universal, i.e. it is governed by a single, dimensionless coupling constant, and it is of paramount importance that the numerical value of this constant is much smaller than unity. It is only this circumstance that makes possible the quantization of the free radiation field, separately from that of the other fields, as a meaningful approximation. The two other types of coupling are, respectively, much stronger and much weaker than the electromagnetic one, and they correspond to a sharp separation of the fermion fields into two families, the leptons and the baryons: the leptons stand out as being only coupled between themselves and with all other fields by the weak (and electromagnetic) interactions; the baryons are coupled between themselves and with the mesons by the strong interactions, and for this reason baryons and mesons are conveniently grouped together under the name of hadrons.

The lepton family consists only of fermions of spin 1/2: the electrons, the muons (which seem to behave exactly like electrons of larger mass) and two distinct fields of vanishing mass, the neutrinos (which always occur in association with either electrons or muons, respectively). The weakness of the interactions to which the leptons are subject again confers their quantized free field description the character of a reliable first approximation. The weak interaction seems to have a universal form and a universal strength. In contrast with the other interactions, it has the remarkable property of not being separately invariant for spatial reflexion and particle-antiparticle conjugation, but only – except for a slight deviation in one case, which is not yet understood – with respect to the product of the two transformations. All three types of interaction seem to be invariant for time reversal, as well as for the product of the three transformations just mentioned.

It deserves emphasis that the conception of quantized fields interacting by mutual couplings eliminates the dualism of matter and force of classical physics and substitutes for it a complementarity relation between the particle and field aspects of each type. Each physical agency plays a double role: in its particle aspect, it acts as source of other fields and is acted upon by them; in its field aspect, it transmits interactions between other agencies which emit and absorb it. The classical dualism, as was pointed out by Maxwell, led to an infinite regress: each field of force needed a medium for its propagation between the particles it coupled, and the constituent particles of this medium had to be coupled by some other field, which required some other medium again for its propagation. Hertz attempted to circumvent this difficulty by eliminating the concept of force altogether from mechanics and describing the coupling between particles as a convection process effected by another kind of hidden particles; he could not complete this scheme, however, and had no successor.

The complementarity between field and particle, which solves this dilemma, is no mere formal relation. Even in non-relativistic quantum mechanics, the field aspect of the electrons has conspicuous effects, such as providing the homopolar chemical bonds. In the relativistic domain, it is especially the electron-antielectron pair field that gives rise to peculiar phenomena, such as scattering of light by light, resulting from an electron pair interaction between photons. Owing to their strong interactions, hadrons exhibit many effects of this type: thus, a nucleon is surrounded by a 'meson cloud' (to use a somewhat misleading terminology), in the sense that the spatial distribution of its properties is essentially determined by its coupling with the meson fields and thus appreciably differs from the idealized point distribution resulting from its own "naked" field operator. In particular, the observed masses of the hadrons must be regarded as effective masses embodying contributions due to their coupling with the other fields.

2.4. *Hadron Symmetries*

Whereas the description of the hadrons by field operators associated with the spatial rotation group is a certainly valid representation of their spin properties, there is no justification for assuming the validity of free field equations for these operators, and the true mode of description of the hadron fields and their mutual strong couplings is still very much in doubt. The great variety of production and decay processes suggests the existence of some underlying structure, the fundamental constituents of which, however, are still unknown: from this point of view, the observed unstable particles would be regarded as metastable, or resonant, states of various assemblies of these fundamental constituents. In this situation, a phenomenological analysis is helpful in establishing general symmetry properties of the hypothetical structures, which represent so many conditions restricting the choice of the properties which have to be ascribed to the fundamental constituents in order to account for the phenomena. In spite of its mainly descriptive character, this analysis has proved remarkably fruitful in pointing to a missing state which has subsequently been found, and to some quantitative relations which could be verified.

If one neglects all weak couplings with leptons, there are two stable hadrons, one representative of the fermion family, the nucleon, the other of the boson type, the pion; the latter, in its field aspect, forms the main component of the nuclear interactions which bind nucleons into the stable structures of atomic nuclei. The finite range of the nuclear interaction transmitted by the meson field is directly related to the mass of the meson in its particle aspect as a consequence of the complementarity between the two aspects. The strict stability of the proton is of course essential to ensure the permanence of matter; this requires moreover that all the atoms are built up of nucleons and electrons of the same kind, and not of antinucleons and antielectrons. Atoms of the latter kind would be indistinguishable from their conjugates, and it is conceivable that the universe could have separate parts consisting of matter and antimatter, respectively. Such worlds and anti-worlds could co-exist so long as they are prevented by some mechanism (such as appropriate magnetic fields) from colliding and annihilating each other.

The concept of particle-antiparticle conjugation is immediately extended to all baryons; it can also be defined for mesons, by means of suitable pairs of field operators belonging to the same state (of positive energy). If a particle is charged, its antiparticle carries an equal charge of opposite sign, and is therefore always distinct from the particle. A neutral baryon has an antiparticle that is also neutral, but exhibits opposite electromagnetic behaviour of higher multipole order and is thus again distinguishable from the particle. Neutral mesons may or may not be distinct from their antiparticles; we shall presently see according to which criterion. Whereas mesons and antimesons can be created in any number in strong interaction processes, only creations or annihilations of baryon-antibaryon pairs are found to occur; if we define a baryonic number in such a way that its negative values refer to antibaryons, and the value zero to mesons, we may state a general law of strict conservation of the baryonic number in all strong interaction processes.

If we only take the stable hadrons into consideration, the characterization of their state of electric charge or electric neutrality exhibits a duplicity for the nucleons and a triplicity for the pions, suggesting an association with the two lowest irreductible representations of a rotation group in some abstract space, in formal analogy with the case of the intrinsic spin; this abstract space is called isospace and the corresponding abstract angular momentum isospin. If one tries, however, to extend this representation to the other hadrons, one finds that while the successive values of a component of the isospin give in each case the correct sequence of positive, zero and negative values of the electric charge that the corresponding hadron may take, the starting point of this sequence may vary and must be fixed by another parameter assigned to the hadron, and called its hypercharge. In particular, the hypercharge determines whether or not a neutral meson is its own antiparticle. The isospin group, therefore, does not provide a sufficient basis for the classification of the hadrons with respect to their charge; the necessary extension – giving the mathematical significance of the hypercharge – proves to be remarkably simple. The spin and isospin groups are completely characterized by the duplicity expressed by the lowest representation and corresponding to physical properties of the stable fermions; the further representations are derived by reducing those obtained by forming the direct products of increasing numbers of the lowest double-valued representation. The latter forms the group of unitary transformations of Hermitian operators with two eigenvalues – the direct mathematical expression of the physical duplicity of states. Now, the group of transformations of the hadron charge states turns out to be that obtained by a similar reduction of direct products of the unitary transformations of triplicity operators – Hermitian operators with three eigenvalues (of zero sum). Each irreducible representation of this group is indeed characterized by a pair of integers, which fixes the range of variation of a component of isospin and of the hypercharge.

2.5. *The Problem of Hadron Structure*

What physical feature does the triplicity underlying hadron classification correspond to? It is tempting to bring it into relation with the structure of the hadrons and assume that it indicates the number of independent constituents of this structure. The experimental data do not contradict this hypothesis, but do not seem sufficient so far to give it a sufficiently definite form; moreover, no hadron constituents have hitherto been observed in the free particle state. The dynamical aspect of hadron structure raises a peculiar problem, reminiscent of the fundamental difficulty affecting the classical matter-force dualism: are we going to ascribe the forces holding the hadron constituents together to some field of a different nature, and thus introduce into the description a dualism (not now of matter and force, but of structural composition) which might require further reduction? or must we rather aim at a closed, self-contained description not involving any other element than the assumed hadron constituents themselves? Strictly speaking, such a question can only be decided by observation; meanwhile, it is on the latter alternative that the greatest efforts are being concentrated. Here, however, another undecided issue presents itself: is it possible to uphold a space-time description of the course of interaction processes, or must the account of these processes, embodying the formation and decay of bound structures, be limited to the asymptotic description offered by a general collision operator?

The first approach must be based on the conception of a self-interaction of a single basic constitutive field, expressed by a non-linear coupling term in the fundamental field equation. The structural features must then be introduced into the state-vector representing the vacuum state, and from which all physical states are generated by application of creation operators of the basic field; this can be done in principle by a formal procedure inspired by various examples occurring in the theory of the solid state (such as the theory of ferromagnetism or superconductivity – phenomena exhibiting an asymmetry not present in the law of interaction, but arising from condition imposed on the state of the system).

In the second approach, to which the picturesque name of 'bootstrap' has been given (in order to recall the fact that no hadron plays a more fundamental structural role than any other), one has to extract all structural properties from the analytical properties of the collision operator, and one hopes that the general requirements imposed upon such an operator could be sufficiently stringent to lead to an essentially unique determination of its analytical behaviour.

Adhuc sub judice lis est.

PART IV

SOCIAL RELATIONS OF SCIENCE

CHAPTER 1

THE ORGANIZATION OF SCIENTIFIC RESEARCH*

[1948d]

It is not without hesitation that I have acceded to the request of the Council of the World Federation of Scientific Workers to present a few observations on the organization of science within the framework of the Perrin-Langevin commemoration. The Federation Council, I am not precisely sure why, seems to feel that I have the custom of treating problems from a theoretical point of view and has asked me, as Mr. [J. G.] Crowther has explained with his well-known humor, to discuss the *philosophy* of the organization of science. But for someone who has had the privilege, as I have, of hearing Langevin develop his ideas on this subject, who still remembers that warm and persuasive voice, that luminous clarity with which he develops his arguments, it truly seems that there is nothing more to be said on the subject; and anything I could say would be but a pale reflection of what Langevin expressed so well, and more importantly, put into practice so well. Moreover, is it not superfluous to discuss the need for organization of science or the forms it should take in this city where the first university, the first institute for scientific research (this Collège de France where we are meeting), the first Academy of Science, the first scientific journal and the first Polytechnic School have all been founded.

These examples show sufficiently well that the problem of scientific organization is not new, and indeed, that in the course of time it has had quite varied solutions. At the present moment, we are witnessing, on one hand, a very marked new awareness on the part of some scientists of the need to set up the organization of scientific work on new bases; and on the other hand, a lively opposition to this point of view coming from those who stress what they call the *freedom* of the scientist. But the spectacle of this opposition is not new either. Each stage of scientific development has produced similar conflicts in outlook. On this subject I shall mention an anecdote concerning Joule, the Manchester physicist whose work contributed to the establishment of the law of conservation of energy. His fundamental first paper had been refused by the Royal Society. Many years later – when the law he had discovered was universally accepted and he was acclaimed as one of the greatest scientists of the period – he was asked what his impression had been when he had learned of this rejection of his first work. His answer was: "Well, I was not at all surprised. I could very well imagine those gentlemen sitting around a table saying: 'how could anything good come from a city where one dines at noon?' "[1] If Joule did indeed say those words, it shows that he had very perceptively grasped the conflict of attitude, relating even to scientific problems, which existed between those who, like himself, came from the industrial bourgeoisie of Manchester (which still dined at noon) and the members of the Royal Society who, during that period, belonged to a completely different class, the old English aristocracy or that portion of the bourgeoisie which had been assimilated to this aristocracy. From the scientific point of view, these people lived according to the ideas and traditions of the eighteenth century and were not capable of grasping the novelty in the viewpoint Joule suggested.

There is, in reality, no eternal precept which would govern the organization of science or, inversely, which would tend to show that any organization of science would be incompatible with some sort of freedom essential to scientific work. Stating the question thus, in abstract, metaphysical terms by separating it from concrete historical reality is condemning ourselves to understanding nothing about it. In the present case, as for any other social problem – for it is a social problem that is at stake here – the only rational method which allows some light to be shed on the subject is the dialectical method. However, I shall certainly refrain from inflicting on you a dialectical analysis of the question of scientific organization going back to the flood. I shall limit myself to a few examples relating to the modern period of scientific development, that is, to the period beginning around the end of the Renaissance in the sixteenth century and going up to the middle of the nineteenth century. It is during this period, in fact, that the modern scientific spirit was really formed.

If one examines the attitudes of the principle scientists during the course of this epoch of creation of modern science, – not of all, obviously, for they represent a remarkable gallery of unusual individuals, but of the most representative by virtue of the universality of the contribution they made to science – one is struck, I think, by the profound harmony prevailing between their properly scientific activity and their relationships with the societies in which they lived. And as a corollary, one finds perfect harmony between what is called pure science (which I would prefer to call basic science) and its applications to the various technical and social needs of the period. These scientists are not anchorites developing their thoughts in isolation; on the contrary, they are, in various ways, actively involved in the social life of their period. They themselves attached great importance to this social role of science. They constantly speak of it in their works. I could cite abundant proof of this. Many examples can be found in the works of Galileo as well as in those of Newton or Laplace and in the writings of the Encyclopedists; the Encyclopedia itself is a monumental example of this intellectual attitude. Even satire directed against science – I am thinking principally of Swift's mordant satire in Gulliver's voyage to Laputa – takes particular aim at the tendency of science to isolate itself from society; what Swift ridicules is the lack of practical sense shown by certain members of the Royal Society, or which he imputes to them.

I shall content myself with recalling in greater detail one example which seems to me particularly striking: It is that of one of the most outstanding figures of French and of world science; I mean Sadi Carnot, this young man who, towards 1824, alone developed the fundamental laws of thermodynamics. However, though he was working without guidance, inspired by his own genius, he was not at all isolated: On one hand, he drew his scientific inspiration from daily contact with the scientists of that brilliant school of French physics then in full bloom, and on the other hand, he had perceived a nascent social need of the period to which his research was to respond. His basic work (in fact, the only one he published) is entitled *Reflections on the Motive Power of Fire and on the Proper Means for Developing This Power*; and Carnot arrived at his famous theorem, that principle of universal scope which is one of the great natural laws, by studying the output of a steam engine, drawing up a balance sheet of the quantity of coal burned in the engine and the quantity of work that could be obtained from it. It is quite characteristic of Carnot that he was profoundly aware of the significance of his work and the conditions under which it could become productive. One of his friends reports that when he was engaged in drafting his paper, Sadi Carnot frequently visited him to submit the parts already written to him and to ask him to criticize them; Carnot wanted to reassure himself that his ideas were expressed in a sufficiently simple manner so that everyone could understand them. He was thinking especially of the artisans to whom, he felt, his work should be directly useful and who did not have the academic education which would have allowed them to understand complicated formulas.

On the other hand, one has the very clear impression that this intense social activity of the scientists who contributed to the establishment of modern science was not at all of a sort to impede their properly scientific work, rather the contrary. Galileo spent enormous amounts of time calculating projectile trajectories because that was work which was directly useful to his *patrons*, for those who gave him a living. But that did not prevent him from treating the problem in such a superlative manner that he was able to deduce from it the very foundations of mechanics. There is another case, perhaps even more interesting because it presents the situation from a rather more complex aspect. Oersted was prevented by his practical occupations from pursuing the consequences of his discovery of one of the great principles of electromagnetism. He published only a few pages on the subject, which suffice, however, to show that he had clearly perceived the majority of these consequences. But it is to others, to Biot and Savart and to Ampère, that we owe the development of these consequences in depth. One might ask whether it is not a shame that Oersted was thus prevented from pursuing his scientific research by these social activities. For if he was not able to exploit his discovery, it is because he was actively engaged in creating conditions favorable to the development of science in Denmark: It was he who founded a Polytechnic School in Copenhagen on the model of the one in Paris, who renewed science education at the university, who created a Society for the Diffusion of Science and who introduced a very profound reform in the secondary education of his country. But we see, in short, that this discovery, though it was not exploited by Oersted himself, immediately found other physicists, such as a Biot, an Ampère, to develop it and deduce all its consequences. In fact, a truly useful scientific idea – that is to say, useful in the general conditions of social development at the moment when it is produced – never lies fallow for long; it always finds people to cultivate it. Undoubtedly, one may regret that Oersted did not harvest all the fruits of his discovery; but this is a personal question with no bearing on the fundamental fact of the development of this discovery, which is a social fact; for a scientific discovery is an essentially collective activity and destined for the collectivity. And moreover, Oersted's organizational work, this work to which he devoted so many years that he could doubtless have employed in purely scientific work – was it not this which in the long run permitted the vigorous expansion of scientific culture in Denmark which gave us Niels Bohr?

I have tried to evoke for you some examples of scientists who have made real contributions to the establishment of modern science and to show you that this scientific activity developed in close contact with the whole social evolution of the period. Nonetheless, we must not look only toward the heights. There are, in the same period, other attitudes toward science, which correspond to other social milieux. I do not wish to enumerate all of them. There is, first, the resolutely anti-scientific attitude, the one which results from religious or political intolerance. But there is another which is perhaps more interesting for our purposes; I will simply illustrate it by the situation in England in the eighteenth century: It is the attitude of a certain part of the English aristocracy which preferred scientific research, which was perhaps considered a more distinguished pastime, to foxhunting or to the debauchery so well depicted by Hogarth's caricatures. Moreover, this aristocracy produced some truly astonishing geniuses: Cavendish, for example, as was found after his papers were studied, discovered quite by himself several fundamental laws of electrostatics which were not rediscovered and published by other physicists until much later. Thus, if he alone had made these discoveries, they would have benefited no one until the day when chance exhumed them from his archives.

This leads us to examine the origins of that individualistic attitude that we observe in some contemporary scientists, an attitude which might be termed that of *Science for Science's Sake*, a counterpart of the school of *Art for Art's Sake.* It is currently extolled by some as being the attitude which suits the so-called *entirely independent* scientist who lives only in his scientific thought and accepts no organization, no cooperation, no constraint which could injure the free flowering of his work and hence the progress of science. If we attempt to analyze the origin of this attitude – which appears as a new element in the evolution of the scientist's outlook, since it was absent during the centuries which saw the birth of modern science – we can find, I believe, two major streams of influence.

As for the first, should I call it snobbery? I mean by that that it refers to an attitude which the bourgeoisie adopted when it developed in the nineteenth century, because it was not capable, it would seem, of creating its own culture, and so it tried to remedy this lack by imitating the aristocracy it had just supplanted. From the scientific point of view, it adopted this attitude of the aristocratic amateurs I alluded to previously. A really striking example was brought out by Crowther in the second edition of his book on British scientists of the nineteenth century; it is that of Perkin, the creator of the English chemical industry. Perkin retired at the age of 35 after having made his fortune and devoted the rest of his existence primarily to music. Moreover, as an object of scientific research, he had chosen a problem of only mediocre interest which was not inspired by any great idea, but which had the distinction of being related to one of Faraday's numerous discoveries in the area of magnetism. From that moment on, he led the aristocratic existence which he evidently took as his model. Naturally, the consequences for the English chemical industry were disastrous. England lost the lead which Perkin's first efforts had given it, and the monopoly of the chemical industry passed to Germany whose researchers were not encumbered by this aristocratic tradition.

Nevertheless, I do not believe that this first circumstance is the most important one. I think that the true cause of this individualistic attitude on the part of the nineteenth century scientists, which offers such a striking contrast with that of their predecessors, lies in the growing complexity of science. This complexity led to a more extensive division of labor, not only among scientists working in different disciplines, but also among scientists concerned with basic science, on one hand, and those more interested in developing applications of scientific principles. Here we have the testimony and the example of Faraday who is certainly the type most representative of the scientists of this period. When he had attained a certain maturity in his scientific work, Faraday consciously and voluntarily gave up pursuing the practical applications of his discoveries, which could obviously have very quickly procured him a fortune comparable to that of Perkin. He preferred to concentrate on the search for the fundamental principles of science because he was probably aware that he was particularly suited for such research and because it offered him more profound satisfaction. There is, perhaps, a certain amount of asceticism in this attitude; however, I would not like to exaggerate this out of proportion: for from the industrial class to whom he left the exploitation of his discoveries, Faraday received – if I can use a somewhat brutal expression – as the price for this decision, marks of respect and financial support which enabled him to enjoy high social prestige and to lead a life exempt from any other care than that of pure scientific research.

In my opinion, what is important about Faraday's attitude is that by making this decision, he voluntarily renounced any control over the use that would be made of his discoveries, thus giving one of the first examples of an attitude which has subsequently spread widely among men of science. Nonetheless, I do not wish to blame Faraday himself too much; for prior to making this decision he had shown that he understood perfectly the close ties which exist between scientific research and its technical and social applications. He also belonged to that line of scientists, which I described earlier, who make no distinction of principle between their attitude as researchers and their social activity. In Faraday's mind, the choice he made was dictated only by purely practical considerations. But the others, who after him developed this new attitude of the scientist with regard to science, neglected, consciously or unconsciously, this fine distinction which Faraday still maintained. We then witnessed, especially in Germany, the development of this new social type of scientist shut up in the world of pure science, and even in an extremely narrow area of science, to the point of taking pride in knowing nothing of what was happening outside of his laboratory. Particularly in Germany we see another aspect, another consequence, of this attitude develop, even more dangerous than the one I have just pointed out. We see flourishing there that type of the Herr Geheimrat who reigns in his laboratory as an absolute master, tyrannizes his assistants and bristles at his colleagues. These scientists think of themselves as living in a manner entirely separated from the social activity of their period. But we know very well what to think of that illusion. We have seen at work those chemists, those geographers, those anthropologists, all those who laid the bases of the Nazi ideology; we have seen them, under the guise of an objective activity in the ethereal spheres of pure science, in reality put all their talent and all their energy at the service of German imperialism. Hence this process does not result at all in a purification of science, as some declare, but, in fact, in its prostitution.

Another aspect which perhaps has already struck you in the development of this new attitude of the scientists of the nineteenth century is the fundamental contradiction which is its basis. For, indeed, if the point of departure for this separation between pure science and applied science is the division of labor, what does that mean: If there is a division of labor, it is because science has become so complex in its various ramifications that it requires a more developed organization, and this division of labor is precisely a process belonging to this organization. It exists in science as in industry, and this similarity between the conditions of development in science and industry is not fortuitous. Basically what took place in science in the nineteenth century is the same process of increasingly complex organization which unfolded in industry as the latter assumed greater and greater proportions. So much so, that we can say that these representatives of the individualistic attitude of *Science for Science's Sake* are the scientific counterparts of the industrial leaders, the great capitalist entrepreneurs. At the present moment, following their traditional political tendencies, these are either somewhat belated representatives of the laissez-faire of a century ago or, a more serious matter, active participants in imperialist expansion.

Thus I see in this individualistic attitude a fundamental contradiction which corresponds to the very structure of the social system in which this attitude developed. This is why I do not think it suffices to lay stress on this attitude, but rather think it necessary to examine the means of overcoming this contradiction. Of course, the only possible way of attaining this goal is to seek a form of scientific organization which is harmoniously adapted to the current needs of the society we live in, a society which is involved in a process of rapid transformation. We are doing nothing new in stating the problem in this fashion. It has always been so stated. When the *Collège de France* was founded here, the motives which inspired its founders were perhaps not very conscious, but it was clearly a question of a new organization of scientific research made necessary by the defects of the university of that time. When the *École Polytechnique* was founded, it was quite consciously intended to serve as a basis, a scientific instrument for a new social development which was then coming to life.

What are the current problems of scientific organization? They result precisely from this division of scientific work made necessary by the complexity of science. The goal of scientific organization is to overcome the contradictions resulting from this division of labor. The fact is that specialists who work in different domains and who no longer establish enough contacts between their research and their respective ideas run the risk of going off into dead ends; for it is a commonly observed fact that great scientific discoveries often result from contacts between different points of view. It is important then – and it is toward this goal that current trends in the organization of science are oriented – to establish organic contacts among specialists in different branches who each have their contributions to make to the common tasks.

We have an initial example of this in those great machines which are used in physics today: They represent in themselves veritable little factories, in the functioning of which teams collaborate whose members are, indubitably, all physicists, but belonging to very different specialties in physics. Still more important is the correlation among different sciences applied to the same problem, whose various aspects call for the collaboration of specialists from different branches of science. I am thinking, for example, of the problems of biophysics or of biochemistry. There are naturally a multitude of examples to cite in this respect. Finally, there are long-term systematic research efforts which necessitate enormous amounts of detailed work for which a single person could not suffice. Here again there is need for a team of numerous collaborators whose activities are coordinated.

Of course, I will not insist on the purely technical problems raised by this organization of science, but there is a sort of human aspect of the problem to which I would like to direct your attention. To help you understand the matter, I can do nothing better than to tell you very briefly what happened in Professor Powell's laboratory in Bristol when he put into practice his new technique for studying cosmic rays by the tracks that these rays leave in photographic emulsions. There was, first of all, a technical problem to resolve, the preparation of photographic emulsions with a particular composition and texture. This technical problem, let us say in passing, was able to be resolved thanks to the support of the British government, but especially thanks to the exemplary manner in which the government gave its support. This manner simply consists of generously providing the necessary financial means to those who request them with a very clear idea of how they want to use them, and then to allow these people complete freedom in the manner of using the funds. In the case which concerns us, the result of this farsighted policy was a marvellous development of the technique of photographic emulsions, which permitted Powell and Occhialini and their collaborators to make an absolutely remarkable series of discoveries in the domain of cosmic rays in a very short time. I can even say that this development has just begun and that new discoveries must be expected in this area which will soon open new perspectives into the most fundamental problems of the structure of matter. Well, there was not only a technical problem to solve in order to apply Powell's method; there was also a human problem: very simply, that of scrutinizing these photographic plates. Indeed, it was a problem of discovering, by means of improved microscopes, tracks exhibiting various interesting peculiarities, and for this it was necessary to systematically examine hundreds of plates. Powell had the idea of using for this work a group of young women with no special qualifications in physics but whom he intended simply to initiate into the routine of microscopic inspection of the plates. But then a very curious phenomenon took place. These young women were not content with a purely routine task; they expressed the desire to understand what they were doing. They took an interest in the aim of their work and in all its modalities. Thus in Powell's laboratory a quite extraordinary atmosphere was created which shows the enormous potential of humanly conceived scientific organization.

It is scarcely necessary to add that scientific organization must not just respond to the solution of immediate problems, but that there are also major questions requiring long-term research whose possibilities for future application cannot presently be foreseen. Clearly, such long-term undertakings merit support equally as much as those whose benefits are immediately apparent. This point is very strongly advanced by the adepts of the individualistic attitude to whom I alluded earlier. They seem to fear that greater development of scientific organization will imperil the progress of this type of work. It seems to me that they are quite wrong to be alarmed. I think, on the contrary, that these problems too can only be resolved within the framework of a more extensive organization, and that consequently the real way to promote this type of investigation is to encourage measures for coordinating scientific work on an increasingly large scale. The Académie des Sciences of Paris at the outset of its activity offered a rather interesting example of this type: the encouragement it gave to the perfection of measurement methods in geodesy. Here, too, the starting point for the work of Abbé Picard may appear somewhat futile with the perspective of history: It was simply a question of the best way to bring water into the park of Versailles. But their end result in the eighteenth century was those magnificent expeditions to Lapland and Peru whose findings allowed one of the most fundamental problems then confronting science to be solved: to confirm Newton's theory of gravity by the shape of the earth.

Of course, it is essential that questions of organization be left entirely in the hands of scientists themselves. Powell's example shows quite well that, from the moment they know what they want, scientists are quite capable of making good use of the resources they are given and do not need the advice of people who are not directly involved in scientific research. Undoubtedly the ever more extensive organization of science presents its own problems; the greater its extent, the less we may lose sight of the dangers of bureaucratization. But I believe that it is enough to think about these dangers and to be firmly resolved to fight them where they arise and to avoid them where they have not yet developed in order to be definitively protected from them. Hence I believe it can be concluded that it is within the framework of this new organization of science, created by the scientists themselves in the manner most appropriate to satisfy the needs of our epoch, that these scientists will find the true freedom which is dear to all and whose essential importance all understand. This question of freedom within the framework of a collective organization has been raised, moreover, on the general level of society: True freedom can only exist when a solid material base has been established for a rational organization of scientific work.

NOTES

* Collège de France, session of November 18th, 1948, in a Symposium on the title subject; taken from the published transcript.

[1] [The quotation is retranslated from the French. – Ed.]

CHAPTER 2

THE ATOMIC RESEARCHER

*The Atomic Physicist's Tasks, Goals, and Methods**†

[1968j]

When we hear what atomic physicists have to relate about the properties and behaviour of atoms and atomic nuclei, and when we see what they are able to achieve by ingeniously exploiting atomic phenomena, we may well wonder how on earth they set to work exploring this strange world and manage to penetrate so deeply into it. In the following I shall attempt to answer this question, to lift the veil a little and initiate you into the methods that atomic physicists have developed and which they use in their work.

What is so singular about atomic research is that we have no means of getting into direct contact with atoms; for we cannot see them, not even with the most powerful microscopes, and we have absolutely no possibility of interfering with their behaviour. We have, therefore, to use indirect methods: our procedure may be compared to detective work. Atomic particles may leave traces that we can examine, and we can then speculate about the way in which these traces have arisen and what they may mean in terms of the properties of the particles and laws of their behaviour. First and foremost, then, the art lies in producing these traces with the aid of suitable apparatus and in interpreting them correctly, so that on the basis of the sporadic information they give us we can form an overall picture of the atomic processes themselves. But, you may well ask, how can we produce traces if we do not already know something about the processes by which they are produced, which are, after all, the very processes we are trying to find out? Are we not facing a contradiction here?

Well, it is indeed a difficult situation, but it is a matter of using common sense and of not acting like the licentiate in Poul Møller's charming and profound short story 'The Tale of a Danish Student': you will recall the wretched licentiate, who was quite unable to make up his mind to begin writing his dissertation before he had studied how a pen is made. The way out of the difficulty is, of course, to proceed by stages. We have to start with some supposition as to how an atomic particle might conceivably behave in given circumstances; such suppositions are often suggested by analogy with the behaviour of large bodies, and they serve as a provisional starting-point for the setting up of the first experiments. From the results of these we see how far our guess was right; we correct our first picture and prepare new experiments from which we hope to be able to draw further conclusions. In this way the original rough sketch gradually turns into a fuller and more correct picture: for it is obvious that as our ideas become more and more detailed they offer more and more reliable starting-points for subsequent steps that are to carry us further.

You will see that this method consists of two different but indissolubly linked components: on the one hand the experiments, which are the only source of our knowledge of the laws of nature, and on the other the theoretical reflexion upon the results of the experiments and upon the preparation of new experiments, a reflexion just as indispensable as the experiments themselves. These two sides of the work, the experimental and the theoretical, assist each other. The preparation of an experiment requires the use of theory in order to form in advance as clear an idea as possible of how a given experimental set-up will work, and the interpretation of the experimental results is also impossible without the help of the theory. On the other hand, the theory must take precise account of the experimental results: there is no use fancying some picture of the behaviour of atoms from arbitrary analogies; if the theory is to be fruitful, it must be based on facts derived from experiments.

When mention is made of experiments in the realm of atomic physics, one immediately thinks of huge machines and complicated apparatus, and when mention is made of theories, one immediately thinks of abstract mathematics, and then it becomes hard to discern any coherent relationship between the two. I hope that it will now be easier for you to understand that it can make sense to use these mighty material and intellectual tools when you know that they form the final stage in a long development whose beginnings were extremely modest and straightforward. And it is important to observe that every step in this development was plain and necessary and dictated simply by common sense without anything mysterious about it.

In this connexion it is of great interest, as a matter of principle, to realize that also in the domain of atoms or even of atomic nuclei, very simple experiments may sometimes lead to the discovery of quite fundamental facts and laws of nature. An experiment is, after all, a question that we ask of nature, and if we succeed in asking the question in such a way that the answer must be yes or no, this kind of answer can be obtained by simple experiments, designed so as to establish that a particular effect either does or does not occur: there is then no need for any accurate measurement of the effect, but merely for the detection of its occurrence. Precisely because of this simplicity, the regularities that we get to know by such yes-and-no experiments are among the most assured that we have, and they are, moreover, of very great generality: they belong to the main features of our view of the world.

When these main features have been drawn in, we attempt to bring out more and more details, and this of course requires a more and more sophisticated art of experimentation. Here the technical aids are of quite decisive importance. The difficulty is that the processes we are looking for have very small effects: they can only be discovered by amplifying these effects, so that they become more accessible to observation. A well-known example of the ingenious amplifying devices that now belong to the arsenal of the nuclear physicist is the bubble chamber, in which an atomic particle forms a track of its path that can be photographed and measured. Now comes another great difficulty: many interesting processes to which an atomic particle gives rise, for instance the splitting of an atomic nucleus with which it collides, are so rare that millions of photographs have to be taken before one is found on which the process has left its easily recognized trace. The working of the bubble chamber must therefore be made fully automatic, and the stage has even been reached at which the viewing of the many photographs and the detection of interesting traces on them is completely automatized.

Seen from another angle our investigations aim at analysing the structure of all the combinations of various kinds of particle which have been found in the world of atoms: this began of course with those that have the weakest binding, the atoms themselves. The next step was the investigation of the structure of the atomic nucleus, where the forces involved are a thousand times greater than those that hold atoms together, and now work is going on breaking down the very building-blocks of the nuclei themselves, the nucleons; the forces needed here are again a thousand times greater than the nuclear forces. This will explain to you why nuclear physicists are driven to build more and more gigantic machines that can supply these large energies in a form suitable for this deeper and deeper penetration into the structure of matter.

In our present context we are not interested so much in machines as in people. But I had to tell you a little about the machines in order to make the human aspect of the matter comprehensible. For now you will have no difficulty in understanding that the age of the lone individual discoverer has passed. These great machines, these delicate automatic contrivances, have to be looked after and kept in running order by highly specialized technicians, and every experiment that employs them is so complicated and has to be so meticulously thought out and prepared that it simply exceeds the powers of the individual physicist; only large groups of specialists are able to carry it out. It demands experts in the art of experimentation and experts in theory as well as experts in calculation or the use of computers, and often a score of people have to work intensely for months before a single result can be achieved that permits us to fill in perhaps one small gap in our knowledge, to answer one modest question about nature.

This remarkable development in the organization of research, its transition from individual to collective form, is on the same lines as the general tendency in the social evolution of out time. But it has perhaps a special significance, for it shows how a free association of men – in spite of the ambitions of the individuals (to which physicists are just as prone as anybody else!) – can, if these people are united in a common ideal, prosper and give fruitful results. As used to happen around trading centres and factories, there now arise around the great atomic machines whole cities whose population consists of scientists and technicians: is this perhaps a presage of the rising of a new form of society whose activity will be in the service of enlightenment and peace?

But apart from these remote and uncertain prospects, how can we give an account of the yield that we have already derived from these investigations carried out with such powerful means? Here again the theoretical view helps us, for it is after all its purpose to gather the experience that we gain through the experiments into a clear and coherent whole. It is quite natural to begin by comparing atomic particles to the large bodies, whose behaviour is well known to us: one forms, as one says, a model of these particles. Then it is a matter of refining these models as far as this is possible: and just as more sophisticated experimental techniques are required, so the theoretical work requires the more sophisticated tools that mathematics offers us. But it should be remembered that mathematics has here no other function than to bring about a great economy of thinking, which alone makes it possible to master the long and involved chains of conclusions necessitated by the indirect nature of atomic research. For in the last resort we have, of course, to build our ideas of the atomic world on the simple experiences that are common to us all: how could we otherwise understand each other when we are speaking about atoms?

And now comes the last great surprise: it turns out that in describing atomic phenomena one particular picture cannot suffice: according to the circumstances we are obliged to make use of several quite different pictures; one can only describe the atomic world, as it were, from one point of view at a time, and the prospects it offers from various angles are so different that they cannot be fused into one single picture in the usual sense.

This situation has given atomic physicists occasion for fascinating reflections upon the very essence of human cognition, which I shall not enter upon here. I just wanted to mention it as a concluding touch in the picture of the atomic physicist that I have attempted to outline. I scarcely venture to hope that I have succeeded in giving an adequate insight into his methods of work, but perhaps I have managed to persuade you that the secret weapons that have secured him such brilliant triumphs are persevering toil and common sense.

NOTES

* Speech on the Danish radio.

† Translation by David Stoner, revised by L. R.

CHAPTER 3

TECHNICAL AND SOCIAL ASPECTS OF THE DEVELOPMENT OF THE EUROPEAN SCIENTIFIC RESEARCH ORGANIZATIONS

[1970e]

The development of large scientific research bodies on the European scale present new aspects, full of lessons for the future. The subject of this report is to assemble some considerations well-known to all who have contact with this development, with a view more specifically to its significance and its consequences for the progress of technology and the relations between science and technology.

The creation of European research bodies in fields such as high-energy physics and cosmic ray physics could be superficially justified by the fact that the amount of equipment necessary for this research surpasses the financial resources of a single country (even those of the big industrial nations of Europe). It is clear, however, that the trend towards co-operation of European research workers responds to a deeper need: this trend manifests itself, for instance, in the creation of a similar body for molecular biology – a field in which the financial pressure is not so great – or in the creation of a European Physical Society, the activity of which seems to be directed towards the improvement of the means of communications and of discussion of the research results either by means of their publication or by their presentation at conferences at a European level. In all these cases, it is a matter of the research worker overcoming the obstacles to the co-ordination and concentration of their efforts which are artificially created by the political and economic fragmentation of Europe.

1. Social phenomena of a new kind

Thus, the trend to association in the scientific field meets the preoccupations of political and economic forces which, on their side, are trying to lower or remove the same barriers and to create a European community. This identity of aim has not escaped governments, which are including scientific research in their general programmes for European integration. It is not surprising that in doing so they emphasize the importance of scientific research for all aspects of economic development: for no one at present contests the fact that economic development dominates all the others. However, one must be careful not to measure the relative importance of scientific development exclusively in economic terms, as certain politicians have a tendency to do. It is advisable to realize that capital investments in scientific research are mostly long-term investments from which it is unreasonable to expect an immediate return; moreover, the financing of this research, however high it may appear taken in isolation, is only a trifling part of the total budget of a state participating in the European effort (especially if one compares it with the extravagant wastage represented by military expenditure). In the narrow limits of budgetary considerations, one should rather stress the very real and by no means negligible profits which the building, equipment and the supply of energy and other products, of a large European laboratory such as the Nuclear Research Centre (CERN) bring to the most varied industries, from building to the different branches of electrotechnics, of electronics, of the photographic industry, of optics and many others. However, that is not the main point; in order to measure the importance of the European research undertaking, one has to consider them as social phenomena of a new kind and to note the repercussions of their existence on the whole process of forming the European community.

The first fact to command attention is that, on the human level, the foundation, the setting in motion and the functioning of these undertakings has not been hampered by any of the national traditions or prejudices which raise so many obstacles in other fields. Quite the contrary, in the vast laboratories of CERN (which we shall take as an example) where hundreds of scientific workers of all European nationalities are assembled, there reigns a spirit of co-operation and friendship that cannot be affected by such antagonisms. Moreover, this understanding spreads beyond the confines of the central laboratory and extends most successfully to the various participating countries: this is due to the very organization of the work at the European level. National teams which co-operate in research, the starting point of which is the central laboratory, must, in effect, proceed to analyse and make an abstract of the results in their national laboratories, in accordance with a jointly prepared plan; in this way active and lasting relations are created between these national laboratories. The feeling of participating in work with a clearly defined aim, the success of which demands co-ordination of efforts within a willingly accepted discipline, suffices to create this atmosphere of cordial solidarity, while not excluding a healthy competition which foreshadows the sort of human relation which will have to be established in a European community to make it viable. It is important to stress that the nature of the great scientific experiments undertaken on a European scale requires the close co-operation of scientific workers of all types: experimenters, theoreticians, instrumentation specialists, engineers, technicians; so that the solidarity which unites them overcomes any social distinction and excludes any artificial separation between science and technology.

Fig. 1. The CERN 2 metre hydrogen buble chamber minus its magnet. In September 1969 it reached the total of ten million photographs, taken since it came into operation less than five years earlier.

(Photo: CERN)

2. New types of apparatus

From another point of view the research being carried out in European laboratories is exercising a key influence on the progress of many branches of industry. The type of problem on which study is concentrated in these laboratories must of necessity use the most advanced experimental techniques, and further leads to increasing improvement and even to the consideration and study of entirely new techniques. This aspect is considered to be sufficiently important by the firms manufacturing measuring apparatus and other specialized equipment for them to establish offices in the immediate vicinity of the CERN laboratories. The ordering of new types of apparatus, made to precise specifications, is stimulating competition at the highest level between industrialized concerns of various countries, accent being placed on quality as much as, or more, than, on quantity. There are numberless examples of these relationships of a new type between science and industry. Even the foundations of the installation of high-energy accelerators raise technical problems beyond the traditional industrial personnel; only to mention one example, the construction of superconducting electromagnets is being studied at present in close collaboration between teams of technicians belonging to scientific laboratories (in Europe and the U.S.) and industrial firms wishing to develop this entirely new technology. In addition, it is the precision industries, above all in the field of electronics, but also of optics and precision engineering, which are benefiting from the fields of activity which the European laboratories are opening up to them. In this sector, the construction of large bubble chambers is raising all sorts of technical problems, and the study of millions of tracks of individual events that these chambers supply requires the development of automatic systems of scanning and of measurement of the films containing these tracks; finally, the analysis of the results of measurement and the theoretical questions that they raise can be worked out only with the help of the most powerful electronic computers. It is scarcely necessary to stress the repercussions that this technological progress, raised by purely scientific problems, can have in totally different and very much more extensive fields; for example, the study of methods of electronic calculation being carried out at CERN has given rise to the gradual development of an independent section embracing the whole field of this new technique, with multiple applications, which we call informatics.

3. A NEW TYPE OF SCIENTIFIC WORKER

Thus, we are witnessing, starting from an effort to concentrate purely scientific research in a sufficiently large network of industrially developed European countries, the spontaneous formation of new methods of organization of that research, finishing up with the removal of artificial barriers which the fragmentation of the diverse activities which research demands had put up between them: this fragmentation is giving place to a harmonious synthesis between theory and experiment, and above all between scientific activity and technical activity. Obviously, it is the collective methods of organization, the teamwork, which make possible this going beyond the individualist stage; but it is noticeable that the preoccupation with synthesis penetrates, as it were, from the team to the individuals who comprise it: even if each one of these remains a 'specialist' and plays the part of specialist in the common work, he nevertheless is conscious of the whole of the problem and interests himself inevitably in all its aspects. The creation of a new type of scientific worker, combining in himself the competence of the specialist and the spirit of synthesis of the true man of science, there indeed is the most engaging prospect which the rise of the large laboratories opens on the future.

CHAPTER 4

SOCIAL AND INDIVIDUAL ASPECTS OF THE DEVELOPMENT OF SCIENCE

[1971f]

To Igor Tamm, as a token of old friendship

The questions I propose to discuss in this essay are often the object of considerations that do not particularly contribute to their clarification, inasmuch as they are inspired by preconceived or unfounded opinions about the meaning of science to society and to the individual scientist. I shall try to approach the matter from a different angle, by the only method on which we can fully rely, namely the very method we apply in science: we first try to formulate the problem in a sensible way, then we look for evidence from experience that may throw light on it and perhaps lead to its solution or at any rate to well-founded conclusions.

In our case the difficulties start with the very formulation of the problems. Certainly, we have the two aspects – social and individual – in the development of science: science is part of the activities of society, and on the other hand it definitely requires individual effort. Now starts the dispute about which of the two components is decisive in this development: is the development of science essentially determined by forces and influences coming from society or, on the contrary, by the individual contributions of great scientists? Thus, it has been alleged that physics would have been exactly what it is today even if Newton, Einstein and Bohr had never lived: if Newton had not discovered the law of gravitation, somebody else would have made the discovery, and if Bohr had never lived, we should have had atomic theory all the same, only (some people think) without all this nonsense about complementarity. This is the school of thought called 'historical materialism'. The opposite school maintains that physics, as known to us, is entirely the work of great men like Newton, Einstein and Bohr: if it has not been for the contributions from these men, physics would not only be different, but would simply not be. I may have exaggerated the contrast to some extent, just to emphasize how artificial it must appear.

We know very well that the individual scientist – and this applies even to the greatest – never faces the problems alone, but builds upon a scientific tradition, established by his predecessors and from which science receives its character of continuity. Individual efforts may certainly lead to fundamental changes in this tradition, but never to a complete break. There are several examples of new discoveries or original thoughts which find no response in the society of the time, because somehow they do not conform with its general attitude; hence they are not appreciated and, for a while, remain unproductive. In such cases these ideas will simply have to be rediscovered at a later stage when, as one puts it, the time is ripe for society to take them up and proceed with them. The problem must therefore be formulated quite differently: it is actually a matter of interaction between the individual scientist and society, an interaction presenting many aspects that are worth while investigating. Apart from its intrinsic interest, such an investigation may be of practical importance for the elucidation of the problems connected with the present organisation of scientific activity.

I may perhaps illustrate the situation by a simple analogy with atomic physics. We may regard the large bodies as atomic systems, and by statistical considerations try to derive their properties from the behaviour of the individual atoms; but we can also formulate the fundamental laws obeyed by these bodies in a way directly referring to the observation of the phenomena. Now it is by no means justified to state that this last, 'macroscopic', description is made superfluous by the atomic representation of the course of the phenomena: the epistemological problem of atomic physics rather consists in establishing a harmony between the two aspects.

Now, when we consider some development in a society, we look at it by definition from a statistical point of view, just as we do when considering the properties of a large body from the atomic point of view: for as soon as we use the term society, we imply that individual peculiarities are left out of account. It is therefore not surprising that if one enquires about the part science plays in the development of society, one isolates the very features which do not depend on the efforts of the individual scientists but only reveal the general attitude of society to science. It is then also to be expected that it is the application of science to problems useful for society which will take a prominent place. By only paying attention to this side of the matter, one is easily led to the widespread opinion that science is only important by its application, its usefulness to society, while questions of principle, the pursuit of so-called pure research, are a luxury in which society can only indulge if its economy is sufficiently prosperous.

If, on the contrary, one is interested in the contribution of the individual to the manifold activities of society – just as in atomic physics, when one studies the individual atomic processes – the attention will be concentrated on the research work itself, at the risk of losing sight of its relation to other social tasks. It is essential, therefore, to keep the whole situation in mind, and find a mode of description which takes due account of the interaction between the individual scientist and the society in which he works.

We are here dealing, in fact, with a complementary situation. In atomic physics it is the relationship of complementarity between the atomic and macroscopic descriptions that allows us to achieve full harmony between the two. Likewise, as I have tried to indicate, the features of the historical development that are prominent when one considers it from a social point of view are complementary to the problems that arise when one tries to analyse the situation of the individual scientist and the circumstances which determine both the trend of his research and the response it receives from his contemporaries.

So far I have only presented general considerations, and general considerations do not carry us far beyond the trivial. Excitement starts when one tries to go deeper into the problem and to investigate in detail the two complementary points of view. To this end one must look for facts that can elucidate these various situations. In the first place, we have the very growth of science, which we know best from the history of our own society, where science has reached its greatest development. We also know this development (although not so well) from other societies, particularly from China, which has, fortunately enough for our present purpose, remained isolated from Western Europe for a long time; for this provides us with some material which may serve as a basis for comparison and offer examples of different possibilities of development.

But of course history is not our only source of information. Scientific thinking, the activity of the scientist, is open for observation and analysis: this is a science by itself, which is called epistemology. It would seem natural that the object of study in epistemology should be the fully-fledged scientist, the adult person. Immediately, however, a problem of quite decisive importance presents itself: whence does the adult acquire the ability to think in the way we call rational? It is clear that the reply to this question must colour our whole view of the meaning of science: just think of the far-reaching consequences that have been drawn from the belief that rational thinking would be, so to speak, a gift of God, that man is a being who distinguishes himself from animals just by the innate ability of thinking by means of abstract concepts.

Here, clearly, is a problem which cannot be solved by mere guesswork, but only by seeking guidance in experience. One can, at any rate, approach the problem by watching children's reactions when they are confronted with various tasks adapted to their age and stage of development. Such investigations, which require of course a strict method of experimentation, were initiated by Piaget in Geneva and continued over the past forty years under his leadership. They have led to extremely instructive results, bearing upon all the problems of epistemology and offering a firm basis for a reply to the question of the formation of our abstract thinking, and to what extent it originates in the individual or in society.

By exploiting the historical facts as well as the results of the experiments on the psychogenesis of thinking, we can try, on a purely empirical basis, without arbitrary assumptions, to build up a really scientific epistemology. Let us consider in particular what we may conclude from such considerations regarding the relations between science and society. Anything approaching a systematic presentation would lead us too far; we must be content with a few examples picked out from an abundant material.

The first example will illustrate how a scientist of flesh and blood (and not the idealized picture scientists sometimes give of themselves) will regard his activity and his relationship to the community in which he lives. I choose it from the period of formation of modern science and of the powerful method by which it could make such great and rapid progress. The men who took part in this process were keenly conscious of opening new adventurous paths, and they were very open-hearted about their purposes and ambitions. I might have adduced one of the adherents of Bacon's programme in England, Hooke for instance, but I prefer to take Descartes, whose influence went deeper and – although indirectly – lasted longer. His theories had serious defects, but also contained fruitful thoughts, which the following generation – Huygens and Newton – took over and carried further: the mechanical picture of the world – the idea that all phenomena originate from the motions of material particles acting up on each other by collisions – was predominant in physics until the end of the 19th century. Certainly, Newton did oppose some of Descartes' theories, but fundamentally his attitude was thoroughly Cartesian. Curiously enough, the same applies to d'Alembert, who considered himself one of Newton's disciples and actually played an important part in the transmission of Newton's ideas (his pupils Lagrange and Laplace were among the founders of the French school, which was the leading one in mathematical physics during the greater part of the 19th century); a closer examination of d'Alembert's studies and papers shows, however, that he was equally influenced by Cartesian ideas, which he received from Descartes' sagacious follower Malebranche.

Well, what does Descartes himself say in his book on method, which appears as the manifesto of the new philosophy, exclusively based on human reason? He wants the reader to believe that he has invented his method merely by his own thinking; he tells us also, however, that he had been travelling about to watch the behaviour of society. As we know, he was strongly influenced by the Dutch community at a time when it was the most advanced in both technical and scientific respects. But this extreme individualist, who equates existence with consciousness, also asks the question why he should concern himself with science, and answers it by emphasizing the usefulness of science to the community. He concludes by saying that the noblest science is medicine, because health is man's most precious possession, and that the study of medicine should therefore give the scientist the highest satisfaction.

This clear recognition of the social role of science is not the same, however, as saying that scientific research would find its inspiration or motivation exclusively in direct social tasks. The question how far social requirements have exerted a furthering influence on the progress of science is one of the most difficult in historical research, and the answer to it in individual cases demands an extremely critical analysis in order to avoid faulty evaluations. In this connexion I might recall a particularly instructive example of such a mistaken interpretation, which was even announced with some rather superfluous dramatization. It happened at a conference in London in the thirties. A Russian delegation turned up too late, as usual, and the somewhat stiff Englishmen did not allow them any time to present their contributions. Thereupon, they had these printed in record time and distributed among the participants. One of their papers caused a sensation: it purported to show, in the spirit of historical materialism, that Newton's discovery of the law of gravitation was prompted by his endeavours to satisfy one of the most pressing social requirements of the time – the determination of longitude at sea, a problem that had arisen from the expansion of long distance sailing and which was undoubtedly very important in connexion with the remunerative trade in spices. It can indeed be documented that Newton took an early interest in geography (he edited an issue of Varenius' manual) and even industriously studied books on navigation.

That such a thesis, not very original and (to say the least) very superficial, could then impress the English, only evidences the sad neglect of Newton studies in England, as a result of which they were not prepared to take a more critical view of the matter. Since then one has digged enough in Newton's papers to form a clear and consistent picture of the motivation that dominated his whole activity, not only in science, but also in history and theology: Newton, whose attitude was at once rationalistic and deeply religious, regarded as his life's task to arrive by rational means at the fullest possible knowledge of God's plan of creation – reason was indeed the faculty God had endowed us with for this very purpose. He tried to achieve it partly by disclosing the laws established by God for the motion and constitution of matter, partly by deciphering the rational code which God had used to inform mankind of his intentions in the prophetic books.

A close study of Newton's decisive contribution to the theory of planetary motion makes it appear as the natural completion of a long evolution, which unfolds itself – from the Ptolemaic tradition via Copernicus and Kepler to Newton – according to its intrinsic problem formulation, its own dialectic. Certainly, each step of this development did bear practical fruit: the Copernican tables of planetary motions were better than the previous ones, Kepler's tables even more so; but it is much more significant that each step corresponded to the elimination of assumptions that had proved too narrow, to a widening of the whole picture of the world, whose epistemological importance went far beyond the practical requirements of society, even beyond society's capacity to assimilate unfamiliar ideas. This intrinsic power of reason, tending to the full unfolding of logical trains of thought, has been as efficient for the advance of science as the external incitements which arise from the unfolding of social forces.

Most often the two complementary motives occur in complicated interplay, and it then hardly makes sense to emphasize the one at the exclusion of the other.

However wrong it was to associate the problem of longitude with Newton, it is nevertheless quite true that this problem had a considerable influence on the progress of physics. When Galilei discovered Jupiter's moons, he saw at once that the problem of longitude could be solved by using the motions of the moons as a celestial clock, and he spent many nights patiently observing their changing configurations; later he even tried to sell the method to the Dutch. Then came Rømer, who carried on these observations with improved instruments; he did not bring the practical problem nearer to a solution, but made his famous discovery of the finite velocity of propagation of light – a discovery without any logical connexion with the original purpose of the observations. How many examples could be given of such unexpected turns in the process of discovery, exhibiting how loosely it is related to its social motivation!

Eventually, it turned out that astronomical observations were not at all suited for the practical determination of longitude. The final solution of the problem was a purely technical achievement: the construction of a sufficiently strong and stable chronometer. This was the way Huygens – with his sound practical common sense – had directly recognized as the right one, but the technology of his time could not cope with the task; it took as much as a whole century from Huygens' experiments of 1659 until Harrison managed to make the first chronometer which satisfied all requirements. This aspect of the story of the longitude problem is a clear case of direct social influence upon an extremely important branch of science; it was in the course of his work with this problem that Huygens developed his profound dynamical ideas about the motion of the pendulum. Here again, however, we must guard us against too simple a schematization. If one enquires why the technical development took so long, one finds that it was dependent on advances in entirely different domains, namely in our knowledge of the elastic and thermal properties of metals. Only very thorough studies in widely different directions – and not smart generalities – may elucidate so complex an interplay between research and society.

Let us now see what we can learn from the psychological investigations I mentioned as the other important source of information available to us. In the first place, we can definitely conclude that rational thinking is no innate faculty: at birth the infant's nervous system is only provided with a few indispensable reflex mechanisms, but is otherwise quite blank and susceptive of imprint of a variety of sensori-motoric schemes, which the child diligently collects and starts coordinating. At the age of 18 months the infant has got so far as to distinguish between the perceptions belonging to its interior world and those coming from outside, and may thus be said to have recognized the untenability of idealism. It may be more surprising that the logic we use as adults is not natural to the child's way of thinking, but is only acquired after a long period of development, during which the child goes by his own logic, not based on abstract concepts but on concrete representations of operations which are imagined to be carried out with a definite purpose in view; this means that at this stage of development action predominates over contemplation, and finalistic causality over physical causality. Decisive for the transition to rational thinking is the recognition of the specific power of language, which resides in the property of words, abstract concepts, to serve as substitutes for the corresponding representations, and in the fact that such concepts detached from concrete representations may be combined and related to each other far more readily than these representations. Now, language is no spontaneous invention of the individual child: it is a method of communication that the child learns from society – one of the richest social traditions which give a distinctive mark to human behaviour. Only very few societies, however, have developed language to an instrument for rational thinking: this requires an incitement which can only be derived from a complex economic activity, like the modern industrial mode of production, or the abortive start to such a production mode in Hellenistic times. Therefore, it is very instructive to find that even children of our highly developed society only discover the logical use of language at the latest stage of the formation of their reasoning faculty.

On the whole, there is a wide-going parallelism between the reactions of the individual child to natural phenomena in the successive stages of his mental development and the historical development of science. The comparisons which can be made in this respect are so striking that the points of conformity can hardly be accidental. During the first stage of development, between the ages of about 3 to 7 years, the child lives in a mythical, animistic world of imagination just like the so-called 'primitive' peoples; at the same time, for the child as well as for the primitive hunter, animism or magic by no means exclude acquisition of realistic bodies of knowledge and skills. The latter become predominant in the following stage, when the child operates with concrete representations of his actions, and on this basis constructs, with uninhibited fantasy, explanations of the natural processes which bear the mark of the dynamical and finalistic tendencies of his thinking. Now, these explanations are often exactly like those we find in the writings of the Greek physicists: thus a little boy actually proposed to explain the continued motion of a stone after it had left the hand that had thrown it, in exactly the same way as Aristotle – the air pushed by the stone streams back behind it and pushes it further on. So it is more than a way of speaking to refer to the Greek view of the world as the 'childhood' of science. For Aristotle, rational thinking was the special mode of thinking of the philosophers, as opposed to that of the common people – a mode of thinking so new that it was a great task for him to formulate its laws.

I shall limit myself to these disconnected indications; they can only give a glimpse of the prospects that an attack on the epistemological problems by scientific method may open.

BIBLIOGRAPHY OF THE WRITINGS OF LÉON ROSENFELD

1922: (a) 'Sur l'homogénéité d'une définition', Univ. de Liège, *Bull. Scient. de l'Association des Elèves des Ecoles Spéciales* (AEES) **20** 9–11 (1922)

(b) 'Sur la trisection de l'angle', Univ. de Liège, *Bull. Scient. de l'AEES* **20** 86–89 (1922)

1923: (a) 'De quelques "quadrateurs" du cercle', Univ. de Liège, *Bull. Scient. de l'AEES* **20** 129–144 (1923)

(b) 'Problèmes de tracés continus', Univ. de Liège, *Bull. Scient. de l'AEES* **21** 37–58 (1923)

(c) Notices Biographiques, 'Eugène Catalan', Univ. de Liège, *Bull. Scient. de l'AEES* **21** 65–70 (1923)

(d) 'Questions à résoudre', Univ. de Liège, *Bull. Scient. de l'AEES* **21** 113 (1923)

1924: (a) 'Sur quelques paralogismes mathématiques', Univ. de Liège, *Bull. Scient. de l'AEES* **21** 135–150 (1924)

(b) 'La biréfringence électrique', Univ. de Liège, *Bull. Scient. de l'AEES* **22** 158–178 (1924)

1927: (a) 'L'électron magnétique en mécanique ondulatoire', *Bull. Acad. Roy. Belgique* (5) **13** 326–328 (1927)

(b) 'L'univers à cinq dimensions et la mécanique ondulatoire', *Bull. Acad. Roy. Belgique* (5) **13** 304–325, 447–458, 573–579, 661–682 (1927)

(c) 'L'électron magnétique et la mécanique ondulatoire', *Compt. Rend.* **184** 1540–1541 (1927)

(d) 'La théorie des couleurs de Newton et ses adversaires', *Isis* **9** 44–65 (1927)

(e) 'Le problème logique de la définition des nombres irrationnels', *Isis* **9** 345–358 (1927) (For English translation see [1977c])

1928: (a) (*with E. E. Witmer*) 'Über die Hohlraumstrahlung und die Lichtquantenhypothese', *Z. Phys.* **47** 517–521 (1928)

(b) (*with E. E. Witmer*) 'Über die Beugung der de Broglieschen Wellen am Krystallgitter', *Die Naturwissenschaften* **16** 149 (1928)

(c) (*with E. E. Witmer*) 'Über die Beugung von de Broglieschen Wellen an Krystallgittern', *Z. Phys.* **48** 530–540 (1928)

(d) 'Brechungsindex der Elektronen und Diamagnetismus', *Die Naturwissenschaften* **17** 49–50 (1929)

(e) (*with E. E. Witmer*) 'Über den Brechungsindex der Elektronenwellen', *Z. Phys.* **49** 534–540 (1928)

(f) 'Quantenmechanische Theorie der Natürlichen optischen Aktivität von Flüssigkeiten und Gasen', *Z. Phys.* **52** 161–174 (1928)

(g) 'René-François de Sluse et le problème des tangentes', *Isis* **10** 416–434 (1928)

(h) 'Le premier conflit entre la théorie ondulatoire et la théorie corpusculaire de la lumière', *Isis* **11** 111–122 (1928)

1929: (a) 'Zur Theorie des Faradayeffekts', *Z. Phys.* **57** 835–854 (1929)

(b) 'Über die longitudinalen Eigenlösungen der Heisenberg-Paulischen elektromagnetischen Gleichungen', *Z. Phys.* **58** 540–555 (1929)

(c) 'Edmund Hoppe (1854–1928)', *Isis* **13** 45–50 (1929)

1930: (a) 'Bemerkung über die Invarianz der kanonischen Vertauschungsrelationen', *Z. Phys.* **63** 574–575 (1930)

(b) 'Zur Quantelung der Wellenfelder', *Ann. der Phys.* (5) **5** 113–152 (1930)

(c) 'Über die Gravitationswirkungen des Lichtes', *Z. Phys.* **65** 589–599 (1930)

1931: (a) (*with J. Solomon*) 'Sur la théorie quantique du rayonnement', *J. de Phys. et Rad.* (7) **2** 139–147 (1931)

(b) (*with J. Solomon*) 'Zur Theorie der Hohlraumstrahlung', *Die Naturwissenschaften* **19** 376 (1931)

(c) 'Zur Kritik der Diracschen Strahlungstheorie', *Z. Phys.* **70** 454–462 (1931)

(d) 'Zur korrespondenzmässigen Behandlung der Linienbreite', *Z. Phys.* **71** 273–278 (1931)

(e) 'Bemerkung zur korrespondenzmässigen Behandlung des relativistischen Mehrkörperproblems', *Z. Phys.* **73** 253–259 (1931)

(f) 'Über die quantentheoretische Behandlung der Strahlungsprobleme', Estratto dagli *Atti del Convegno di Fisica Nucleare* della 'Fondazione Alessandro Volta': 5–9 (1931)

1932: (a) 'Über eine mögliche Fassung des Diracschen Programms zur Quantenelektrodynamik und deren formalen Zusammenhang mit der Heisenberg-Paulischen Theorie', *Z. Phys.* **76** 729–734 (1932)

(b) 'La théorie quantique des champs', *Annales de l'Institut H. Poincaré* **2** 25–91 (1932)

(c) 'Marcus Marcis Untersuchungen über das Prisma und ihr Verhaltnis zu Newtons Farbentheorie, *Isis* **17** 325–330 (1932) (For English translation see [1978e])

1933: (a) (*with Y. Cambresier*) 'On the Dissociation of Molecules in the Atmospheres of the Stars of the Main Sequence', *Monthly Notices of the Roy. Astron. Soc.* **93** 710–723 (1933)

(b) (*with N. Bohr*) 'Zur Frage der Messbarkeit der elektromagnetischen Feldgrössen', *Mat.-fys. Medd. Dan. Vid. Selsk.* **12** no. 8 (1933) (For English translation see [1978l])

(c) 'The Dissociation of Molecules in the Atmospheres of the Carbon Stars', *Monthly Notices of the Roy. Astron. Soc.* **93** 724–729 (1933)

1935: (a) 'Kvanteteori og Feltfysik', *Fysisk Tidsskrift* **33** 109–121 (1935)

(b) (*with S. Chandrasekhar*) 'Production of Electron Pairs and the Theory of Stellar Structure', *Nature* **135** 999–1000 (1935)

1936: (a) 'Sur l'enseignement au Danemark', *Bull. Assoc. Amis de l'Univ. de Liège* **8** 135–149 (1936)

(b) 'La première phase de l'évolution de la théorie des quanta', *Osiris* **2** 149–196 (1936) (For English translation see [1978i])

1937: (a) (*with P. Swings*) 'Considerations Regarding Interstellar Molecules', *Astrophys. J.* **86** 483–486 (1937)

(b) 'Le dualisme entre ondes et corpuscules', *Archeion* **19** 74–77 (1937)

(c) 'C'est un animateur, un apôtre (Langevin 65 years old)', *L'Humanité*, 22 Jan. 1937

1938: (a) 'Remarques sur la question des précurseurs', *Archeion* **21** 74–77 (1938)

1939: (a) (*with C. Moller*) 'Theory of Mesons and Nuclear Forces', *Nature* **143** 241–242 (1939)
(b) (*with C. Moller*) 'The Electric Quadrupole Moment of the Deuteron and the Field Theory of Nuclear Forces', *Nature* **144** 476–477 (1939)
(c) (*with C. Møller and S. Rozental*) 'Connection between the Life-Time of the Meson and the Beta-Decay of Light Elements', *Nature* **144** 629–630 (1939)
1940: (a) (*with C. Møller*) 'On the Field Theory of Nuclear Forces', *Mat.-fys. Medd. Dan. Vid. Selsk.* **17** no. 8 (1940)
(b) 'Sur le tenseur d'impulsion-énergie', *Mémoires Acad. Roy. de Belgique* **18** no. 6: 1–30 (1940) (For English translation (see [1978q])
1941: (a) 'La genèse des principes de la thermodynamique', *Bull. Soc. Roy. Sc. Liège* **10** 199–212 (1941) (For English translation see [1978g])
1942: (a) 'Ontwikkeling van de causaliteitsidee', *Inaugurale rede Univ. Utrecht* (Amsterdam: North-Holland Publ. Co., 1942) pp. 1–27
(b) (*with J. K. Lubanski*) 'Sur la représentation des champs mésiques dans l'espace à cinq dimensions', *Physica* **11** 117–134 (1942)
(c) 'Meson Theories in Five Dimensions', *Proc. of Nederl. Akad. van Wetenschappen* (Amsterdam) **45** no. 2: 155–158 (1942).
(d) 'L'évolution de l'idée de causalité', *Mém. Soc. Sc. Liège* **6** 59–87 (1942) (For English translation see [1978m])
(e) 'Sur la définition du spin d'un champ de rayonnement', *Bull. Acad. Roy. Belg.* **28** 562–568 (1942) (For English translation see [1978r])
(f) 'Sur le comportement d'un ensemble canonique lors d'une transformation adiabatique', *Proc. of Nederl. Akad. van Wetenschappen* (Amsterdam) **45** 970–972 (1942) (For English translation see [1978s])
1943: (a) 'Polarisatieverschijnselen van electronenstralen', *Nederlands Tijdschrift voor Natuurkunde* **10** 53–70 (1943)
(b) (*with C. Møller*) 'Electromagnetic Properties of Nuclear Systems in Meson Theory', *Mat.-fys. Medd. Dan. Vid. Selsk.* **20** no. 12: 1–66 (1943)
(c) 'Elementaire kernprocessen en energieproductie in sterren', *Nederlands Tijdschrift voor Natuurkunde* **10** 365–375 (1943)
1944: (a) 'Utviklingen av Årsaksbegrepet', *Fra Fysikkens Verden* **49** 49–61, 113–122 (1944)
(b) 'Wisselwerkingen tussen elementaire deeltjes', *Nederlands Tijdschrift voor Natuurkunde* **11** 124–133 (1944)
1945: (a) (*with J. K. Lubanski*) 'On the Disintegration of the Deuteron by Electron Impact', *Experientia* **1** no. 6 (1945)
(b) 'Penetration of Fast Nucleons into Heavy Atomic Nuclei', *Nature* **156** 141–143 (1945)
(c) 'Non-Central Coupling in the Mixed Meson Theory of Nuclear Forces', *Mat.-fys. Medd. Dan. Vid. Selsk.* **23** no. 13: 3–17 (1945)
(d) 'Niels Bohr, naar aanleiding van zijn 60e verjaardag', *De Vrije Katheder* **5** no. 28: 268 (19 October 1945)
(e) *Niels Bohr, An Essay Dedicated to Him on the Occasion of His Sixtieth Birthday 1945* (Amsterdam: North-Holland Publ. Co., 1945, sec. ed. 1961), (Nordita reprint No. 57) (Reprinted in [1978a], pp. 313–326)
(f) 'My Initiation', *J. of Jocular Physics* **2** 7 (Institute of Theoretical Physics, Copenhagen, October 1945) (Reprinted in [1978a], pp. xxxi–xxxiv)

1946: (a) *De ontsluiting van de atoomkern* (Amsterdam: De Arbeiderspers, 1946) 132 pp.
(b) 'Ierse indrukken', *De Vrije Katheder* **6** no. 16: 245 (1946)
(c) 'Physische verschijnselen in gasnevels', *Nederlands Tijdschrift voor Natuurkunde* **12** 156–167 (1946)

1947: (a) 'The Two-Nucleon Problem', *Phys. Soc. Cambridge Conf. Rep.* 133–142 (1947)
(b) 'On the Method of History of Science', *Archives Intern. d'Histoire des Sciences* **1** 1–6 (1947) (Reprinted in [1978a], pp. 3–6)

1948: (a) 'Phenomenological Description of Nuclear Forces and Meson Properties', *Research* (supplem. vol. Colston papers) 135–140 (1948)
(b) *Nuclear Forces* (Amsterdam: North-Holland Publ. Co., 1948) xx + 540 pp.
(c) 'J. H. Lubanski' (*obituary*), *Acta Phys. Pol.* **9** 63 (1948)
(d) 'Remarques sur l'organisation de la science', *L'organisation de la recherche scientifique* (Hommage national à Paul Langevin et Jean Perrin) (Paris, 1948) pp. 50–60 (For English translation see [1978t])
(e) 'Dynamisch denken in de wetenschap', *De Vrije Katheder* **7** no. 42: 668 (1948)

1949: (a) (*with K. J. le Couteur*) 'Canonical Transformations of the Hamiltonian in Meson Field Theory', *Phil. Mag.* (7) **11** 151–157 (1949)
(b) 'Some Impressions of University Life in Belgium and Holland', *Universities Quarterly* **3** 593–599 (1949)

1950: (a) (*with N. Bohr*) 'Field and Charge Measurements in Quantum Electrodynamics', *Phys. Rev.* **78** 794–798 (1950) (Reprinted in [1978a], pp. 401–412)
(b) 'Stationary States of Light Nuclei', *Helv. Phys. Acta.* **23** suppl. 3: 211–224 (1950)
(c) 'Meson Fields and Nuclear Forces', *Progr. Theor. Phys.* **5** 519–522 (1950)
(d) 'Problems of Nuclear Forces', *Les particules élémentaires*, Rapports et discussions du 8e conseil de physique Solvay: 179 (1950)
(e) 'Early History of Quantum Mechanics', *Nature* **166** 883–884 (1950)
(f) 'Professor Einstein's Dilemma', *The Listener* **44** 823 (1950)
(g) 'View of the Universe, Einstein's Two Equations', *Manchester Guardian* (25 March 1950)

1951: (a) *Theory of Electrons* (Amsterdam: North-Holland Publ. Co., 1951, and New York: Dover Publications Inc., 1965) 119 pp.
(b) 'Electromagnetic Properties of Nuclei and Nuclear Structure', *Physica* **17** 461–483 (1951)
(c) 'Outlook of the Philosophy of Science', Convocation address, Patna University (1951) pp. 1–11
(d) 'Achievement of a Nineteenth Century Physicist: the Joule Museum Established', *Manchester Guardian* (4 July 1951)
(e) 'The Dome of Discovery, Science Advertised not Portrayed', *Manchester Guardian* (12 May 1951)

1952: (a) 'Joule's Scientific Outlook', *Bull. Brit. Soc. Hist. Sc.* **1** 169–176 (1952) (Reprinted in [1978a], pp. 112–122)
(b) 'Prof. H. A. Kramers' (*obituary*), *Nature* **170** 99–100 (1952)
(c) *L'exploration du noyau atomique* (Paris: Hermann, 1952)
(d) 'The Ghost of the "Aether" ', *Manchester Guardian* (19 Jan. 1952)

1953: (a) (*with N. Bohr*) 'Mesure des champs et de la charge en electrodynamique quantique', *Particules Fondamentales et Noyaux* (*Colloques Internationaux du Centre Nat. de la Rech. Scient.*, vol. 38) (Paris, CNRS, 1953) p. 87 (French translation of [1950a])

(b) 'Collisions entre nucléons et propriétés nucléaires', *ibid.*, pp. 117–125
(c) 'Strife about Complementarity', *Science Progress* **163** 393–410 (1953) (Reprinted in [1977a], pp. 465–483; English translation of [1953g])
(d) 'Problems of Interpretation of Quantum Electrodynamics', *Physica* **19** 859–868 (1953)
(e) 'Introduction to *Essays on the Social History of Science*', *Centaurus* **3** no. 1–2: 7–11 (1953)
(f) 'Problèmes d'interpretations en electrodynamique quantique', *J. de Phys. Radium* **14** no. 12 (1953)
(g) 'L'évidence de la complementarité', in *Louis de Broglie, physicien et penseur* (Paris: Albin Michel, 1953) (For English translation see [1953c])
(h) 'The Philosophy of Atomic Physics', *The Listener* **49** 215–216, 253–254 (1953)

1954: (a) 'Causalité statistique et ordre en physique et en biologie', *Sep. Anais da Acad. Brasileira de Ciências* **26** 47–50 (1954)
(b) 'A filosofia da fisica atômica', *Ciência e Cultura* **6** 67–72 (1954)
(c) 'Rationalism in Antiquity', *Bull. Brit. Soc. Hist. Sc.* **1** 255–257 (1954) (Reprinted in [1978a], pp. 29–31)
(d) *La structure du noyau atomique*, Rencontres 130–167 (Hainaut, Belgique, 1954)
(e) 'La responsabilité sociale de l'homme de science', *Comprendre*, no. 12 (1954)
(f) 'Social Conditions and the Development of Science', *Nature* **173** 1102 (1954)
(g) 'Marie Curie Celebrations in Warsaw', *Nature* **174** 959–960 (1954)

1955: (a) 'On the Foundations of Statistical Thermodynamics', *Acta Phys. Polon.* **14** 3–39 (1955) (Reprinted in [1978a], pp. 762–807)
(b) 'On Quantum Electrodynamics', in *Niels Bohr and the Development of Physics* (London: Pergamon Press, 1955) pp. 70–95 (Reprinted in [1978a], pp. 413–441)
(c) 'J. Podolanski' (*obituary*), *Nature* **175** 795 (1955)
(d) 'A Voyage to Laplacia', *J. of Jocular Physics* **3** (Institute of Theoretical Physics, Copenhagen, 1955) (Reprinted in [1978a], pp. 704–708)

1956: (a) 'The Velocity of Light and the Evolution of Electrodynamics', *Suppl. Nuovo Cim.* **4** 1630–1669 (1956) (Reprinted in [1978a], pp. 134–177)
(b) 'Conclusions', Amsterdam Nuclear Reactions Conference, *Physica* **22** 1115–1123 (1956)
(c) 'Remarks on the Charge Distribution in Nuclei', *Nucl. Phys.* **2** 450–453 (1956)
(d) 'O rasseyanii bystrykh chastits yadrami' (On the Scattering of Fast Particles by Nuclei), *Usp. Fiz. Nauk* **60** 565–572 (1956)
(e) (*with A. P. Hatton*) 'An Analysis of Joule's Experiments on the Expansion of Air', *Centaurus* **4** 311–318 (1956) (Reprinted in [1978a], pp. 123–133)
(f) 'J. D. Bernal, Science in History' (*book review*), *Centaurus* **4** 285–296 (1956) (Reprinted in [1978a], pp. 7–15)

1957: (a) 'Misunderstandings about the Foundations of Quantum Theory', *Observation and Interpretation*, ed. S. Körner (*Colston Papers* **9**) (London: Butterworths Scientific Publications, 1957) pp. 41–61 (Reprinted in [1978a], pp. 495–502)
(b) 'Komplementarność', *Kosmos* (Warsaw) **B3** 257–262 (1957)
(c) 'Rozkład ładunku w jadrach' (Charge Distribution in Nuclei), *Postepy Fizyki* (Warsaw) **8** 407–415 (1958)

1958: (a) 'Nuclear Orientation', *Physica* **24** Suppl. 63 (1958)
(b) 'Complementarity and Statistics. I and II', *Det Kgl. Norske Vid. Selsk. Forh.* **31** nos. 9–10 (1958) (Nordita reprint No. 2) (Reprinted in [1978a], pp. 484–494)

1959: (a) 'Interactions nucléaires aux basses énergies et structure des noyaux'. Conclusions, *Comptes Rendus du Congrès International de Physique Nucléaire*, Paris 1958 (Paris: Dunod, 1959) pp. 319–332 (Nordita reprint No. 22)

(b) 'Max Planck et la définition statistique de l'entropie', *Max-Planck-Festschrift* (*1958*) (Berlin: VEB Deutscher Verlag der Wissenschaften, 1959) pp. 203–211 (Nordita reprint No. 25) (For English translation see [1978j])

(c) 'Jacques Solomon' (*obituary*), in *A la mémoire de 15 savants français assassinés par les Allemands 1940–1945* (Paris: Comité à la mémoire des savants français victimes de la barbarie allemande 1940–1945, 1959) pp. 25–28 (For English translation see [1978k])

(d) 'The Statistical Character of Atomic Processes', *Current Science* (Bangalore) **28** 227–229 (1959)

1960: (a) 'Introductory Considerations on Elementary Constituents and Their Interactions', *Interazioni Deboli*, ed. L. A. Radicati (*Rendiconti della Scuola Intern. di Fisica "Enrico Fermi"*), Corso 11, Varenna, 1959 (Bologna: Zanichelli, 1960) pp. 1–8 (Nordita reprint No. 44)

(b) 'Oriented Nuclei', *Interazioni Deboli*, ed. L. A. Radicati (*Rendiconti della Scuola Intern. di Fisica "Enrico Fermi"*), Corso 11, Varenna, 1959 (Bologna: Zanichelli, 1960) pp. 197–250 (Nordita reprint No. 45)

(c) 'The Correspondence of Isaac Newton, vol. 1: 1661–1675, ed. by H. W. Turnbull' (*book review*), *Nature* **187** 537–538 (1960) (Nordita reprint No. 39)

(d) 'Heisenberg, Philosophy and Science' (*book review:* W. Heisenberg, *The Revolution in Modern Science*), *Nature* **186** 830–831 (1960)

(e) 'Hasard et nécessité dans la nature', *Bull. de l'Institut Français de Copenhague* **76** 109–112 (1960)

1961: (a) (*with J. Humblet*) 'Theory of Nuclear Reactions (I). Resonant States and Collision Matrix', *Nucl. Phys.* **26** 529–578 (1961) (Nordita reprint No. 62)

(b) 'Theory of Nuclear Reactions (II). The Foundation of the Optical Model', *Nucl. Phys.* **26** 579–593 (1961) (Nordita reprint No. 63)

(c) 'On the Isolated and Adiabatic Susceptibilities', *Physica* **27** 67–78 (1961) (Nordita reprint No. 55) (Reprinted in [1978a], pp. 747–761)

(d) 'Ryosiron no Kiso to Sohosei', *Nippon Buturigakkai Si* (*J. Phys. Soc. Japan*) **16** 218–226 (1961) (Nordita reprint No. 58)

(e) 'Foundations of Quantum Theory and Complementarity', *Nature* **190** 384–388 (1961) (Nordita reprint No. 59) (Reprinted in [1978a], pp. 503–516)

(f) 'Marginalia to Newton's Correspondence', *Isis* **52** 117–119 (1961) (Nordita reprint No. 56)

(g) 'The Correspondence of Isaac Newton, vol. 2: 1676–1687, ed. by H. W. Turnbull' (*book review*), *Nature* **189** 171–172 (1961) (Nordita reprint No. 49)

(h) 'Atomforskeren, Om atomfysikerens opgaver, mål og arbejdsmetoder', *Perspektiv* (Copenhagen), **8** 31–34 (1961)

1962: (a) 'Phénoménologie des forces nucléaires', *Colloque sur la physique nucléaire des basses énergies* (Brussels: Institut Interuniversitaire des Sciences Nucléaires, 1962) pp. 355–375 (Nordita reprint No. 95)

(b) 'Questions of Irreversibility and Ergodicity', *Ergodic Theories*, ed. P. Caldirola (*Rendiconti della Scuola Intern. d. Fisica "Enrico Fermi"*), corso 14, Varenna, 1960 (Bologna: Zanichelli, 1962) pp. 1–20 (Nordita reprint No. 69) (Reprinted in [1978a], pp. 808–829)

(c) 'Complementarity', *ICSU Review* **4** 42–47 (1962) (Nordita reprint No. 66)

(d) 'Le conflit épistémologique entre Einstein et Bohr', *Revue de Métaphysique et de Morale* **67** 147–151 (1962) (Nordita reprint No. 92) (For English translation see [1963c])

(e) 'Einstein i razvitiye fiziko-matematicheskoi mysli' (Einstein and the Development of Physical–Mathematical Thought) (sbornik statei) (Moscow: Izd. Akad. Nauk, 1962) pp. 89–93

(f) 'Nogle minder om Niels Bohr', *Fysisk Tidsskrift* **60** 65–75 (1962) (Nordita reprint No. 105)

(g) 'Niels Bohr 7. oktober 1885–18 november 1962, Niels Bohrs indsats i epistemologien', *Overs. Dan. Vid. Selsk.* 90–97 (1962–1963) (Nordita reprint No. 109)

(h) 'The Correspondence of Isaac Newton, vol. 3: 1688–1694, ed. by H. W. Turnbull' (*book review*), *Nature* **195** 414–416 (1962) (Nordita reprint No. 82)

(i) 'Miss Margrethe Have (*in memoriam*)', *Nucl. Phys.* **31** 689 (1962)

1963: (a) 'On Quantization of Fields', *Nucl. Phys.* **40** 353–356 (1963) (Nordita reprint No. 97) (Reprinted in [1978a], pp. 442–445)

(b) 'The Foundation of the Optical Model', *Proc. of the Conference on direct interactions and nuclear reaction mechanisms, Padua, 1962* (New York: Gordon and Breach, 1963) pp. 69–79 (Nordita reprint No. 112)

(c) 'The Epistemological Conflict between Einstein and Bohr', *Z. Phys.* **171** 242–245 (1963) (Nordita reprint No. 91) (Reprinted in [1978a], pp. 517–521; English translation of [1962d])

(d) 'Matter and Force after Fifty Years of Quantum Theory', *Physics in the Sixties*, ed. S. K. Runcorn (London: Oliver and Boyd, 1963) pp. 1–22 (Nordita reprint No. 120) (Reprinted in [1978a], pp. 247–265

(e) Niels Bohr, *On the Constitution of Atoms and Molecules*, Reprint of 1913 papers with an Introduction by L. Rosenfeld (Copenhagen: Munksgaard Ltd., 1963) pp. xi–lii

(f) 'Niels Bohr, 7 October 1885–18 November 1962', *Nucl. Phys.* **41** 1–16 (1963) (Nordita reprint No. 100)

(g) 'Niels Bohr (7 October 1885–18 November 1962)', *Bull. Soc. Belg. de Physique* Ser. III, No. 4: 252–256 (1963) (Nordita reprint No. 110)

(h) 'Niels Bohr's Contribution to Epistemology', *Physics Today* **16** no. 10: 47–52 and 54 (1963) (Nordita reprint No. 121) (Reprinted in [1978a], pp. 522–535)

(i) 'Nogle minder om Niels Bohr', *Fysisk Tidsskrift* **60** 65–75 (1963) (Nordita reprint No. 105)

1964: (a) 'Newton as a Historian' (*book review:* F. E. Manuel, *Isaac Newton, Historian*), *Nature* **202** 43 (1964) (Nordita reprint No. 125)

(b) 'Niels Bohr', *Voprosy Istorii Estestvoznaniia i Tekhniki*, no. 17 (1964)

1965: (a) 'Theory of Nuclear Reactions (VIII). Time Evolution of the Scattering Process', *Nucl. Phys.* **70** 1–27 (1965) (Nordita reprint No. 167)

(b) 'On the Foundations of the Theory of Nuclear Reactions', *Proc. Int. Conf. on Elementary Particles, Kyoto, 1965* (Kyoto, 1966) pp. 182–205

(c) 'The Measuring Process in Quantum Mechanics', *Suppl. Prog. Theor. Phys., Commemoration Issue for the 30th Anniversary of the Meson Theory by Dr. H. Yukawa* 222–231 (1965) (Nordita reprint No. 186) (Reprinted in [1978a], pp. 536–546)

(d) 'Newton and the Law of Gravitation'. *Arch. Hist. Exact Sc.* **2** 365–386 (1965) (Nordita reprint No. 173) (Reprinted in [1978a], pp. 58–87)

(e) 'Niels Bohr et sa contribution à l'épistémologie', *Nucleus* **6** 167–172 (1965) (Nordita reprint No. 185)

1966: (a) 'Quantentheorie und Gravitation', *Entstehung, Entwicklung und Perspektiven der Einsteinschen Gravitationstheorie: Einstein-Symposium vom 2. bis 5. November 1965 in Berlin* (Berlin: Akademie-Verlag, 1966) pp. 185–197 (Nordita reprint No. 200) (For English translation see [1978n])

(b) 'Tale ved Universitetets Minnehøjtidelighed for Professor Niels Bohr torsdag d. 7. oktober 1965', *Fra Fysikkens Verden* **28** 70–72 (1966) (Nordita reprint No. 221)

1967: (a) 'Descartes i Uppsala, Striderna om "nya filosofien" 1663–1689, Rolf Lindborg' (*book review*), *Isis* **58** 136 (1967) (Nordita reprint No. 243) (For English translation see [1978f])

(b) 'Birth of the Calculus' (*book review: The Mathematical Papers of Isaac Newton*, vol. 1 (1664–1666), ed. D. T. Whiteside), *Nature* **216** 305 (1967) (Nordita reprint No. 248)

(c) (*with Erik Rüdinger*) 'The Decisive Years 1911–1918', *Niels Bohr, His Life and Work as Seen by His Friends and Colleagues*, ed. S. Rozental (Amsterdam: North-Holland Publ., 1967) pp. 38–73

(d) 'Niels Bohr in the Thirties. Consolidation and Extension of the Conception of Complementarity', *ibid.*, pp. 114–136

1968: (a) 'Prospects of the Theory of Nuclear Structure and Nuclear Reactions', *Proc. of the Int. Conf. on nuclear structure, Tokyo, 1967, Suppl. J. Phys. Soc. Jap.* **24** 9–13 (1968) (Nordita reprint No. 264)

(b) 'Dynamical Theory of Nuclear Resonances', in *Spectroscopic and Group Theoretical Methods in Physics, Racah memorial volume* (Amsterdam: North-Holland Publ. Co., 1968) pp. 203–230 (Nordita reprint No. 278) (Reprinted in [1978a], pp. 830–860)

(c) 'Coupling between Compound and Single-Particle Resonances', Addendum to 'Dynamical Theory of Nuclear Resonances', in *Spectroscopic and Group Theoretical Methods in Physics, Racah memorial volume* (Amsterdam: North-Holland Publ. Co., 1968) pp. 203–230 (Nordita reprint No. 278a) (Reprinted in [1978a], pp. 861–865)

(d) 'The Structure of Quantum Theory', Lecture given at 12e Congrès Int. d'Histoire des Sciences, Paris, 1968, *Revue de Synthèse* **3** nos. 49–52 (1968) (Reprinted in [1978a], pp. 866–878)

(e) 'The Conception of the Meson Field: Some Reminiscences and Epistemological Comments', *Suppl. Progr. Theor. Phys.* **41** C1–C7 (1968) (Nordita reprint No. 298); 257–267 (1968) (Nordita reprint No. 307) (Reprinted in [1978a], pp. 327–334)

(f) 'Questions of Method in the Consistency Problem of Quantum Mechanics', *Nucl. Phys.* **A108** 241–244 (1968) (Nordita reprint No. 254) (Reprinted in [1978a], pp. 609–613

(g) *The Method of Physics* (Unesco Report, 1968) © 1968 by Unesco. (Reprinted in [1978a], pp. 614–636)

(h) 'For Scientists and Historians' (*book review: The Mathematical Papers of Isaac

Newton, vol. 2 (1667–70), ed. D. T. Whiteside), *Nature* **218** 989–990 (1968) (Nordita reprint No. 269)

(i) 'The Correspondence of Isaac Newton, vol. 4 (1694–1709), ed. by J. F. Scott' (*book review*), *Nature* **217** 485 (1968) (Nordita reprint No. 255)

(j) 'The Atomic Researcher: The Atomic Physicist's Tasks, Goals and Methods', *Perspektiv* **8** 31 (1968) (Reprinted in [1978a], pp. 892–896)

1969: (a) 'Summary Talk', *Int. Conf. on Clustering Phenomena in Nuclei, Bochum, 1969* (Vienna: IAEA, 1969) pp. 197–201 (Nordita reprint No. 339)

(b) 'Celestial and Terrestrial Physics in Historical Perspective', in *The Application of Modern Physics to the Earth and Planetary Interiors*, ed. by S. K. Runcorn (London: Wiley-Interscience, 1969) pp. 1–6 (Nordita reprint No. 304) (Reprinted in [1978a], pp. 346–353)

(c) 'Newton's Views on Aether and Gravitation', *Arch. Hist. Exact. Sc.* **6** 29–37 (1969) (Nordita reprint No. 313) (Reprinted in [1978a], pp. 88–98)

(d) 'Newton's thinking' (*book review: The Mathematical Papers of Isaac Newton*, vol. 3, ed. D. T. Whiteside), *Nature* **224** 89 (1969) (Nordita reprint No. 312)

(e) 'Flicker in the Darkness' (*book review: Nicole Oresme and the Medieval Geometry of Qualities and Motions*, ed. M. Clagett), *Nature* **222** 197 (1969) (Nordita reprint No. 377) (Reprinted in [1978a], pp. 45–48)

(f) 'Atombegrebets forvandlinger gennem tiderne', *Fysisk Tidsskrift* **67** no. 2–3; 78–91 (1969) (Nordita reprint No. 314) (For English translation see [1978d])

1970: (a) 'Unitarity of the Collision Matrix and Interdependence of Resonance Parameters', *Acta Physica Pol.* (*Niewodniczanski Festschrift*) **A38** 603–619 (1970) (Nordita reprint No. 380)

(b) 'L'évolution des conceptions scientifiques d'Oersted', *Overs. Dan. Vid. Selsk.* 1–17 (1970–1971) (For English translation see [1978h])

(c) 'Berkeley *Redivivus*' (*book review:* W. Heisenberg, *Natural Law and the Structure of Matter*), *Nature* **228** 479 (1970) (Nordita reprint No. 372) (Reprinted in [1978a], pp. 686–687)

(d) 'Aspects techniques et sociaux du développement des organismes européens de recherche scientifique', *Le Monde Scientifique* **14** 13–15 (1970)

(e) 'Technical and Social Aspects of the Development of the European Scientific Research Organizations', *Scientific World* **14** 9–10 (1970) (Reprinted in [1978a], pp. 897–901: English translation of [1970d])

1971: (a) 'Erkendelsesteori pa naturvidenskabeligt grundlag', *Physica Norvegica* **5** 319–326 (1971) (Nordita reprint No. 424) (For English translation see [1978p])

(b) 'Unphilosophical Considerations on Causality in Physics', in *Perspectives in Quantum Theory, Essays in Honor of Alfred Lande*, eds., Yourgrau and van der Merwe (Cambridge, Mass.: MIT Press, 1971) pp. 219–235 (Nordita reprint No. 434) (Reprinted in [1978a], pp. 666–680)

(c) 'Quantum Theory in 1929: Recollections from the First Copenhagen Conference' (Copenhagen: Rhodos, 1971) 15 pp. (Nordita reprint No. 387) (Reprinted in [1977a], pp. 302–312)

(d) 'Men and Ideas in the History of Atomic Theory', *Arch. Hist. Exact. Sc.* **7** 69–90 (1971) (Nordita reprint No. 382) (Reprinted in [1978a], pp. 266–296)

(e) 'Quelques réflections sur la connaissance', *Volume Jubilaire Louis d'Or, Mémoires de la Soc. Roy. Sc. Liège*, 6me série, vol. I, fasc. 2: 187–190 (1971) (Nordita reprint No. 400) (For English translation see [1978o])

(f) 'Sociale og individuelle aspelter af naturvidenskabens udvikling', *Fysisk Tidsskrift* **69** no. 4–5: 97–106 (1971) (Nordita reprint No. 412) (For English translation see [1972g])

1972: (a) (*with C. George and I. Prigogine*) 'The Macroscopic Level of Quantum Mechanics', *Mat.-fys. Medd. Dan. Vid. Selsk.* **38** no. 12: 1–44 (1972) (Nordita reprint No. 466) (Reprinted in part in [1978a], pp. 571–574 and 581–597)

(b) (*with C. George and I. Prigogine*) 'The Macroscopic Level of Quantum Mechanics', *Nature* **240** 25–27 (1972) (Nordita reprint No. 477) (Reprinted in [1978a], pp. 574–581

(c) 'Nuclear Reminiscences', in *Cosmology, Fusion and Other Matters, George Gamow Memorial Volume*, ed. Frederick Reines (Colorado Associated University Press, 1972) pp. 289–299 (Nordita reprint No. 469) (Reprinted in [1978a], pp. 335–345)

(d) 'General Introduction to Irreversibility', in *Irreversibility in the Many-Body Problem*, ed. by J. Biel and J. Rae (New York: Plenum Publ. Corp., 1972) (Nordita reprint No. 527) (Partially reprinted in [1978a], pp. 547–570)

(e) 'Introduction to Niels Bohr, *Collected Works* Vol. 1' (Amsterdam: North-Holland Publ. Co., 1972)

(f) 'Condillac's Influence on French Scientific Thought', in *The Triumph of Culture: 18th Century Perspectives*, ed. by P. Fritz and D. Williams, (Toronto: A. M. Hakkert Ltd., 1972) pp. 157–168 (Nordita reprint No. 497) (Reprinted in [1978a], pp. 655–665)

(g) 'Social and Individual Aspects of the Development of Science', in *Problems of Theoretical Physics* [Igor Tamm Festschrift, in Russian but Rosenfeld's paper appeared in English – Ed.] (Moscow: Nauka, 1972) (Reprinted in [1978a], pp. 902–910. English translation of [1971f])

1973: (a) (*with I. Prigogine, C. George, and F. Henin*) 'A Unified Formulation of Dynamics and Thermodynamics', *Chemica Scripta* **4** 5–32 (1973) (Nordita reprint No. 504)

(b) 'The Wave-Particle Dilemma', in *The Physicist's Conception of Nature*, ed. by J. Mehra (Dordrecht: D. Reidel Publ., 1973) pp. 251–263 (Nordita reprint No. 528) (Reprinted in [1978a], pp. 688–703

1974: (a) 'Statistical Causality in Atomic Theory', in *The Interaction Between Science and Philosophy*, ed. Y. Elkana (The Van Leer Jerusalem Foundation Series) (New York: Humanities Press, 1974) pp. 469–480 (Reprinted in part in [1978a], pp. 547–570)

1977: (a) *Selected Papers of Léon Rosenfeld*, eds. R. S. Cohen and J. Stachel (*Boston Studies in the Philosophy of Science* volume XXI) (Dordrecht: D. Reidel Publ., 1978)

(b) 'Irreversibility – a Lay Sermon', dedicated to K. Bleuler, in [1978a], pp. 681–685.

(c) 'The Logical Problem of the Definition of Irrational Numbers' in [1978a], pp. 16–28; English translation of [1927e]

(d) 'The Transformations of the Atomic Concept through the Ages' in [1978a], pp. 32–44; English translation of [1969f]

(e) 'Marcus Marci's Investigations of the Prism and Their Relation to Newton's Theory of Color" in [1978a], pp. 49–54; English translation of [1932c]

(f) 'R. Lindborg, *Descartes at Uppsala*' (*book review* in [1978a], pp. 55–57; English translation of [1967a]

(g) 'The Genesis of the Laws of Thermodynamics' in [1978a], pp. 99–111; English translation of [1941a]

(h) 'The Evolution of Oersted's Scientific Concepts' in [1978a], pp. 178–192; English translation of [1970b]

(i) 'The First Phase in the Evolution of the Quantum Theory' in [1978a], pp. 193–234; English translation of [1936b]

(j) 'Max Planck and the Statistical Definition of Entropy' in [1978a], pp. 235–246; English translation of [1959b]

(k) 'Jacques Solomon' in [1978a], pp. 297–301; English translation of [1959c]

(l) (*with N. Bohr*) 'On the Question of the Measurability of Electromagnetic Field Quantities' in [1978a], pp. 357–400; English translation of [1933b]

(m) 'The Evolution of the Idea of Causality' in [1978a], pp. 446–464; English translation of [1942d]

(n) 'Quantum Theory and Gravitation' in [1978a], pp. 599–608; English translation of [1966a]

(o) 'Some Reflections on Knowledge' in [1978a], pp. 637–642; English translation of [1971e]

(p) 'Epistemology on a Scientific Basis' in [1978a], pp. 643–654; English translation of [1971a]

(q) 'On thè Energy-Momentum Tensor' in [1978a], pp. 711–735; English translation of [1940b]

(r) 'On the Definition of Spin for a Radiation Field' in [1978a], pp. 736–741; English translation of [1942e]

(s) 'On the Behavior of a Canonical Ensemble during an Adiabatic Transformation in [1977a], pp. 742–746; English translation of [1942f]

(t) 'The Organization of Scientific Research' in [1978a], pp. 881–891; English translation of [1948d]

In addition: numerous book reviews in *Centaurus, Proc. Phys. Soc., Nature, Nucl. Phys., Science Progress, Arch. Hist. Exact. Sc.*, a.o.

INDEX OF NAMES

SYNTHESE LIBRARY

Studies in Epistemology, Logic, Methodology,
and Philosophy of Science

1. J. M. Bocheński, *A Precis of Mathematical Logic.* 1959, X + 100 pp.
2. P. L. Guiraud, *Problèmes et méthodes de la statistique linguistique.* 1960, VI + 146 pp.
3. Hans Freudenthal (ed.), *The Concept and the Role of the Model in Mathematics and Natural and Social Sciences. Proceedings of a Colloquium held at Utrecht, The Netherlands, January 1960.* 1961, VI + 194 pp.
4. Evert W. Beth, *Formal Methods. An Introduction to Symbolic Logic and the Study of Effective Operations in Arithmetic and Logic.* 1962, XIV + 170 pp.
5. B. H. Kazemier and D. Vuysje (eds.), *Logic and Language. Studies Dedicated to Professor Rudolf Carnap on the Occasion of His Seventieth Birthday.* 1962, VI + 256 pp.
6. Marx W. Wartofsky (ed.), *Proceedings of the Boston Colloquium for the Philosophy of Science 1961-1962,* Boston Studies in the Philosophy of Science (ed. by Robert S. Cohen and Marx W. Wartofsky), Volume I. 1963, VIII + 212 pp.
7. A. A. Zinov'ev, *Philosophical Problems of Many-Valued Logic.* 1963, XIV + 155 pp.
8. Georges Gurvitch, *The Spectrum of Social Time.* 1964, XXVI + 152 pp.
9. Paul Lorenzen, *Formal Logic.* 1965, VIII + 123 pp.
10. Robert S. Cohen and Marx W. Wartofsky (eds.), *In Honor of Philipp Frank,* Boston Studies in the Philosophy of Science (ed. by Robert S. Cohen and Marx W. Wartofsky), Volume II. 1965, XXXIV + 475 pp.
11. Evert W. Beth, *Mathematical Thought. An Introduction to the Philosophy of Mathematics.* 1965, XII + 208 pp.
12. Evert W. Beth and Jean Piaget, *Mathematical Epistemology and Psychology.* 1966, XII + 326 pp.
13. Guido Küng, *Ontology and the Logistic Analysis of Language. An Enquiry into the Contemporary Views on Universals.* 1967, XI + 210 pp.
14. Robert S. Cohen and Marx W. Wartofsky (eds.), *Proceedings of the Boston Colloquium for the Philosophy of Science 1964-1966, in Memory of Norwood Russell Hanson,* Boston Studies in the Philosophy of Science (ed. by Robert S. Cohen and Marx W. Wartofsky), Volume III. 1967, XLIX + 489 pp.

15. C. D. Broad, *Induction, Probability, and Causation. Selected Papers.* 1968, XI + 296 pp.
16. Günther Patzig, *Aristotle's Theory of the Syllogism. A Logical-Philosophical Study of Book A of the Prior Analytics.* 1968, XVII + 215 pp.
17. Nicholas Rescher, *Topics in Philosophical Logic.* 1968, XIV + 347 pp.
18. Robert S. Cohen and Marx W. Wartofsky (eds.), *Proceedings of the Boston Colloquium for the Philosophy of Science 1966-1968*, Boston Studies in the Philosophy of Science (ed. by Robert S. Cohen and Marx W. Wartofsky), Volume IV. 1969, VIII + 537 pp.
19. Robert S. Cohen and Marx W. Wartofsky (eds.), *Proceedings of the Boston Colloquium for the Philosophy of Science 1966-1968*, Boston Studies in the Philosophy of Science (ed. by Robert S. Cohen and Marx W. Wartofsky), Volume V. 1969, VIII + 482 pp.
20. J.W. Davis, D. J. Hockney, and W. K. Wilson (eds.), *Philosophical Logic.* 1969, VIII + 277 pp.
21. D. Davidson and J. Hintikka (eds.), *Words and Objections: Essays on the Work of W. V. Quine.* 1969, VIII + 366 pp.
22. Patrick Suppes, *Studies in the Methodology and Foundations of Science. Selected Papers from 1911 to 1969.* 1969, XII + 473 pp.
23. Jaakko Hintikka, *Models for Modalities. Selected Essays.* 1969, IX + 220 pp.
24. Nicholas Rescher *et al.* (eds.), *Essays in Honor of Carl G. Hempel. A Tribute on the Occasion of His Sixty-Fifth Birthday.* 1969, VII + 272 pp.
25. P. V. Tavanec (ed.), *Problems of the Logic of Scientific Knowledge.* 1969, XII + 429 pp.
26. Marshall Swain (ed.), *Induction, Acceptance, and Rational Belief.* 1970, VII + 232 pp.
27. Robert S. Cohen and Raymond J. Seeger (eds.), *Ernst Mach: Physicist and Philosopher*, Boston Studies in the Philosophy of Science (ed. by Robert S. Cohen and Marx W. Wartofsky), Volume VI. 1970, VIII + 295 pp.
28. Jaakko Hintikka and Patrick Suppes, *Information and Inference.* 1970, X + 336 pp.
29. Karel Lambert, *Philosophical Problems in Logic. Some Recent Developments.* 1970, VII + 176 pp.
30. Rolf A. Eberle, *Nominalistic Systems.* 1970, IX + 217 pp.
31. Paul Weingartner and Gerhard Zecha (eds.), *Induction, Physics, and Ethics: Proceedings and Discussions of the 1968 Salzburg Colloquium in the Philosophy of Science.* 1970, X + 382 pp.
32. Evert W. Beth, *Aspects of Modern Logic.* 1970, XI + 176 pp.
33. Risto Hilpinen (ed.), *Deontic Logic: Introductory and Systematic Readings.* 1971, VII + 182 pp.
34. Jean-Louis Krivine, *Introduction to Axiomatic Set Theory.* 1971, VII + 98 pp.
35. Joseph D. Sneed, *The Logical Structure of Mathematical Physics.* 1971, XV + 311 pp.
36. Carl R. Kordig, *The Justification of Scientific Change.* 1971, XIV + 119 pp.
37. Milič Čapek, *Bergson and Modern Physics*, Boston Studies in the Philosophy of Science (ed. by Robert S. Cohen and Marx W. Wartofsky), Volume VII. 1971, XV + 414 pp.

38. Norwood Russell Hanson, *What I Do Not Believe, and Other Essays* (ed. by Stephen Toulmin and Harry Woolf). 1971, XII + 390 pp.
39. Roger C. Buck and Robert S. Cohen (eds.), *PSA 1970. In Memory of Rudolf Carnap*, Boston Studies in the Philosophy of Science (ed. by Robert S. Cohen and Marx W. Wartofsky), Volume VIII. 1971, LXVI + 615 pp. Also available as paperback.
40. Donald Davidson and Gilbert Harman (eds.), *Semantics of Natural Language.* 1972, X + 769 pp. Also available as paperback.
41. Yehoshua Bar-Hillel (ed.), *Pragmatics of Natural Languages.* 1971, VII + 231 pp.
42. Sören Stenlund, *Combinators, λ-Terms and Proof Theory.* 1972, 184 pp.
43. Martin Strauss, *Modern Physics and Its Philosophy. Selected Papers in the Logic, History, and Philosophy of Science.* 1972, X + 297 pp.
44. Mario Bunge, *Method, Model and Matter.* 1973, VII + 196 pp.
45. Mario Bunge, *Philosophy of Physics.* 1973, IX + 248 pp.
46. A. A. Zinov'ev, *Foundations of the Logical Theory of Scientific Knowledge (Complex Logic)*, Boston Studies in the Philosophy of Science (ed. by Robert S. Cohen and Marx W. Wartofsky), Volume IX. Revised and enlarged English edition with an appendix, by G. A. Smirnov, E. A. Sidorenka, A. M. Fedina, and L. A. Bobrova. 1973, XXII + 301 pp. Also available as paperback.
47. Ladislav Tondl, *Scientific Procedures*, Boston Studies in the Philosophy of Science (ed. by Robert S. Cohen and Marx W. Wartofsky), Volume X. 1973, XII + 268 pp. Also available as paperback.
48. Norwood Russell Hanson, *Constellations and Conjectures* (ed. by Willard C. Humphreys, Jr.). 1973, X + 282 pp.
49. K. J. J. Hintikka, J. M. E. Moravcsik, and P. Suppes (eds.), *Approaches to Natural Language. Proceedings of the 1970 Stanford Workshop on Grammar and Semantics.* 1973, VIII + 526 pp. Also available as paperback.
50. Mario Bunge (ed.), *Exact Philosophy – Problems, Tools, and Goals.* 1973, X + 214 pp.
51. Radu J. Bogdan and Ilkka Niiniluoto (eds.), *Logic, Language, and Probability. A Selection of Papers Contributed to Sections IV, VI, and XI of the Fourth International Congress for Logic, Methodology, and Philosophy of Science, Bucharest, September 1971.* 1973, X + 323 pp.
52. Glenn Pearce and Patrick Maynard (eds.), *Conceptual Change.* 1973, XII + 282 pp.
53. Ilkka Niiniluoto and Raimo Tuomela, *Theoretical Concepts and Hypothetico-Inductive Inference.* 1973, VII + 264 pp.
54. Roland Fraïssé, *Course of Mathematical Logic* – Volume 1: *Relation and Logical Formula.* 1973, XVI + 186 pp. Also available as paperback.
55. Adolf Grünbaum, *Philosophical Problems of Space and Time.* Second, enlarged edition, Boston Studies in the Philosophy of Science (ed. by Robert S. Cohen and Marx W. Wartofsky), Volume XII. 1973, XXIII + 884 pp. Also available as paperback.
56. Patrick Suppes (ed.), *Space, Time, and Geometry.* 1973, XI + 424 pp.
57. Hans Kelsen, *Essays in Legal and Moral Philosophy*, selected and introduced by Ota Weinberger. 1973, XXVIII + 300 pp.
58. R. J. Seeger and Robert S. Cohen (eds.), *Philosophical Foundations of Science. Proceedings of an AAAS Program, 1969*, Boston Studies in the Philosophy of

Science (ed. by Robert S. Cohen and Marx W. Wartofsky), Volume XI. 1974, X + 545 pp. Also available as paperback.

59. Robert S. Cohen and Marx W. Wartofsky (eds.), *Logical and Epistemological Studies in Contemporary Physics,* Boston Studies in the Philosophy of Science (ed. by Robert S. Cohen and Marx W. Wartofsky), Volume XIII. 1973, VIII + 462 pp. Also available as paperback.
60. Robert S. Cohen and Marx W. Wartofsky (eds.), *Methodological and Historical Essays in the Natural and Social Sciences. Proceedings of the Boston Colloquium for the Philosophy of Science 1969-1972,* Boston Studies in the Philosophy of Science (ed. by Robert S. Cohen and Marx W. Wartofsky), Volume XIV. 1974, VIII + 405 pp. Also available as paperback.
61. Robert S. Cohen, J. J. Stachel and Marx W. Wartofsky (eds.), *For Dirk Struik. Scientific, Historical and Political Essays in Honor of Dirk J. Struik,* Boston Studies in the Philosophy of Science (ed. by Robert S. Cohen and Marx W. Wartofsky), Volume XV. 1974, XXVII + 652 pp. Also available as paperback.
62. Kazimierz Ajdukiewicz, *Pragmatic Logic,* transl. from the Polish by Olgierd Wojtasiewicz. 1974, XV + 460 pp.
63. Sören Stenlund (ed.), *Logical Theory and Semantic Analysis. Essays Dedicated to Stig Kanger on His Fiftieth Birthday.* 1974, V + 217 pp.
64. Kenneth F. Schaffner and Robert S. Cohen (eds.), *Proceedings of the 1972 Biennial Meeting, Philosophy of Science Association,* Boston Studies in the Philosophy of Science (ed. by Robert S. Cohen and Marx W. Wartofsky), Volume XX. 1974, IX + 444 pp. Also available as paperback.
65. Henry E. Kyburg, Jr., *The Logical Foundations of Statistical Inference.* 1974, IX + 421 pp.
66. Marjorie Grene, *The Understanding of Nature: Essays in the Philosophy of Biology,* Boston Studies in the Philosophy of Science (ed. by Robert S. Cohen and Marx W. Wartofsky), Volume XXIII. 1974, XII + 360 pp. Also available as paperback.
67. Jan M. Broekman, *Structuralism: Moscow, Prague, Paris.* 1974, IX + 117 pp.
68. Norman Geschwind, *Selected Papers on Language and the Brain,* Boston Studies in the Philosophy of Science (ed. by Robert S. Cohen and Marx W. Wartofsky), Volume XVI. 1974, XII + 549 pp. Also available as paperback.
69. Roland Fraïssé, *Course of Mathematical Logic* – Volume 2: *Model Theory.* 1974, XIX + 192 pp.
70. Andrzej Grzegorczyk, *An Outline of Mathematical Logic. Fundamental Results and Notions Explained with All Details.* 1974, X + 596 pp.
71. Franz von Kutschera, *Philosophy of Language.* 1975, VII + 305 pp.
72. Juha Manninen and Raimo Tuomela (eds.), *Essays on Explanation and Understanding. Studies in the Foundations of Humanities and Social Sciences.* 1976, VII + 440 pp.
73. Jaakko Hintikka (ed.), *Rudolf Carnap, Logical Empiricist. Materials and Perspectives.* 1975, LXVIII + 400 pp.
74. Milič Čapek (ed.), *The Concepts of Space and Time. Their Structure and Their Development,* Boston Studies in the Philosophy of Science (ed. by Robert S. Cohen and Marx W. Wartofsky), Volume XXII. 1976, LVI + 570 pp. Also available as paperback.

75. Jaakko Hintikka and Unto Remes, *The Method of Analysis. Its Geometrical Origin and Its General Significance*, Boston Studies in the Philosophy of Science (ed. by Robert S. Cohen and Marx W. Wartofsky), Volume XXV. 1974, XVIII + 144 pp. Also available as paperback.
76. John Emery Murdoch and Edith Dudley Sylla, *The Cultural Context of Medieval Learning. Proceedings of the First International Colloquium on Philosophy, Science, and Theology in the Middle Ages – September 1973*, Boston Studies in the Philosophy of Science (ed. by Robert S. Cohen and Marx W. Wartofsky), Volume XXVI. 1975, X + 566 pp. Also available as paperback.
77. Stefan Amsterdamski, *Between Experience and Metaphysics. Philosophical Problems of the Evolution of Science*, Boston Studies in the Philosophy of Science (ed. by Robert S. Cohen and Marx W. Wartofsky), Volume XXXV. 1975, XVIII + 193 pp. Also available as paperback.
78. Patrick Suppes (ed.), *Logic and Probability in Quantum Mechanics*. 1976, XV + 541 pp.
79. Hermann von Helmholtz: *Epistemological Writings. The Paul Hertz/Moritz Schlick Centenary Edition of 1921 with Notes and Commentary by the Editors.* (Newly translated by Malcolm F. Lowe. Edited with an Introduction and Bibliography, by Robert S. Cohen and Yehuda Elkana), Boston Studies in the Philosophy of Science (ed. by Robert S. Cohen and Marx W. Wartofsky), Volume XXXVII. 1977, XXXVIII+204 pp. Also available as paperback.
80. Joseph Agassi, *Science in Flux*, Boston Studies in the Philosophy of Science (ed. by Robert S. Cohen and Marx W. Wartofsky), Volume XXVIII. 1975, XXVI + 553 pp. Also available as paperback.
81. Sandra G. Harding (ed.), *Can Theories Be Refuted? Essays on the Duhem-Quine Thesis*. 1976, XXI + 318 pp. Also available as paperback.
82. Stefan Nowak, *Methodology of Sociological Research: General Problems*. 1977, XVIII + 504 pp.
83. Jean Piaget, Jean-Blaise Grize, Alina Szeminska, and Vinh Bang, *Epistemology and Psychology of Functions*, Studies in Genetic Epistemology, Volume XXIII. 1977, XIV+205 pp.
84. Marjorie Grene and Everett Mendelsohn (eds.), *Topics in the Philosophy of Biology*, Boston Studies in the Philosophy of Science (ed. by Robert S. Cohen and Marx W. Wartofsky), Volume XXVII. 1976, XIII + 454 pp. Also available as paperback.
85. E. Fischbein, *The Intuitive Sources of Probabilistic Thinking in Children*. 1975, XIII + 204 pp.
86. Ernest W. Adams, *The Logic of Conditionals. An Application of Probability to Deductive Logic*. 1975, XIII + 156 pp.
87. Marian Przełęcki and Ryszard Wójcicki (eds.), *Twenty-Five Years of Logical Methodology in Poland*. 1977, VIII + 803 pp.
88. J. Topolski, *The Methodology of History*. 1976, X + 673 pp.
89. A. Kasher (ed.), *Language in Focus: Foundations, Methods and Systems. Essays Dedicated to Yehoshua Bar-Hillel*, Boston Studies in the Philosophy of Science (ed. by Robert S. Cohen and Marx W. Wartofsky), Volume XLIII. 1976, XXVIII + 679 pp. Also available as paperback.
90. Jaakko Hintikka, *The Intentions of Intentionality and Other New Models for Modalities*. 1975, XVIII + 262 pp. Also available as paperback.

91. Wolfgang Stegmüller, *Collected Papers on Epistemology, Philosophy of Science and History of Philosophy*, 2 Volumes, 1977, XXVII + 525 pp.
92. Dov M. Gabbay, *Investigations in Modal and Tense Logics with Applications to Problems in Philosophy and Linguistics.* 1976, XI + 306 pp.
93. Radu J. Bogdan, *Local Induction.* 1976, XIV + 340 pp.
94. Stefan Nowak, *Understanding and Prediction: Essays in the Methodology of Social and Behavioral Theories.* 1976, XIX + 482 pp.
95. Peter Mittelstaedt, *Philosophical Problems of Modern Physics*, Boston Studies in the Philosophy of Science (ed. by Robert S. Cohen and Marx W. Wartofsky), Volume XVIII. 1976, X + 211 pp. Also available as paperback.
96. Gerald Holton and William Blanpied (eds.), *Science and Its Public: The Changing Relationship*, Boston Studies in the Philosophy of Science (ed. by Robert S. Cohen and Marx W. Wartofsky), Volume XXXIII. 1976, XXV + 289 pp. Also available as paperback.
97. Myles Brand and Douglas Walton (eds.), *Action Theory. Proceedings of the Winnipeg Conference on Human Action, Held at Winnipeg, Manitoba, Canada, 9-11 May 1975.* 1976, VI + 345 pp.
98. Risto Hilpinen, *Knowledge and Rational Belief.* 1979 (forthcoming).
99. R. S. Cohen, P. K. Feyerabend, and M. W. Wartofsky (eds.), *Essays in Memory of Imre Lakatos*, Boston Studies in the Philosophy of Science (ed. by Robert S. Cohen and Marx W. Wartofsky), Volume XXXIX. 1976, XI + 762 pp. Also available as paperback.
100. R. S. Cohen and J. J. Stachel (eds.), *Selected Papers of Léon Rosenfeld*, Boston Studies in the Philosophy of Science (ed. by Robert S. Cohen and Marx W. Wartofsky), Volume XXI. 1978, XXX + 927 pp.
101. R. S. Cohen, C. A. Hooker, A. C. Michalos, and J. W. van Evra (eds.), *PSA 1974: Proceedings of the 1974 Biennial Meeting of the Philosophy of Science Association*, Boston Studies in the Philosophy of Science (ed. by Robert S. Cohen and Marx W. Wartofsky), Volume XXXII. 1976, XIII + 734 pp. Also available as paperback.
102. Yehuda Fried and Joseph Agassi, *Paranoia: A Study in Diagnosis*, Boston Studies in the Philosophy of Science (ed. by Robert S. Cohen and Marx W. Wartofsky), Volume L. 1976, XV + 212 pp. Also available as paperback.
103. Marian Przełęcki, Klemens Szaniawski, and Ryszard Wójcicki (eds.), *Formal Methods in the Methodology of Empirical Sciences.* 1976, 455 pp.
104. John M. Vickers, *Belief and Probability.* 1976, VIII + 202 pp.
105. Kurt H. Wolff, *Surrender and Catch: Experience and Inquiry Today*, Boston Studies in the Philosophy of Science (ed. by Robert S. Cohen and Marx W. Wartofsky), Volume LI. 1976, XII + 410 pp. Also available as paperback.
106. Karel Kosík, *Dialectics of the Concrete*, Boston Studies in the Philosophy of Science (ed. by Robert S. Cohen and Marx W. Wartofsky), Volume LII. 1976, VIII + 158 pp. Also available as paperback.
107. Nelson Goodman, *The Structure of Appearance*, Boston Studies in the Philosophy of Science (ed. by Robert S. Cohen and Marx W. Wartofsky), Volume LIII. 1977, L + 285 pp.
108. Jerzy Giedymin (ed.), *Kazimierz Ajdukiewicz: The Scientific World-Perspective and Other Essays, 1931 - 1963.* 1978, LIII + 378 pp.

109. Robert L. Causey, *Unity of Science*. 1977, VIII+185 pp.
110. Richard E. Grandy, *Advanced Logic for Applications*. 1977, XIV + 168 pp.
111. Robert P. McArthur, *Tense Logic*. 1976, VII + 84 pp.
112. Lars Lindahl, *Position and Change: A Study in Law and Logic*. 1977, IX + 299 pp.
113. Raimo Tuomela, *Dispositions*. 1978, X + 450 pp.
114. Herbert A. Simon, *Models of Discovery and Other Topics in the Methods of Science*, Boston Studies in the Philosophy of Science (ed. by Robert S. Cohen and Marx W. Wartofsky), Volume LIV. 1977, XX + 456 pp. Also available as paperback.
115. Roger D. Rosenkrantz, *Inference, Method and Decision*. 1977, XVI + 262 pp. Also available as paperback.
116. Raimo Tuomela, *Human Action and Its Explanation. A Study on the Philosophical Foundations of Psychology*. 1977, XII + 426 pp.
117. Morris Lazerowitz, *The Language of Philosophy. Freud and Wittgenstein*, Boston Studies in the Philosophy of Science (ed. by Robert S. Cohen and Marx W. Wartofsky), Volume LV. 1977, XVI + 209 pp.
118. Tran Duc Thao, *Origins of Language and Consciousness*, Boston Studies in the Philosophy of Science (ed. by Robert S. Cohen and Marx. W. Wartofsky), Volume LVI. 1979 (forthcoming).
119. Jerzy Pelč, *Semiotics in Poland, 1894 - 1969*. 1977, XXVI + 504 pp.
120. Ingmar Pörn, *Action Theory and Social Science. Some Formal Models*. 1977, X + 129 pp.
121. Joseph Margolis, *Persons and Minds, The Prospects of Nonreductive Materialism*, Boston Studies in the Philosophy of Science (ed. by Robert S. Cohen and Marx W. Wartofsky), Volume LVII. 1977, XIV + 282 pp. Also available as paperback.
122. Jaakko Hintikka, Ilkka Niiniluoto, and Esa Saarinen (eds.), *Essays on Mathematical and Philosophical Logic. Proceedings of the Fourth Scandinavian Logic Symposium and of the First Soviet-Finnish Logic Conference, Jyväskylä, Finland, 1976*. 1978, VIII + 458 pp. + index.
123. Theo A. F. Kuipers, *Studies in Inductive Probability and Rational Expectation*. 1978, XII + 145 pp.
124. Esa Saarinen, Risto Hilpinen, Ilkka Niiniluoto, and Merrill Provence Hintikka (eds.), *Essays in Honour of Jaakko Hintikka on the Occasion of His Fiftieth Birthday*. 1978, IX + 378 pp. + index.
125. Gerard Radnitzky and Gunnar Andersson (eds.), *Progress and Rationality in Science*, Boston Studies in the Philosophy of Science (ed. by Robert S. Cohen and Marx W. Wartofsky), Volume LVIII. 1978, X + 400 pp. + index. Also available as paperback.
126. Peter Mittelstaedt, *Quantum Logic*. 1978, IX + 149 pp.
127. Kenneth A. Bowen, *Model Theory for Modal Logic. Kripke Models for Modal Predicate Calculi*. 1978, X + 128 pp.
128. Howard Alexander Bursen, *Dismantling the Memory Machine. A Philosophical Investigation of Machine Theories of Memory*. 1978, XIII + 157 pp.
129. Marx W. Wartofsky, *Models: Representation and Scientific Understanding*, Boston Studies in the Philosophy of Science (ed. by Robert S. Cohen and Marx W. Wartofsky), Volume XLVIII. 1979 (forthcoming). Also available as a paperback.
130. Don Ihde, *Technics and Praxis. A Philosophy of Technology*, Boston Studies in

the Philosophy of Science (ed. by Robert S. Cohen and Marx W. Wartofsky), Volume XXIV. 1979 (forthcoming). Also available as a paperback.

131. Jerzy J. Wiatr (ed.), *Polish Essays in the Methodology of the Social Sciences*, Boston Studies in the Philosophy of Science (ed. by Robert S. Cohen and Marx W. Wartofsky), Volume XXIX. 1979 (forthcoming). Also available as a paperback.
132. Wesley C. Salmon (ed.), *Hans Reichenbach: Logical Empiricist.* 1979 (forthcoming).
133. R.-P. Horstmann and L. Krüger (eds.), *Transcendental Arguments and Science.* 1979 (forthcoming). Also available as a paperback.

SYNTHESE HISTORICAL LIBRARY

Texts and Studies
in the History of Logic and Philosophy

Editors:

1. M. T. Beonio-Brocchieri Fumagalli, *The Logic of Abelard.* Translated from the Italian. 1969, IX + 101 pp.
2. Gottfried Wilhelm Leibniz, *Philosophical Papers and Letters.* A selection translated and edited, with an introduction, by Leroy E. Loemker. 1969, XII + 736 pp.
3. Ernst Mally, *Logische Schriften,* ed. by Karl Wolf and Paul Weingartner. 1971, X + 340 pp.
4. Lewis White Beck (ed.), *Proceedings of the Third International Kant Congress.* 1972, XI + 718 pp.
5. Bernard Bolzano, *Theory of Science,* ed. by Jan Berg. 1973, XV + 398 pp.
6. J. M. E. Moravcsik (ed.), *Patterns in Plato's Thought. Papers Arising Out of the 1971 West Coast Greek Philosophy Conference.* 1973, VIII + 212 pp.
7. Nabil Shehaby, *The Propositional Logic of Avicenna: A Translation from al-Shifā: al-Qiyās,* with Introduction, Commentary and Glossary. 1973, XIII + 296 pp.
8. Desmond Paul Henry, *Commentary on De Grammatico: The Historical-Logical Dimensions of a Dialogue of St. Anselm's.* 1974, IX + 345 pp.
9. John Corcoran, *Ancient Logic and Its Modern Interpretations.* 1974, X + 208 pp.
10. E. M. Barth, *The Logic of the Articles in Traditional Philosophy.* 1974, XXVII + 533 pp.
11. Jaakko Hintikka, *Knowledge and the Known. Historical Perspectives in Epistemology.* 1974, XII + 243 pp.
12. E. J. Ashworth, *Language and Logic in the Post-Medieval Period.* 1974, XIII + 304 pp.
13. Aristotle, *The Nicomachean Ethics.* Translated with Commentaries and Glossary by Hypocrates G. Apostle. 1975, XXI + 372 pp.
14. R. M. Dancy, *Sense and Contradiction: A Study in Aristotle.* 1975, XII + 184 pp.
15. Wilbur Richard Knorr, *The Evolution of the Euclidean Elements. A Study of the Theory of Incommensurable Magnitudes and Its Significance for Early Greek Geometry.* 1975, IX + 374 pp.
16. Augustine, *De Dialectica.* Translated with Introduction and Notes by B. Darrell Jackson. 1975, XI + 151 pp.

17. Arpád Szabó, *The Beginnings of Greek Mathematics.* 1979 (forthcoming).
18. Rita Guerlac, *Juan Luis Vives Against the Pseudodialecticians. A Humanist Attack on Medieval Logic.* Texts, with translation, introduction and notes. 1978, xiv + 227 pp. + index.

SYNTHESE LANGUAGE LIBRARY

Texts and Studies
in Linguistics and Philosophy

1. Henry Hiż (ed.), *Questions.* 1977, xvii + 366 pp.
2. William S. Cooper, *Foundations of Logico-Linguistics. A Unified Theory of Information, Language, and Logic.* 1978, xvi + 249 pp.
3. Avishai Margalit (ed.), *Meaning and Use. Papers Presented at the Second Jerusalem Philosophical Encounter, April 1976.* 1978 (forthcoming).
4. F. Guenthner and S. J. Schmidt (eds.), *Formal Semantics and Pragmatics for Natural Languages.* 1978, viii + 374 pp. + index.
5. Esa Saarinen (ed.), *Game-Theoretical Semantics.* 1978, xiv + 379 pp. + index.
6. F. J. Pelletier (ed.), *Mass Terms: Some Philosophical Problems.* 1978, xiv + 300 pp. + index.